T0173036

Nanofabrication
Handbook

Nanofabrication Handbook

Edited by
Stefano Cabrini
Satoshi Kawata

CRC Press
Taylor & Francis Group
Boca Raton London New York

CRC Press is an imprint of the
Taylor & Francis Group, an **informa** business

CRC Press
Taylor & Francis Group
6000 Broken Sound Parkway NW, Suite 300
Boca Raton, FL 33487-2742

First issued in paperback 2019

© 2012 by Taylor & Francis Group, LLC
CRC Press is an imprint of Taylor & Francis Group, an Informa business

No claim to original U.S. Government works

ISBN-13: 978-1-4200-9052-9 (hbk)
ISBN-13: 978-0-367-38165-3 (pbk)

This book contains information obtained from authentic and highly regarded sources. Reasonable efforts have been made to publish reliable data and information, but the author and publisher cannot assume responsibility for the validity of all materials or the consequences of their use. The authors and publishers have attempted to trace the copyright holders of all material reproduced in this publication and apologize to copyright holders if permission to publish in this form has not been obtained. If any copyright material has not been acknowledged please write and let us know so we may rectify in any future reprint.

Except as permitted under U.S. Copyright Law, no part of this book may be reprinted, reproduced, transmitted, or utilized in any form by any electronic, mechanical, or other means, now known or hereafter invented, including photocopying, microfilming, and recording, or in any information storage or retrieval system, without written permission from the publishers.

For permission to photocopy or use material electronically from this work, please access www.copyright.com (http://www.copyright.com/) or contact the Copyright Clearance Center, Inc. (CCC), 222 Rosewood Drive, Danvers, MA 01923, 978-750-8400. CCC is a not-for-profit organization that provides licenses and registration for a variety of users. For organizations that have been granted a photocopy license by the CCC, a separate system of payment has been arranged.

Trademark Notice: Product or corporate names may be trademarks or registered trademarks, and are used only for identification and explanation without intent to infringe.

Library of Congress Cataloging-in-Publication Data

Nanofabrication handbook / editors Stefano Cabrini and Satoshi Kawata.
    p. cm.
  Includes bibliographical references and index.
  ISBN 978-1-4200-9052-9 (alk. paper)
  1. Nanostructured materials--Handbooks, manuals, etc. 2. Nanostructures--Handbooks, manuals, etc. I. Cabrini, Stefano. II. Kawata, Satoshi, 1966-

TA418.9.N35N2523 2012
620.1'15--dc23

2011037490

Visit the Taylor & Francis Web site at
http://www.taylorandfrancis.com

and the CRC Press Web site at
http://www.crcpress.com

To my father Antonio and my mother Ninni.

—Stefano Cabrini

To my wife Megumi for her everlasting support.

—Satoshi Kawata

Contents

PART I   Standard Lithography

PART II   New Lithographic Techniques

PART III   Nanofabrication Applications

Preface

Nanofabrication is a relatively new field in the world of science—it is known as an innovative and interdisciplinary field that is expanding rapidly. This handbook is a unique collection of various approaches to nanofabrication. Since the field is ever changing and expanding, the challenges put upon the editors are considerable; we are well aware that we are describing a field that is constantly broadening its definition. However, it is of the utmost importance, because of the expanding nature of the field, to address new and different approaches to this fascinating and important field of science. In this handbook, you will find an introduction to "nanofabrication" based on various approaches from different scientists. Each contributor has attempted to mark boundaries, define subfields, offer practical instructions and examples, and pave the way for future generations to travel. The book is composed of many chapters by various authors chosen for their expertise in their given field of study.

The motivation for this kind of handbook arose when one of us (SC) was interim director of the Nanofabrication Facility at The Molecular Foundry, a nanoscience and nanotechnology user facility at the Lawrence Berkeley National Laboratory in Berkeley, California. I was in charge of the design of the laboratory, planning the scientific strategies, and starting new programs. But one of the many duties was also to act as a "tour guide" to the many visitors and potential users of the facility. Chemists, biologists, electronic engineers, physicists, journalists, and many others were coming to see and to learn more about this new field of study that the Department of Energy was initiating. Some of them were already experts and some were complete novices, so for every tour I was telling a different story; each time I had to explain the nanofabrication process using different examples, different objectives, and different perspectives. Although all of my explanations were valid, they were also just small pieces of a much larger puzzle. Moreover, when some of the "tourists" asked for a reference, I could not point them to anything. I could think of many books dedicated to the individual aspects of the field, but no single source had yet to really define, describe, and explain the total vision of nanofabrication. Because of this experience, I realized that a book was necessary, not only to fill the large gap in the world of scientific

publications, but also to give some practical information for those just entering the field. The first big challenge was the definition of nanofabrication. What exactly is nanofabrication? Is it the scaling down of microfabrication? Is it engineering support for nanosciences? Or is it more complex?

We are developing methods, techniques, and knowledge to reproduce and control matter at a nanometer scale. To do this, we need to understand the physical phenomena that occur in that regime by controlling the environment (to a nanometer size) as we perform our experiments and validate our theories. Matter in that size regime presents all sorts of astonishing properties that can be applied to many fields, thereby bringing tremendous benefits to science, society, and ultimately, our lives. Nanofabrication devices can contribute by the transfer of such properties from the mesoscale to the real world. In short, we apply all our macroskills to manipulate the nanoworld, better understand its properties, and control them for applications that will enhance daily life on a global scale.

This handbook is the first attempt to clarify these aspects and provide the most complete resource for this rapidly emerging interdisciplinary field. Each chapter focuses on a particular method or aspect of study. For every method, the contributors discuss the underlying theoretical basis, resolution, patterns and substrates used, and applications. This builds upon our understanding of microfabrication, while moving into the "nano" dimension and showing that its use at this scale requires a different process and understanding. For each experiment, it shows how to approach key problems in the decision-making process related to materials, methods, surface considerations, and so on.

Nanofabrication Handbook encourages readers to consider fabricating techniques that better fit their particular research needs. It is designed to serve as an essential resource tool for consulting in the experimental design phase of any project utilizing nanofabrication techniques. It is written for scientists and students from different disciplines, who want to approach a nanoscience experiment using nanofabrication. We have endeavored to provide a broad vision of the most critical problems and how to approach and solve them. The contents include basic definitions and introduce the main underlying concepts of nanofabrication. The core material offers a wide variety of illustrative examples of cutting-edge applications with discussion of major advantages and disadvantages of each approach. We believe that it is an excellent resource for those who will plan experiments in a team, whether in industry or academia.

In summary, it is our hope that this book gives the reader a foundation to enter the complex world of nanofabrication, and also inspires the scientific community-at-large to push the limits of nanometer resolution to make full use of that "plenty of room at the bottom."

Editors

Stefano Cabrini is the director of the Nanofabrication Facility at Molecular Foundry (Lawrence Berkeley National Laboratory), and is a member of the Inter-facility Nanophotonics Group. Molecular Foundry among the five Department of Energy Nano-Scale Research Centers that involve user-oriented facilities dedicated both to user projects and to internal research. He earned his "laurea" degree in physics from the University of Rome in 1991. In 1996, he received a European postdoc fellowship at the Institut d'Optique Théorique et Appliquée at Orsay, France. After that he became a researcher at the National Research Council (CNR) at the Institute of Solid State Electronics in Rome, Italy. In 2001, he worked as a senior scientist at Elettra Synchrotron Light Source in Trieste, Italy, in the Laboratory for Interdisciplinary Lithography. In 2006, he joined the Molecular Foundry just before it opened and he had the opportunity to organize the new nano-fabrication facility, with a special focus on the "Single Digit (sub-ten nanometer) Nanofabrication." Cabrini has over 90 publications in the field of nanofabrication. His research interests include nanophotonics and meta-materials (metallic resonator and photonic crystals), semiconductor device fabrication, NEMS–MEMS, and the development of new lithographic tools and processes.

Satoshi Kawata received his BS, MS, and PhD in physics from Osaka University, Japan in 1974, 1976, and 1979, respectively. After performing postdoctoral research at the University of California, Irvine, he joined the Department of Applied Physics, Osaka University as an assistant professor, where later he became professor of applied physics. He is also the chairman of Nanophoton Corp., and chief scientist of the Nanophotonics Laboratory, RIKEN. His research interests lie in the fields of nanophotonics, biophotonics, plasmonics, metamaterials, photonic crystals, and nanofabrication through two-photon absorption techniques. He has published over 200 journal papers and is the recipient of many awards, including the Medal of Purple Ribbon from the Japanese Emperor in 2007; the Science and Technology Award from the Japanese Ministry of Education, Culture, Sports, Science and Technology in 2005; the Da Vinci Excellence, Moët Hennessy Louis Vuitton (LVMH) prize in 1997; and the Japan IBM Science Award in 1996. He has been the editor of the *Journal of Optics Communications* since 2000 and is a Fellow of the Optical Society of America, SPIE, and Institute of Physics.

Contributors

Arata Aota
Institute of Microchemical Technology
Kanagawa, Japan

Sergey Babin
Abeam Technologies
Castro Valley, California

Stefano Cabrini
Molecular Foundry
Lawrence Berkeley National Laboratory
Berkeley, California

Patrizio Candeloro
BIONEM Lab
University of Magna Graecia
Catanzaro, Italy

Harish Chandran
Department of Computer Science
Duke University
Durham, North Carolina

Allan Chang
Lawrence Livermore National Laboratory
Livermore, California

Stephen Y. Chou
Department of Electrical Engineering
Princeton University
Princeton, New Jersey

Maria Laura Coluccio
NanoBioScience Department
Fondazione Istituto Italiano di Tecnologia
Genova, Italy

and

BIONEM Lab
University of Magna Graecia
Catanzaro, Italy

Gordon S. W. Craig
Nanoscale Science and Engineering Center
University of Wisconsin—Madison
Madison, Wisconsin

Gobind Das
NanoBioScience Department
Fondazione Istituto Italiano di Tecnologia
Genova, Italy

Francesco De Angelis
NanoBioScience Department
Fondazione Istituto Italiano di Tecnologia
Genova, Italy

and

BIONEM Lab
University of Magna Graecia
Catanzaro, Italy

Enzo Di Fabrizio
NanoBioScience Department
Fondazione Istituto Italiano di Tecnologia
Genova, Italy

and

BIONEM Lab
University of Magna Graecia
Catanzaro, Italy

Hiroyuki Fujita
CIRMM
Institute of Industrial Science
University of Tokyo
Tokyo, Japan

Jacques Gierak
Laboratoire de Photonique et de
  Nanostructures
CNRS
Marcoussis, France

Nikhil Gopalkrishnan
Department of Computer Science
Duke University
Durham, North Carolina

Peter Hawkes
CEMES–CNRS
Toulouse, France

Koji Ishibashi
Advanced Device Laboratory
RIKEN—The Institute of Physical and
    Chemical Research
Saitama, Japan

Ralf Jede
Raith GmbH
Dortmund, Germany

Linke Jian
Singapore Synchrotron Light Source
National University of Singapore
Singapore

Satoshi Kawata
Department of Applied Physics
Osaka University
Osaka, Japan

Dan Kercher
Hitachi GST Research
San Jose, California

Takehiko Kitamori
Institute of Microchemical Technology
Kanagawa, Japan

and

Department of Applied Chemistry
The University of Tokyo
Tokyo, Japan

Roman Krahne
Istituto Italiano di Tecnologia
Genova, Italy

Thomas LaBean
Department of Computer Science and
    Department of Chemistry
Duke University
Durham, North Carolina

Carlo Liberale
NanoBioScience Department
Fondazione Istituto Italiano di Tecnologia
Genova, Italy

Liberato Manna
Istituto Italiano di Tecnologia
Genova, Italy

Shinji Matsui
Graduate School of Science, LASTI
University of Hyogo
Hyogo, Japan

Yoshio Mita
School of Engineering
The University of Tokyo
Tokyo, Japan

Herbert O. Moser
Singapore Synchrotron Light Source

and

Department of Physics
National University of Singapore
Singapore

Omkar A. Nafday
Department of Biological Sciences
Florida State University
Tallahassee, Florida

Patrick Naulleau
Center for X-Ray Optics
Lawrence Berkeley National Laboratory
Berkeley, California

Paul F. Nealey
Department of Chemical and Biological
    Engineering
University of Wisconsin—Madison
Madison, Wisconsin

John H. Reif
Department of Computer Science
Duke University
Durham, North Carolina

and

Faculty of Computing and Information
    Technology
King Abdulaziz University
Jeddah, Saudi Arabia

Aditi Risbud
Molecular Foundry
Lawrence Berkeley National Laboratory
Berkeley, California

Thomas Schenkel
Accelerator and Fusion Research Division
Lawrence Berkeley National Laboratory
Berkeley, California

P. James Schuck
Molecular Foundry
Lawrence Berkeley National Laboratory
Berkeley, California

Steven Shannon
Department of Nuclear Engineering
North Carolina State University
Raleigh, North Carolina

T.-C. Shen
Department of Physics
Utah State University
Logan, Utah

Satoru Shoji
Department of Applied Physics
Osaka University
Osaka, Japan

Takuo Tanaka
Metamaterials Laboratory
RIKEN Advanced Science Institute
Saitama, Japan

Stephen Thoms
School of Engineering
University of Glasgow
Glasgow, Scotland, United Kingdom

Yuan Wang
Nano-Scale Science and Engineering Center
University of California
Berkeley, California

Alexander Weber-Bargioni
Molecular Foundry
Lawrence Berkeley National Laboratory
Berkeley, California

Xiang Zhang
Department of Mechanical Engineering
University of California
Berkeley, California

Standard Lithography

1. Introduction to Nanofabrication

Stefano Cabrini

Molecular Foundry, Lawrence Berkeley National Laboratory, Berkeley, California

Aditi Risbud

Molecular Foundry, Lawrence Berkeley National Laboratory, Berkeley, California

Nanoscale materials provide a fertile ground between molecular and bulk systems. Capitalizing on the fundamental differences in physical, chemical, and biological behavior of materials at dimensions between 1 and 100 nm, nanoscience examines the compelling and singular properties found neither in atoms and molecules nor at macroscopic dimensions. Such unique properties have already improved routine commercial products such as sunscreen and stain-resistant clothing, and hold promise for uncovering new materials with applications in energy, security, agriculture, and healthcare.

In general, materials fall into three categories: metals or conductors, insulators, and semiconductors. At nanometer-length scales (a nanometer is one billionth of a meter), a material's properties can be altered by minute changes in its size, which affect how its electrons behave (delocalized, localized, or somewhere in between). As a metal shrinks in size, for example, its electrons no longer roam in bands of energy, and instead are restricted to discrete energy levels. This quantization of energy—commonly called quantum confinement—ultimately determines a material's properties. In addition, defects drive a material's behavior in insulating materials such as transition metal oxides—an effect that is enhanced in nanoscale structures, which have high surface areas and/or interfacial areas. In semiconductors, electron–hole pairs (called excitons) are crucial to processes such as light emission; in nanoscale materials, the exciton size is determined by the material's size or arrangement with respect to other nanostructures, rather than the electron–hole interaction [1].

Ideally, nanoscale materials combine advantageous features of bulk and molecular systems into materials that can be readily designed, manufactured, and integrated into functional devices. It is evident that the fabrication methods currently applied by the microelectronics industry will not translate directly to the nanoscale. Therefore, the design and manufacture of structures at nanometer dimensions—what we call nanofabrication—requires innovations in material synthesis, physical patterning, structural characterization, and theory, encompassing disciplines such as biology, chemistry, physics, and engineering. As such, scientists around the world are actively seeking and testing "top-down" and "bottom-up" strategies to engineer surface nanostructures [2] that can interface with macroscopic materials and can be scaled up to manufacturing devices for real-world applications.

"Top-down" strategies make use of advanced lithography, electron-beam writing, and nanoimprinting techniques to directly generate a pattern into a substrate. These techniques, due to their historical origin, are especially fruitful in developing metal- and semiconductor-based nanostructures. "Bottom-up" methods exploit atoms and molecules that spontaneously assemble into organized structures, without human intervention [3]. Commonly employed by nature, self-assembly of materials occurs at length scales ranging from cellular components to star-studded galaxies. This tactic is particularly useful for constructing systems, especially biological, organic, and other soft materials, with nanoscale precision, particularly for structures below the optical diffraction limit [4].

By combining "bottom-up" and "top-down" strategies, we can integrate nanosystems and tailor materials to achieve a specific application. However, as the dimensions

Nanofabrication Handbook. Edited by Stefano Cabrini and Satoshi Kawata © 2012 CRC Press/Taylor & Francis Group, LLC. ISBN: 978-1-4200-9052-9

of a material decrease, special consideration must be taken to understand the size-dependent properties of a material within the context of a given application, while optimizing molecular and material design [5]. For example, optical properties originate from the bandgap of a nanoscale material, while magnetic properties are dominated by the exchange interactions between spin states [1]. To effectively generate nanostructures with resolutions below 10 nm, it is likely that both "top-down" and "bottom-up" methods will need to work in concert with larger-scale components in functional devices [2]. This also requires examining nanomaterials with high precision to tease out structure–property relationships. The advent of techniques for atomic scale imaging, such as scanning tunneling microscopy and atomic force microscopy, along with advances in traditional electron microscopy, has greatly aided efforts to view nanoscale structures and their modification under external perturbation [6].

The famous lecture "There's plenty of room at the bottom," presented by Richard P. Feynman to the American Physical Society in Pasadena at the California Institute of Technology in December 1959 [7], and the fabrication of the first solid-state transistor [8] unveiled a new era in the scientific and technological world. Feynman's landmark lecture previewed the possible arrangement of a material atom by atom to find new properties and a great range of potential applications. This was a new way to see matter and it opened a new scientific challenge we now call nanoscience. The advent of the transistor effectively launched the entire semiconductor revolution, bringing advanced technology to a greater part of the world. Although the impact of Feynman's words and the transistor's discovery have been exhaustively discussed, one of the main aspects of interest is the development of the science and technology developed during this period that allowed the microelectronic industry to manufacture devices at submicron dimensions with incredible control and speed.

Moore's law [9] describes a long-term trend in the history of computing hardware: the number of transistors that can be placed inexpensively on an integrated circuit doubles approximately every 2 years. This trend continues today and is now pushing the fabrication and manufacturing worlds toward the nanoscale. This has opened a Pandora's box of nanoscience in the quest to find more room at the bottom, giving researchers a new opportunity to understand the scientific principles and properties of a few atoms or molecules. Worldwide,

there is a common effort to study, understand, and control matter at the nanoscale and, subsequently, use this knowledge to reveal macroscopic phenomena, for astonishing new applications in healthcare, technology, and energy sustainability.

The United States, in particular, has invested significant resources into nanoscience and technology research and development. This can be evidenced by the 2001 National Nanotechnology Initiative, which pooled the collective efforts of 15 U.S. federal agencies broadly involved in nanoscience and nanotechnology, from fundamental research to regulation and commercialization (FY 2011 budget: $1.8 billion) [10]. In support of the National Nanotechnology Initiative, the U.S. Department of Energy's Scientific User Facilities Division, for example, has developed and launched five new nanoscale science research centers to support the synthesis, processing, fabrication, and analysis of materials at the nanoscale.

These centers are designed to be premier user centers for interdisciplinary research at the nanoscale, serving as the basis for a national program that encompasses new science, new tools, and new computing capabilities. Together, the nanoscale science research centers provide a gateway to other major user facilities for x-ray, neutron, and electron scattering. Each nanoscience user facility contains clean rooms, laboratories for nanofabrication, one-of-a-kind signature instruments, and other tools (e.g., nanopatterning instruments and research-grade probe microscopes). These facilities provide researchers from around the world free access to state-of-the-art instrumentation, computational methods, and expert scientific staff—a novel and standalone setting designed to promote collaborative work among scientists from varied disciplines including chemistry, biology, physics, materials science, engineering, and computer science. These centers exemplify the type and scale of interdisciplinary settings needed to tackle large-scale scientific challenges, such as our current and future needs for carbon-neutral and renewable energy sources. Indeed, these centers have accelerated efforts in developing nanotechnology for solar energy collection and conversion, sustainable manufacturing of nanoscale materials, and nanoelectronics [10].

This is just one example of the broad efforts made by governments in Europe, Australia, Canada, and Japan to push nanoscience development: for example, institutes such as the National Institute of Nanotechnology in Canada, Nanotechnology Network in Japan, Centre for Nanoscience and Nanotechnology in Australia, and Nanoscience Foundries and Fine Analysis in Europe are involved in similar programs. These countries dedicate significant resources to investigate nanoscience and address larger-scale science and technology problems.

In all these efforts, nanofabrication is one of the most important components. But what exactly is nanofabrication? Frequently, the name causes some misunderstanding as it was derived historically from the microfabrication discipline [11]. This discipline was developed and enhanced by the microelectronic "revolution" that occurred in the second part of the twentieth century. Large competition in the semiconductor industry pushed the development of new techniques to write circuits at successively smaller dimensions, but with incredible stability and reproducibility. Today, we are at a 45 nm node made by 193 nm immersion optical lithography [12]. This allows the development of new manufacturing techniques to realize new devices and discover materials not previously accessible. As a result, new lithographic tools for "writing" new circuits, new etching tools to transfer a pattern on a suitable substrate, and new tools to deposit ever-improving materials were developed during this period. New sophisticated technology for optimized device fabrication, as well as new tools, have resulted in better performance and resolution. Although applications outside the semiconductor industry were pursued in parallel, the main driver for controlling the dimensions of devices was the "transistor" business. However, it became clear there were many other applications that required control of many other parameters or the use of different materials. Hence the ability to fabricate at a "small" scale was used for microfludic devices, microscopy, microelectromechanics, and many other fields. Every time a new technological improvement was available, a new scientific and technological frontier opened.

With this continuous evolution, the microfabrication community systematically arrived at nanofabrication. As is often the case, the final result of this outcome was quite different from its initial status. Although lithography, etching, and deposition are still important, several other components became part of the game. Since matter demonstrated different behavior at nanoscale dimensions (below 100 nm), it became important to study theoretically and experimentally what this confinement could induce. As a result, new calculation methods were developed, along with new

tools for nanoimaging, more refined spectroscopies, and new lithographic tools. Chemistry played an essential part in bringing a huge variety of new materials as well as new mechanisms to control and study their behavior at the nanoscale. A huge variety of scientific fields are now interested in all the capabilities of nanoscience as well as of nanofabrication: with new problems to investigate, they stimulated new solutions and new ways to investigate the results, consequently opening new frontiers and opportunities even in totally different fields. Potential applications also provide feedback on the quality of fabrication, suggesting a way to investigate properties of matter at the nanoscale.

A "nanofabricator" should prepare an experiment to control as best as possible all nanoscale properties such that a process may be used to obtain reproducible results in an application. Not only will this provide a significant contribution to the basic understanding of all the physical phenomena occurring at the "bottom," it could also give scientists a path to effectively use these properties in more controllable microscale devices that can fundamentally contribute to the macroscale world. The "nanofabricator" needs dedicated spaces or laboratories, some specific tools, some basic techniques, and—perhaps most importantly—a stimulating and interdisciplinary cultural environment to guide one in the daily challenges of exploring the "room at the bottom."

Controlling every parameter during different steps of nanofabrication processes requires a specific environment. The temperature should be kept constant: chemical and physical properties of materials can change drastically, and thermal expansion of tools can compromise nanometer-scale stability (and consequently resolution). Uncontrolled humidity can create the same kind of problems. Particles of dust might approximate an enormous mountain with respect to the dimensions of fabricated devices. As these particles will negatively impact device performance, it is important to work in a dust-free environment, such as a "clean room" laboratory [13].

In contrast to a traditional microfabrication clean room in which the protocol does not allow any "dirty stuff" to be used (e.g., metals, organics, or biopolymers), for a nanofabrication clean room, some of the materials important for nanoscience fall into this category. All of these materials (that can be considered dust pollution) should be handled as not to compromise the dust-free environment, so a higher class of clean room should be maintained in the specific areas in which these materials can be handled. This laboratory should have

- A very clean (Class 100) and yellow light area (to avoid exposing photosensitive materials) dedicated to lithographic techniques such as electron beam, optical, nanoimprinting, and focused ion beam
- An area for imaging and microscopy with very low levels of environmental and vibrational noise
- An acoustically isolated space for "noisy" tools, such as etchers, sputtering equipment, and other tools requiring high vacuum chambers
- A space for all wet chemical manipulation not requiring acoustic and vibrational noise isolation
- A dedicated area for working with nanoparticles
- An area for initial characterization before a sample is removed from the clean room and contaminated with dust

Some of the typical tools and techniques used in nanofabrication are described in the first two sections of this book. As this field is so young and dynamic, it is almost impossible to capture all the aspects that make this discipline so intriguing and fascinating in a single book. In the first section, the basic techniques derived from classical microfabrication linked to basic lithographic processes and pattern transfer concepts (represented in Figure 1.1) are described. Here, we will try to enhance the nanoscale aspects of these techniques and the most recent developments. Some aspects,

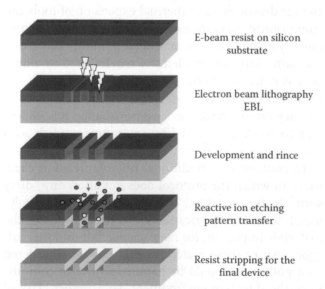

E-beam resist on silicon substrate

Electron beam lithography EBL

Development and rince

Reactive ion etching pattern transfer

Resist stripping for the final device

FIGURE 1.1 Schematic representation of a typical lithographic nonofabrication process.

which did not change significantly from the microscale to the nanoscale, will not be treated in this section. For example, the science of the photoresists and material deposition have been extensively detailed elsewhere and cannot be covered in a few chapters on nanofabrication [14].

Although there are several new interesting techniques in the world of thin-film deposition, it is important to highlight at least one recent technique that is getting a great deal of attention for its versatility and high resolution: atomic layer deposition or ALD [15,16].

ALD is based on the sequential use of gas-phase chemicals, called precursors, deposited on a surface or substrate. The majority of ALD reactions use two precursors that react with a surface one after another in a sequential manner. By exposing precursors to a growth surface repeatedly, a thin film is deposited. This is a self-limiting and conformal technique, keeping the amount of material deposited in each reaction cycle constant. In addition, a uniform monolayer is created due to diffusion of the gas precursors absorbed on a sample surface. ALD is similar to chemical vapor deposition (CVD), but, in ALD, the CVD reaction is broken into two half-reactions, keeping the precursor materials separate during growth. Due to these characteristics, ALD makes atomic-scale deposition control possible. Consequently, both the thickness and composition of a thin film may be controlled with very high resolution. This technique brings incredible advantages to nanofabrication.

The second section of this book will review new techniques developed more recently in research laboratories but now appearing in many industrial applications, such as nanoimprinting, directed self-assembly, scanning probe lithography, and nanoparticle fabrication. These are all very promising techniques opening new and exciting possibilities for manipulating matter at molecular and even atomic levels. These methods represent a real road to nanofabrication needed to match the level of precision and reproducibility of classical techniques described in the first section. It will also become clear how interdisciplinary the field of nanoscience truly is: biological, physical, and chemical considerations are equally important. This necessitates an engineering approach to control tools and experimental conditions to make nanofabrication techniques as reliable as possible.

The interdisciplinary nature of nanofabrication will also become obvious during the third section of this book, in which some practical examples of the applica-

FIGURE 1.2 Scanning electron micrograph of a plasmonic gold antenna built on and AFM tip apex by focused ion beam lithography. The antenna can squeeze the visible light in the small gap and this hot spot can be used as a near field optical probe.

tion of nanofabrication will be discussed. The application of nanofabrication in diverse fields of science and engineering will show how different techniques are useful in solving practical problems, and how fine control of each step in nanofabrication can help create an ideal experimental environment in many areas of nanoscience. It will also become clear how the requirements of various applications can create new ways to fabricate devices and provide feedback to scientists and researchers studying nanofabrication (Figure 1.2). All of these examples will reveal how different aspects of nanofabrication are intimately linked to one another and how deep knowledge of the effects of fabrication can significantly impact understanding of our world at the molecular and atomic level. We will also learn how nanofabrication can make this knowledge useful in the everyday world.

References

1. G. D. Scholes and G. Rumbles, Excitons in nanoscale systems, *Nature Materials* **5**:683–696, 2006.
2. J. V. Barth, G. Costantini, and K. Kern, Engineering atomic and molecular nanostructures at surfaces, *Nature* **437**:671–679, 2005.
3. G. M. Whitesides and B. Grzybowski, Self-assembly at all scales, *Science* **295**:2418–2421, 2002.
4. S. R. Quake and A. Scherer, From micro- to nanofabrication with soft materials, *Science* **290**:1536–1540, 2000.
5. A. P. Alivisatos, P. F. Barbera, A. W. Castleman, J. Chang, D. A. Dixon, M. L. Klein, G. L. McLendon. et al., From molecules to materials: Current trends and future directions, *Advanced Materials* **10**:1297–1336, 1998.
6. K. J. Klabunde and R. M. Richards (eds), *Nanomaterials in Chemistry*. Hoboken, NJ: John Wiley & Sons, 2009.
7. R. P. Feynman, Plenty of room at the bottom, *Caltech Engineering and Science* **23**(5):22–36, 1960.
8. W. Shockley (inventor) and Bell Telephone Labor, Inc. (applicants), Solid-state transistor. U.S. Patent US2569347 (A), issued September 25, 1951.
9. G. E. Moore, Cramming more components onto integrated circuits, *Electronics Magazine* **38**(8):114–117, 1965.
10. United States. *The National Nanotechnology Initiative—Research and Development Leading to a Revolution in Technology and Industry (Supplement to the President's FY 2011 Budget)*. Arlington, VA: National Science and Technology Council, 2010, p. 18.
11. P. Rai-Choudhury (ed.), *Handbook of Microlithography, Micromachining, and Microfabrication. Volume 1: Microlithography, Volume 2: Micromachining, and Microfabrication (SPIE Press Monograph, Vol. PM39)*. Bellingham, WA/Herts, UK: SPIE / IEEE, 1997.
12. Y. Wei and R. L. Brainard, *Advanced Processes for 193-nm Immersion Lithography*. Bellingham, WA: SPIE Press Book, 2009.
13. W. Whyte, *Cleanroom Technology: Fundamentals of Design, Testing and Operation*. Hoboken, NJ: John Wiley & Sons, 2010.
14. M. Ohring, *The Materials Science of Thin Films: Deposition and Structure*. San Diego, CA: Academic Press, 2002.
15. A. Sherman, *Atomic Layer Deposition for Nanotechnology: An Enabling Process for Nanotechnology Fabrication*. USA: Ivoryton Press, 2008.
16. M. Ritala and M. Leskelä, Atomic layer epitaxy—A valuable tool for nanotechnology? *Nanotechnology* **10**(1):19–24, 1999.

Chapter 1

2. Electron Beam Lithography

Stephen Thoms

School of Engineering, University of Glasgow, Glasgow, Scotland, United Kingdom

Nanofabrication is engineering on a tiny scale where the number of atoms across a structure becomes countable, and rather interesting effects appear. Conventional engineering deals with objects on the scale of a ship or car, or perhaps an MP3 player, but here one is engaged in a construction exercise where a pinhead is enormous and even a hair's breadth—about 100,000 nm—is large. At the outset it is worth pausing to think about this scale, which is very different from our everyday experience. Imagine, if you can, that you are nanowoman (or perhaps

Nanofabrication Handbook. Edited by Stefano Cabrini and Satoshi Kawata © 2012 CRC Press / Taylor & Francis Group, LLC. ISBN: 978-1-4200-9052-9

nanoman), and 200 nm tall (Figure 2.1). Most surfaces such as paper or even the keys on the keyboard in front of me are rough on the scale of microns or more, and would present a formidably difficult terrain for nanoman to get about on. Indeed it is more or less impossible to carry out nanopatterning on anything other than highly polished surfaces. But even here life is different. Even on dry days you can expect to get wet feet since the layer of surface water is typically 0.1–1 nm deep, and possibly up to 10 nm. It is not just water that covers the landscape; many processes leave surface residues. You should also consider the crowds: 100 nanofolk could gather in a square micron; 100 million could gather in a square millimeter. This is part of the power of miniaturization. Finally, nanofolk

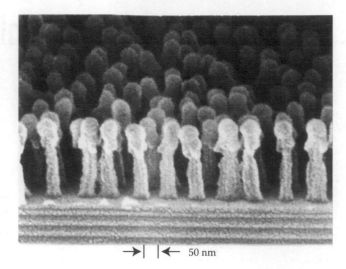

→| |← 50 nm

FIGURE 2.1 "At the Debutante's Ball, young nanowomen gather in their finery at the edge of the stage to weep because the nanoboys won't dance with them" from http://www.zyvexlabs.com/EIPBNuG/EIPBN1995/1995.html. This image was awarded "Most Bizarre Micrograph" at the *1995 International Conference on Electron Ion and Photon Beam Technology and Nanofabrication's Micrograph Contest*. The artist is Tim Savas from Massachusetts Institute of Technology, who took the image with an initial magnification of 100,000× on a Zeiss DSM982 Gemini scanning electron microscope.

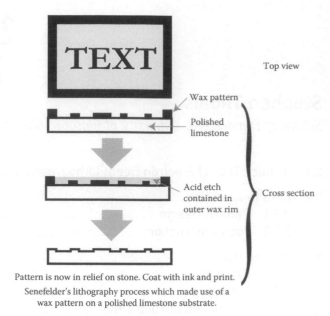

Pattern is now in relief on stone. Coat with ink and print.
Senefelder's lithography process which made use of a wax pattern on a polished limestone substrate.

FIGURE 2.2 Senefelder's first lithography process.

almost never come across a straight smooth edge—typically any straight line along the surface is rough on a scale of a few nanometers which would be like us finding no edge smoother than a pebble-dashed wall.

Roughly speaking, we can divide nanofabrication into two areas. The first is lateral—or horizontal—engineering where we are concerned with making two-dimensional (x, y) patterns on a surface; and second, depth engineering where we work in the third, or z, dimension. The first is the art of *lithography*; the second is that of *pattern transfer*, which includes deposition and etching where we add and subtract material, respectively.

Nanolithography is therefore a major part of nanofabrication, and so a vital skill to master if you are to be a nanofabricator. The nanopart is easy enough to understand since it is just a length scale, albeit a very small one. But what is meant by lithography? According to the *Oxford English Dictionary*, lithography is the "process of obtaining prints from stone or metal surface so treated that what is to be printed can be inked but the remaining area rejects ink." It comes from the Greek: lithos is stone; grapho is to write. Its history can be traced back a long way, but perhaps a good place to start for our purposes is with Alois Senefelder (1771–1834) who came up with the scheme shown in Figure 2.2.

Senefelder used wax to protect certain regions of the limestone from the acid etch, and after etching removed the wax to obtain the desired print pattern, in relief, on the surface. In the crudest method, the wax pattern could be obtained simply by drawing with a wax crayon. This process and its derivatives became the foundation for much of the modern printing industry. The process is also remarkably similar to the lithography process used today in microfabrication labs across the world, hence the use of the same name. In microfabrication, the wax is replaced by a material called *resist*, so-called because it resists the etching process, thereby protecting the material beneath it. A silicon wafer typically replaces the limestone but other substrates are often used such as quartz, GaAs, or more complex layers of different materials such as silicon on sapphire.

There is a useful distinction between patterning the resist (applying the wax in Senefelder's approach) and the subsequent etching process. In the micro- and nanofabrication world, the former is known as lithography and the latter as pattern transfer. Electron beam lithography is concerned with producing a fine pattern, typically measured in tens of nanometers, in resist on the substrate. This resist pattern is then usually transferred to a more useful material such as a metal or dielectric, which can be used for building electrical or optical devices. An electron beam can be focused much smaller than the finest wax crayon, and is computer controlled to produce the desired pattern almost perfectly. Nevertheless, there are similarities in the approach. In particular, in both cases the writing process is a serial one, which means it can be fairly slow to cover large areas. But just as Senefelder used his patterned stone to print many copies rapidly, in the same way there are methods for today's lithographer to rapidly reproduce a pattern once written.

2.1 A Quick Tour of Electron Beam Lithography

First let us have a brief diversion to talk about how electrons can be manipulated. Because electrons are charged particles, they can be steered by electric or magnetic fields. Usually magnetic fields are employed, using electromagnets, and two important parts of an electron column are lenses and deflectors. The lenses are used to focus the electrons and in many ways behave like the more familiar light lenses. For instance, they obey the lens law, and just as many glass elements are put together to make a light microscope, so multiple electron lenses are used to form a versatile electron microscope. They have the advantage over light lenses in that it is very easy to vary their strength simply by varying the current passing through the electromagnets. Likewise deflectors can be constructed which can scan a beam from side to side, again controlled by the current passing through the deflection coils.

Now back to electron beam lithography. Electron beam tools can be constructed which focus a beam down to very small sizes, round about 1 nm in the case of a typical scanning electron microscope (SEM). Under computer control, this beam is scanned across the surface of a wafer. As it does, so it induces chemical changes in a thin layer of polymer, known as a resist, with which we have previously coated the surface. When the patterning is complete, we place the wafer in a chemical bath, which removes some of the resist revealing the scanned pattern. There are two types of resist: positive tone in which the electron beam exposed regions are removed, and negative tone, in which the nonexposed regions are removed.

A simple picture of the electron beam lithography process is a sort of ultimate inkjet printer where the ink droplets are replaced by electrons and the paper by a resist-coated wafer. A computer controls the pattern being written in both cases. Of course there are many subtleties to lithography, which are discussed in the following sections.

The resist should possess a number of important properties. It must be able to withstand the pattern transfer processes such as etching. It also needs to be sensitive to the electron beam so that it can be patterned, and of course it needs to be removable once we have finished with it. You will find that some resists are very stubborn to remove once the pattern transfer process is finished, and this limits their usefulness.

Obtaining sub-20 nm lithography is not straightforward, even using a good tool with a sub-5 nm beam size. In this section, a brief overview of the various topics that will be covered in more depth in the rest of the chapter is given.

You will need to understand something of the physics of electron–solid interactions. Solid matter is, of course, mostly empty space and an electron beam can penetrate many microns before coming to rest, or possibly emerging back into the vacuum chamber of the electron beam tool. There are two types of scattering event. In *elastic* events, the electron maintains its energy but changes direction; this is typically an electron–nucleus interaction. Electron–electron scattering is generally *inelastic* and slows down the electron, which is why it will eventually come to rest if it does not leave the substrate. Most of the resist exposure is caused by *secondary* electrons that arise from inelastic scattering events and typically have energies in the range 10–100 eV and a range of 2–3 nm. The secondary electrons give rise to a fundamental limit on the resolution that can be attained. These scattering processes are described in more detail in Box 2.1.

The electron beam diameter is another resolution limiting factor. A good SEM can have a 30 kV beam with a 0.5 nm probe size. Other optimizations, such as maintaining focus across a large field, are important for a purpose-built lithography tool and minimum beam diameters may be around 3–4 nm, comparable with the secondary electron range. Of course, the beam needs to be in focus on your resist, and this depends on having a flat substrate and making good use of the available technology.

Such considerations show that even for a hypothetical "zero-width" design, a finite linewidth will be achieved, typically 10 nm or greater. This growth for any small feature you attempt to write is called *bias*. Bias needs to be accounted for in the design stage, either in the computer-aided design editor, or else at the postprocessor stage. This has a profound effect on dose assignment for very high-resolution features.

You also need some understanding of the chemistry of the available resist and development processes. polymethyl methacrylate (PMMA) is a common high-resolution positive tone resist, which comes in a variety of molecular weights, and there are a bewildering array of different processes reported in the literature. Its resolution and line edge roughness depend strongly on the developer used. Hydrogen silsquioxane (HSQ) is a negative tone resist with a similar resolution but a much smaller line edge roughness. Again the development regime is critical. There is a class of resists which are so-called *chemically amplified*, which have benefits in terms of writing speed but generally lower resolution. The highest commonly available resists available at the moment are PMMA, HSQ, and calixerenes.

The concept of process latitude or process window is of considerable importance, especially since variations are inherent in all processing despite best efforts to keep everything the same. The development temperature will vary slightly from day to day, and development time probably cannot be controlled better than to the nearest second, at least when carried out with beakers and tweezers. Thus, we expect a 5 s development process to have six times the variability of a 30 s process and these will have similar effects to dose variations. The beam current will always vary with time and position—typically variations are of the order of 1%. This means that you can expect around 1% dose variations in any pattern. The dose window is the range of doses for which a process works. An easy process would have a big window— say a dose between 800 and 1200 μC cm^{-2} which is

BOX 2.1 ELECTRON SCATTERING

In 1909, Geiger and Marsden noticed that α-particles passing through a thin gold foil were occasionally scattered through large angles close to 180° (Geiger 1909). From this Rutherford deduced the nuclear picture of the atom, which we recognize today, and the scattering process experienced by the α-particles is known as Rutherford scattering. This same scattering process affects electrons as they travel through matter. It is an elastic process and so electrons do not loose energy in such collisions; they simply experience a large change in direction. Electron–electron scattering events, on the other hand, are generally inelastic and gradually slow the electron down.

The cross section for Rutherford scattering of electrons into solid angle Ω is given by

$$\frac{d\sigma}{d\Omega} = \sigma_0(1 - \cos\theta + 2\mu)$$

with

$$\sigma_0 = \frac{Z^2 e^4}{4E^2}$$

where Z is the atomic number, e the electronic charge, and E the energy per electron (Shimizu and Ze-Jun 1992). The scattering angle is θ and μ is a screening parameter usually taken as a constant. Note that the scattering mean free path is inversely

proportional to the scattering cross section, and so increases with the square of the energy, and decreases with the square of the atomic number. Thus electrons travel in comparatively straight lines through resist, and suffer more scattering in Si (atomic number 14), and more again in GaAs (atomic numbers 31 and 33).

The second of the two important scattering mechanisms is electron–electron scattering, which is described by the continuous slowing down approximation of Bethe (Shimizu and Ze-Jun 1992):

$$\frac{dE}{ds} = -\left(\frac{2\pi e^4 N_A \rho Z}{A}\right)\frac{\ln(\gamma E/J)}{E} \quad (2.1)$$

where N_A is Avogadro's number, ρ is the material density, and A the atomic weight; s is distance traveled in the solid; γ is a constant (=1.1658) and J is the mean excitation energy of the solid, assumed to be a constant for a given material. For our purposes, it is important to note that this equation represents the rate of energy (E) loss per distance, and that this increases approximately linearly with density, and decreases inversely with energy. This is assuming that both the log term and Z/A are constant. What this tells us is that electrons travel further through low-density solids, like polymer resists, than they do through denser materials like Si or GaAs. Also we can see that the rate of energy loss increases as the electron slows down. For example, the rate of energy loss will double if the initial beam energy is halved, say from 100 to 50 kV. This is important because this energy loss from the incoming electron is precisely the energy that exposes the resist, and so we see that higher energy beams need a higher dose to expose the same resist. PMMA requires a dose of about 200 µC/cm² at 50 kV, but 400 µC/cm² at 100 kV, and only 40 µC/cm² at 10 kV.

Let us consider a 100 kV primary electron passing through PMMA, a typical resist. The mean free path between elastic electron scattering events is about 250 nm; by way of comparison, the elastic mean free path is 75 nm in Si, and 21 nm in GaAs. Since 250 nm is considerably greater than the typical resist thicknesses used for high-resolution exposures, many of the electrons will pass through the resist undeflected. Thus there will be little undercut caused by elastic scattering for the 100 kV electrons. For lower energy beams the number of scattering events increases markedly, giving a tapered cross-sectional profile with larger widths near the wafer surface. This is generally speaking a bad thing, but the increasing undercut in positive tone resist is beneficial for a lift-off processes.

For the same 100 kV primary beam striking a 100-nm-thick PMMA layer, there are about 53 inelastic scattering events producing secondary electrons per 1000 primary electrons. About 80% of these secondary electrons have an energy of below 200 eV and a range <5 nm. Some higher-energy secondary electrons are produced and a 1 keV secondary electron has a range of about 45 nm. Despite their small number, secondary electrons are important in the resist exposure process, causing about 80% of the resist exposure in the example above. This is a consequence of Equation 2.1. A proportion of the energy loss results in chain scission in PMMA giving rise to resist exposure. The energy required for a chain scission event in PMMA is 4.9 eV and the chain scission event cross section peaks at a secondary electron energy of about 15 eV.

An important electron scattering effect to consider is the backscattering of electrons. This is actually Rutherford scattering again, but viewed in a different light. Although few of the electrons scatter in passing through the resist, almost all will suffer multiple scattering events in the bulk substrate. Some will scatter more than 90° from the original trajectory and emerge from the substrate back into the vacuum. These so-called backscattered electrons pass through the resist again with lower energy than they did the first time, and over a wider area. The reduced energy is important since it means they will be more effective at exposing the resist than they were on their first passage through the resist. The range of the backscattered electrons increases with energy and decreases with the atomic number of the substrate. At 100 kV, the range is about 31 µm for a silicon substrate and 11 µm for GaAs. By range here is meant the standard deviation of the distribution which is approximately Gaussian.

not too hard to hit. If the pattern was only acceptable for doses between 1000 and 1050 µC cm⁻², then failures are much more likely, if not inevitable. Sadly, as we push the resolution down, dose windows have a habit of shrinking.

This all means that you cannot rely on a single result to establish a process. A little knowledge of statistics will be very useful when trying to figure out the dose window with as few runs as possible. When developing a new process, it is also vital to carry the process through to completion before coming to any conclusions. A simple example is a liftoff process. Do not rely on electron micrographs of the resist to figure out the optimum dose. Not only does the electron beam cause the resist to shrink and move, this inspection will tell you little about whether the resist profile is suitable for liftoff.

Next a word about inspection. There are published papers describing the fabrication of sub-10 nm lines with unconvincing looking images of narrow lines. It is difficult to acquire a good scanning electron micrograph without contamination or other damages to the pattern. But even after acquiring this skill, you still will not really know if your lines are 8, 10, or 15 nm since the interaction between the electrons and the substrate can make a difference of a few nanometers. You should also be aware of the potential and limitations of other measurement techniques. Atomic force microscopy, AFM, is not much good for measuring narrow line widths but excellent for thicknesses. Transmission electron microscopy gives excellent linewidth resolution, but sample preparation is complicated, and each image is of a single cross-section, and so good statistical data are hard to acquire. Electrical linewidth measurements are a bit fiddly but are a good way to collect lots of data rapidly.

Finally you should aim for the lithographer's equivalent of "green" fingers which is a slightly elusive qual-

2.2   Data Preparation

2.2.1   Pattern Design

This subject is described elsewhere in this book. Here, we mention just a few points that particularly affect electron beam lithography. The first is bias. There is almost always a difference between the digital line width in a pattern and the actual developed feature size after processing. This comes about from electron scattering in the resist discussed in Box 2.2, from the finite size of the beam, and from intrinsic resolution limitations in the resist. This effect can be appreciated by considering Figure 2.3, which shows the variation in linewidth obtained with dose when writing lines with 10, 20, 40, and 80 nm digital linewidths in PMMA and then carrying out liftoff. Results are also shown for different beam currents. For a well-characterized process, there is generally a fixed difference between the digital size and the actual size, known as bias. This can be implemented by altering all the sizes in the design as appropriate. However, this is not very convenient, especially if the process and thus the bias changes, and so many design packages enable the user to define a bias, which means that you can design shapes at the correct size, and leave the resizing to later.

Another important concern is the writing of patterns larger than a single writing field of the electron beam tool. How you tackle this depends very much on what sort of tool you have at your disposal. These can range from an SEM with a lithography attachment to a full commercial electron beam lithography tool. See Box 2.3. The former often has a stitching performance or around a micron or so. The latter will have vastly better stitching accuracy, typically of the order of 10–50 nm. You can almost ignore stitching effects when designing patterns for a good tool, but not quite: a 10 nm line will still suffer a distinct kink at a stitch boundary even if it is not altogether broken.

Most vector scan e-beam tools used for nanolithography have two types of shapes available: rectangles and trapezia. All the other shapes available to you in the design package will at some point in the process be converted in to these primative shapes. It

ity. The green-fingered lithographer is meticulous, has keen powers of observation and leaves as little to chance as possible. One works with deliberation rather than haste since one good result in a week is better than three failures.

is important to keep this in mind when designing shapes which will be written at the extreme of the tool resolution. For instance, when writing curves the actual shapes written by the e-beam tool will be a series of small rectangles and trapezia that approximate the original curve. This leads to a large increase in the number of written shapes, and occasionally results in tool-specific errors.

The pixel size used by the tool in writing also has to be considered at this stage. Generally speaking, there are two important parameters here, namely the beam step size and the placement grid. Usually the beam step size is restricted to being an exact multiple of the placement grid size. The placement grid defines the smallest adjustment in beam position which can be achieved in a field. So any shape, for examples, a square, can be placed to this accuracy. A typical example would be for a 16 bit pattern generator and a 200 μm field size. In this case, there are 2^{16} pixels across the 200 μm field, leading to a placement grid of

$$200/65, 536 \sim 0.003\,\mu m = 3\,nm.$$

As a rule of thumb the beam step size should be a little smaller than the beam diameter in order to give smooth edges to shapes (see Figure B.2.4.1 in Box 2.4 later in this chapter). This is not always possible to arrange, for instance when using a fast resist and a field emission gun, which produces a high beam current density. Aim to have all shape sizes an exact multiple of the beam step size. It is easy to find yourself in the position where rounding errors in an array mean that nominally identically sized shapes turn out to be a different number of beam step sizes across. This is especially important when dealing with curves such as arrays of small circles typically used for photonic crystals. The end result is that after writing and development the supposedly identical circles have a small but unwanted variety of shapes and sizes. Sometimes with small circles it is better to hand craft a single circle in

BOX 2.2   THE PROXIMITY EFFECT

This arises from electron scattering effects. Imagine a point beam striking a resist-coated surface and consider the dose received by the surrounding area of resist. Without scattering, all the dose would be concentrated at one point; in practice, it is smeared out as shown in the point spread function shown in Figure B.2.2.1. A real beam of course has finite size, usually considered to be of a Gaussian distribution, and the point spread function needs to be convolved with the primary beam distribution to give the actual electron dose received at adjacent points.

Two models are shown, the upper curve including secondary electrons is a better reflection of reality. This is for a resist thickness of 300 nm, a 100 kV beam and a Si substrate. Note that the axes are logarithmic so that the deposited energy density drops by two orders of magnitude in the first 100 nm, and there are many electrons present up to the backscatter radius of about 50 μm.

The point spread function after convolution with the Gaussian primary beam is often approximated by the double Gaussian approximation

$$f(r) = \frac{1}{\pi(1+\eta)}\left[\frac{1}{\alpha^2}\exp\left(\frac{-r^2}{\alpha^2}\right) + \frac{\eta}{\beta^2}\exp\left(\frac{-r^2}{\beta^2}\right)\right] \quad (2.2)$$

where α is the radius of the primary beam and β the radius of the backscattered approximation.

Sometimes a third term is added to account for fast secondary electrons in the intermediate range, roughly, 10 nm–1 μm. Typical values for these constants are shown in Table B.2.2.1. Note that the term η is the ratio of total integrated electron dose from the primary and backscattered contributions, and is roughly equal to unity for a range of substrates. This indicates that approximately half of the electron exposure comes from long-range backscattered electrons. This is the origin of the proximity effect which describes the fact that adjacent shapes written by the e-beam tool affect each other and that account must be taken of this when preparing data for electron beam exposure. Also note that the value of η varies considerably from author to author. This arises from different measurement techniques and the fact that the double Gaussian approximation is only an approximate fit to the real distribution.

Proximity effect correction software is available which will calculate the actual doses required to compensate for the proximity effect. It is often useful to be able to estimate the dose required for simple shapes without the use of software, and approximate techniques for doing this are described in Box 2.4 "Estimating Dose."

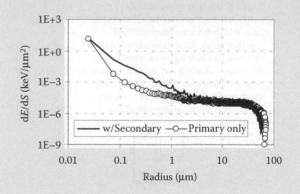

FIGURE B.2.2.1 Point spread function for 100 kV beam. (From Ivin, V. V. et al. 2002. *Microelectronic Engineering* **61–62**: 343–349. With permission.)

Table B.2.2.1   Double Gaussian Proximity Effect Parameters for Various Substrates

Substrate	kV	β (μm)	η (μm)	Reference
Si	20	2.0	0.74	Owen (1990)
Si	50	9.5	0.74	Owen (1990)
Si	100	31.2	0.74	Owen (1990)
Si	50	10.2	0.51	Boere et al. (1990)
C	50	17.5	0.19	Boere et al. (1990)
GaAs	50	3.9	0.88	Boere et al. (1990)
GaAs	100	11.1	0.89	Boere et al. (1990)
InP	50	4.1	0.99	Boere et al. (1990)
Si	50	8.8	0.75	Rishton and Kern (1987)
GaAs	50	3.28	1.07	Rishton and Kern (1987)

Chapter 2

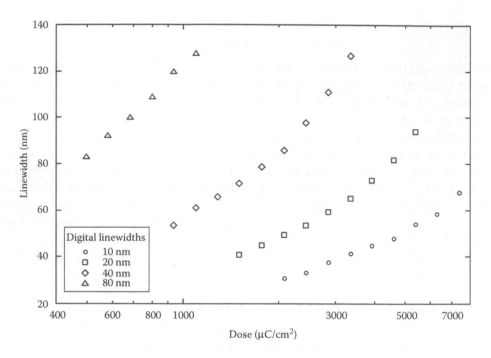

FIGURE 2.3 Linewidth against dose for PMMA thickness of 150 nm; results are shown for 10, 20, 40, and 80 nm digital linewidths.

BOX 2.3 TOOL CONSTRUCTION: SEM OR PURPOSE-BUILT E-BEAM LITHOGRAPHY TOOL

A generic electron beam tool is shown in Figure B.2.3.1; such a tool can either start life as an SEM, or else be designed from the ground up as a lithography tool. A number of companies will modify an SEM to be used as a lithography tool. Such modified microscopes can do some excellent lithography, but there are some differences which mean that purpose-built tool can do a much better job. We consider a few of the more important differences here. You should also bear in mind that there are different levels of modification from simply adding a scanning unit to almost a complete rebuild.

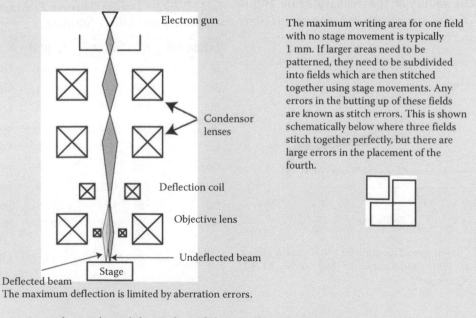

The maximum writing area for one field with no stage movement is typically 1 mm. If larger areas need to be patterned, they need to be subdivided into fields which are then stitched together using stage movements. Any errors in the butting up of these fields are known as stitch errors. This is shown schematically below where three fields stitch together perfectly, but there are large errors in the placement of the fourth.

FIGURE B.2.3.1 A generic electron beam lithography tool showing the main components.

First, consider the scan coils. An SEM takes images relatively slowly: 20 s is a fairly fast image collection time, and so the scan coils do not have to operate quickly. They do have TV rate scanning, but this mode typically comes with a fairly low number of lines, say 512, and some image distortion compared to the slower scan rates. This means that the scan coils can be built with large numbers of turns, giving a high inductance, and making them slow to respond to large positional shifts of the beam. By way of contrast, a purpose-built tool typically has separate main and subfield deflection coils, which enables faster patterning while maintaining positional accuracy. In addition, the scan field of an SEM will have greater distortions than are present in a commercial tool. For instance, an SEM probably has no more than 2000 lines in the image, so does not require precision greater than 1 part in 2000. A purpose-built tool will have precision of a few 10 s of nm across a 1 mm field, or 1 part in 30,000, or better.

The stage on a purpose-built tool will have better specifications for position, speed of motion, and drift, all of which are more important for lithography than for image acquisition. The position is measured by a laser interferometer and the error between the desired position and the measured position is dynamically corrected using beam deflection. Thus, the stage, interferometer, and dynamic beam deflection can be considered as a single system giving positional accuracy of a few nm over the full stage size of typically 150–300 mm.

A purpose-built tool will have a height measurement system which measures the distance of the wafer from the pole piece and enables the focus and field size to be adjusted automatically as the wafer is moved.

An SEM probably wins out in outright probe diameter, with round about 1 nm being available. The downside is that deflection aberrations are not very important to an SEM, so this small probe diameter will only be maintained over a small field size, say <50 μm. A purpose-built tool, on the other, may be limited to a probe diameter of 3 nm, but it can maintain this over a field of several hundred microns and automatically adjust the focus as the wafer moves, which the SEM cannot. If you want to write the single smallest structure in the world, then an SEM is ideal; if you want to write it many times over across a sample (say a 100 μm square array), then the advantage swings to the lithography tool.

the design package rather than drawing a circle and relying on shape conversion to turn it into a circle on the e-beam tool. This is often the case when designing small arrays of 100 nm or so sized circles for photonic crystal applications.

One exception to the general rule of pixel and beam diameter sizing arises when writing arrays of dots. Usually the fastest way to write an array of dots is to set the beam step size to equal the dot period and use a beam diameter a little smaller than the desired dot size. Then simply write a large rectangle which will end up as an array of dots if you choose the dose correctly. This saves the shape overheads associated with each dot, which for an array of 10^9 dots can amount to several hours and dominate the writing time when the pattern is designed as an array of small squares, or, even worse, circles.

2.2.2 Job Construction

Pattern design is important, but is only part of the overall design process. The e-beam tool will have a number of parameters that will need to be set to the correct values if your jobs are to succeed. There are different aspects to a writing job, and depending on what you wish to achieve, different ones will be to the fore. In particular, the following may be of interest to you:

1. Achieving a certain feature size
2. Achieving a certain periodicity
3. Accurate pattern placement relative to a previous lithographic layer (alignment)
4. Large area pattern integrity, for example, a uniform grating over many mm (stitching)
5. Accurate placement of two parts of the pattern written with different beam sizes (overlay)
6. Writing a prescribed area of the pattern for a planned experiment

Let us consider various tool parameters and how they bear upon these issues.

A suitably small beam size is clearly essential for writing nanoscale patterns. If smallness were the only thing that mattered, the choice would be easy: simply select the smallest spot available with your tool. But life is never so straightforward and there are a number of reasons why this might not be a good idea, the main one being the resulting job time. If you need a large area of pattern, you may find that it is just not possible to use the smallest beam and you have to consider using something larger with an increased beam current. Another issue is that often small beams are provided on SEMs that provide a marginal improvement in resolution for a considerable expenditure of effort. For instance, with 10 pA rather than 20 pA, you may achieve a 1.0 nm beam rather than a 1.2 nm beam. However, in practice, you may well achieve a worse diameter, say 1.5 nm if you do not have a suitable substrate for focusing a small beam. Moreover, your resist process may be limited to 10 nm, and you might achieve equally good lithography with a 100 pA beam and a 2.5 nm spot.

A useful rule of thumb is to initially choose a beam diameter one-fifth of your minimum feature size assuming square features, and about one-third of the minimum feature size for straight lines, where you need not sharp corners. So if you need 20 nm lines you might initially consider using a 7 nm beam as this would make the processing comparatively easy. If this is too slow, you can trade ease of processing for writing time and increase the beam diameter almost up to the line width, say up to 15 nm in this case, and take the hit in significantly harder process control.

The placement grid size, beam step size, and dose are closely linked parameters. The first two were discussed in some detail under the design heading. You will need to set them when using the tool, and it is probably here that you will make your final check that the writing frequency is within range. The writing frequency is determined by the beam step size, dose, and beam current according to the equation

$$f = \frac{1}{\tau} = 10^5 \frac{i_b}{\text{dose} \times (\text{bss})^2} \qquad (2.3)$$

where f is the frequency in MHz, τ is the dwell time in μs, i_b is the beam current in nA, the dose is in μC/cm^2 and bss is the beam step size in nm. If the frequency is beyond the capability of your hardware, you will need to reduce it by either decreasing the beam current or increasing the beam step size. Changing the dose is not usually an option unless it is a very minor change. Either of the two main options will reduce the beam

size relative to the beam step size which might move the pattern to the domain where the beam is much smaller than the beam step size and the edges become noticeably ragged. On some systems, it is possible to reduce this effect by deliberately introducing a small amount of defocus. Most tools allow you to set the dose and then automatically adjust the frequency depending on the beam current and beam step size.

The field size is often closely linked to the placement grid size. This is an important parameter that affects both stitching performance and writing speed. Generally speaking, if your whole pattern will fit into one field, that is a good option because there will be no stitching. Of course, pattern placement will still not be perfect because electronic beam deflection always has some errors, leading to in-field distortions. Large fields are good because they reduce the number of stage movements required and this will increase the writing speed. It will also reduce the number of stitch boundaries. Conversely, large fields have the drawback that beam aberrations increase away from the axis so that the beam diameter may begin to grow significantly toward the edge of the field, adversely affecting your patterning. In addition when the field size is large the stitch errors will tend to be larger than for smaller field sizes. The maximum useable field size varies considerably between tools, and so you need to be aware of the specifications of your instrument.

Some more subtle parameters that can be set at the job stage are various delays that affect the writing speed and accuracy. For instance, when loading a holder onto the stage, there will inevitably be a period of relatively rapid thermal drift until thermal equilibrium has been reached. This can take up to 6 h or longer. Of course if the system does not have good thermal management, for instance, if its environment changes temperature by a degree or so over a period of a day, then there will be continuous drift due to thermal effects. To some extent the effect of thermal drift can be compensated for by making regular position adjustments by measuring a fixed point on the holder (or wafer) at regular intervals, although such adjustments themselves always lead to small additional placement errors. Other delays that may be settable are the delays between stage movement and the start of writing, and also main field settling intervals. In general, all these involve a compromise between speed of writing and the positional accuracy of the lithography.

At this stage, you also need to decide how many times the pattern will be written across the wafer. Sometimes, a large pattern is simply written once in the center of the wafer. Other times, the pattern is smaller and can be repeated across the wafer at regular intervals—often known as step and repeat. One important variant of this is to vary the dose at every step. This is sometimes called a dose wedge, and after processing can be used to determine which is the best dose for the given pattern and process conditions. Most people are tempted to use a linear dose scale, going say from 100 to 500 μC cm^{-2} in steps of 20. Resist the temptation! A geometric progression is much more useful, and steps around 10% are typical. In the above example, this would mean a dose step of 10 at the bottom end, and almost 50 at the top

2.3   Writing the Job

Once the data are prepared and the resist layer is deposited on the substrate, the writing process can begin. The resist deposition process is described in a later section.

2.3.1   Calibration

The first step is to calibrate the e-beam tool. Using a modified SEM, this needs to be done manually, whereas with a purpose-built tool, this may be accomplished by issuing a single command. But it is still important to understand what is being carried out.

Gun alignment. A correctly aligned gun will maximize both beam current and beam current stability. For some stable field emission, systems, the gun alignment only has to be carried out on an occasional basis, for example, once a month, but typically this is done for every job.

Focus. The beam must be in focus on the resist-coated surface if good lithography is to be obtained. With an electron beam tool the focus control includes the stigmators, which means effectively that there are three knobs to adjust, not just the one. This can be either manual or automatic, but in either case the focusing requires a high-contrast object, or mark, to focus on which cannot be immediately adjacent to the pattern region. This is because focusing requires the beam turned on which exposes the local area. There needs to be some way to make sure that the beam remains in focus when the stage is moved from the focus mark to the pattern region, and this varies from tool to tool. A purpose-built tool will have a method of measuring the wafer height and automatically adjusts the focus as appropriate. A modified SEM may provide a method to focus on the four corners of the area to be patterned,

end. The whole range of doses will be covered with 18 patterns using the geometric progression, as opposed to 21 with the linear progression. What is more, the geometric progression provides a constant 10% resolution, whereas the linear progression gives a very coarse 20% resolution at the bottom end, and an almost unusably fine 4% resolution at the top. If your dose lattitude is as small as 4%, then your process will be very hard to maintain—see the section on PMMA processing for an example. Box 2.4 gives more details on how to determine the correct dose for a given pattern.

and then use linear interpolation to adjust the focus for points in between.

Pull-in. Mechanical stage movement is typically limited in resolution to a few microns, whereas the absolute positioning required in a lithography tool is a few nm. The stage positioning system is usually supplemented by an electronic deflection system that works in tandem with the laser interferometer which measures the stage position. This electronic deflection is called by different names in different tools, but we will call it pull-in. For system purposes, the combined mechanical/electrical positioning system is regarded as a single entity used to move the stage position. The point of this calibration is to make sure that the pull-in deflection is exactly correct, in other words that 1000 μm of electronic shift is equivalent to 1000 μm of stage shift. This option will only be available on higher-end tools.

Field size adjustment. This also determines the resolution and beam step size. For a tool with good stitching performance, this needs to be done very accurately so that the writing field size (electronic deflection) is precisely calibrated to the step size when moving from field to field (mechanical motion). As with focus, this needs to be adjusted for the height difference between the calibration mark and the pattern region.

Beam current. This requires a Faraday cup which is simply a hole such that the primary beam enters the hole, but very few backscattered or secondary electrons are able to escape it. This is often a small aperture (~50 μm) placed over a blind hole machined into a holder. Box 2.5 gives a schematic diagram of a typical Faraday cup.

Chapter 2

BOX 2.4   DOSE ESTIMATION

When writing a large area, the dose is simply the charge per unit area, which by rearranging Equation 2.3 can be expressed as

$$\text{dose} = 10^5 \frac{\tau i_b}{bss^2} = \frac{\text{Charge per pixel}}{\text{Pixel area}}$$

To first order the dose required to expose resist is independent of the beam step size. All that is required when varying the beam step size, for a fixed current, is that the dwell time be adjusted accordingly.

When writing a narrow line the situation is different. Assume for the moment that the line is exposed by a single pass as in (b) or (c) of Figure B.2.4.1. To expose the resist with the correct dose always requires the same deposited charge, and this clearly requires the writing frequency to scale inversely with beam step size. But this is quite different to Equation 2.3 where the frequency varies inversely to the square of the beam step size. Thus it is appropriate to define a separate *line dose* for exposing lines.

You can also understand the difference by noting that the area enclosed by the pixels in (b) is exactly one half of the area enclosed by the pixels in (c), and yet in both cases the actual area written is the same, namely one beam width multiplied

by the length of the line. The dose referred to in Equation 2.3 is sometimes called the area dose.

The line dose is defined as the charge per unit length of line, usually its units are nC/cm and with this definition the dose required to write a line does not vary with beam step size. Some e-beam tools allow the line dose to be set directly for single pass lines, and the concept is useful for quoting the dose required for writing narrow lines since the dose does not then depend on the beam step size. In a similar way, it is often useful to express the dose required for writing a single point or dot as simply the amount of charge per dot, measured in nC. Often the only dose which can be set for an e-beam tool is the area dose, and you must remember to vary it with beam step size for single pass lines.

Often it is useful to estimate the dose required for a given pattern even though a proximity correction program can calculate this for you. If nothing else it gives a feel for what is going on. The easiest starting point is the large area clearing dose which we shall call D_0. This is for an area much larger than the range of the backscattered electrons; for instance the center of a 200 μm square. For various reasons—for instance beam current drift—you should have a minimum dose of a little bigger

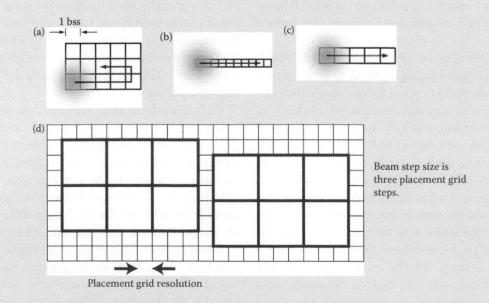

FIGURE B.2.4.1   A Gaussian beam being scanned (a) throughout a rectangle and (b, c) along a line. The beam step size in (b) is one half of the beam step size in (a) and (c). Finally, (d) shows the difference between beam step size and the placement grid.

than D_0, say $1.1D_0$. The resist at this point receives roughly speaking a dose of twice what you give it, since for typical substrates the total backscattered dose is roughly equal to the forward dose. (Computer programs of course use exact numbers, but we will not bother here.) So the actual dose the resist receives is $2.2D_0$, though you only specify a value of $1.1D_0$.

Next consider what happens at a corner of the large square. The resist at this point only receives backscattered electrons from one quarter of all possible directions. So the actual dose is now only $1.1D_0 + 1.1D_0/4 \sim 1.4D_0$. This means you need to give the corners a bigger dose by a factor of $2.2/1.4 \sim 1.6$, and so the design dose given to the corner regions becomes $D = 1.7D_0$.

Let us now consider an isolated 1-µm-wide line. This is fairly wide, but well under the backscattered electron range. So to a first approximation none of the backscattered electrons end up in the line, and so the design dose has to be $D = 2.2D_0$. If we wish to shrink the line much further, we need to begin to think about bias. So for instance a 100 nm linewidth may be written as a digital line of 60 nm. This means that all the electrons required to expose the 100-nm-wide line must be supplied in the 60-nm-wide line, and so the dose is multiplied by a further 100/60 and becomes $3.7D_0$. This increase in dose becomes more pronounced as the widths decrease; and so for instance a 20-nm-wide line may be written using a 5 nm digital linewidth which would require a dose of round about $8.8D_0$.

Other situations can be considered by the same kind of reasoning. So, for instance, consider a 1:1 grating of 100 nm lines on a 200 nm period. If the grating covers a large area, the 50% fill factor means that each point receives 50% of the backscattered dose. Taking the bias into account, we find that the required dose D is given by $D = (1 + 0.5)1.1 \times 100/60 \approx 2.8D_0$.

Frequency. Once the beam current is known, the scanning frequency can be calculated from the required dose and the beam step size.

2.3.2 Exposure

After calibration we can finally expose the wafer with our pattern. This stage consists first of a move to the correct position followed by a determination of the height of the wafer at this point. Next the focus and field sizes are adjusted as appropriate and finally the tool can write the pattern. One thing to remember is that mechanical movements are slow compared to electronic deflections. This means that your control program should aim to minimize stage movements. So, when writing a pattern, use a zig-zag-style stage movement rather than a raster-scan approach since the latter has a long and unnecessary fly-back after each line. The zig-zag style is sometimes called boustrophedonically, and is exactly as an inkjet printer prints a page of text.

One way to reduce stitch errors is to write the pattern multiple times, but each time with the stitch boundaries at different positions in the file. For instance, if a 200 µm field is being used, stage movements of 100 µm could be used which would give stitch boundaries every 100 µm rather than every 200 µm. At the same time, the pattern is exposed four times, and so you need to give the pattern one-quarter the dose which of course requires four times the writing frequency. The stitch boundary positions will also be in the middle of the field three out of four times, where there is no stitching, and hence the stitch errors will be less pronounced. This of course needs a higher-end system with a laser interferometer on the stage.

Intermediate calibrations are needed for long exposures. Both focus and field size will drift over periods of 15–60 min and therefore need to be recalibrated from time to time. You should also expect some holder drift, even in a temperature-stabilized system. Expansion coefficients for metals are about $10^{-5}/°C$, so, for 1°C temperature change and 0.1 m holder size, the expansion is about 1 µm. So, for a fairly small 0.25°C temperature shift from load lock to stage, there could be 250 nm of motion over several hours, and quite likely more. Motions of 100 nm/h are quite common immediately after a holder exchange, and if this is too much, then some delay time is necessary. One way to correct for this is by using a drift calibration. This simply means checking the position against fixed reference point at regular intervals and making the necessary corrections.

BOX 2.5 TOOL CONSTRUCTION: BEAM CURRENT MEASUREMENT AND DEPTH OF FOCUS

Beam currents are measured using a Faraday cup as illustrated in Figure B.2.5.1. In essence this is very simple, and can be made simply by milling a blind hole and placing a small aperture, say 20 μm, over the top. It works by ensuring that all the outgoing electrons, both secondary and backscattered, are trapped within the cup. In contrast, if the beam is placed outside the cup, the current measurement is the difference between the primary beam current and the current carried away by backscattered and secondary electrons. This will generally underestimate the beam current and therefore lead to overexposure.

One feature of an SEM as compared to an optical microscope is the large depth of focus, and this also applies to an electron beam lithography tool. For a field emission gun and a 60 μm final aperture size 30 mm away (typical numbers), the depth of focus can simply be estimated from the geometry by a 60 μm increase over 30 mm, or 2 nm over 1 μm. For a 3 nm probe diameter this yields a depth of focus of round about 2 μm. This means that it is possible to write high-resolution features over topography, especially if the height differences present are known and compensated for. Figure B.2.5.2 shows a cross-shaped aperture written on top of a 5 μm high pyramid using e-beam (Zhang et al. 2010).

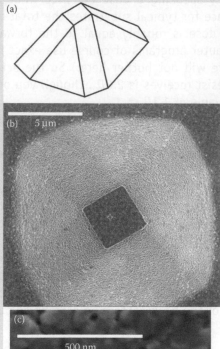

FIGURE B.2.5.2 A small cross-shaped aperture written at the apex of a pyramid to form a Scanning Near Field Microscopy probe. The pyramid was coated with Al and the cross was etched through the metal. (a) Schematic view of pyramid. The top is about 10 μm above the base. (b) SEM micrograph looking down from above showing the whole pyramid. (c) Detail showing the cross written by electron beam lithography. (Courtesy of Y. Zhang and J.M.R. Weaver.)

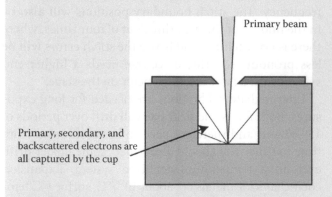

FIGURE B.2.5.1   A Faraday cup, used to measure beam current.

2.3.3 Estimating the Job Time

To make the best use of time, the ideal e-beam job would spend most of its time actually writing the pattern. This is called beam time, which, as well as being the dominant contributor to total writing time, should also be as short as possible. In practice, there are a number of significant overheads that contribute to the total job time, and which in certain circumstances dominate the writing time. The details depend on the e-beam tool, but will include calibration time, stage movement time, shape time, and alignment time. Let us consider each of these times in some detail.

Beam time is the time it takes to expose a given area of pattern at a set dose with a given beam current. There are two different ways of looking at this. In the first, we simply consider the area to be written, the average dose, and the beam current. Then simply by considering units we see that

$$\text{dose} = \frac{i\,t}{A}, \qquad (2.4)$$

where i is the beam current, t is the exposure time, and A is the area. Since dose is usually expressed in $\mu C\,cm^{-2}$, and beam current in nA, we can rearrange this to give

$$t = 1000\,\frac{\text{dose}(\mu C\,cm^{-2})\,A(cm^2)}{i(nA)}. \qquad (2.5)$$

So, if we wish to write a 1 cm^2 array of 30 nm dots on a 60 nm pitch using PMMA as the resist and a 2 nA beam current we can reason as follows. The clearing dose for PMMA is (depends on the exact conditions) 300 $\mu C\,cm^{-2}$. We have a fill factor of 0.25, since one-quarter of the area is covered in dots and the other three-quarters are empty; this gives a writing area of 0.25 cm^2. The writing time is then

$$t = 1000 \times 300 \times 0.25/2 \approx 40{,}000\ \text{s}$$

This is over 11 h, and so it is a long writing time, even neglecting overheads. This shows incidentally why PMMA is considered a slow resist, and why faster resists that require lower doses are desirable. If we were to double the size of the dots, but keep the mark–space ratio the same, then all the numbers in the above equation remain the same apart from the beam current. For 60 nm dots, we can to a first approximiation use double the beam diameter which means four times the beam current (see Box 2.6). In this case, the writing time reduces to about 3 h which becomes much more manageable, though it can be greatly decreased by using a faster resist.

The other way to look at the same issue is to evaluate the writing frequency, f, and count the number of pixels N which are to be exposed. The beam time is then simply given by

$$t = N/f.$$

The frequency can be calculated using Equation 2.3 and the number of pixels can be calculated from a knowledge of the area to be patterned and the beam step size. Often this is somewhat more complicated than the first method, but it does have the advantage that the intermediate step gives the writing frequency. This is important because your e-beam tool will have a range of possible writing frequencies, and if the calculated value lies outside this range, then you have to change some of your assumptions. Moreover, once you are familiar with a particular resist you will have a good idea of the writing frequencies you use for typical applications, and so you can simply plug in a value of f without having to calculate it.

In the above example, you may wish to write the dots as small octagons, with a combined area of 0.25 cm^2. With a beam diameter of 6 nm, a beam step size of 5 nm and a beam current of 2 nA, and using the same clearing dose, we find that

$$f = 10^5\,\frac{2}{300 \times 5^2} \approx 27\ \text{MHz}.$$

Depending on your tool, this may or may not be possible. If not, the simplest remedy is to increase the beam step size. Assuming it is, we can next figure out the area in pixels which is given by the area divided by the square of the beam step size:

$$N = \frac{0.25\ cm^2}{(5 \times 10^{-7})^2} = 10^{12}.$$

And so, finally, the writing time is given by $t = N/f = 10^{12}/(26 \times 10^6) \approx 40{,}000$ s, which is the same as before of course.

Let us next consider shape time. This is the time overhead for writing a single shape and is typically of the order of 5–50 μs. This time includes the shape loading time from disk to pattern generator, the delay times built into the hardware to account for main- and subfield settling times, blanking time, and possibly other contributions too, depending on the particular hardware available. The settling times come about because the beam cannot be moved instantaneously from one position to another. The deflection mechanism used is based on magnetic deflectors that involve an electromagnet with many turns and a high inductance. Hence there must always be a settling time, otherwise the beginning of each shape will be distorted.

So, in the example above, we have a large number of dots, each being written as an octagon. Pattern generators cannot typically write an octagon as a single shape and the dots must therefore be written as two trapezoids each. So the total number of shapes, N_s, is given by $N = 2\,(10^7/60)^2 \approx 5 \times 10^{10}$. Even if we assume a blisteringly fast 1 μs shape delay time, this corresponds to

BOX 2.6 TOOL CONSTRUCTION: BEAM DIAMETER AND CURRENT

A fundamental principal of charged particle optics is the conservation of brightness (Grivet 1972). In terms of an electron beam lithography tool, this means that for a constant aperture size the current density of the final beam is approximately constant as the beam current is varied by adjusting the condenser lenses. For large beams and sources such as LaB$_6$ or tungsten hairpin, the relationship holds very closely and implies that the beam current varies as the square of the beam diameter. So if you double the beam current, the beam size will increase by a factor of 1.4. For smaller beams, the beam diameter becomes dominated by aberrations, either spherical or chromatic, and the decrease of beam diameter with current begins to tail off. For field emission sources where the brightness is much greater, aberrations become dominant at larger beam currents. Figure B.2.6.1 shows a typical relationship between beam diameter and current for a thermal field emission gun, a common source in modern e-beam tools. Decreasing the final aperture will reduce the effect of aberrations, in particular spherical aberration, but the current decreases as the square of the aperture diameter, which limits the minimum aperture size which is practical. In addition, for very small apertures electron diffraction effects become important. The beam diameter d_p is given by

$$d_p = \sqrt{d_g^2 + d_c^2 + d_s^2 + d_d^2}$$

where d_g is the Gaussian probe size given by demagnification of the virtual source size by the electron optics, which can be expressed as

$$d_q = \left(\frac{J}{I}\right)^{1/2} \frac{1}{\pi\alpha}.$$

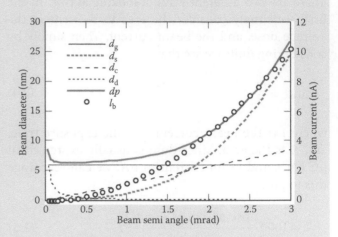

FIGURE B.2.6.1 Calculated beam diameter vs beam semi-angle for a typical lithography tool with a field emission gun. Various contributions to the final probe diameter are shown, which are finally added in quadrature. The electron probe current is also shown.

The spherical aberration, d_s, and the chromatic aberration, d_c, are given by

$$d_s = \frac{1}{2}C_s\alpha^3; \quad d_c = C_c\alpha\frac{\Delta E}{E}$$

and the blur due to diffraction, d_d, is given by

$$d_d = \left(\frac{0.6\,\lambda}{\alpha}\right)$$

J is the probe current, I the source brightness, α the beam semi-angle at the focus point, C_s the spherical aberration coefficient, C_c the chromatic aberration coefficient, E the energy spread of the source, E the beam energy, and λ the electron wavelength.

another 50,000 s, and probably about 10 times this. As a result, the shape time dominates the beam time, and indeed is the major contributor to writing time. This is why dots are generally written as a single large rectangle with a large beam step size. In this way, the shape time is eliminated and beam time again becomes the dominant term.

The next item to consider is stage time. This is the time taken for the stage to move between writing fields, and usually includes a stage settling time. This number will vary considerably from tool to tool, but is typically of the order 0.1–1.0 s. This is a big time, but fortunately there are not usually too many stage movements. In the example above the writing field may be 0.5 mm, in which case there are 20 × 20 fields, and so 400 stage movements. This makes a total time of something like 200 s. Generally speaking, stage time is not a major issue unless you have a sparse pattern covering a large area which requires movement to every field. On a 4″ wafer for instance, with the 0.5 mm field size mentioned above, there are something like 24,000 stage movements, which even for a fast stage represents something like an hour's overhead in the job time.

For alignment jobs (see Section 2.3.4), there is also a time overhead for finding markers. Ideally, you should arrange the job so that marker searches—typically four per cell—take place every 15 min or so. Allowing 5 s for each stage movement and marker search, the overhead here is 20 s in 900, or about 2%, which is not a major issue. A badly posed job can have marker searches much more often if the cell size is small, and it is possible to have the writing time dominated by marker search time. This is a BAD THING, and generally something to avoid when planning a job. When planning a job, a good target is for most of the elapsed time to be beam time which is the time which really counts.

Finally, calibration and setup time should be considered. This will typically be something like 2–5 min, and should be repeated every half hour or so to keep the tool finely tuned throughout the writing period. The exact times here are of course very much tool dependent, and in particular you should make sure you know how frequently recalibrations should take place. This is clearly less often for a more stable tool.

2.3.4 Alignment

Almost all practical devices need several lithography and pattern transfer steps. For instance, consider a transistor which at the very least needs two types of material, one for the gate and another for the source and drain contacts. These two materials must be arranged on the substrate with their own separate patterns, and this requires two lithography and pattern transfer steps. These different lithography steps are often known as layers. Figure 2.4 shows a schematic view of these two layers and how they need to be placed on the substrate. It also shows different ways in which the alignment can go wrong. A practical transistor circuit will need many other layers besides, for instance, to provide electrical isolation between transistors, to make resistors and capacitors, and to wire circuit elements together. These general principles apply to other types of devices such as photonic structures and bioelectronics. And so the ability to align one layer to another is an essential part of the lithographer's toolkit.

A number of different errors that may occur when aligning one layer to another are shown in Figure 2.4. A translation error means that the pattern is simply shifted from where it is supposed to be. A rotation error

means that only one point on your pattern can be in the correct location, and the placement error in x and y increases linearly with distance away from this point. To understand how important this can be, consider a rotational error of only 0.1°, which is a typical rotational accuracy attainable by mechanical rotation of the substrate. At a distance of 1 mm from the center of rotation, the positional error is 1000 sin 0.1 ≈ 1.7 μm, which is a fatal error for most nanofabrication. Gain—or scale—errors come from various sources but can, for example, be temperature related. A temperature change of 1°C for a Si wafer means an expansion of 2.6 ppm or 2.6 nm per mm. This is not very much over a small distance, but for some patterns that extend over 10s of millimeters the 10s of nanometers of error can become significant. Larger temperature shifts can easily occur if different layers are written using different lithography tools; moreover, many materials, such as GaAs (Ignatova et al. 2010), have higher linear expansion coefficients. Another source of a gain error is the issue of writing different layers on different tools, for instance, lower resolution using photolithography and higher

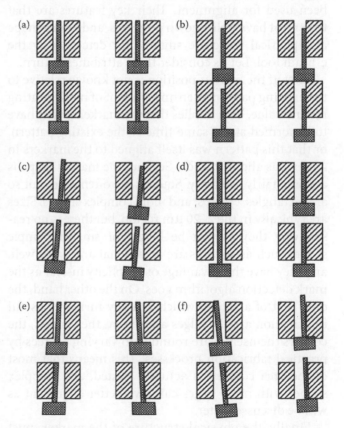

FIGURE 2.4 Schematic showing alignment of source/drain and gate layers for a transistor. (a) Shows correct alignment, (b) shows a translation error, (c) shows a rotation error, (d) shows a gain error, (e) shows keystone error, and (f) shows all types of error combined.

resolution by e-beam. In this example, the micron on the mask plate used for the photolithography may not be calibrated to the micron on your e-beam tool; differences of 10s of ppm are quite common.

Keystone errors arise from variations in x gain in the y direction and vice versa. This sort of error is typical in any tool which has to position accurately in two dimensions and is often seen, for instance, in video projectors. The result is that a large rectangle is actually somewhat trapezoidal, just the shape the keystone at the top of an archway in architecture. Normally the effect is small and not too important, but with alignment this is not the case. A change in width of 50 nm between the top and bottom of a 500 μm field is hardly visible normally, but would lead to significant alignment errors on the nanometer scale if uncorrected.

In order to correct for all these errors, the e-beam tool has to detect the shape and position of the existing pattern and adapt its own field shape and position to match. It does this typically by finding alignment markers that are usually placed at the corners of a rectangle surrounding the pattern. Different markers have been used for alignment. Their key features are that they must have well-known positions, and be of a shape and physical structure suitable for detection by the e-beam tool. Let us consider these attributes in turn.

First, if the marker position is not known relative to the existing pattern, then any amount of mark locating is of no value. This implies that the markers either have to be formed at the same time as the existing pattern, or that this pattern was itself aligned to the markers in a previous alignment job. Second, the marker shape is actually fairly arbitrary. Squares are often used, but so are rectangles, crosses, and more complex shapes. Sizes are typically in the 2–20 μm region, but there is no reason why they cannot be bigger or smaller. Simple shapes such as squares are traditional and work well, and they have the advantage of simplicity insofar as the mark detection algorithm goes. On the other hand, the only part of a square which actually furnishes useful information are the edges away from the corners; the corners themselves are rounded to varying degrees by practical fabrication processes. This means that most of the area covered is actually wasted, and complex shapes with more edges can give better alignment as will be discussed later.

Finally, the physical structure of the marker must be suitable for detection by the e-beam tool. Generally speaking, the detection is carried out using a backscattered detector since the structure is covered by resist and so secondary electrons tell you little about the underlying structure of the marker. Some secondary electron detectors can pick up a significant fraction of backscattered electrons and be used for mark detection. Typically, the markers are made of a metal with a high atomic number, such as gold or tungsten. This is because the backscattered yield increases rapidly with atomic number. Aluminum, for instance, has almost the same atomic number as silicon, and so provides no contrast under backscattered electrons and is entirely useless as a marker. The thickness of the markers is usually in the 50–100 nm range. Another sort of marker is an etched structure. A deep pit will appear dark under backscattered illumination, but as the depth decreases the center of the pit becomes brighter and only the edges provide good contrast. Either sort can work, depending on the detection algorithm available. To obtain a uniform contrast within the pit, an aspect ratio of about 0.3 is required (e.g., 3 μm deep for a 10 μm wide-marker). This is not always easy to achieve. In an analogous way, the marker can be a raised mesa. Finally, sometimes markers are formed by crystallographic KOH etching (or the like) of silicon, yielding edges sloped along the (111) plane. Box 2.7 gives some further details about detectors in an e-beam tool and some of their other uses.

You should be aware that the very process of locating markers exposes the resist around the marker. This is inevitable since the location is carried out by the electron beam itself. It is possible that by using an image correlation technique to detect the marker, the exposure dose experienced by the marker is sufficiently low to prevent exposure of the resist. But this is very much the exception rather than the rule. In general, you should expect your marker regions to be subjected to whatever pattern transfer process the wafer will experience, and therefore quite likely be destroyed. Therefore, it is useful to have multiple markers at each corner of the pattern so that if one set is destroyed, backups are available. The exception of course is when you use negative resist. In this case, exposure around the markers protects them from subsequent patterning, and you may well wish to deliberately expose a region around the markers to make sure they are preserved for future use.

The mathematics of the alignment process is fairly straightforward, and usually handled automatically by system software. Each marker search and locate operation yields two numbers: an x and y position. Including keystone, there are eight corrections to be made, namely x and y offset, rotation x and y, gain x and

BOX 2.7 TOOL CONSTRUCTION: DETECTORS AND THEIR USES

All e-beam tools are equipped with an electron detector of some kind, either secondary electron or backscattered, or possibly both. These have two main uses as far as e-beam lithography is concerned. First, they enable the column to be aligned and the beam to be focused; second they can be used to detect registration markers on the wafer being exposed and thus align one pattern to another. For conventional e-beam tools with beam energy 10 kV or above, a backscattered detector is generally more useful for picking up markers which are buried in resist.

Beam set up is usually fully automatic for purpose-built e-beam lithography tools, whereas for modified SEMs these set ups will generally be done manually before commencing the exposure. For automated systems a calibration mark giving high contrast is provided, often a square or a cross, which the tool will use for its calibrations.

The on-wafer registration markers can come in various forms, both in terms of shape and in terms of structure. The backscattered signal increases with atomic number, and so high atomic number metals such as gold or tungsten are commonly used as markers. Typically, a 50–100-nm-thick layer of these metals makes an easy-to-detect marker. Another approach is to use an etched pit or raised area on the substrate. The backscattered signal from a pit is less than that from the surrounding area.

y, and keystone x and y. Rotation errors are generally analyzed and corrected separately for the two axes x and y because there is no such thing as perfectly orthogonal axes in real life. This applies to both the e-beam tool writing the new pattern, and whatever tool was used to fabricate the old pattern. Therefore, it is best to consider the x- and y-directions separately and correct by adding a little x to y, and vice versa. The same applies to gain corrections. Thus there are eight degrees of freedom to correct, which requires four markers. A simple matrix operation will produce the correct transformations from the postions of the four located markers. Note carefully that the markers are best placed at the corners of a rectangle, in other words, around the pattern. If the markers are arranged in a cross arrangement, say at (1, 0), (–1, 0), (0, 1), and (0, –1), then the required matrix becomes singular and a solution cannot be found. This is because such a distribution of markers produces no information on keystone correction.

It is possible to obtain alignment with fewer markers. For instance, by omitting keystone correction only three markers need to be found. If it is assumed that the x and y rotations are the same, and that the x and y gains are also the same, then only two markers will suffice. When only one marker can be found, the situation is essentially hopeless since rotation cannot be corrected for.

Figure 2.5 shows some common types of markers along with marker defects which are often encountered and about which you should be aware. It is important to realize that the accuracy of the alignment achieved is limited by the accuracy of the marker search, and so it is vital to make high-quality markers. With a sharp, defect-free edge, marker locate accuracies using square shapes can be as good as 10 nm leading to alignment errors below 15 nm. Better alignment can be achieved by using correlation techniques with more complex shapes. For instance, Docherty et al. (2008, 2009) show alignment of 0.63 nm using Penrose tiles as markers. The advantages of this kind of marker pattern consist in (a) better mark locate because of the large number of edges; (b) this kind of mark is tolerant of fabrication defects unlike simple rectangular shapes; and (c) the dose required to obtain the image can be kept sufficiently low to prevent resist exposure during the marker locate.

When planning an alignment job, the main causes of errors need to be kept in mind. Alignment errors come from a number of sources, chief among which are the accuracy with which the marks can be located, the accuracy of the correction which can be applied in terms of how well it fits the original pattern distortions, and drift during writing. We have considered the first carefully. The second depends very much on the exposure tool in question—or tools if the first layer was fabricated using a different lithography tool. When aiming for sub-20 nm alignment accuracy, the stability with time of pattern placement across a field must be better than this, but also absolute positional accuracies can become important—especially if two different tools are being used. Finally, positional drift should not be neglected, especially if the exposure takes a long time. Typically, positional drift may be of the order of 100 nm over an hour, though it can be better or worse. So, for

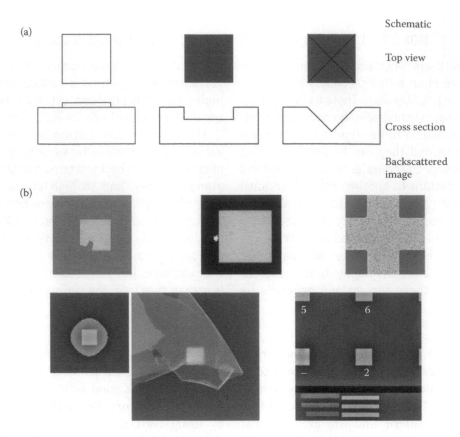

FIGURE 2.5 Different types of alignment markers and some typical marker defects which can arise. (a) Types of markers and (b) common marker defects.

an exposure period of an hour or more the alignment errors can easily be dominated by drift. The easiest way to prevent this is to ensure that the alignment procedure is repeated at regular intervals, say every 15 min.

One option that can be useful is to make sure that the marker positions themselves are as accurate as possible by writing them as quickly as possible, so as to prevent drift.

2.4 Processing Issues

2.4.1 Spinning and Developing PMMA, a Common Resist

Here, we concentrate on PMMA because it is a very common resist, easy to use, and has very high resolution. PMMA is therefore a good choice of resist for nanofabrication. It is the highest resolution available positive tone resist with a number of groups claiming to have achieved sub-10 nm lines using it. It is available with a wide range of molecular weights, from around 50,000 to 1,000,000, of which more later.

Whatever resist you use, it is important to choose the correct thickness for carrying out lithography. A useful rule of thumb is to aim for an aspect ratio of 3:1, so for 10 nm features the resist thickness should be about 30 nm. Thicker resist is almost always a good thing after writing the pattern since it enables easier pattern

transfer: in particular it lasts longer in a dry etch process. It is also easier to see in a microscope. Thinner resist has the benefit of being easier to pattern at high resolution. For features larger than about 30 nm and a high beam energy this is partly because there is more electon scattering in the thicker resist, but by the time the resist is below 100 nm in thickness the majority of 100 kV electrons pass through without any interactions and so scattering is less important. Of course for lower beam energies, there is significant scattering through a 100 nm layer of resist. The other reason to prefer thin resist is the question of resist stability. A tall resist structure is more likely to loose adhesion to the surface and collapse; for high-density patterns the capillary forces during drying can be strong enough to completely ruin the pattern as shown in Figure 2.6. So the 3:1 aspect ratio is not a fixed rule: for isolated trenches a larger aspect ratio is perfectly possible, whereas for dense gratings a lower aspect ratio may be required.

For this reason, you must be able to control the thickness of your resist to suit your patterning requirements. For this it is useful to know the formula

$$\text{thickness} = k\frac{\sqrt{ss}}{c^2}, \qquad (2.6)$$

where k is a constant, ss is the spin speed, and c the concentration. Varying the spin speed is the easiest way to control thickness, but since good films can only be obtained for a narrow range of spin speeds, say 2–8 krpm, this gives a variation of a factor of 2 in thickness which is often not sufficient. One solution is therefore to buy many bottles of different thickness from the manufacturer. This can be very expensive, so another approach is to dilute the resist yourself. This should be done by weight for Equation 2.6 to apply, and care needs to be taken to ensure that the correct solvent is used for the dilution.

The next issue is that of development. See Box 2.8. Most resists have a number of different developers that can be used, and PMMA is no exception. One of the more common is the IPA: methyl isobutyl ketone (MIBK) solvent system in different ratios. Other developers that can yield high resolution are xylene, IPA:water, and cellusolve:methanol. The IPA:MIBK system consists of a strong solvent (MIBK) and a poor solvent (IPA) mixed in different ratios. Larger concentrations of MIBK give greater sensitivity and reduced contrast and resolution. For high resolution, a suitable ratio is 2.5 parts IPA and 1 part MIBK. A ratio of 3:1 is more commonly used but suffers from lower sensitivity than 2.5:1 with no additional resolution. It also has a significant "tail" on the contrast curve leading to more resist residues. Typical development conditions are 23°C for 30 s in a beaker, generally with gentle agitation, followed by an IPA rinse of 20 s and a blow dry. A perusal of the literature suggests a plethora of different ways in which to improve this process and obtain greater resolution. Some of the suggestions are the use of ultrasonic agitation (Chen and Ahmed 1993), cold development (Hu et al. 2004), and rapid thermal processing (Arjmandi et al. 2009). The standard process will get you down to 10 nm; some of the bells and whistles will improve this to around 7 nm, but since there is no standard way to measure resist resolution at these length scales (see Section 2.4.2) it is not entirely clear which process, if any, is the best.

One issue which should be emphasized is consistency. By writing a dose test, it is relatively easy to obtain very high-resolution features at some dose. If you want to repeat this, for instance to make a transistor, then you will need to take care of the process control. The most critical parameter here is developer temperature which very strongly affects the sensitivity. Greeneich (1975) shows that for the IPA:MIBK system the development process is controlled by an activation energy E_A which varies with the developer concentration. He gives E_A values of 1.04 and 2.43 for 1:1 and 3:1 ratios, respectively. For the 2.5:1 developer, we find E_A is 1.66 eV, which means that the development rate is given by

$$\text{rate} \propto \exp\left(-\frac{E_A}{kT}\right),$$

and so

$$\frac{\text{rate}(T_1)}{\text{rate}(T_2)} = \frac{e^{-E_a/kT_1}}{e^{-E_a/kT_2}} = \exp\left(-\frac{E_a}{k}\frac{T_1 - T_2}{T_1 T_2}\right) = 1.365,$$

when $T_1 = 296$ K and $T_2 = 297$ K. This corresponds to a 37% change in sensitivity per °C change in development temperature. From the linewidth–dose plot shown in Figure 2.3, this corresponds to a change in linewidth from 38 to 45 nm. Although this change in linewidth may seem small, it is for isolated lines at 100 kV, which is the best of all possible worlds for linewidth independence of dose. For lower beam energies, or denser patterns, the effect can be much more dramatic. It also shows fairly clearly that there is

300 nm

FIGURE 2.6   Resist line collapse when the period is small and the aspect ratio large. The resist here was UVN.

BOX 2.8   RESIST CONTRAST

If the thickness of resist after development is plotted against the log of the exposure dose, a graph is obtained as shown in Figure B.2.8.1; the data cure from the work of Chen et al. (1999). This is known as a contrast curve and it is generally summarized using two figures of merit: sensitivity and contrast. For a positive tone resist, the sensitivity is the dose where the remaining thickness becomes zero. This happens at a lower dose for a more sensitive resist, and generally speaking higher sensitivity is better since the writing time reduces. In Figure B.2.8.1, UVIII shows the highest sensitivity and PMMA the lowest. The contrast is the slope of the linear portion of the curve and for most purposes a high contrast is desirable.

Another useful graph is that of development rate versus dose on a log–log scale. This is known as a Hurter–Driffield (H–D) plot and an example is shown in Figure B.2.8.2. The resist contrast is the slope of the linear region.

The two definitions of contrast can be seen to be the same. The fraction of thickness remaining can be written in terms of the development rate as

$$T = 1 - A \times \text{rate}$$

for some constant A. But the rate is proportional to dose^γ, for contrast γ as defined by the H–D plot. So for some constant k we can write

$$T = 1 - k \times \text{dose}^\gamma.$$

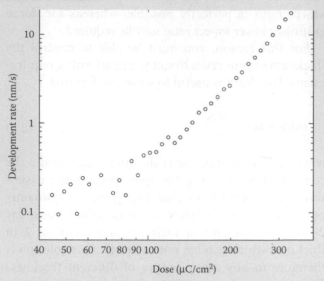

FIGURE B.2.8.2   A 270 nm thick PMMA baked at 180°C and developed in 2.5:1 IPA:MIBK for 30 s. The thickness loss in the developed areas was then measured and the development rate in nm/s obtained, which was plotted against log dose. This graph gives rate 3×10^{-8} dose$^{3.48}$.

Now the conventional e-beam contrast plot is T against D, where $D = \ln \text{dose}$, and we can write T as a function of D

$$T = 1 - ke^{\gamma D}. \tag{2.7}$$

This equation has a similar form to the curves shown in Figure B.2.8.1, and the slope of the curve is given by

$$\frac{dT}{dD} = -k\gamma e^{\gamma D}.$$

The conventional contrast, from the thickness versus log Dose plot is the slope of the linear region. There is no region of constant slope according to this equation, but it is approximately linear near $T = 0$. At this point, from Equation 2.7 we have $ke^{\gamma D} = 1$, and so

$$\frac{dT}{dD} = -\gamma$$

as required.

The two types of plot can give complementary information. It is often useful to have the log–log H–D curve since it gives information on the

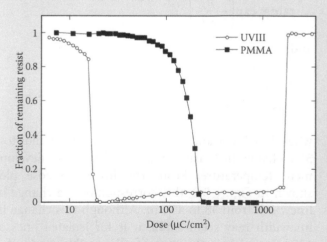

FIGURE B.2.8.1   Contrast curves for UVIII and PMMA resists. The UVIII resist was developed using CD26 and the PMMA using pure O-xylene.

resist removal rate at comparatively low doses. This information is hard to see from the lin–log plot. For example, consider writing a small gap between large squares with PMMA as the resist on a GaAs substrate. By considering the contribution from backscattered electrons, the dose in the center is about 55% of the clearing dose, and this can easily be seen to give a resist dissolution rate in PMMA of 13% of the clearing rate—in other words 13% of the resist is lost (probably more since we have to do more than just clear the resist from the patterned areas). This is a significant rate of resist loss indicating that higher contrast resists would be better for this application. On the other hand, consider Figure B.2.8.1, where the full lin–log plot for UVIII shows useful information which would not be immediately apparent from an H–D plot.

no point in reading off a dose wedge to 5% accuracy if your temperature is not under control. A temperature variation of ±0.1°C is a reasonable target, which gives an effective dose stability of around 4% under these conditions.

Another parameter which needs to be controlled is the development time; but by using a longer development time (30 s), the typical error of ±1 s is only about 3% in developer time. Assuming that the beam deposits energy uniformly through the depth of the resist (a good assumption for high kV and thin resists), this corresponds to a 3% change in development rate. For a typical PMMA contrast of 3.4 (see Box 2.8), this corresponds to a 1% dose variation sensitivity which is not of great importance. When using ultrasonically assisted development with development times of around 5 s, the errors in development time become much more significant.

After development there is often a thin scum of resist left on the surface where the resist has been developed away. This can vary in thickness and is often not only a continuous film, but also a series of granules as shown in the AFM micrograph in Figure 2.7. See Macintyre et al. (2009) for details on how this image was obtained.

It is a good to remove this residue by means of an oxygen plasma before the next process step.

2.4.2 Inspection

It is important to know what you are looking for when inspecting your patterns after development, or indeed after subsequent processing. This section is not intended to teach you how to use a microscope; rather it is to help you to relate what you see to the electron beam lithography process so that you can more easily gauge whether your lithography was successful, and determine what improvements can be made.

2.4.2.1 Optical Microscope

Although this is a chapter on nanolithography there is a great deal which can be learned about your e-beam lithography process from the comparatively low power of a good optical microscope.

The first thing to be aware of is what a well-exposed pattern should look like when using a positive resist. Figure 2.8a shows a dose wedge pattern which is simply a series of rectangles written at different doses in

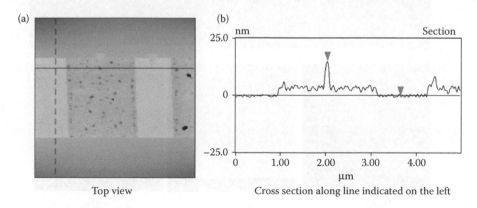

(a) Top view

(b) Cross section along line indicated on the left

FIGURE 2.7 Resist residues in PMMA after developing using 2.5:1 IPA:MIBK developer. The resist was entirely removed on the left and right sides of the image area, and so the granular residues can be seen to lie atop a continuous film of about 3 nm thickness. The granules themselves are about 10 nm in height. (See Macintyre et al. (2009)).

300-nm-thick PMMA. This is relatively a thick resist that allows patterning down to about 100 nm for dense features and isolated trenches of about 50 nm. This thickness of resist has the advantage of being easy to see optically so that the effects are much clearer here. If you were to write the same dose wedge but on thinner, say 100-nm-thick PMMA, you would find it harder to see a good dose, but still possible. If the resist is much thinner than that, correct dose clearout becomes difficult to see. Note first that the color seen through the optical microscope corresponds to the thickness of the remaining resist, and so the image is a good indicator of resist clearout. When the resist is fully cleared out, the patterned area is typically a uniform white color for a semiconductor or glass surface. A 10 nm variation in resist thickness often can fairly easily be seen by eye, and so it is easily possible to see

subfield stitching in the partially exposed resist areas. These are the horizontal lines across the patterned areas. There is also a vertical line at dose 0.800. A very small gap between subfields leads to a slight thickening of the resist at that point, which can be seen under the microscope. Likewise there is a resist thinning for a slight overlap between fields.

Note next that the pattern clears first in the middle (dose 1.042) and then at the ends. For a square, it is the corners that clear last. This is a consequence of the proximity effect which means that for a rectangle which is given uniform dose, the corners have the least effective dose because they receive the fewest backscattered electrons. Hence when inspecting a pattern optically, always check that the corners have cleared out correctly since this is where problems are likely to lie. Observe the faint shadow of resist in the corners for

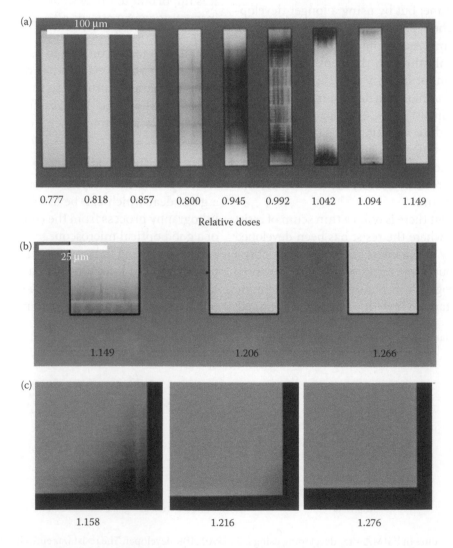

FIGURE 2.8   Optical micrograph of a dose wedge in 300 nm PMMA. (a) Shows a wide range of doses; (b) shows a detail for a few doses; and (c) shows doses for the corners of 150 μm squares rather than the rectangles in (a) and (b).

dose 1.216 in Figure 2.8c. This is something you need to watch for. Clearly, when you are using proximity correction, extra dose is given to the corners so that the effect will not be the same, but still look carefully at the edges of large features. Also, be aware that you will need a high magnification (50× or 100× objective) when looking for pattern clearout in the corners.

It is advisable to write a dose wedge with large features on every substrate you pattern since this can be used as a simple check of the processing conditions. If the dose readings differ markedly from your standard values, then you can expect any nanoscale patterns to also have problems.

Structures that are smaller than the resolution of the optical microscope can still be seen, and you will quickly become used to the appearance of a well-processed feature. In particular, gratings with a subresolution period can be seen as a uniform color, and this is particularly useful since any deviation from uniformity alerts you to a possible flaw in the patterning. In particular, it is very easy to spot a small stitch error along a grating. A stitch error of around 20–50 nm can show up very clearly optically and be very hard to find with an electron microscope.

2.4.2.2 Scanning Electron Microscope

The SEM is the tool par excellence for inspecting nanopatterns. When using it, you should bear in mind that resist layers are sensitive to electron exposure and so the features you are looking at may well vary under the beam. In particular, PMMA and other resists tend to shrink under intense electron irradiation, which means, for instance, that isolated trenches will appear to get larger during inspection. A good test is to take two consecutive images and compare them: ideally they should give identical sizes. Another related issue is that of contamination which can make features appear larger than they really are (or smaller, if you are looking at a gap).

It is often a good idea to carry out pattern transfer, for example, liftoff, before SEM inspection, since then you are looking at the features you actually want rather than an intermediate step. This is particularly relevant when trying to establish the dose for a desired linewidth. Not only may the SEM strongly affect the resist as you take the image, but there is also the real chance that the pattern transfer technique will impart a bias to the structure. For instance, an etch may enlarge or shrink a resist feature, and so it is much more useful to inspect the final feature rather than an intermediate step. Of course, there will be times when looking at the

intermediate steps is important. For instance, if your pattern looks a mess after etching then the etcher is likely to blame the lithographer and vice versa. Having some images of the resist is invaluable in that case.

2.4.2.3 Atomic Force Microscope

This is an excellent tool for measuring resist thicknesses; it is not so useful for lateral dimensions since there will always be the offset of the probe size in any linewidth measurements, unless you are using a very expensive critical dimension atomic force microscope (CD-AFM). Knowledge of resist thicknesses in electron beam lithography is often useful, for instance if you are fabricating a grating. The resist thickness before lithography is presumably known, but in a large area grating the unexposed resist receives roughly half the clearing dose and will thin somewhat, depending on the resist contrast. The AFM is the best tool in this situation to measure the thickness.

2.4.3 The Process Window

This describes the range of conditions in which a process works. A process here comprehends all the steps required to go from a bare surface to patterned resist of correct thickness and feature size. Some of the more important conditions are dose, bake temperatures [critical for chemically amplified (CA) resist], developer, development time and temperature, spin conditions, and beam size/current. The process window is the range of conditions in which the process works. Clearly, the bigger the window the easier it is to keep the process working. At the extreme limit a process window may be so small that it is more or less impossible to meet, and the results on your test wafer can never be transferred to your actual devices.

A process window is a multidimensional object which most of us find hard to visualize. See Box 2.9 for some further details. It is easiest to picture the process window in one dimension (e.g., dose) or two dimensions (e.g., dose and bake temperature).

2.4.4 How to Establish a Process

This is never done in a vacuum: you will have some idea of what you are trying to achieve. For instance, your goal may be 25 nm resist lines on an 80 nm period. Moreover, you may well know that you wish to use this as an etch mask (for instance) and that the etch resistance of the resist is going to be crucial. Possibly linewidth control may not be too important, but you require line edge roughness better than 1 nm. So the first stage of developing a new process is to crystallize exactly what it is you wish to achieve, and to write this down in order of priority.

The next stage is to come up with an initial guess of a process. This can come from various sources. It may be something you have read in the literature; it may be a suggestion from a colleague, or a minor improvement of a process you already have running. Often you may find a number of possible processes, which you can then screen by asking questions such as, "Which would be the simplest to develop?,"

BOX 2.9 A LITTLE MATHS AND STATISTICS

Statistical process control is very important in manufacturing, but it is not very appropriate to lab-based work where the number of runs required is usually quite small, say of the order of one a month. That having been said, it is important to get some handle on the errors involved in your process or else you can spend your life chasing your tail.

Variability in lithography results is difficult to quantify with a small number of samples. Suppose you process a number of samples and measure the linewidths. Assume the individual samples really are independently processed—see later for more details—then your linewidth measurements might well have a Normal distribution. By calculating the mean (w_0) and standard deviation (σ) of the measurements, we can say that 95% of future lithography runs will lie within $w_0 \pm 2\sigma$. But there is a snag with this reasoning inasmuch as it only applies to the real mean (w_r) and standard deviation (σ_r) of the system. Something like 100 measurements from independently processed samples are required for w_0 and σ to be a good approximation to w_r and σ_r.

Even for a fairly small number of samples w_0 is a reasonably good estimate of w_r. For n samples the 95% confidence interval is $w_0 \pm 2\sigma_r/\sqrt{n}$ (so 95% of the time the real mean will lie within this band). The real killer is the confidence interval for the standard deviation which is much larger than most people expect, as shown in Table B.2.9.1. So for example, if you do a single run you have no idea of what the process variablity is. Next suppose you carry out two runs and calculate the standard deviation to be 10 nm. From Table B.2.9.1, you see that the actual standard deviation could actually be anywhere between 4.5 and 319 nm (with a 95% confidence level). This clearly is not much better than with just a single run.

Even with five samples which are quite a lot for most lab-based work, the 95% confidence interval would be between 6 and 29 nm in this case, which is still quite a big spread, but at least it is manageable. For an unknown standard deviation, what you really need is the t-distribution rather than the normal distribution; you will find information on this easily in standard textbooks.

If you are doing repeated measurements to measure process variability, you need to be careful to make sure the results are independent. If you run them consecutively through the e-beam tool, you must be careful to reset all the parameters between each sample; this means for instance deliberately destroying the focus and astigmatism, and resetting the lenses and so forth. And of course the measurements must be on different substrates which are processed independently.

Another aspect of measuring variability is to look at the variance from each of a number of important factors. The idea here is to identify the major causes of variation which then helps keep variability under control. We can use the Taylor-series approximation, keeping only terms of the first order. For example, if we measure the linewidth for two post exposure bake (PEB) temperatures we can find the variation of linewidth as a function of PEB temperature.

In general, given that the linewidth varies as a function of dose (d), focus (f), and so on, then the

Table B.2.9.1 Confidence Intervals for the Standard Deviation, σ, from a Number, N, of Measurements

N	95% Confidence Interval of σ
2	$0.45–31.9\sigma$
3	$0.52–6.29\sigma$
5	$0.60–2.87\sigma$
10	$0.69–1.83\sigma$
100	$0.88–1.16\sigma$

variation of w with each of its dependent variables can be expressed as follows:

$$w = w\left(d, f, t, d_t, \cdots\right)$$

$$\Delta w = \frac{\partial w}{\partial d}\Delta d + \frac{\partial w}{\partial f}\Delta f + \frac{\partial w}{\partial t}\Delta t + \frac{\partial w}{\partial d_t}\Delta d_t + \cdots$$

if we know the variability of dose (Δd in the above equation) and $\delta w/\delta d$, we can calculate the first term and so forth. A good estimate of the Δx terms is important, and these will come from an understanding of your tool and process equipment. For instance you may have measured the temperature variations caused by a particular hotplate to be ±2°C, in which case Δt would be 4°C.

Usually we are only interested in the sources of variation which give the biggest errors. Typically some of these might be the developer temperature, dose, and perhaps the focus (if the tool does not have a height meter). Then the variations can be estimated by evaluating the derivatives, $\delta w/\delta x$, and using your estimates for the Δx terms. The derivatives can be measured by measuring w for large and small values of the parameter x, while holding all the other factors constant. A relatively large variation has to be used for this to give good results.

If we now assume that the errors are independent, then each of the error terms, $(\delta w/\delta x)\Delta x$, can be added quadrature to obtain an estimate of the total error given by

$$\Delta w = \sqrt{\left(\frac{\partial w}{\partial d}\Delta d\right)^2 + \left(\frac{\partial w}{\partial f}\Delta f\right)^2 + \left(\frac{\partial w}{\partial t}\Delta t\right)^2 + \left(\frac{\partial w}{\partial d_t}\Delta d_t\right)^2 + \cdots}$$

This means that the biggest errors are by far the most important and should be sought out.

"Which would provide the best platform for future developments?," and "Which has the greatest chance of success?" So, in the end you have an initial guess at a process, which will include the resist, the required thickness, the estimated dose, and the development and bake conditions.

The next stage is range checking in which you seek to establish the approximate range of each important parameter for further study. If you are simply extending an existing process, you may skip this step and move on to fine tuning. Each substrate should contain a dose wedge which covers a large range of doses, and it is better to err on the side of too large a range rather than too small. Generally speaking, you will have a rough estimate of the correct dose and you should aim to go both up and down from that by a factor of at least 3, preferably more. By increasing the beam step size, you are able to lower the bottom end of the dose range; increasing the beam current enables higher doses to be reached without excessive exposure times. The downside for both these adjustments is a poorer quality of lithography, but good lithography is not the aim here. Rather you want to home into an approximate process as quickly as possible. Major parameters often come in groups such as developer concentration/temperature and time; or bake temperature and time. Often for initial range checking it suffices to vary only one parameter from each group, such as developer concentration and bake temperature.

Finally, you will need to fine tune the process, which involves smaller changes in parameters and slowly homing into the best processing conditions. A very useful method of attacking this sort of multidimensional problem is to use design of experiments (DOE) methodology, for instance factorial experiments (there are many textbooks available). These characterize a number of dimensions at once instead of the more traditional one variable at a time approach. As well as being more time-efficient in covering multiple dimensional parameter space, the DOE methodologies can uncover relationships between parameters, which would not otherwise be apparent.

As part of this process, you should measure the variability of the process so that you have a good idea of the size of the process window. This sort of information drops easily out of the DOE methodology. Without this information, you can easily become frustrated by a process which always appears to work when tested, but never on real devices. This is a common problem for processes on the edge of working when the dose range for the process to work is very small—say less than ±2%.

2.4.5 Liftoff

Liftoff is a well-established e-beam process for making metallic features. In general, compared with an etch process its reliability is poorer, but it is less expensive, and good enough for many purposes. It works best with an undercut resist profile which can be obtained very easily in positive tone resists for lower beam energy (say ≤50 kV). It is good for feature sizes down to 10 nm or even smaller.

It has been used with both single layers of PMMA (Beaumont et al. 1981) and bilayers of PMMA (Beaumont et al. 1980). A layer of LOR resist under PMMA (Chen et al. 2004), or indeed ZEP520 (Tu et al. 2007), has been shown to enhance liftoff in some circumstances; other approaches such as ZEP520–PMMA bilayer have also been successfully used (Maximov et al. 2002).

Generally speaking, lift-off works best with positive tone resist, but with some extra processing it can be used with negative tone resists. For instance, with an HSQ/PMMA bilayer good negative liftoff has been achieved (Yang et al. 2008).

2.4.6   Troubleshooting When Things Go Wrong

Very often problems arise within e-beam lithography, and the following notes give some hints on how to solve them. There are three major sources of faults: the resist processing procedure, the e-beam tool itself, and the data preparation. To assist the tracing of problems, you should take time to make sure your exposure sequences on the tool are well logged; you should also always write a standard test pattern on your substrate.

Every tool will have different logging capabilities, but as a general rule log as much as you can. If log files get too detailed, however, it is difficult sometimes to extract the relevant information, and so some common sense is required here. Certainly log files should record all the tool parameters such as beam energy, lens currents, beam current, and vacuum levels; if your tool measures sample height it is useful to log the height at each exposure field. Most of this information can be ignored when things go well, but it can be invaluable when there are problems.

Standard test patterns are also invaluable. They will need to be tailored slightly for different resist systems, which may just be a matter of varying the beam settings and the dose. A dose wedge consisting of a series of rectangles written at different doses is a basic diagnostic tool. By running a profilometer along with it, the contrast of the resist system can be established, and a simple visual inspection using an optical microscope will reveal the clearing dose, which should be the same from run to run. Small lines to test the resolution of the process and verniers to measure field stitching are

other useful additions to a test pattern. In the event of the main pattern failing in some way, the test pattern can give information about whether the process has shifted, and by how much.

Here are some common problems with notes on what steps you can take to solve them:

- *No pattern on the substrate after development.* If you have a test pattern, look to see if that is present. Typical causes are
  o The pattern is there but is very small and you cannot find it. A single 100 nm line is hard to find on a 4 in. wafer. Make sure you have some strategy for finding small patterns, for instance writing a larger block nearby with a bigger beam current.
  o If the test pattern is missing as well, then was your wafer upside down in the tool? (It has been known, especially if the backside is shiny with a thin-film coating.)
  o Check the log file. If the job had a measured beam current of 1 nA and it wrote for 2 h you would expect something to be there, but maybe the writing time was 2 s, or the beam current dropped off during the writing.
  o Check the data preparation. You may have written the wrong (empty) layer.
- *The patterns appear bloated.*
  o This is either overexposure or overdevelopment. The test pattern should help to distinguish. Typical causes are the use of the wrong (or badly made up) developer, or possibly the pattern preparation has left multiple shapes on top of each other, or simply used the wrong dose. Remember too the proximity effect, which will make a dense pattern overexposed if given the same dose as a sparse pattern.
  o Again the use of a standard test pattern would distinguish between pattern preparation issues and processing errors.
- *The patterns are not cleared out.*
  o This is essentially the reverse situation of the previous problem with correspondingly inverted causes.
- *Linewidth inconsistency.*
  o Here you need to know your process: what parameters most strongly affect the linewidth? See under "how to establish a process."
- *Grating collapse/resist adhesion.*
  o Resist collapse typically looks as shown in Figures 2.5 and 2.9. It can be caused either by

FIGURE 2.9 HSQ lines on silicon; (a) without a Ti layer; (b) using a 2 nm Ti layer as an adhesion promoter. Note that the line collapses without the Ti layer.

capillary forces on small resist gaps during drying, or else because of poor adhesion between the resist and the substrate. Important variables here are the substrate, resist, grating period, linewidths, resist thickness, and method of drying.

o Adhesion can be improved by paying careful attention to drying the wafer–bake up to 220°C for 30 min to drive off water; an oxygen plasma treatment immediately prior to resist application often helps. Adhesion promoters are useful, for example, Hexamethyldisilazane (HMDS) for silicon substrates, but note that these are substrate specific, and so HMDS is not suitable for III–V materials. A thin, 2 nm, Ti layer can also substantially improve adhesion (Macintyre et al. 2006), as shown in Figure 2.9.

o Even with good adhesion, the resist can collapse in dense line–space patterns. The easiest fix is to reduce the resist thickness, but this is not always easily possible. Drying from a lower surface tension liquid can help—for example, IPA instead of water. Note, however, that this is only suitable for some resists. The use of critical point dryers is also beneficial (Wahlbrink et al. 2006).

• *Proximity related: small gaps.*

o Look out for proximity-effect-related issues, such as bowing in narrow gaps. This is easily fixed by using proximity correction software, although you should note that some patterns cannot be corrected. A case in point is submicron gap in very large (>100 μm) pads using a low contrast resist, for example PMMA. Although resist will remain in the gap as required, because of the low contrast of the resist there will inevitably be some resist thinning in such regions which can be problematic for subsequent processing. A higher contrast resist is the answer here.

• *Machine related: missing shapes/unwanted shapes/focus.*

o There are a host of things that can go wrong with an e-beam tool, which are specific to a particular tool. For instance, the pattern generator can miss shapes, or parts of shapes. Height measurement errors can be caused by a transparent substrate which lead to stitch and focus errors. The blanker could fail leaving telltale traces of exposed resist across the wafer. In all such cases, a thorough understanding of your own e-beam tool is invaluable.

When diagnosing problems with an e-beam tool, it is often useful to use a simple resist like PMMA rather than a chemically amplified one since resist issues are less likely to be present.

2.5 Some Resists

Resists can be divided into different categories. One helpful division is between positive and negative tone resists; this is particularly important for e-beam lithography where using the wrong tone of resist can drastically increase the writing time. Another useful distinction is between chemically amplified and nonchemically amplified resists. The former are usually much faster, but do not have the ultimate resolution of some of the nonchemically amplified resists PMMA and HSQ. Important issues to consider when choosing a resist are speed, etch resistance, resolution, processing ease, thickness, availability, and cost.

Nonchemically amplified resists typically have sensitivities in the range 100–500 $\mu C/cm^2$ at 50 kV. Almost certainly, the best-known e-beam resist in this category is PMMA, which is positive tone, very high resolution and slow. Linewidths below 7 nm have been reported using it (Chen and Ahmed 1993). Its other drawback is its poor dry etch resistance. ZEP520 resist is about five times faster but has resolution of about 20 nm (Namatsu et al. 1995), and is very expensive. It is based on a similar chemistry to PMMA but does have a significantly better dry etch resistance.

HSQ is a negative tone with similar resolution to PMMA and which gives very small linewidth roughness (Namatsu et al. 1998; Kupper et al. 2006; Grigorescu and Hagen 2009). It also has a similar speed to PMMA. HSQ has very different chemistry to polymer resists consisting mainly of silicon oxide after exposure rather than being an organic polymer. This can often be very useful, but it does also mean that its dry etch resistance in fluorine chemistry is quite poor. Micro resist technology's ma-N is lower-resolution negative tone resist without much speed advantage over HSQ but uses a more conventional polymer chemistry which is sometimes useful.

Chemically amplified resists are generally about 10 times faster than nonchemically amplified resists, with typical sensitivities in the range 10–50 $\mu C/cm^2$, which provides a major improvement in writing time. Generally speaking, the resolution attainable is not as good as with PMMA or HSQ, but can be below 50 nm, which is often adequate. Two examples are the positive tone resists UVIII or UV5 (Thoms et al. 1999), and the negative tone resists NEB-31 and NEB-22. NEB has a slightly higher ultimate resolution than the UVx family, but both can get down to around 50 nm. The UVx resists are primarily deep ultraviolet (DUV) rather than e-beam resists which in many ways is an advantage. Because there is a much bigger market, the price is lower, and the supply is more certain than for resists manufactured purely for the e-beam lithography industry. Many other DUV resists can also be used for e-beam.

2.6 Future Outlook

Compared to other laboratory equipment, electron beam lithography tools are long lasting and it is not uncommon to find 20-year-old tools still producing good work. This is not to say that e-beam tools have not improved over this time period, but rather that the improvements have been steady and incremental, so that the key components of today's tools are recognizably the same as those of yesteryear. Often one of the hardest components of an e-beam tool to maintain is the control computer since computers and their interfaces can change very markedly over this time period.

With this in mind, it is relatively easy to predict some likely tool evolution over the coming decade. The probe size will continue to fall slowly due to improvements in electron optics. There are no revolutionary new source designs on the horizon and so it is unlikely that a major improvement in gun brightness will occur, as happened in the advance from LaB_6 to thermal field emitters over the past 10 years. Incremental improvements in all areas such as beam positioning and scan frequency are very likely.

Resist changes and improved understanding of existing ones. PMMA has been the resist of choice for many years because of its relatively simple processing and high resolution. There has been a gradual increase in understanding its chemistry and processes such as ultrasonic and cold development have been developed which have increased its resolution to the sub-10 nm domain in the past 10 years. Further advances are likely (e.g., MNE09) and PMMA is likely to remain an important resist for many years to come. The advent of field emission sources has meant that a large beam current can be focused into a sub-10 nm beam, which means that the slowness of PMMA is not such an issue as in the past. For feature sizes, around 10 nm shot noise can become a major issue and faster resists are not needed at this resolution. HSQ has established itself over the past decade as the high-resolution negative resist of choice and is likely to remain popular, but other high-resolution resists may well be developed. Because the scattering range of secondary electrons is 2–3 nm, we are unlikely to see individual linewidths dropping much below 5 nm, but there is room for reducing the minimum patterned pitch which currently stands at about 20 nm. Some new resists will inevitably be developed over the coming years, these will be used in the 10–100 nm regime where there is scope for more speed. This may come from the world of extreme ultraviolet (EUV) lithography, which is being developed for wafer-scale 22 nm production and needs suitable resists. Any resists developed for EUV are likely to be also suitable for electron beam lithography.

Integration with self-assembly. One area growing in importance is the interface between top-down and bottom-up approaches to nanopatterning. Electron beam lithography is in many ways the standard form of top-down patterning where the user can easily define an arbitrary pattern. Bottom-up patterning involves the use of carefully chosen or designed molecules, which combine spontaneously to form patterns on the surface. By appropriately prepatterning the surface using other lithographies, for example, e-beam, the self-assembled nanostructures can be given more regularity giving in some ways the best of both worlds: the speed, low cost, and high resolution of self-assembly with the controlled placement of top-down lithography. This is an area that looks set to expand over the coming years.

E-beam no longer possesses significantly higher resolution than photolithography: its advantage is cost and flexibility, which make it suitable for research in a growing number of areas. Twenty, or even 10, years ago one of the main advantages of electron beam lithography over photolithography was its higher resolution. This advantage is still there but has been very considerably eroded to the point where 22 nm lithography is now within the sights of photolithography and will almost certainly be achieved. In that time, electron beam lithography has scarcely improved beyond the 10 nm mark where it has been throughout that time. There are reports of isolated lines of around 5 nm (Chen and Ahmed 1993; Grigorescu et al. 2007; Thoms and Macintyre 2010), but in terms of lines and spaces,

10 nm is about as good as is possible (Grigorescu et al. 2007; Yang and Berggren 2007). So electron beam lithography will have to look elsewhere for its advantage over photolithography. The main advantage at the moment is flexibility and cost. Not only are the tools required for 20 nm photolithography priced at hundreds of millions of dollars; the running costs per year in terms of mask plates run into many millions. This is far too expensive for most small research institutes. Moreover, the high throughput of photolithography is not required in research and so e-beam has a very solid place. In addition, it is much faster to move from concept to reality using an e-beam tool since no intermediate mask step is required, and e-beam is also able to write on nonplanar surfaces, which cannot be done with photolithography.

Multibeam and projection optics. Finally, a word on other forms of electron beam lithography. This chapter has concentrated on single Gaussian beam lithography. There are a number of tools in development that aim to increase the throughput of electron beam tools to rival photolithography. These use various approaches but typically consist of multibeams which are focused using one set of electron optics, but blanked by multiple blankers, for example, refer Pease (2009) and Petric et al. (2009). If these developments succeed to the point of becoming used for mainstream semiconductor fabrication, there may well be beneficial knock-on effects for lab-based nanofabricators using e-beam tools. The most obvious one would be the extra research effort going into suitable e-beam resists with potential improvements in both performance and cost.

2.7 Symbols and Abbreviations

A	writing area (square centimeters)		N	number of pixels
bss	beam step size		PMMA	polymethyl methacrylate
c	resist concentration (g/L)		SEM	scanning electron microscope
f	writing frequency (MHz)		ss	spin speed (rpm)
HSQ	hydrogen silsquioxane		t	writing time (s)
i	beam current (nA)		τ	dwell time (μs)

References

Arjmandi, N., L. Lagae, and G. Borghs. 2009. Enhanced resolution of poly(methyl methacrylate) electron resist by thermal processing. *Journal of Vacuum Science and Technology B* **27**(4): 1915–1918.

Beaumont, S. P., P. G. Bower, T. Tamamura, and C. D. W. Wilkinson. 1981. Sub-20nm-wide metal lines by electron-beam exposure of thin poly(methyl methacrylate) films and liftoff. *Applied Physics Letters* **38**(6): 436–439.

Beaumont, S. P., T. Tamamura, and C. D. W. Wilkinson. 1980. A two layer resist system for efficient liftoff in very high resolution electron beam lithography. In: Kramer, R. P., editor, *Microcircuit Engineering*, p. 381, Vol. 80, Delft. Delft University Press.

Boere, E., E. van der Drift, J. Romijn, and B. Rousseeuw. 1990. Experimental study on proximity effects in high voltage e-beam lithography. *Microelectronic Engineering* **11**: 351–354.

Chen, W. and H. Ahmed. 1993. Fabrication of 5-7 nm wide etched lines in silicon using 100 keV electron-beam lithography and polymethylmethacrylate resist. *Applied Physics Letters* **62**(13): 1499–1501.

Chen, Y., D. Macintyre, and S. Thoms. 1999. Fabrication of T-shaped gates using UVIII chemically amplified DUV resist and PMMA. *Electronics Letters* **35**(4): 338–339.

Chen, Y., K. Peng, and Z. Cui. 2004. A lift-off process for high resolution patterns using PMMA/LOR resist stack. *Microelectronic Engineering* **73–74**: 278–281.

Docherty, K. E., S. Thoms, P. Dobson, and J. M. R. Weaver. 2008. Improvements to the alignment process in a commercial vector scan electron beam lithography tool. *Microelectronic Engineering* **85**(5–6): 761–763.

Docherty, K. E., K. A. Lister, J. Romijn, and J. M. R. Weaver. 2009. High robustness of correlation-based alignment with Penrose patterns to marker damage in electron beam lithography. *Microelectronic Engineering* **86**(4–6): 532–534.

Geiger, H. 1909. On a diffuse reflection of the alpha-particles. *Proceedings of the Royal Society of London Series a-Containing Papers of a Mathematical and Physical Character* **82**(557): 495–500.

Greeneich, J. S. 1975. Developer characteristics of poly-(Methyl Methacrylate) electron resist. *Journal of the Electrochemical Society* **122**: 970–976.

Grigorescu, A. E. and C. W. Hagen. 2009. Resists for sub-20-nm electron beam lithography with a focus on HSQ: State of the art. *Nanotechnology* **20**(29), doi: 10.1088/0957-4484/20/29/292001.

Grigorescu, A. E., M. C. van der Krogt, and C. W. Hagen. 2007. Sub-10 nm structures written in ultra-thin HSQ resist layers, using electron beam lithography. *Proceedings of SPIE* **6519**: 65194A–65191–65194A–65112.

Grigorescu, A. E., M. C. van der Krogt, C. W. Hagen, and P. Kruit. 2007. 10 nm lines and spaces written in HSQ, using electron beam lithography. *Microelectronic Engineering* **84**(5–8): 822–824.

Grivet, P. (1972). *Electron Optics*. Oxford and New York: Pergamon Press.

Hu, W., K. Sarveswaran, M. Lieberman, and G. H. Bernstein. 2004. Sub-10 nm electron beam lithography using cold development of poly methylmethacrylate. *Journal of Vacuum Science and Technology B* **22**(4): 1711–1716.

Ignatova, O., S. Thoms, W. Jansen, D. S. Macintyre, and I. Thayne. 2010. Lithography scaling issues associated with III-V MOSFETs. *Microelectronic Engineering* **87**(5–8): 1049–1051.

Ivin, V. V., M. V. Silakov, D. S. Kozlov, K. J. Nordquist, B. Lu, and D. J. Resnick. 2002. The inclusion of secondary electrons and Bremsstrahlung x-rays in an electron beam resist model. *Microelectronic Engineering* **61–62**: 343–349.

Kupper, D., T. Wahlbrink, W. Henschel, J. Bolten, M. C. Lemme, Y. M. Georgiev, and H. Kurz. 2006. Impact of supercritical CO_2 drying on roughness of hydrogen silsesquioxane e-beam resist. *Journal of Vacuum Science & Technology B* **24**(2): 570–574.

Macintyre, D. S., O. Ignatova, S. Thoms, and I. G. Thayne. 2009. Resist residues and transistor gate fabrication. *Journal of Vacuum Science & Technology B* **27**(6): 2597–2601.

Macintyre, D. S., I. Young, A. Glidle, X. Cao, J. M. R. Weaver, and S. Thoms 2006. High resolution e-beam lithography using a thin titanium layer to promote resist adhesion. *Microelectronic Engineering* **83**(4–9): 1128–1131.

Maximov, I., E. L. Sarwe, M. Beck, K. Deppert, M. Graczyk, M. H. Magnusson, and L. Montelius. 2002. Fabrication of Si-based nanoimprint stamps with sub-20 nm features. *Microelectronic Engineering* **61–62**: 449–454.

Namatsu, H., M. Nagase, K. Kurihara, K. Wadate, and K. Murase. 1995. 10-nm silicon lines fabricated in (110) silicon. *Microelectronic Engineering* **27**(1–4): 71–74.

Namatsu, H., Y. Takahashi, K. Yamazaki, T. Yamaguchi, M. Nagase, and K. Kurihara. 1998. Three-dimensional siloxane resist for the formation of nanopatterns with minimum linewidth fluctuations. *Journal of Vacuum Science & Technology B* **16**(1): 69–76.

Owen, G. 1990. Methods for proximity effect correction in electron lithography. *Journal of Vacuum Science & Technology B* **8**(6): 1889–1892.

Pease, F. R. 2009. To charge or not to charge: 50 years of lithographic choices. *Journal of Vacuum Science & Technology B* **28**(6): C6A1–C6A6.

Petric, P., C. Bevis, A. Carroll, H. Percy, M. Zywno, K. Standiford, A. Brodie, N. Bareket, and L. Grella. 2009. REBL: A novel approach to high speed maskless electron beam direct write lithography. *Journal of Vacuum Science & Technology B* **27**(1): 161–166.

Rishton, S. A. and D. P. Kern. 1987. Point exposure distribution measurements for proximity correction in electron-beam lithography on a sub-100 nm scale. *Journal of Vacuum Science & Technology B* **5**(1): 135–141.

Shimizu, R. and D. Ze-Jun. 1992. Monte Carlo modelling of electron–solid interactions. *Reports on Progress in Physics* **55**(4): 487–531.

Thoms, S. and D. S. Macintyre. 2010. Linewidth metrology for sub-10-nm lithography. *Journal of Vacuum Science & Technology B* **28**(6): C6H6–C6H10.

Thoms, S., D. S. Macintyre, and Y. F. Chen. 1999. Evaluation of Shipley UV5 resist for electron beam lithography. *Lithography for Semiconductor Manufacturing* **3741**: 138–147.

Tu, D. Y., M. Liu, L. W. Shang, C. Q. Xie, and X. L. Zhu. 2007. A ZEP520–LOR bilayer resist lift-off process by e-beam lithography for nanometer pattern transfer. *7th IEEE Conference on Nanotechnology, 2007 (IEEE-NANO 2007)*, Hong Kong, pp. 624–627.

Wahlbrink, T., D. Kupper, Y. M. Georgiev, J. Bolten, M. Moller, D. Kupper, M. C. Lemme, and H. Kurz. 2006. Supercritical drying process for high aspect-ratio HSQ nano-structures. *Microelectronic Engineering* **83**(4–9): 1124–1127.

Yang, H. F., A. Z. Jin, Q. Luo, J. J. Li, C. Z. Gu, and Z. Cui. 2008. Electron beam lithography of HSQ/PMMA bilayer resists for negative tone lift-off process. *Microelectronic Engineering* **85**(5–6): 814–817.

Yang, J. K. W. and K. K. Berggren. 2007. Using high-contrast salty development of hydrogen silsesquioxane for sub-10-nm half-pitch lithography. *Journal of Vacuum Science & Technology B* **25**(6): 2025–2029.

Zhang, Y., K. E. Docherty, and J. M. R. Weaver. 2010. Batch fabrication of cantilever array aperture probes for scanning near-field optical microscopy. *Microelectronic Engineering* **87**(5–8): 1229–1232.

3. Nanofabrication with Focused Ion Beams

Jacques Gierak
LPN–CNRS, Marcoussis, France

Ralf Jede
Raith GmbH, Dortmund, Germany

Peter Hawkes
CEMES–CNRS, Toulouse, France

Nanofabrication Handbook. Edited by Stefano Cabrini and Satoshi Kawata © 2012 CRC Press / Taylor & Francis Group, LLC. ISBN: 978-1-4200-9052-9

Chapter 3

3.1   Introduction

In this chapter, the potential of focused ion beam (FIB) technology and its ultimate applications are reviewed. After an introduction to the FIB technology and to the operating principles of liquid metal ion sources (LMIS), to ion optics and to instrument architectures, several advanced applications are described and discussed. Finally, we introduce and illustrate some future trends for FIB nanofabrication schemes combining top-down and bottom-up processing capabilities.

3.1.1   Background

The patterning of samples using the FIB technique is a very popular approach in the field of inspection of integrated circuits and electronic devices manufactured by the semiconductor industry or research laboratories. This is the case mainly for prototyping devices. The FIB technique, with which materials can be engraved on a very fine scale, complements other lithographic techniques, such as optical lithography. The main difference is that FIB allows direct patterning and therefore does not require an intermediate sensitive medium or process (resist, metal-deposited film, etching process). FIB allows three-dimensional (3D) patterning of target materials using a finely focused pencil of ions with speeds of several hundreds of km/s at impact. As far as the choice of ion is concerned, most metals can be used in FIB technology as pure elements or in the form of alloys but in practice, gallium (Ga+ ions) is preferred in

most cases. FIB patterning can be achieved by local surface defect generation, by ion implantation, or by local sputtering. These modes of operation are obtained very easily by varying the locally deposited ion fluence as a function of the sensitivity of the target and of the selected FIB processing method.

Some brief historical reminders are discussed here. Around 1959, Feynman, who was later awarded the Nobel Prize, first raised the question of the limits of miniaturization in a visionary speech that was to become very well known (Feynman 1959). In this speech, he described the basis of a new research area that is today known as nanoscience. He also reviewed most of the applications that are nowadays hot topics in nanotechnology, mentioning information storage at a reduced size, biological systems, and computer miniaturization. The main question was not whether this was possible but rather how to write small enough. At that time, the idea of developing integrated circuits was still in the distant future. Therefore, when Feynman came to the point of proposing a method of writing small, it was without any *a priori* knowledge that he proposed the use of ion beams to sculpt matter on the required nanoscale.

Why did it take so long to put this idea into practice? In the microelectronic and microengineering area, the use of ions having energies between 1000 and some thousands of electronvolts has always been very common. Certainly, ion etching and ion implantation are widely used. But here, in most cases, it is a question of processes based on ion beams having a large section, and not focused. The most popular application of ion beams is to achieve large-scale and uniform mechanical or reactive material sputtering. A gas introduced in a chamber is subjected to an ionization process caused by electrons emitted by a cathode. Grid electrodes are then used to extract the resulting neutralized ion beam and guide it down to a target. The gas used in most cases is argon, accelerated under a voltage around 500 V, with a current density of some mA/cm². Another application is the reactive ion etching process where ion bombardment is simultaneously generated, thus allowing a kinetic energy transfer to the sample surface combined with a chemical reactivity of the ionized gas (fluorated or chlorinated) toward the target material. The chemically reactive species are selected to allow volatile species to be created and efficiently evacuated by the chamber pumping system. In this case, we speak of reactive ion etching or "dry etching" because the etching process takes place within a plasma process.

Ion implantation consists in modifying the properties of a given material by inserting selected ions within the volume of a target. In microelectronics, this technique is widely used, since it enables semiconductor materials to be doped. This technique is also used for some other applications like surface treatments, for example, hardening. Within a vessel containing the element to be implanted, a plasma is ignited. Then, an electric field is generated to extract the ion beam, which is filtered and accelerated to energies ranging from some keV to several hundreds of keV and finally directed onto a sample surface. The penetration depth of the ions is usually adjusted to be in the micrometer range. Commonly used ions are boron, arsenic, phosphorus, germanium, and silicon implanted in materials such as silicon and gallium arsenide. The implantation doses vary between 10^{11} and 10^{16} ions/cm². One should also note that this technique has also been successfully employed to improve mechanical properties of materials (reducing wear and friction), to reduce corrosion of surfaces, or to improve surface properties (surface energy and absorptivity).

Up to the middle of the 1970s, no ion source bright enough for the ion beam to be refocused within a spot having a current density as high as 1 A/cm² was available. This became possible only recently after the discovery of LMIS, which have a high brightness and a very small emitting area. This discovery opened new perspectives by allowing the development of instruments capable of generating and using highly focused ion beams for direct patterning of materials. These FIB systems, all based on LMIS and coupled

with dedicated optics, are capable of generating ion beams with energies between one and some hundreds of keV. These LMIS based on a field emission process are the descendents of the field ionization sources discovered in the 1950s by the German physicist Erwin Müller (Müller and Tsong 1969). In these field ionization sources, which deliver very low ion currents, the electrostatic field applied to a tip having an end radius as small as 10 nm allows ionization of gas atoms that are injected in the device. This principle is still central in field ion microscopy, the first-ever technological setup capable of displaying atoms and of determining the nature of the atoms constituting the tip material.

The first investigations made on LMIS were initiated with the aim of developing ion thrusters for spacecraft. Then, in the early 1970s, Krohn and his colleagues, seeking emitters capable of producing small electrically charged droplets, observed that metals having high-surface tension were emitting ions rather than droplets when they were placed in a sufficiently intense electric field (Krohn and Ringo 1975). This discovery of a high-brightness ion source using a liquid metal was to generate considerable interest. During the same period, Clampitt, who was also investigating this kind of emitter (Clampitt et al. 1975), and the team of Sudraud also succeeded in developing several kinds of LMIS (Sudraud et al. 1978).

Later, in the year 1978, Seliger and his collaborators at the Hughes Research Laboratories were the first to employ this emitter for direct patterning; the first-ever scanning ion microscope was realized by coupling an LMIS to charged particle optics (Seliger et al. 1979). The era of FIB development had begun. In the early 1980s, many FIB systems were developed and tested with different results, which we will recall and analyze in the following paragraphs. For further details of the early history, see Prewett and Mair (1991), Orloff et al. (2003), Giannuzzi and Stevie (2005), and Utlaut (2009), and for an analysis of the physics of the LMIS, see Forbes and Mair (2009).

3.1.2 Scanning Ion Microscopy

One of the early application domains for the FIB was related to scanning ion microscopy (SIM) and to chemical analysis via secondary ion mass spectroscopy (SIMS). The SIM technique (Figure 3.1a) was mainly investigated by Levi-Setti (Levi-Setti et al. 1985), who in the early 1980s developed a scanning ion microscope capable of resolving objects as small as some tens

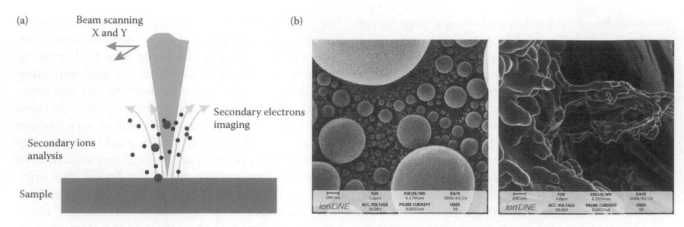

FIGURE 3.1   (a) Schematic representation of the interaction of a pencil of ions scanned over the surface of a target. (i) The collection of the secondary electrons allows reconstruction of an image of the surface while (ii) collection and analysis of the secondary ions allow building chemical mapping of the sample. (b) Example of images obtained in scanning ion microscopy (SIM) for image resolution evaluation. (Adapted from Raith. 2009. Raith GmbH, ionLiNE lithography, nanofabrication and engineering workstation, http://www.raith.com/) (Left) Surface of a test sample consisting of tin (Sn) balls having various sizes. (Right) Test sample.

of nanometers. The main interest of this method, when coupled with a secondary ion analyzer, is to perform SIMS. This technique is widely used to analyze the composition of solid surfaces and thin films. Sputtering the surface of the specimen is achieved using a primary ion beam and the ejected secondary ions are analyzed. This technique allows local mapping of the surface constituents with a very low detection limit around 10^{12}–10^{16} atoms/cm^3. Then, because the sample surface is eroded during the analysis, this technique also allows a depth profile of the target composition to be made. This 3D mapping is possible down to depths of several micrometers, making the technique irreplaceable in microelectronics. Among its limitations are the limited lateral resolution and the fact that this analysis is destructive since the imaged/analyzed surface is indeed slowly etched.

Besides abundant information about the physico-chemical properties of the material surfaces imaged, originating from the very small interaction volume of the impinging ions with material that allows high-emission yield, high-topographic sensitivity, ion-channeling contrast, and an overall better material contrast than scanning electron microscopy (Sakai et al. 1999; Suzuki et al. 2004), SIM is frequently used to quantify the ultimate resolution of FIB systems. Figure 3.1b represents such an SIM image where the "imaging resolution" may be deduced from the image sharpness produced by the instrument, that is, the ability to distinguish two objects spatially. This resolution is expressed in terms of contrast difference according to the Rayleigh criterion, which can be expressed as follows:

$$C = \frac{I_{max} - I_{min}}{I_{max} + I_{min}}$$

In practice, FIB systems delivering low probe currents, of the order of a fraction of a picoampere to limit surface-engraving processes, allow high-resolution images to be acquired. This is facilitated by the use of image acquisition systems having integration and smoothing capabilities. Generally, specific test samples are used, which are made of two different elements having very different secondary electron emission yields such as gold (Au) structures deposited onto a graphite (C) substrate. From the resulting images, a physical distance, the resolution limit, is deduced for a signal contrast variation around 10%. At this point, it is important to insist that this measure does not necessarily indicate the practical capability of the instrument in question to modify any target material at the same scale (Orloff et al. 1996; Rue 2009). The main limitation encountered arises first from the very low probe currents that are used. Certainly, this helps preserve the sample from too intense sputtering effects but on the other hand, the probe currents remain too weak (<1 pA) to allow reproducible patterning.

3.1.3   FIB Lithography on Sensitive Media

This other promising application was intensively investigated in the early 1980s by several teams, notably those of Seliger and later of Kubena (Kubena et al. 1991). This technique initially used sensitive media or resists, very similar to those used in lithography techniques employing photon or electron irradiation. The results obtained by Kubena in particular very soon reached a level of performance almost comparable to the best results ever obtained in electron beam lithography (EBL). The realization of features fabricated using FIB lithography with lateral widths as small as 8 nm in a poly(methyl methacrylate) (PMMA) resist layer was in particular demonstrated (Kubena et al. 1991).

Some years ago, we investigated the potential of FIB lithography using the classical organic resist PMMA (Gierak et al. 1997b). Our objective was to evaluate the resolution and the shot noise effects caused by statistical fluctuations found by Matsui et al. (1986) when using this resist in very high-resolution FIB lithography. For this purpose, several arrays of lines were written using a 30 keV gallium probe focused in a 10 nm spot size and transporting a current of about 10 pA. After FIB exposure, the features were developed using methyl isobutyl ketone-propanol solution (1:3) for 30 s followed by a rinse in pure propanol (10 s). A 10-nm-thick aluminum layer was then deposited before lift-off in hot trichloroethylene (Figure 3.2). Reproducible arrays of aluminum lines were observed for irradiation doses per unit length ranging from 3×10^7 to 3×10^9 ions/cm. For doses exceeding 3×10^8 ions/cm, the influence of secondary electrons causes a lateral parasitic exposure and cross-linking of the PMMA occurs. The exposure latitude was thus found to be very narrow, limited between 3×10^7 and 3×10^8 ions/cm. The minimum linewidth we have obtained using PMMA was found to be close to 20 nm.

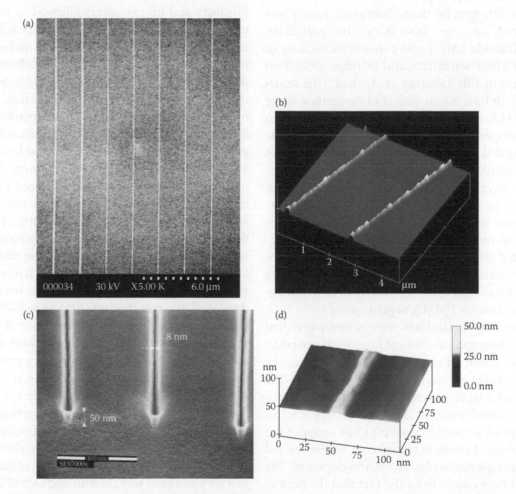

FIGURE 3.2 (a) Scanning electron microscopy image of an array of aluminum lines obtained using FIB lithography and lift-off using a 50-nm-thick PMMA layer. The minimum measured width is 20 nm. (b) Atomic force microscopy (AFM) image of an array of aluminum lines obtained using FIB lithography on the same sample after aluminum deposition and lift-off (ions, Ga+; energy, 30 keV; ion dose, 3×10^7 ions/cm, silicon substrate) (Gierak et al. 1997b). (c) SEM image of an array of FIB-etched lines on a multilayer sample AlF$_3$ (50-nm-thick)/ GaAs, for an incident ion dose of about 92 nC/cm. The sample is tilted at 45°, the edges of the resist-etched layer appear transparent, and the lines in GaAs have a very reproducible width of 8 nm. (d) AFM image of a single line obtained by FIB lithography in the negative mode on a 50-nm-thick AlF$_3$ layer (ion dose 1.97 nC/cm, Si substrate, Ga+ ions, 30 keV). The linewidth is 10 nm and the total height is nearly 20 nm.

Chapter 3

A further important limitation, which we also encountered with EBL, is related to the extremely thin resist layers that must be used (thickness around 50 nm). This limitation rendered the lift-off and final transfer of the features painted in the resist quite complicated and difficult. The reasons for this are related to the mass of the ions (gallium) and to their limited energy (usually some tens of keV). As a consequence, the penetration depth of such ions is much smaller than that of other particles and hence only very thin resist films were satisfactorily processed using FIB. An alternative that was investigated later consisted in using ions of lower atomic mass and higher energy (>100 keV), but the benefits, obtained at the price of increasing instrumental complexity, never demonstrated substantial improvements, in particular with respect to the EBL technique.

Ion beam lithography using inorganic resists was also explored in our laboratory. In particular, aluminum fluoride (AlF_3) was considered because of its relatively lower sensitivity and intrinsic resolution of about 2 nm in EBL (Murray et al. 1984). The sensitivity of AlF_3 in EBL is four orders of magnitude lower than for PMMA, and flux effects have been reported in the self-developed positive regime. Following this, we have investigated the applicability and performances of AlF_3 thin films (50 nm thick) used as negative inorganic resist layers for FIB nanolithography (Gierak et al. 1997b). In comparison with PMMA resist, we obtained better resolution limits. Indeed, we demonstrated that 10-nm-wide lines could be very reproducibly fabricated using a Ga^+ beam of 30 keV incident energy. The resist sensitivity was found to tolerate around 10^{10} ions/cm, a value that is two orders of magnitude lower than for PMMA organic resist.

FIB lithography capabilities were recently revisited with a new lithographical concept proposed by Arshak (Arshak et al. 2004; Gilmartin et al. 2007). Here, the aim is to organize a bilayer resist on a sample. The superficial resist layer is first irradiated with gallium ions to paint a mask pattern. Then, using a dry reactive ion etching process using oxygen (O_2), the image of the gallium-implanted mask is transferred to the second resist layer using oxygen plasma dry development. The main benefit here comes from the fact that the performances are significantly improved in terms of writing speed and of final structure aspect ratio. Furthermore, the writing process can be achieved for very low ion fluences (of the order of 10^{12} ions/cm²) and the final development is achieved in a single step. Finally, this technique, which allows deep sub-100 nm feature fabrication and high-aspect-ratio resist features (up to 21:1), is suitable for use in applications requiring high-resolution and high-aspect-ratio lithography. Indeed, the International Technology Roadmap for Semiconductors (ITRS) describes the development of new lithographical methods as one of the biggest challenges facing the semiconductor industry as it progresses to smaller geometry nodes (ITRS 2007). FIB lithography has significant advantages over alternative nanolithography techniques, particularly when resist sensitivity, topographical effects, proximity effects, and backscattering are compared.

3.1.4 Ion Implantation

The direct implantation of doping species, achieved via a focused beam in semiconductors, was first studied by Miyauchi and his coworkers followed by several industrial companies, which invested considerable efforts and resources in this field. The objective was to employ FIB implantation techniques for direct realization of complex integrated circuits (Miyauchi and Hashimoto 1986). In particular, in one interesting project, an attempt was made to implement a maskless ion implantation capability inside a molecular beam epitaxy chamber allowing growth of a GaAs or AlGaAs epitaxial layer and *in situ* 3D pattern doping during crystal growth. For this purpose, specific ion sources were developed to combine p- or n-doping species while retaining the conventional FIB local engraving and imaging capabilities. Nevertheless, this technique, which seemed promising at first sight, never reached a productivity level compatible with the anticipated production requirements. It turned out quite soon that the ion emitters (LMIS) that were specifically developed to accommodate alloys (Au–Si–Be) capable of delivering (after filtering) beams of p- or n-type doping species or even insulating material were not emitting enough current to ensure a satisfactory production. The development of FIB systems operating at voltages from 100 to 200 kV was then quickly abandoned. However, in the field of research on electronic components, this technique made it possible to obtain properties that were inaccessible by other standard manufacturing techniques by adjusting exactly the doping of certain regions in a very localized way (Matsui and Ochiai 1996).

3.1.5 Localized Engraving

This domain of application is nowadays the most widely used application area for FIB technology. This application consists in elaborating structures via direct ion etching well below the micrometer scale. In the typical energy range for the ions used (gallium ions having 5–50 keV energies), the general principle is that the kinetic energy of the incident ion and its momentum are transferred to the target through elastic and inelastic interactions. The penetration depth of such a gallium ion into a target is of the order of 10 nm (Ziegler 2008), a distance that is much shorter than for electrons of comparable energy (Figure 3.3a). In the case of inelastic interactions (electron–ion collisions), part of the ion energy is transferred to the electrons of the target material and this allows ionization to take place. In the case of elastic interactions (ion–nucleus collisions), the energy of the ion is mainly transferred to the atoms of the target in the form of displacements. This causes damage and subsequently sputtering phenomena at the surface of the irradiated material. The mechanism most generally used to describe this ion–solid interaction is the model of collisional cascades. If the energy received by a target atom exceeds a critical value, the atom will be moved from its initial site and, in a crystalline material, a pair—interstitial atom/vacancy—will be created. This primary collision can then be followed by others and gives rise to multiple secondary collisions as long as the energy communicated to the target atoms exceeds their own displacement energies. As an example, the critical displacement energy for materials is of the order of 5–20 eV, a value clearly exceeding the binding energy for the materials usually used in microelectronics, which is of the order of 1 eV. The consequence is that effective engraving processes for most of the materials traditionally used in microelectronics are obtained via controlled FIB irradiation.

This mechanical sputtering process ratio varies according to the nature of the target and the nature and energy of the incident ion. The total number of atoms sputtered when a pencil of ions is scanned over a target can be expressed as follows:

$$N = \frac{V \rho N_A}{A} = Y N_i N_x N_y$$

where V denotes the affected volume, ρ the target density, A the atomic mass of the sputtered material, and N_A Avogadro's constant. Y is the sputtering ratio (number of sputtered atoms from the target/number of incident ions), N_i the number of incident ions during the irradiation time, N_x the number of impact points in the x direction, and N_y the number of impact points in the y direction (Figure 3.3b). The total volume removed in a predefined digitized pattern consisting of N_x points in the x direction and N_y in the y direction can then be written as

$$V = \frac{Y N_i N_x N_y A}{\rho N_A}$$

The sensitivity of the target material to the sputtering effects denoted by S (typically in $\mu m^3/nC$) can be expressed as

$$S = \frac{YA}{\rho N_A e}$$

This sensitivity was measured (Orloff et al. 1996; Stark et al. 1995) in the case of a pencil of gallium ions at 30 keV. This parameter was found to vary from 0.08 for alumina, a material having low sensitivity to the

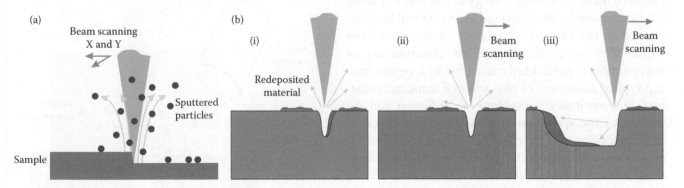

FIGURE 3.3 (a) Schematic representation of the FIB engraving process. Here, the target material sputtered by the ion beam is shown to redeposit randomly. (b) Schematic principle of an FIB etching process with a probe scanned from the left (i) to the right (iii). Prior to local erosion effects, the ion bombardment causes damage and redeposition of sputtered material occurs. These effects have to be considered and can be controlled.

ion erosion, up to 0.61 for gallium arsenide (GaAs) and peaks around 1.5 for gold, which is one of the most common and most sensitive materials for FIB engraving processes. The average sputtering ratio for various elements is about 2 atoms/incident gallium ion. It is then clear that the engraving speeds are limited, with respect to practical ion current densities, to some cubic micrometers per minute (1 $\mu m^3 = 10^{-9}$ mm^3).

Besides the etching speed, the homogeneity of the FIB engraving process is strongly affected by the properties of the target material (crystalline, polycrystalline, amorphous, doping, etc.) as well as by the angle of incidence of the ion beam. The bottom planes of FIB-etched structures are difficult to keep flat and usually exhibit residual roughness up to some micrometers. Also, ion irradiation generates electric charges that must be evacuated at the risk of causing drifts of the machining process or causing damage to the device structure. Finally, as a result of the ion impacts, ion bombardment on a target surface generates defects, which are crippling when the target material is a highly crystalline material. For some patterning applications such as quantum wells (QWs) and two-dimensional electron gases (2DEG) made of very sensitive III–V materials (GaAs materials and related compounds), the FIB technique remains difficult to use without extreme caution (Gierak et al. 2006).

3.1.6 The Injection of Reactive Gases and Metallic Precursors

As we have just seen, ion sputtering is a mechanical process, essentially governed by physical mechanisms of energy transfer. These mechanisms allow us to remove atoms from the target surface and to communicate to them an average energy of the order of some eV. Consequently, these ejected elements will be redeposited in the immediate vicinity of the zone from which they were extracted. This phenomenon of redeposition is particularly sensitive in a region having typical diameters of the order of some micrometers. It constitutes a problem of pollution for the surface of the specimen and in some cases such as deep engraving may generate partial refilling of previously processed structure. Finally, there is a risk of short circuit, between interconnects for example. To bypass this limitation and increase the productivity, a solution was proposed by Gamo (Kosugi et al. 1991), who suggested associating the FIB technique with reactive gas injection technology. The aim was to

benefit from the reactive ion etching improvement, allowing the etching speeds for many materials to be increased dramatically and eliminating the effect of redeposition (Figure 3.4).

Chemically assisted FIB engraving processes are based on the local injection in the working chamber of a reactive gas, introduced by means of a microactuated capillary positioned in the immediate vicinity of the working area. The surface of the sample adsorbs a part of the injected gas that readily reacts with sputtered atoms of the target by forming volatile compounds. These etching volatile products are not redeposited since they are evacuated by the pumping system of the chamber. Thanks to this, an appreciable improvement of the engraving speed by factors varying from 10 to 100 according to the material and the gas mixture is achieved (Matsui and Ochiai 1996; Reyntjens and Puers 2001). Nowadays, a plurality of etching processes and recipes have been developed for improving etching selectivity between given materials, reducing redeposition, and allowing significant improvement of the aspect ratio (width/depth) of the engraved features. This functionality is a parameter of the highest importance in modern device-editing applications where lateral resolution and deep reconnects are simultaneously required. As limitations of this method, we can mention contamination effects of the sample surface and limitations in ultimate resolution due to scattering effects of the incoming ions in the adsorbed gas layers.

Another very important requirement in device-editing applications is the possibility of depositing locally conductive materials by means of an FIB system. A first approach was explored in which a noble

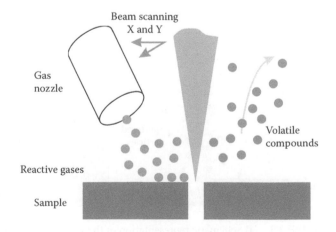

FIGURE 3.4 Schematic representation of a chemically assisted FIB engraving process allowing etching speeds to be improved and redeposition effects minimized.

metal LMIS (gold) emitted molecular ions or charged clusters, which were then deposited on the surface of a sample (D'Cruz et al. 1985). The lateral resolution of the deposited metal structures was found to be limited by the large energy dispersion of the emitted cluster beams (several tens of eV) and was never improved to values better than some micrometers. This disappointing result was found to be a definitive limitation of this technique, which was, on the other hand, capable of creating high-purity deposited films.

More recently, a technique for depositing conductive media by FIB-induced deposition has been developed: ion-beam-induced deposition. This technique is based on a mechanism of energy transfer between the secondary electrons generated by the impact of the FIB probe onto a sample surface. This energy transfer causes local "cracking" of gas precursor molecules injected and adsorbed on a target surface (Figure 3.5). The nonvolatile compounds of this reaction, which are generally selected for their metallic or insulating character, form a solid-state deposit adhering to the surface of the target. The materials generally deposited are platinum (Pt) or tungsten (W). In the case of tungsten, the organometallic precursor gas is the compound $W(CO_6)$. It is also possible to depose silica locally by using as precursor gas the compound 1,3,5,7-tetramethylcyclotetrasiloxane (TMCTS) in the presence of oxygen (O_2), or even in the presence of steam. A recurring criticism of these

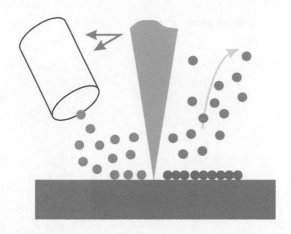

FIGURE 3.5 Schematic representation of local deposition assisted by FIB. The gas precursor introduced by the capillary is decomposed by the secondary electrons generated by the interaction of the primary ion beam with the target.

very popular methods concerns the chemical composition of the conductive deposits obtained, which incorporate a high percentage of carbon and therefore some traces of gallium. In consequence, relatively important resistivities are observed for the deposited conductive structures (Matsui and Ochiai 1996). Nevertheless, as this technique can be easily combined with direct engraving, it is widely used since it becomes possible using the same tool to engrave or deposit structures locally both with 3D control.

3.2 The Liquid Metal Ion Source: Still the Best Choice for FIB?

Since the early developments (Seliger et al. 1979), the FIB technology has been continuously and intensively improved in terms of lateral patterning resolution and of processing capabilities. A faithful illustration of this tendency may be found in the etching speeds and the transported current values that have been improved by factors exceeding 100 in some cases. The resolution of the patterned structures is a second illustration of this progress. FIB technology developments were started with the aim of achieving submicron patterning; nowadays, recent FIB machines are all targeting nanometer-size patterning.

3.2.1 Principle and General Properties of Liquid Metal Ion Sources

The sources of ions emitted from a molten metal, more commonly called LMIS, constitute a particular class of

field-effect emitters. These emitters are compact, easy to produce and use, and have general properties that make them ideal for integration in dedicated FIB research instruments intended for micro- or nanoprototyping:

- Moderate emitted current typically 2–3 μA (minimum energy width)
- Point emitter with a narrow emission area (smaller than 10 nm)
- Emission current distributed over a relatively broad solid angle (more than 0.1 sr)

The LMIS emitters are point-like ion sources and their intrinsic brightness remains among the highest known to date (Melngailis 2001; Utlaut 2009). The operation of an LMIS, schematically represented in Figure 3.6a and b, is obtained by the diffusion of a supply metal film maintained in a liquid phase on the

Chapter 3

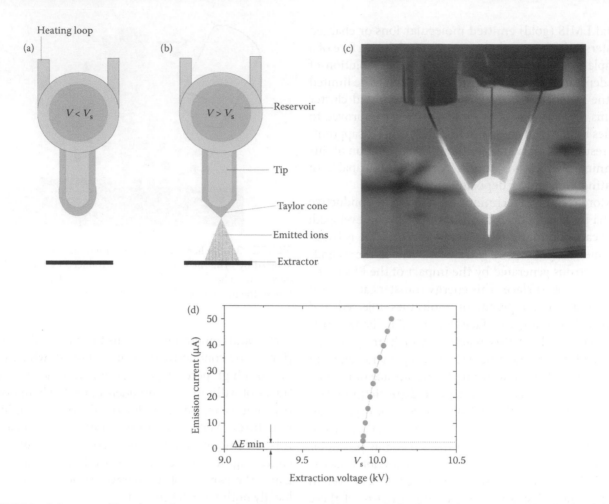

FIGURE 3.6 Schematic of liquid metal ion source (LMIS) operation. (a) Tip-applied voltage is below critical threshold voltage V_s. (b) Tip-applied voltage exceeds extraction voltage V_s; the Taylor cone and ion emission appear. (c) *In situ* picture of an LMIS heated to around $T \sim 900°C$ during emission tests into a high-vacuum chamber. (d) Current/voltage (I/V) characteristics of the LMIS.

surface of a refractory metal tip. By subjecting this tip, which is covered with the liquid metal film and placed under vacuum, to a positive potential difference V of several kilovolts, the metal film at the apex of the tip will become distorted by the local electric field and will take the form of a stable conical structure. This equilibrium, where electric and mechanical forces balance, was studied around the beginning of the 1960s by the British physicist Geoffrey Taylor and is therefore known as the Taylor cone (Taylor 1964).

Three forces determine the equilibrium of this conical structure: the electrostatic force, the surface tension, and the pressure inside the molten metal film. The electric forces tend to draw the surface away from the molten metal film according to the lines of the applied electrostatic field, while the surface stress tends to preserve the cohesion of the liquid. The two forces are inversely proportional to the square of the radius of curvature of the liquid surface: the more the

free surface of the metal is curved (the radius of curvature is estimated to be about some nanometers), the stronger will be the centrifugal attraction due to the local electric field, but the larger will be the centripetal cohesive force, due to the surface tension. In experiments, one observes, for a critical tension V_s (threshold voltage) applied to the point, an ejection of matter in the form of ions that occurs at the apex of the tip (Figure 3.6b). By increasing the voltage gradually, the emission current increases with a slope that is directly governed by the tip radius and the supply function of the needle that supports the ion emission (Figure 3.6c). When the emission current reaches a value of several tens of microamperes, an emission of charged clusters is observed as well as the ion emission.

3.2.1.1 Energy Dispersion of the Emitted Beam

This parameter is a particularly important factor because it governs the magnitude of the spreading caused by the chromatic aberration of the lenses; as we will see, this constitutes the principal limitation for obtaining ion probes in the nanometer range. Unlike sources of electrons operated in field emission (FE), for which the energy dispersion is at worst about 0.2–0.3 eV, the energy dispersion of the emitted ions for an LMIS is much greater. This dispersion has been measured experimentally (Swanson and Bell 1989), and was found to increase with the emission current following a model proposed by Knauer (1981). According to this model, the variation of the kinetic energy of the ion around its median value eV_o would be due to a relaxation of the Coulomb potential energy of the beam leading to a dispersion of energy ΔE of the form

$$\Delta E = 5.8\pi \frac{e}{4\pi\varepsilon_o}\left(\frac{m}{V_o}\right)^{1/3} r_s j_o^{2/3}$$

where V_o is the acceleration potential of the beam, e and m are the charge and the mass, respectively, of the particles, ε_o the dielectric constant, r_s the effective source radius, and j_o the emitted current density.

In the case of gallium ions, this dispersion has been the subject of several experimental studies. A minimum value of 4.5 eV for an emission current of about 2 μA is found by several authors (Swanson et al. 1979; Prewett et al. 1982; Mair et al. 1983; Nakayama and Makabe 1993; Beckman et al. 1996). It was quickly verified that this high value of the energy dispersion could not be explained on the basis of the usual theory involved in field evaporation mechanisms (Mair et al. 1983). For a gallium surface subjected to an electrostatic field E, a good approximation of the full-width at half-maximum (FWHM) of the energy distribution of simply charged ions is given by the following expression (Mair et al. 1983):

$$\text{FWHM} = eE\Delta x = 0.076 \cdot eV^{3/2}E\phi^{-1/2}$$

In the case of a ϕ value around 4.5 eV and an evaporation field $E = 15$ V/nm, this expression yields $\Delta E = 0.5$ eV, a value that is almost 10 times lower than the experimental results. To try to explain this discrepancy between the experimental values, all of the order of 4.5 eV, and this calculation, more sophisticated models (Suvorov and Forbes 2004; Van Es et al. 2004) or better understanding of the physical phenomena involved is necessary (Forbes and Mair 2009).

3.2.1.2 Angular Intensity

The system LMIS + extractor electrode forms a so-called "weak" lens; the angular growth m of the beam will vary with the extraction voltage applied and with the geometry of the system LMIS + extractor electrode. This can be expressed as follows:

$$m = \frac{\alpha}{\theta_o} = M_c^{-1}\left(\frac{V_o}{V}\right)^{1/2} \tag{3.1}$$

where θ_o and α are the initial and final angular apertures, M_c is the linear magnification of the optics relative to the virtual image plane, V_o the applied voltage in the "object" plane where the particles are emitted (a value ranging between 0.1 and 1 eV), and V the voltage applied to the particle in the image plane (final acceleration voltage).

One can then write the angular intensity of the emitted beam as $dI/d\Omega$ using the geometry illustrated in Figure 3.7a:

$$\frac{dI}{d\Omega} = \frac{Jr^2(1-\cos\theta_o)}{(1-\cos\alpha)} \tag{3.2}$$

Then, combining Equations 3.1 and 3.2, we obtain

$$\frac{dI}{d\Omega} \cong \frac{Jr^2}{m^2}$$

In the case of a source of gallium ions as described in the literature (Bell and Swanson 1986), the nominal angular density on the axis is about 20 μA/sr for an emitted current of about 2 μA. In addition, we have seen that the angular magnification m varies with V, the tension applied to the particle in the image plane (final acceleration voltage). It is thus possible to vary the angular density of an ion source over a relatively large range. Thus, within the framework of the project NanoFIB (NanoFIB 2007), this effect was investigated and angular density values as high as 80 μA/sr were obtained (Van Es et al. 2004).

3.2.1.3 Diameter of the Virtual Source

From the physical point of view, as explained above, an LMIS is a point-like source of ions with an emissive zone some nanometers in diameter. But from the geometrical optics point of view, the only parameter to be considered is the virtual source diameter d_v, that is, the zone from which the emitted ions seem to come when

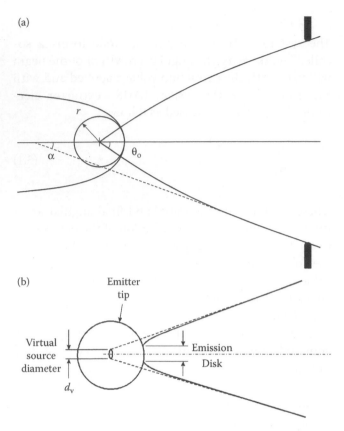

(a)

(b)

FIGURE 3.7 (a) Illustration representing the initial trajectories of the ions emitted in the object plane (emitter) and in the exit plane: the diaphragm of the electrode of extraction. (b) Principle of the determination of the virtual source size of an LMIS by extrapolation of the tangents to the trajectories of the emitted ions, which define the diameter of the virtual source d_v. In this diagram, the physical apex of the tungsten supporting needle is represented as a sphere.

they enter the optical system after acceleration. This virtual source diameter d_v is larger than the emissive zone. This widening effect can be explained on the basis of the study of Knauer (Knauer 1981) and by interactions between particles (Boersch effect). The latter occurs in the zone of very high charge density in the immediate vicinity of the emissive surface, where ion densities can reach values as high as 10^{22}–10^{23} ions/cm³.

The diameter of the virtual source has been estimated by extrapolating the tangents of the trajectories of the ions obtained at the entry of the electrode that fixes the acceleration energy back to a plane behind the emissive surface (Figure 3.7b). This virtual source diameter was initially measured in experiments using well-calibrated transfer optics (Komuro et al. 1983). Later, the virtual source diameter has been calculated (Ward 1985) using a Monte Carlo-type program taking into account the Coulomb interactions between

particles in the immediate vicinity of the emissive zone (Boersch effect), and values ranging between 50 and 100 nm were estimated. It was also shown that the virtual source diameter increases with the current of emitted ions. For $I = 2\ \mu A$ and a gallium LMIS, it is generally accepted that d_v falls between 30 and 50 nm.

3.2.1.4 Brightness

The brightness of the source of ions is the last important parameter since it will fix the value of the current in each plane crossed by the beam. One generally uses the concept of unnormalized brightness or the term "reduced brightness," which has the accelerating voltage in the denominator. The brightness can be expressed as

$$B = \frac{dI}{d\Omega \cdot dS} \quad \text{or} \quad B_{\text{red}} = \frac{dI}{d\Omega \cdot dS \cdot V}$$

The (unnormalized) B is expressed in A/m² sr and we find

$$B = \frac{4 \cdot dI/d\Omega}{\pi \cdot d_v^2}$$

Writing $dI/d\Omega = 20\ \mu A/sr$ and $d_v = 50$ nm, we finally obtain $B \sim 10^6$ A/cm² sr.

The brightness of the LMIS is one of the highest values ever reported for an ion source, making it almost ideal for FIB applications.

3.2.2 Realization: Example of an LMIS

The number of ionic species available with the LMIS does not cover all metals. Although it is possible, in certain cases, to use some of these elements in the form of alloys and then "sort" the desired ionic species, the following constraints govern the realization of an LMIS capable of emitting ions from a given metal:

- The vapor pressure of the metal to be ionized must be sufficiently low so that the supply metal has a vacuum liquid phase under high vacuum. For example, one cannot use magnesium or zinc since these metals sublime when one attempts to melt them under high-vacuum conditions.
- The wettability of tungsten (tip and filament constituent material) by the supply metal to be ionized must be good. The supply metal must diffuse into grooves etched in the surface of the tungsten tip. This condition is not fulfilled for the couples silver/tungsten and copper/tungsten.

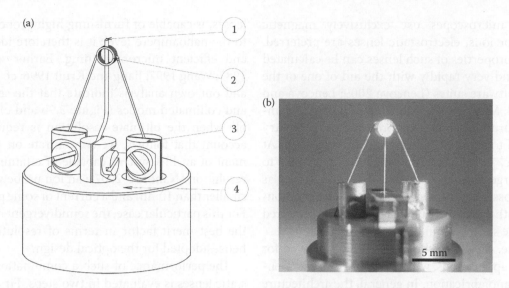

FIGURE 3.8 (a) Geometry of a liquid metal ion source for gallium ions developed at the LPN-CNRS. (1) Tungsten tip (0.38 mm in diameter) terminated with a conical apex. (2) Filament heating reservoir consisting of a 0.125 mm tungsten wire rolled up in the form of jointed loops. (3) Grip contacting support. (4) Mounting base. (b) Source assembled after loading with the supply metal (gold in this case).

- Alloy formation between the tip, heating filament, and the supply metal is forbidden. Rapid corrosion of the filament would ensue and thus destruction of the source. This is the case, for example, with the couple aluminum/tungsten.
- A crucial point for the applications of other supply elements is their melting temperature. Many materials are therefore used in the form of alloys, mostly in the eutectic composition (Bischoff 2008) allowing operation at reduced temperatures.

A source of ions is composed of two main parts, both fabricated starting from polycrystalline tungsten wires. The tip, drawn up in the shape of a stem, is mechanically polished with a conical shape at its end (Reference mark 1 in Figure 3.8a). The radius of curvature of the final apex strongly depends on the method of polishing used and on the application considered. This value is generally between a few tens and a few

hundreds of nanometers. A filament rolled up in the form of a series of circular loops is then inserted around the tip to constitute a reservoir holding the supply metal to be ionized (Reference mark 2 in Figure 3.8a). The reservoir is finally loaded by immersion in a crucible containing the desired metal or alloy maintained in the liquid phase by Joule heating, preferably in ultrahigh-vacuum conditions (UHV). The result of this last operation makes it possible to fill the reservoir with a drop of metal having a mass of about 1/3 of a gram that contains ~2.5 × 10^{21} gallium atoms.

The theoretical autonomy, for an emission current of a few microamperes, corresponding to a flow of about 1.25 × 10^{13} gallium ions per second should thus reach several thousands of hours. Unfortunately, in practice, LMIS lifetimes do not exceed a value ranging between 500 and 1500 h; this is mainly for reasons of contamination of the emitting tip by "back sputtering" effects (Galovich 1988).

3.3 Basics of Charged Particle Optics: Application to an FIB Column

3.3.1 Elementary Optics

The focusing and dispersive properties of charged particle optics devices that use static or quasi-static electric or magnetic fields can be described in the terminology of geometrical optics. Many detailed studies of lenses, deflectors, mirrors, and less common optical elements for charged particles have been made since 1927, when Hans Busch showed that rotationally

symmetric magnetic or electrostatic fields act as lenses for electrons. Those first studies were made in an attempt to characterize the concentrating effect of such fields in cathode ray tubes but they soon led to the development of the first electron microscopes by Ernst Ruska and Max Knoll. Today, such instruments have reached such a high degree of perfection that the microscopist can reach atomic resolution (Hawkes 1995; Kirkland et al. 2007; Cockayne et al. 2009).

Chapter 3

Electron microscopes use exclusively magnetic lenses but for ions, electrostatic lenses are preferred. The optical properties of such lenses can be calculated accurately and very rapidly with the aid of one of the dedicated software suites (Lencova 2009; Lencová and Zlámal 2008; Munro 2009; Munro et al. 2006). A family of configurations can be explored and the geometry best adapted to a given application can be selected. At one extreme, a high current may be desirable, to remove a large volume of material from a target as quickly as possible; for imagery and for exact positioning, on the other hand, weak currents will be preferred to protect the sample from unwanted erosion.

In practice, the FIB technique is now employed for very diverse applications, which extend from microengraving to nanofabrication. In general, the architecture of an FIB column is first developed to deliver as small a spot as possible for a given current, after which the suitability of this basic geometry for the many tasks it may be called on to perform is investigated. In particular, a very important requirement for the semiconductor industry is flexibility, for the current transported may vary between 0.1 pA and some nanoamperes. Most FIB columns therefore consist of two electrostatic lenses, which are operated in one of the four modes shown in Figure 3.9. The crossover mode (Figure 3.9a), in which a beam-defining aperture is situated between the two

lenses, is capable of furnishing high probe currents up to the nanoampere level; it is therefore ideal for rapid and efficient micromachining. Earlier work (Orloff 1987; Wang 1997; Jiang and Kruit 1996; cf. Utlaut 2009) and our own analysis indicate that the semidivergent and collimated modes (Figure 3.9b and c) are preferable when the ultimate resolution is required. In the account that follows, we concentrate on the development of an FIB system capable of attaining very high resolution (NanoFIB 2007): an ion probe with FWHM smaller than 10 nm and a current of some picoamperes. For this particular case, the semidivergent mode offered the best merit factor in terms of resolution and was hence adopted for the optical design.

The performance of such a combination of electrostatic lenses is evaluated in two steps. First, the distribution of the electrostatic field is established, typically by the finite-element method. Trajectories through this field are then calculated and the size of the spot and the current distribution within it are then determined. If the optical system were perfect, the spot size, the image of the source, would be obtained simply by multiplying the source size δ by the magnification factor M of the optical system. Like all optical systems, however, an ion optics column suffers from various defects or aberrations, some of which are severe. Such defects generate additional contributions to the spot

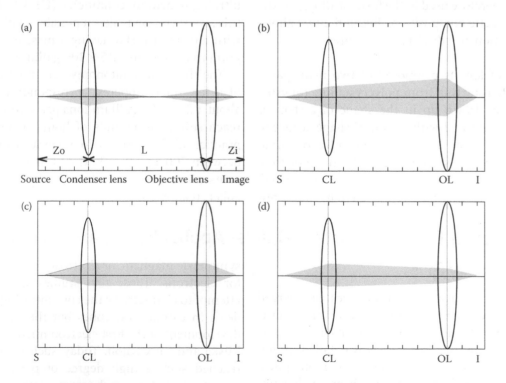

FIGURE 3.9 Main charged particle optics modes used in conventional FIB systems. (a) Mode known as "crossover" capable of conveying a high-current density, (b) semidivergent mode allowing a high resolution to be achieved, (c) parallel, and (d) semiconvergent modes.

size, thus broadening the image of the source produced at the target.

Several terms contribute to the resulting spot diameter. Before examining these in detail, however, we need to examine the definition of spot size more closely. Although the FWHM is widely used and very convenient, it is not the only possible measure and has the inconvenience that it is not independent of assumptions about the probe profile (see e.g., Reimer 1998). A better choice is the diameter of the disk containing a specified fraction of the total current, typically $d(0.5)$, the diameter containing 50% of the probe current. For extensive discussion, see the authoritative paper by Bronsgeest et al. (2008), which should be read in conjunction with an earlier paper by Barth and Kruit (1996). We return to this after examining the individual contributions to the spot size.

3.3.1.1 Magnification Term

$$d_G = M \cdot d_v$$

where M denotes the linear magnification of the optical system and d_v is the size of the virtual source. For a two-lens system, we have $M = M_1 \cdot M_2$, where M_1 and M_2 are the magnifications of the two lenses. In the case that interests us here, namely an ion probe size smaller than the diameter of the point source, M must be <1. In practice, the magnification must often be well below one; in our case, we have set $M \approx 1/10$ so that the term $M \cdot d_v$ contributes little to the final spot size. It is important to remember that the probe current and the spot size are not independent for a given source. The various parameters are related as follows:

$$d_G = M \cdot d_v = \frac{2}{\pi \alpha_p} \left(\frac{I_p}{B_{red} V_p} \right)$$

in which α_p denotes the semiangle of the beam at the probe, I_p is the current, B_{red} is the reduced brightness, and eV_p is the energy of the beam at the probe, where e is the electron charge.

3.3.1.2 Chromatic Aberration Term

In the case of ion beams extracted from an LMIS, this term is one of the most important contributions, if not the most important. Chromatic aberration is the direct consequence of the large energy spread of the emitted ions. For such a source, and in the case of gallium, the most favorable supply element, this energy spread is

found to have a minimum around 5 eV for an emission current of ~2 μA. This value increases rapidly with the emission current. The focusing action of the electrostatic lens will hence not be uniform for all the particles but will vary according to their velocities. The focused spot becomes blurred to a "disk of chromatic aberration," the diameter of which can be expressed as

$$d_c = \alpha_p C_c \frac{\Delta E}{E}$$

where α_p is again the semiangle of the beam at the probe, C_c the chromatic aberration coefficient, ΔE the energy spread, and E the acceleration energy of the gallium ions. As explained above, we need to say exactly how the various quantities are defined. If we adopt the definition $d(0.5)$, we find

$$d_c(0.5) = 0.6 C_c \frac{\Delta E(0.5)}{eV_p} \alpha_p$$

where $\Delta E(0.5)$ denotes the energy spread within which 50% of the current lies.

The coefficient C_c is found to be characteristic of the geometry and excitation of the electrostatic lens used. It is therefore very important to select a lens configuration and excitation parameters giving the lowest possible value of C_c, keeping in mind that the coefficient C_c has a tendency to decrease when convergence increases. Asymmetric lens designs are usually considered to be the best choice here. For an extremely thorough study of electrostatic lens properties, see Lencová (2008) or Hawkes and Kasper (1996).

3.3.1.3 Geometrical Aberrations

All electrostatic lenses suffer from geometrical aberrations, which reflect the fact that such lenses provide linear point-to-point imaging only in the first approximation. The aberrations are classified as spherical aberration, coma, astigmatism, field curvature, and distortion. For probe-forming systems, only the spherical aberration is of real importance, being the only aberration that does not vanish or become negligible for points on or close to the optic axis. Its effect may be understood from the fact that particles traveling far from the optic axis experience a stronger focusing effect than those closer to the axis. As a result, ions emerging from a point on the axis will be focused not to a point but to a circular disk. The diameter of this "disk of least confusion" is least in a plane slightly closer to the lens than the Gaussian image plane and is given by

$$d_s = \frac{1}{2}\alpha_p^3 C_S$$

The corresponding $d(0.5)$ value is

$$d_s(0.5) = \left(\frac{1}{2}\right)^{5/2} \alpha_p^3 C_S = 0.18\alpha_p^3 C_S$$

where C_s is the spherical aberration coefficient. It is important to note that the spherical aberration term varies as α_p^3. For a small α_p, as in our case, the contribution from this aberration will be small.

Diffraction must also be considered in general but for ions the wavelength

$$\lambda = \frac{h}{p} = \frac{h}{\sqrt{2M_I eV}}$$

is short. Here, h is Planck's constant, M_I is the mass of the ion in question, and V is the accelerating voltage. For gallium ions with an energy of 30 keV, the wavelength is only 0.544 pm. Any contribution to the spot size is given by

$$d_D \approx \frac{\lambda}{\alpha_p}$$

which gives $d_D = 0.5$ nm for $\alpha_p = 1$ mrad. This is negligible compared with the other contributions.

3.3.1.4 Addition of the Various Contributions

It has been common in the past to estimate the resulting spot size by adding the individual contributions in quadrature:

$$d = \sqrt{d_G^2 + d_C^2 + d_S^2}$$

but this is known to be a crude estimate. A better estimate is obtained from the expression (Barth and Kruit 1996)

$$d_p = \left[\left\{ \left(d_D^4 + d_s^4\right)^{\frac{1.3}{4}} + d_G^{1.3} \right\}^{\frac{2}{1.3}} + d_c^2 \right]^{1/2}$$

In the absence of any diffraction contribution and a negligible spherical aberration term, this collapses to the quadrature value.

3.3.2 Example of an Optimization Project: An FIB Column Optimized for Very High Resolution (NanoFIB)

The FIB system examined here was developed in the LPN-CNRS exclusively for nanofabrication, that is, patterning at the sub-10 nm scale. We designed our ion optics column to allow the routine generation of ion probes with FWHM well below 10 nm. Our second requirement was to maintain the transported current in the range 5–10 pA depending on the size and position of the beam-defining aperture. In this FIB column, the latter (typically 5 μm in diameter giving a probe angle around 0.1 mrad) is placed in front of the first lens (Figure 3.10), just at the entrance of the ion optics. This arrangement is not appropriate for operating the system with an intermediate crossover. Unlike conventional FIB systems, therefore, it is not possible to modulate the transported current by varying the crossover position through a fixed aperture. On the other hand, accelerated ions having divergent or perturbed trajectories can be easily rejected, so that only emitted ions with paraxial trajectories can enter the optics and reach the target. The main benefit is that the so-called tails of the probe current distribution at the target level are strongly truncated.

Furthermore, our setup has the advantage that the ion gun where the ions are extracted and accelerated is

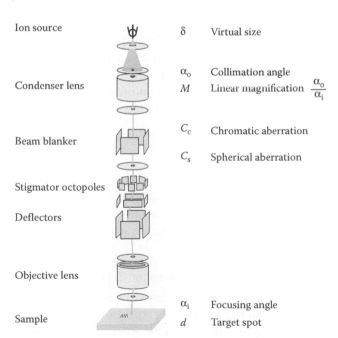

FIGURE 3.10 Illustration of the FIB optics concept detailed in this study and definition of the main parameters used for the calculation of the properties of the double-lens optics.

clearly separated from the ion optics where the beam is transported, focused, and scanned. This is particularly important because the extraction/accelerating region plays a major role in governing the size of the virtual source and the value of the angular distribution of the current in the emitted beam. This combination allows us to vary the source emission parameters widely without influencing the optical settings of the ion optics itself.

The performance of the ion optics part was optimized for an ion energy range between 30 and 40 keV. As illustrated in Figure 3.10, our optical column uses two asymmetric lenses working in the decelerating mode and for high-demagnification conditions. In this column, the deflection plates are located between the lenses, allowing a reduction of the final lens working distance (WD), a necessary condition to achieve a strong demagnification of the virtual source size.

For the attainment of very high resolution, the machining tolerances have to be in the micrometer range for the lenses and great care must be taken to avoid introducing dust or scratching any surfaces during assembly. Such tolerances are indispensable, for any residual imperfections will generate parasitic aberrations, which will contribute to the final spot size (Grivet 1972; Zworykin et al. 1945; Hawkes and Kasper 1996). These aberrations can arise for several reasons: first, from poor machining of the aperture of the central electrodes of the lenses, or from a tilt of the lens, or from off-center displacements with respect to the optic axis. The ensuing contributions, which are listed here by order of decreasing influence, can now be very precisely evaluated numerically but are difficult to predict or identify unambiguously. As reported long ago by Zworykin (Zworykin et al. 1945), poor tolerance of the column can introduce additional defects greater than the classical aberrations of the designed optics. This argument justifies the extreme attention and care that must be taken during the mechanical realization and assembly of high-resolution ion optics.

The performance that we present in Figure 3.11 and below is that attained with the NanoFIB instrument developed within the framework of the European project of the same name (NanoFIB 2007). Thus, for the following conditions: ion energy = 40 keV; source size (d_v) = 30 nm; E = 5 eV; α_o = 0.1 mrad, α_i = 0.84 mrad; and distance from condenser lens to target ~85 mm, the various contributions have the following values:

Magnification:

$$d_G = M \cdot d_v, \quad d_G = 3.58 \text{ nm}$$

Aberrations:

$$d_c = C_c \frac{\Delta V}{V} \alpha_p, \quad d_c = 2.93 \text{ nm}$$

$$d_s = \frac{1}{2} C_s \alpha_p^3, \quad d_s = 0.03 \text{ nm}$$

$$d^2 = d_G^2 + d_c^2 + d_s^2, \quad d \approx 5 \text{ nm}$$

This calculated performance level was then checked and validated by comparison with experimental results, presented below. This is of particular importance since the question of FIB practical resolution and overall performance is a sensitive item. For the reasons we have already listed and in the light of the application parameters we summarize below, it is important to quantify experimentally the resolution of a given FIB instrument and to verify it regularly. There are many parameters that influence directly and profoundly the performance of FIB machines. From a strictly applied point of view, FIB technology based on gallium ions is a technique of localized structuring and not a technique of destructive imaging microscopy. In this sense, the relevant criterion for quantifying the resolution of any given FIB instrument is its ability to transfer structures having lateral dimensions as small as possible. This can be measured in terms of a number of features or points per unit length, for example. This resolution is, however, found to be dependent on the

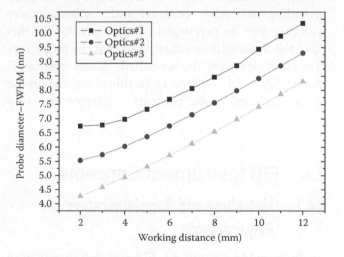

FIGURE 3.11 Evolution of the diameter of a gallium ion probe calculated for different generations and designs of nano-FIB optics. The improvements are mainly related to the design and the improvement of the properties of the electrostatic lenses and specifically of the C_c term.

physicochemical characteristics of the target materials (surface roughness, structural morphology, conducting or insulating characteristics, etc.) and also of the pattern geometry itself.

The preferred alternative consists in defining the effective FIB resolution as the width of the smallest feature (point, line) that the system is able to transfer via the given probe. This may be established by local pulverization (high fluences ~10^{16}–10^{18} ions/cm²), by injection of defects, local disorder, or by modification of the physicochemical properties of the surface of the substrate (low or very low fluences ~10^{12}–10^{14} ions/cm²). If this criterion is adopted, it is clear that the size of the probe itself is no longer the only parameter involved and the "selectivity" becomes an important factor here. The selectivity is defined as the ability to modify a precise region while preserving the adjacent areas.

The relevant factors to consider are thus the size and the profile of the current distribution delivered by the optical system at the target surface. Another important quantity is the density of current available in the ion probe. As an example, to allow the reader to appreciate the required level of accuracy for FIB focus fine adjustments, we present in Figure 3.12 a schematic representation of the envelopes of the trajectories calculated at the target level, in the planes of the "Gauss optimum" and of the "circle of least confusion" (Figure 3.12—planes 1 and 2, respectively). When adjusting the focus of a scanning ion microscope, the operator will instinctively optimize the "image sharpness" or image information content. The Rayleigh criterion providing maximum contrast and resolving power will therefore be privileged by the operator. In this case, the focus will be adjusted so that the target surface coincides with the so-called "Gauss optimum" plane (Figure 3.12—plane 1). In this plane, the profile of the incident probe presents a narrow and very

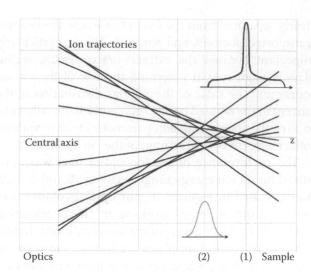

FIGURE 3.12 Schematic representation of the trajectories of ions at the exit of an electrostatic lens with aberration. (1) Geometrical plane of the "Gauss optimum" with a profile presenting not only a narrow peak but also side wings. (2) Optimum plane for FIB writing experiments ("circle of least confusion") giving a bell-shaped profile for the current distribution.

intense central part but with lateral tails that are by no means negligible.

However, our calculations of the ion probe diameters (FWHM) show that the probe diameter is smallest in a different plane, requiring a stronger excitation of the lenses. This is known as the plane of the "circle of least confusion" (Figure 3.12—plane 2) and is the preferred plane in high-resolution electron microscopy (in the absence of aberration correction). Here, the profile of the probe is bell shaped. The lateral extension of the ion distribution is much smaller and the current density is highest. Hence, this mode is to be preferred when attempting to achieve maximum FIB patterning resolution and selectivity. From this result, it is clear that FIB optics have different optima for (1) best imaging resolution and (2) highest patterning resolution.

3.4 FIB Instrument Concepts

3.4.1 Metrology and Nanofabrication Requirements

As discussed in Section 3.1, FIB systems already have a history of more than 30 years, starting with early instruments that were mainly intended for thin film analysis and early exploratory lithography applications. From the beginning, almost all systems were based on Ga LMIS emitters owing to their excellent performance and handling characteristics. Alloy emitters were employed in the 1980s for exploitation in laboratory applications but have never been developed to an acceptable usability level. The beam energies of these tools were typically set between 30 and 100 kV, yielding spot sizes from around 50 nm to a few 100 nm.

A major breakthrough in FIB technology and applications appeared in the 1990s in connection with the requirements of transmission electron microscopy (TEM) sample preparation and for 3D inspection—mainly assigned to semiconductor devices. This was the era of the successful dual/cross-beam instruments combining FIB for milling and scanning electron microscopy (SEM) for *in situ* observation. A solid equipment market established itself, driven by companies like FEI, Orsay Physics, Seiko Instruments, JEOL, and Zeiss. The beam resolution was reduced down to the 10 nm level and below at beam energies converging on 30 kV—all with Ga ions for the applications mentioned above plus the advancing nanotechnologies and materials science in general. The applications all involve milling—removing material in a controlled way—and depositing material in combination with injected precursor gases.

As FIB-based nanofabrication allows us to address many application fields ranging from surface modification at the nanometer scale up to milling large holes, state-of-the-art FIB instruments allow the beam current to be selected, typically from 0.1 pA up to a few tens of nA. In most cases, this is done by switching between different mechanical apertures, sometimes in combination with modified lens operation, but the beam emission current always remains constant in the range of a few µA. Depending on the beam current selected, spot sizes vary from below 5 nm in the lowest beam current regimes up to a few 100 nm in high-current regimes.

The great majority of FIB instruments now available on the market are based on the very successful combined FIB/SEM architecture often referred to as "cross-" or "dual-" beam systems. In such a setup, the FIB column is situated on a platform together with an SEM, the two optic axes converging at the same point of the sample surface. The "cross-" or "dual-" beam architecture was developed in the early 1980s (Sudraud et al. 1988) (Figure 3.13a) and allows efficient *in situ* observation, nondestructive control, and real-time monitoring of FIB processes. Typical examples are microetching and microdeposition tasks. Because of its unrivalled performances, this combination immediately became very popular and was intensively used by the microelectronics industry. Nowadays, most FIB manufacturers propose this type of instrument (Figure 3.13b) (Morgan et al. 2006; FEI 2008; Orsay Physics 2009).

The combined FIB/SEM instruments generally consist of the following components and properties:

- The main system platform includes a high-vacuum chamber, which carries an electron beam column and an ion beam column. In most systems, the electron column is mounted in a vertical orientation whereas the ion beam column is inclined.

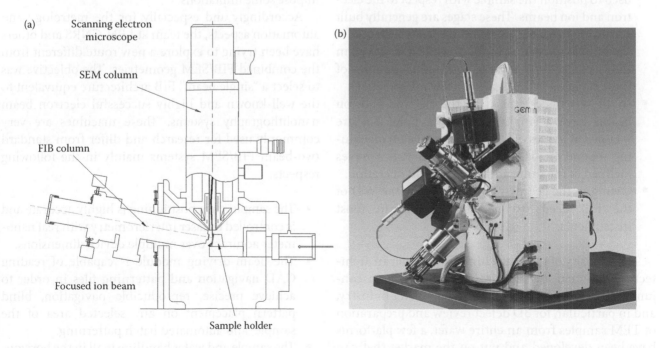

FIGURE 3.13 (a) The combined FIB/SEM initially developed by Sudraud. Here, the two beams cross at an angle of 83° on the sample. (b) Example of a commercial FIB/SEM station from Carl Zeiss NTS equipped with an FIB column and gas injection systems from Orsay Physics (From Sudraud, P., Ben Assayag, G., and Bon, M. 1988. *J. Vac. Sci. Technol. B* 6, 234; Orsay physics. 2009. Orsay Physics, Products FIB, http://www.orsayphysics. com/)

- The electron beam is mainly used for imaging. Electrons generated in the gun unit at the top of the column are focused onto the sample. By means of an electron deflection system that is typically of a magnetic type, the beam is scanned across the sample. The primary beam interaction with the sample results in a variety of secondary sources producing secondary electrons, x-rays, and backscattered electrons. All these signal sources originate at the sample at or close to the location of beam impact. The signals contain sample and surface information and are collected by suitable detectors; the corresponding images are then displayed to the operator.

- The ion beam is used to modify the sample, laterally and in depth. The principle of ion column operation is similar to that of the electron column but works with heavy ions, providing a much more violent impact at the sample (= sputtering). The ion–sample interaction is mostly used to drill holes in a well-defined geometry so that, for example, a thin cross section can be made, extracted, and analyzed in a TEM. Other applications include modifications on semiconductor wafers for defect inspection or circuit edit applications in device engineering (Melngailis et al. 1985).

- Inside the vacuum chamber is a sample stage that is used to position the sample with respect to the electron and ion beams. These stages are generally built according to an eucentric tilt design, which allows the sample to be kept in focus while it is moved in the *xy* plane. Typically, the positioning accuracy of these stages is on the scale of a few micrometers.

- In all systems of this type, the electron and ion beams have a single coincidence point but are inclined at angles of 50–55°; the ion beam machining operation is hence executed on samples inclined at the same angle by stage tilt operation.

- Owing to the geometrical restrictions, the WD of the FIB needs to be not <15 mm in most instruments.

The majority of these machines is based on an architecture designed for small samples; however, in conjunction with the needs of the semiconductor industry, and in particular, for 3D defect review and preparation of TEM samples from an entire wafer, a few platforms have been developed and put on the market that can handle entire 200- or even 300-mm-sized wafers (Seiko 2009; FEI 2009; Applied Materials 2009). The key characteristics for these machines and architectures are automation and operational stability—at the expense of some loss of ultimate performance in terms of patterning resolution and flexibility.

The combined FIB/SEM design concept is attractive for many applications as it allows the action of the ion beam to be observed with the electron beam in or close to real time.

However, with the advent of nanofabrication, new constraints have appeared in terms of positioning accuracy of the FIB spot and related specimen motion and positioning. For example, FIB manufacture of structures wider than the elementary patterning field (squares a few tens of micrometers in length) may require the initial pattern to be split into subpatterns that fit in individual elementary writing fields. To achieve this, field stitching via repositioning of each elementary subpattern with the writing resolution (some tens of nm) is often proposed. Concomitantly, FIB patterning of a large number of features in an automated way onto a preexistent structure also requires high-precision positioning. Such tasks are difficult to control without adequate metrological positioning tools. Indeed, the requirements here in terms of metrology are considerably more severe. In particular, in combined FIB/SEM systems, the sample position at high-tilt angles and the complex geometrical arrangements between sample and FIB column, which in some cases result in nonsymmetrical electromagnetic fields, may impose some limitations.

Accordingly and especially for the metrology and automation aspects, the team at LPN-CNRS and others have been trying to explore a new route different from the combined FIB/SEM geometries. The objective was to select a "single beam" FIB architecture equivalent to the well-known and highly successful electron beam nanolithography systems. These machines are very commonly used for research and differ from standard two-beam FIB/SEM systems mainly in the following respects:

- The sample stage positioning is highly accurate and is controlled by laser interferometry with real nanometer accuracy over multiple device dimensions.

- The beam-driving module is capable of reading CAD navigation and patterning files in order to achieve precise, reproducible navigation, blind pattern placement on any selected area of the sample, and automated batch patterning.

- The sample and wafer handling is all in the horizontal plane and allows careful planar sample mounting and leveling, thus minimizing artifacts such as positional or beam focus drifts during operation.

- The position of the beam–sample interaction is highly symmetrical and virtually field-free, thereby reducing beam-scanning distortions and beam-shape effects (astigmatism) as found in FIB/SEM machines with electron beam columns and/or complex mechanical arrangements in the vicinity of the sample.
- The WD for the ion beam can be selected over a wide range for optimized operation both for extremely high-resolution work (few mm WD) and large area patterning (10–20 mm WD).
- In order to keep the ion beam in focus over the entire sample/wafer area, all the techniques known from electron or laser beam lithography machines, such as sample height sensors and focus feedback loops, are readily available.
- In order to provide longer stability, the specifications of the dedicated control electronics and power supplies, such as high-voltage power supplies, scan amplifiers, and scan generators, are typically better by at least one order of magnitude than in conventional FIB/SEM instruments. This makes it possible to provide and specify excellent beam current and position stabilities.

Since the requirements were very similar to those of the EBL machines, the LPN-CNRS laboratory has selected this architecture for its new-generation FIB instruments; this choice should match the research objectives, namely, exploration of the ultimate performance achievable using an FIB (Raith 2009).

The beauty of this approach is that it allows (i) automated "no-imaging" wafer-scale applications, (ii) fabrication with automated field stitching for large continuous patterns, (iii) step and repeat overlay with local autoalignment or alignment based on pure laser interferometer accuracy. In connection with a special exposure mode, which operates with a continuous path-controlled stage motion, nanofabrication for extended (mm-sized) structures even without stitching (Raith 2008) can be added. Moreover, this ion beam lithography approach and system architecture allow FIB-based nanofabrication to be combined with EBL, optical prepatterning, and nano-imprint lithography processes.

In a unique and very complementary approach to these areas, a gas field-ion (He) system has recently been introduced by ALIS/Carl Zeiss NTS (Morgan et al. 2006), allowing beam resolutions down to the sub-nm level and medium energies; owing to the physics of this particular source technology delivering an extremely high

brightness but very low currents; however, the applications are focused on microscopy (Hill et al. 2012) and metrology at high magnification (and hence small fields).

Applications around ion beam lithography, implantation, and direct surface modification based on the use of alloy emitters and in conjunction with a dedicated large-area patterning platform have been approached in a very limited way as only a few commercial instruments had been made available (from Microbeam, Inc., Nanofab, Inc., and JEOL Ltd.). Typically, these instruments operated at high-beam energies (100–150 kV) and were limited to about 50 nm beam diameter. Most of them used various alloy sources in combination with an $E \times B$-type mass filter. However, these instruments historically did not achieve a technological or commercial breakthrough.

In recent years, advances in nanotechnology research, materials science, and semiconductor device engineering combined with stringent demands to limit the focused beam interaction volume in three dimensions to the smallest possible values and to manipulate the local chemistry have renewed interest in having a choice of ion species other than Ga. These demands are now spurring the development of a new style of nanofabrication platforms, combining multi-ion species, low energies, and high-patterning resolution.

3.4.2 Gas Delivery Systems for Enhanced Etching and Deposition

In order to extend the direct surface modification and milling capabilities of FIBs, many instruments employ additional gas delivery systems to allow chemically enhanced and selective etching and/or material deposition. These gas delivery systems provide a range of precursor gases and direct those at high-flux rates to the ion beam interaction regime while maintaining the high-vacuum level required for safe operation. Commercial instruments use either individual needles or a common nozzle block with multiple needles to supply the gases. Generally, they provide programmable control of one to five gas sources; the user can create simple or complex process flows (Figure 3.14).

In milling and etching tasks, the gas precursors may increase the etching rate, eliminate redeposition, and/or allow the evaporation of volatile species. Other precursor gases allow ion-beam-assisted material deposition (metal or insulator), which involves the following steps: (i) adsorption of the gas molecules on the

Chapter 3

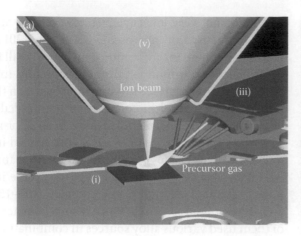
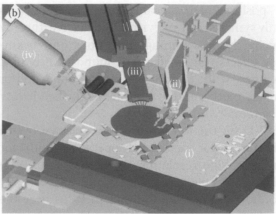

FIGURE 3.14   Schematic view of the sample geometry in a single-beam FIB nanopatterning instrument showing (i) sample platen, (ii) two nanomanipulators, (iii) gas injection system, and (iv) secondary electron detector and FIB column. (a) Side view with FIB and its gas delivery system in operation; (b) top view (From Raith. 2009. Raith GmbH, ionLiNE lithography, nanofabrication and engineering workstation, http://www.raith.com/)

substrate, (ii) dissociation of the gas molecules using the energy brought by the ion beam, and (iii) deposition of the atoms and removal of the organic ligands.

3.4.3   *In Situ* Manipulation for Sample Preparation and Materials Characterization

In many applications, it is not only desirable to machine and potentially image with the ion and/or electron beam, but also to place probes interactively on selected nanoscale objects or to manipulate or extract objects created or modified by the FIB. To accomplish these tasks, a set of micro- or nanomanipulators can be added to modern FIB instruments. Typically, these manipulators utilize high-resolution piezoelectric drives for three independent axes of motion. In most cases, these devices are mounted in a top-plate arrangement fixed in the vacuum chamber but in some cases they are side-mounted on the sample stage. Depending on the selection of the end-effectors used in conjunction with the manipulator, the most common applications are (i) TEM lift-out preparation, (ii) *in situ* four-point probing measurements, and (iii) *in situ* mechanical testing.

3.4.4   System and Patterning Control

Flexible control of the ion beam scanning across the sample is mandatory for all FIB systems for use in micro- and nanofabrication. The control needs to provide flexibility far beyond pure raster scanning as

provided in any SEM operation. In most applications, the patterning systems are PC-based and transfer simple or more complex CAD data into sequences of lines. These lines can be drawn by a dedicated pattern generator sending analog signals to the scan generators, which are linked to electrostatic deflection systems inside the FIB columns.

Modern FIB instruments thus support various beam writing strategies. Raster scan (Figure 3.15b) is one basic method. Here, the beam is swept across the entire field, sequentially pixel by pixel, with the beam being turned off and on (blanked and unblanked) as needed to expose the desired pattern. This strategy is based on a relatively simple but effective architecture. The disadvantage is that because the beam is scanned across the entire writing field, sparse patterns take just as long to write as dense patterns. An additional drawback is that dose adjustment within the pattern, for creating 3D dose profiles and patterns, is inherently more difficult. In contrast, vector scan (Figure 3.15c) utilizes the method of jumping from one patterned area to the next, skipping over all areas where no patterns are located. This method makes the vector scan approach faster than raster scan for sparse patterns. There is little difference in writing time for dense patterns. Adjustments to dose can be accomplished during the jumps between adjacent patterns. A disadvantage is that some beam settling time is required to ensure the required placement accuracy of the individual patterns.

Unlike EBL, where only the total local dose is important, these modes have to be applied in multiple loops

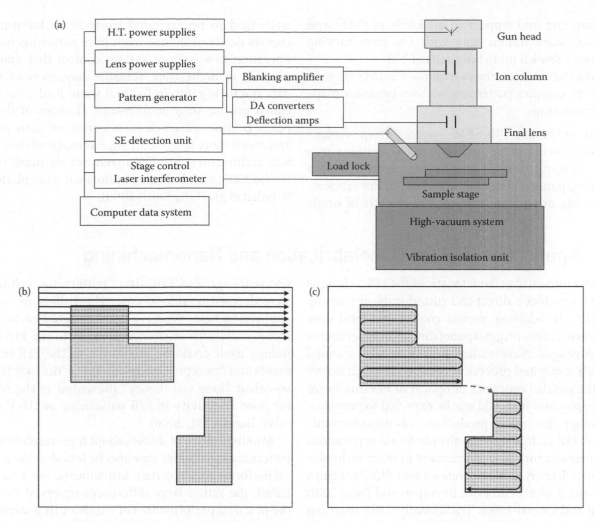

FIGURE 3.15 (a) System architecture of a dedicated FIB nanofabrication instrument. Right side: scan modes in ion beam patterning, showing raster scan operation in (b) and vector scan mode in (c) (single loop indicated, only).

in ion beam direct patterning in order to control and minimize artifacts such as redeposition. Moreover, in combination with gas injection systems for deposition or enhanced etching, refresh and beam "revisiting" times have to be controlled carefully.

Beyond the pure physical sputtering, FIB patterning results in a number of side effects, some desirable and others parasitic, such as implantation, (re)deposition of the sputtered material, local amorphization, swelling, and so on. In order to find optimized patterning mechanisms that allow for these effects, and moreover, to control localized deposition and enhanced etching in combination with gas injection systems, modern pattern generator systems provide functions like fast loops, flexible control of scan directions, intelligent use of the beam blanker, and much more. As an example, flexible step size, dwell time, and loop settings are critical for

gas-assisted processes, particularly deposition, where incorrect parameters can lead to unexpected results.

Typical performance characteristics of pattern generator systems for FIB nanofabrication include

- 16-Bit scanning along two axes with flexible beam blanking control for variable dwell times down to the range of few tens of nanoseconds
- Sequential or parallel patterning of shapes each with its associated parameter sets (beam current, dwell time, dwell spacing, scan direction, fast scan loops, etc.)
- Integrated imaging and alignment capabilities for precise write-field size control and for pattern placement within preexisting structures
- Integration of sample stage, gas injection, and beam control

Chapter 3

- Handling and import of all kinds of CAD and pixel-based design data with file sizes varying from a few kB up to hundreds of MB
- Intuitive user interfaces to allow a variety on simple or complex patterning without extensive preparation times

State-of-the-art FIB/SEM systems from various vendors provide these functions for operations in a single imaging field or can be complemented by a dedicated pattern generator attachment. For fabricating devices or patterns that require the size of single write-field to be extended from 20 to 100 μm, the control needs to change from pure pattern-generator operation to a form of system control that makes it possible to orchestrate versatile sequences of CAD data fracturing into individual write-fields, the associated precise stage movements, changes of the ion beam column parameters, selection of scan modes and much more (Figure 3.15a). Examples of these system architectures are FIB machines for mask repair (Seiko 2008) and dedicated nanofabrication platforms (Ghaleh et al. 2007; Raith 2009).

3.5 Application: FIB for Nanofabrication and Nanomachining

The most interesting characteristic of the FIB technique is that it provides a direct and customizable patterning capability. In addition, precise control and local dose adjustments allow original nanofabrication experiments to be envisaged. Nevertheless, patterning with scanned FIBs is a sequential process. It is therefore much slower than the parallel processes of optical or electron beam lithography and it should not be expected to provide a technology for mass production of nanoelectronic devices. FIB technology has already found applications in IC manufacture as a complement to other technologies, for relatively small operations on individual chips on a wafer (Orloff et al. 1996; Reyntjens and Puers 2001; Matsui and Ochiai 1996). Traditionally, FIB patterning is based on material removal by ion sputtering at the micrometer scale or just below. A focused gallium ion beam having an energy typically around 30 keV is scanned over the sample surface to create a pattern through topographical modification, deposition, or sputtering. The first consequence is that, mainly because of the high-ion doses required (~10^{18} ions/cm^2) and of the limited beam particle intensity available in the probe, FIB etching-based processes remain relatively slow. We recall that for most materials, the material removal rate for a 30 keV gallium ion is around 1–10 atoms per incident ion, corresponding to a machining rate of around 0.1–1 μm^3 per nC of incident ions.

The second consequence is that, for most applications, the spatial extension of the phenomena induced by FIB irradiation constitutes a major drawback. This damage generation was immediately pointed out, some time ago, as the main limitation for the realization of highly localized structures by FIB (Yamamoto et al. 1993; Laruelle et al. 1990). As a result, the idea of using FIB-related processes for nanofabrication was regarded as a pure dream only a few years ago and FIB direct patterning techniques were deemed unable to enter the ambitious "nano" application field. We have demonstrated earlier that some limitations attributed globally to the FIB technology itself could be attributed to the FIB instruments and concepts employed. This is the case for the so-called "long tail theory" presented as the reason for poor selectivity in FIB patterning of III–V crystals (Gierak et al. 2006).

Another source of doubts about high-resolution FIB patterning capabilities may also be found in the practical performances of certain instruments. For a nonspecialist, the rather large differences observed between the practical performances of current FIB instruments are difficult to understand. Indeed, between the imaging mode with probe currents ranging from 0.1 to 0.5 pA and the patterning mode where the probe current can usually be raised up to the nA regime, the performances seem to be contradictory. The consequence is that high-spatial imaging resolution close to or below 5 nm may be obtained in the SIM mode but such values may remain inaccessible in the patterning mode, mostly because of limitations in the ion emission process occurring in the LMIS (Rau et al. 1998). On the other hand, there is a clear benefit: a low probe current helps preserve the sample. Subsequent switching of the system to the high-probe current regime allows higher erosion speeds. These two regimes are absolutely essential and were optimized accordingly by FIB manufacturers to comply with the customers' specifications. There is no doubt that most existing FIB systems were developed with these specifications in mind. As a result of this analysis, some years ago we came to the conclusion that for our research purposes a dedicated FIB concept optimized solely to deliver the highest possible patterning resolution was necessary (Figure 3.16).

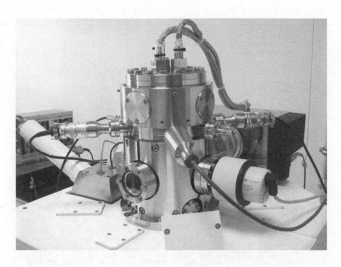

FIGURE 3.16 Picture of the 2008-generation single-beam architecture FIB machine developed at LPN-CNRS (From Gierak, J. et al. 2010b. *Microelectron. Eng.* 87, 1386–1390.)

Nevertheless, FIB technology exhibits the interesting ability of allowing direct and local surface modifications with dimensions that are characteristic of nanoscience. The structuring of materials via etching or deposition may be achieved efficiently and in a controlled way down to the scale of some nanometers, and in our opinion this avenue remains incompletely explored. Consequently, the exact potential of the FIB technique in nanotechnology still requires to be assessed, after which new routes, applications, or limits may be found. Among the examples of applications that we have selected for presentation below, some are perfect examples of the difficult task of supporting an innovative application idea along the critical route to its acceptance.

For nanometer-scale FIB patterning, the appropriate level of interaction between ions and solids is often translated toward low fluence effects. Collisional defects in the vicinity of the surface-irradiated layers play the major role in our approach. This is due to the shrinkage of the lateral dimensions, to the very small thickness of the active layers used, and to the extreme sensitivity of these materials to ion bombardment. Indeed, most of the target materials involved in nanofabrication experiments (III–V crystals, thin magnetic crystalline films, or inorganic compounds with weak bonds) exhibit very high-ion sensitivity, well below the 10^{14} ions/cm^2 range (30-keV Ga$^+$ ions). As a direct consequence, local chemical reactivity or crystal modification of these materials induced by local ion bombardment becomes a high-speed process, offering new possibilities for localized structuring of materials or selective deposition of nanograins, for example.

3.5.1 Local Defect Injection or Smoothing in Magnetic Thin Film Direct Patterning (Ferré and Jamet 2007)

The rapid increase of areal storage densities in magnetic hard disk drives (HDDs) may in the future be limited by the thermal instability of small magnetic domains, which is known as superparamagnetism. A patterned storage medium featuring independent bit cells consisting of single magnetic domains would make it possible to relax this limitation. Ion-beam irradiation has been shown to modify the magnetic properties of Co/Pt multilayers (Co/Pt MLs) with large perpendicular magnetic anisotropy (PMA) (Chappert et al. 1998) and patterning of magnetic properties can be achieved by local ion beam exposure (Devolder et al. 1999; Terris and Rettner 2002). Ion patterning techniques are basically resistless structuring processes and are therefore attractive for manufacturing the magnetic storage media of the future. Complex patterns can be easily defined. Ultrahigh-density recording using nanopatterned magnetic media is obviously also a major issue. The challenge here is to obtain significant density enhancement by FIB patterning without the signal-to-noise ratio degradation entailed by surface roughness modification and nanostructuring jaggedness between information bits.

For this example, we have used magnetic nanostructures fabricated on virgin, high-quality, ferromagnetic Pt/Co ultrathin films and multilayer structures. The samples were produced by sputtering (Chappert et al. 1998).

In this application, the goal is to isolate and investigate collective behavior of magnetically stable domains organized in a matrix. To achieve this, the pencil of gallium ions is scanned over a set of horizontal and vertical lines having fixed separations. One initial idea was to use the sputtering effect of the gallium beam to cut a boundary line. This idea was tested but immediately discarded. The low patterning speeds measured (a few µm/s or less) were not compatible with the reproducible patterning of a sufficient number of magnetic elements (10^4–10^6). Moreover, redeposition of sputtered contaminating materials and, in some cases, charging effects causing shifts were evidenced. Furthermore, the initially flat surface morphology of the film was to be preserved, in order to ensure a maximum compatibility with potential magnetic reading/recording processes. Another patterning route was therefore explored for magnetic layers, based on an ion beam mixing process of ultrathin interfaces.

In order to pattern the lines, the ion probe is digitally scanned with a pixel time and pixel-to-pixel distance set to allow linear fluence values ranging from 1 to 0.01 nC/cm. These fluences, allowing writing speeds of about 0.08 and 80 mm/s, respectively, are used for patterning Pt(3.4 nm)/Co(1.4 nm)/Pt(4.5 nm) samples (Figure 3.17). For a magnetic multilayer consisting of a six-period Pt(2.8 nm)/[Pt(0.6 nm)/Co(0.3 nm)]$_6$/Pt(6.5 nm) ferromagnetic multilayer, an even higher sensitivity has been evidenced and writing speeds up to 200 mm/s ($\sim 10^{12}$ ions/cm^2) were reached (Hyndman et al. 2001). To our knowledge, this is the present record for a sequential patterning technique. The writing strategy we selected was based on independent patterns designed to fit in a single writing field (50 $\times$ 50 μm). For the probe scanning process a meander mode or S mode was selected and no beam blanking was activated inside scanned lines or complete patterns. This choice was made for reasons of accuracy arising from beam stabilization not compatible with the very high sensitivity of the patterned media.

With this patterning method and the FIB technology described earlier, we were able to demonstrate that Pt/Co(1.4 nm)/Pt nanodots with sizes down to 70 nm remain ferromagnetic. This a very good point for the FIB processing technique since the spatial localization is very high and this is encouraging for the design of ultrahigh-density recording media up to the superparamagnetic limit. Further investigations were also made possible such as

- Competition between exchange and dipolar interactions

We have shown that FIB-patterned lines at very low fluences cannot fully isolate the designed magnetic nanodots. The magnetic interaction (from parallel to antiparallel) can be manipulated by increasing the linear fluence, which induces a transition from a collective to a more individual behavior (Figure 3.17b) (Ferré and Jamet 2007). This can be exploited to realize magnetic logic devices. Isolated type of elements may also be designed for targeted fundamental research.

- Creation of magnetic nanocircuits and magnetic logic circuits

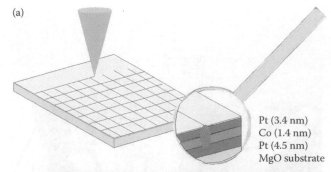

(a)

Pt (3.4 nm)
Co (1.4 nm)
Pt (4.5 nm)
MgO substrate

(b)

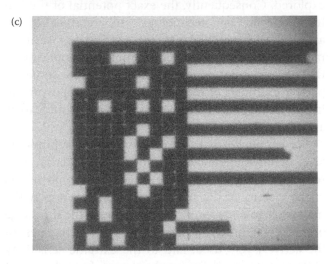

(c)

FIGURE 3.17 (a) Illustration of the low fluence irradiation method for defining an array of dots on an ultrathin magnetic film Pt(3.4 nm)/Co(1.4 nm)/Pt(4.5 nm). (b) High-resolution magneto-optical microscopy images of demagnetized Pt/Co(1.4 nm)/Pt FIB-patterned structures. Square (1 $\times$ 1 μm) dot arrays of magnetic monodomains. (c) 2 $\times$ 2 μm dots coupled with channels. Note the clear separation lines where the high-resolution FIB probe was scanned (From Ferré, J. and Jamet, J.P.: In *Handbook of Magnetism and Advanced Magnetic Materials*. 2007. Wiley, Chichester.)

Single narrow magnetic nanotracks may be created between two close parallel lines. Information can then be transmitted by fast wall motion in such nanotracks; this is favored by an irradiation-induced wetting effect at track edges (Jamet et al. 2001; Ferré and Jamet 2007).

3.5.2 Surface-Induced Defects for Local Guided Self-Organization (Bardotti et al. 2002)

The design and fabrication of quantum dot (QD) systems and the study of their properties are playing an increasingly important role, mainly because of the large number of potential applications in fields such as high-density information storage (Tbits/in.^{2}), nanoelectronics, nanooptics, magnetics, and magnetoelectronics devices (Orlov et al. 1997; Sun et al. 1989; Andres et al. 1996). Several ways of producing such systems have been developed, including top-down, bottom-up, or combined top-down–bottom-up technologies. However, in most cases, the individual nanosize dots that must exhibit specific structure/morphology and properties are difficult to obtain using the existing preparation techniques. In this second example, we try to exploit the effects of local defect injection induced by FIB irradiation as a fundamental process causing localized chemical or

crystal modification of the target. The creation of defect cascades and other irradiation-induced defects has often been considered as a limiting factor in the application of FIB technology. This is especially true for high-dose ion-milling applications when typical doses exceed 10^{18} ions/cm^2 and for targets that are very sensitive to ion bombardment. However, for FIB fabrication of nanometric devices, the appropriate level of interaction between the ions and the solid usually requires lower doses. Our aim here was to create localized surface modifications (i.e., FIB-induced artificial defects) on a very highly crystalline material (highly oriented pyrolitic graphite or HOPG) in order to trap subsequently deposited metal nanoparticles. The basic pattern design is a matrix of points (dots) defined with a fixed step size and constant point fluence. Initially, the nature of the required ion-induced modifications was unknown. For this reason, we started with high-ion point doses (3×10^5 ions/point) to promote ion sputtering and create nanoholes (Figure 3.18a and b) in the graphite material. The lower limit, achieved with very low ion point doses (~50 ions/point), was also tested in order to investigate the influence of crystal amorphization effects.

In these experiments, the beam-writing strategy was designed to generate matrices of isolated points with horizontal and vertical intervals between impacts set to 1 µm, 300 nm (Figure 3.18c), and 100 nm. The pixel

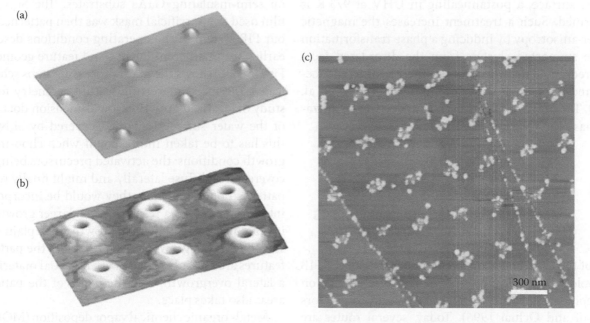

FIGURE 3.18 Tapping mode atomic force microscopy (TMAFM) images of FIB-induced defects defined with a 300 nm interval distance. (a) Nanobump in HOPG (10^4 ions/point; height Z_{max} = 1 nm). (b) Nanocrater in HOPG (10^5 ions/point; Z_{max} = 5 nm). (c) Tapping mode atomic force microscopy images (2×2 µm) of Co$_{50}$Pt$_{50}$ deposited on FIB-patterned HOPG after annealing. The equivalent surface density in a data storage application here would be around 10 Gbit/in.^{2} (From Mélinon, P. et al. 2008. *Nanotechnology* 19, 235305 (9pp); Prével, B. et al. 2004. *Appl. Surf. Sci.* 226, 173.)

dwell time for a 5 pA probe current was varied from 10 ms to 1.6 μs, corresponding to 50 and 3×10^5 ions/point, respectively. The writing procedure for the point matrix patterns was set to a meander mode. The beam blanking is hence activated only at the end of a complete pattern during specimen repositioning. This scheme was made possible by the excellent performance of our pattern generator for which the ramp-up time between two adjacent pixels is only a few nanoseconds.

$Co_{50}Pt_{50}$ magnetic clusters (2 nm mean diameter) are deposited in a soft landing regime, using the LECBD (low-energy cluster-beam deposition) technique (Perez et al. 2001), onto functionalized HOPG substrates with 2D-organized arrays of nanodefects. Such defects act as nucleation centers for the deposited clusters, which diffuse on the HOPG surface leading to the formation of stable 2D-organized cluster-assembled dots (Bardotti et al. 2002). In Figure 3.18c, we present a tapping mode atomic force microscopy (TMAFM) image of 0.2-nm-thick $Co_{50}Pt_{50}$ clusters deposited onto an FIB prepatterned HOPG substrate (10^4 ions/point and step distance = 300 nm) after annealing. The diffusion of the clusters and their sensitivity to tiny FIB defects (nanoprotrusions) lead to the formation of 2D-organized arrays of cluster-assembled dots on the surface-induced defects in a guided self-organization process.

To improve the quality of the organization on the HOPG surface, a postannealing in UHV at 973 K is performed. Such a treatment increases the magnetic cluster anisotropy by inducing a phase transformation of the supported clusters from the disordered face-centered-cubic (fcc) structure to the $L1_0$-ordered face-centered-tetragonal structure (fct) (Hannour et al. 2005). The positive effect of annealing on the organization has been clearly demonstrated.

3.5.3 Local and Selective Epitaxy of III–V Semiconductor Quantum Dots (Sagnes et al. 2006)

One of the early and most exciting promises of the FIB technology was related to *in situ* structure generation and epitaxial growth of high-quality semiconductors (Matsui and Ochiai 1996). Today, several routes are being explored in an attempt to combine local doping (FIB) or local etching with 3D control (FIB) via the molecular beam epitaxy technique (Sugimoto et al. 1991; Reuter et al. 2003).

The new photonics materials are key elements in a wide area of optical communications, data storage and processing, as well as in sensing, display, and lighting technologies. The discrete nature of the electronic states of In(Ga)As/GaAs QDs is very promising for the realization of high-performance optoelectronic devices (Zhukov et al. 1999; Park et al. 1999). In particular, In(Ga)As/GaAs QD-based lasers (which can be designed for operation at 1.3 μm) present low thresholds and high-characteristic temperatures. The main limitation of these lasers is their rather low maximal modal gain, which could be improved by increasing the QD density (Saint-Girons et al. 2003). The control of the QD nucleation using FIB-textured substrates could allow an increase of their density as well as a reduction of their size dispersion, thus leading to a considerable improvement in the related laser performance. The control of the QD size dispersion could also facilitate the fabrication of broadband emitters or amplifiers for telecommunication applications.

We used our nano-FIB system to pattern nanostructures into a Si_3N_4 thin film to be followed by localized epitaxial regrowth. This technique is of considerable interest since the feature size is very small and should allow the localized growth of QDs.

The technique we have developed is similar to lithography with a thin Si_3N_4 layer of 10 nm deposited on semi-insulating GaAs substrates. The Si_3N_4 thin film used as a sacrificial mask was then patterned with our FIB system in the operating conditions described earlier, using different fluence and feature geometries. Figure 3.19a shows the main processing steps schematically. The most interesting feature geometry for our study of localized growth is low-dimension dots. Most of the wafer surface remained covered by Si_3N_4 and this has to be taken into account when choosing the growth conditions: the activated precursors being on a covered area diffuse laterally and might finally reach a patterned opening where they would be incorporated into the crystal. As a result, a much larger growth rate is obtained than for planar growth on a plain GaAs substrate. If the growth rate is too high, the patterned features are not only "filled" with epitaxial material but a lateral overgrowth over the edges of the patterned areas also takes place.

Metal–organic chemical vapor deposition (MOCVD) was used to regrow GaAs within the patterned holes. The growth was carried out at a pressure of 60 Torr and at a temperature of around 500°C. Tri-methyl gallium (TMGa) and arsine (AsH_3) are used as precursors.

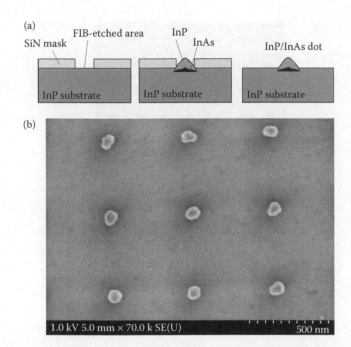

FIGURE 3.19 (a) Schematic representation of the main processing steps used for the selective epitaxy experiments. FIB nanopatterning of the Si_3N_4 masking layer, epitaxial regrowth using the patterned mask to localize the growth, and HF removal of the mask. (b) Scanning electron micrograph of a dense field of round features shows a homogeneous and well-localized growth (From Sagnes, I. et al. 2006. Selective epitaxy of patterned III–V semiconductors surfaces using FIB technology, *E-nano Newslett.* December, 11.)

The growth conditions—mainly the temperature and the flow of the TMGa precursors—were carefully chosen to obtain a very low growth rate (Sagnes et al. 2006). Figure 3.19b shows the structures after the regrowth process. Here, the Si_3N_4 mask was removed by hydrofluoric acid (HF) prior to observation in an SEM. Homogeneous growth within the patterned areas can be observed and these results are very promising for the growth of localized QDs (Kitslaar et al. 2006).

3.5.4 FIB Engineering of the Optical Properties of Apertures and Microcavities

3.5.4.1 3D Near-Field Experimental Study of the Light Transmitted by a Hole Array (Wang et al. 2009)

Using a scanning near-field optical microscope (SNOM), we have investigated the electromagnetic field distribution on an array of subwavelength holes pierced by FIB machining into a 100 nm-thick gold layer. The apertures have been made with a sub-10 nm probe of 30 keV gallium ions. The geometry described in

Figure 3.20 consists of an arrangement of circular holes separated by 900 nm and whose diameter is 340 nm.

The goal of the study is to understand the role of surface plasmon polaritons (SPP) in the spectacular optical properties observed on nanostructured surfaces such as the extraordinary optical transmission or the beam collimation effect (Lezec et al. 2002). When illuminated in a transmission mode, the holes may diffract the light in the far field but also generate surface waves that will propagate on the metal/dielectric interface and interact with other nanostructures. The direct observation of these surface waves with an SNOM provides a microscopic description of the phenomena and may explain many far-field experimental results.

The particular SNOM technique we have employed uses a small fluorescent particle settled at the end of an atomic force microscope tip as a nanodetector of light (Aigouy et al. 2007). The sample is illuminated in transmission by the 975 nm line of a laser diode and the fluorescence is collected in the green with a high-numerical aperture objective situated above the tip. For these particular experiments, the SNOM images were obtained by moving the particle above the structure, in two scanning modes:

1. In the first mode (Figure 3.20c), the tip is moved in the xy plane giving a map of the electromagnetic field distribution on the surface. The light that emerges from the apertures is visible, as well as a background signal between the holes. This background signal is due to surface waves created by the apertures and propagating on the metallic surface.
2. In the second mode (Figure 3.20c), the tip is moved in the yz plane above a line of holes. Now, we observe the diffraction of light in free space above the holes as well as the vertical extension of the local electromagnetic field above the surface.

We are currently studying these effects in detail by performing experiments on hole arrays that have different characteristics (hole diameter and separation) and on isolated apertures (holes and slits) surrounded by subwavelength grooves.

3.5.4.2 FIB Engineering of the Optical Properties of Microcavities (Coll. I. Robert, I. Abraham, LPN-CNRS)

In solid-state physics, two main mechanisms for light confinement are used. The first exploits the total internal reflection at the interface between two media with different refractive indices: total internal reflection bounces

Chapter 3

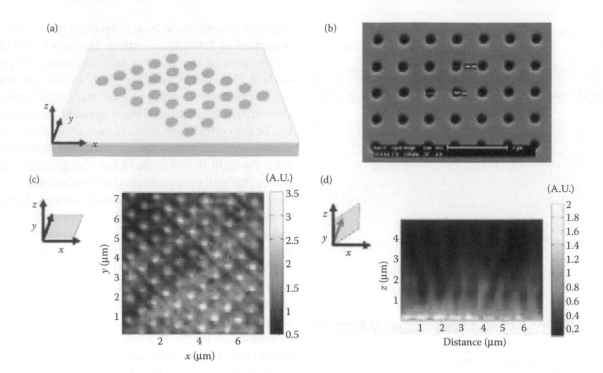

FIGURE 3.20 (a) Schematic description of the structure; (b) SEM image of the FIB-patterned hole array; (c) corresponding SNOM images of the field distribution on the sample and (d) in a plane perpendicular to the surface. (From Wang, B. et al. 2009. *Appl. Phys. Lett.* 94, 011114.)

light toward the region of higher refractive index when the light strikes the interface at a steep angle. The second exploits the destructive interference between optical waves impinging on a medium with a periodically modulated index of refraction, so-called photonic crystals.

One strategy to confine light in the three spatial directions and form a microcavity combines total internal reflection in one or two spatial directions and interference effects in the other direction(s). Such cavities are more amenable to fabrication than 3D photonic crystal resonators, while retaining or approximating many of their desirable properties, such as 3D confinement of light. One example is the 2D photonic crystal cavity in a slab waveguide, consisting of a 2D array of holes perforated through a thin membrane. In these cavities, the photonic-bandgap effect is used for strong light confinement in the transverse directions, and total internal reflection at the air–slab interface ensures light confinement in the longitudinal direction. The cavity is formed by the introduction of a point defect in the 2D photonic crystal. The simplest cavity geometry is created by removing a finite number of holes from a perfect array of holes. For instance, the so-called H1 cavities are formed by one missing hole in a regular triangular photonic crystal with ΓM-type boundaries.

The resonant modes (leaky and guided) of a symmetric dielectric slab waveguide separate out into TE and TM polarizations that also have even or odd spatial symmetry with respect to the center of the waveguide. For the TE modes, the electric field is polarized in the plane of the waveguide and for the TM modes, the magnetic field is polarized in the plane of the waveguide. The resonant modes of a 2D patterned dielectric slab surrounded by air are, however, not purely TE or TM but rather what we designate TE-like and TM-like. Nevertheless, emitters located in the middle of the membrane and with dipoles oriented in the plane of the membrane (such as the fundamental transitions of self-assembled QDs) will only be coupled to TE-like modes. For suitable geometries defined by the slab thickness d, the radius of the holes r and the period of the photonic crystal a (in our experience, $a \sim 0.7\,\lambda/n$, $d/a \sim 0.7$; and $r/a \sim 0.3$, with $\lambda \sim 950$ nm and $n \sim 3.5$), the cavity supports localized TE-like modes. The fundamental one is a twofold degenerate dipole mode with a maximal electric field at the center of the cavity (Painter et al. 1999). At the cavity center, one of these modes is polarized in the x direction while the other is polarized in the y direction (Figure 3.21a). However, fabrication imperfections split the desired mode degeneracy to engender two linearly orthogonal polarized modes with wavelengths (and hence energies) split by 4.5 nm on average (see Figure 3.21b). Moreover, the implementation of the cavity effect with single QDs requires a precise spectral overlap

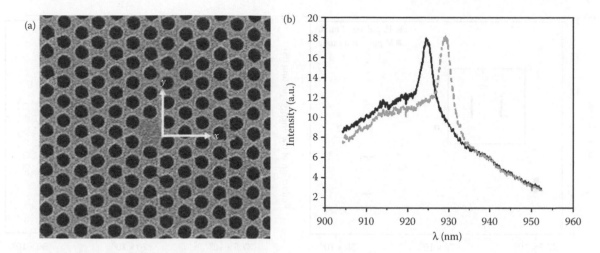

FIGURE 3.21 (a) SEM image of the H1 photonic crystal defect region with x and y directions of the twofold degenerate fundamental dipole mode. (b) Typical polarization-resolved optical spectra of an H1 cavity embedding three arrays of quantum dots nonresonantly optically pumped at 4 K (solid line and dashed line correspond to the spectra of the cavity obtained with two orthogonal linear polarizations). The cavity displays two linearly polarized modes with orthogonal polarization (H and V) and split by 4 nm.

between the cavity mode and the optical transition of the emitter. Nonetheless, because it is very difficult to predetermine the exact resonance energies of the QD and of the cavity, the difficulty in tuning the resonance energy of a fabricated nanocavity has so far limited the application of these solid-state cavity quantum electrodynamics nanostructures.

In order to tune this unintentional splitting and the cavity resonance, one strategy consists of correcting spectral offsets and split degeneracies after the fabrication process. One postfabrication technique to blue-shift the cavity resonance uses oxidation–wet etching cycles (Badolato et al. 2005). Another approach uses atomic force microscope nano-oxidation of the cavity surface. Relative tuning between two nanocavity modes is achieved through careful choice of the oxide pattern (Hennessy et al. 2006). In our experiments, we plan to replace the atomic force microscope nano-oxidation of the cavity surface by FIB nanoscale patterning. This patterning induces a nanoscale modification of the refractive index and red-shifts the energies of modes.

The impact of FIB nanoscale patterning on the optical properties of the H1 cavities has been studied on three H1 cavities. Two patterns have been designed according to the insets in Figure 3.22, where the patterns are depicted as thick black lines. Both patterns induce a red-shift of the two polarized modes of roughly 7.5 nm for the lower-energy mode and 9.5 nm for the higher-energy mode (Figure 3.22a). This red-shift increases when the FIB ion density is increased. It also indicates that the FIB process does not locally remove matter, since local etching would reduce the average index overlapping the

mode and induce a blue-shift. The red-shift is likely to result from a local deposit of Ga ions out of the top surface causing a local increase of the average refractive index. The larger detuning of the higher-energy mode induces a decrease of the energy splitting of the two modes of roughly 40% (Figure 3.22b). The reduction of this energy splitting also increases with the FIB ion density. We do not observe a significant reduction of the quality factors Q of the cavities after FIB processing, which caused only a 20% reduction on average of Q.

This demonstration indicates that FIB nanopatterning is very encouraging for engineering the optical properties of photonic crystal slab cavities. The FIB low-dose irradiation has allowed us to modify the near-field cavity geometry directly and thus to fine-tune the nanocavity resonance wavelengths. Further investigations on the FIB irradiation pattern and ion dose control should allow us to tune a single mode over several nanometers deterministically and nearly continuously. Such a technique is in principle applicable to all photonic crystal-based cavity structures and is therefore very promising.

3.5.5 FIB–Patterned Solid–State Nanopores for the Manipulation and Analysis of Biological Molecules (Schiedt et al. 2010)

For biophysics, a topic of the highest importance is the fabrication of isolated single-nanometer-scale holes, or

Chapter 3

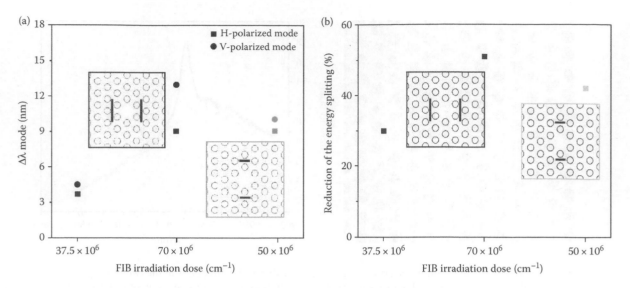

FIGURE 3.22 (a) Amplitude of the red-shift of the two linearly polarized modes with modal wavelength initially around 930 nm for two different (black and gray) FIB irradiated patterns consisting of the red lines and different FIB ion doses. (b) Relative reduction of the energy splitting between the two linearly polarized modes for two different (black and gray) FIB irradiated patterns consisting of the thick black lines and different FIB ion doses.

"nanopores" in thin insulating solid-state membranes for use at the heart of devices capable of detecting and manipulating individual biomolecular DNA engines. Synthetic nanopores are extremely sensitive single-molecule sensors, the surface properties of which can be tailored in such a way as to govern interactions with various analytes, resulting in "smart" nanopore sensors. With nanopores, the confinement, dynamics, and transport properties of single macromolecules can be studied at the nanometer scale with a temporal resolution of 10 μs for the electrical method of detection and perhaps even better with optical detection methods. Nanopores are fantastic tools for mimicking biological objects and functions, in particular membrane proteins. Molecular biology is a source of inspiration in this field of research: living cells can synthesize a wide variety of macromolecules with atomic precision, each of which has a specific function in the cell. Of particular importance are the spatial separation between two compartments with different environments and the "unidimensionality" that opens the way to sequential re- or denaturation experiments, macromolecular synthesis, or destruction. In recent years, the passive aspects of the pore physics has been extensively studied, in particular, the passive transport of macromolecules, but much remains to be done in this field.

The detection of the translocation of biological molecules such as proteins or DNA through a pore of a few nanometers by electrical methods (Figure 3.23) has been shown to be an efficient way of understanding some biological mechanisms (Kasianowicz et al. 1996), such as DNA unzipping (Mathé et al. 2005) or protein denaturation (Oukhaled et al. 2007). The nanopores can be fabricated by the insertion of a protein channel in a lipid bilayer or by a nanopatterning technique (Storm et al. 2003). Artificial or solid-state nanopores have many attractive features, notably excellent resolution while offering a longer lifetime and a higher resistance to environmental conditions than biological nanopores. Furthermore, the passage of individual molecules through nanosized pores in membranes is at the center of a host of biological processes. The underlying principle is that one can detect the passage of a single molecule through a single protein channel inserted in a plane lipidic membrane subjected to a potential difference by measuring the variation of the electric current induced by the presence of the molecule in the pore. Similarly, a solid-state membrane containing an artificial nanopore can serve as a dividing wall in an electrolytical cell and a translocating molecule can be detected via the modification of the ionic conduction of the pore.

This technique is extremely sensitive and opens interesting prospects for the physics of confined polymers, the development of nanosensors, and the sequencing and handling of macromolecules (in particular DNA) (Auvray 2000). It was shown that the translocation time of passage of a DNA molecule through a nanopore is a function of the length of this molecule but also of the affinities that it establishes

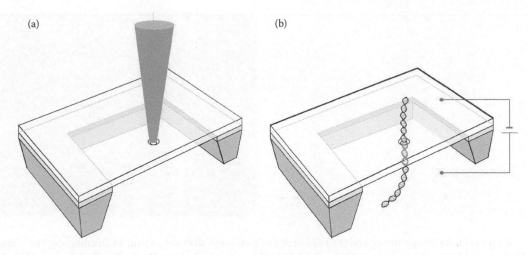

FIGURE 3.23 (a) Schematic representation of the fabrication of the nanopore and (b) after insertion in an electrophoretic cell translocation of a DNA molecule through the FIB-prepared synthetic nanopore.

with the walls of the pore. The translocation of one single molecule through a nanopore can be detected if the latter has a suitable size and thickness. When a macromolecule (transported by electrophoresis) enters a nanopore, it will prevent the ions from conducting the current in the pore and will result in a current blockade. The main technological aspect here is to define a nanopore size below 5 nm and then to ensure the integration of the membrane in a suitable setup with which the measurement can be made. Because of the application potential of these devices, the use of artificial membranes is now privileged (Stein et al. 2002; Biance et al. 2006; Gierak et al. 2007; Zhao et al. 2007; Dekker 2007; McNally et al. 2008).

Among the most promising approaches for the manufacture of such nanopores, there are two principal techniques: local erosion using an FIB instrument and intense irradiation in a high-voltage transmission electron microscope. To date the most reproducible technique seems to be that using the FIB technology, which consists in etching a hole with an initial diameter of a few tens of nanometers and then partially filling it by means of low-energy ion bombardment (Stein et al. 2002). For example, by using this technique, nanopores have been manufactured with sizes as small as 1.8 nm and used successfully to achieve voltage-driven DNA translocations through the nanopore (Li et al. 2001).

In a collaborative application project, we have decided to explore the possibility of fabricating solid-state nanopores with diameters below 5 nm directly; this would make it possible to access molecules so that DNA or proteins, for example, could be manipulated and studied at the single-molecule level in native conditions. For these experiments, we use dielectric films with thicknesses down to a few nm. The FIB-processed membranes were drilled with a 35 keV Ga^+ beam, focused to a 5 nm FWHM probe that carries around 2 pA. A specific processing methodology was developed with two main steps carried out sequentially on the same membrane batch:

1. Calibration of the critical fluence giving open nanopores. This first step was carried out with a relatively wide fluence range. The irradiation pattern selected here was a matrix of dots (Figure 3.24a) with the point fluence varying continuously between 10 and 200 ms within a single writing field, giving point doses of $2 \times 5 \ 10^6$ up to 5×10^7 ions/point. After SEM or TEM observation (Figures 3.24b and 3.25), the critical fluence yielding an open nanopore for a given membrane thickness and material (SiC, Si_3N_4, SiO_2, etc.) was identified and set as the reference parameter for the second processing step.

2. Nanopore engraving in a batch process. The final and decisive advantage of our FIB processing technique is to allow the realization of a large number of identical nanopores within a single processing batch. Figure 3.26a shows a 2 in. wafer containing about 200 identical membrane devices that have been processed in a single step using the highly accurate navigation and placement facility of our high-resolution FIB nanowriter (Figure 3.26b) (Schiedt et al. 2009).

Chapter 3

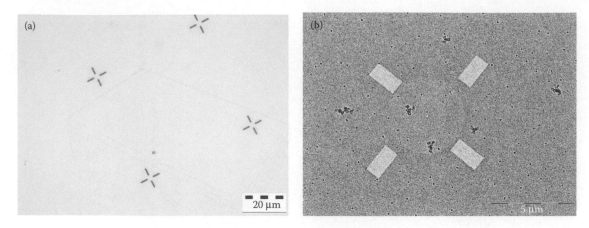

FIGURE 3.24 (a) Optical microscopy image and (b) TEM image of nanopores drilled in a thin SiC membrane. The alignment crosses deliberately added around each nanopore are perfectly visible in both imaging techniques (From Biance, A.L. et al. 2006. *Microelectron. Eng.* 83, 1474–1477.)

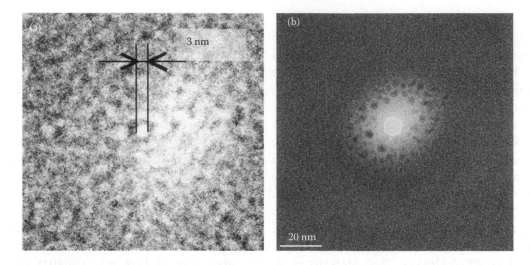

FIGURE 3.25 TEM images of pores of different sizes drilled in a 20-nm-thick SiC film (a) and 50-nm-thick Si_3N_4 film (b) with a dose ~10^6 ions/pt. The smaller pores have a similar aspect ratio: length/diameter ~7.

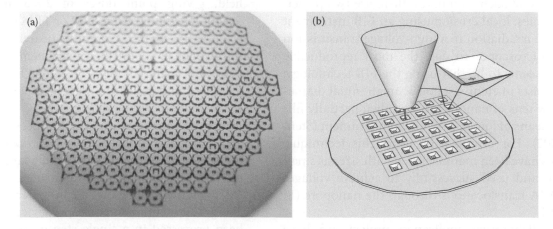

FIGURE 3.26 Image of a silicon wafer (2 in.–50 mm diameter) processed via lithography in an array of individual 3 mm disks supporting a 20-nm-thick SiC membrane. (a) View of the 3 mm disks with prethinned areas for cleavage and the transparent window in the middle. (b) Optical microscopy image of the array of membranes after FIB engraving. Each pore is inspected individually and then marked on the image with a red circle. (From Schiedt, B. et al. 2010. *Microelec. Eng.* 87, 1300–1303.)

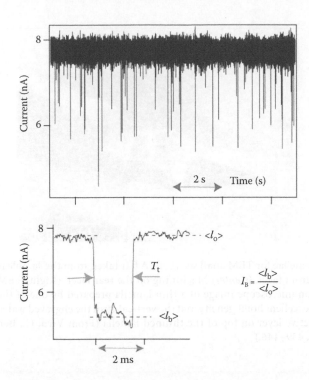

FIGURE 3.27 Current recording of double-strand DNA translocation events through a 10 nm artificial nanopore fabricated at LPN-CNRS. Inset showing a typical individual event with important blocking current $I_B \sim$ 20%. (Solution KCl 1 M, Tris, EDTA 10 mM, pH 7.6, DNA concentration: 0.7 nM.) (From Oukhaled, G. et al. 2011. *ACS Nano* 5, 3628–3638.)

In this last example, the advantage of combining the advanced FIB processing with a parallel fabrication technique such as conventional lithography and etching processes appears clearly. Such an efficient combination demonstrates that FIB processing can join the advanced technologies, thus allowing new fabrication routes to be opened. In this particular application field, we believe that the FIB technique will bring artificial nanopores into more widespread use for nanobiology research and nanomedicine (Figure 3.27).

3.5.6   FIB Patterning of Thin Foils from TEM Lamellae to Graphene

3.5.6.1   Preparation of Thin Membranes for Transmission Electronic Microscopy (Vieu et al. 1994)

The idea of preparing thin lamellae using an FIB engraving process for subsequent high-resolution crystal structure analysis in the transmission electron microscope would have been considered daring and

provocative some years ago. However, during the manufacture of microelectronic devices, developers often run up against the problem of observation of their detailed structures. Toward the end of the 1980s, these developers were looking for practical solutions and tools that would provide more accurate information for circuit editing. Within this framework, the TEM was used as the technique of choice, near-atomic resolution being possible (Hawkes 1995; Reimer and Kohl 2008), but a preparation technique had to be found that would allow thinned zones to be obtained from very precisely selected areas. For this, an FIB-based technique was developed with which thinned zones or cross sections could be located and prepared with great precision for subsequent TEM analysis. This FIB thinning process, schematically represented in Figure 3.28a, has rapidly attracted considerable attention. In this process, the FIB is used to trim thin walls (<50 nm) that are then transparent to the 100 kV electrons and their internal structure can then be inspected in a TEM (Vieu et al. 1994).

Thus, although it was originally regarded as controversial, this method has unanimously imposed itself today. The results obtained are very reproducible, localized, and obtained rapidly. FIB preparation of thin crystalline membranes that are homogeneous and have very few artifacts is very popular. This method makes it possible to manufacture walls at selected places with a thickness of <100 nm, a figure that could reach 10 µm on widely different materials such as semiconductors, metals, and even insulators. The success rate can reach 100%. It is interesting to note that all the commercial FIB machines have now been optimized for this precise area of application. Additional capabilities such as the possibility of lifting out the thinned membrane directly from a precise zone of the sample onto a TEM specimen holder (Stevie et al. 1998) have been added. However, in the light of growing difficulties with observation artifacts and FIB damage, it seems that the optimization of the radial profile of the ion probe and in particular the reduction of the lateral tails of the probe current distribution still require improvement or at least a better compromise needs to be found. The race toward maximum transported currents seems to have reached its limits here.

The profile for the ion probe current distribution must be as abrupt as possible for this application. Here, the lateral "tails" have deleterious effects that can only be compensated with a protective layer deposited beforehand on the device. On the other hand, the FWHM of the ion probe and therefore the resolution

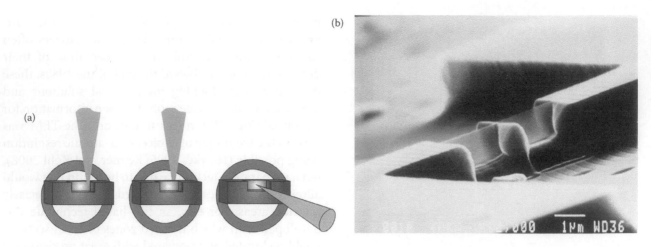

FIGURE 3.28 (a) Principle of FIB preparation of thin high-quality lamellae for TEM analysis. (Left) A bar taken from the sample is glued on a copper ring grid and then FIB-machined, starting on its front facet. (*Center*) Machining on the rear facet. (Right) TEM inspection and analysis of the thinned lamella. (b) Scanning electron microscope image of a thin lamella prepared by one of the authors (JG) using the method described above with an FIB. Note the excellent homogeneity and the verticality of the engraved walls, their transparency to the 40 kV electrons, and the absence of protective layer on top of the thinned lamella (From Vieu, C., Ben Assayag, G. and Gierak, J. 1994. *Nucl. Instrum. Meth. Phys. Res. B* 93, 439–446.)

have very little importance here and, together with the ion probe current value, these parameters must be specifically optimized (Gierak et al. 1997b). It is clear that maximum engraving speed is not a guarantee for preserving the structure quality of the manufactured membrane and for the absence of residual contamination (redeposition). To machine thin lamellae with good crystallographic quality (Figure 3.28b), the current density may be limited to around 10 A/cm²; on materials such as GaAs, engraving speeds of about 10 μm³/min can then be achieved.

Beyond these values, the constraints (thermal, mechanical) generated in the machined material and the redeposition of the engraved species have detrimental consequences for the quality of the lamellae. With reference to the limits of this method, one must remember that the ultimate thickness of the specimens cannot be reduced below 50 nm. The diffraction patterns of such thin lamellae reveal the completely amorphous nature of the residual material. In practice, thicknesses of about 100 nm are satisfactory. Recently, it was demonstrated that the depth of ion-induced damage in the lamellae can be decreased strongly by reducing the energy of the beam down to around 5 keV in a final polishing step (Giannuzzi and Garrison 2007). As an example of the ultimate performance, we mention the visualization of stress fields on a nanowire array made of silicon nitride (Si_3N_4) and deposited on multilayer GaAs/GaAlAs and encapsulated in SiO_2 (Pépin et al. 1997).

3.5.6.2 Patterning of Graphene Atomic Thin Layers (Lucot et al. 2009)

Graphene is generating considerable interest in materials science and condensed-matter physics. One crucial technological problem, which will govern future applicability of this material, is related to the patterning of graphene while preserving the exceptionally high crystallinity and electronic properties of this material. We have already shown earlier in this chapter that ultrathin membranes may be the ideal templates for FIB nanoengraving (Gierak et al. 2007). This is even more interesting if the membrane is highly crystalline, conductive, and thin enough, that is, with a thickness well below the projected range of the incoming ions. SRIM simulations (Ziegler 2008) of the interaction of a 35 keV Ga⁺ beam with different materials predict a typical projected range around 10 nm. In practice, because the lattice parameters are not taken into account in these simulations, one can expect channeling effects to become predominant for crystalline membranes with nanometer thickness.

In addition, FIB processing of ultrathin membranes exhibits some interesting technical features since there is a clear separation between the upper face bombarded by the incoming ions and the lower face that remains preserved. These advantages are illustrated in Figure 3.29:

- Most of the energy deposited by incoming ions remains located at the upper side of the patterned thin foil, thus preserving the other side. Lateral

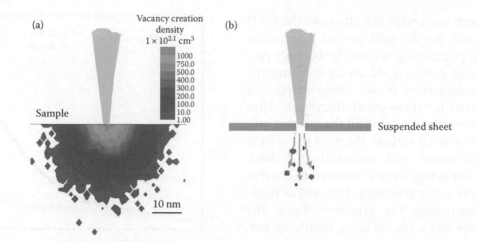

FIGURE 3.29 Schematic comparison of FIB patterning processes. (a) Bulk sample irradiated with 35 keV gallium ions. The damage intensity profile is calculated using the SRIM model. (b) A suspended sheet with a thickness below the calculated projected range of the incoming ions (~10 nm) is engraved primarily by knock-on effects of the incoming ions while scattering effects are negligible. (From Ziegler, J. 2008. *SRIM—The Stopping and Range of Ions in Matter*, http://www.srim.org/)

scattering effects are negligible in the membrane. Finally, the engraved material is efficiently ejected by knock-on effects; very good preservation of the remaining material can therefore be expected.

- There is almost no possibility for sputtered material and for gallium ions to be deposited on the far side of the foil since these particles are ejected via a scattering phenomenon with highly directional transfer of kinetic energy.

In the light of all these arguments, it appears that a suspended crystalline nanometer-thick medium may be the ideal target for exploring the ultimate resolution achievable using an FIB.

The discovery of graphene has generated intense experimental and theoretical activity in materials science and condensed-matter physics (Geim and Novoselov 2007). Graphene is a 2D lattice of carbon atoms arranged in a honeycomb crystal structure with a zero (or near-zero) bandgap and electron or hole energy dispersion resembling that of relativistic massless Dirac fermions. The unique electronic and optical properties of graphene make it a promising material for the development of high-speed electron devices, including field-effect transistors, pn-diodes, terahertz oscillators, and electronic and optical sensors. But realization of graphene-based devices requires new fabrication techniques that preserve the unique properties of graphene. In particular, the edges should be controlled, because their electronic states largely depend on the edge structures (armchair or zigzag) (Tapaszto et al. 2008).

Most of the research on graphene has been conducted on micromechanically cleaved samples on SiO_2 (Geim and Novoselov 2007). Transport measurements show that graphene has remarkably high- electron mobility at room temperature, with reported values in excess of 40,000 cm^2/V s (Chen et al. 2008). The mobility is nearly independent of temperature between 10 and 100 K, which implies that the dominant scattering mechanism is defect scattering. The exact nature of the scattering mechanism remains unclear but interactions with optical phonons of the substrate are likely to play a major role. Surface charge traps, substrate-stabilized ripples, and fabrication residues, on or under the graphene sheet, may all contribute. Consequently, eliminating the substrate by suspending graphene over a trench seems a promising strategy toward higher-quality samples. Recently, dramatically reduced carrier scattering was reported in suspended graphene devices (Du et al. 2008; Bolotin et al. 2008). Mobility as high as 120,000 cm^2/V s at 240 K was obtained. In this work, we report on the fabrication of electrically contacted suspended graphene nanoribbons.

Suspended graphene sheets are fabricated with a peeling process similar to that reported previously (Lucot et al. 2009). In our case, the graphene sheets are mechanically exfoliated over predefined gold structures patterned on a SiO_2 surface. We have patterned the structures directly on the suspended graphene sheets. After careful position monitoring of the graphene sheets with suitable characteristics (size, thickness, shape) using optical microscopy, the FIB pattern placement was coded directly onto the CAD file used previously for the gold contact lithography process. For the positioning, we use deliberately predefined alignment marks (gold crosses). Using the scanning ion microscope mode, these marks are retrieved and used for three-point alignment. After this, a batch patterning process of all the selected graphene sheets was started without the need to perform structure identification and localization. Indeed, using a nanowriter setup, there is no need to observe the structures prior to patterning. This avoids damaging or contaminating the graphene films. The engraving process was achieved using multiscan patterning strategies with a total dose around 3×10^6 ions/pt. This method avoids beam drifts and charging effects due to the proximity of the silicone dioxide surface. Very reproducible suspended graphene nanoribbons (Lucot et al. 2009) were made with this accurate alignment, positioning, and patterning method.

To probe the possible lateral extent of damage induced by the FIB patterning (Gierak et al. 1997a), we have measured the electrical transport characteristics of the FIB-fabricated graphene ribbons. Two-probe electrical measurements on suspended graphene devices are performed in a liquid helium cryostat using standard low-frequency lock-in techniques with an excitation current of 10 nA. In Figure 3.30, we observe semiconducting behavior of the nanoribbons with a strong temperature dependence down to ~4 K. There are two distinct regimes below and above ~100 K for the larger ribbon and ~50 K for the thicker. For all the fabricated samples (between 20 and 50 nm), the conductivity σ systematically exhibits this strong dependence with a factor as large as 1.5–10 for temperatures ranging from 230 to 4 K. This suggests that the high-structural quality of the graphene nanostructures is preserved since in our experiments the transport at the Dirac point is dominated by extrinsic scattering. The temperature dependence cannot be understood within existing theoretical models: further work is needed to elucidate how the different scattering mechanisms, interactions, and boundary conditions affect the temperature behavior and how the unbonded edge configurations (zigzag or armchair) affect the electrical properties.

3.5.6.3 TEM Observation of the Graphene Nanoribbon Structure

Using the same method, graphene sheets were then directly exfoliated onto an electron microscope grid and

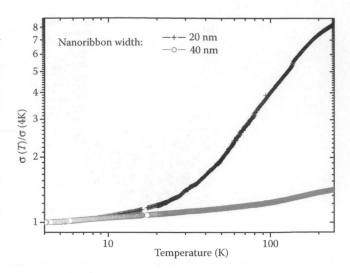

FIGURE 3.30 Conductivity σ normalized to its value at $T = 4$ K as a function of T for FIB-sculpted graphene nanoribbons having lengths around 1 μm and widths of 40 nm (red line) or 20 nm (black line). These nanoribbons were both etched into a ~1-nm-thick suspended graphene sheet using a 35 keV gallium ion beam focused to a 5 nm spot transporting an 8 nA probe current (From Lucot, D. et al. 2009. *Microelectron. Eng.* 86, 882.)

FIB-patterned to allow TEM observation of the edges of the graphene nanoribbon structure (Figure 3.31). These observations have not revealed the existence of gallium ion implantation effects and the lateral extension of

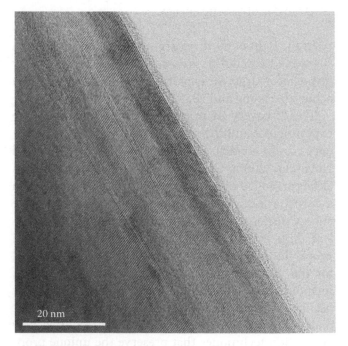

FIGURE 3.31 TEM image of the edge of a nanoribbon sculpted by the high-resolution FIB system developed by the authors. Note the extremely low lateral extension of damages (2 nm) at the interface. This damage layer exactly matches the crystal lattice (Gierak et al. 2010a).

amorphization is strikingly low. This is a clear demonstration that our method and instrumentation allow a state-of-the-art control of the quality of preservation of

3.6 Future Trends

At the heart of the highly successful FIB technology stands the LMIS that was pioneered in the late 1970s (Krohn and Ringo 1975; Sudraud et al. 1978). This emitter was rapidly shown to be a remarkably high-performance source for a large number of ionic species. Thus, even if an extremely wide variety of LMIS have been developed since the middle of the 1970s, it is the gallium ion source that comes first to mind today. It may be interesting to recall that, in an early attempt to understand the physics of the ion emission from liquid metal, one favorite supply metal was gold (Sudraud et al. 1978) and other ion species were also produced. The most important to date are pure elements, including Al, As, Au, B, Be, Bi, Cs, Cu, Ga, Ge, Fe, In, Li, Pb, P, Pd, Si, Sn, U, and Zn. In addition, elements such as As, B, Be, and Si may be produced from alloys (Orloff 1993; Bischoff 2008). Nowadays, even if the LMIS remains the brightest ion source ever fabricated with a very wide spectrum of applications, elements other than gallium are not yet very common.

As an illustration of the capabilities of nongallium FIB beams, we will describe the development of a new FIB approach based on a pure gold-ion source that has allowed us to obtain GaAs nanowires with very

the graphene structural properties as indicated by electrical transport measurements.

high-aspect ratios (length/diameter), difficult to obtain by any other technique (Gierak et al. 2010a). In this last example, we show that a focused gold-ion beam allows high-resolution gold-implanted dots to be created. The interest here is to use ultrahigh purity gold (5 N purity grade) that is preserved down to the target level and to benefit from a well-controlled localized implantation phenomenon (size and dose). Implanted dots (Figure 3.32a) were defined with a 20 keV gold-ion beam on <111> GaAs samples to form gold islands at the surface. Then the sample was transferred into an MBE III–V growth system equipped with a gallium effusion cell and a cracking source of arsenic. GaAs nanowires were then grown on the gold-implanted islands in the MBE III–V chamber. The growth rate was monitored allowing us to obtain the very high aspect and highly crystalline structures. These gold dots were found to promote growth of nanowires with record aspect ratio (Figure 3.32b). Their lengths reach several hundred nanometers and their diameters range from some 10s of nanometers down to less than 10 nm. In our opinion, this result is a clear indication that the use of nongallium FIB beams in emerging nanoscience applications has a very promising future.

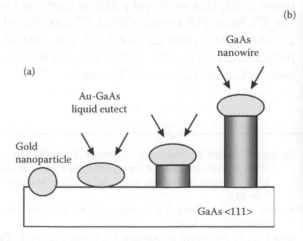

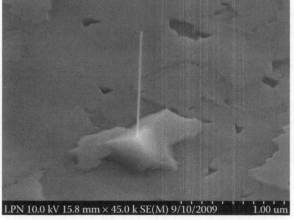

FIGURE 3.32   (a) Schematic principle of the Au-assisted VLS growth process. (b) SEM image of a GaAs single nanowire having a very high-aspect ratio (diameter <10 nm and length around 500 nm). This nanowire was grown using the VLS process on FIB-implanted gold islands (Au+, 20 keV). (From Gierak, J. et al. 2010b. *Microelectron. Eng.* 87, 1386–1390.)

Chapter 3

3.7   Conclusion

The development of the FIB technology is one among many examples of how research results may find unexpected applications in totally different application areas. This is particularly true of the FIB technology development itself, which has benefited from all the earlier advances in field emission physics, charged particle optics theory and modeling, and fundamental instrumentation and applied metrology. All these advances were very quickly and efficiently integrated into FIB instruments, so that in less than one decade these have moved out from a few specialist laboratories to enter almost every modern laboratory, research institute, or processing environment. This is also true for the semiconductor industry, which almost immediately adopted FIB systems for device inspection failure analysis and reverse engineering with a roaring success. A glance at the present field of applications shows how much progress has been made. Consider, for example, the TEM specimen fabrication method that is regarded today as a workhorse for the FIB technology. Who could have predicted back in the beginning of the 1990s, before the development of gas injection, nanomanipulation, and *in situ* welding, that this technique would be so widespread today? The perceived and established limits of the FIB technology are not intrinsic to it but rather to our understanding of these limits and to the restrictions of the present FIB constituents (source, optics, etc.). By improving some of these, we have been able to demonstrate that direct, efficient, and controlled nanopatterning of materials is possible through local nanoengraving, ion-induced modifications, or damage engineering. The FIB processing methods we have developed now appear to be well suited and very promising for several diverse nanotechnology applications. This is especially true when we explore the promising direction of "bottom-up" or "organization" processes achievable via FIB templating and subsequent nanostructure formation. These advances were made possible by optimizing a specific gallium-based FIB instrument, together with innovative patterning schemes and associated processing techniques. This may be of major interest for future applications to nanoelectronics, nanomagnetism, spin-electronics, or nanooptics.

In support of these opinions, we draw attention to one of the latest developments of FIB technology, revealed recently (Notte et al. 2007). This remarkable development concerns a high-resolution ion microscope using a cryogenic helium ion source (Hill et al. 2012). The brightness of this source is even higher than that of an LMIS and the helium ions can therefore be focused in a spot smaller than a nanometer in diameter. Analysis and ion microscopy are possible with this device, which seems a rival to the electron microscope. This newest development, taken together with the results presented earlier in this work, clearly shows than the FIB technique constitutes a technology with a strong potential and experimental limits that have not yet been attained.

Acknowledgments

The authors wish to acknowledge all their colleagues who contributed to the different achievements presented here for their enthusiastic and noteworthy contributions. Jacques Gierak would like to express thanks to A. Septier for his advice, guidance, and support over many years. This work was partially supported under the EC Nano-FIB project G5RD-CT2000-00344, the EC (AMMARE contract G5RD-CT 2001-00478), and by SESAME contract No. 1377, the Région Ile de France and the Conseil Général de l'Essonne.

References

Aigouy, L., Lalanne, P., Hugonin, J.P., Julié, G., Mathet, V., and Mortier, M. 2007. Near-field analysis of surface waves launched at nanoslit apertures, *Phys. Rev. Lett.* 98, 253902.

Andres, R.P., Bein, T., Dorogi, M., Feng, S., Henderson, J.L., Kubiak, C.P., Mahoney, W., Osifchin, R.G., and Reifenberger, R. 1996. "Coulomb Staircase" at room temperature in a self-assembled molecular nanostructure, *Science* 272, 1323.

Applied Materials, Inc. 2009. Applied Materials, Inc., Model SEMVision G3 FIB Star, see http://www.appliedmaterials.com/products.

Arshak, K., Mihov, M., Nakahara, S., Arshak, A., and McDonagh, D. 2004. A novel focused-ion-beam lithography process for sub-100 nanometer technology nodes, *Superlattices Microstruct.* 36, 335–343.

Auvray, L. 2000. Polymers at interfaces, *C. R. Acad. Sci. Paris* t. 1, Série IV, 1123–1124.

Badolato, A., Hennessy, K., Atature, M., Dreiser, J., Hu, E., Petroff, P.M., and Imamoglu, A. 2005. Deterministic coupling of single quantum dots to single nanocavity modes, *Science* 308, 1158.

Bardotti, L., Prével, B., Jensen, P., Treilleux, M., Mélinon, P., Perez, A., Gierak, J., Faini, G., and Mailly, D. 2002. Organizing nano-clusters on functionalized surfaces, *Applied Surface Science* 191, 205–210.

Barth, J.E., and Kruit, P. 1996. Addition of different contributions to the charged particle probe size, *Optik* 101, 101–109.

Beckman, J.C., Chang, T.H.P., Wagner, A., and Pease, R.F.W. 1996. Energy spread in liquid metal ion sources at low currents, *J. Vac. Sci. Technol. B* 14, 3911.

Bell, A.E. and Swanson, L.W. 1986. The influence of substrate geometry on the emission properties of a liquid metal ion source, *Appl. Phys.* A41, 335–346.

Biance, A.L., Gierak, J., Bourhis, E., Madouri, A., Lafosse, X., Patriarche, G., Oukhaled, G. et al. 2006. Focused ion beam sculpted membranes for nanoscience tooling, *Microelectron. Eng.* 83, 1474–1477.

Bischoff, L. 2008. Application of mass-separated focused ion beams in nano-technology, *Nucl. Instrum. Meth. Phys. Res. B* 266, 1846–1851.

Bolotin, K.I., Sikes, K.J., Hone, J., Stormer, H.L., and Kim, P. 2008. Temperature dependent transport in suspended graphene, *Phys. Rev. Lett.* 101, 096802.

Bronsgeest, M.S., Barth, J.E., Swanson, L.W., and Kruit, P. 2008. Probe current, probe size, and the practical brightness for probe-forming systems, *J. Vac. Sci. Technol. B* 26, 949–955.

Chappert, C., Bernas, H., Ferré, J., Kottler, V., Jamet, J.P., Chen, Y., Cambril, E. et al. 1998. Planar patterned magnetic media obtained by ion irradiation, *Science* 280, 1919.

Chen, J.H., Jang, C., Xiao, S., Ishigami, M., and Fuhrer, M.S. 2008. Intrinsic and extrinsic performance limits of graphene devices on SiO_2, *Nature Nano.* 3, 206.

Clampitt, R., Aitken, K.L., and Jefferies, D.K. 1975. Intense field emission ion source of liquid metal, *J. Vac. Sci. Technol.* 12, 1208.

Cockayne, D., Kirkland, A.I., Nellist, P.D., and Bleloch, A., eds. 2009. New possibilities with aberration-corrected electron microscopy, *Phil. Trans. Roy. Soc. London* 367(1903), 3631–3870.

D'Cruz, C., Pourrezaei, K., and Wagner, A. 1985. Ion cluster emission and deposition from liquid gold ions sources, *J. Appl. Phys.* 58, 2724–2730.

Dekker, C. 2007. Solid-state nanopores, *Nat. Nanotechnol.* 2, 209–215.

Devolder, T., Chappert, C., Chen, Y., Cambril, E., Bernas, H., Jamet, J.P., and Ferré, J. 1999. Magnetic properties of He⁺-irradiated Pt/Co/Pt ultrathin films, *Appl. Phys. Lett.* 75, 403.

Du, X., Skachko, I., Barker, A., and Andrei, E.Y. 2008. Approaching ballistic transport in suspended graphene, *Nat. Nano.* 3, 491.

FEI. 2008. FEI Company, Focused Ion Beam (FIB) Tools, http://www.fei.com.

FEI. 2009. FEI Company, Expida series, see http://www.fei.com/products/dualbeams/expida.aspx.

Ferré, J., and Jamet, J.P. 2007. Alternative patterning techniques: magnetic interactions in nanomagnet arrays. In: Kronmüller, H. and Parkin, S., (eds.) *Handbook of Magnetism and Advanced Magnetic Materials*, Vol. 3, pp. 1710–1735. Wiley, Chichester.

Feynman, R.P. 1959. There's plenty of room at the bottom—An invitation to enter a new field of physics, *Caltech's Engineering and Science* http://www.zyvex.com/nanotech/feynman.html.

Forbes, R.G., and Mair, G.L.R. 2009. Liquid metal ion sources. In *Handbook of Charged Particle Optics*, J. Orloff, ed. pp. 29–86. CRC Press, Boca Raton, FL.

Galovich, C.S. 1988. Effects of backsputtered material on gallium liquid metal ion source behavior, *J. Appl. Phys.* 63, 4811.

Geim, A.K., and Novoselov, K.S. 2007. The rise of graphene, *Nat. Mater.* 6, 183.

Ghaleh, F., Köster, R., Hövel, H., Bruchhaus, L., Bauerdick, S., Thiel, J., and Jede, R. 2007. Controlled fabrication of nanoptic patterns on a graphite surface using focused ion beams and oxidation, *J. Appl. Phys.* 101, 044301.

Giannuzzi, L.A., and Garrison, B.J. 2007. Molecular dynamics simulations of 30 and 2 keV Ga in Si, *J. Vac. Sci. Technol. A* 25, 1417.

Giannuzzi, L.A., and Stevie, F.A., eds. 2005. *Introduction to Focused Ion Beams*, Springer, New York.

Gierak, J., Schneider, M., Vieu, C., Ben Assayag, G., and Marzin, J.Y. 1997a. 3D defect distribution induced by focused ion beam irradiation at variable temperatures in a GaAs/GaAlAs multi quantum well structure, *Microelectron. Eng.* 30, 253–256.

Gierak, J., Vieu, C., Schneider, M., Launois, H., Ben Assayag, G., and Septier, A. 1997b. Optimization of experimental operating parameters for very high resolution focused ion beam applications, *J. Vac. Sci. Technol. B* 15, 2373.

Gierak, J., Septier, A., and Vieu, C. 1999. Design and realization of a very high-resolution FIB nanofabrication instrument, *Nucl. Instrum. Meth. Phys. Res. A* 427, 91.

Gierak, J. et al. 2006. Exploration of the ultimate patterning potential of focused ion beams. *Journal of Microlithography, Microfabrication, and Microsystems* 5(1), 011011.

Gierak, J., Madouri, A., Biance, A.L., Bourhis, E., Patriarche, G., Ulysse, C., Lucot, D. et al. 2007. Sub-5nm FIB direct patterning of nanodevices, *Microelectron. Eng.* 84, 779.

Gierak, J., Lucot, D., Ouerghi, A., Patriarche, G., Bourhis, E., Faini, G., and Mailly, D. 2010a. Nano-patterning of graphene structure using highly focused beams of gallium ions, *Mater. Res. Soc. Symp. Proc.* 1259, 47–58.

Gierak, J., Madouri, A., Bourhis, E., Travers, L., Lucot, D., and Harmand, J.C. 2010b. Focused gold ions beams for localized epitaxy of semiconductor nanowires, *Microelectron. Eng.* 87, 1386–1390.

Gierhart, B.C., Howitt, D.G., Chen, S.J., Zhu, Z., Kotecki, D.E., Smiths, R.L., and Collins, S.D. 2007. Nanopore with transverse nanoelectrode for electrical characterisation and sequencing of DNA, *Transducers Nanosensors Proc.* 399–402.

Gilmartin, S.F., Arshak, K., Collins, D., Korostynska, O., and Arshak, A. 2007. Fabricating nanoscale device features using the 2-step NERIME nanolithography process, *Microelectron. Eng.* 84, 833–836.

Grivet, P. 1972. *Electron Optics (2nd Edition)*, Pergamon, Oxford.

ITRS. 2007. International Technology Roadmap for Semiconductors, 2007 Edition. http://www.itrs.net/reports.html.

Hannour, A., Bardotti, L., Prével, B., Bernstein, E., Mélinon, P., Perez, A., Gierak, J., Bourhis, E., and Mailly, D. 2005. 2D arrays of CoPt nanocluster assemblies, *Surface Sci.* 594, 1–11.

Hawkes, P.W., ed. 1995. *Electrons Et Microscopes. Vers Les Nanosciences*, CNRS éditions & Belin, Paris.

Hawkes, P.W., and Kasper, E. 1996. *Principles of Electron Optics, Volume 1 and 2.* Academic Press, London.

Hennessy, K., Högerle, C., Hu, E., Badolato, A., and Imamoglu, A. 2006. Entangled photons from a strongly coupled quantum dot-cavity system, *Appl. Phys. Lett.* 89, 041118.

Hill, R., Notte, J.A., and Scipioni, L. 2012. Scanning helium ion microscopy, *Adv. Imaging Electron Phys.* 170.

Hyndman, R., Warin, P., Gierak, J., Ferré, J., Chapman, J.N., Jamet, J.P., Mathet, V., and Chappert, C. 2001. Modification of Co/Pt

Chapter 3

multilayers by gallium irradiation—Part 1: The effect on structural and magnetic properties, *J. Appl. Phys.* 90, 3843.

Jamet, J.P., Ferré, J., Meyer, P., Gierak, J., Vieu, C., Rousseaux, F., Chappert, C., and Mathet, V. 2001. Giant enhancement of the domain wall velocity in patterned ultrathin magnetic nanowires, *IEEE Trans. Mag.* 37, 2120–2122.

Jiang, X.R., and Kruit, P. 1996. Comparison between different imaging modes in focussed ion beam instruments, *Microelectron. Eng.* 30, 249–252.

Kasianowicz, J.J., Brandin, E., Branton, D., and Deamer, D.W. 1996. Characterization of individual polynucleotide molecules using a membrane channel, *Proc. Nat. Acad. Sci.* 93, 13770–13773.

Kirkland, A.I., Chang, S.L.-Y., and Hutchison, J.L. 2007. Atomic resolution transmission electron microscopy. In *Science of Microscopy*, P.W. Hawkes and J.C.E. Spence, eds, pp. 3–64. Springer, New York and Heidelberg.

Kitslaar, P., Strassner, M., Sagnes, I., Bourhis, E., Lafosse, X., Ulysse, C., David, C., Jede, R., Bruchhaus, L., and Gierak, J. 2006. Towards the creation of quantum dots using FIB technology, *Microelectron. Eng.* 83, 811–814.

Knauer, W. 1981. Energy broadening in field emitted electron and ion beams, *Optik* 59, 335.

Komuro, M., Kanayama, T., Hiroshima, H., and Tanoue, H. 1983. Measurement of virtual crossover in liquid gallium ion source, *App. Phys. Lett.* 42, 908.

Kosugi, T., Gamo, K., Namba, S., and Aihara, R. 1991. Ion beam assisted etching of GaAs by low energy focused ion beam, *J. Vac. Sci. Technol. B* 9, 2660.

Krohn, V.E., and Ringo, G.R. 1975. Ion source of high brightness using liquid metal, *Appl. Phys. Lett.* 27, 479.

Kubena, R.L., Ward, J.W., Stratton, F.P., Joyce, R.J., and Atkinson, G.M. 1991. A low magnification focused ion beam system with 8 nm spot size, *J. Vac. Sci. Technol.* B9, 3079.

Laruelle, F., Hu, Y.P., Simes, R., Robinson, W., Merz, J., and Petroff, P.M. 1990. Optical study of GaAs/GaAlAs quantum structures processed by high energy focused ion beam implantation, *Surf. Sci.* 228, 306.

Lencová, B. 2008. Electrostatic lenses. In *Handbook of Charged Particle Optics*, J. Orloff, ed. pp. 161–208. CRC Press, Boca Raton, FL.

Lencova, B. 2009. *Software for particle optics computations*, http://www.lencova.com.

Lencová, B., and Zlámal, J. 2008. A new program for the design of electron microscopes, *Phys. Procedia.* 1, 315–324.

Levi-Setti, R., Crow, G., Wang, Y.L., Parker, N.W., Mittleman, R., and Hwang, D.M. 1985. High resolution scanning-ion-microprobe study of graphite and its intercalation compounds, *Phys. Rev. Lett.* 54, 2615.

Lezec, H.J., Degiron, A., Devaux, E., Linke, R.A., Martin-Moreno, L., Garcia-Vidal, F.J., and Ebbesen, T.W. 2002. Beaming light from a subwavelength aperture, *Science* 297, 820–822.

Li, J., Stein, D., McMullan, C., Branton, D., Aziz, M.J., and Golovchenko, J.A. 2001. Ion-beam sculpting at nanometre length scales, *Nature* 412, 166.

Lucot, D., Gierak, J., Ouerghi, A., Bourhis, E., Faini, G., and Mailly, D. 2009. Focused ion beam sculpted membranes for nanoscience, *Microelectron. Eng.* 86, 882.

Mair, G.L.R., Forbes, R.G., Latham, R.V., and Mulvey, T. 1983. Energy spread measurements on a liquid metal ion source, *Microcircuit Eng.* 83, 171.

Mathé, J., Aksimentiev, A., Nelson, D.R., Schulten, K., and Meller, A. 2005. Orientation discrimination of single stranded DNA inside the α-hemolysin membrane channel, *Proc. Nat. Acad. Sci.* 102, 12377–12382.

Matsui, S., and Ochiai, Y. 1996. Focused ion beam applications to solid state devices, *Nanotechnology* 7, 247–258.

Matsui, S., Mori, K., Saigo, K., Shiokawa, T., Toyoda, K., and Namba, S. 1986. Lithographic approach for 100 nm fabrication by focused ion beam, *J. Vac. Sci. Technol. B* 4, 845.

McNally, B., Wanunu, M., and Meller, A. 2008. Electromechanical unzipping of individual DNA molecules using synthetic sub-2 nm pores, *Nano Lett.* 8, 3418–3422.

Mélinon, P., Hannour, A., Bardotti, L., Prével, B., Gierak, J., Bourhis, E., Faini, G., and Canut, B. 2008. Ion beam nanopatterning in graphite: Characterization of single extended defects, *Nanotechnology* 19, 235305 (9pp).

Melngailis, J. 2001. Ion sources for nanofabrication and high resolution lithography, *Proceedings of the 2001 Particle Accelerator Conference*, Chicago.

Melngailis, J., Musil, C.R., Stevens, E.H., Utlaut, M., Kellog, E.M., Post, R.T., Geis, M.W., and Mountain, R.W. 1985. The focused ion beam as an integrated circuit restructuring tool, *J. Vac. Sci. Technol. B* 4, 176–180.

Morgan, J., Notte, J., Hill, R., and Ward, B. 2006. An introduction to the helium ion microscope, *Microsc. Today*, 14(4), 24–31.

Müller, E.W., and Tsong, T. 1969. *Field Ion Microscopy*, Elsevier, Amsterdam.

Munro, E. 2009. *MEBS-computer aided design of electron and ion optical system*, http://www.mebs.co.uk/.

Munro, E., Rouse, J., Liu, H.-N., Wang, L.-P., and Zhu, X. 2006. Simulation software for designing electron and ion optical equipment. *Microelectron. Eng.* 83, 994–1002.

Murray, A., Isaacson, M., and Adesida, I. 1984. AIF3—A new very high resolution electron beam resist, *Appl. Phys. Lett.* 45, 589.

Miyauchi, E., and Hashimoto, H. 1986. Application of focused ion beam technology to maskless ion implantation in a molecular beam epitaxy grown GaAs or AlGaAs epitaxial layer for three-dimensional pattern doping crystal growth, *J. Vac. Sci. Technol.* A4, 933.

Nakayama, Y., and Makabe, T. 1993. Investigation of the ion emission mechanism of gallium liquid metal ion sources, *J. Phys. D: Appl. Phys.* 26, 1769.

NanoFIB. 2007. *Nanofabrication with focused ion beams*, http://www.nanofib.com/. See also ftp://ftp.cordis.europa.eu/pub/nanotechnology/docs/n_s_nanofib_27052002.pdf.

Notte, J., Ward, B., Economou, N., Hill, R., Percival, R., Farkas, L., and McVey, S. 2007. An introduction to the helium ion microscope. *AIP Conf. Proc.* 931, 489.

Orloff, J. 1987. Comparison of optical design approaches for use with liquid metal ion sources, *J. Vac. Sci. Technol. B* 5, 175–177.

Orloff, J. 1993. High-resolution focused ion beams, *Rev. Sci. Instrum.* 64, 1105–1130.

Orloff, J., Swanson, L.W., and Utlaut, M. 1996. Fundamental limits to imaging resolution for focused ion beams, *J. Vac. Sci. Technol.* B14, 3759.

Orloff, J. Utlaut, M., and Swanson, L.W. 2003. *High Resolution Focused Ion Beams*, Plenum, New York.

Orlov, A.O., Amlani, I., Bernstein, G.H., Lent, C.S., and Snider, G.L. 1997. Realization of a functional cell for quantum-dot cellular automata, *Science* 277, 928.

Orsay Physics. 2009. Orsay Physics, Products FIB, http://www.orsayphysics.com/.

Oukhaled, G., Mathé, J., Biance, A.-L., Bacri, L., Betton, J.-M., Lairez, D., Pelta, J., and Auvray, L. 2007. Unfolding of proteins and long transient conformations detected by single nanopore recording, *Phys. Rev. Lett.* 98, 158101.

Oukhaled, G., Auvray, L., Bacri, L. et al. 2011. Dynamics of completely unfolded and native proteins through solid-state nanopores as a function of electric driving force, *ACS Nano* 5, 3628–3638.

Painter, O., Vuckovic, J., and A. Scherer, A. 1999. Defect modes of a two-dimensional photonic crystal in an optically thin dielectric slab, *J. Opt. Soc. Am. B* 16, 275.

Park, G., Huffaker, D.L., Zou, Z., Shchekin, O.B., and Deppe, D.G. 1999. Temperature dependence of lasing characteristic for long-wavelength (1.3-μm) GaAs-based quantum-dot lasers, *IEEE Photon. Technol. Lett.* 11, 301–303.

Pépin, A., Vieu, C., Schneider, M., Launois, H., and Nissim, Y. 1997. Evidence of stress dependence in SiO_2/Si_3N_4 encapsulation-based layer disordering of GaAs/AlGaAs quantum well heterostructures, *J. Vac. Sci. Technol. B* 15, 142–153.

Perez, A., Bardotti, L., Prével, B., Jensen, P., Treilleux, M., Mélinon, P., Gierak, J., Faini, G., and Mailly, D. 2001. Quantum-dot systems prepared by 2D organization of nanoclusters preformed in the gas phase on functionalized substrates, *New J. Phys.* 4, 76.1–76.12.

Prével, B., Bardotti, L., Fanget, S., Hannour, A., Mélinon, P., Perez, A., Gierak, J., Faini, G., Bourhis, E., and Mailly, D. 2004. Gold nanoparticle arrays on graphite surfaces, *Appl. Surf. Sci.* 226, 173.

Prewett, P.D., and Mair, G.L.R. 1991. *Focused Ion Beams from Liquid Metal Ion Sources*, Wiley, New York.

Prewett, P.D., Mair, G.L.R., and Thompson, S.P. 1982. Some comments on the mechanism of emission from liquid metal ion sources, *J. Phys. D: Appl. Phys.* 15, 1339.

Purcell, S.T., Vu Thien Binh and Thevenard, P. 2001. Atomic-size metal ion sources: Principles and use, *Nanotechnology* 12, 168–172.

Raith. 2008. TNT 2008, see http://www.tntconf.org/2008/Files/Presentaciones/, pp. 18–22.

Raith. 2009. Raith GmbH, ionLiNE lithography, nanofabrication and engineering workstation, http://www.raith.com/.

Rau, N., Stratton, F., Fields, C., Ogawa, T., Neureuther, A., Kubena, R., and Willson, G. 1998. Shot-noise and edge roughness effects in resists patterned at 10 nm exposure, *J. Vac. Sci. Technol. B* 16, 3784.

Reimer, L. 1998. *Scanning Electron Microscopy*, Springer, Berlin.

Reimer, L., and Kohl, H. 2008. *Transmission Electron Microscopy*, Springer, Berlin.

Reuter, D., Riedesel, C., Schafmeister, P., Meier, C., and Wieck, A.D. 2003. Fabrication of high quality two-dimensional electron gases by overgrowth of focused-ion-beam-doped AlxGa1-xAs, *Appl. Phys. Lett.* 82, 481.

Reyntjens, S., and Puers, R. 2001. A review of focused ion beam applications in microsystem technology, *J. Micromech. Microeng.* 11, 287–300.

Rue, C. 2009. *Methods for Quantifying "Milling Acuity"*, European Focused Ion Beam Users Group 2009 Meeting, October 5, 2009, http://www.imec.be/efug/. Arcachon, France.

Sagnes, I., Stassner, M., Bouchoule, S., Kistlaar, P., Bourhis, E., and Gierak, J. 2006. Selective epitaxy of patterned III–V semiconductors surfaces using FIB technology, *E-nano Newslett.* December, 11.

Saint-Girons, G., Garnache, A., Patriarche, G., and Sagnes, I. 2003. *Proc. IEEE Indium-Phosphide-and-Related-Materials 2003 (Santa Barbara CA)*, Paper TuB2.2.

Sakai, Y., Yamada, T., Suzuki, T., and Ichinokawa, T. 1999. Contrast of scanning ion microscope images compared with scanning electron microscope images for metals, *J. Anal. At. Spectrom.* 14, 419–421.

Schiedt, B., Gierak, J., Madouri, A., Biance, A.L., Bourhis, E., Patriarche, G., Ulysse, C. et al. 2009. Sub-5 nm FIB direct patterning of nanopores, *Mater Res. Soc. Sympos. Proc.* 1191.

Schiedt, B., Auvray, L., Bacri, L., Oukhaled, G., Madouri, A., Bourhis, E., Patriarche, G., Pelta, J., Jede, R., and Gierak, J. 2010. Direct FIB fabrication and integration of "single nanopore devices" for the manipulation of macromolecules, *Microelec. Eng.* 87, 1300–1303.

Seiko. 2008. Seiko Instruments Inc., Model SIR5000, see http://www.siint.com/en/products/mask_repair/SIR5000.html.

Seiko. 2009. Seiko Instruments Inc., Model XVision300, see http://www.siint.com/en/products/fib/XVision300.html.

Seliger, R.L., Kubena, R.L., Olney, R.D., Ward, J.W., and Wang, V. 1979. High-resolution, ion–beam processes for microstructure fabrication, *J. Vac. Sci. Technol.* 16, 1610.

Stark, T., Shedd, B., Bitarelli, J., Griffis, D., and Russell, P. 1995. H_2O enhanced focused ion beam micromachining, *J. Vac. Sci. Technol. B* 13, 282.

Stein, D., Li, J., and Golovchenko, J.A. 2002. Ion-beam sculpting time scales, *Phys. Rev. Lett.* 89, 276106.

Stevie, F.A., Irwin, R.B., Shofner, T.L., Brown, S.R., Drown, J.L., and Giannuzzi, L.A. 1998. Plan view TEM sample preparation using the focused ion beam lift-out technique, *AIP Conf. Proc.* 449, 868.

Storm, A.J., Chen, J.H., Ling, X.S., Zandbergen, H.W., and Dekker, C. 2003. Fabrication of solid-state nanopores with single-nanometre precision, *Nat. Mater.* 2, 537.

Sudraud, P., van de Walle, J., Colliex, C., and Castaing, R. 1978. Contribution of field effects to the achievement of higher brightness ion sources, *Surface Sci.* 70, 392–402.

Sudraud, P., Ben Assayag, G., and Bon, M. 1988. Focused-ion-beam milling, scanning-electron microscopy, and focused-droplet deposition in a single microcircuit surgery tool, *J. Vac. Sci. Technol. B* 6, 234.

Sugimoto, Y., Akita, K., Taneya, M., Kawanishi, H., Aihara, R., and Watahiki, T. 1991. A multichamber system for *in situ* lithography and epitaxial growth of GaAs, *Rev. Sci. Inst.* 62, 1828–1835.

Sun, S, Murray, C.B., Weller, D., Folks, L., and Moser, A. 1989. Monodisperse FePt nanoparticles and ferromagnetic FePt nanocrystal superlattices, *Science* 287, 1989–1992.

Suvorov, V.G., and Forbes, R.G. 2004. Theory of minimum emission current for a non-turbulent liquid-metal ion source, *Microelectron. Eng.* 73–74, 126–131.

Suzuki, T., Endo, N., Shibata, M., Kamasaki, S., and Ichinokaw, T. 2004. Contrast differences between scanning ion and scanning electron microscope images, *J. Vac. Sci. Technol. A* 22, 49.

Swanson, L.W., and Bell, A.E. 1989. Liquid metal ion sources. In *The Physics and Technology of Ion Sources*, I.G. Brown, ed., Wiley, Chichester.

Swanson, L.W., Schwind, A.E., Bell, A.E., and Brady, J.E. 1979. Emission characteristics of gallium and bismuth liquid metal ion sources, *J. Vac. Technol. B* 16, 1864.

Tapaszto, L., Dobrik, G., Lambin, P., and Biro, L.P. 2008. Tailoring the atomic structure of graphene nanoribbons by scanning tunnelling microscope lithography, *Nature Nano.* 3, 397.

Chapter 3

Taylor, G.I. 1964. Disintegration of water drops in an electric field, *Proc. Roy. Soc. Lond.* A280, 383–397.

Terris, B.D., and Rettner, C. 2002. Ion beam patterning of magnetic media, *NanoTech.* 6(4), 79–81.

Utlaut, M. 2009. Focused ion beams. In *Handbook of Charged Particle Optics*, J. Orloff, ed. pp. 523–600. CRC Press, Boca Raton, FL.

Van Es, J.J., Gierak, J., Forbes, R.G., Suvorov, V.G., Van den Berghe, T., Dubuisson, Ph., Monnet, I., and Septier, A. 2004. An improved gallium liquid metal ion source geometry for nanotechnology, *Microelectron. Eng.* 73–74, 132–138.

Vieu, C., Ben Assayag, G., and Gierak, J. 1994. Observation and simulation of focused ion beam induced damage, *Nucl. Instrum. Meth. Phys. Res.* B 93, 439–446.

Wang, L. 1997. Design optimization for two lens focused ion beam columns, *J. Vac. Sci. Technol.* B 15, 833–839.

Wang, B., Aigouy, L., Bourhis, E., Gierak, J., Hugonin, J.P., and Lalanne, P. 2009. Efficient generation of surface plasmon illumination under highly oblique incidence. *Appl. Phys. Lett.* 94, 011114.

Ward, J.W. 1985. A Monte Carlo calculation of the virtual source size for a liquid metal ion source, *J. Vac. Technol.* B3, 207.

Yamamoto, T., Yanagisawa, J., Gamo, K., Takaoka, S., and Murase, K. 1993. Estimation of damage induced by focused Ga ion beam irradiation, *Jpn. J. Appl. Phys.* 32, 6268.

Zeiss. 2008. Carl Zeiss NTS GmbH Nano Technology Systems Division, http://www.smt.zeiss.com/.

Zhao, Q, Sigalov, G., Dimitrov, V., Dorvel, B., Mirsaidov, U., Sligar, S., Aksimentiev, A., and Timp, G. 2007. Detecting SNPs using a synthetic nanopore, *Nano Lett.* 7, 1680–1685.

Zhukov, A.E., Kovsh, A.R., Maleev, N.A., Mikhrin, S.S., Ustinov, V.M., Tsatsul'nikov, A.F., Maximov, M.V. et al. 1999. Long-wavelength lasing from multiply stacked InAs/InGaAs quantum dots on GaAs substrates, *Appl. Phys. Lett.* 75, 1926–1928.

Ziegler, J. 2008. *SRIM—The Stopping and Range of Ions in Matter*, http://www.srim.org/.

Zworykin, V.K., Morton, G.A., Ramberg, E.G., Hillier, J., and Vance, A.W. 1945. *Electron Optics and the Electron Microscope*, Wiley, New York. pp. 600–601.

4. Focused Ion Beam and Electron Beam Deposition

Shinji Matsui

Graduate School of Science, University of Hyogo, Hyogo, Japan

4.1   Introduction

The recent years have witnessed a number of investigations concerning nanostructure technology. The objective of research on nanostructure technology is to explore the basic physics, technology, and applications of ultrasmall structures and devices with dimensions in the sub-10 nm regime. Electron beam (EB) lithography is the most widely used and versatile lithography tool in fabricating nanostructure devices and photomasks. EB can be focused to a diameter <5 nm. The minimum beam diameter of scanning electron microscope (SEM) and scanning transmission electron microscope (STEM) is 1.5 and 0.5 nm, respectively. Focused ion beam (FIB) can be focused close to 5 nm diameter.

Nanofabrication Handbook. Edited by Stefano Cabrini and Satoshi Kawata © 2012 CRC Press/Taylor & Francis Group, LLC. ISBN: 978-1-4200-9052-9.

Chapter 4

Today, the minimum size of Si production devices is down to 50 nm or less. EBs and FIBs have been used to fabricate various two-dimensional (2D) nanostructure devices such as single-electron transistors and metal oxide semiconductor (MOS) transistors with nanometer gate lengths. Ten-nanometer structures can be formed by using a commercially available EB or FIB system with 5–10-nm-diameter beams and high-resolution resist (Matsui 1997). Two-dimensional nanostructure fabrication is therefore already an established process. There are approaches to three-dimensional (3D) fabrication: using a laser, an EB, or an FIB to perform chemical vapor deposition (CVD). FIB- and EB-CVD are superior to laser-CVD in terms of spatial resolution and a beam-scan control.

In this chapter, deposition characteristics and the applications of FIB- and EB-CVD are described. The deposition rate of FIB-CVD is much higher than that of EB-CVD due to factors such as the difference in mass between an electron and an ion. Furthermore, a smaller penetration depth of the ions compared to the electrons makes it easier to create complicated 3D nanostructures. For example, when we attempt to make a coil nanostructure with a linewidth of 100 nm, 10–50 eV electrons pass through the ring of coil and reach the substrate because of the large range of electrons (at least a few microns), which makes it difficult to create a coil nanostructure using EB-CVD. On the other hand, since the range of the ions is a few tens of nanometers or less, the ions are deposited inside the ring. As a result, FIB-CVD is superior to EB-CVD in making complicated 3D nanostructures.

4.2 Focused-Ion-Beam Chemical Vapor Deposition

Seliger et al. (1979) reported Ga ion beams down to 100 nm diameter with current density in the focal spot of 1.5 A/cm^2. Wargner et al. (1990) demonstrated pillars and walls with high aspect ratios achieved using FIB-CVD. Matsui et al. (2000) demonstrated 3D nanostructure fabrication by FIB-CVD.

We used two commercially available FIB systems (SMI9200, SMI2050, SII Nanotechnology Inc., Tokyo, Japan) with a Ga$^+$ ion beam operating at 30 keV. The FIB-CVD used a precursor of *phenanthrene* ($C_{14}H_{10}$) as the source material. The beam diameter of SMI9200 was about 7 nm and that of SMI2050 was about 5 nm. The SMI9200 system was equipped with two gas sources in order to increase the gas pressure. The top of the gas nozzles faced each other and was directed at the beam point. The nozzles faced each other and were directed at the beam point. The nozzles were set at a distance of 40 μm from each other and positioned about 300 μm above the substrate surface. The inside diameter of the nozzle was 0.3 mm. The phenanthrene gas pressure during pillar growth was typically 5×10^{-5} Pa in the specimen chamber, but the local gas pressure at the beam point was expected to be much higher. The crucible of the source was heated to 85°C. The SMI2050 system, on the other hand, was equipped with a single gas nozzle. The FIB is scanned in order to be able to write the desired pattern via computer control, and the ion dose is adjusted to deposit a film of the desired thickness. The experiments were carried out at room temperature on a silicon substrate.

The deposited film was characterized by observing it with a transmission electron microscope (TEM). A thin film of carbon (200 nm thick) was deposited on a silicon substrate by 30 keV Ga$^+$ FIB using phenanthrene precursor gas. The cross sections of the structures created and its electron diffraction patterns were observed by using a 300 kV TEM. There were no crystal structures in the TEM images and diffraction patterns. It was therefore concluded that the deposited film was amorphous carbon (a-C).

Raman spectra of the a-C films were measured at room temperature with 514.5 nm line of an argon ion laser. The Raman spectra were recorded using a monochromator equipped with a charge coupled device (CCD) multichannel detector. Raman spectra were measured at 0.1–1.0 mW to avoid thermal decomposition of the samples. A relatively sharp Raman band at 1550 cm^{-1} and a broad-shouldered band at around 1400 cm^{-1} were observed in the spectra excited by the 514.5 nm line. Two Raman bands were plotted after Gaussian line shape analysis. These Raman bands, located at 1550 and 1400 cm^{-1}, originate from the trigonal (sp^2) bonding structure of graphite and tetrahedral (sp^3) bonding structure of diamond. This result suggests that the a-C film deposited by FIB-CVD is diamond-like carbon (DLC), which has attracted attention due to its hardness, chemical inertness, and optical transparency.

The coordination of carbon atoms in the carbon-based material formed by CVD of phenanthrene assisted by Ga FIB was investigated by the measurement of near-edge x-ray absorption fine structure spectra of the carbon K-edge over the excitation energy range 275–320 eV (Kanda et al. 2006). Novel peak observed at 289.0 eV was assigned to the $1s \rightarrow \sigma^*$ transition of carbon neighboring to the residue gallium. The material formed by this method was found to be Ga-doped DLC, which consists of a high sp^3-hybridized carbon.

The atomic fraction of the FIB-CVD DLC film has been determined as C:Ga:H = 87.4 at%:3.6 at%:9.0 at% using Rutherford backscattering spectrometry (RBS) and elastic recoil detection analysis (ERDA) (Igaki et al. 2007). The hydrogen content of FIB-CVD DLC film was relatively lower than that of DLC films formed by other CVD methods.

4.2.1 Three-Dimensional Nanostructure Fabrication

Beam-induced CVD is widely used in the electrical device industry in the repair of chips and masks. This type of deposition is mainly done on 2D pattern features, but it can also be used to fabricate a 3D object. Koops et al. (1994) demonstrated a nanoscale 3D structure construction by applying EB-induced amorphous carbon deposition to a microvacuum tube. However, FIB-induced CVD seems to have many advantages for the fabrication of 3D nanostructures (Matsui et al. 2000). The key issue to realizing such 3D nanostructures is the short penetration depth of the ions (a few tens of nanometers) into the target material, where the penetration depth of the ions is much shorter than that of electrons (a few microns). This short penetration depth reduces the dispersion area of the secondary electrons, and so the

deposition area is restricted to roughly several tens of nanometers. A 3D structure usually contains overhang structures and hollows. Gradual position scanning of the ion beam during the CVD process causes the position of the growth region around the beam point to shift. When the beam point reaches the edge of the wall, secondary electrons appear at the side of the wall and just below the top surface. The DLC then starts to grow laterally; the width of the vertical growth is also about 80 nm. Therefore, combining the lateral growth mode with rotating beam scanning, it is possible to obtain 3D structures with rotational symmetry like a wine glass.

The process of fabricating 3D structures by FIB-CVD is illustrated in Figure 4.1 (Matsui et al. 2000). In FIB-CVD processes, the beam is scanned in digital mode. First, a pillar is formed on the substrate by fixing the beam position (position 1). After that, the beam position is moved to within a diameter of the pillar (position 2) and then fixed until the deposited terrace thickness exceeds the range of the ions (a few tens of nanometers). This process is repeated to make 3D structures. The key point to making 3D structures is to adjust the beam-scan speed so that the ion beam remains within the deposited terrace, which means that the terrace thickness always exceeds the range of the ions. The growth in the x and y directions are controlled by both beam deflectors. The growth in the z direction is determined by the deposition rate, that is, the height of the structure is proportional to the irradiation time when the deposition rate is constant.

We intend to open up a new field called microstructure plastic arts using FIB-CVD. To demonstrate the possibilities of this field, a "microwine glass" was

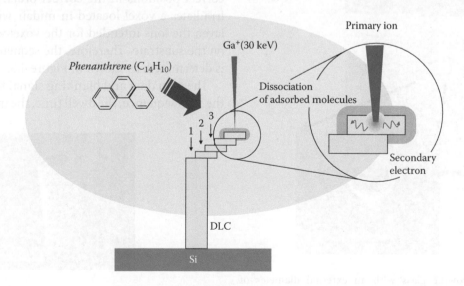

FIGURE 4.1 Fabrication process for 3-D nanostructure by FIB-CVD.

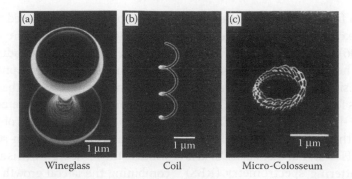

FIGURE 4.2   (a) Micro wine glass with an external diameter of 2.75 μm and a height of 12 μm. (b) Microcoil with a coil diameter of 0.6 μm, a coil pitch of 0.7 μm, and a linewidth of 0.08 μm. (c) "Micro-Colosseum."

created on a Si substrate and a human hair as a work of microstructure plastic arts as shown in Figures 4.2a and 4.3, respectively. A microwine glass with an external diameter of 2.75 μm and a height of 12 μm was formed. The fabrication time was 600 s at a beam current of 16 pA. This beautiful microwine glass shows the potential of the field of microstructure plastic art. A micro-Colosseum and a micro-Leaning Tower of Pisa were also fabricated on a Si substrate as shown in Figures 4.2c and 4.4, respectively.

Various microsystem parts have been fabricated using FIB-CVD. Figure 4.2b shows a microcoil with a coil diameter of 0.6 μm, a coil pitch of 0.7 μm, and a linewidth of 0.08 μm. The exposure time was 40 s at a beam current of 0.4 pA. It was formed by reducing the diameter of the microcoil. The diameter, pitch, and height of the microcoil were 0.25, 0.20, and 3.8 μm, respectively. The exposure time was 60 s at a beam

current of 0.4 pA. The results show that FIB-CVD is a highly promising technique for realizing parts of a microsystem, although their mechanical performances must be measured.

4.2.1.1   Three-Dimensional Pattern-Generating System

We used ion-beam-assisted deposition of a source gas to fabricate 3D structures. The 3D structure is built up as a multilayer structure. In the first step of this 3D pattern-generating system, a 3D model of the structure, designed using a 3D computer-aided design (CAD) system (3D DXF format), is needed. In this case, we realized a structure shaped like a pendulum. The 3D CAD model, which is a surface model, is cut into several slices, as shown in Figure 4.5. The thickness of the slices depends upon the resolution in the z direction (the vertical direction). The x and y coordinates of the slices are then used to create the scan data (voxel data). To fabricate the overhanging structure, the ion beam must irradiate the correct positions in the correct order. If the ion beam irradiates a voxel located in midair without a support layer, the ions intended for the voxel will be deposited on the substrate. Therefore, the sequence of irradiation is determined, as shown in Figure 4.5.

The scan data and blanking signal therefore include the scan sequence, the dwell time, the interval time, and

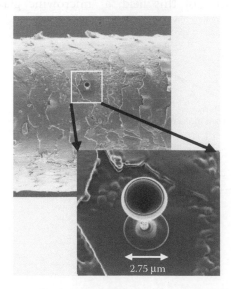

FIGURE 4.3   Microwine glass with an external diameter of 2.75 μm and a height of 12 μm on a human hair.

FIGURE 4.4   Micro-Leaning Tower of Pisa.

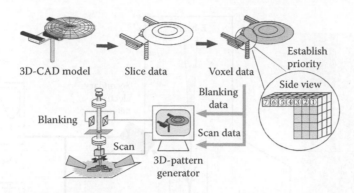

FIGURE 4.5   Data flow of 3D pattern-generating system for FIB-CVD.

the irradiation pitch. These parameters are calculated from the beam diameter, x–y resolution, and z resolution of fabrication. The z resolution is proportional to the dwell time and inversely proportional to the square of the irradiation pitch. The scan data are passed to the beam deflector of the FIB-CVD as are the blanking data. The blanking signal controls the dwell time and interval time of the ion beam.

Figure 4.6 shows a 3D CAD model and a scanning ion microscope (SIM) image of the starship Enterprise NCC-1701D (from the television series Star Trek), which was fabricated by FIB-CVD at 10~20 pA

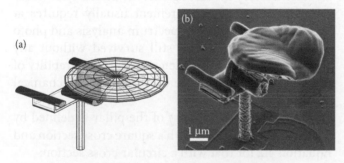

FIGURE 4.6   "Micro-Starship Enterprise NCC-1701D," 8.8 µm long. (a) 3D CAD model. (b) SIM image (tilt 45 deg).

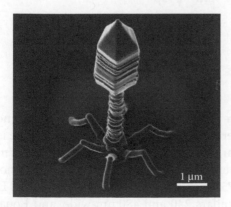

FIGURE 4.7   T-4 bacteriophage.

(Hoshino et al. 2003). The nanospaceship is 8.8 µm long and was realized at about a 1:100,000,000 scale on silicon substrate. The dwell time (t_d), interval time (t_i), irradiation pitch (p), and total process time (t_p) were 80 µs, 150 µs, 2.4 nm, and 2.5 h, respectively. The horizontal overhang structure was fabricated successfully.

Figure 4.7 shows a "nano-T4 bacteriophage," which is an artificial version of the virus fabricated by FIB-CVD on silicon surface. The size of the artificial "nano-T4 bacteriophage" is about 10 times that of the real virus.

4.2.2   Nano–Electromechanics

4.2.2.1   Measuring Young's Modulus

An evaluation of the mechanical characteristics of such nanostructures is needed for material physics. Buks and Roukes reported a simple but useful technique (Buks and Roukes 2001) for measuring the resonant frequencies of nanoscale objects using an SEM. The secondary electron detector in the SEM can detect frequencies up to around 4 MHz, so the sample vibration is measured as the oscillatory output signal of the detector. Buks and Roukes used this technique to evaluate the Casimir attractive force between the two parallel beams fabricated on a nanoscale. We evaluated the mechanical characteristics of DLC pillars in terms of the Young's modulus, determined using resonant vibration and the SEM monitoring technique (Fujita et al. 2001; Ishida et al. 2002).

The system setup for monitoring mechanical vibration is shown in Figure 4.8b. There were two ways of measuring the pillar vibrations. One was active measurement, where the mechanical vibration was induced by a thin piezoelectric device 300 µm thick and 3 mm² square. The piezo device was bonded to the sidewall of the SEM's sample holder with silver paste. The sample holder was designed to observe cross sections in the SEM (S5000, Hitachi) system. Therefore, the pillar's

Chapter 4

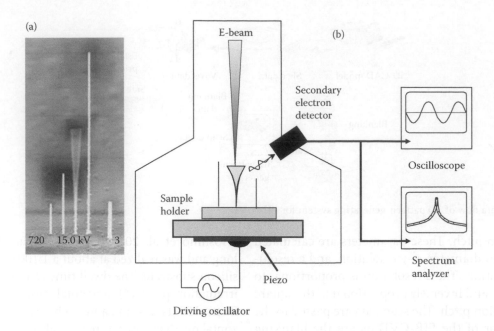

FIGURE 4.8 (a) SEM image of the vibration. The resonant frequency was 1.21 MHz. (b) Schematic diagram of the vibration monitoring system.

vibration was observed as a side-view image, as shown in Figure 4.8a. The range of vibration frequencies involved were 10 kHz up to 2 MHz, which is much faster than the SEM raster scanning speed. Thus, the resonant vibrations of the pillars can be taken as the trace of the pillar's vibration in the SEM image. The resonant frequency and amplitude were controlled by adjusting the power of the driving oscillator.

The other way to measure pillar vibrations is passive measurement using a spectrum analyzer (Agilent, 4395A), where most of the vibration seemed to derive from environmental noise from rotary pumps and air conditioners. Some parts of the vibration result from the spontaneous vibration associated with thermal excitations (Buks and Roukes 2001). Because of the excitation and residual noise, the pillars on the SEM sample holder always vibrated at a fundamental frequency, even if noise isolation is enforced on the SEM system. The amplitude of these spontaneous vibrations was on the order of a few nanometers at the top of the pillar, and high-resolution SEM can easily detect it at a magnification of 300,000.

We arranged several pillars that had varying diameters and lengths. The DLC pillars with the smallest diameter of 80 nm were grown using point irradiation. While we used two FIB systems for pillar fabrication, slight differences in the beam diameters of the two systems did not affect the diameters of the pillars. Larger-diameter pillars were fabricated using an area-limited raster scan mode. Raster scanning a 160 nm² region

produced a pillar with a cross section of about 240 nm², and a 400 nm² scan resulted in a pillar with a cross section of 480 nm². The typical SEM image taken during resonance is shown in Figure 4.8a. The FIB-CVD pillars seemed very durable against the mechanical vibration. This kind of measurement usually requires at least 30 min, including a spectrum analysis and photo recording, but the pillars still survived without any change in resonance characteristics. This durability of the DLC pillars should be useful in nanomechanical applications.

The resonant frequency f of the pillar is defined by Equation 4.1 for a pillar with a square cross section and Equation 4.2 for that with a circular cross section:

$$f_{square} = \frac{a\beta^2}{2\pi L^2}\sqrt{\frac{E}{12\rho}} \tag{4.1}$$

$$f_{circular} = \frac{a\beta^2}{2\pi L^2}\sqrt{\frac{E}{16\rho}} \tag{4.2}$$

where a is the width of the square pillar and/or the diameter of the circular-shaped pillar, L is the length of the pillar, ρ is the density, and E is the Young's modulus. The coefficient β defines the resonant mode and $\beta = 1.875$ for the fundamental mode. We used Equation 4.1 for pillars 240 and 480 nm wide, and Equation 4.2 for pillars grown by point-beam irradiations.

The relationship of the resonant frequency to the Young's modulus, which depends on the ratio of the pillar diameter to the squared length, is summarized in Figure 4.9. All the pillars evaluated in this figure were fabricated using the SMI9200 FIB system under rapid growth conditions. Typical growth rates were about 3–5 μm/min for the 100-nm-diameter and 240-μm-wide pillars, and 0.9 μm/min for the 480-nm-wide pillars. When calculating the line of Figure 4.9, we assumed that the density of the DLC pillars was about 2.3 g/cm³, which is almost identical to that of graphite. The inclination of the line in Figure 4.9 indicated Young's modulus for each pillar. Young's moduli of the pillars were distributed over a range from 65 to 140 GPa, which is almost identical to that of normal metals. Wider pillars tended to have larger Young's moduli.

We found that the stiffness increases significantly as the local gas pressure decreases, as shown in Figure 4.10. While the absolute value of the local gas pressure at the beam point is very difficult to determine, we found that the growth rate can be a useful parameter for describing the dependence of pressure on Young's modulus. All data points indicated in Figure 4.10 were obtained from pillars grown using point irradiation. Therefore, the pillar diameters did vary slightly from 100 nm but did deviate by more than 5%. A relatively low gas pressure, with a good uniformity, was obtained by using a single gas nozzle and gas reflector. We use a cleaved sidewall of Si tips as the gas reflector, which was placed 10–50 μm away from the beam point so as to be facing the gas nozzle. The growth rate was controlled by changing the distance to the wall. While there is a large distribution of data points, the stiffness

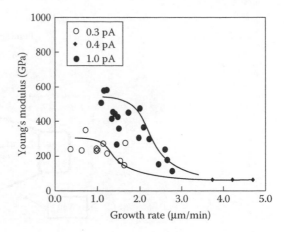

FIGURE 4.10 Dependence of Young's modulus dependence on the growth rate.

of the pillar tended to increase as the growth rate decreased. The two curves in Figure 4.10 represented data points obtained for a beam current of 0.3 pA (open circles) and 1 pA (solid circles), respectively. Both curves show the same tendency; the saturated upper levels of Young's modulus are different for each ion current at low gas pressure (low growth rate). It should be noted that some of the pillar's Young's moduli exceeded 600 GPa, which is of the same order as that of tungsten carbide. In addition, these estimations assume a pillar density of 2.3 g/cm³, but a finite amount of Ga was incorporated with the pillar growth. If the calculation takes the increase in pillar density due to the Ga concentration into account, Young's modulus exceeds 800 GPa. Such a high Young's modulus reaches that of carbon nanotubes and natural diamond crystals. We think that this high Young's modulus is due to surface modification caused by direct ion impact.

In contrast, when the gas pressure was high enough to achieve a growth rate of more than 3 μm/min, the pillars became soft but the change in Young's modulus was small. The uniformity of Young's modulus (as seen in Figure 4.9) presumably results from the fact that the growth occurred in this insensitive region where low levels of source gas limit pillar growth.

4.2.2.2 Free Space Nanowiring

All experiments were carried out in a commercially available FIB system (SMI9200, SII Nanotechnology Inc.) using a beam of 30 kV Ga⁺ ions. The beam was focused to a spot size of 7 nm at a beam current of 0.4 pA, and it was incident perpendicular to the surface. The pattern drawing system [computer pattern generator (CPG)-1000, Crestec Co., Tokyo, Japan] was added

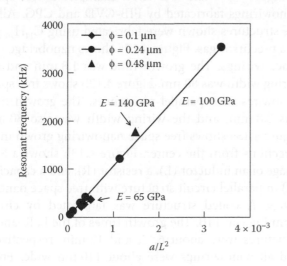

FIGURE 4.9 Dependence of resonant frequency on the pillar length.

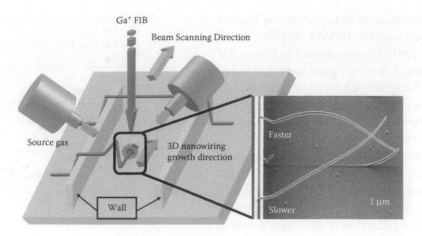

FIGURE 4.11   Fabrication of DLC free space wiring using both FIB-CVD and CPG.

to the FIB apparatus to draw any patterns. Using the CPG, it is possible to control beam-scan parameters such as scanning speed, x–y direction, and blanking of the beam, and so 3D free space nanowiring can be performed (Morita et al. 2003).

Figure 4.11 illustrates the free space nanowiring fabrication process using both FIB-CVD and CPG. When phenanthrene gas or tungsten hexacarbonyl [$W(CO)_6$] gas, which is a reactive organic gas, is evaporated from a heated container and injected into the vacuum chamber by a nozzle located 300 μm above the sample surface at an angle of about 45° with respect to surface, the gas density of the $C_{14}H_{10}$ or $W(CO)_6$ molecules increases on a substrate near a gas nozzle. The nozzle system creates a local high-pressure region over the surface. The base pressure of the sample chamber is 2×10^{-5} Pa and the chamber pressure upon introducing $C_{14}H_{10}$ and $W(CO)_6$ as a source gas was 1×10^{-4} and 1.5×10^{-3} Pa, respectively. If a Ga^+ ion beam is irradiated onto the substrate, $C_{14}H_{10}$ or $W(CO)_6$ molecules adsorbed on the substrate surface are decomposed, and carbon (C) is mainly deposited onto the surface of the substrate. The direction of deposition growth can be controlled through the scanning direction of the beam. The material deposited using $C_{14}H_{10}$ gas was DLC, as confirmed by Raman spectra, and it had a very large Young's modulus of 600 GPa (Matsui et al. 2000; Fujita et al. 2001).

After the two walls were formed in Figure 4.11, free space nanowiring was performed by adjusting the beam scanning speed. The ion beam was used at 30 kV Ga^+ FIB, and the irradiation current was 0.8–2.3 pA. The x and y scanning directions and the beam scanning speed were controlled by the CPG. The height in the z

direction was proportional to the irradiation time. Deposition is made to occur horizontally by scanning the beam at a certain fixed speed in the direction of a plane. However, if the beam scanning speed is faster than the nanowiring growth speed, it grows downward or drops, and conversely if the scanning speed is too slow, the deposition grows slanting upward. Therefore, it is very important to carefully control the beam scanning speed when growing a nanowire horizontally. It turns out that the optimal beam scanning speed to realize a nanowire growing horizontally, using two $C_{14}H_{10}$ gas guns, is about 190 nm/s. The expected pattern resolution archived using FIB-CVD is around 80 nm because both the primary Ga^+ ion and secondary-electron scatterings occur over distances around 20 nm (Fujita et al. 2001, 2002).

Figures 4.12 and 4.13 show examples of free space nanowirings fabricated by FIB-CVD and CPG. All of the structures shown were fabricated using $C_{14}H_{10}$ gas as a precursor gas. Figure 4.12a shows nanobridge free space wirings. The growth time was 1.8 min and the wiring width was 80 nm. Figure 4.12b shows free space nanowires with parallel resistances. The growth time was 2.8 min, and the wiring width was also 80 nm. Figure 4.13a shows free space nanowiring grown in 16 directions from the center. Figure 4.13b shows a SIM image of an inductor (L), a resistor (R), and a capacitor (C) in parallel circuit structure with free space nanowirings. A coiled structure was fabricated by circle-scanning Ga^+ FIB. The growth times of the L, R, and C structures were about 6, 2, and 12 min, respectively, and all nanowirings were about 110 nm wide. From these structures, it is possible to fabricate nanowiring at an arbitrary position using FIB-CVD and CPG. These

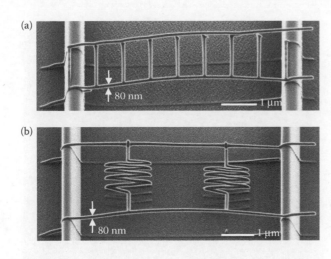

FIGURE 4.12 (a) DLC free space wiring with a bridge shape. (b) DLC free space wiring with parallel resistances.

results also indicate that various circuit structures can be formed by combining L, C, and R.

The free space wiring structures were observed using 200 keV TEM. The analyzed area was 20 nm diameter. Figure 4.14a and b shows TEM images of DLC free space wiring and a pillar, respectively. It became clear from these energy dispersive x-ray spectroscopy (EDX) measurements that the dark part (A) of Figure 4.14a corresponds to the Ga core, and the outside part (B) of Figure 4.14a corresponds to amorphous carbon. This free space wiring therefore consists of amorphous carbon with a Ga core. The center position of the Ga core is actually located below the center of the wiring. However, in the case of the DLC pillar, the Ga core is located at the center of the pillar. To investigate the difference between these core positions, the Ga core distribution in free space wiring was observed in detail by TEM. The center position of the

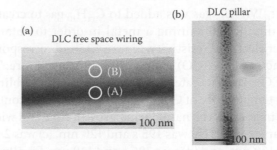

FIGURE 4.14 TEM images of (a) DLC free space wiring and (b) DLC pillar.

Ga core was about 70 nm from the top, which was 20 nm below the center of the free space wiring. We calculated an ion range of 30 kV Ga ions into amorphous carbon, using TRIM (transport of ions in matter), which was 20 nm. The calculation indicates that the displacement of the center of the Ga core in the nanowiring corresponds to the ion range.

The electrical properties of free space nanowiring fabricated by FIB-CVD using a mixture of $C_{14}H_{10}$ and $W(CO)_6$ were measured. Nanowirings were fabricated on an Au electrode. These Au electrodes were formed on a 0.2-μm-thick SiO_2 on Si substrate by an EB lithography and lift-off process. Two-terminal electrode method was used to measure the electrical resistivity of the nanowiring. Figure 4.15a shows the nanowiring fabricated using only $C_{14}H_{10}$ source gas. The growth time here was 65 s and the wiring width was 100 nm.

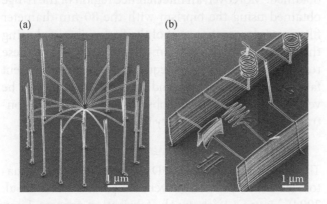

FIGURE 4.13 (a) Radial DLC free space wiring grown into 16 directions from the center. (b) Scanning ion microscope (SIM) micrograph of inductor (L), resistor (R), and capacitor (C) structure.

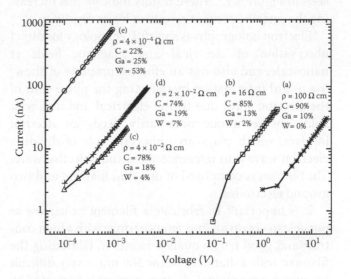

FIGURE 4.15 Electrical resistivity measurement for nanowiring. Electrical resistivity ρ was calculated by *I–V* curve. Elemental contents C, Ga, and W were measured by SEM-EDX. (a) Only $C_{14}H_{10}$ gas, (b)–(d) a mixture gas of $C_{14}H_{10}$ and $W(CO)_6$, and (e) only $W(CO)_6$ gas.

Chapter 4

Next, $W(CO)_6$ gas was added to $C_{14}H_{10}$ gas to create a gas mixture containing a metal in order to obtain a lower electrical resistivity. Figure 4.15b–d corresponds to increasing $W(CO)_6$ contents in the gas mixture. The $W(CO)_6$ content rate was controlled by the sublimation temperature of $C_{14}H_{10}$ gas. As the $W(CO)_6$ content was increased, the nanowiring growth time and width became longer: (b) was 195 s and 120 nm, (c) was 237 s and 130 nm, and (d) was 296 s and 140 nm. Finally, we tried to fabricate free space nanowiring using only $W(CO)_6$, but did not obtain continuous wiring because the deposition rate for a source gas of just $W(CO)_6$ was very slow.

The electrical resistivity of Figure 4.15a fabricated using only $C_{14}H_{10}$ source gas was 1×10^2 Ω cm. The elemental contents were 90% C and 10% Ga, which were measured using a SEM-EDX spot beam. The *I–V* curves in Figure 4.15b–d correspond to increasing $W(CO)_6$ content in the gas mixture. On increasing the $W(CO)_6$ content, the electrical resistivity decreases as shown in *I–V* curves (Figure 4.15b–d). Moreover, the Ga content also increased because the growth of nanowiring slowed down; the irradiation time of Ga+ FIB became longer. The electrical resistivities of the *I–V* curves in Figure 4.15b–d were 16×10^{-2}, 4×10^{-2}, and 2×10^{-2} Ω cm, respectively. The electrical resistivity of Figure 4.15e, which was fabricated by using only $W(CO)_6$ source gas, was 4×10^{-4} Ω cm. Increasing the Ga and W metallic content decreases the electrical resistivity, as shown by the SEM-EDX measurements seen in Figure 4.15. These results indicate that increasing the metallic content results in lower resistivity.

Electron holography is a useful technology for direct observation of electrical and magnetic fields at nanoscale, and also has an efficient property of showing useful information by detecting the phase shift of the electron wave due to the electrical and magnetic field. The technique necessarily needs an electron biprism, which plays an important role of dividing electron wave into reference wave and objective wave. The biprism is composed of one thin filament and two ground electrodes.

It is important to fabricate a filament as narrow as possible to obtain an interference fringe with a high contrast and good fringe quality. However, fabricating the filament with a diameter below 500 nm is very difficult because a conventional electron biprism is fabricated by pulling a melted glass rod by hand. To overcome this problem, we introduce a new fabrication technique of the electron biprism using FIB-CVD and evaluate the characteristics of the new biprism (Nakamatsu et al. 2008).

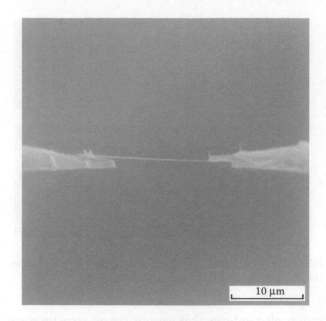

FIGURE 4.16 Electron biprism fabricated by FIB-CVD.

Figure 4.16 shows a SEM micrograph of the FIB-CVD biprism. We successfully fabricated DLC wiring with smooth surface in between W rods by free space wiring fabrication technology of FIB-CVD. The 80 nm DLC thin wiring works as the filament of the biprism. The diameter and length of the filament are 80 nm and 15 μm, respectively.

Figure 4.17 shows the interference fringes obtained using the biprism with a filament of (a) 80 nm diameter and (b) 400 nm diameter, and corresponding fringe profiles. The applied prism voltage was 20 V. The filament with 400 nm diameter, close to the standard size used in the conventional electron biprism, was fabricated by Pt-sputter coating onto the 80-nm-diameter filament. The interference fringes were successfully obtained. Moreover, an interference region of the fringe obtained using the biprism with the 80-nm-diameter filament is larger than that of the fringe obtained using the biprism with the 400-nm-diameter filament. These results demonstrate an adequacy of the thin filament fabricated by FIB-CVD, and the new biprism will be very useful for an accurate observation with a high contrast and good fringe quality in electron holography.

4.2.2.3 Nano-Electrostatic Actuator

The fabrication process of 3D nano-electrostatic actuators (and manipulators) is very simple (Kometani et al. 2004). Figure 4.18 shows the fabrication process. First, a glass capillary (GD-1, Narishige Co., East Meadow, NY, USA) was pulled using a micropipette puller (PC-10, Narishige Co.). The dimensions of the glass

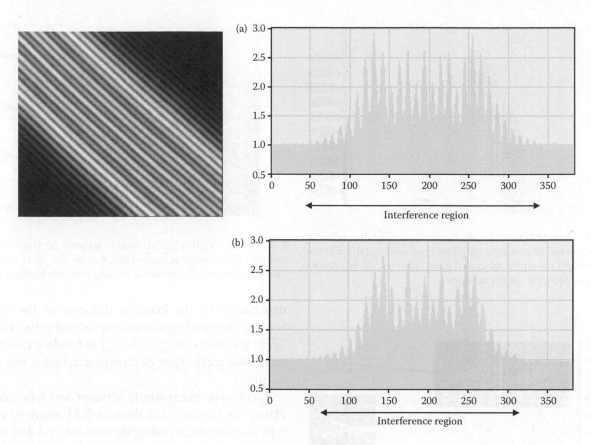

FIGURE 4.17 Interference fringes and corresponding fringe profiles obtained using the biprism with a diameter of (a) 80 nm and (b) 400 nm.

capillary were 90 mm in length and 1 mm in diameter. Using this process, we obtained a glass capillary tip with a 1 μm diameter. Next, we coated the glass capillary surface with Au by DC sputtering. The Au thickness was ~30 nm. This Au coating serves as the electrode that controls the actuator and manipulator.

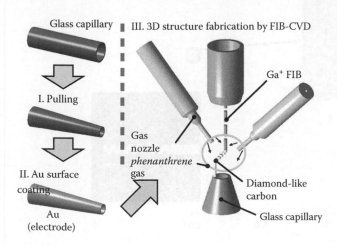

FIGURE 4.18 Fabrication process of 3D nano-electrostatic actuators.

Then, the 3D nano-electrostatic actuators and manipulators were fabricated by FIB-CVD. This process was carried out in a commercially available FIB system (SIM9200, SII Nanotechnology Inc.) with a Ga$^+$ ion beam operating at 30 keV. FIB-CVD was carried out using a precursor of phenanthrene as the source material. The beam diameter was about 7 nm. The inner diameter of each nozzle was 0.3 mm. The phenanthrene gas pressure during growth was typically 5×10^{-5} Pa in the specimen chamber. The Ga$^+$ ion beam was controlled by transmitting CAD data on the arbitrary structures to the FIB system. A laminated pleats-type electrostatic actuator was fabricated by FIB-CVD. Figure 4.19a shows a SIM image of a laminated pleats-type electrostatic actuator fabricated at 7 pA and 60 min exposure time. Figure 4.19b shows the principle behind the movement of this actuator. The driving force is the repulsive force due to the accumulation of electric charge. This electric charge can be stored in the pleats structures of the actuator by applying a voltage across the glass capillary. The pillar structure of this actuator bends due to charge repulsion, as shown in Figure 4.19b. Figure 4.20 shows the

Chapter 4

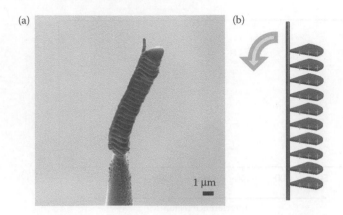

FIGURE 4.19 Laminated pleats-type electrostatic actuator. (a) SIM image of a laminated pleats-type electrostatic actuator fabricated on the tip of Au-coated glass capillary. (b) Illustration of moving principle for the actuator.

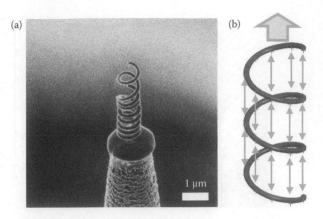

FIGURE 4.21 Coil-type electrostatic actuator. (a) SIM image of a coil-type electrostatic actuator fabricated on the tip of Au-coated glass capillary. (b) Illustration of moving principle for the actuator.

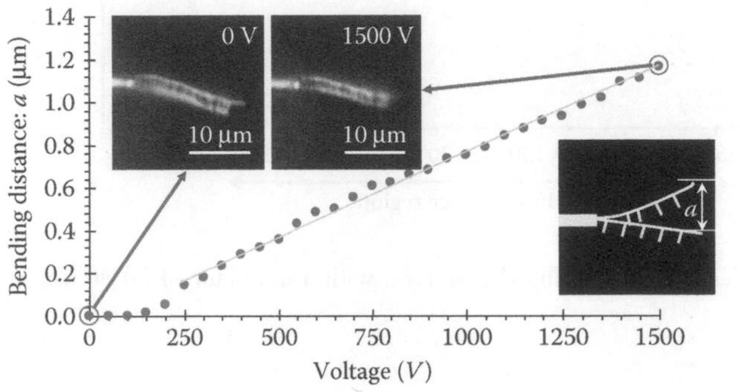

FIGURE 4.20 Dependence of bending distance on the applied voltage.

dependence of the bending distance on the applied voltage. The bending distance is defined as the distance "a" in the inset of Figure 4.20. The bending rate of this laminated pleats-type electrostatic actuator was about 0.7 nm/V.

A coil-type electrostatic actuator was fabricated by FIB-CVD. Figure 4.21a shows a SIM image of a coil-type electrostatic actuator fabricated at 7 pA and 10 min of exposure time. Figure 4.21b shows the principle behind the movement of this actuator, which is very simple. The driving force is the repulsive force induced by electric charge accumulation; the electric charge can be stored in this coil structure by applying a voltage across the glass capillary. This coil structure expands and contracts due to charge repulsion, as shown in Figure 4.21b. Figure 4.22 shows the dependence of the

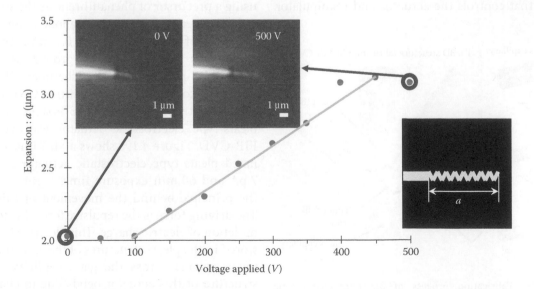

FIGURE 4.22 Dependence of coil expansion on the applied voltage.

coil expansion on the applied voltage. The length of the expansion is the distance "*a*" in the inset of Figure 4.22. The result revealed that the expansion could be controlled in the applied voltage range from 0 to 500 V.

4.2.3 Nano-Optics: Brilliant Blue Observation from a *Morpho*-Butterfly-Scale Quasi-Structure

The *Morpho*-butterfly has brilliant blue wings, and the source of this intense color has been an interesting topic of debate for a long time. Due to an intriguing optical phenomenon, the scales reflect interfered brilliant blue color for any angle of incidence of white light. This color is called a structural color, meaning that it is not caused by pigment reflection (Vukusic and Sambles 2003). When we observed the scales with a SEM (Figure 4.23a), we found 3D nanostructures 2 μm in height, 0.7 μm in width, and with a 0.22 μm grating pitch on the scales. These nanostructures cause a similar optical phenomenon to the iridescence produced by a jewel beetle.

We duplicated the *Morpho*-butterfly-scale quasi-structure with a commercially available FIB system (SMI9200, SII Nanotechnology Inc.) using a Ga⁺ ion beam operating at 30 kV (Watanabe et al. 2005). The

beam diameter was about 7 nm at 0.4 pA. The FIB-CVD was performed using a precursor of phenanthrene.

In this experiment, we used a computer-controlled pattern generator, which converted 3D CAD data into a scanning signal, which was passed to an FIB scanning apparatus in order to fabricate a 3D mold (Hoshino et al. 2003). The scattering range of the Ga primary ions is about 20 nm and the range of the secondary electrons induced by the Ga ion beam is about 20 nm, so the expected pattern resolution of the FIB-CVD is about 80 nm.

Figure 4.23b is a SIM image of the *Morpho*-butterfly quasi-structure fabricated by FIB-CVD using 3D CAD data. This result demonstrates that FIB-CVD can be used to fabricate the quasi-structure.

We measured the reflection intensities from *Morpho*-butterfly scales and the *Morpho*-butterfly-scale quasi-structure optically; white light from a halogen lamp was directed onto a sample with angles of incidence ranging from 5° to 45°. The reflection was concentrated by an optical microscope and analyzed using a commercially available photonic multichannel spectral analyzer system (PMA-11, Hamamatsu Photonics K.K., Hamamatsu City, Japan). The intensity of incident light from the halogen lamp had a peak at a wavelength close to 630 nm.

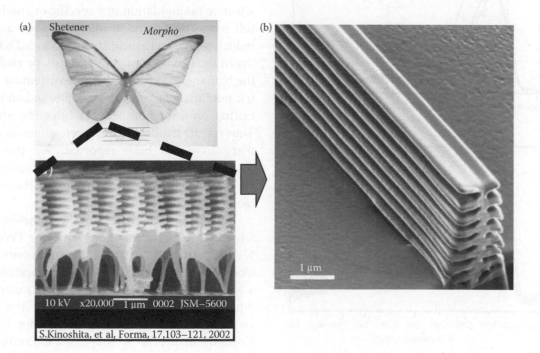

FIGURE 4.23 *Morpho*-butterfly scales. (a) Optical microscope image showing top view of *Morpho*-butterfly. SEM image showing a cross-sectional view of *Morpho*-butterfly scales. (b) SIM images showing an inclined view of *Morpho*-butterfly-scale quasi-structure fabricated by FIB-CVD.

Chapter 4

The *Morpho*-butterfly-scale quasi-structure was made of DLC. The reflectivity and transmittance of a 200-nm-thick DLC film deposited by FIB-CVD, measured by the optical measurement system at a wavelength close to 440 nm (the reflection peak wavelength of the *Morpho*-butterfly), were 30% and 60%, respectively. Therefore, the measured data indicated that the DLC film had high reflectivity near 440 nm, which is important for the fabrication of an accurate *Morpho*-butterfly-scale quasi-structure.

We measured the reflection intensities of the *Morpho*-butterfly scales and the quasi-structure with an optical measurement system, and compared their characteristics. Figure 4.24a and b respectively shows the reflection intensities from *Morpho*-butterfly scales and the quasi-structure. Both gave a peak intensity near 440 nm and showed very similar reflection intensity spectra for various angles of incidence.

We have thus successfully demonstrated that a *Morpho*-butterfly-scale quasi-structure fabricated using FIB-CVD can give almost the same optical characteristics as real *Morpho*-butterfly scales.

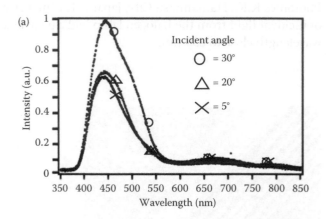

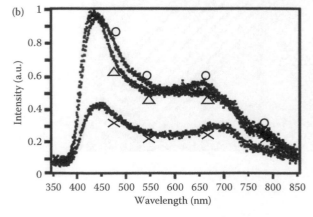

FIGURE 4.24 Intensity curves of the reflection spectra for (a) *Morpho*-butterfly scales and (b) *Morpho*-butterfly-scale quasi-structure.

4.2.4 Nanobiology

4.2.4.1 Nano-Injector

Three-dimensional nanostructures on a glass capillary have a number of useful applications, such as manipulators and sensors in various microstructures. We have demonstrated the fabrication of a nozzle nanostructure on a glass capillary for a bioinjector using 30 keV Ga$^+$ FIB-assisted deposition with a precursor of phenanthrene vapor and etching (Kometani et al. 2003). It has been demonstrated that nozzle nanostructures of various shapes and sizes can be successfully fabricated. An inner tip diameter of 30 nm on a glass capillary and a tip shape with an inclined angle have been realized. We reported that DLC pillars grown by FIB-CVD with a precursor of phenanthrene vapor have very large Young's moduli that exceed 600 GPa, which potentially makes them useful for various applications (Fujita et al. 2001). These characteristics are applicable to the fabrication of various biological devices.

In one experiment, nozzle nanostructure fabrication for biological nano-injector research was studied. The tip diameters of conventional bioinjectors are >100 nm and the tip shapes cannot be controlled. A bionano-injector with various nanostructures on the top of a glass capillary has the following potential applications (shown in Figure 4.25): (1) injection of various reagents into a specific organelle in a cell, (2) selective manipulation of a specific organelle outside a cell by using the nano-injector as an aspirator, (3) reducing the mechanical stress produced when operating in the cell by controlling the shape and the size of the bionano-injector, and (4) measurement of the electric potential of a cell, an organelle, and an ion channel exiting on a membrane, by fabricating an electrode. Thus far, 3D nanostructure fabrications on a glass capillary have not been reported. We present nozzle nanostructure fabrication on a glass capillary by FIB-CVD and etching in order to confirm the possibility of bionano-injector fabrication.

The nozzle structures of the nano-injector were fabricated using a function generator (Wave factory, NF Electronic Instruments, Yokohama, Japan). Conventional microinjectors are fabricated by pulling a glass capillary (GD-1, Narishige Co.) using a micropipette puller (PC-10, Narishige Co.). The glass capillary was 90 mm in length and 1 mm in diameter. Conventionally, the tip shape of a microinjector made by pulling a glass capillary, and which is used as an injector into a cell, is controlled by applying mechanical grinding (or not). However, the reliability of this

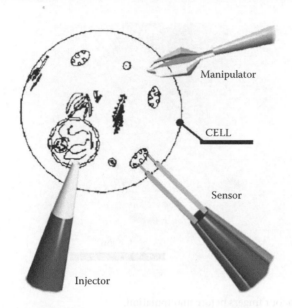

FIGURE 4.25 Potential uses for a bionano-injector.

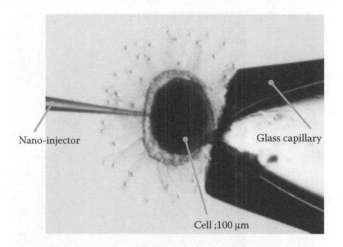

FIGURE 4.27 Injection into an egg cell (*Ciona intestinalis*) using a bionano-injector.

technique for controlling tip shape is very poor and requires experienced workers.

A bionano-injector tip was fabricated on a glass capillary by FIB-CVD, as shown in Figure 4.26a–c. First, FIB etching made the tip surface of the glass capillary smooth. Then, a nozzle structure was fabricated at the tip by FIB-CVD. Figure 4.26a shows the surface of a chip smoothed at 120 pA and after 30 s exposure time by FIB etching, with an inner hole diameter of 870 nm. A nozzle structure fabricated by FIB-CVD with an inner hole diameter of 220 nm is shown in Figure 4.26b. Figure 4.26c is a cross section of Figure 4.26b. These results demonstrate that a bionano-injector could be successfully fabricated by 3D nanostructure fabrication using FIB-CVD. The bionano-injector was used to inject dye into an egg cell (*Ciona intestinalis*) as shown in Figure 4.27.

4.2.4.2 Nanomanipulator

An electrostatic 3D nanomanipulator that can manipulate nanoparts and operate on cells has been developed by FIB-CVD. This 3D nanomanipulator has four fingers so that it can manipulate a variety of shapes. To move the nanomanipulator, electric charge is accumulated in the structure by applying voltage to the four-fingered structure, and electric charge repulsion causes them to move. Furthermore, we succeeded in catching a microsphere (made from polystyrene latex) with a diameter of 1 μm using this 3D nanomanipulator with four fingers (Kometani et al. 2005a).

The glass capillary (GD-1, Narishige Co.) was pulled using a micropipette puller (PC-10, Narishige Co.). A tip diameter of about 1.0 μm could be obtained using this process. Then, the glass capillary surface was coated with Au to fabricate an electrode for nanomanipulator control. The thickness of the Au coating was about 30 nm. Finally, a 3D nanomanipulator structure with

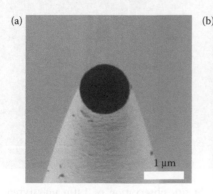
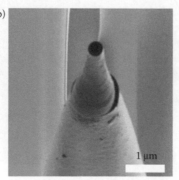
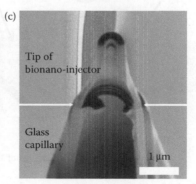

FIGURE 4.26 SIM images of bionano-injector fabricated on a glass capillary by FIB-CVD (a) before FIB-CVD, (b) after FIB-CVD, and (c) a cross section of (b).

Chapter 4

(a)

(b)

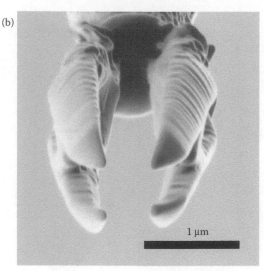

1 µm

FIGURE 4.28 SIM image of the 3D electrostatic nanomanipulator with four fingers before manipulation.

four fingers (Figure 4.28) was fabricated by FIB-CVD on the tip of the glass capillary with an electrode.

Microsphere (a polystyrene latex ball with a diameter of 1 µm) manipulation was carried out using the 3D nanomanipulator with four fingers. An illustration of this manipulation experiment is shown in Figure 4.29. By connecting the manipulator fabricated by FIB-CVD to a commercial manipulator (MHW-3, Narishige Co.), the direction and movement along the x-axis, y-axis, and z-axis could be controlled. The microsphere target was fixed to the side of a glass capillary, and the

manipulation was observed from the top with an optical microscope.

The optical microscope image of Figure 4.30 shows the situation during manipulation. First, the 3D nanomanipulator was made to approach the microsphere; no voltage was applied. Next, the four fingers were opened by applying 600 V in front of the microsphere and the microsphere could be caught by turning off the voltage when the microsphere was in the grasp of the nanomanipulator. The 3D nanomanipulator was then removed from the side of the glass capillary. Note that the action of catching the microsphere occurs due to the elastic force of the manipulator's structure. We

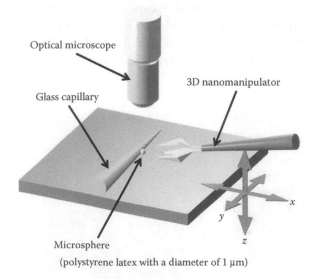

FIGURE 4.29 Illustration of 1 µm polystyrene microsphere manipulation by using 3D electrostatic nanomanipulator with four fingers.

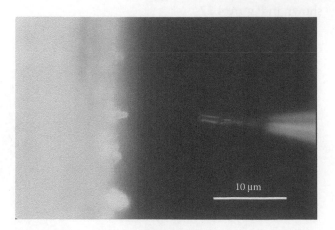

10 µm

FIGURE 4.30 *In situ* observation of 1 µm polystyrene microsphere manipulation by using a 3D electrostatic nanomanipulator with four fingers.

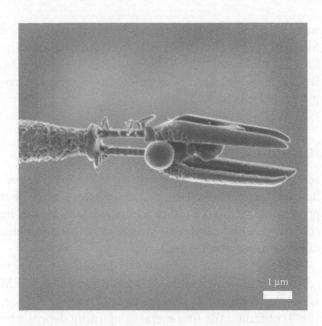

FIGURE 4.31 SIM image of the 3D electrostatic nanomanipulator with four fingers after manipulation.

succeeded in catching the microsphere, as shown in the SIM image in Figure 4.31.

4.2.4.3 Filtering Tool with Nano-Net Structure

It is necessary to operate and analyze single cell and organelle with high accuracy to understand the biological phenomenon more. For this reason, bio-nanotools are very useful and important tools to make unknown biological phenomenon clear. Therefore, we have been studying about bionano-tools with 3D structures by using FIB-CVD. Three-dimensional nanostructures on a glass capillary have a number of useful applications such as manipulators and sensors in the various microstructures. Thus far, bionano-injector or nano-net mounted on the top of a glass capillary as 3D nanotools were reported. The bionano-injector is a very useful tool because the shape and size of its tip can be freely designed for various functions (Kometani et al. 2003). And we succeeded in capturing polystyrene microspheres with a diameter of 2 μm by using the nano-net (Kometani et al. 2005b).

Nano-net structure has wide applications such as manipulation and filtering of subcellular organelles. The filtering tool was developed as one of the applications of a nano-net. The purposes of the filtering tool are a speedy filtering and capture of an arbitrary-size organelle. If the organelle can be quickly operated, an energetic organelle that is necessary to obtain accurate biological information can be obtained.

The filtering tool was fabricated on the tip of the glass capillary by FIB-CVD (Kometani et al. 2006). First, the glass capillary of 90 mm length and 1 mm diameter (GD-1, Narishige Co.) was pulled by a micro-pipette puller (PC-10, Narishige Co.). Second, an outer diameter of the glass capillary tip was adjusted to 13 μm in the diameter by using FIB etching, and an inner diameter of 7 μm was obtained. Third, the glass capillary surface was coated with Au for avoiding the charging by Ga ion beam. In the final fabrication process, the filtering tool structure was fabricated by FIB-CVD.

Figure 4.32 shows the fabrication result of the filtering tool with a nano-net structure. This filtering tool has a penetration needle in the cell, the organelle retention table, and a nano-net structure to filter organelles. The wire size of a nano-net was 200 nm. These parts of a filtering tool were made of DLC.

Capillary phenomenon was used as the aspiration force of the filtering tool. For this reason, organelles such as the chloroplast in the cell are aspirated to the direction of the nano-net. And then, the arbitrary-size organelle is selected by the nano-net because large organelles are caught in the nano-net but small organelles are inhaled into the glass capillary by passing through the nano-net.

By using a filtering tool, we carried out the experiment of chloroplast capturing under the optical microscope. In this experiment, we chose chloroplasts that were in the cells of *Egeria densa*'s leaf as a target. Optical microscope images as shown in Figure 4.33a–d were obtained. First, a filtering tool was approached to the cell wall (Figure 4.33a). And we shoved the cell wall by using the needle of the filtering

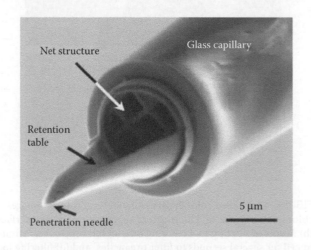

FIGURE 4.32 SEM image of filtering tool fabricated by FIB-CVD.

Chapter 4

tool (Figure 4.33b). We then maintained the filtering tool for a few seconds in the cell to filter organelles (Figure 4.33c). Finally, a filtering tool containing the chloroplast was put out from the cell (Figure 4.33d). The total operation time for capturing the chloroplast was about 7 s. After this experiment, we observed the filtering tool by scanning electron microscopy. The filtering tool had a chloroplast as shown in Figure 4.34. In this way, we succeeded in capturing the chloroplast with this tool.

4.3   Electron Beam Chemical Vapor Deposition

Nanometer structures can be fabricated by electron-beam-induced surface reaction because beam diameters, as small as 1 nm, can be formed with conventional electron optical equipment. Computer-controlled direct writing is also possible for EB-CVD. Metals, semiconductors, and inorganic metals can be deposited by this means. Two major advantages are that processing methods in microelectronic devices can be simplified and new device structure with nanometer dimensions can be realized.

Broers et al. (1976) were able to deposit nanometer-scale carbon contamination structures in a STEM with electrons of 45 keV energy. Matsui and Mori (1984) obtained the first metal-containing deposits by introducing a metal–organic gas ($Cr(C_6H_6)_2$) vapor in the vacuum chamber of a SEM. Since the first experiments, many EB-CVD metal-containing structures have been deposited under a wide range of experimental conditions in the past 30 years (Matsui and Mori 1986; Koops et al. 1988, 2001a; Matsui et al. 1989; Ochiai et al. 1996; Utke et al. 2000; Floreani et al. 2001; Mitsuishi et al. 2003; Shimojo et al. 2004; Brintlinger et al. 2005; Tanaka et al. 2005; Barry et al. 2006; Perentes et al. 2007; Botman et al. 2008a,b). The first metal-containing 3D nanostructures were deposited by Koops et al. (1988), using the focused electron beam of a conventional SEM. Koops et al. demonstrated some applications such as a field emitter that were realized using EB-CVD (Floreani et al. 2001).

FIGURE 4.33   **(See color insert.)** Optical microscope images of chloroplast capture. (a) Filtering tool approaching cell wall surface, (b) needle of filtering tool entering cell, (c) filtering tool remaining in cell for several seconds to filter organelles, and (d) filtering tool with chloroplast removed from cell.

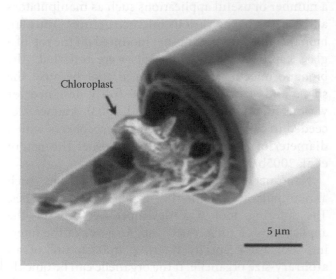

FIGURE 4.34   SEM image of filtering tool with chloroplast obtained after experiment.

4.3.1 EB–CVD by SEM

4.3.1.1 *In Situ* Observation by Auger Electron Spectroscopy

When an electron beam is irradiated onto a substrate in EB-CVD, source gas molecules adsorbed on a substrate are dissociated into nonvolatile and volatile materials. Nonvolatile materials are deposited on a substrate, while volatile materials are evacuated. When source gas molecules adsorbed on a substrate are dissociated by electron beam irradiation, Auger electrons are emitted from deposited materials at the same time. Therefore, the deposited material growth process can be observed *in situ* by Auger electron spectroscopy (AES) (Matsui and Mori 1987).

Figure 4.35 shows an *in situ* observation system for electron-beam-induced surface reaction by AES. This system has been constructed by modifying the conventional AES apparatus. Source chamber and mass flow controller (MFC), which controls the source gas flow rate, were added to the conventional AES apparatus. The system was evacuated by an ion pump with a 2×10^{-9} Torr base pressure after bake-out. This system also contains an electron beam gun, a cylindrical mirror analyzer (CMA), and an Ar ion gun. The source gas is introduced through the MFC, and the source gas molecules are adsorbed onto the substrate. The source gas molecules adsorbed onto the substrate are dissociated by a primary electron beam from the AES system and nonvolatile materials are deposited onto the substrate. At the same time that the primary electron beam is irradiated onto the substrate, Auger electrons are emitted from the deposited material surface. Therefore, the growth process of deposition of materials can be observed *in situ* by detecting Auger electrons using CMA. In this experiment,

in situ AES observation has been carried out for W deposition using WF_6 source gas on Si(100) substrate. The electron beam voltage and current were 10 keV and 1 µA, respectively. This experiment has been carried out by point-beam exposure. The AES apparatus minimum beam diameter is ~1 µm. However, ~30 µm beam diameter, which was obtained by defocusing, has been applied to avoid effects of beam profile and to obtain a uniform deposited film. Therefore, the beam current density was ~0.14 A/cm². The electron beam was irradiated at room temperature. It is estimated by calculation that temperature increase of the sample during electron beam exposure was <10°C.

Figure 4.36 shows Auger intensity variations during electron beam irradiation at 3.5×10^{-7} Torr WF_6 source gas pressure. W Auger intensity increases exponentially as the electron beam exposure time increases. W Auger intensity saturates at a 10 min exposure time because W deposited film thickness reaches the escape depth. On the other hand, Si Auger intensity decreases because W deposited film was formed on the Si substrate. The contents (at. %) calculated from the AES spectrum, as shown in Figure 4.36, were 85%W, 7.5%F, and 7.5%O at a 15 min exposure time. It is believed that O originates in residual oxygen in the source and sample chambers. If ultrahigh vacuum with 10^{-10} Torr order is achieved in both chambers, a higher-quality film will be deposited. A growth rate for EB-CVD has been calculated on the assumption that EB-CVD is a layer growth. The calculated growth rate for 3.5×10^{-7} Torr is 0.15 nm/min.

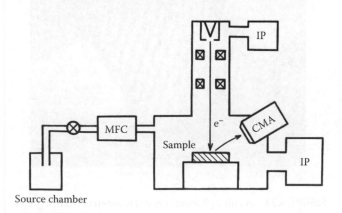

FIGURE 4.35 Experimental AES system arrangement to make *in situ* observations on EB-CVD process.

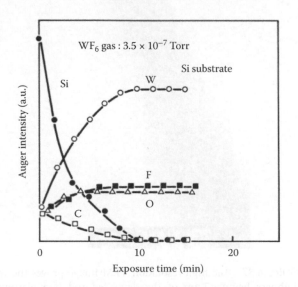

FIGURE 4.36 Relations between Auger intensity variations and electron exposure time for W deposition on Si substrate.

4.3.1.2 Fabrication of Photonic Crystal and Field Emission Electron Source

Koops et al. (2001a) reported the experiments using 3D EB-CVD. This technique is very well suited for rapid prototyping of photonic crystals. It is used to grow arrays of rods from nanocrystalline material having a very high aspect ratio over 15 under computer control. The nanolithography tool in use is based on a SEM Jeol JSM 6300 F with beam control system VIDAS (Kontron). All possible precautions have been taken to compensate for thermal drifts, mechanical vibrations, and magnetic interference by external compensating systems (Hübner et al. 2001). These measures allow the high-precision fabrication of 2D arrays of dielectric rods with the dimensions required for photonic crystals. For the precursor material supply to the waveguide sample, a specially designed environmental chamber is used (Koops et al. 2001b). The EB-CVD process renders 2 nm edge roughness of the deposited rods. This is a surface quality that is unreachable by standard lithography or beam-assisted etching techniques. It allows minimizing scattering of photons. Figure 4.37 proves with a high-resolution TEM image the wall roughness being <2 nm of the rod grown by EB-CVD using cyclopentadienyl-platinum-trimethyl as a precursor material.

EB-CVD constructs the rods from nanocrystals of 2 nm diameter embedded into a carbonaceous matrix. This results in an edge roughness of below 2 nm, which is an outstanding feature of this process, as is demonstrated in Figure 4.37.

Photonic crystals with cubic, hexagonal, or honeycomb elementary cells are fabricated under computer control. A VIDAS macrolanguage allows to quickly adjust the design parameters of the crystals that are to be grown. There are no other process steps required to generate the crystal by the time-consuming growth process. Using 25 keV electrons and 20 pA current, a crystal of 100 rods is grown in 40 min. For this time, the sample in the microscope is not allowed to drift by more than 20 nm.

Figure 4.38 shows a cubic photonic crystal, which is constructed into a prefabricated gap in a poly(methyl methacrylate) (PMMA) waveguide for investigation of the transmission properties of the photonic crystal. The pitch is 580 μm, the rod diameter is 150 nm, and the height is 2.5 μm.

The characterization of photonic crystals and related devices is a very challenging task. This is due to the small size of the devices, having diameters of 3–6 λ, which is 4.5–9 μm at λ = 1.5 μm. Monomode fibers have a core diameter of 9 μm. Therefore, a well-designed optical beam converter is required to couple light into photonic crystals. To measure the polarization-dependent transmission of photonic crystals, a sophisticated metrology setup was used. The transmission measurement did prove a band gap filter in the regime from 1250 to 1650 nm.

EB-CVD is a rapid prototyping technology with a very high resolution and accuracy to investigate the

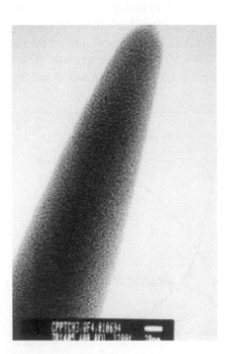

FIGURE 4.37 The high-resolution TEM image proves the wall roughness being <2 nm of the deposited rod if a platinum-containing precursor material is used. (After Koops, H. W. P. et al. 2001a. *Microelectron. Eng.* **57–58**: 995–1001.)

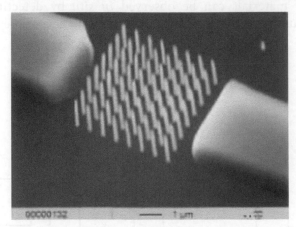

FIGURE 4.38 A cubic photonic crystal is constructed in a prefabricated gap in a PMMA waveguide for investigation of the transmission properties of the phonic crystal. (After Koops, H. W. P. et al. 2001a. *Microelectron. Eng.* **57–58**: 995–1001.)

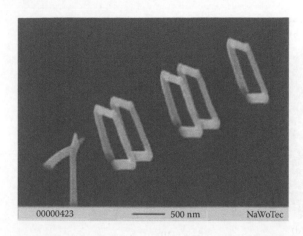

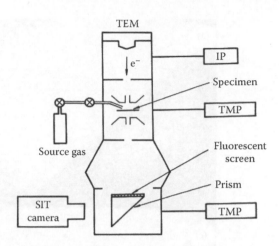

FIGURE 4.39 Smith–Purcell electron optics realized with EB-CVD. (After Floreani, F., Koops, H. W., and Elsäßer, W. 2001. *Microelectron. Eng.* **57–58**: 1009–1016.)

FIGURE 4.40 Experimental arrangement diagram of TEM *in situ* observation. Gas introducing system and VTR system are included.

activity and characteristics of photonic crystals and PC devices. The initial promising results prove the applicability of the technology and the optical action of the devices.

EB-CVD can build similar structures in the sub-μm dimensions with wires of 80–200 nm width. Figure 4.39 shows a 3D buildup with conducting material resembling a field emission electron source, having an emitter tip and an extractor ring around it (Floreani et al. 2001). The structure has a capacitance of 24 pF between the tip and the ring.

4.3.2 EB-CVD by STEM

4.3.2.1 *In Situ* Observation by TEM

The deposited material growth process can be observed *in situ* by TEM, if WF_6 source gas molecules are adsorbed on a TEM specimen (Matsui and Ichihashi 1988). The employed electron microscope was EM-002A (Akashi beam Technology Corp.), equipped with a real-time TV monitor system, specially improved for *in situ* observation as shown in Figure 4.40. The microscope resolution was 0.23 nm at 120 keV, which allowed the imaging of individual rows of atom columns in W crystals. To reduce problems in regard to specimen contamination, the instrument was operated under ultrahigh vacuum pumping system with turbo-molecular pumps and ion pumps. The gas injection tube had a 3 mm inner diameter and was about 5 mm from the sample surface. Fine and spherical Si particles, about 60 nm in diameter, were used as TEM specimens. The small particles were made by a gas evaporation method in argon under a reduced atmosphere, where

the Si vapor was condensed into small particles. They were <100 nm in diameter and were usually covered with 1–3-nm-thick SiO_2. The gas molecules were adsorbed on the fine Si particles, which were on the TEM specimen grid. The gas molecules were excited by the TEM electron beam and dissociated into W metal and F_2 gas. W metal was deposited on the Si surface and growth occurred. The growth process was observed *in situ* at a single-atom resolution level by an electron microscope equipped with a TV monitor system. All micrographs were reproduced from single frames of the video tape recorder (VTR) tape. Two types of experiments were performed. First, the electron beam was irradiated on a WF_6 adlayer formed on the fine Si particle surface, in order to clarify an initial growth process of EB-CVD. Second, the electron beam was irradiated on the Si fine particle surface while WF_6 was flowing on the surface, to study layer growth process of EB-CVD.

Figure 4.41 shows a typical series of electron micrographs of *in situ* TEM observation variations during electron beam irradiation for the WF_6 adlayer. These micrographs were selected from a VTR tape that ran over a 30 min span. Electron beam irradiation time for Figure 4.41a–d were 0, 3, 15, and 30 min, respectively. This result indicates that W atoms, dissociated by electron beam irradiation from the WF_6 adlayer, coalesced and grew under electron beam irradiation. According to the real-time observation on a TV screen, moving clusters often collided with each other, causing coalescence.

Figure 4.42 shows TEM *in situ* observation variations during electron beam irradiation under 5×10^{-7} Torr

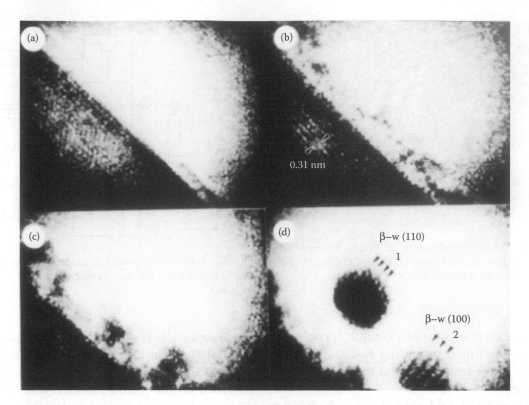

FIGURE 4.41   Electron beam exposure time dependence of W cluster growth on a Si particle. Exposure times: (a) 0, (b) 3, (c) 15, and (d) 30 min.

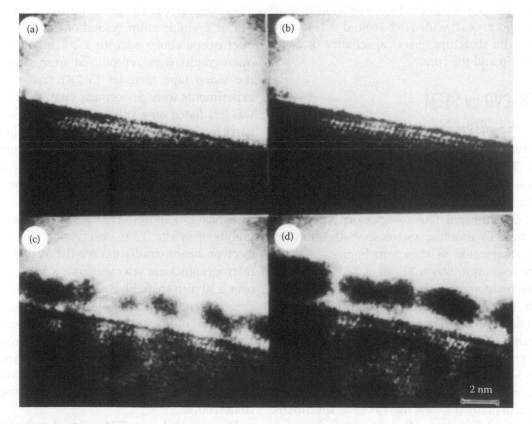

FIGURE 4.42   Electron beam exposure time dependence of W layer growth on fine Si particles at 5×10^{-7} Torr WF_6 gas pressure. Exposure times: (a) 0, (b) 15, (c) 25, and (d) 30 s.

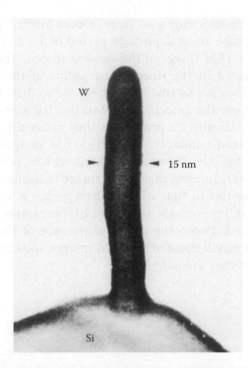

FIGURE 4.43 Electron micrograph for a W rod with 15 nm diameter. The W rod was fabricated using a 120 kV electron beam with 3 nm diameter.

WF$_6$ gas pressure. Electron beam irradiation times were 0, 15, 25, and 30 s for Figures 4.42a–d, respectively. These micrographs indicate that the W clusters, as the initial growth state, were formed, and then the W layer was formed by W clusters coalescing due to electron beam irradiation.

A nanometer-scale W rod was fabricated by a focused electron beam (3 nm diameter) moved manually under 1×10^{-6} Torr WF$_6$ gas atmosphere. Electron beam energy was 120 keV and its current density was ~100 A/cm^2. The moving speed of the beam was 1 nm/s. The electron beam was parallel to the substrate and its scanning direction was normal to the surface. An electron micrograph of the fabricated W rod is shown in Figure 4.43. The rod radius was ~15 nm. Bright contrast for the inner rod shows that the center of the rod was probably tungsten silicide. The beam conditions for forming the rod were not the same as those for taking electron micrographs. The electron beam was spread to about 1 μm diameter for taking this photograph.

4.3.2.2 Resolution of EB-CVD by STEM

Tanaka et al. (2005) reported the resolution of EB-CVD by STEM. The TEM instrument in this study is a JEM-2000VF with a field emission gun,

and the base pressure of the chamber is 1×10^{-7} Pa. The deposition apparatus is attached to a vacuum port of the TEM column. It consists of a variable leak valve, a gas nozzle ~1 mm in diameter, and heating system. The substrates were Si(111), conventional dimpled and ion milled. Most of the EB-CVD experiments were performed on the thinnest parts of the substrate, mostly amorphized during the ion milling. The precursor gas was W(CO)$_6$, and the EB-CVD was done at room temperature. The microscope is operated at 200 kV. The electron beam intensity was ~5×10^3–5×10^4 A cm^{-2} and the probe size was ~1–2 nm.

By using a 2 nm probe, arrays of dots are fabricated successfully from W(CO)$_6$. Figure 4.44 shows a TEM image of one of the arrays. The dots were fabricated by changing the irradiation time from 15 s (top left), 10 s (top right and bottom right), and 5 s (bottom left), respectively. The pressure during the deposition was kept at 2×10^{-6} Pa. The distance between the dots in the horizontal direction is ~12 nm and in the vertical direction is ~14 nm. Their structure is W–C mixed amorphous because they contain carbon from the precursor gas. It is recognized that the size of the dots becomes smaller as the irradiation time becomes shorter. By measuring the size, the 15 s dot is ~3 nm in diameter, the 10 s dots are ~2.5 nm, and the 5 s dot is ~2 nm, respectively.

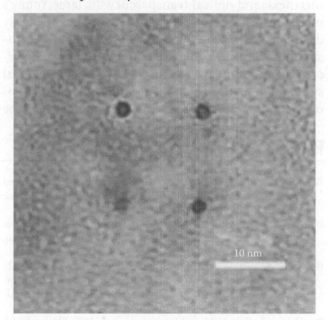

FIGURE 4.44 TEM image of W nanodots deposited for different times. The top left dot was fabricated for 15 s, the ones on the right were fabricated for 10 s, and the bottom left nanodot was fabricated for 5 s. (After Tanaka, M. et al., 2005. *Surf. Interface Anal.* **37**: 261–264.)

Chapter 4

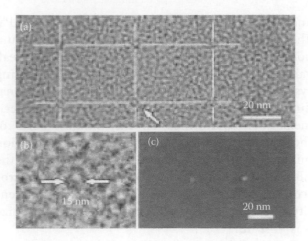

FIGURE 4.45 An array of the smallest W nanodots ever achieved by EB-CVD: (a) TEM image of the array; (b) magnified image of one of the dots shown by the arrow in (a); (c) the corresponding HAADF-STEM image. (After Tanaka, M. et al., 2005. *Surf. Interface Anal.* **37**: 261–264.)

Figure 4.45 shows an array of dots formed with a 1 nm probe and a deposition period of 5 s. In Figure 4.45a, a TEM image of the array is shown. The dots are located in the intersecting points of the white lines. They are so small that it is hard to distinguish them from the amorphous substrate. The size of the dot is ~1.5 nm. To prove that they are really tungsten-mixed nanodots, HAADF-STEM images were taken for the same substrate, which produces Z-contrast images, that is, the image is atomic number-sensitive in that a heavy atom makes a brighter contrast. Figure 4.45c shows one of these images. The distance between the dots and the size of the dots almost match those of the TEM images, thus indicating that they are nanodots.

4.4 Summary

The 3D nanostructure fabrication has been demonstrated by 30 keV Ga+ FIB-CVD and a phenanthrene source as a precursor. The film deposited on a silicon substrate was characterized using a transmission microscope and Raman spectra. This characterization indicated that the deposited film is DLC, which has attracted attention due to its hardness, chemical inertness, and optical transparency. Its large Young's modulus, which exceeds 600 GPa, makes it highly desirable for various applications. A nano-electrostatic actuator and 0.1 μm nanowiring were fabricated and evaluated as parts of nanomechanical system. Furthermore, nano-injector and nanomanipulator were fabricated as novel nanotool for the manipulation and analysis of subcellular organelles. These results demonstrate that FIB-CVD is one of the key technologies needed to make 3D nanodevices that can be used in the field of electronics, mechanics, optics, and biology.

Two- and three-dimensional nanostructure fabrications have been described by EB-CVD using SEM and STEM. As an application in two and three dimensions by EB-CVD using a 10 keV SEM, nano-optics with the rapid prototyping of photonic crystals and nano-electronics with microtriodes have been described. The smallest particle size with 1.5 nm in diameter was fabricated by EB-CVD with $W(CO)_6$ precursor using a 200 keV STEM.

References

Barry, J. D., Ervin, M., Molstad, J. et al. 2006. Electron beam induced deposition of low resistivity platinum from Pt(PF₃)₄. *J. Vac. Sci. Technol. B* **24**: 3165–3168.

Botman, A., de Winter, D. A. M., and Muders, J. J. L. 2008a. Electron-beam-induced deposition of platinum at low landing energies. *J. Vac. Sci. Technol. B* **26**: 2460–2463.

Botman, A., Hesselberth, M., and Mulders, J. J. L. 2008b. Investigation of morphological changes in platinum-containing nanostructures created by electron-beam-induced deposition. *J. Vac. Sci. Technol. B* **26**: 2464–2467.

Brintlinger, T., Fuhrer, M. S., Melngailis, J. et al. 2005. Electrodes for carbon nanotube devices by focused electron beam induced deposition of gold. *J. Vac. Sci. Technol. B* **23**: 3174–3177.

Broers, A. N., Molzen, W. W., Cuomo, J. J., and Wittels, N. D. 1976. Electron-beam fabrication of 80-Å metal structures. *Appl. Phys. Lett.* **29**: 596.

Buks, E. and Roukes, M. L. 2001. Stiction, adhesion energy, and the Casimir effect in micromechanical systems. *Phys. Rev. B* **63**: 033402.

Floreani, F., Koops, H. W., and Elsäßer, W. 2001. Operation of high power field emitters fabricated with electron beam deposition and concept of a miniaturized free electron laser. *Microelectron. Eng.* **57–58**: 1009–1016.

Fujita, J., Ishida, M., Ochiai, Y., Ichihashi, T., Kaito T., and Matsui, S. 2002. Focused ion beam-induced fabrication of tungsten structures. *J. Vac. Sci. Technol. B* **20**: 2686.

Fujita, J., Ishida, M., Sakamoto, T., Ochiai, Y., Kaito, T., and Matsui, S. 2001. Graphitized wavy traces of iron particles observed in amorphous carbon nano-pillars. *J. Vac. Sci. Technol. B* **19**: 2834.

Hoshino, T., Watanabe, K., Kometani, R. et al. 2003. Development of three-dimensional pattern-generating system for focused-ion-beam chemical-vapor deposition. *J. Vac. Sci. Technol. B* **21**: 2732.

Hübner, U., Plontke, R., Blume, M., Reinhardt, A., and Koops, H. W. P. 2001. On-line nanolithography using electron beam-induced deposition technique. *Microelectron. Eng.* **57–58**: 953–958.

Igaki, J., Saikubo, A., Kometani, R. et al. 2007. Elementary analysis of diamond-like carbon film formed by focused-ion-beam chemical vapor deposition. *Jpn. J. Appl. Phys.* **46**: 8003.

Ishida, M., Fujita J., and Ochiai, Y. 2002. Density estimation for amorphous carbon nanopillars grown by focused ion beam assisted chemical vapor deposition. *J. Vac. Sci. Technol. B* **20**: 2784.

Kanda, K., Igaki, J., Kato, Y., Kometani, R., Saikubo, A., and Matsui S. 2006. Surface modification of PTFE by synchrotron radiation under the O_2 gas atmosphere. *Radiat. Phys. Chem.* **75**: 1850.

Kometani, R., Funabiki, R., Hoshino, T. et al. 2006. Cell wall cutting tool and nano-net fabrication by FIB-CVD for subcellular operations and analysis. *Microcircuit Eng.* **83**: 1642.

Kometani, R., Hoshino, T., Kondo, K. et al. 2004. Characteristics of Nano-electrostatic actuator fabricated by focused ion beam chemical vapor deposition. *Jpn. J. Appl. Phys.* **43**: 7187.

Kometani, R., Hoshino, T., Kondo, K. et al. 2005a. Performance of nanomanipulator fabricated on glass capillary by focused-ion-beam chemical vapor deposition. *J. Vac. Sci. Technol. B* **23**: 298.

Kometani, R., Hoshino, T., Kanda K. et al. 2005b. Three-dimensional high-performance nano-tools fabricated using focused-ion-beam chemical-vapor-deposition. *Nucl. Instr. Meth. Phys. Res. B* **232**: 362.

Kometani, R., Morita, T., Watanabe, K. et al. 2003. Nozzle-nanostructure fabrication on glass capillary by focused-ion-beam chemical vapor deposition and etching. *Jpn. J. Appl. Phys.* **42**: 4107.

Koops, H. W., Kertz, J., Rudolph, M., Weber, M., Dahm, G., and Lec, K. L. 1994. Characterization and application of materials grown by electron-beam-induced deposition. *Jpn. J. Appl. Phys.* **33**: 7099.

Koops, H. W., Weiel, R., Kern, D. P., and Baum, T. H. 1988. High-resolution electron-beam induced deposition. *J. Vac. Sci. Technol. B* **6**: 477.

Koops, H. W. P., Hoinkis, O. E., Honsberg, M. E. W. et al. 2001a. Two-dimensional photonic crystals produced by additive nanolithography with electron beam-induced deposition act as filters in the infrared. *Microelectron. Eng.* **57–58**: 995–1001.

Koops, H. W. P., Reinhardt, A., Klabunde, F., Kaya, A., and Plontke, R. 2001b. Vapour supply manifold for additive nanolithography with electron beam induced deposition. *Microelectron. Eng.* **57–58**: 909–913.

Matsui, S. 1997. Nanostructure fabrication using electron beam and its application to nanometer devices. *Proc. IEEE* **85**: 629.

Matsui, S. and Ichihashi, T. 1988. *In situ* observation on electron-beam-induced chemical vapor deposition by transmission electron microscopy. *Appl. Phys. Lett.* **53**: 842.

Matsui, S., Ichihashi, T., and Mito, M. 1989. Electron beam induced selective etching and deposition technology. *J. Vac. Sci. Technol. B* **7**: 1182.

Matsui, S., Kaito, T., Fujita, J., Komuro, M., Kanda, K., and Haruyama, Y. 2000. Nanoelectromechanical device fabrications by 3-D nanotechnology using focused-ion beams. *J. Vac. Sci. Technol. B* **18**: 3181.

Matsui, S. and Mori, K. 1984. New selective deposition technology by electron beam induced surface reaction. *Jpn. J. Appl. Phys.* **23**: L706.

Matsui, S. and Mori, K. 1986. New selective deposition technology by electron-beam induced surface reaction. *J. Vac. Sci. Technol. B* **4**: 299.

Matsui, S. and Mori, K. 1987. *In situ* observation on electron beam induced chemical vapor deposition by Auger electron spectroscopy. *Appl. Phys. Lett.* **51**: 646.

Mitsuishi, K., Shimojo, M., Han, M., and Furuya, K. 2003. Electron-beam-induced deposition using a subnanometer-sized probe of high-energy electrons. *Appl. Phys. Lett.* **83**: 2064.

Morita, T., Kometani, R., Watanabe, K. et al. 2003. Free-space-wiring fabrication in nano-space by focused-ion-beam chemical vapor deposition. *J. Vac. Sci. Technol. B* **21**: 2737.

Nakamatsu, K., Yamamoto, K., Hirayama, T., and Matsui, S. 2008. Fabrication of fine electron biprism filament by free-space-nanowiring technique of focused-ion-beam + chemical vapor deposition for accurate off-axis electron holography. *Appl. Phys. Express* **1**: 117004.

Ochiai, Y., Fujita, J., and Matsui, S. 1996. Electron-beam-induced deposition of copper compound with low resistivity. *J. Vac. Sci. Technol. B* **14**: 3887.

Perentes, A., Sinicco, G., Boero, G., Dwir, B., and Hoffmann, P. 2007. Focused electron beam induced deposition of nickel. *J. Vac. Sci. Technol. B* **25**: 2228–2232.

Seliger, R. L., Kubena, R. L., Olney, R. D., Ward, J. W., and Wang, W. 1979. High-resolution, ion-beam processes for microstructure fabrication. *J. Vac. Sci. Technol.* **16**: 1610.

Shimojo, M., Takeguchi, M., Tanaka, M., Mitsuishi, K., and Furuya, K. 2004. Electron beam-induced deposition using iron carbonyl and the effects of heat treatment on nanostructure. *Appl. Phys. A* **79**: 1869–1872.

Tanaka, M., Shimojo, M., Han, M., Mitsuishi, K., and Furuya, K. 2005. Ultimate sized nano-dots formed by electron beam-induced deposition using an ultrahigh vacuum transmission electron microscope. *Surf. Interface Anal.* **37**: 261–264.

Utke, I., Hoffmann, P., Dwir, B., Leifer, K., Kapon, E., and Doppelt, P. 2000. Focused electron beam induced deposition of gold. *J. Vac. Sci. Technol. B* **18**: 3168–3171.

Vukusic, P. and Sambles, J. R. 2003. Photonic structures in biology. *Nature* **424**: 852–855.

Wargner A., Levin, J. P., Mauer, J. L., Blauner, P. G., Kirch, S. J., and Long, P. 1990. X-ray mask repair with focused ion beams. *J. Vac. Sci. Technol. B* **8**: 1557.

Watanabe, K., Hoshino, T., Kanda, K., Haruyama, Y., and Matsui, S. 2005. Brilliant blue observation from a morpho-butterfly-scale quasi-structure. *Jpn. J. Appl. Phys.* **44**: L48.

Chapter 4

5. Plasma–Assisted Pattern Transfer at the Nanoscale

Steven Shannon

Department of Nuclear Engineering, North Carolina State University, Raleigh, North Carolina

5.1 Overview

This chapter presents a very general overview of pattern transfer technologies for nanomanufacturing that employs plasma discharges. The author recognizes the following: (1) although this field is vital for pattern transfer, it is decidedly decoupled from the field of pattern generation, and therefore most of the subject matter of this text, and (2) plasma etching is a very broad field, and so this chapter touches on everything from gas delivery to radio frequency (RF) heating to particle acceleration. Therefore, the material in this chapter is kept intentionally broad in nature. If the reader has a deeper interest in plasma processing, the author highly recommends the preeminent text in the field, *Principles of Plasma Discharges and Material Processing* by Michael Leiberman and Alan Lichtenburg.

5.2 Definition of Key Terms

- Ambipolar diffusion—Charged particle transport in quasineutral plasmas driven in part by internal fields formed due to the difference in ion and electron mobilities.
- Anisotropy (*A*)—The rate of material removal as a function of direction, typically defining the rate of vertical removal over the rate of horizontal removal.

- Atomic layer etching (ALE)—Etch processes that mimic atomic layer deposition processes in that process is self-terminating on an atomic monolayer level, enabling an atomic level of pattern transfer control.
- Black silicon—Micromasked silicon nanowires formed during etch processes.
- Bosch process—Bimodal etch process where material is sequentially etched and deposited via step-by-step process modification.
- Capacitively coupled plasma (E-mode heating)—Plasma heating via electric field coupling at the

Nanofabrication Handbook. Edited by Stefano Cabrini and Satoshi Kawata © 2012 CRC Press / Taylor & Francis Group, LLC. ISBN: 978-1-4200-9052-9

Chapter 5

periphery of the plasma discharge, typically on plasma boundaries facing electrically driven surfaces. Capacitively coupled systems typically have lower electron densities accompanied with moderate-to-high sheath voltages at the periphery.

- Cross section ($\sigma(E)$)—The energy-dependent relative probability of specific collisional events occurring in a plasma including ionization, energy transfer, and excitation.
- Debye length (λ_{de})—The characteristic scale length of a plasma, defined as the distance over which the charged species in a plasma screen out an electrostatic potential via charge separation.
- Direct current heating—Heating mechanism where a plasma is sustained via DC current from a biased surface to ground; the plasma is heated via ohmic power dissipation.
- Electron cyclotron resonance (ECR) heating—Heating mechanism where free electrons in a plasma resonate with an externally applied electromagnetic source (typically at microwave frequencies) as well as a DC magnetic field, generating an orbital trajectory for the electron that promotes heating and confinement.
- Electron density (n_e)—The number of free electrons per unit volume in a plasma. For pattern transfer process plasmas, this typically ranges from 10^8 cm^{-3} to 10^{11} cm^{-3}.
- Electron energy distribution function ($f(E)$)—The distribution of electrons in a plasma with respect to energy that plays a principal role in electron impact events such as ionization and molecular dissociation and therefore play a key role in process chemistry and surface interaction.
- Electron temperature (T_e)—The average energy of electrons in a plasma discharge; this number typically assumes a Maxwell–Boltzmann electron energy distribution function.
- Faceting—Corner rounding due to ion sputtering-induced erosion of the mask layer.
- Inductively coupled plasma (H-mode heating)—Plasma heating via current coupling between an external RF-driven coil and the free electrons in the bulk plasma. Inductively coupled plasmas tend to have higher density and lower sheath voltages than other heating methods.
- Ion energy (E_i)—Kinetic energy of ions in a plasma system, typically referring to the energy of the ion when it is incident on the pattern substrate, typically proportional to sheath voltage.

- Ion flux (I_i)—Number of ions incident on the patterned substrate per unit time per unit area, typically proportional to electron density.
- Ion sputter etch—Material removal process by which the surface atoms are removed via ballistic collisions with ions sans chemical reaction or other enhancements.
- Mask trim—Process by which the mask layer is isotropically etched in order to reduce a positive pattern's overall width and thereby reducing the minimum pattern transfer size. Although trim does enable smaller pattern transfer, it cannot increase pattern pitch or density.
- Passivation—Utilizing plasma chemistry to modify surfaces to conditions that do not react with dissociated species in the plasma to form volatile by-products, thereby protecting the surface from etching processes.
- Pattern distortion—Modification of pattern layers including pattern roughening and buckling that can then be transferred to the underlying etch layer.
- Plasma etch (dry etch)—Utilization of plasma processes, primarily electron impact collision events, to generate the necessary conditions for material removal in pattern transfer processes.
- Radiofrequency (RF) heating—Plasma heating process by which electrons are heated via collisional and stochastic heating driven by time-varying electromagnetic fields, typically in the RF region.
- Reactive ion etching—Process by which reactive chemistry is combined with ion bombardment to drive surface reactions that produce volatile by-products and thereby enable material removal enhanced in the direction of ion bombardment.
- Residence time (τ)—The average time it takes for feedgas molecules to travel through the plasma volume.
- Reactive ion etching (RIE) Lag—Aspect ratio-dependent etching of features in reactive ion etch processes caused by the decreased transport of reactive and energetic species as aspect ratio increases.
- RIE twist—Phenomena where etch profiles deviate direction randomly resulting in randomly oriented high aspect ratio structures.
- Selectivity—The rate of etch layer removal normalized to the removal of either mask material (mask selectivity) or etch stop layer (underlayer selectivity).
- Sheath—Periphery of the plasma that retains a net positive charge due to electron transport to

the walls, forming an electric field that acts to equilibrate negative particle flux with positive particle flux to maintain bulk plasma quasineutrality; this phenomena also provides the mechanism for ion acceleration in ion sputtering and RIE.

- Sheath voltage (V_s)—The voltage that forms between a plasma and its containing vessel surfaces in order to maintain quasineutrality in the bulk plasma.
- Templating—Process by which a spare template pattern is transferred to a material to act as a template for smaller patterned features formed by nonoptical patterning techniques (such as di-block copolymer patterning) in order to establish longer range order and directionality.
- Tilting—High aspect ratio etching phenomena where etch profiles are etched off-normal relative to a substrate surface due to angular deviations in ion and reactant fluxes.
- Top-down—Term referring to process flow for pattern transfer where a mask material is first deposited over a material of interest and used to transfer the pattern onto an underlayer material via etch processes.
- Undercutting—Etch phenomena where the pattern transfer process removes material underneath the mask, resulting in pattern transfer distortion.
- Quasineutrality—Fundamental requirement for a plasma discharge that the total amount of free negative charge (electrons and negative ions) be balanced by the total amount of free positive charge (ions).
- Wet etch—Pattern transfer process where a mask pattern is transferred to an underlayer material via chemical bath.

5.3   Introduction

Selective material removal at the nanoscale requires extraordinarily fine control of chemical and energy deposition on material surfaces to ensure a one-to-one transfer of nanoscale patterns to relevant materials. In order to achieve this, nonequilibrium chemistries and directional energy deposition are typically needed in order to preferentially remove or passivate material depending on its composition or spatial orientation. Fundamentally, this is an identical problem to the one that has been addressed in top-down manufacturing of integrated circuit devices for decades, where lithographic patterns formed using photosensitive coatings and sacrificial hard masks are used to transfer detailed patterns onto a plurality of materials via selective chemical removal. In fact, similar process limitations such as ion beam erosion of mask patterns have been observed for nanoscale patterning solutions such as nanoimprint lithography and copolymer micelle lithography (Figure 5.1).[1,2] A simplified schematic of this is shown in Figure 5.2, which depicts the formation of a nanoscale hole defined by some patterning layer over a substrate material where one desires to transfer the pattern. A source of chemistry and energy is incident on surfaces A (the horizontal surfaces of the hole), B (the vertical surfaces of the hole), and C (the surfaces of the nanopattern structure). If the source is purely isotropic (i.e., energy and chemistry constituents are incident on the surfaces with equal isotropic distribution), then reactions on surfaces A and B cannot be independently controlled, resulting in undercutting of the mask and expansion of the nanoscale features that were originally desired. This isotropic undercutting is illustrated in Figure 5.3a. However, if the chemistry and the energy can be controlled with respect to direction of incidence on the substrate, it would logically follow that one would be able to control the type and rate of chemical reaction that occurred at the surface. For example, if all reactive species were to only reach surface A and not reach surface B, then material removal would only occur on horizontal surfaces and undercutting would be minimized. This directional chemical removal has been the principal mechanism for pattern transfer. For the most part, RIE has been used to provide removal rate directionality due to its uniquely anisotropic energy deposition via ion acceleration at the plasma periphery, and is the central technology that will be reviewed in this chapter. This unique energy deposition technique also carries with it process limitations such as the enhanced erosion of the mask material C and undesired damage to material B, and presents the potential mechanism that the predicted scale length limitation for plasma-assisted pattern transfer. This too will be reviewed in this chapter.

Another mechanism that can be utilized in pattern transfer that can extend its capability and plays a central role in current state-of-the-art nanoscale pattern transfer technologies is anisotropy enhancement via

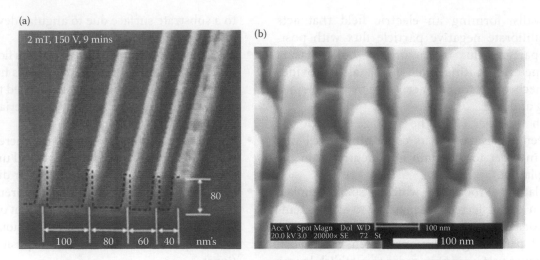

FIGURE 5.1 (a) Etched nanoscale structured formed using nanoimprint lithography, highlighting mask erosion effects due to ion bombardment. (Reprinted from Morecroft, D., Yang, J. K. W., Schuster, S. et al. 2009. *J. Vac. Sci. Technol. B* 27(6), 2837. With permission.) (b) Nanopillars etched into silicon using copolymer micelle lithography and chlorine plasma etching—taken from copolymer micelle lithography. (Reprinted from Krishnamoorthy, S., Gerbig, Y., Hilbert, C. et al. 2008. *Nanotechnology* 19, 285301 (6pp). With permission.)

selective surface passivation. For example, if the hypothetical problem illustrated in Figure 5.2 is maintained at a temperature that is low enough to make reactive by-products nonvolatile, and energy is directionally deposited on surface A only, one could theoretically passivate surface B with residual by-product while providing a constant reactive surface on A. One or both of these mechanisms can promote anisotropic material removal, and are the key elements in current

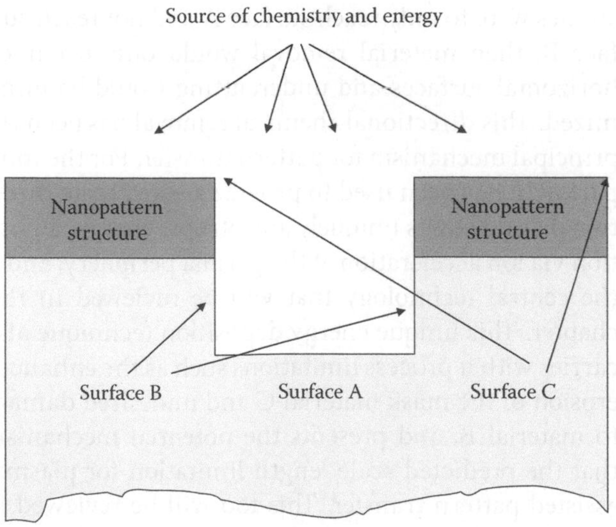

FIGURE 5.2 Formation of a nanoscale hole defined by some patterning layer over a substrate material. A source of chemistry and energy is incident on surfaces A (the horizontal surfaces of the hole), B (the vertical surfaces of the hole), and C (the surfaces of the nanopattern structure).

state-of-the-art pattern transfer, such as is illustrated in Figure 5.3b. More advanced etch processes that utilize a bimodal etch/dep process, such as Bosch processing, rely heavily on this passivation technique, and promising results have been demonstrated using this technique to etch features down to <40 nm and over 1.6 μm deep.[3]

Transitioning these chemically driven, directionally dependent systems down to the nanoscale presents fundamental challenges. In addition, the unique techniques for pattern generation at the nanoscale present new opportunities for pattern transfer to contribute to the overall fabrication process (such as template synthesis for self-organized structures or mask tapering). What is required for nanoscale pattern transfer is one or both of these anisotropic mechanisms that provides 1:1 transfer of patterns down to the nanometer feature scale while minimizing erosion or modification of surrounding material. This requires directional delivery of either chemistry or energy to drive a plurality of reactions that will depend on the orientation of the surface relative to the incidence of reactants or energy. This system needs to be manufacturing worthy and economically viable to enable economic commercialization of nanopatterned systems. Finally, this system needs to be robust enough to provide consistent delivery of reactants and energy to ensure repeatable transfer of nanoscale patterns. Fortunately, these challenges closely mirror those faced in the top-down fabrication of larger devices, presenting a promising roadmap for pattern transfer at these scale lengths.

(a)

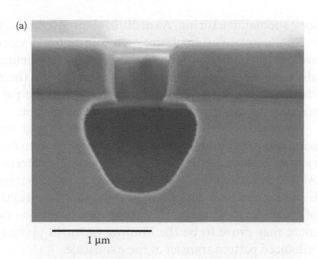

(b)

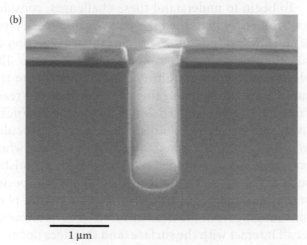

FIGURE 5.3   Isotropic and anisotropic etch profiles obtained from the same mask at different process conditions. (Reprinted from Wu, B., Kumar, A., and Pamarthy, S. 2010. *J. Appl. Phys.* 108, 051101. With permission.)

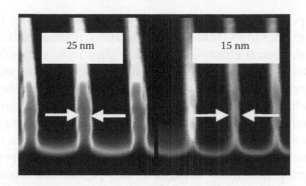

FIGURE 5.4   Nanoscale structures etched into silicon using cryogenic plasma etch processing. (Reprinted from Olynick, D. L., Liddle, J. A., Harteneck, B. D. et al. 2007. *Proceedings of SPIE* 6462, 64620J-1. With permission.)

The predominant process for anisotropic pattern transfer for the past 30 years has been plasma-assisted material removal, commonly referred to as plasma etch, dry etch, or RIE. Plasma etch utilizes nonequilibrium chemistry generated in a plasma glow through various mechanisms including ballistic particle collisions, photon interaction, and gas phase chemical reactions to produce a combination of reactive species that can interact with the material to be patterned. The mechanisms for the formation of these reactive species allow these chemistries to be formed at

relatively low temperatures, opening the window for an array of processes that otherwise would not be possible due to the thermal limitations of the devices being fabricated. Along with this unique, low temperature chemistry, plasmas provide directional energy deposition via ion acceleration perpendicular to the surface of the substrate due to the formation of a space charge sheath at the plasma periphery. These ions provide directional energy deposition within a few monolayers of the surface and modify the near surface temperature and chemical equilibrium of the exposed surface. Additionally, these ions can be chemically reactive with the substrate material, providing a level of directionality with respect to chemistry as well. The result is a system capable of transferring patterns with minimal distortion onto an array of materials down to the nanoscale. As of the writing of this chapter, pattern transfer down to 15 nm has been demonstrated by several groups, including the Molecular Foundry at Lawrence Berkeley National Laboratory.[4] Figure 5.4 shows a high-resolution SEM image of a 25 and 15 nm structure generated using RIE-based pattern transfer technology.

This chapter details the overarching mechanisms and challenges in plasma etching, with emphasis on mechanisms that present unique challenges for nanoscale pattern transfer. Specifically, this chapter explores methods for controlling both the chemistry and energy delivery of these systems in order to refine pattern transfer processes at the nanoscale.

5.4   Brief History of Plasma Etching

Plasma-assisted removal of material layers for top-down manufacturing can be traced back to the late

1970s, when researchers were studying new ways to control patterning that would extend the range of operation of the current state-of-the-art technology of wet etching.[5] These first-generation plasma etchers were barrel etchers, designed to chemically etch substrates in batches and basically mimic the processes that were at the time dominated by wet chemistry. Initial studies not only concluded that the primary etch mechanism was chemical, but also ironically suggested that ion bombardment was not preferred due to the surface damage that could be caused by ion sputtering (the irony being that this phenomena would drive plasma etching as the dominant pattern transfer technology for the next 30 years, and that now as we approach the nanoscale may very well be the limiting factor in the extension of this technology). During these preliminary studies, reduced undercutting and an improved level of process anisotropy was still observed, and the mechanism for undercutting reduction initially hypothesized to be due to ion acceleration across the plasma sheath.[6]

Since this initial effort focused on reproduction of wet chemistry effects, the synergistic effects of plasma chemistry and directional ion energy deposition were not fully characterized until 1979, when Colburn and Winters published their now famous work detailing the enhanced removal of silicon using a combination of XeF_2 gas and Ar ion bombardment.[7] In this work, it was shown that the individual contributions of either XeF_2 gas or Ar ion bombardment would only result in a silicon removal rate of <1 nm/min, but that the combined influence of both XeF_2 gas and Ar ion bombardment would increase the removal rate by almost an order of magnitude, suggesting a synergistic removal rate that was much greater than the sum of the parts. Although the theory that this synergistic effect existed prior to the publication of this article, this was the first rigorous demonstration of this benefit, and this publication is often credited as the starting point for the field of plasma etching.

The primary beneficiary of this work has been microelectronic fabrication. No other industry has stretched the capabilities of plasma-assisted material removal to the extent that this industry has, and the result has been an amazing pace of innovation in the field to keep up with pattern generation technology through optical lithography. The combined maturation of plasma etch and optical lithography has been the primary mechanism for the adherence of the industry to the now famous Moore's law, which states that the density of devices on an integrated circuit will double roughly every 18 months.[8] This marriage of plasma etch to Moore's law has positioned it uniquely for nanomanufacturing. As of 2010, 45 nm technology is commonplace using plasma etch, and 32 nm devices are on the horizon. The technology has therefore already demonstrated the ability to approach these challenging scale lengths and accommodate the pattern sizes possible with advanced nanolithography.

The principle challenges in plasma etching at the nanoscale are profile control, repeatability, and maintaining the integrity of the remaining material. Although each of these challenges exist at much larger feature scales than 10s of nanometers, the scale length of these devices exasperates these issues and one or more may prove to be the limiting factor in plasma-enhanced pattern transfer at the nanoscale.

To begin to understand these challenges, consider a silicon wafer patterned with an organic mask layer of thickness d (photoresist) that exposes a section of silicon with radius R to the etching plasma. The plasma introduces a multitude of components to the surface including neutrally charged species (reactants, nonreactants, particles), charged species (positively and negatively charged atomic and molecular ions as well as electrons), and photons. Even when one considers a simplified reactive plasma chemistry such as chlorine, the number of contributing species is substantial: Cl_2, Cl, Cl_2^+, Cl^+, Cl_2^-, Cl^-, e^-, and photons ranging from the near infrared to deep ultraviolet all interact with the surface (and this does not even account for the impact of etch by-products such as sputtered Si, $SiCl$, $SiCl_2$, $SiCl_3$, and $SiCl_4$ as well as mask material that are introduced into the gas phase during the etch process). More complicated chemistries, particularly those involving fluorocarbon feedgases, can have as many as 40 coupled reactions just to account for the gas phase, with an additional 36 mechanisms that account for surface reactions.[9] This multitude of species presents a very challenging (and fortuitous) environment for the selective removal of this material. The primary challenge comes from the fact that the plasma itself is a highly nonequilibrium system, where the relative temperatures of all of these species are extremely different from each other (ranging from room temperature to nearly 100,000 K) and are often not represented by a Maxwellian distribution (particularly the charged species), making accurate prediction and modeling of these systems a significant challenge. The fortuitousness of this environment is that this nonequilibrium collection of charged particles presents a mechanism for directed energy transfer due to the sheath formation on the periphery of the plasma discharge.

5.5   Plasma Chemistry

The formation of reactive species in plasma is primarily through electron impact dissociation of a molecular feedgas. The relative rates of production for these dissociated species are primarily determined by the gas phase density of the feedgas n_g, the density of free electrons n_e, and the distribution of these electrons in energy space, $f(E)$. These parameters are linked together through the energy-dependent interaction cross section $\sigma(E)$ that defines the relative target size that the energetic electron encounters for a specific reaction. Together, these parameters provide a relationship for the production rate of a dissociated specie from a specific electron impact interaction:

$$\text{Rate} = \int_0^\infty n_e \cdot f(E) \cdot \sigma(E) \cdot \sqrt{\frac{2 \cdot E}{m_e}} \, dE \qquad (5.1)$$

where m_e is the mass of an electron. The total rate of production of a dissociated specie from the background feedgas is the sum of all possible dissociative mechanisms:

$$\text{Rate} = \sum_{\text{reactions}} \int_0^\infty n_e \cdot f(E) \cdot \sigma_{\text{reaction}}(E) \cdot \sqrt{\frac{2 \cdot E}{m_e}} \, dE \qquad (5.2)$$

These reactions can be numerous. For example, fluorine production through electron impact dissociation of C_4F_8 can follow two separate paths.[9]

Loss mechanisms are also a vital mechanism that determines the relative concentration of dissociated species, and are primarily driven by surface reactions, gas-phase recombination, and the gas residence time τ in the plasma reactor. Surface reactions and gas residence time effects are the mechanisms that are most directly influenced by process conditions and are therefore the most commonly used means for controlling dissociative specie loss in the reactor. Surface reactions are most typically controlled via surface temperature control (not only of the substrate material, but also the plasma-facing components that make up the plasma reactor). The residence time influences the rate of loss of species as it defines how long gas species can be modified via plasma interaction (electron collision, etc.) and therefore the rate of subsequent breakdown of dissociated species into more elementary components due to subsequent electron collisions, and is most conveniently controlled via feedgas flow rate and reactor geometry. This residence time in comparison with the rate of dissociation of the feedgas presents a mechanism for subsequent dissociation of already decomposed species. This complicates the calculation of production rate in that the integral now depends on the concentration of a dissociated specie and not the feedgas in the reactor, and therefore depends on the rate of production and loss of the dissociated specie itself. The result is that the total rate of production of a specific dissociated specie now becomes a solution derived from a system of coupled differential equations, where the production rates of species via specific reactions via Equation 5.1 are balanced with the loss of dissociated species due to recombination, surface reactions, and loss to the system due to residence time effects. This calculation is even further complicated by the fact that the electron energy distribution function, $f(E)$, is in large part determined by the relative rates of these reactions due to their inelastic nature, and therefore changes with respect to the rates of formation of various species. Finally, the calculation of cross sections for different electron impact reactions for dissociated species is extremely challenging, and these cross sections tend to vary significantly from reference to reference, lending an inherent source of error to many of these models. Despite these challenges, very detailed global models of reactive plasma chemistries have been performed by several groups, and these models have been incorporated into several plasma reactor models to study industrial plasma processes, most notably the Hybrid Plasma Equipment Modeler developed by Mark Kushner's groups at the University of Illinois, Iowa State, and the University of Michigan, respectively.

There are many ways to generate a plasma discharge for materials processing applications. Plasmas can be driven by energy sources ranging from direct current drives all the way to microwave frequencies. The most common heating mechanism, however, is using RF power coupling to the plasma reactor. This is due to the relative simplicity of the power coupling system, the capability to transmit RF frequencies using both conducting and insulating materials (and thereby providing a wide range of material options to control surface chemistry in plasma reactors), and the ability to couple power to either the sheath (for ion heating to promote anisotropic energy deposition) or the bulk plasma (to promote feedgas dissociation) through antenna design or frequency selection. This heating can be further augmented by using external magnetic fields or additional power supplies that can provide the user with some level of independent control of electron heating versus ion heating.

The power that is dissipated into the plasma promotes electron heating for plasma chemistry control. The steady-state electron density and energy produced by this auxiliary heating is largely determined by the balance of particle loss versus particle production as well as detailed energy balance between dissipated power and power lost at the periphery of the system. On a first order, particle loss is determined by the equating of boundary conditions at the plasma edge (typically particle density at the boundary or particle flux at the boundary) with the diffusive properties of the plasma generated by the combination of collisional effects and ambipolar diffusion effects generated by the collective response of the plasma to maintain quasineutrality within the scale length of the discharge itself. Since the motion of charged species in the bulk plasma are largely governed by parameters that have dependence on both electron energy and ion energy, and boundary conditions define the geometric impact on the steady-state solution of the plasma discharge, there is a synergy

between the plasma geometry and the electron energy distribution function that drives the reactions in the bulk plasma. These, along with the residence time effects discussed earlier, are the primary mechanisms that are behind adaptive geometry designs for plasma reactors, particularly moving substrate holders that can effectively change the volume to surface ratio of the plasma, thereby influencing the steady-state electron energy distribution function. Externally applied magnetic fields are used in a similar way by producing a diffusion tensor for charged species that can impact charge specie diffusion and mobility and thereby change the steady-state conditions for a constant surface-to-volume ratio. Energy balance in the plasma system determines electron density by balancing the amount of power dissipated in the plasma with the amount of power lost per electron–ion pair formed. This loss term accounts for energy loss due to ionization potential, energy transfer collisions, and acceleration of ions to the surface, which in turn is due to the electric potentials that are formed on the periphery of the plasma discharge due to the formation of plasma sheaths.

5.6   Plasma Sheaths

Plasma sheaths are formed on the periphery of a plasma discharge in order to equilibrate the loss of negatively charged and positively charged species in the plasma. This allows a plasma of finite size to maintain quasineutrality, which is the condition where the net charge of the entire plasma discharge is zero, and is a necessary condition for low-temperature plasmas.

Sheaths are formed due to the relative masses and velocities of ions and electrons. Consider a simplified case where a plasma made up entirely of positively charged ions and negatively charged electrons is contained in a grounded, conducting chamber that is grounded at some time $t = 0$ (Figure 5.5a). Due to the lower mass and comparable energy of the electrons, their velocities are much greater than the ions, and for short time periods compared to the ratio of Debye length to electron velocity (where the Debye length is the characteristic scale length of a plasma defined as $\lambda_{de} = \sqrt{\varepsilon_0 T_e / e n_e}$ where T_e is the average electron energy in volts), the electrons are more readily lost to the grounded periphery while the ions in this short time scale remain relatively stationary (Figure 5.5b). This produces an electron depletion region at the periphery of the plasma where the electron density drops to near zero while the ion density is maintained at roughly the

same order of magnitude as the original charges specie density (Figure 5.5c). This volume positive charge produces an electric field normal to the surface of the plasma vessel. This internally generated electric field's magnitude is such that the flux of positive and negative charge to the walls is equal, thereby maintaining quasineutrality in the bulk plasma (Figure 5.5d). This electric field, while confining the electrons, accelerates the ions perpendicular to the chamber and provides an anisotropic energy component to the surface of the chamber. This is the primary mechanism for ion acceleration for RIE.

These sheaths can be manipulated by controlling the potential of the plasma-facing surface, thereby affecting the potential that the ions are accelerated through. Typically, this manipulation is brought about by electrically floating part of the chamber (typically referred to as an electrode or cathode) and biasing it either with a DC potential or a rectified RF waveform. DC potentials require a conducting path from the plasma-facing surface to ground, and are therefore not typically used since many materials of interest for both processing and chamber design have insulating properties. Some examples of this would be anodization of the chamber wall to prevent chemical attack

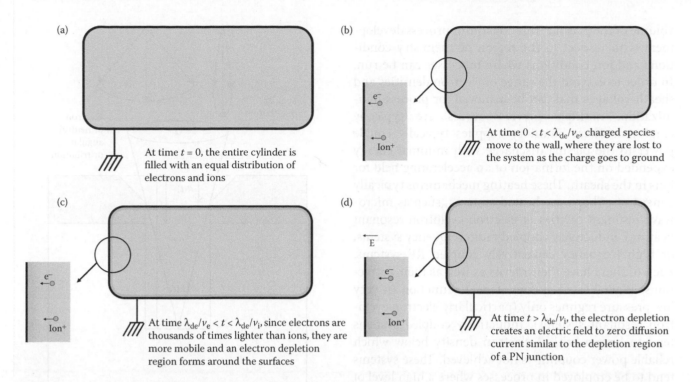

FIGURE 5.5 Schematic diagram of a simplified case when plasma that is made up entirely of positively charged ions and negatively charged electrons is contained in a grounded, conducting chamber (a), the loss of charge carriers at the periphery of the plasma (b), the electron depletion formed due to higher electron velocity (c), and formation of an internal field in the plasma due to the formation of a positive space charge (d). The region where this periphery field exists is commonly referred to as the dark space or plasma sheath.

and subsequent metal contamination, etching of insulating films such as silicon dioxide, and electrode coatings with insulating materials for electrostatic clamping of substrates for temperature control. RF-driven sheaths do not require a conducting path to ground. The RF waveform produces a time averaged DC potential across the sheath about which the RF waveform oscillates. The sheath effectively expands and collapses as the instantaneous voltage oscillates about this DC potential. Again, the amplitude of both the DC field and the RF field is determined by the fundamental need to maintain quasineutrality in the plasma. However, the additional variable of electrode biasing does provide the user with some level of ion energy control, which is a vital component in plasma etch process development. In addition, the proper

selection of one or more frequencies to drive this electrode system can also provide the user with not only control of the time average ion energy, but also the ion energy distribution function that is brought about due to the combination of the oscillating field and the inherent inertia of the ions as they respond to these fields.[10] RF sheaths are capable of providing controllable ion energies ranging from the 10s of electron volts to over 1 keV. It must be noted, however, that there is a lower limit to ion energy control due to the eventual need for a baseline potential to meet the quasineutral requirements of the bulk plasma. This makes some amount of ion heating an eventuality in all plasma reactors, and establishes a lower limit for ion benefit as well as ion damage that cannot be overcome by sheath control.

5.7 Controlling Chemistry and Ion Energies in Critical Plasma Applications

Splitting power between sheath acceleration and chemistry production determines the level at which these two components contribute to surface reactions. Typically, for a given process condition (gas composition, pressure, etc.), a plasma driven by a single power

source will follow a specific trajectory with respect to electron density n_e and sheath voltage V_s, determined by the particle and energy balance conditions that are established by energy and particle losses on the surface of the plasma compared to production rates in the volume of the plasma. This constrains process development with respect to the region of chemistry conditions and ion conditions where processes can be run. In order to expand the range of electron densities and sheath voltages that can be achieved for process optimization, multiple power sources are typically employed. These auxiliary supplies typically couple power to the plasma discharge with minimal energy expended on the formation of an accelerating field for ions in the sheath. These heating mechanisms typically consist of either wave heated systems (such as microwave resonant cavities or electron cyclotron resonant heating), inductively coupled radio frequency systems, or high frequency capacitively coupled RF systems. Each of these have their merits as well as their detractions. Wave-heated systems tend to function in very low pressure regimes only (particularly electron cyclotron resonant systems). Inductively coupled systems tend to have a critical electron density below which reliable power coupling is not achieved. These systems tend to be employed in processes where a high level of dissociation is required, such as chlorine-based or oxygen-based processes. High-frequency capacitive systems tend to have a larger dynamic range of electron density control. However, these systems also tend to suffer from standing wave effects and edge electrode H-mode heating that both can adversely affect process uniformity for large area processing. These systems are then combined with either DC or low-frequency coupling over the substrate-holding electrode that couple power to the sheath for ion heating, thus providing decoupled control of electron heating and ion sheath heating.

This combination of unique chemistry and ion sheath heating is incident on a substrate patterned with a mask material that is typically more impervious to this environment than the underlayer that the pattern is to be transferred to. The underlying material is removed layer by layer at a rate determined by the flux of reactant and energetic ions incident on the surface. Etch profile evolution is a dynamic system, with the erosion of etch features continuously changing the flux of these species to the surface. Heil and Anderson demonstrated a simple geometric argument that accurately predicted the slowing of pattern transfer processes due to time-dependent geometry considerations, and noted that by accounting for three principle pattern characteristics: mask thickness L_0, facet depth $R(t)$, and feature depth $L(t)$, as illustrated in Figure 5.6, that the removal rate as a function of geometry could be determined.[11] Typically, reaction rates are deter-

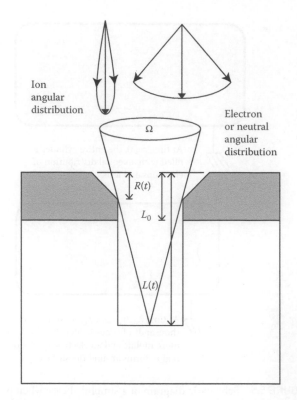

FIGURE 5.6 Geometry considerations for calculating time-dependent etch parameters. (Reprinted from Keil, D. and Anderson, E. 2001. *J. Vac. Sci. Technol. B* 6(19), 2082–2089. With permission.)

mined by calculating the steady-state surface coverage of the plurality of reactive species on the surface of the substrate and the removal rate of volatile by-products as determined by these relative surface coverages. The complexity of these models varies from system to system, but typically amount to nothing more than a system of coupled differential equations, similar to those used to determine the steady-state dissociated species in the bulk plasma. Several groups have demonstrated the ability to predict the nuances of profile evolution, including phenomena such as tapering, notching, faceting, and twisting. Figure 5.7 depicts one feature level simulator that captures many of the nuances of high aspect ratio etching of sub-100 nm features, validated by cross-sectional SEM analysis of the modeled process.

The evolution of plasma reactors for pattern transfer beyond the micron scale and into the nanoscale has primary focused on four areas: plasma heating, gas delivery, system materials, and closed loop process control. As discussed earlier in this chapter, plasma heating, primarily through RF heating, influences the nonequilibrium chemistries that make plasmas so unique. Additionally, these heating systems provide the energy for ion acceleration and energy

Feature tilt and
etch depth
(a)   (b)

FIGURE 5.7 Characterization of RIE lag measured (a) and predicted (b) profiles for sub-100 nm HAR structures highlighting capability to predict challenging profile features, in this case HAR via twisting due to mask aberrations. (Reprinted from Welsh, S., Keswick, K., Stout, P. et al. 2009. *Semi. International* February, 18–22. With permission.)

deposition to the wafer surface. Most advances in this area that has enabled the extension of plasma etching to sub-100 nm feature sizes has focused on how the plasma responds to power sources with different characteristic frequencies and combining multiple power sources driven at multiple frequencies to tune specific process conditions. These drive frequencies range from the kHz range to 100s of MHz. These power sources are driven both at constant power delivery, and pulsed on and off to provide even larger dynamic range with respect to chemistry. Gas delivery advances have enabled spatially variant gas composition control and fast time response gas modulation, enabling Bosch-type and gas chop processes that are currently the most widely demonstrated technologies for sub-20 nm processing. System materials enable economic fabrication of these devices without significant preventative maintenance, as well as provide a passive mechanism for power delivery control and chemistry modification. Closed-loop control of various process subsystems ranging from commonplace (pressure control, gas flow control, load power RF delivery) to advanced (multivariate process monitoring, fast transition control) has reduced the variance in these processes as the reduction in critical dimensions has demanded tighter variance from run to run.

5.8 New Plasma Etch Processes for Nanostructures

The limits of scaling of electronic devices and the death of Mark Twain are similar in that both have been famously prematurely announced. With respect to current predictive studies of the limitations of plasma etch for pattern transfer, current studies predict a practical limit of approximately 5 nm.[12] The reason for this limitation is due to the damage layer that is produced due to chemical and ionic attack of the device sidewalls during the etch process. The thickness of this damage layer presents the predicted limitation of plasma etching, even in the presence of ideal ion beam incidence and mask structure. The repeatability of these processes also presents a very challenging limitation with respect to critical dimension scaledown.

As devices transition into the nanoscale, the concept of "pattern transfer" also begins to transcend the idea of simply transferring a pattern onto an underlayer that has prevailed the technology for the past 30 years. Plasma processing for pattern transfer at the nanoscale now enables the patterning itself through structures that provide long range order in self-assembled patterns. Plasmas are used to modify the patterns themselves by trimming critical dimensions down through carefully engineered mask erosion. The concept of etch selectivity has now transcended the idea of only selectively etching the underlying material from the mask structure. Now, pattern transfer demands the selective removal of specific nanostructured phases, effectively etching away the undesired material phases in order to enhance the performance characteristics of remaining materials. Finally, in what on the surface appears to be a complete misnomer, plasma etch is used to "grow" nanoscale features such as nanowires and nanocrystalline materials.

One-to-one nanoscale pattern transfer has two key challenges. The first is the ability to etch anisotropic, high aspect ratio features such as trenches and via's with nanoscale pitch. The second is the ability to stop pattern transfer processes within monolayers of the etch stop layer reliably. Bimodal plasma processing consisting of alternating etch steps and deposition steps have shown very promising results in sub-20 nm processing. This process is typically referred to as Bosch processing. Although it focuses on larger dimension features, one of

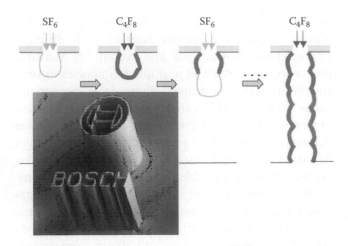

FIGURE 5.8　Bosch process flow and Bosch etch in silicon (inset). (SEM photograph reprinted from Tillocher, T., Dussart, R., and Overzet, L. J. 2008. *J. Electrochem. Soc.* 155(3), D187–D191. With permission.)

the most complete parametric studies of Bosch-type processing provided in Reference.[13] The basic idea behind Bosch processing is designing a process that can reliably switch from a depositing regime to an etching regime and back again. A simplified Bosch process flow is shown in Figure 5.8. This cyclical etch/dep/etch/dep process provides consistent sidewall passivation and a high degree of anisotropic process capability that makes it the current state of the art for advanced processes

such as through-wafer etching and HAR via processes. These processes have been demonstrated for nanoscale pattern transfer for pitches <20 nm. A similar methodology is employed to control etch depth for critical processes as well. Atomic layer etching also employs a bimodal process sequence. However, this sequence serves to provide a self-limiting surface activation process that enables material removal with monolayer control. A simplified atomic layer etching flow is illustrated in Figure 5.9. Precision etch depth of nanoscale insulating structures for transistor gates has been demonstrated with cycle control of etch rate down to 1.2 Å/cycle using reactive chemistries in conjunction with ion beam exposure for anisotropic material removal.[14]

Two key challenges for the reliable integration of these bimodal processes are transition stability between steps and fast, reliable feedgas transition. Process stability is a particular challenge during the transition between these steps as the gas phase chemistry can present some very unstable transients in a plasma discharge when electronegative gases such as SF_6 or oxygen are employed.[15] These instabilities can produce inconsistent transitions from cycle to cycle, and in some cases result in total or intermittent extinguishing of the plasma discharge. Efforts to passively and actively control these instabilities have had some success.[16] Specifically, the incorporation of Faraday shields on inductive systems to limit the amount of capacitive coupling has demon-

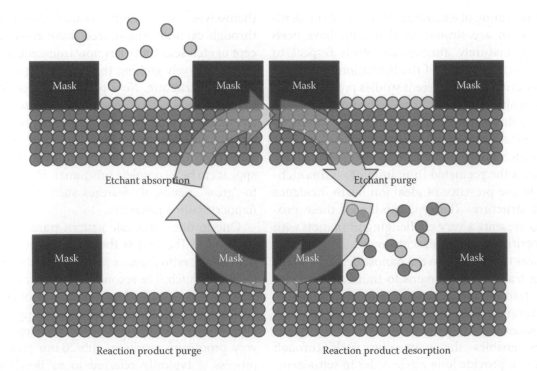

FIGURE 5.9　Simplified process flow for atomic layer etch processes.

strated an ability to greatly reduce or remove these instabilities.[17] However, the best-known-method for dealing with these phenomena to this day continues to be avoidance of conditions where they manifest. Gas delivery presents the second challenge in reliable bimodal processing for nanomanufacturing for both Bosch and ALD processes. Since these cycles consist of steps that tend to only last a few seconds, it is important to be able to quickly transition feedgases from one mix to another. However, the trapped volume of gas between the chamber valve and the flow controller, combined with the response time of the flow controller to down-stream transients, makes reliable gas delivery challenging, as system lag and overshoot can produce flow and pressure instabilities that can increase the process variability of a pattern transfer recipe. Currently, this issue is resolved with a combination of pressure feedback flow control, fast valve gas injection, and close proximity design where the flow control and valve system are maintained with minimal gas line length and maximum conductance to the process chamber so as to provide a fast transition process with minimal overshoot and sta-bilization times of <1 s.[18]

Nanoscale pattern transfer often requires nonopti-cal lithographic techniques that employ self-organiza-tion of patterned structures. Close-packed nanoparticles and diblock copolymers are currently the primary self-organization-based nanoscale pat-terning techniques. Although these techniques dem-

onstrate extraordinary capability to generate features with pitch down to 10s of nanometers, the orientation and long range order of these structures is difficult to control, and presents one of the key challenges in employing these technologies in volume manufactur-ing. One technique that enables control of orientation and long range order is substrate templating. In sub-strate templating, a patterned template that is much larger than the actual device pitch is established on the substrate with more controllable lithographic tech-niques. These larger pitch structures then "pin" the self-organized patterns and allow for longer range order.[19] Figure 5.10 shows a top-down simplified flow for orienting high-density nanoscale patterns using templating.

Mask trim is a relatively old technology that is being used to extend older patterning technologies to the nanoscale. In mask trim processes, the physical size of pattern features is reduced prior to pattern transfer through lateral etching of the mask layer. Although etch technology has provided an excellent means for vertical anisotropy, horizontal anisotropy methods are not achievable through plasma-based pattern transfer. Therefore, trim processes typically employ an isotropic (or relatively isotropic) mask etch process. A typical mask trim flow is shown in Figure 5.11. Note that a single step trim process using mask removal does not reduce pitch, but only reduces the size of the mask feature. Also, due to the isotropic nature of a mask trim process, mask etch

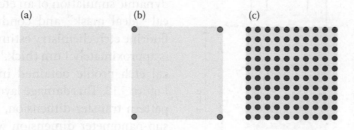

FIGURE 5.10 Pattern templating simplified flow. A substrate (a) is patterned with a sparse template (b) that is used to pin a self-organized structure that provides a more dense pitch (c) that can be etched using standard pattern transfer techniques.

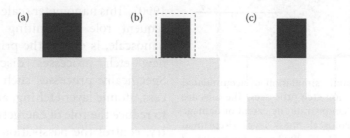

FIGURE 5.11 Mask trim simplified flow. A mask with larger feature size (a) is exposed to an isotropic etch process, shrinking the fea-ture size (b), followed by a main etch step (c).

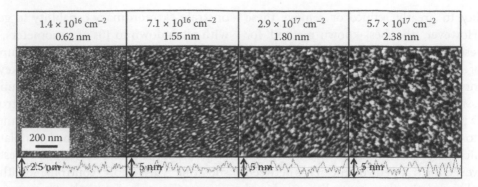

| 1.4×10^{16} cm^{-2}
0.62 nm | 7.1×10^{16} cm^{-2}
1.55 nm | 2.9×10^{17} cm^{-2}
1.80 nm | 5.7×10^{17} cm^{-2}
2.38 nm |

FIGURE 5.12 Top down SEM image of blanket photoresist illustrating mask roughness due to ion and photon fluence during plasma processing. (Reprinted from Titus, M. J., Nest, D. G., Chung, T. -Y. et al. 2009. *J. Phys. D: Appl. Phys.* 42, 245205 (13pp). With permission.)

selectivity in the subsequent main etch step also tends to suffer when trim techniques are employed. It is also worth noting that similar processes that deposit monolayers of mask material on an existing mask (such as atomic layer deposition) can "fatten" masks and provide smaller feature sizes when negative pattern transfer (via's instead of pillars as shown in Figure 5.11) is desired.

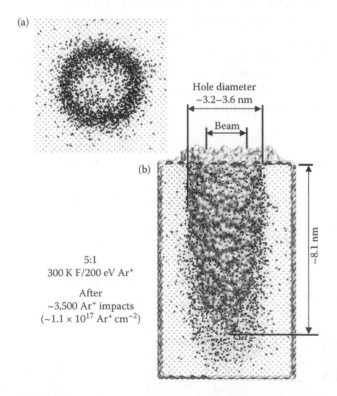

(a)

Hole diameter
~3.2–3.6 nm

Beam

(b)

5:1
300 K F/200 eV Ar$^+$

After
~3,500 Ar$^+$ impacts
(~1.1 × 10^{17} Ar$^+$ cm^{-2})

~8.1 nm

FIGURE 5.13 Molecular dynamic simulation of accumulated silicon damage during reactive ion etch processes. The ions are incident at a central point to demonstrate the extent of damage in the presence of an ideal nanoscale mask layer. (Reprinted from Vegh, J. J. and Graves, D. B. 2008. *1st International Conference on Laser and Plasma Applications in Materials Science*, Vol. 1047, pp. 74–78. With permission.)

Mask distortion presents another significant challenge for the transfer of nanoscale patterns. Plasmas for etching, although designed to minimally impact the mask layer, can distort or roughen the pattern layer due to ion and photon bombardment. This can result in inconsistent patterning across a substrate and tremendous variability in etched structures. Figure 5.12 shows the progression of mask roughness due to plasma exposure of 193 nm photoresist.[20]

As feature sizes continue to shrink, plasma interaction with feature sidewalls are one of the proposed mechanisms that will present a lower limit for plasma enhanced pattern transfer. Specifically, ion-collision induced surface damage introduces a lower limit defined by the depth of the damage layer. Molecular dynamic simulation of an etch profile using a theoretical "ideal mask" and conditions typically found in fluorine etch chemistry estimate that this damage layer is approximately 1 nm thick.[11] A cross section of a typical etch profile obtained in this study is shown in Figure 5.13. This damage layer present a lower limit for pattern transfer dimension, as even an ideal mask of sub-nanometer dimension will effectively be "undercut" by this damage layer, presenting a similar challenge with respect to pattern transfer geometry that isotropic wet etching presented prior to the early 1990's. This nanometer scale damage layer, and its subsequent role in limiting pattern transfer at the nanoscale, is one of the primary inspirations for the new etch processes discussed in this chapter. Specifically, processes such as bimodal (Bosch) process, atomic layer etching, and cryogenic etching seek to reduce the role of energetic ions in sidewall chemistry, control the passivation that protects these sidewalls, and extend plasma enhanced pattern transfer beyond this proposed dimensional limit.

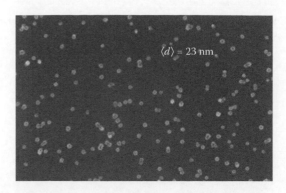

$\langle d \rangle = 23$ nm

FIGURE 5.14 Nanoparticles formed in an acetylene/argon plasma. (Reprinted from Schulze, M., von Keudell, A., and Awakowicz, P. 2006. *Plasma Sources Sci. Technol.* 15, 556–563. With permission.)

One plasma chemistry mechanism that presents a challenge for nanoscale pattern transfer is the formation of nanoscale particles in process plasmas. Depending on process conditions, particles with diameters down to the 10s of nanometers can be grown in the gas phase, presenting a mechanism for pattern screening (Figure 5.14).[21] These particles have been observed in micron-scale device processing as well, and have been attributed to the silicon "grass" effect that is generated when these particles provide nanoscale screening of plasma etch processes (effectively acting as an uncontrolled mask layer) resulting in a grass like structure on the silicon surface commonly referred to as "black silicon."[22]

References

1. Morecroft, D., Yang, J. K. W., Schuster, S., Berggren, K., Xia, Q., Wu, W., and Williams, R. 2009. Sub-15 nm nanoimprint molds and pattern transfer. *J. Vac. Sci. Technol. B* 27(6), 2837.
2. Krishnamoorthy, S., Gerbig, Y., Hilbert, C., Pugin, R., Hinderling, C., Brugger, J., and Heinzelmann, H. 2008. Tunable, high aspect ratio pillars on diverse substrates using copolymer micelle lithography: An interesting platform for applications. *Nanotechnology* 19, 285301 (6pp).
3. Morton, K. J., Nieberg, G., Bai, S., and Chou, S. Y. 2008. Wafer-scale patterning of sub-40 nm diameter and high aspect ratio (>50:1) silicon pillar arrays by nanoimprint and etching. *Nanotechnology* 19, 345301 (6pp).
4. Olynick, D. L., Liddle, J. A., Harteneck, B. D., Cabrini, S., and Rangelow, I., 2007. Nanoscale pattern transfer for templates, NEMs, and nano-optics. *Pro. SPIE* 6462, 64620J-1.
5. Abe, H. 1974. The application of gas plasma to the fabrication of MCIS LSI. *Proc. 6th Con. Solid State Devices, Tokyo 1974, Supp. to Oyo Buturi (J. Japan SOC. Appl. Phys.)*, 44, 287–251.
6. Suzuki, K., Okudaira, S., Sakudo, N. et al. 1977. Microwave plasma etching. *Jpn. J. Appl. Phys.* 16, 1979–1984.
7. Colburn, J. W. and Winters, H. F. 1979. Ion and electron assisted gas surface chemistry—An important effect in plasma etching. *J. Appl. Phys.* 50, 3189–3197.
8. Moore, G. E. 1965. Cramming more components onto integrated circuits. *Electronics* 38, 114–117.
9. Kokkoris, G., Goodyear, A., Cooke, M., and Gogolides, E. 2008. A global model for C_4F_8 plasmas coupling gas phase and wall surface reaction kinetics. *J. Phys. D: Appl. Phys.* 41, 195211.
10. Shannon, S., Hoffman, D., Yang, J. G., Paterson, A., and Holland, J. 2004. The impact of frequency mixing on sheath properties: Ion energy distribution and V_{dc}/V_{rf} interaction. *J. Appl. Phys.* 97, 103304.
11. Keil, D. and Anderson, E. 2001. Characterization of reactive ion etch lag scaling. *J. Vac. Sci. Technol. B* 6(19), 2082–2089.
12. Vegh, J. J. and Graves, D. B. 2008. Molecular dynamics simulations of nanometer-scale feature etch. *1st International Conference on Laser and Plasma Applications in Materials Science*, Vol. 1047, pp. 74–78, Algiers, Algeria, 23–26 June 2008.
13. Jansen, H. V., de Boer, M. J., and Unnikrishnan, S. 2008. Black silicon method: X. A review on high speed and selective plasma etching of silicon with profile control: An in-depth comparison between Bosch and cryostat DRIE processes as a roadmap to next generation equipment. *J. Micromech. Microeng.* 19, 033001 (41pp).
14. Park, J. B., Lim, W. S., Park, B. J., Park, I. H., Kim, Y. W., and Yeom, G. Y. 2009. Atomic layer etching of ultra-thin HfO_2 film for gate oxide in MOSFET devices. *J. Phys. D: Appl. Phys.* 42, 055202 (5pp).
15. Chabert, P., Lichtenburg, A. J., Lieberman, M. A., and Marakhtanov, A. 2001. Instabilities in low-pressure electronegative inductive discharges. *Plasma Sources Sci. Technol.* 10, 478.
16. Goodman, D. L. and Benjamin, N. M. P. 2003. Active control of instabilities for plasma processing with electronegative gases. *J. Phys. D: Appl. Phys.* 36, 2845.
17. Marakhtanov, A. M., Tuszewski, M., Leiberman, M. A. Lichtenburg, A. J., and Chabert, P. 2003. Stable and unstable behavior of inductively coupled electronegative discharges. *J. Vac. Sci. Technol. A* 21, 1849–1864.
18. Morishita, S., Goto, T., Nagase, M., and Ohmi, T. 2009. Precise and high-speed control of partial pressures of multiple gas species in plasma process chamber using pulse-controlled gas injection. *J. Vac. Sci. Technol. A* 27(6), 423–430.
19. Ogawa, T., Takahashi, Y., Yang, H., Kimura, K., Sakurai, M., and Takahashi, M. 2006. Fabrication of Fe_3O_4 nanoparticle arrays via patterned template assisted self-assembly. *Nanotechnology* 17, 5539–5543.
20. Titus, M. J., Nest, D. G., Chung, T.-Y., and Graves, D. 2009. Comparing 193 nm photoresist roughening in an inductively coupled plasma system and vacuum beam system. *J. Phys. D: Appl. Phys.* 42, 245205 (13pp).
21. Schulze, M., von Keudell, A., and Awakowicz, P. 2006. Characterization of a rotating nanoparticle cloud in an inductively coupled plasma. *Plasma Sources Sci. Technol.* 15, 556–563.
22. Oehrlein, G. S., Rembetski, J. F., and Payne, E. H. 1990. Study of sidewall passivation and microscopic silicon roughness phenomena in chlorine-based reactive ion etching of silicon trenches. *J. Vac. Sci. Technol. B* 8, 1199–1212.

Chapter 5

23. Welsh, S., Keswick, K., Stout, P., Lee, W. S., and Doan, K. 2009. Advanced DRAMs Drive High-AR Etch Advances. *Semi. International* February, 18–22.

24. Tillocher, T., Dussart, R., and Overzet, L. J. 2008. Two cryogenic processes involving SF_6, O_2, and SiF_4 for silicon deep etching. *J. Electrochem. Soc.* 155(3), D187–D191.

25. Wu, B., Kumar, A., and Pamarthy, S. 2010. High aspect ratio silicon etch: A review. *J. Appl. Phys.* 108, 051101.

6. Optical Lithography

Patrick Naulleau

Lawrence Berkeley National Laboratory, Berkeley, California

6.1 Background

Optical lithography is a photon-based technique comprised of projecting, or shadow casting, an image into a photosensitive emulsion (photoresist) coated onto the substrate of choice. Today, it is the most widely used lithography process in the manufacturing of nanoelectronics by the semiconductor industry, a $200 billion industry worldwide.

Optical lithography's ubiquitous use is a direct result of its highly parallel nature allowing vast amounts of information (i.e., patterns) to be transferred in a very short time. For example, considering the specification of a modern leading edge scanner (150–300 mm wafers per hour and 40 nm two-dimensional pattern resolution), the pixel throughput can be found to be approximately

Nanofabrication Handbook. Edited by Stefano Cabrini and Satoshi Kawata © 2012 CRC Press / Taylor & Francis Group, LLC. ISBN: 978-1-4200-9052-9

1.8T pixels per second. This capability has arguably enabled the computing revolution we have undergone over the past 50 years.

Within the realm of optical lithography, there exists a wide diversity of implementation both in wavelength and optical configuration. Wavelengths range from the traditional visible and ultraviolet ranges down to extreme ultraviolet (EUV) and even soft x-ray. Optical configurations range from the simplest case of direct shadow casting to complex multi-element refractive and/or reflective imaging systems. Additionally, diffractive systems can be used for applications such as interference and scanning probe lithography.

6.1.1 History

The earliest optical lithography tools used in the manufacturing of semiconductor devices were of the type

Chapter 6

classified as contact printers. In these systems, a mask is placed in direct contact with the photoresist-coated wafer and light is shined through the mask. Patterned areas on the mask served to block the light causing the negative of the pattern to be transferred to the wafer. The problem with the contact approach, however, was the rapid generation of defects on the mask, which are subsequently replicated in all exposures. The industry addressed this problem with the introduction of proximity lithography which is essentially the same as contact lithography but with a small air gap maintained between the surface of the mask and the wafer. This mitigated the defect problem but at the cost of resolution limitations arising from diffraction, or spreading of the light, upon propagation of the light through the free-space gap between the mask and wafer.

The free-space diffraction problem was eventually solved by introducing an imaging system between the mask and the wafer. The gap can now effectively be eliminated since the function of the imaging system is to replicate the electric field present in its object plane to its image plane. Any focus error in this optical system can be thought of simply as equivalent to the gap present in the proximity tool with the further benefit of allowing the gap to effectively become negative thereby expanding the acceptable gap or focus operating range. In addition to solving the proximity diffraction problem, using an imaging system enables demagnification from the mask to the wafer. This is beneficial since it greatly relaxes mask requirements both in terms of feature quality and defects. The demagnification cannot be made too large, however, since mask size would become an issue. Modern projection optical lithography tools use a demagnification of 4.

6.1.2 Today's Optical Lithography Tools

Projection lithography tools now come in two variations: step and repeat, and step and scan. In the step and repeat system (a stepper), the entire mask is illuminated and projected onto the wafer exposing one "die" (approximately 25×25 mm in size at the wafer). The light is then turned off and the wafer shifted (stepped) and the exposure process repeated. This cycle is continued until the entire wafer is exposed. In a step and scan system (a scanner), the imaging field size is reduced to a slit (typically on the order of 6×25 mm at the wafer) greatly facilitating the design and fabrication of the optical system. The mask and wafer stages are then scanned in opposite directions at proper speeds such that the entire mask pattern is replicated in one scan thereby creating an exposed die this time with a typical size of approximately 25×32 mm at the wafer. As with the stepper, the light is then turned off and the wafer shifted over to an unexposed region where the die scan process is repeated.

Since the advent of the scanner, further changes/improvements to the technology have come in the form of increases in numerical aperture, decreases in wavelength, and the introduction of immersion fluids between the projection optic and the wafer. One of the most significant developments currently underway is the reduction of the wavelength from 193 to 13.5 nm. This quantum leap in wavelength comes with many additional changes including high-vacuum operation and the requirement for all reflective components including both the optics and the mask. Reflective imaging systems, however, are not new to lithography; in fact many of the earliest systems were based on reflective optics due to their achromatic characteristics which was crucial before line-narrowed lasers were developed.

It is interesting to note that while contact lithography represents the dawn of the technology, one could argue that it has made a resurgence in the form of nanoimprint lithography. Nanoimprint uses direct contact between the mask and the wafer and for the case of "step and flash," light is shined though the mask to "cure" the resist. The difference, however, is that the light itself does not transport the pattern but rather simply cross-links the photoresist material. It is the mask that transports the pattern by physically displacing the photoresist in the patterned area before cross-linking. For this reason, unlike any of the optical lithography methods described so far, the illumination wavelength has no effect on the resolution of the process. Thus, although the nanoimprint process does use photons, we do not classify it as an optical lithography technique.

The topic of optical lithography is by far too vast to be covered in one small chapter. The goal here is simply to provide an introduction of the topic with the hope of making the reader aware of the various optical lithography options available, as well as to provide some basic understanding of the capabilities and limits of the technology. Since resolution is typically of paramount concern for nanofabrication, an attempt is made to provide a fundamental understanding of resolution limits and depth of focus in various optical systems. Next the issue of coherence is addressed, and again with particular focus on resolution and depth of focus. Finally, the future of optical lithography is explored, ending with a brief discussion of practical considerations for lab-based use. For much more detailed discussions of opti-cal lithography, the reader is referred to several exhaustive texts on the topic [1–3].

6.2 Resolution

Wavelength is the fundamental limiting factor in determining the resolution of optical lithography systems. However, wavelength alone does not provide the entire picture, also crucial to understanding the resolution limits in optical lithography systems is the concept of diffraction. Diffraction occurs as light is passed through a limiting aperture. Although beyond the scope of this chapter, the phenomenon of diffraction can be readily predicted using Maxwell's equations and heuristically explained using Huygen's Principle, which itself can be derived from Maxwell's equations [4,5]. Using these techniques, one can show that the diffraction half angle θ introduced to a plane wave of wavelength λ upon propagation through an aperture of width W is

$$\theta = \sin^{-1}(\lambda / W) \tag{6.1}$$

Figure 6.1 shows an example of the diffraction process where we assume a wavelength of 13.5 nm and a slit aperture size of 300 nm with propagation distances of up to 25 μm. The spread of the beam is clearly observed.

The aforementioned simple diffraction equation can be directly applied to predict the maximum allowable gap in a proximity lithography tool. Given a target resolution W, and a wavelength λ, and setting the maximum allowable diffraction blur to be equal to the target resolution, the required gap L can be written as

$$L = W \sqrt{\frac{W^2}{\lambda^2} - 1} \tag{6.2}$$

Given a target resolution of 50 nm and a wavelength of 13.5 nm, the gap would have to be smaller than 178 nm.

Although the aforementioned diffraction equation is also directly responsible for the resolution limit of an imaging lithography tool, the connection is less evident. In heuristically understanding the relationship between resolution and diffraction in this case, it is useful to think of the lens as a component that simply inverts the diffraction caused by the mask. Taking for granted that the propagation of light is reversible, to produce an image of an aperture of width W at some wavelength λ, we are required to generate a converging

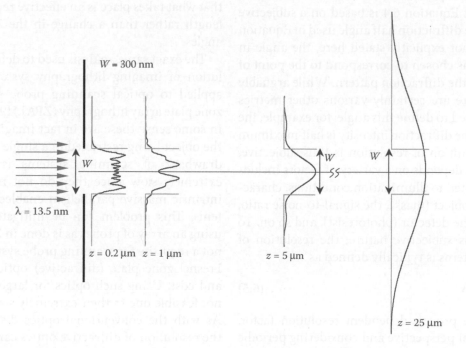

FIGURE 6.1 Example of the diffraction process where we assume a wavelength of 13.5 nm and a slit aperture size of 300 nm with propagation distances of up to 25 μm.

wave where the convergence angle is simply equal to the divergence angle that would be produced upon diffraction in the forward case. Thus the minimum image size a lens can produce depends on the range of input angles that can be inverted by the lens and the wavelength,

$$W = \lambda / \sin(\theta_c) \tag{6.3}$$

where θ_c is the maximum half angle accepted by the lens. The *sine* of the collection half angle (assuming a medium index of unity) is referred to as the numerical aperture (NA) of the lens allowing the minimum feature size equation to be rewritten as

$$W = \lambda / NA \tag{6.4}$$

It is important to note that the magnification has been assumed to be unity in the discussion here. Typical lithography optics are demagnifying and we are generally interested in the resolution on the image side (unlike microscopy systems), thus the NA of interest is the image-side NA rather than the acceptance NA. The two, however, are simply related by the magnification. For example, a so-called 4× lithography system, which demagnifies the object by a factor of 4 will have an image-side NA that is four times larger than the object-side NA.

Using the image-side NA, Equation 6.4 now can be thought of representing the resolution limit of the projection tool. The problem with this interpretation, however, is that Equation 6.4 is based on a subjective definition of the diffraction half angle used in Equation 6.1. Although not explicitly stated here, the angle in Equation 6.1 was chosen to correspond to the point of the first null in the diffraction pattern. While arguably reasonable, there are certainly various other metrics that could be used to define this angle: for example, the angle at which the diffraction intensity is half-maximum or $1/e$. The definition of resolution is thus subjective, and in practice, depends on a variety of factors including but not limited to illumination conditions, characteristics of the object (mask), the signal-to-noise ratio, capabilities of the detector (photoresist), and so on. To account for this subjective nature, the resolution of lithography systems is typically defined as

$$R = k_1 \lambda / NA \tag{6.5}$$

where k_1 is the process-dependent resolution factor. From the optical perspective and considering periodic structures, physics can be shown to set the lower limit of k_1 to 0.25. For isolated structures of relatively loose

pitch, however, the ultimate limitation of what can be printed depends more on the process than the optics. Thus, effective k_1 factors of smaller than 0.25 can be achieved. Examples of this are now routine in the integrated circuit (IC) industry, for example, 22 nm devices will be commercialized in the near future using 193 nm lithography. Assuming a numerical aperture of 1.35 (the highest currently available), this would correspond to a k_1 factor of 0.15.

Given the definition of NA explained here, one might ask how it is possible to achieve an NA that is greater than unity. The answer to this apparent dilemma is that we had assumed a medium index of unity. The complete definition of NA is in fact

$$NA = n \sin(\theta) \tag{6.6}$$

where n is the index of refraction of the medium between the lens and the image plane. This fact has long been used by microscopists in the form of oil-immersion lenses. The IC industry has recently adapted this technology to produce water-immersion lithography tools [6,7] with NAs of up to 1.35. In principle, even higher NA tools could be developed; however, materials issues have halted progress on that front [8]. Heuristically, it is instructive to consider the NA to be defined with an n of unity and instead note that the effective wavelength in the medium is λ/n, where λ is the vacuum wavelength. From this perspective, we see that what takes place is an effective reduction in wavelength rather than a change in the actual collection angles.

The exact same analysis used to determine the resolution of imaging lithography systems can also be applied to optical scanning probe systems such as zone plate array lithography (ZPAL) [9]. This is because in some sense these are in fact imaging systems with the object being restricted to a single point. The main drawback of scanning systems is that they are extremely slow since they do not make use of the intrinsic massive parallelism enabled by optical systems. This problem can be mitigated, however, by using an array of probes as is done in ZPAL. Although not a requisite for scanning probe systems, ZPAL uses Fresnel zone plate (diffractive) optics for simplicity and cost. Using such optics for large field systems is not feasible due to their extremely small field of view. As with the conventional optics discussed hitherto, the resolution of diffractive optics can also be characterized by the NA. Although beyond the scope of this chapter, it can also be shown that the resolution limit of a diffractive lens is simply defined by the size of the smallest zone width on the lens [10]. This should not be a surprise since the zone width will determine the converging diffraction angle.

Another important class of optical lithography tools, especially for lab use, is the interference tool, where two mutually coherent beams are combined at an angle (Figure 6.2). The mechanism used to create the two beams can vary and includes refractive, diffractive, and reflective methods. Ultimately, all that matters is the combining angle and the wavelength. Using Fourier optics [4], a plane wave traveling at some angle can be expressed by its spatial frequency

$$f_x = \sin(\theta)/\lambda \qquad (6.7)$$

As shown in Figure 6.2, the interfering frequency becomes the difference between the two or $2 f_x$. From the perspective of resolution, it is instructive to instead consider one-half the period of the interference term. The interference pattern period (T) and resolution can be written as

$$T = 1/(2f_x) = \frac{1}{2}\lambda/\sin(\theta) \qquad (6.8a)$$

$$R = \frac{1}{2}T = \frac{1}{4}\lambda/\sin(\theta) \qquad (6.8b)$$

Maximum resolution is achieved when the interfering beams travel in opposite directions ($\theta = 90°$), enabling the patterning of $\lambda/4$ features, not coincidentally matching the $k_1 = 0.25$ limit discussed here. In addition, immersion methods can be used to further push the resolution by reducing the effective wavelength [11].

In the aforementioned resolution discussions, the implicit assumption was made that the image is observed at the ideal focal plane. In practice, however, it may be difficult to maintain the wafer in that ideal

plane and even more fundamentally, the photoresist being imaged into will have some finite thickness. Consequently, the longitudinal distance over which the resolution of an optical system is preserved, or its depth of focus (DOF) becomes extremely important. In practical terms, the DOF can be defined as the longitudinal distance over which the change in size of a single image point is less than or equal to the minimum size of the image point as set by the diffraction limit discussed above.

We again begin by considering the simple proximity lithography case. Recall that when determining the maximum allowable gap, as described here, a similar criterion was used requiring the diffraction blur to be less than or equal to the actual feature size on the mask. Thus one can think of this maximum gap (Equation 6.2) as being equivalent to the DOF since mechanical constraints limit the minimum size of the gap to 0.

For projection systems (refractive, reflective, and/or diffractive), the DOF can be elucidated with simple geometry (Figure 6.3). Using geometric optics, the single-sided blur as a function of defocus (d) can be expressed as

$$\text{Blur} = d\,\text{NA} \qquad (6.9)$$

Setting the maximum allowable blur to be equal to the resolution limit of the system yields

$$\text{Blur}_{max} = d_{max}\text{NA} = \lambda/\text{NA} \qquad (6.10)$$

Now solving for d_{max} yields

$$d_{max} = \lambda/\text{NA}^2 \qquad (6.11)$$

Again applying a process-dependent factor, the DOF becomes

$$\text{DOF} = k_2\,\lambda/\text{NA}^2 \qquad (6.12)$$

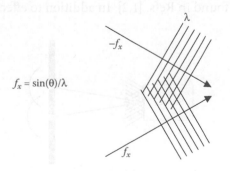

FIGURE 6.2 Schematic of a lithographic interference tool where two mutually coherent beams are combined at an angle.

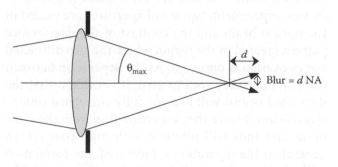

FIGURE 6.3 Schematic describing the geometry leading to depth of focus limits in optical systems.

Chapter 6

where k_2, similar to k_1, is a constant representative of the lithographic process conditions. A typical value for k_2 for a conventional process is 0.5.

Turning to interference tools (Figure 6.2), and again assuming full spatial coherence (coherence issues will be discussed in more detail in Section 6.3), it is evident that the generated interference will be independent of longitudinal position. Thus, in an ideal interference tool the DOF is effectively infinite. In practice, however, the DOF in the perfect coherence case will be limited by the finite overlapping footprint of the two interfering beams. Since the two beams are crossing each other as shown in Figure 6.2, this overlapping footprint will be maximized at only one longitudinal point and will decrease linearly from that point in either direction. This constraint, however, is certainly not very restrictive and can be represented mathematically as

$$DOF = D / [2 \tan(\theta)] \qquad (6.13)$$

where D is the beam diameter and 2θ is the angle between the two interfering beams.

In conclusion, it is noted that alternative and certainly more complete discussions of resolution and depth of focus can be found in the literature [4,5,12].

6.3   Coherence

Although not explicitly addressed in the discussions so far, illumination coherence plays an important role in the achievable resolution/DOF of optical systems. However, by and large, modern day steppers employ extremely narrowband sources allowing temporal coherence effects to be ignored leaving only spatial coherence. As discussed later, the exception to this are diffractive methods in which case temporal coherence can play an important role. Detailed discussions of coherence theory in general can be found in the literature [5,13]. In the parlance of lithography, coherence is almost universally described in terms of the partial coherence factor or σ. Most fundamentally, σ can be thought of as the ratio of the diffraction-limited resolution to the coherence width. When the coherence width is larger than the resolution limit, σ is <1. In most practical cases, σ is <1 and typically falls in the range of 0.2–0.8, with 0.2 being close to coherent and 0.8 being close to incoherent.

Implicit in the definition discussed earlier is the assumption that the concept of coherence width is understood, thus it is quickly reviewed here. In a practical sense, coherence width is best explained by recalling Young's double-slit experiment (Figure 6.4). In this experiment, two small apertures are placed in the optical beam and the contrast of the interference pattern created in the region where the two diffracted waves overlap is observed. As the separation between the two apertures goes to zero, it is evident that the diffracted beams will become fully correlated (mutually coherent) since they are emanating from the same point and thus will interfere with high contrast. In general, as the separation is increased, the correlation will decrease and so will the interference contrast. The coherence width can be defined as the separation between the two apertures where the interference contrast drops to 50%.

A basic understanding of the effect of coherence on imaging performance is perhaps best achieved from the perspective of the optical system transfer function [4]. Given a unity contrast sinusoidal object, the transfer function describes the contrast of the resulting sinusoidal image for all possible frequencies. For illustrative purposes, Figure 6.5 shows the transfer function of an ideal projection system for three difference values of σ: 0.1, 0.5, and 1. The frequency axis in Figure 6.5 is normalized to λ/NA, which represents the coherent cutoff of the system. The plot shows that, in terms of ultimate resolution capabilities, larger values of σ are preferable, but they come at the cost of performance at more moderate feature sizes. It is important to note that these results assume on axis illumination and a pure amplitude mask; letting these variables float can significantly change the results allowing the imaging performance to be optimized for specific feature types. These technologies are generally referred to as resolution enhancement techniques. Although beyond the scope of this chapter, detailed information on this topic can be found in Refs. [1,2]. In addition to effecting res-

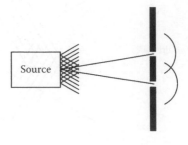

FIGURE 6.4   Schematic depiction of Young's double-slit experiment.

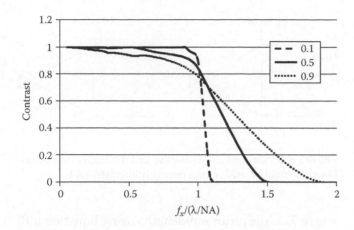

FIGURE 6.5 Transfer function of an ideal projection system for three difference values of σ: 0.1, 0.5, and 1. The frequency axis is normalized to λ/NA, which represents the coherent cutoff of the system.

olution, these various parameters also have significant impact on DOF. It is important to note that these techniques do not change the fundamental resolution limits set by λ and NA but rather provide a mechanism for reducing k_1 and/or k_2.

As mentioned in the previous section, coherence plays an important role in the determination of DOF for interference lithography systems. In that scenario, the concept of partial coherence factor is less useful. More insight can be gained by instead using the unnormalized coherence width W_c. The requirement for DOF now becomes that the shear (lateral displacement) between the two interfering waves be smaller than W_c. Determining the shear depends on the type of interference tool being used, wavefront or amplitude division. Examples of the two different types (both based on gratings) are shown in Figure 6.6. In wavefront division, very large coherence width is required since by design, the two interfering beams are extracted from different lateral locations in the incident wavefront. The coherence width is required to be significantly larger than the total printed width and thus W_c plays little role in the DOF. Rather the DOF is determined by the coherent equation, that is, Equation 6.13.

Amplitude division, on the other hand, can be applied in cases where there is significantly less coherence because full copies of the input wavefront are created and then recombined allowing zero shear to be obtained at the cross-over point. As the beams propagate away from the cross-over point, a shear is introduced that is directly proportional to the interference angle. In such a system, the DOF can be shown to be

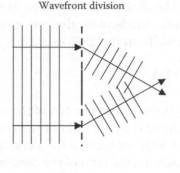

Wavefront division

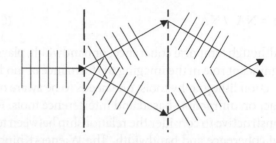

Amplitude division

FIGURE 6.6 Schematics depicting the distinctions between wavefront and amplitude division interference tools.

$$\text{DOF} = W_c / [2 \tan(\theta)] \tag{6.14}$$

where 2θ is again the angle between the two interfering beams.

Finally, the lateral coherence (W_c) itself is considered. Although a full discussion of this vast topic is certainly beyond the scope of this chapter, valuable insight can be gained by considering a simple limiting case. The simple case we consider is a fully incoherent source that is re-imaged to the entrance plane of our system, be it the mask in a projection lithography tool or the beamsplitter in an interference tool. The optic used to re-image the source is often referred to as the "condenser" and such an illumination system is commonly referred to as "critical" [13] (Figure 6.7). In this case, the coherence width is simply determined by the resolution of the condenser lens. Heuristically, this can be explained by noting that source variations, which lead to incoherence, that are

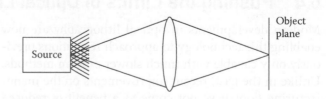

FIGURE 6.7 Schematic of a "critical" illumination system.

finer than the condenser resolution cannot be reproduced. Thus, mathematically we can express the coherence width in this case as

$$W_c = \lambda / NA_c \tag{6.15}$$

where NA_c is the image-side NA of the condenser lens. Having defined σ as the ratio of the resolution to the coherence width, we see that for a critical illuminator, σ simply becomes the ratio of the condenser NA to the object-side imaging NA.

$$\sigma \approx NA_c / NA_o \tag{6.16}$$

Although temporal coherence, or bandwidth, plays an insignificant role in the imaging performance of modern projection lithography tools, one needs to be aware of its impact on diffraction-based and interference tools. First it is instructive to consider the relationship between temporal coherence and bandwidth. The Wiener–Khinchin theorem [13] teaches us that there exists a Fourier transform relationship between the two. Thus, to first order, the coherence time becomes the inverse of the bandwidth, making short temporal coherence equivalent to large bandwidth. This is important because the effect of bandwidth on a system is much easier to visualize than the effect of temporal coherence.

Beginning with the diffraction-based zoneplate system such as ZPAL, the highly chromatic nature of the Fresnel zoneplate makes bandwidth a concern. The focal length of a zoneplate can be shown to be [10]

$$f \approx D\, \Delta r / \lambda \tag{6.17}$$

where D is the diameter and Δr is the outer zone width. The focal length is inversely proportional to the wavelength, thus resolution will be adversely affected by increased bandwidth. From Equation 6.17 the change in focal length as a function of bandwidth, $\Delta\lambda$ can be written as

$$\Delta f = \Delta\lambda\, D\, \Delta r / \overline{\lambda}^2 \tag{6.18}$$

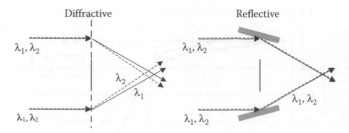

FIGURE 6.8 Schematics of reflective and diffractive beam combiners as could be used in interference lithography tools.

where $\overline{\lambda}$ is the mean wavelength. Using Equation 6.10 to determine the resolution based on a defocus of $1/2\Delta f$ (the single-sided focal change), the resolution can now be expressed as

$$R = \frac{\Delta\lambda}{\overline{\lambda}} \frac{D}{4} \tag{6.19}$$

Conversely, when considering interference tools, diffractive properties are in fact beneficial in terms of tolerance to bandwidth. This can be explained with the help of Figure 6.8 showing both reflective and diffractive beam combiner cases. For simplicity, wavefront division and spatial coherence are assumed. In the diffractive case, the longer wavelength (λ_2) is bent more. The frequency of the interference fringes produced will then depend on the angle of interference and the wavelength. It can be shown that the increased diffraction angle is exactly balanced by the increased wavelength to produce the identical frequency fringes as produced by λ_1. Thus, the system can produce high-quality fringes with broadband light. In the reflective system, on the other hand, the interference angles are identical for both wavelengths owing to the achromatic nature of reflection. This causes the two different wavelengths to produce fringes of two different frequencies thereby in sum generating poor-quality fringes in the presence of broadband light. The tolerable bandwidth depends on the lateral range over which one desires to produce high-contrast fringes [13].

6.4   Pushing the Limits of Optical Lithography

Modern developments in optical lithography are now enabling this technology to approach resolutions previously only capable with much slower e-beam methods. Unlike in the past, recent improvements on the manufacturing floor have not come as a benefit of reduced wavelength. The last attempt to change wavelength was the shift to 157 nm which was abandoned due to the birefringence problem. Currently, leading edge industry is still operating at a wavelength of 193 nm but as described in the resolution section the numerical aperture has been increased to beyond unity using an immersion process akin to oil-immersion microscopy.

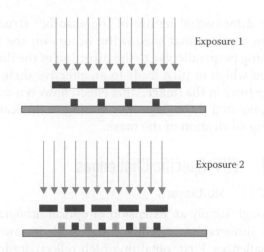

FIGURE 6.9 Schematic depicting a double-patterning process.

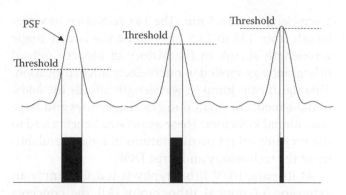

FIGURE 6.10 Schematic showing how dose can be used to shrink feature size at a fixed pitch apparently circumventing the λ/NA limit, but only for isolated features.

An alternative of the immersion process is to keep the numerical aperture fixed and define the effective wavelength as the vacuum wavelength divided by the index of the medium. At a wavelength of 193 nm, the refractive index of water is $n_w = 1.437$, thus water immersion can be used to operate at an effective wavelength of 134 nm. This wavelength corresponds to a single exposure periodic structure resolution limit of 33.5 nm.

The key qualifiers in the resolution limit just mentioned are "single exposure" and "periodic." Industry has recently been pushing beyond even this limit using a concept that can simply be described as interlacing. Referred to by the industry as "double patterning," these techniques rely on printing at a looser pitch and then shifting and printing again, thereby decreasing the effective pitch as shown in Figure 6.9. The process works because optical systems fundamentally limit pitch and not isolated feature size. The "size" of an isolated feature after going through a threshold process (which ideally is what a photoresist does) is determined by the threshold level and can become

arbitrarily small, ultimately being limited by noise and process control. Traditional resolution descriptions of isolated features are determined by the width of the point-spread function (PSF) which is customarily defined as the full-width at half-maximum. Thus the definition assumed a threshold value of 50%, but as we allow the threshold to get higher and higher, the feature width becomes even smaller (Figure 6.10). Although it is fine for isolated features, such a process cannot arbitrarily shrink the minimum resolvable separation between two features, thus the need for double patterning.

The aforementioned simplified description represents only one of the many potential double-patterning approaches including an approach called "spacer" [14] requiring only a single exposure and relying on processing to double the pattern density. In principle, these methods can be extended beyond double patterning to triple and even more to further push down feature sizes. However, even double patterning comes at a prohibitive cost compared to single exposure methods raising concerns about the commercial viability of extending these techniques to triple and more.

6.5 Taking Optical Lithography to the Extreme

The escalating costs and difficulties associated with subquarter-wavelength lithography using double patterning and beyond provide great impetus for the return of a viable single-patterning optical technique. As shown in Section 6.2, the only way to drive the optical resolution down in terms of half pitch is to increase the NA and/or decrease the wavelength. As mentioned here, increases in NA have been pushed to the limit and are now constrained by the availability of adequate high-index materials. Materials limitations in terms of

suitable high-quality low birefringence glass have also put an end to wavelength reduction, at least for refractive systems. For wavelength reduction, however, one can turn to reflective optics to get around the materials issues, but reflective solutions come with restrictions in NA due to geometric limitations. The NA restrictions, however, are readily overcome through even greater reductions in wavelength.

This train of thought has led to the development of extreme ultraviolet (EUV) lithography which relies on a wavelength of 13.5 nm. The 14× reduction in wavelength from 193 to 13.5 nm will be the largest single wavelength shrink in the history of modern optical lithography as applied to microelectronics fabrication. This significant jump in wavelength affords the additional benefits of small numerical apertures and large operational k_1 factors. These associated benefits lead to the very important manifestations of long extendability of the technology and large DOF.

At its core, EUV lithography is indeed simply an extension of optical lithography. All the concepts presented up to this point in this chapter remain directly applicable to EUV. The difference in implementation is simply that we now rely on reflective components for both the mask and the imaging optics instead of the refractive components. Figure 6.11 schematically depicts the reflective EUV configuration in a more simplified implementation. It should be noted that the use of reflective optics is not a fundamental concern since such optics have long been utilized in the ultraviolet and deep ultraviolet lithography regime in the form of catadioptric exposure lenses [15–18]. In addition to the obvious change to all reflective components, geometry requires the mask to be tilted and the light to come in at an angle compared to the on-axis condition for the transmission case. The angle of the mask is readily compensated for by also tilting the wafer at an angle that is scaled down by the magnification of the optical system. A more subtle impact of the mask tilt and the

three-dimensional nature of the absorber structure on the mask is that shadowing occurs on the lines running perpendicular to the direction of the illumination which in turn leads to an effective shrinking of the lines in the image. This effect, however, can be compensated through proper biasing of the features during fabrication of the mask.

6.5.1 EUV-Specific Challenges

6.5.1.1 Multilayers

Although simply an extension of optical lithography, EUV lithography certainly comes with its own set of challenges. First, obtaining high reflectivity in the EUV wavelength regime is not trivial. The development of EUV lithography has been enabled by the invention of high reflectivity near-normal incidence Bragg coatings [19] allowing high NA EUV optics with reasonable throughputs to be fabricated. One such coating is the molybdenum–silicon multilayer providing peak reflectivity near 13.5 nm. It is typically comprised of 40 bilayers with a bilayer thickness of approximately 7 nm. This coating now serves as the basis for all EUV lithography optics. Although tremendous improvements have been made in this area allowing near-theoretical limit reflectivities to be achieved on a routine basis, the theoretical limit is only 70%. This relatively low reflectivity places significant constraints on the number of mirrors that can be used in both the illuminator used to transport and shape the light from the source to the mask as well as in the imaging optic itself. Current manufacturing class EUV tools operating at 0.25 NA utilize six mirrors in the imaging system and approximately four in the illuminator. Further adding the mask, a typical tool might have 11 multilayer surfaces which in the ideal case would give you a total reflectivity of 2%. If reduction in bandwidth is also considered, one can find the effective throughput of the tool to shrink even faster as mirrors are added.

6.5.1.2 Source

In addition to putting constraints on the optical design, the throughput issue places higher demands on source power and source power has long been viewed as the biggest challenge facing EUV. Current lithography tools use high-power excimer lasers, however, direct scaling of laser technology to EUV wavelengths is not feasible from the perspective of use in exposure tools. Two major classes of sources are under development for use in production scale EUV

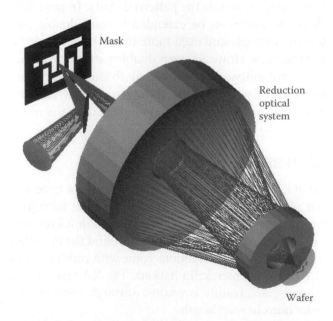

FIGURE 6.11 Schematic depiction the reflective lithography configuration required at EUV wavelengths.

exposure tools: laser-produced [20,21] and discharge-produced [22] plasmas. Not only must EUV sources produce high power, but they must do it in a clean manner making debris mitigation another crucial issue for the source. High-power EUV sources typically generate large amounts of energetic debris which, if allowed to strike mirrors, will quickly deteriorate them. Various methods exist for controlling debris [23–28]. It is important to note that debris mitigation typically comes at the cost of optical throughput and thus is an important consideration in tool design. Despite these challenges, tremendous improvements have been made on both types of sources allowing good results to be obtained from alpha tools and supporting pilot tool operation in the near future.

6.5.1.3 Vacuum

Because EUV light is strongly absorbed by all materials, including atmosphere, EUV systems operate under high vacuum. Moreover, the purity of the vacuum is also crucial due to the fact that the high energy of the EUV photons has the ability to dissociate molecules leading to contamination of optical surfaces and subsequent loss of reflectivity. One common potential source of contamination is residual hydrocarbon in the vacuum. Hydrocarbons dissociated by the EUV radiation are highly reactive and lead to carbon growth on the multilayer surfaces. A 1% loss in reflectivity requires only 0.8 nm of carbon growth. For a system with a total of nine multilayer reflections, such contamination would lead to a throughput loss of 10%. Carbon contamination, however, is generally accepted to be a reversible process. Oxidation of multilayers is another potential problem. This can occur when oxygen-containing molecules, such as water, are split into radicals by means of photoemission leading to oxidation of the multilayer surface and again reflectivity loss. Oxidation is of greater concern than carbon growth because its reversal is considerably more complicated [29].

6.5.1.4 Mask Defects

As noted earlier, in EUV lithography, the mask must also be reflective. The most daunting challenge this reflective architecture poses to the mask is the possibility of defects embedded underneath or within the multilayer stack. If these embedded defects lead to surface indentations on the order of 3 nm or even smaller, they will act as strong phase-shifting defects with potentially considerable impact on the printed image. The detection, mitigation, and repair of such defects are crucial engineering challenges facing the commercialization of EUV lithography.

6.5.1.5 Flare

Another issue particularly relevant to EUV is flare. Flare in optical systems in general is simply scattered light leading to a DC background in the image and subsequent loss of contrast. The only major contributor to flare in EUV systems is projection optics roughness and the resulting scatter [30]. EUV lithography's short wavelength renders it very vulnerable to surface roughness. Although atomic level long-range accuracy as required for wavefront control is now readily achieved [31], such precision cannot be obtained at the shorter spatial scales relevant to flare. Early EUV production tools are expected to have flare levels on the order of 10% with potentially significant variations across the field. The resulting CD variation, however, can be controlled using mask-based flare compensation techniques [32–33].

6.5.2 From the Lab to the Factory

EUV lithography has been in development since the mid-1980s but has recently entered the early manufacturing integration study phase [34–36]. Currently operational are full-field "alpha" tools with NAs of 0.25 [37–39] and microfield exposure tools with NAs of 0.3 [40–43]. Assuming a k_1 factor of 0.4, the corresponding half-pitch resolution limits are 22 and 18 nm for these two sets of tools, respectively. Figure 6.12 shows recent results from the SEMATECH Berkeley Microfield Exposure Tool used for advanced development of resists, processes, and masks.

The next phase in commercialization will be the deployment of so-called "pilot line" tools used to establish high-volume manufacturing processes. These tools are currently (2010) under development and slated to be delivered to several key integrated circuit manufacturers [44]. As with the alpha tools, the pilot tools will also have an NA of 0.25, however, the first production tools are expected to have NAs of approximately 0.32 being suitable for half pitches down to 16 nm. As with conventional optical lithography systems in the past, extension of EUV to even finer resolutions is expected with increasing NA and decreasing wavelength. Designs with NAs of up to 0.7 have been presented [45] as well as multilayers with reflectivities of 41% at 6.8 nm wavelength [46]. Using these parameters and an assumed k_1 limit of 0.25, EUV would be extendable to a half-pitch resolution of 2.4 nm.

Chapter 6

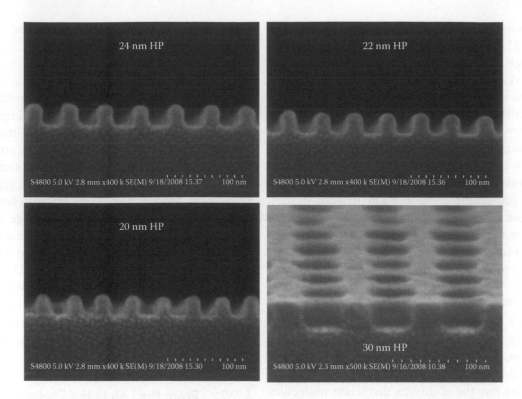

FIGURE 6.12 Recent results from the SEMATECH Berkeley Microfield Exposure Tool showing patterning capability down to 20 nm lines and spaces and 30 nm contacts.

6.6 Practical Considerations for Lab-Based Nanofabrication

While arguably the most widely used and highest throughput lithography technique for patterning of nanostructures, leading edge optical lithography is not particularly well suited for low-volume lab-based applications. This is due to the extremely high capital cost of leading edge tools and masks. Nevertheless, optical lithography can play an important role in this regime. For example, older generation depreciated and/or contact lithography tools could be used in conjunction with slower high-resolution methods such as e-beam to pattern large area "support" structures, alleviating write time burden on the e-beam tool.

In terms of actually using optical lithography for very fine patterns, if access to projection tools is available, one must also be aware of mask requirements both in terms of lead time and costs. If needs are limited to strictly periodic structures, either lines or contacts, interference lithography is an excellent option. Interference lithography can also be combined with other lithography techniques, optical or otherwise, to customize the periodic pattern for the generation of more complex structures.

References

1. H. J. Levinson, *Principle of Lithography*, Second Edition, SPIE Press, Bellingham, Washington, 2005.
2. C. Mack, *Fundamental Principles of Optical Lithography: The Science of Microfabrication*, Wiley-Interscience, New York, 2007.
3. K. Suzuki and B. W. Smith, *Microlithography: Science and Technology*, Second Edition, CRC Press, Boca Raton, Florida, 2007.
4. J. W. Goodman, *Introduction to Fourier Optics*, Second Edition, McGraw-Hill, New York, 1996.
5. M. Born and E. Wolf, *Principles of Optics: Electromagnetic Theory of Propagation, Interference and Diffraction of Light*, Seventh Edition, Cambridge University Press, Cambridge, New York, 1999.
6. H. Kawata, J. M. Carter, A. Yen, and H. I. Smith, Optical projection lithography using lenses with numerical apertures greater than unity, *Microelectron. Eng.* 9, 31–36, 1989.
7. S. Owa and H. Nagasaka, Immersion lithography; its potential performance and issues, *Proc. SPIE* 5040, 724, 2003.

8. T. Miyamatsu, Y. Wang, M. Shima, S. Kusumoto, T. Chiba, H. Nakagawa, K. Hieda, and T. Shimokawa, Material design for immersion lithography with high refractive index fluid (HIF), *Proc. SPIE* 5753, 10, 2005.

9. D. Carter, D. Gil, R. Menon, I. Djomehri, and H. Smith, Zoneplate array lithography (ZPAL): A new maskless approach, *Proc. SPIE*, 3676, 324, 1999.

10. D. Attwood, *Soft X-Ray and Extreme Ultraviolet Radiation: Principle and Applications*, Cambridge University Press, New York, 1999.

11. J. A. Hoffnagle, W. D. Hinsberg, M. Sanchez, and F. A. Houle, Liquid immersion deep-ultraviolet interferometric lithography, *J. Vac. Sci. Technol.* B 17, 3306–3309, 1999.

12. F. Jenkins and H. White, *Fundamentals of Optics*, Fourth Edition, McGraw-Hill, New York, 1976.

13. J. W. Goodman, *Statistical Optics*, Wiley-Interscience, New York, 1985.

14. Y-K. Choi and T-J. King, A spacer patterning technology for nanoscale CMOS, *IEEE Trans. Electron. Dev.* 49, 436–441, 2002.

15. Y. Ohmura, M. Nakagawa, T. Matsuyama, and Y. Shibazaki, Catadioptric lens development for DUV and VUV projection optics, *Proc. SPIE* 5040, 781–788, 2003.

16. D. Williamson, Catadioptric optical reduction system with high numerical aperture, US Patent #5537260, 1996.

17. C. Rim, Y. Cho, H. Kong, and S. Lee, Four-mirror imaging system (magnification +1/5) for ArF excimer laser lithography, *Opt. and Quant. Electron.* 27, 319–325, 1994.

18. Y. Zhang, D. Lu, H. Zou, and Z. Wang, Excimer laser photolithography with a 1:1 broadband catadioptric optics, *Proc. SPIE* 1463, 456–463, 1991.

19. J. H. Underwood and T. W. Barbee, Jr., Layered synthetic microstructures as Bragg diffractors for x-rays and extreme ultraviolet: Theory and predicted performance, *Appl. Opt.* 20, 3027–3034, 1981.

20. D. Brandt et al., LPP source system development for HVM, *Proc. SPIE* 7271, 727103–727103-10, 2009.

21. A. Endo et al., Laser-produced plasma source development for EUV lithography, *Proc. SPIE* 7271, 727108–727108-7, 2009.

22. M. Yoshioka et al., Xenon DPP source technologies for EUVL exposure tools, *Proc. SPIE* 7271, 727109–727109-8, 2009.

23. S. Harilal, B. O'Shay, and M. Tillack, Debris mitigation in a laser-produced tin plume using a magnetic field, *J. Appl. Phys.* 98, 036102, 2005.

24. S. Fujioka et al., Properties of ion debris emitted from laser-produced mass-limited tin plasmas for extreme ultraviolet light source applications, *Appl. Phys. Lett.* 87, 241503, 2005.

25. S. Namba, S. Fujioka, H. Nishimura, Y. Yasuda, K. Nagai, N. Miyanaga, Y. Izawa, K. Mima, and K. Takiyama, Spectroscopic study of debris mitigation with minimum-mass Sn laser plasma for extreme ultraviolet lithography, *Appl. Phys. Lett.* 88, 171503, 2006.

26. E. Vargas López, B. Jurczyk, M. Jaworski, M. Neumann, and D. Ruzic, Origins of debris and mitigation through a secondary RF plasma system for discharge-produced EUV sources, *Microelectron. Eng.* 77, 95–102, 2005.

27. E. Antonsen, K. Thompson, M. Hendricks, D. Alman, B. Jurczyk, and D. Ruzic, Ion debris characterization from a z-pinch extreme ultraviolet light source, *J. Appl. Phys.* 99, 063301, 2006.

28. V. Sizyuk, A. Hassanein, and V. Bakshi, Modeling and optimization of debris mitigation systems for laser and discharge-produced plasma in extreme ultraviolet lithography devices, *J. Micro/Nanolithography, MEMS and MOEMS* 6, 043003, 2007.

29. H. Oizumi, A. Izumi, K. Motai, I. Nishiyama, and A. Namiki, Atomic hydrogen cleaning of surface Ru Oxide formed by extreme ultraviolet irradiation of Ru-capped multilayer mirrors in H_2O ambience, *Jap. J. Appl. Phys.* 46, L633–L635, 2007.

30. J. M. Bennett and L. Mattson, *Introduction to Surface Roughness and Scattering*, Optical Society of America, Washington, DC, 1989.

31. T. Miura, K. Murakami, H. Kawai, Y. Kohama, K. Morita, K. Hada, and Y. Ohkubo, Nikon EUVL development progress update, *Proc. SPIE* 7271, 72711X, 2009.

32. D. Tichenor, D. Stearns, J. Bjorkholm, E. Gullikson, and S. Hector, Compensation of Flare-Induced CD Changes EUVL, U.S. Patent #6815129, 11/9/2004.

33. F. Schellenberg, J. Word, and O. Toublan, Layout compensation for EUV flare, *Proc. SPIE* 5751, 320–329, 2005.

34. H. Meiling et al., First performance results of the ASML alpha demo tool, *Proc. SPIE* 6151, 615108-1–12, 2006.

35. O. Wood et al., Integration of EUV lithography in the fabrication of 22-nm node devices, *Proc. SPIE* 7271, 727104-727104-10, 2009.

36. J. Park et al., The application of EUV lithography for 40 nm node DRAM device and beyond, *Proc. SPIE* 7271, 727114-727114-8, 2009.

37. G. Vandentop et al., Demonstration of full-field patterning of 32 nm test chips using EUVL, *Proc. SPIE* 7271, 727116-727116-9, 2009.

38. H. Meiling et al., Performance of the Full Field EUV Systems, *Proc. SPIE* 6921, 69210L-69210L-13, 2008.

39. B. LaFontaine et al., The use of EUV lithography to produce demonstration devices, *Proc. SPIE* 6921, 69210P–69210P-10, 2008.

40. P. Naulleau et al., Status of EUV micro-exposure capabilities at the ALS using the 0.3-NA MET optic, *Proc. SPIE* 5374, 881–891, 2004.

41. A. Brunton et al., High-resolution EUV imaging tools for resist exposure and aerial image monitoring, *Proc. SPIE* 5751, 78–89, 2005.

42. P. Naulleau, C. Anderson, K. Dean, P. Denham, K. Goldberg, B. Hoef, B. La Fontaine, and T. Wallow, Recent results from the Berkeley 0.3-NA EUV microfield exposure tool, *Proc. SPIE* 6517, 65170V, 2007.

43. N. Nishimura, G. Takahashi, T. Tsuji, H. Morishima, and S. Uzawa, Study of system performance in SFET, *Proc. SPIE* 6921, 2008.

44. H. Meiling et al., EUVL system: Moving towards production, *Proc. SPIE* 7271, 727102-727102–15, 2009.

45. J. Benschop, ASML EUV Program: Status and Prospects, 2009 International Symposium on Extreme Ultraviolet Lithography, Prague, Czech Republic, October 18–October 23, 2009, *Proceedings Available from SEMATECH*, Austin, TX.

46. T. Tsarfati, E. Zoethout, E. Louis, R. van de Kruijs, A. Yakshin, S. Müllender, and F. Bijkerk, Improved contrast and reflectivity of multilayer reflective optics for wavelengths beyond the extreme UV, *Proc. SPIE* 7271, 72713V, 2009.

Chapter 6

7. Soft X-Ray Lithography

Herbert O. Moser

Singapore Synchrotron Light Source, National University of Singapore, Singapore
Department of Physics, National University of Singapore, Singapore

Linke Jian

Singapore Synchrotron Light Source, National University of Singapore, Singapore

Nanofabrication Handbook. Edited by Stefano Cabrini and Satoshi Kawata © 2012 CRC Press / Taylor & Francis Group, LLC. ISBN: 978-1-4200-9052-9

Chapter 7

7.1   Introduction

The significance of lithography in manufacturing, in contrast to primary pattern generation, stems from its giving access to a fast repeatable coverage of large areas with wafer sizes up to 30 cm. Using synchrotron radiation, the lateral extension of the sample can be bigger than the largest present-day wafers up into the meter range (Hahn et al. 2010; Utsumi and Kishimoto 2005). Processing many structures on the wafer in parallel, lithography is faster and cheaper than serial primary pattern generation.

From the 1970s, soft x-ray lithography (SXRL) was considered a potential candidate for semiconductor manufacturing should the industry need processes suitable to produce finer details than photolithography at that time. Originally, for microelectronics, resist layers could be as thin as possible to maintain safe processing in etching steps. An important extension was the consideration of high aspect ratio and tall resist structures when SXRL and deep x-ray lithography (DXRL) were applied to other manufacturing fields than microelectronics (Becker et al. 1986). The spectral range of SXRL is situated between hard x-rays and ultraviolet (UV). SXRL is similar to hard x-ray lithography with differences such as a possible exploitation of C and O K edge resonance or the necessity to perform it under vacuum.

The current development is characterized by the absence of a strong driver like the microelectronics community previously. However, research labs at universities and synchrotron light sources are pursuing work on devices and processes. The major ones are ANKA, BESSY, CAMD, CNTech, LASTI, LILIT, MIT, NSRL/USTC, and SSLS.

7.1.1   History

X-ray lithography took off with Feder (1970) writing an IBM Technical Report on x-ray projection printing of electrical circuits. Spears and Smith (1972) described an SXRL technique for contactless replication of sub-μm linewidth planar-device patterns that used specially developed soft x-ray masks for the 4–14 Å wavelength range. Elastic-surface-wave-transducer patterns with 1.3 μm electrode spacings were written onto such masks by electron-beam techniques and successfully replicated. Feder et al. (1975) demonstrated the replication of 0.1 μm geometries, and synchrotron radiation came on stage when Spiller et al. (1976) performed x-ray lithography at DESY (http://hasylab.desy.de). Using soft x-rays with $\lambda > 10$ Å, they reached 50 nm spatial resolution.

McGowan et al. (1979) used the technique for micro(chemical) imaging of biological cells in polymethylmethacrylate (PMMA) with a resolution down to 10 nm. A marked difference in the relief replica in PMMA resulting from the differential absorption by the dried cells of carbon K_α radiation at 4.48 nm just below the carbon K edge and broad band synchrotron radiation with $\lambda > 1.5$ nm demonstrated the high-resolution chemical identification of the cell constituents. The biological sample served as the mask itself to be replicated into PMMA and was the first example of exploiting resonance absorption at the carbon K edge.

In the mid-1980s, the technological exploitation for semiconductor lithography set in, eventually leading to SXRL's heydays when large companies pushed the development including IBM (Wilson 1993), Mitsubishi Electrical Corporation (MELCO) (Kitayama and Itoga 2002; Nakanishi et al. 1995), Sumitomo Heavy Industries (SHI) (Takahashi 1987, 1990), Sumitomo Electrical Industries (SEI) (Tomimasu et al. 1985; Tomimasu 1987), and Oxford Instruments (Wilson et al. 1990, 1993; Wood 2001). Monographs and conference proceedings may be found in Thompson et al. (1994), Valiev (1992), Madou (1997), Gentili et al. (1994), and Suzuki and Smith (2007).

As (deep) UV lithography (DUVL) continued to meet the equipment development requirements set by semiconductor industry under Moore's law, there was no point of insertion of SXRL into the chip manufacturing process and the development efforts weakened. Eventually, IBM, MELCO, and others ceased pursuing the industrial use of SXRL following commercial considerations as DUVL was making steady progress. Eventually, MELCO, CNTech, SHI, and others promoted a second-generation x-ray lithography (PXL-II) in the early 2000s (Feldman et al. 2001; Khan et al. 2001a,b; Kitayama and Itoga 2002; Marumoto et al. 2003; Toyota et al. 2001; Watanabe et al. 2002), but activities waned around 2005.

Initially, SXRL was a proximity lithography where the mask and the substrate are separated by a small gap and the patterns are transferred via shadow casting. However, in the second half of the 1980s, researchers developed the projection lithography aspect so successfully that it led to the massive effort toward extreme ultraviolet lithography (EUVL) represented by the EUV Limited Liability Co. Therefore, EUVL is treated separately in this handbook and elsewhere (Bakshi 2009). For proximity SXRL, the present development is mostly driven by device fabrication in universities and research labs. With growing nanoscience and technology efforts, part of the nanofabrication community is pursuing the use of SXRL for nanomanufacturing.

In the following, we shall review the field of SXRL. Physical principles of the whole process chain are outlined, and technological embodiments and processes described.

7.1.2 "Soft" Definition of SXRL

SXRL may be defined as a *proximity lithography using photons in the soft x-ray range* shifting the open question to defining the soft x-ray range. Another statement, disguised within EUVL, can be found in Mack (2006) saying "EUV lithography: lithography using light of a wavelength in the range of about 5 to 50 nm, with about 13 nm being the most common. Also called soft x-ray lithography" (p. 99). From the energy–wavelength product for electromagnetic waves, 1239.9 nm eV, the corresponding photon energies are 247.9, 24.79, and 95.4 eV, respectively.

However, this definition includes projection lithography such as EUVL. To distinguish EUVL from SXRL, the lower bound for SXRL should exceed 90 eV, but include 284 eV of the carbon K edge. The upper bound may be seen at the Si K edge (1.839 keV) because Si-based mask membranes are frequently used. While Attwood (1999) considers that SXRL utilizes nominally 0.7–1.2 nm wavelength radiation (1.0–1.8 keV photon energy), we prefer defining *SXRL as the proximity lithography in the spectral range including the carbon K edge* (284 eV, 4.37 nm) *up to the silicon K edge* (1839 eV, 0.67 nm).

Finally, we clearly distinguish SXRL from the so-called soft lithography that is a direct replication method with many variants (Xia and Whitesides 1998).

7.1.3 Need and Motivation

SXRL is aiming at ever-smaller structures. With progressing nanotechnology, the further reduction of sizes

7.2 Theoretical Background

7.2.1 Physical Processes

The basic mechanism leading to the structuring of resists by either chain scission or cross-linking is the absorption of radiation leading to cascades of processes

Table 7.1 Basic Features of SXRL

Spectral range	250–1800 eV, 5–0.7 nm Band-pass controllable via filters (high pass) and mirrors (low pass)
Source	Synchrotron radiation: conventional and compact storage ring Discharge plasma Laser plasma Bremsstrahlung (MIRRORCLE-driven storage rings) X-ray tube
Resist	PMMA (positive) SU-8 (negative, chemically amplified)
Mask membrane and absorber	Si_3N_4, Au
Environment	Fine vacuum

down to the nanoscale calls for shorter wavelengths than used in DUVL. Achieving sub-10 nm resolution may be crucial to join the scale of self-assembly technique, thus enabling building electrode structures for molecular electronics. The obvious way of improving resolution is to reduce wavelength as resolution scales with $\sqrt{\lambda \cdot g}$ in case of proximity where λ is the wavelength and g is the proximity gap. Hard x-rays may be too violent for nanostructures. Side effects such as photo and Auger electron generation in both the resist and the substrate, as well as mask and substrate fluorescence, may degrade the quality of nanostructures. Moreover, hard x-rays are not much absorbed in thin resist layers, and so they are more suitable for DXRL which is routinely performed at labs such as ANKA, BESSY, CAMD, ELETTRA, and SSLS.

Thus, soft x-rays seem to be a viable compromise for manufacturing nanostructures, particularly, if the focus is also on comparably tall and high-aspect-ratio structures. As a proximity lithography, SXRL is much simpler in terms of equipment, it has a good working depth as compared to projection systems, and it is characterized by a very good absorption, little scattering, and comparably small secondary effects (Smith and Schattenburg 1994). Table 7.1 shows some basic features of SXRL.

such as photoionization, photodissociation, excitation, and fluorescence, caused by the primary photons, and similar processes in which the energy is further dissipated by the photoelectrons, Auger electrons, and so on, until ending up as phonons or heat. The resist is commonly a long-chain polymer such as PMMA (Feder et al. 1975; Spears and Smith 1972). Monomolecular resists were studied as well (Klauser et al. 2004). In the polymer case, we distinguish between positive resists (PMMA) that respond by a reduction of the molecular weight of their constituents, mainly by chain scission (Feder et al. 1975; Spears and Smith 1972), or negative resists (SU-8) that are based on cross-linking of chains and usually have chemical amplification (Deguchi et al. 1999; Dellmann et al. 1998; Engelke et al. 2004; Microchem 2001; Lee et al. 1995).

The linear absorption coefficient of a material for x-rays is proportional to the cubes of the atomic number Z and the wavelength λ.

$$\mu_{lin} \propto Z^3 \lambda^3$$

Heavier atoms absorb more strongly than lighter ones and absorption decreases for shorter wavelengths or harder photons (Maid et al. 1989). A parameterization including absorption edges is due to Victoreen (1943, 1948). Convenient diagrams of the mass absorption coefficient that is related to the linear absorption coefficient via the density and of the atomic scattering

$$\mu_m(cm^2 g^{-1}) = \mu_{lin}(cm^{-1}) / \rho(cm^{-3} g)$$

factors were distilled by Attwood (1999, pp. 428), from which we derive Table 7.2 that gives an overview of the relevant cross sections of C, O, H, and N, the most common elements in resists. The basic tendency of cross sections is their falling off with increasing photon energy beyond their maximum, superposed by the resonances when the photon energy becomes large enough to excite a transition from a lower state in the atom to a higher one or the ionization continuum. It is very interesting to compare cross sections

of the main resist constituents to see that depending on photon energy, the contribution of the various atoms to the exposure-induced processes may be completely different.

Photoionization cross sections for C, N, and O are about 10^{-19} cm^2 for 1 keV photon energy. Details may be found in Veigele (1973) and Verner (http://www.pa.uky.edu/~verner/photo.html). Brundle et al. (1992) describe that for carbon the cross section for ejecting a 1 s electron is much bigger than for either 2 s or 2 p, and for the Auger process as well. Atomic oxygen (Angel and Samson 1988) has a maximum cross section of 1.34×10^{-17} cm^2 at 20 eV, 1.65×10^{-18} cm^2 at 100 eV, 3.3×10^{-19} cm^2 at 200 eV, and 1.6×10^{-19} cm^2 at 280 eV.

Photodissociation is treated in Zaikov (1995) with special emphasis on PMMA degradation (pp. 172, 193, in the chapter on polyacrylates by Melnikov et al.) They also mention the formation of free radicals, that is, neutral fragments that have unpaired electrons that can cause chain scission.

Chain scissions occur as a fraction of all possible dissipation channels, a quantum yield. For positive resists, quantum yield in terms of numbers of main-chain scissions per incident photon is an essential issue as they are usually less sensitive than negative resists. With negative and chemically amplified resists (CARs), the emphasis is not so much on quantum yield, but on other properties that may affect cross-linking (Wong et al. 2006).

Much work was published on direct cross sections for chain scission in positive resists in the UV. Zweifel et al. (2009, p. 192) reviewed the UV degradation of PMMA including main- and side-chain scission processes. Gupta et al. (1980) analyzed the photon-induced chain scission at 254 and 266 nm showing that the excitation of the monomer preferably affects the oxygen-containing ester group as expected since the cross section of oxygen is more than twice as large as that of carbon, let alone hydrogen.

Shultz et al. (1985) found the quantum yield for PMMA chain scission by 214–229 nm wavelength UV to be $\phi_d = 0.03$ scissions per absorbed photon. Postbaking of the irradiated films at 150°C for 1 h under reduced-pressure flowing nitrogen increased the quantum yield to 0.04.

Torikai et al. (1989) irradiated PMMA with monochromatic light of 260, 280, and 300 nm wavelength finding photodegradation for $\lambda < 260$–300 nm, but not for $\lambda > 320$ nm. The number of main-chain scissions (N_{cs}) had a maximum at $\lambda = 280$ nm. Under the same conditions, a longer-term irradiation at lower

Table 7.2   Cross Section $\sigma_p/10^{-18}$ cm^2 for Photoabsorption versus Photon Energy for Selected Elements

	E/eV			
Element	100	300	700	1000
H	0.01923	0.00059	3.71E−05	1.13E−05
C	0.48638	0.91391	0.117266	0.044229
N	1.01274	0.07117	0.196756	0.076898
O	1.80436	0.13146	0.305024[a]	0.122621

[a] Factor of 10 typo in Attwood (1999) corrected.

intensity yielded a greater amount of main-chain scissions. A linear relationship was found between the number of main-chain scissions and the incident photon intensity. The average quantum yields were 2.1×10^{-4}, 2.4×10^{-4}, and 4.1×10^{-4} for wavelengths of 260, 280, and 300 nm, respectively. The photo-induced side-chain scission was found to initiate the main-chain scission.

For x-rays, Chooi et al. (1988) gave a quantum yield for PMMA chain scission in the spectral range of 0.8–1.8 keV as well as a comparison with the UV (4–6 eV) results claiming that x-rays are 10× more efficient. However, with the number of chain scissions N and exposure dose D, they give

$$N = m \cdot D/(\mathrm{J\,cm^{-2}})$$

with $m = 18.86$ for x-rays and $m = 7.43$ for UV entailing a ratio between x-rays and UV of 2.54, not 10, leaving a discrepancy.

In case of negative and CARs, one of the issues is to avoid cross-linking that is not caused by the incident radiation such as thermally initiated cross-linking. Wong et al. (2006) studied this effect by monitoring the 914 cm^{-1} absorption peak finding the onset at about 120°C.

The next fundamental question asks for the main influences determining pattern transfer accuracy, that is, the spatial resolution and the smoothness of contours, such as line edge roughness (LER) for thin and sidewall roughness (SWR) for thick resists as outlined in the following.

A radiation source produces a wavefield that is spatially modulated after passing through a mask. Shadowing and diffraction by mask features are of predominant importance. Next, the modulated wavefield is transformed by passing the proximity gap and impinges on the interface to the resist layer where reflection and refraction occur. Upon propagation through the resist, there will be refraction at internal inhomogeneities such as density fluctuations of the resist or impurities, scattering from the same density fluctuations, and, of course, absorption which is the desired effect.

For monochromatic radiation, there is only a decrease of the power distribution according to Beer's law when the wave propagates through the resist. For wide-band radiation, dispersive effects occur resulting in a spectral shift to harder photons deeper into the resist, a phenomenon called beam hardening.

The structure definition inside the resist will depend on the three-dimensional (3D) spatial power density distribution that causes the exposure and the aerial image. Although frequently neglected, radiation scattered by inhomogeneities of the resist like density fluctuations or impurities ultimately represents a source of loss of spatial resolution. A concise description may be found in Attwood (1999, p. 24). The transition from high to low dose, the aerial image contrast, is a key factor affecting the sharpness and smoothness of edges. As for the wavefield, the amplitude or power gradient can be estimated from Fraunhofer or Fresnel pattern. Fraunhofer diffraction at a Fresnel number of $N_f = a^2/\lambda D = 0.5$ yields a modulation from maximum to zero over a distance a where λ is the wavelength, D is the proximity gap, and $2a$ is the clear feature width (Saleh and Teich 1991). For 1 nm wavelength, 2 µm gap, and 64 nm clear aperture, we obtain 32 nm. Critical dimensions would benefit from larger Fresnel numbers and exploiting the steepening response of the resist. Shin et al. (2001) experimentally studied correlations between LER and aerial image contrast for SXRL, EUVL, and electron beam lithography (EBL), under identical processing conditions using atomic force microscopy with carbon nanotube tips to image top and bottom of trenches with very high resolution. Results indicated that higher aerial image contrast leads to lower LER, more pronounced for UV-6 photoresist than for PMMA. Top surface roughness results showed similar trends with LER. Higher aerial image modulation also yielded a higher resist sidewall angle, in particular for PMMA.

Next, we consider resist reactions under exposure. As the macromolecules are either downsized or upsized, what are the spatial inhomogeneities that are introduced? This is probably different between positive and negative resists. In the former case, typical size scale lengths range from the bond length between C atoms in an alkane chain backbone 1.554 Å (Beyer and Walter 1991, p. 96), over about 10 Å for the persistence length (local stiffness) to about 100 Å of coil radius (Baschnagel et al. 2000).

Further resolution loss comes from the secondary exposure by mostly Auger electrons. From carbon, for instance, we get KLL Auger electrons at about 260–280 eV. As the diffraction blur decreases with photon energy while the range of Auger electrons is constant, there will be a photon energy at which diffraction blur can be neglected compared to Auger blur. The effect of direct photoelectrons was studied by Smith and Schattenburg (1994) and Carter et al. (1997), and simulated by Ocola and Cerrina (1993, 1994) showing that it is not the mere existence of secondary processes leading to structure degradation, but the dose density they generate as described by the energy-deposition point-spread-function (EDPSF). Minimum blur occurs at about 1.2 keV or 1 nm. Moreover, there might be secondary effects due to fluorescence and photoelectrons from both the mask and the substrate (Ogawa et al. 1989).

Another issue is the selectivity of the developer to distinguish molecules of different size in either positive or negative resists. The development process exposes the irradiated resist to the developer for a given time. The solvability of molecules being a function of their size (molecular weight), a given developer would dissolve molecules up to a certain size. Qualitatively, it amounts to being a time integral of the size-dependent removal rate and the spatial distribution of chain lengths or molecular weights.

Finally, mechanical issues of differential thermal expansion of the mask and the substrate upon heating during exposure, and vibration between the mask and the substrate or between the mask–substrate stack and the radiation source during exposure are also potential causes of structural loss. Appropriate countermeasures may include the selection of materials, damping of mechanical vibrations, and beam stabilization.

7.2.2 Theoretical Description

Schattenburg et al. (1991) analyzed soft x-ray diffraction from 100 nm gold structures to find out the appropriate approximation to solving Maxwell's equations for determining the aerial image. Using a method-of-moments and a finite-difference time-domain approach, the common use of Kirchhoff boundary conditions was shown unsuitable because it introduces unphysically high spatial frequencies in the diffracted fields in typical situations such as 100 nm structure detail, 1.3 nm wavelength, and 11 μm proximity gap.

Guo and Cerrina (1993) modeled x-ray proximity lithography, eventually providing the CNTech Toolset simulation software (http://www.nanotech.wisc.edu/toolset/). Stimulated by Hundt and Tischer (1978) and Murata (1985), Early et al. (1990) found, by exposing a 30 nm Au absorber line into PMMA by contact printing with soft x-rays of wavelengths 4.48 nm (CK),

1.33 nm (CuL), and 0.83 nm (AlK), that the linewidth in PMMA was independent of the wavelength within 5 nm concluding that the exposure blur was not due to photoelectrons, but compatible with Auger electrons. Ocola et al. (1993, 1994) studied in detail the theoretical aspects of the energy deposition by Auger electrons confirming that the spatial dose density is the dominant quantity controlling blur instead of the range of electrons generated in the course of the energy dissipating processes. For short-range Auger electrons, the dose density is much higher than for large-range direct photoelectrons.

The dissolution of resist was studied theoretically and experimentally by various authors. Kim et al. (2006) simulated the full resist process in photolithography. Raptis et al. (2000) studied the dissolution of polymethacrylate-based positive resists, both theoretically and experimentally. Liu et al. (1998) investigated the chemical dissolution rate of PMMA in the case of DXRL with the aim of producing microstructures with inclined side walls. Controlling the dose distributions and related dissolution rates in both exposed and shaded areas, they specialized in the fabrication of pyramidal microstructures using thin mask absorber patterns. Houle et al. (2002) used the critical ionization model to simulate resist line shapes. This model requires a critical level of ionization for a polymer chain to move from the film into solution and can describe polymer dissolution features like the dependence on chain length and solution pH. Developing a reaction scheme that described the coupled reversible ionization–relaxation steps that transform a polymer chain from an unsolvated into a solvated form, they showed that nonlinearities inherent in the dissolution kinetics are responsible for resist imaging leading, for example, to increased roughening as the aerial image contrast is decreased.

7.2.3 Simulation

As far as known to us, two prominent simulation tools are available on the Internet. One is the CNTech Toolset by Cerrina and co-workers (http://www.nanotech.wisc.edu/toolset/), the other one DoseSim (Meyer et al. 2003) from Karlsruhe Institute of Technology (KIT). Toolset focuses more on the thin resist case of microelectronic manufacturing, whereas DoseSim is targeted at high-aspect-ratio DXRL. SHI also generated codes such as BLOD and BLEX to account for the optical transport, aerial image, and resist dissolution (Toyota et al. 2001).

7.3 Process

7.3.1 Mask Making

Basically, we distinguish

- The transmission mask featuring a membrane that supports the absorber
- The stencil mask consisting of absorber structures supporting themselves in free space

The former is most common. It has the pattern-defining absorber supported by a transparent membrane that is held itself by a massive window-frame or ring. A stencil mask may be required when any influence of a membrane must be avoided. However, the pattern geometry is constrained to be singly connected in order to be held safely.

Figure 7.1 shows absorption lengths for mask materials versus wavelength. Membrane materials include Be, polyimide, C (diamond), Mylar, Si_3N_4, and SiC, whereas absorbers include Au, Pt, Ta, and W. As for the mask membranes, high transparency and thermomechanical stability are crucial. Thus, membranes should be thin and have small atomic number Z, whereas absorbers should be thick and have large Z to achieve high x-ray absorption. Although various absorber and membrane materials have been used in the fabrication of x-ray masks including tungsten absorbers (Kola et al. 1991) or diamond membranes (Löchel et al. 1992; Ravet and Rousseaux 1996), the most common materials are a Si_3N_4 membrane and Au absorber. The membrane is

made from a silicon wafer coated on both sides with low-stress silicon nitride thin film made by low-pressure chemical vapor deposition (LPCVD) (Ohta et al. 1994). The membrane pattern is opened in resist with photolithography; the patterned resist serves as a mask for the dry etching of the silicon nitride film with a CF_4/O_2 mixture. The silicon substrate is etched with KOH solution to get the window (Figure 7.2). Nowadays, typical mask blank materials are Si_3N_4, SiC, diamond, or Be. Polyimide or SiON were used earlier (Schattenburg et al. 1987). Si_3N_4 mask blanks are available commercially, for example, from Silson (http://www.silson.com/index.html) or NTT (http://www.ntt-at.com/products_e/x-ray_masks/index.html). NTT is also commercializing complete x-ray masks.

Figure 7.3 shows a generic mask-making process. A mask blank is coated with a thin adhesion layer/plating base of Cr/Au by sputter deposition or evaporation and then spin-coated with a thin e-beam resist such as PMMA. Primary patterns are typically generated by e-beam writing. After development, void patterns are filled with an absorber such as Au by electroplating or sputter deposition. The remaining resist is dissolved to obtain the pure Au pattern. The layer produced by sputter deposition is so lifted off. There are two distinct process routes in use depending on when the window is opened, namely, the WBP route (Window opening by Si etching Before Patterning) (Figure 7.3) and WAP route (Window opening After Patterning). WAP is identical to WBP except that the window in the Si wafer is etched at the very end of the process. In general, WAP has the advantage that all processes are run on a robust Si wafer, unlike WBP in which PMMA patterning and gold absorber plating are carried out on the thin Si_3N_4 film. But WAP makes the last step, the Si etching to open the window, very critical because its failure wipes out the whole fabrication of the x-ray mask.

For primary pattern, mostly electron beam writing is used and, subsequently, either additive or subtractive processes come into play to transfer the resist pattern into the absorber. Electroplating or metal deposition followed by lift-off are the additive processes while dry etching of an absorber layer under the resist is a subtractive process. The required spatial resolution permitting direct laser writing may also be used. For writing into thick resists, proton beam writing was studied (Yue et al. 2008, 2009).

Instead of inorganic membranes such as Si_3N_4, organic membranes were made from Mylar, Kapton,

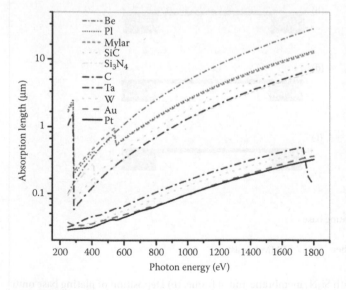

FIGURE 7.1 Absorption lengths for various membranes and absorber materials.

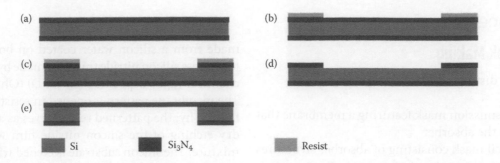

FIGURE 7.2   Fabrication of a Si_3N_4 membrane. (a) Deposition of Si_3N_4 film on both sides. (b) Resist coating and UV pattering. (c) Dry etching Si_3N_4 film with resist pattern as mask. (d) Removal of the resist. (e) Wet etching Si to open the window.

and Pyrolene (Dupont trade names). Polyimide has an order-of-magnitude higher attenuation length than Si_3N_4 (Figure 7.1) enabling relatively thick and stable membranes. To make a polyimide membrane according to Zhao et al. (2008), a 2-μm-thick polyimide film is spin-coated on the front side of a silicon wafer and baked on a hot plate. Then, the wafer is fixed on a clamping apparatus leaving only a 10-mm-diameter circle in the middle of its back side etched by a mixture of hydrofluoric, nitric, and acetic acids to open the window. Next, a Cr/Au plating base is deposited onto the surface of the polyimide membrane and covered by a positive resist layer (ZEP520A by Nippon Zeon). Upon e-beam patterning and development, electroplating yields about 500-nm-thick Au absorber.

Comparing polymer with inorganic membranes, Schattenburg et al. (1987) discarded polyimide for

sub-200 nm linewidth because it needed contact printing which created the risk of mask damage due to distortion. Instead, they used stress-controlled $SiON_{0.5}$ membranes of 6 cm² area attached to a narrow rim etched into a Si substrate. Accounting for penumbra, diffraction, and photoelectron range, they found the trade-off between gap and wavelength at 0.1 μm linewidth, 4 μm gap, and $\lambda = 1.3$ nm. Gap uniformity was ~0.5 μm. High-aspect-ratio PMMA structures were achieved with steep sidewalls, suggesting that 50 nm lines could be faithfully printed.

As in optical lithography, controlling of the phase of the transmitted radiation to improve resolution was considered by Malueg et al. (2004) who discussed the modeling, fabrication, and experimental application of x-ray phase masks showing that three to five times reduction of clear features may be achieved, promising sub-50 nm gate widths.

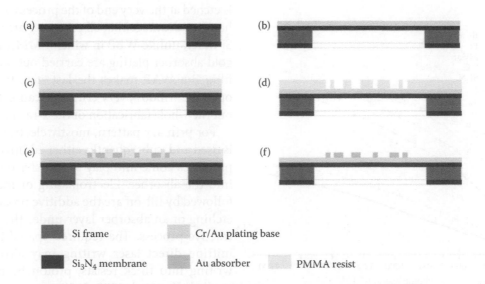

FIGURE 7.3   Generic mask process-WBP route. (a) Mask blank with Si_3N_4 membrane and Si frame. (b) Deposition of plating base onto the front surface of Si_3N_4 membrane. (c) Application of PMMA resist (coating and prebaking). (d) Patterning by e-beam lithography. (e) Deposition of Au into PMMA mould by electroplatng. (f) Removal of PMMA resist.

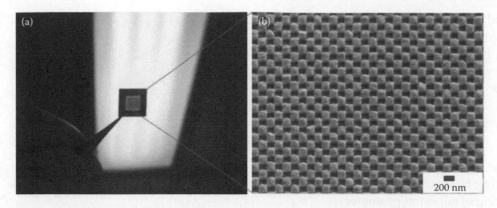

FIGURE 7.4   SXRL mask.PMMA patterned with Sirion/Nabity system on a Si_3N_4-membrane with 5×5 mm^2 usable area (a) and gold microstructures plated through PMMA pattern ((b), scale bar 200 nm).

For primary pattern generation, there are high-end e-beam writers such as ELS-7000 (ELIONIX INC.) that feature up to 100 keV electron energy, 5 nm minimum linewidth and 30 nm overlay and stitching accuracy (http://www.elionix.co.jp/english/products/ELS/ELS7000.html). In addition, there are add-ons to scanning electron microscopy (SEM) systems, such as the Nano Pattern Generation System (NPGS) (http://www.jcnabity.com/12nmline.htm) or the ELPHY kit by Raith. They can reach minimum feature sizes of 20 nm. Lacking positional measurement and feedback for precise stage positioning, they do not provide stitching of adjacent writing fields. An SXRL mask written by means of the Sirion/NPGS is displayed in Figure 7.4.

With the He ion microscope (HIM) claiming a minimum spot size of 0.25 nm, dense sub-10 nm features were written (Sidorkin et al. 2009; Ward et al. 2006; Winston et al. 2009). Advantageously, the absence of proximity effects was shown by dot arrays down to 6 nm dot diameter and 14 nm pitch patterned into a 5-nm-thick hydrogen silsesquioxane (HSQ) electron resist.

Finally, we note maskless SXRL (Smith et al. 2000) that circumvents conventional masks. Termed as zone-plate-array lithography (Carter et al. 1999; Gil et al. 2000), an array of Fresnel zone plates (FZPs) focuses radiation beamlets onto a substrate. Individual beamlets are turned on and off by upstream micromechanics as the substrate is scanned under the array creating patterns of arbitrary geometry with a minimum linewidth equal to the width of the outermost zone. Using 4.5 nm radiation, lines and spaces of 20 nm seem achievable provided zone plates are available.

7.3.2   Substrate Preparation

Substrates support the resist layer to be patterned and auxiliary layers like the plating base or sacrificial layer. Resist–substrate adhesion is critical requiring substrate cleaning to remove contamination, baking to remove moisture, and, eventually, adding a suitable adhesion promoter to keep water away and improve chemical bonding. Usually, substrates are cleaved from Si wafers to the required size, degreased by sonication in acetone, isopropanol, methanol, and rinsed with deionized water. The adsorbed water is removed by baking at 150–300°C for 30–60 min. Popular adhesion promoters are hexamethyldisilazane for PMMA and OmniCoat™ for SU-8. PMMA is the most common resist (Table 7.3) featuring high resolution, high contrast, yet low sensitivity. ZEP-520 claims a higher sensitivity and better etch resistance. Negative resists commonly exhibit much higher sensitivities than positive resists due to chemical amplification. Their resolution has reached sub-70 nm lines (Deguchi et al. 1999).

Table 7.3   Common Resists, Developers, Substrates, and Applications

Resist	Tone	Developer	Applications/Substrates
PMMA	+	MIBK:IPA (1:3); GG developer (Ghica et al. 1982)	DiFabrizio et al. (2004): Photonic crystal/GaAs Liu et al. (2008): Fresnel zone plate/SiN Kato et al. (2007): Grating/Si
ZEP-520	+	Xylene: p-dioxane	Wang et al. (2004): Zone plate/SiN Zhao et al. (2008): QDADG/ PI membrane
SAL601	–	Shipley MF312	Ravet and Rousseaux (1996): x-ray masks/SiC
SU-8	–	SU-8 developer	Reznikova et al. (2008): SU-8 grating/Si

Chapter 7

The typical thickness of resist layers for SXRL nanoapplications is less than a few μm because of the difficulties in achieving higher aspect ratio. Spin coating typically yields single resist layers of a few hundred nm. For thicker layers, multiple spin coating applies.

Substrate preparation typically involves the

- Cleaning of the substrate with IPA, sonication for 5–10 min
- Dehydrating by singe
- Spin coating of adhesion promoter
- Spin coating of resist

PMMA sample preparation involves

- Singe at 150°C for 30 min
- Spin coat at 3000 rpm for 30 s (for 400 nm thickness)
- Bake at 160–180°C for 15–30 min

7.3.3 Irradiation through Masks

An x-ray mask must transmit a high fraction of photon flux for fast processing combined with enough contrast for reliable structuring (Atoda et al. 1983; Matsui et al. 1979). Contrast may be defined as the ratio $T_2/(T_1 \cdot T_2) = 1/T_1$, where T_1 and T_2 are the transmissions of absorber patterns and membrane, respectively (Figure 7.5). The membrane transmission T_2 does not affect contrast, but must be large enough for high throughput and fast processing. Figure 7.6 shows the transmission of a 1000-nm-thick Si_3N_4 membrane and a 500-nm-thick gold layer that is providing sufficient contrast with <1000 eV photons. The spectral contrast is also plotted. It is higher for lower photon energies and vice versa. As an example, for 1050 eV photons and an Au absorber thickness of 500 nm, 99% of the x-ray photons would be absorbed and 50% transmitted through 1 μm Si_3N_4. The contrast would be 100. The mask contrast also

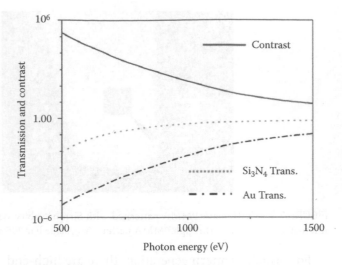

FIGURE 7.6 Spectral transmission and contrast of layers of 1000 nm Si_3N_4 and 500 nm Au.

must take into account the resist systems. A value of 30 gives good exposure for PMMA whereas 10 is required for SU-8.

Mask alignment is of concern mostly for microelectronics manufacturing because integrated circuits need many mask levels. The present-day SXRL for Research & Development (R&D) and prototyping of micro/nanostructures (see Section 7.5) is usually working with one mask only without any need of alignment. Smith et al. (1973) and Moon et al. (1998) studied alignment schemes realizing an interferometric broadband imaging (IBBI) system demonstrating sub-1 nm consistency of independent IBBI measurements and the ability to feedback lock the mask relative to a wafer to within a mean of 0.0 nm and a standard deviation of 1.4 nm.

7.3.4 Exposure of Resists

Soft x-ray exposures need vacuum to prevent absorption by air. Main considerations for choosing a resist include its sensitivity, spatial resolution, and etch resistance (Oizumi et al. 1994). For typical resists, see Table 7.3. A useful analogy assumes that exposure by x-rays is almost the same as by e-beam. The difference is that electrons directly interact with resist molecules whereas x-rays generate photo- and Auger electrons which then interact with resist molecules. Thus, any electron-sensitive resist can also serve for x-rays (Cui 2005).

A single-layer resist is sufficient for most practical uses. Its thickness is controlled by spin-coating parameters and solvent content. The absorption of

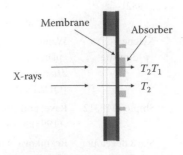

FIGURE 7.5 Definition of x-ray mask contrast.

radiation in the resist and the transformation via either chain scissions or cross-linking have been discussed above. Exposure conditions have to be optimized to have good absorption by the resist, and negligible response from the substrate. Exposure times fully depend on the source type. With 300 mA electron current in the storage ring, typical exposure times at SSLS for PMMA and SU-8 are 40 min and 0.4 min, respectively, for 3 μm resist thickness and a scan range of 100 mm. Resist exposure requires the mask held against the resist surface via a spacer (Kapton foil) (Figure 7.7).

PMMA serves as an x-ray resist since 1970 (sum formula $C_5H_8O_2$). It is normally in white powder form. A quantity of 5–10% of PMMA thoroughly mixed with anisole solvent makes up the PMMA resist. Commercial PMMA resists from, for example, MicroChem (http://www.microchem.com) have two standard relative molecular masses, 495 and 950 kDa. Although a low-molecular-mass PMMA is more sensitive, both contrast and resolution are lower. Therefore, high-resolution SXRL needs 950 kDa PMMA.

First formulated at IBM (Gelorme et al. 1989), photoresist SU-8 is based on an epoxy resin with high functionality (sum formula $C_{24}H_{26}O_4$) that is dissolved in an organic solvent, gammabutyrolactone, with added triaryl sulfonium salt acting as a photoacid generator. Suppliers include MicroChem and Micro Resist Technology.

The calculated absorption lengths for PMMA and SU-8 versus photon energy (http://www-cxro.lbl.gov) show the incident radiation fully absorbed in the thin resist layers for photons close to the C K edge (284 eV), whereas it is only partly absorbed at >1 keV (Figure 7.8).

As for spatial resolution, Chen et al. (1997) achieved 200 nm linewidth with an aspect ratio of >10 using 2–3-μm-thick PMMA 950 kDa. Burkhardt et al. (1995) developed an accurate mask-copying process introducing capacitive gap control between parent and daughter mask to replicate about 30 nm features. They claimed

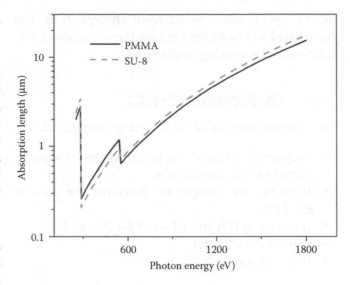

FIGURE 7.8 Absorption lengths of PMMA ($C_5H_8O_2$) and SU-8 ($C_{24}H_{26}O_4$) versus photon energy. Densities: PMMA 1.19 g/cm³, SU-8 1.21 g/cm³.

fabrication of sub-70 nm features when applied to mask-to-substrate exposure. Such minimum features are still larger than the resolution limit of PMMA as shown by Yasin et al. (2001) where <5 nm PMMA resist lines were obtained using a 100 kDa PMMA and ultrasonically assisted development in a 3:7 mixture of water and isopropanol. Using the commercially available chemically amplified UVII-HS, Hoffnagle et al. (2002, 2003) evaluated resist resolution by interference lithography and subsequent analysis of the line spread function claiming a potential resolution of approximately 50 nm.

Chemically amplified negative resists may encounter problems, particularly, at their surface and substrate interface which may reduce resolution (Lenhart et al. 2007) by effects including surface skin formation caused by resist contaminants or airborne amines and footing or undercutting at the bottom interface. Bottom antireflective coatings were also studied.

Direct writing by means of a focused soft x-ray beam into an organic resist was investigated by Klauser et al. (2004) and Ballav et al. (2008). Monomolecular aliphatic and aromatic self-assembled monolayers served as resist which responded by irradiation-induced desorption of molecular fragments. A potential spatial resolution of a few nm was claimed. One might speculate that undulator beams from advanced third-generation light sources might enable practical direct photon beam writing.

No work has been found on what we would call "resonance exposure," for example, at 284 eV for C and

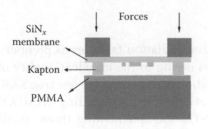

FIGURE 7.7 Schematic setup of the positioning of mask and substrate during exposure. The proximity gap is set by a Kapton foil spacer.

543 eV for O, which would favor absorption in thin resists and so resolution because the influence of diffraction and scattering might be less.

7.3.5 Development of Resists

The essential steps of PMMA development include

- Prepare for 20 mL developer, IPA, and DI water in three beakers, respectively.
- Immerse the sample in developer for 45 s at 20–21°C.
- Immerse in IPA and DI water for 20 s each.
- Rinse with DI water.
- Dry with air.

Similarly, for SU-8 development

- Prepare one beaker of a suitable amount of SU-8 developer for coarse developing, one beaker of SU-8 developer for fine developing and rinsing, and IPA for rinsing.
- Immerse the sample in first beaker for 60 s for developing.
- Immerse the sample in second beaker for 30 s for rinsing.
- Rinse with IPA.
- Dry with air.

In either case, if the resist structure is to be characterized by SEM, then sputter deposit a gold layer of about 50 Å.

The influence of the development on the spatial resolution is a key problem being considered since the 1970s. The basic issue is to find the rate of removal and the main effects influencing the interface and its roughness, given a spatial distribution of average molecular weight varying monotonously in the etch direction. For an overview, see Cowie (1994).

For positive resists such as PMMA, the basic physicochemical process is the dissolution of molecules in an appropriate developer—with tetrahydrofuran being the absolute solvent. Ehrfeld et al. (1986) indicated that the development makes structures sharper than the x-ray dose distribution. It has also been recognized from early literature that the specific polymer is influencing the LER and SWR (Yamaguchi et al. 1997; Yoshimura et al. 1993) by its structure or by particles formed during dissolution. As noted above, scale lengths of polymers lie between 1 and 100 Å. In PMMA, <5 nm lines were achieved by e-beam writing and developing in a water–isopropanol solvent assisted by ultrasonication (Yasin et al. 2001). These authors assumed that the spatial resolution was determined by the solvent-dependent gyration radius of the polymer molecule. Gyration radius and LER are consistently related (Allen et al. 2007).

Yamaguchi et al. (1997, 2000) described the formation of 20–30 nm-large polymer aggregates in resist films as responsible of LER and developed a special resist (suppressed aggregate extraction development resist) to reduce LER. Ma et al. (2003) simulated LER from a percolation model in the UV-6 negative resist. Although there is no full picture yet of the most important factors influencing LER or SWR, the resist and its development seem to be critical. According to Reynolds and Taylor (1999), SWR in a positive-tone, chemically amplified x-ray resist depends on the composition and polydispersity of the base polymer, the effect of reduction of feature size, and acid concentration.

Finally, small high-aspect-ratio structures may be deformed by capillary forces arising from wet-chemical process steps such as development and rinsing, the so-called stiction. Tanaka et al. (1993, 1994a,b) mitigated stiction by heating and/or flood exposure during rinsing. Raccurt et al. (2004) counteracted stiction by working wet-in-wet and, for drying, by covering with a layer of pentane that has less than five times the surface tension of water. Van Spengen et al. (2002) modeled processes to predict stiction, Maboudian and Howe (1997) used supercritical CO_2 drying, and Vora et al. (2005) included reinforcing structures in the design.

7.4 Apparatus

7.4.1 Introduction

An SXRL exposure system must feature a vacuum chamber that houses the key process components to prevent absorption of soft x-rays by air (Figure 7.9). Differences between various SXRL systems stem from the source and type of lithography, that is, proximity or projection. Although it would be natural to use synchrotron radiation facilities as premier sources of soft x-rays owing to the excellent quality of synchrotron radiation, there are only a few true SXRL facilities installed at light sources including CAMD, ELETTRA, NSRL/USTC. Complementing those, small inexpensive laboratory sources are under development (Gaeta et al. 2002; Minkov et al. 2007; Sugano et al. 2008; Verma et al. 2008).

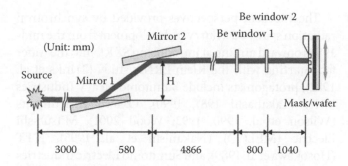

FIGURE 7.9 Generic setup for SXRL at a synchrotron light source. (Courtesy J. Göttert, CAMD.)

7.4.2   Source

Starting with standard x-ray tubes (Feder 1970; Feder et al. 1975), SXRL nowadays is using mostly synchrotron radiation and plasma sources for power and resolution reasons (Brodie and Murray 1992). Sources based on transition radiation from relativistic electrons are also under development. A comparison of the x-ray power for electron bombardment, laser plasma, and synchrotron light sources was given by Cui (2005) (Table 7.4).

In the following, synchrotron light sources and plasma sources will be addressed.

7.4.2.1   Synchrotron Light Source

Since Spiller et al. (1976), synchrotron light is the source of choice due to its small divergence, high flux, and broad continuous spectrum. From 1985 to 1995, work on synchrotron light sources for SXRL was strongly pushed. After an early review (Grobman 1984), literature exploded (Attwood 1999, refs. 59–82, Chapter 10; Heuberger 1985a,b), reflecting the heydays of synchrotron light source development for SXRL.

It is well known that electrical charges emit electromagnetic radiation when accelerated. For charges moving at a speed close to that of light, the spatial and spectral distribution of the radiation has special properties and is called synchrotron radiation as it was discovered at the General Electric synchrotron in 1947. The theory is due to Schott (1912), Iwanenko and Pomeranchuk (1944), Schwinger (1946, 1949), and Sokolov and Ternov (1968). Blewett (1988) gave a concise overview of the discovery of synchrotron radiation.

Table 7.4   X-Ray Sources and Their Irradiation Power

Source Type	Power (mJ cm^{-2} s^{-1})
Electron bombardment	0.01–0.1
Laser plasma	0.01–1
Synchrotron radiation	10–100

The following formulae describe the spatial and spectral power distribution, and the total power radiated per electron, respectively.

$$\frac{dP(t')}{d\Omega} = \frac{1}{\mu_0 c^3}\frac{q^2}{(4\pi\varepsilon_0)^2}\frac{\left[\vec{n}\times((\vec{n}-\vec{\beta})\times\dot{\vec{\beta}})\right]^2}{(1-\vec{n}\vec{\beta})^5}$$

$$\frac{d^2 I}{d\omega d\Omega} = \frac{1}{\mu_0 c^3}\frac{q^2}{(4\pi\varepsilon_0)^2}\left(\frac{\omega\rho}{c}\right)^2\left(\frac{1}{\gamma^2}+\vartheta^2\right)^2$$
$$\left[K_{2/3}^2(\xi)+\frac{\vartheta^2}{(1/\gamma^2)\vartheta^2}K_{1/3}^2(\xi)\right]$$

$$P(t')=\frac{2}{3}r_0 m_0 c\dot{\vec{\beta}}^2\gamma^4$$

Table 7.5 explains the parameters. The first equation shows that the radiation is concentrated in a narrow forward lobe leading to high brightness. The ratio of forward-to-backward power density is $\left((1+\vec{n}\vec{\beta})/(1-\vec{n}\vec{\beta})\right)^5 \approx (4\gamma^2)^5$. For 700 MeV electrons (Helios 2 at SSLS), this equals 2.4×10^{34}. A parameterized form of the half-aperture angle was derived from the second equation by Green (1976) as $\sigma_R = (0.57/\gamma)(\lambda/\lambda_c)^{0.43}$ in radian. Here, the forward lobe is assumed to have a Gaussian spatial distribution with variance σ_R. For SSLS, this value is 0.416 mrad. $\lambda_c = (4\pi/3)(\rho/\gamma^3)$ is the critical wavelength that divides the power spectrum into two equal halves.

Table 7.5   Explanation of Parameters Used in Synchrotron Radiation Theory

$\beta = v/c$	normalized velocity of particle
γ	ratio particle energy over rest energy
ϑ	angle between $\vec{n}$ and mid-plane (latitude)
ρ	local orbit curvature radius
ω	angular frequency
$K_{2/3}(\xi), K_{1/3}(\xi)$	modified Bessel functions with $\xi = \dfrac{\omega\rho}{3c}\left(\dfrac{1}{\gamma^2}+\vartheta^2\right)^{3/2}$
Ω	solid angle
q	electric charge of particle
r	distance particle-observer
$\vec{n}$	unit vector from particle to observer
$r_0 = \dfrac{e^2}{4\pi\varepsilon_0 m_0 c^2}$	classical electron radius
m_0	rest mass

Chapter 7

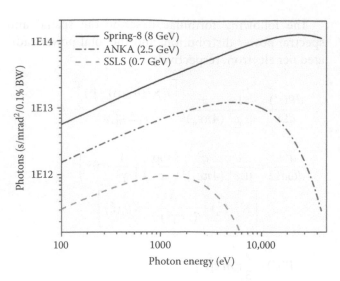

FIGURE 7.10 Spectrum of electrons under a constant radial acceleration for the cases of SSLS (700 MeV, 4.5 T), ANKA (2.5 GeV, 1.5 T), Spring-8 (8 GeV, 0.679 T). An electron current of 100 mA was assumed in all cases.

The total power emitted by an electron (third equation) is also much higher for highly relativistic electrons, for example, about 3.5×10^{12} for 700 MeV. Figure 7.10 depicts the spectrum of a charge moving perpendicularly to the field lines in a uniform magnetic field thus undergoing a radial acceleration. As the electron beam cross section which the synchrotron radiation is emitted from is of the order of 1000 μm² for modern third-generation light sources, it is a highly brilliant source. Moreover, for the same reason of the narrow forward lobe, it features a continuous spectrum that extends from hard x-rays continuously to the fundamental given by the revolution frequency of the electron bunches in the storage ring, 56 MHz for Helios 2. In fact, the lower limit for practical use is the far infrared at about 10 cm⁻¹. Thus, the spectral range of useful synchrotron radiation spans more than seven orders of magnitude. For SXRL, a spectral band pass is selected to optimize the exposure time and resolution.

Magnets producing a uniform field are called dipole or bending magnets. Being part of storage rings as necessary accelerator elements, they may be normal- or superconducting having different field strengths. Moreover, there has been intense work on special magnets to optimize generation of synchrotron radiation including superconducting wavelength shifters, superconducting wigglers, and undulators, both permanent magnet and superconducting. They deliver harder photons, higher flux, brightness, and brilliance. Their potential still remains to be fully exploited.

The excellent perspectives provided by synchrotron radiation spawned a flurry of development from the mid-1980s onward aiming at introducing SXRL into the wafer fab. Starting with the Klein ERNA study (Trinks et al. 1982), protagonists include Sumitomo Heavy Industries (SHI) (Takahashi 1987, 1990), Oxford Instruments (Wilson et al. 1990, 1993; Wood 2001), Mitsubishi Electric (MELCO) (Nakanishi et al. 1995), NTT (Hosokawa et al. 1989), and Sumitomo Electric Industries (SEI) (Tomimasu et al. 1985; Tomimasu 1987).

SHI's AURORA featured the smallest design orbit of 1 m Ø at 650 MeV electron energy (now at Ritsumeikan (Kato et al. 2007)). They used one single axisymmetric superconducting dipole which necessitated innovative resonance injection from a 150 MeV microtron.

A normal conducting racetrack version AURORA 2-D built later has become the source of HiSOR at Hiroshima University. SEI teamed up with Electrotechnical Laboratory (ETL) at Tsukuba to realize four rings dubbed NIJI that also relied on superconducting magnets. NTT built a superconducting ring SuperALIS and a normal conducting NAR at their Atsugi labs as did MELCO at Amagasaki. Oxford Instruments built Helios 1 and 2, meanwhile both at SSLS. In the United States, DoE funded the SXLS project (Murphy et al. 1990) at NSLS/BNL with Grumman as the general contractor.

Later, Yamada (1989) invented the photon storage ring and developed his family of driven storage rings named MIRRORCLE (Minkov et al. 2006, 2007; Toyosugi et al. 2007), currently the most compact relativistic electron machines. 20 MeV MIRRORCLE-20SX uses a 240-nm-thick Be foil as an internal target for transition radiation delivering >150 mW power with photon energy >500 eV. Sugano et al. (2008) designed a compact 6 MeV source using a periodic aluminum microstructure with vacuum gaps to generate resonance transition radiation.

7.4.2.2 Plasma Sources

Depending on temperature and density, hot plasmas radiate line spectra characteristic for the neutral and ionized atomic species present as well as continuous Bremsstrahlung. Nowadays, the line spectrum of Sn or Xe is used as soft x-ray source. To produce the hot plasma, electrical gas discharges and laser heating are used. As the latter focused on EUVL since 2004, it will not be presented here.

Building on nuclear fusion work (Anderson et al. 1958), x-ray production is investigated in gas discharges that carry a pulsed electric current strong enough to magnetically compress and heat the gas to plasma temperatures (pinch). A specific embodiment is the "plasma focus"

(Mather 1965). With their FMPF-1 fast miniature plasma focus, Wong et al. (2004, 2006) and Verma et al. (2008) achieved a soft x-ray yield of about 50 mJ per shot in the spectral range of 0.9–1.6 keV into the full solid angle of 4π from a mixture of 90% D_2 and 10% Kr. Another example may be found in Bogolyubov et al. (1998).

7.4.3 Exposure Station

7.4.3.1 Beamline

Figure 7.11 presents a typical arrangement for SXRL at a synchrotron light source (CAMD, http://www.camd.lsu. edu/beamlines.htm) featuring mirrors upstream the sample that provide a variable low pass at fixed exit beam to select a soft x-ray photon range from the broad synchrotron radiation spectrum. The exposure station is located at about 10 m from the source. The mask and substrate are mounted in parallel with the proximity gap most simply set by a thin Kapton foil. The surface of the mask–substrate stack is generally perpendicular to the photon beam, but can be inclined for special purposes. The beam footprint on the mask–substrate stack is a horizontal ribbon that is typically 10 mm high due to the small divergence of synchrotron radiation. A uniform dose distribution across the whole assembly is achieved by either vertically scanning the mask–substrate assembly, or scanning the mirrors to sweep the light beam, or scanning the electron beam (Moser and Lehr 1989;

Tomimasu et al. 1985). Eventually, the mechanical scanning of the mask–substrate stack prevailed.

SXRL exposure stations are commonly customized. Jenoptik have offered a standard series of scanners. There are only a few exposure stations capable of SXRL at synchrotron light sources including CAMD, LILIT, and NSRL/USTC. Others have the processes available, but feature a harder spectrum like ANKA and SSLS.

At CAMD, the XRLM1 beamline (Figure 7.11) reflects the x-ray beam from two parallel mirrors that are made from Si and coated with a >120-nm-thick Cr layer. The first mirror is water cooled. Cutoff photon energy is about 3 keV. The spectral influence of the various beamline elements is displayed in Figure 7.12.

At NSRL, beamline U1 is dedicated to SXRL in a wavelength range of 0.5–2 nm (http://www.nsrl.ustc. edu.cn). It uses one plane low-pass mirror at an incidence angle of 20 mrad. The mirror scans ±4 mrad to get 30×30 mm^2 exposure area with 90% uniformity. Differential pumping enables windowless operation with $<5 \times 10^{-8}$ Pa pressure at the mirror and $<5 \times 10^{-4}$ Pa in the exposure chamber.

ELETTRA runs the Laboratory for Interdisciplinary Lithography (LILIT) facility that is capable of providing photon energies within 1–2 keV (http://www. elettra.trieste.it/). A spatial resolution of <100 nm can be achieved. The x-ray stepper accommodates up to 6 in. wafers (Romanato et al. 2001).

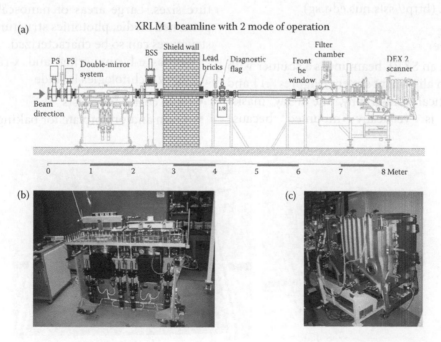

(a) XRLM 1 beamline with 2 mode of operation

(b)

(c)

FIGURE 7.11 (a) Schematic layout of XRLM 1 beamline at CAMD; (b) IRELEC double mirror system; beam incidence from the right; (c) Jenoptik DEX 2 scanner. (Courtesy J. Göttert, CAMD.)

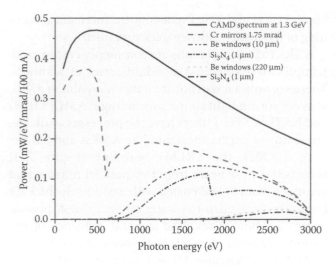

FIGURE 7.12 Spectral distribution after different components in the XRLM1 beamline: Dipole radiation spectrum, double Cr mirrors, Be windows, and Si_3N_4 mask membrane. The case of a 10-μm-thick Be window is fictitious and is included to show the effects more clearly. In reality, the beamline has a total of 220 μm Be from two windows with 100 and 120 μm thickness as represented by the flat-lying curves.

Although the LIGA-1 beamline at ANKA (http://ankaweb.fzk.de) is equipped with a Cr-coated Si mirror with grazing incidence angle of 17.5 mrad, its 1.5–4 keV photon energy range has only 300 eV overlap with SXRL in the present definition. Similarly, SSLS has a low energy cutoff at about 2 keV from its Be windows and, lacking mirrors, a high-energy roll-off at 10 keV, too hard for SXRL (http://ssls.nus.edu.sg).

7.4.3.2 Scanner

The end-station in an SXRL beamline is a scanner with a workstage and an alignment system (Figures 7.11 and 7.13). Unlike optical lithography, the x-ray mask–substrate stack is vertically mounted because synchrotron x-rays commonly propagate horizontally. The workstage must enable the assembly of both the mask and the substrate with a precisely controllable gap between them. As the vertical beam size is typically only 10 mm, the stage must be moved up and down to let the beam expose the whole wafer of 4″ diameter or more. At beamline U1 of NSRL/USTC, the mirror is scanned instead of the stage.

7.4.4 Metrology and Auxiliary Technologies

Metrology is critical for SXRL as feature sizes go down to <10 nm. Dimensions in x-ray masks produced by e-beam writing and in nanostructures copied from the masks by SXRL must be accurately measured to check compliance with specifications. Common metrology tools such as SEM and AFM/STM have been used to measure dimensions and image surfaces and defects. The HIM has become an important add-on to the equipment portfolio due to its expected resolution of <0.25 nm (Ward et al. 2006). It provides topographic features and analytical information about sample surfaces (Scipioni et al. 2009; Sijbrandij et al. 2010).

Among synchrotron radiation methods, grazing incidence small-angle x-ray scattering (GISAXS) was successfully applied to extended areas of regular nanostructure arrays (Hofmann et al. 2009) to reveal the morphology and arrangement of the patterned features and determine average values of pitch and structure sizes. Large areas of nanoscale arrays used for patterned media, photonics structures, and electronics structures can so be characterized.

Besides e-beam writer and x-ray source/scanner, auxiliary technologies include

- Spin coater for resists
- Oven and/or hotplate for baking

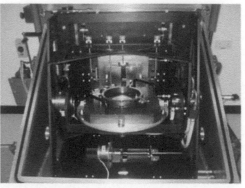

FIGURE 7.13 The Oxford Danfysik scanner attached to the LiMiNT beamline at SSLS.

- Wet bench for sample cleaning, resist development, etching
- Sputtering and/or evaporation tools for metal films
- PECVD for depositing of Si_3N_4, SiC
- Electroplating station for thick metal deposition
- Plasma RIE for cleaning/descum

7.5 Applications

Here, we briefly highlight recent applications. A major advantage of SXRL is the exposure of high-aspect-ratio structures over wide fields. Chen et al. (2007b, 2009) achieved isolated lines/pillars and periodic hole arrays with smallest feature sizes of 200 nm and aspect ratios of 10 with straight and smooth sidewalls (Figure 7.14). X-ray masks were fabricated on 1-µm-thick Si_3N_4 membranes using e-beam writing at 30 keV, electroplating of 250-nm-thick gold layers and soaking in acetone to remove unexposed PMMA. A quantity of 2–3-µm-thick PMMA 950 k was exposed at NSRL in the spectral range of 500–3000 eV. The proximity gap was 13 µm. For a clear mask feature of 200 nm, the Fresnel number ranged from 0.31 to 1.85, just adequate for this size. Mask contrast was estimated to be 15, taking into account the 20 nm Au seed layer. The

Moreover, chemistry laboratory space must be available for handling process chemicals including acetone, isopropanol, methanol, resist, developer, etchants, and electroplating chemicals. The chemistry lab is preferably combined with the clean room that may range in quality between classes 10 and 1000.

exposed resist was developed in an MIBK:IPA (1:3) solution for 30 s. Pattern sizes on the substrate and on the mask differed by <10%. For some pattern, diffraction effects and distortion were observed, especially at large exposure dose. Moreover, in some areas of the substrate, isolated nanostructures collapsed indicating stiction. Thus, further optimization of exposure dose and development time as well as stiction countermeasures is indicated. Applications of such high-aspect-ratio nanostructures include nanofilters, nanosensors, and photonic bandgap lasers (Chen et al. 2007a).

Romanato et al. (2001, 2003) and Di Fabrizio et al. (2003, 2004) pursued a broad micromanufacturing program including 3D photonic crystals from Ni and Au by means of inclined SXRL at 0.5–1.8 keV. The lattice parameter ranged from 1 µm to 300 nm, pillar diameters

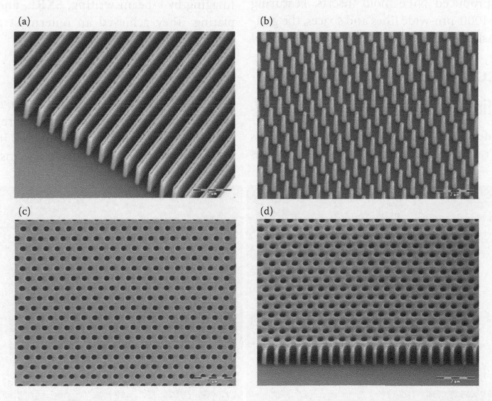

FIGURE 7.14 SEM images of PMMA nanostructures made by SXRL: (a) isolated walls; (b) isolated pillars; (c) hexagonal hole array (perpendicular view); and (d) hexagonal air hole array (inclined view). Scale bar 1 µm for (a) and (b), 2 µm for (c) and (d). (Permission by World Scientific Publishing.)

Chapter 7

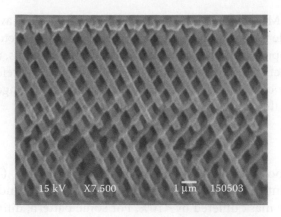

FIGURE 7.15 Yablonovite structure made from crossing pillars and generating up to 14 parallel lattice planes. Scale bar 1 μm. (Permission by IOP Publishing.)

from 300 to 100 nm (Figure 7.15). In one example, pillar diameter and length in a Yablonovite Au structure was 300 nm and about 7 μm, respectively, resulting in an aspect ratio of 23 and eight parallel layers. Masks were made from a 1-μm-thick Si_3N_4 membrane with either 800 nm or 300-nm-thick gold absorber patterns. Malvezzi et al. (2002) also fabricated air/GaAs/AlGaAs photonic-crystal asymmetric waveguides.

Kato et al. (2007) made high-aspect-ratio gratings in PMMA and produced NiFe mold inserts. Featuring either 500- or 1000-nm-wide lines and spaces, the gratings reached aspect ratios of 4.4. The mask consisted of

7.6 Discussion

7.6.1 Limiting Effects

SXRL encounters natural limits when structure sizes approach the range of fundamental physical processes

500-nm-thick Ta absorber on a 2.2-μm-thick Si_3N_4 membrane. Applications of such gratings (Figure 7.16) are related to blue-light DVDs.

Wang et al. (2004) demonstrated sub-μm-scale manufacturing using the KIT 100 keV-e-beam writer for mask making and CAMD for SXRL. The mask featured about 2 μm gold absorbers on a 1-μm-thick Si_3N_4 membrane and was used to expose 6-μm-thick PMMA resist. As an example, Figure 7.17 shows 400 nm diameter 6 μm tall PMMA posts.

Cao et al. (2007) and Zhao et al. (2007, 2008) manufactured quantum dot array diffraction gratings (QDADG) for soft x-ray spectroscopy. Consisting of a bitmap-like distribution of 470-nm-thick Au dots that generate a sinusoidal transmission distribution over the substrate, they are distinguished by the lack of higher-order diffraction and overlap of orders. The QDADG was fabricated by e-beam writing of a soft x-ray mask and a subsequent SXRL step to reach a dot thickness of 470 nm. Dot sizes were either 250 or 500 nm, grating density 250 and 1000 L/mm.

FZPs are important components for x-ray focusing to very small spots. Their critical issue is the width of the outermost ring in relation to its thickness. Liu et al. (2008) fabricated high-aspect-ratio FZPs for hard x-ray imaging by e-beam writing, SXRL, and gold electroplating. They achieved an outermost zone width of 150 nm and a thickness of 660 nm.

including the spatial dose distribution, range of Auger electrons, fluorescence radiation, and resist size scales. Exploration of such limits is going on. In 2001, several protagonists presented extrapolations that 35 nm

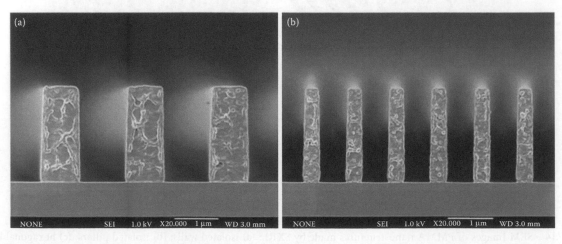

FIGURE 7.16 SEM micrograph of a 2.2-μm-high PMMA grating element and either 1000 nm (a) or 500 nm (b) width. (Permission by Springer Science+Business Media.)

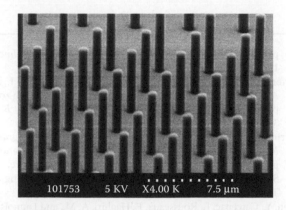

FIGURE 7.17 PMMA posts, 6 µm tall and 400 nm in diameter. (Permission by IOP Publishing.)

should be printable (Khan et al. 2001a) or proposed a technique for 25 nm x-ray nanolithography (Toyota et al. 2001). Others studied focusing x-ray masks for printing very narrow features (Feldman et al. 2001) or high-energy proximity x-ray lithography (Khan et al. 2001b). Near-field x-ray lithography simulations and experiments were pursued (Bourdillon et al. 2003; Kong et al. 2003). The demonstration of <5 nm structures in PMMA (Yasin et al. 2001) shifted the burden to other process steps such as mask making, exposure, and metrology—rather close to the conclusion arrived at by Cerrina (1997).

7.6.2 Ongoing Development

By now, SXRL is no longer in the limelight. Activities have abated after 2005 compared with the heydays in the 1980s/1990s as can be inferred from the disappearance of papers in JVST EIPB conferences. According to Smith et al. (2005), interest in and support for SXRL had waned due to a decision in semiconductor industry and the emergence of the imprint lithography. Literature shows the use of SXRL being motivated predominantly by structures and devices that need SXRL for manufacturing. Laboratories working on SXRL include NSRL/

USTC, LILIT, CAMD, SSLS, PPL, and ANKA. Efforts at MIT and CNTech have diminished.

7.6.3 Future Perspective

There exists an opportunity for SXRL to reinvent itself by new ideas instead of a push by industry. The advantage of SXRL is the nanoscale high-aspect-ratio high-volume manufacturing based on parallel-processing lithography followed, where applicable, by plastic molding (LIGA). Obviously, there is competition with nanoimprint lithography, and future development will show the relative merits of SXRL with LIGA or NIL with mold inserts made from primary pattern generation. More competition will come from interference lithography that has demonstrated nanoscale fabrication (Brueck 2005).

One challenge to SXRL is to achieve 10 nm structures. The reason is that in nanofabrication, there are two basically different approaches, the top-down lithography with the geometry controlled by primary pattern generation, and the bottom-up self-assembly. There is a gap between these approaches as self-assembly goes up to the 1 nm order while SXRL is in the <100 nm range. Considering fan-in fan-out issues in nanotechnology, it seems crucial to bridge this gap at a reasonable value that could be around 10 nm. Critical issues for progress include the reduction of the proximity gap and the tighter specifications of flatness and roughness of mask and resist surfaces. Mask making may also benefit from He ion beam writing.

Facilities for SXRL are around, partly user facilities such as the synchrotron light sources, partly plasma sources or small compact synchrotron light sources. Future applications can be abundant across all of nanotechnology, in particular, nanophotonics with nanoplasmonics (Maier 2007), metamaterials (Engheta and Ziolkowski 2006), nanoelectronics using lumped nanocircuit elements (metatronics) (Engheta 2007), optical metamaterials (Cai and Shalaev 2009), and the meta-foil (Moser et al. 2009).

7.7 Conclusion

SXRL is a valuable and probably critical technique for nanoscale manufacturing despite having lost the limelight of the stormy development in the past when it was driven by anticipated chipmaking. The "10 nm challenge," the development of the demand for cheap

large-area nanostructures, and the competition with other lithographies such as nanoimprint and interference lithography are factors likely to shape the future development of SXRL.

References

Allen, R., Ober, C., Ayothi, R., and Shiota, A. 2007. Macromolecules in lithography. *White Paper, ITRS*. Available at http://www.itrs.net/links/2007ITRS/LinkedFiles/ERM/Macromolecules%20in%20Lithography%20white%20paper_Rev_4ACC.doc

Anderson, O. A., Baker, W. R., Colgate, S. A., Ise, J., Jr., and Pyle, R.V. 1958. Neutron production in linear deuterium pinches. *Phys. Rev.* **110**: 1375.

Angel, G. C. and Samson, J. A. R. 1988. Total photoionization cross sections of atomic oxygen from threshold to 44.3 Å. *Phys. Rev. A* **38**: 5578.

Atoda, N., Kawamatsu, H., Tanino, H., Ichimura, S., Harita, M., and Hoh, K. 1983. Diffraction effects on pattern replication with synchrotron radiation. *J. Vac. Sci. Technol. B* **1**: 1267–1270.

Attwood, D. 1999. *Soft x-rays and Extreme Ultraviolet Radiation, Principles and Applications*. Cambridge: Cambridge University Press.

Bakshi, V. ed. 2009. *EUV Lithography*. Hoboken, New Jersey: SPIE and John Wiley & Sons.

Ballav, N., Chen, C., and Zharnikov, M. 2008. Electron beam and soft x-ray lithography with a monomolecular resist. *J. Photopolym. Sci. Technol.* **21**: 511–517.

Baschnagel, J., Binder, K., and Milchev, A. 2000. Mobility of polymers near surfaces. In *Polymer Surfaces, Interfaces and Thin films*, p. 1. eds. A. Karim and S. Kumar, Singapore: World Scientific Publishing.

Becker, E. W., Ehrfeld, W., Hagmann, P., Maner, A., and Muenchmeyer, D. 1986. Fabrication of microstructures with high aspect ratios and great structural heights by synchrotron radiation lithography, galvanoforming, and plastic moulding (LIGA process). *Microelectron. Eng.* **4**: 35–56.

Beyer, H. and Walter, W. 1991. *Lehrbuch der Organischen Chemie*. Stuttgart: S. Hirzel Verlag.

Blewett. J. P. 1988. Synchrotron Radiation—1873 to 1947. *Nucl. Instr. Meth. A* **266**: 1–9.

Bogolyubov, E. P., Bochkov, V. D., Veretennikov, V. A., Vekhoreva, L. T., Gribkov, V. A., Dubrovskii, A. V., Ivanov, Yu. P. et al. 1998. A powerful soft x-ray source for x-ray lithography based on plasma focusing. *Phys. Scr.* **57**: 488–494.

Bourdillon, A. J., Boothroyd, C. B., Williams, G. P., and Vladimirsky, Y. 2003. Near field x-ray lithography simulations for printing fine bridges. *J. Phys. D: Appl. Phys.* **36**: 2471–2482.

Brodie, I. and Murray, J. J. 1992. *The Physics of Micro/Nano-Fabrication*. New York: Plenum Press.

Brueck, S. R. J. 2005. Optical and interferometric lithography—Nanotechnology enablers. *Proc. IEEE* **93**: 1704–1721.

Brundle, C. R., Evans, C. A., and Wilson, S. 1992. *Encyclopedia of Materials Characterization: Surfaces, Interfaces, Thin Films*. Stoneham, Massachussetts: Butterworth-Heinemann.

Burkhardt, M., Silverman, S., Smith, H. I., Antoniadis, D. A., Rhee, K. W., and Peckerar, M. C. 1995. Gap control in the fabrication of quantum-effect devices using x-ray nanolithography. *Microelectron. Eng.* **27**: 307–310.

Cai, W. and Shalaev, V. 2009. *Optical Metamaterials: Fundamentals and Applications*. Berlin, Germany: Springer.

Cao, L. F., Förster, E., Fuhrmann, A., Wang, C. K., Kuang, L. Y., Liu, S. Y., and Ding, Y. K. 2007. Single order x-ray diffraction with binary sinusoidal transmission grating. *Appl. Phys. Lett.* **90**: 053501.

Carter, D. J. D., Pepin, A., Schweizer, M. R., Smith, H. I., and Ocola, L. E. 1997. Direct measurement of the effect of substrate photoelectrons in x-ray nanolithography. *J. Vac. Sci. Technol. B* **15**: 2509–2513.

Carter, D. J. D., Gil, D., Rajesh M., Mondol M. K., Smith H. I., and Anderson, E. H. 1999. Maskless, parallel patterning with zone-plate array lithography (ZPAL). *J. Vac. Sci. Technol. B* **17**: 3449–3452.

Cerrina, F. 1997. X-ray lithography, In *Handbook of Microlithography, Micromachining, and Microfabrication, Vol. 1: Microlithography*, p. 251, ed. P. Rai-Choudhury. Stevenage, Hertfordshire, UK: SPIE—The International Society for Optical Engineering and The Institution of Electrical Engineers.

Chen, Y., Carcenac, F., Rousseaux, F., Haghiri, A. M., and Launois, H. 1997. High resolution x-ray lithography: Features of two-dimensional patterning. *J. Photopolym. Sci. Technol.* **10**: 619–623.

Chen, A., Chua, S. J., Xing, G. C., Ji, W., Zhang, X. H., Dong, J. R., Jian, L. K., and Fitzgerald, E. A. 2007a. Two-dimensional AlGaInP/GaInP photonic crystal membrane lasers operating in the visible regime at room temperature. *Appl. Phys. Lett.* **90**: 011113.

Chen, A., Liu, G., Jian, L. K., and Moser, H. O. 2007b. Synchrotron-radiation-supported high-aspect-ratio nanofabrication. *COSMOS (Special Issue: Nanoscience & Nanotechnology I)* **3**: 79–88.

Chen, A., Liu, G., Jian, L. K., and Moser, H. O. 2009. Synchrotron-radiation-supported high-aspect-ratio nanofabrication, In *Selected Topics in Nanoscience and Nanotechnology*, ed. Andrew T. S. Wee. Singapore: World Scientific Publishing Company.

Chooi, J. O., Moore, J. A., Corelli, J. C., Silverman, J. P., and Bakhru, H. 1988. Degradation of poy(methylmethacrylate) by deep ultraviolet, x-ray, electron beam, and proton beam irradiations. *J. Vac. Sci. Technol. B* **6**: 2286–2289.

Cowie, J. M. G. 1994. Wet development of polymer resists: A guide to solvent selection. *Adv. Mater. Opt. Electron.* **4**: 155–163.

Cui, Z. 2005. *Micro-Nanofabrication*. Berlin, Germany: Springer.

Deguchi, K., Nakamura, J., Kawai, Y. Ohno, T., Fukuda, M., Oda, M., Kochiya, H., Ushiyama, Y., Hamada, H., and Shimizu. T. 1999. Lithographic performance of a chemically amplified resist developed for synchrotron radiation lithography in the sub-100-nm region. *Jpn. J. Appl. Phys.* **38**: 7090–7093.

Dellmann, L., Roth, S., Beuret, C. Racine, G.-A., Lorenz, H., Despont, M., Renaud, P., Vettiger, P., and de Rooij, N. F. 1998. Fabrication process of high aspect ratio elastic and SU-8 structures for piezoelectric motor applications. *Sens. Actuators, A* **70**: 42–47.

Di Fabrizio, E., Cabrini, S., Cojoc D., Romanato, F., Altissimo, M., Kaulich, B., Kumar, R. et al. 2003. Shaping-rays by diffractive coded nano-optics. *Microelectron. Eng.* **67–68**: 87–95.

Di Fabrizio, E., Romanato, F., Cabrini, S., Kumar, R., Perennes, F., Altissimo, M., Businaro, L. et al. 2004. X-ray lithography for micro- and nano-fabrication at ELETTRA for interdisciplinary applications. *J. Phys. Condens. Matter* **16**: S3517–S3535.

Early, K., Schattenburg, M. L., and Smith, H. I. 1990. Absence of resolution degradation in x-ray lithography for λ from 4.5 nm to 0.83 nm. *Microelectron. Eng.* **11**: 317–321.

Ehrfeld, W., Muenchmeyer, D., and Stutz, A. 1986. Untersuchungen zum Entwicklungsverhalten eines Roentgenresists aus vernetztem Polymethylmethacrylat, Primaerbericht Nr. 02.01.05.P.12C, Forschungszentrum Karlsruhe, unpublished.

Engelke, R., Engelmann, G., Gruetzner, G., Heinrich M., Kubenz, M., and Mischke, H. 2004. Complete 3D UV microfabrication technology on strongly sloping topography substrates using epoxy photoresist SU-8. *Microelectron. Eng.* **73–74**: 456–462.

Engheta, N. 2007. Circuits with light at nanoscales: Optical nanocircuits inspired by metamaterials. *Science* **317**: 1698–1702.

Engheta, N. and Ziolkowski, R. W. eds. 2006. *Electromagnetic Metamaterials: Physics and Engineering Explorations.* Hoboken, New Jersey: Wiley-IEEE Press.

Feder, R. 1970. X-ray projection printing of electrical circuit patterns. *Technical Report TR22.1065.* East Fishkill facility, Hopewell Junction, NY: IBM Components Division.

Feder, R., Spiller, E., and Topalian, J. 1975. Replication of 0.1 μm geometries with x-ray lithography. *J. Vac. Sci. Technol.* **12**: 1332–1334.

Feldman, M., Khan, M., and Cerrina, F. 2001. Focusing x-ray masks for printing very narrow features. *J. Vac. Sci. Technol. B* **19**: 2434–2438.

Gaeta, C. J., Rieger, H., Turcu, I. C. E., Forber, R. A., Campeau, S. M., Cassidy, K. L., Powers, M. J. et al. 2002. High-power compact laser-plasma source for x-ray lithography. *Jpn. J. Appl. Phys.* **41**: 4111–4121.

Gelorme, J. D., Cox, R. J., and Gutierrez, S. A. R. 1989. Photoresist composition and printed circuit boards and packages made therewith, United States Patent 4882245. Available at http://www.freepatentsline.com/4882245.html

Gentili, M., Giovannella, C., and Selci, S. eds. 1994. *Nanolithography: A Borderland between STM, EB, IB, and X-Ray Lithographies.* Dordrecht, Netherlands: Kluwer Academic Publishers.

Ghica, V. and Glashauser, W. 1982. Verfahren fuer die spannungsrissfreie Entwicklung von bestrahlten Polymethylmethacrylat-Schichten. Deutsche Offenlegungsschrift DE 3039110.

Gil, D., Menon, R., Carter D. J. D., and Smith, H. I. 2000. Lithographic patterning and confocal imaging with zone plates. *J. Vac. Sci. Technol. B* **18**: 2881–2885.

Green, G. K. 1976. Spectra and optics of synchrotron radiation, In *Brookhaven National Laboratory Report BNL 50522.* Springfield, Virginia: National Technical Information Service.

Grobman, W. 1984. X-ray lithography. In *Handbook on Synchrotron Radiation,* p. 1131. Vol. 1, ed. E.-E. Koch. North Holland, Amsterdam.

Guo, J. Z. Y. and Cerrina, F. 1993. Modeling x-ray proximity lithography. *IBM J. Res. Dev.* **37**: 331–349.

Gupta, A., Liang, R., Tsay, F. D., and Moacanin, J. 1980. *Macromolecules* **13**: 1696–1700.

Hahn, L., Schwartz, G., Saile, V., and Schulz, J. 2010. First automated production line for x-ray-LIGA (FELIG) is brought on line. *Microsyst. Technol.* **16**: 1287–92.

Heuberger, A. 1985a. X-ray lithography with synchrotron radiation. *Z. Phys. B—Condensed Matter* **61**: 473–476.

Heuberger, A. 1985b. X-ray lithography. *Microelectron. Eng.* **3**: 535–556.

Hofmann, T., Dobisz, E., and Ocko, B. M. 2009. Grazing incident small angle x-ray scattering: A metrology to probe nanopatterned surfaces. *J. Vac. Sci. Technol. B* **27**: 3238–3243.

Hoffnagle, J. A., Hinsberg, W. D., Houle, F. A., and Sanchez, M. I. 2003. Use of interferometric lithography to characterize the spatial resolution of a photoresist film. *J. Photopolym. Sci. Technol.* **16**: 373–379.

Hoffnagle, J. A., Hinsberg, W. D., Sanchez, M. I., and Houle, F. A. 2002. Method of measuring the spatial resolution of a photoresist. *Opt. Lett.* **27**: 1776–1778.

Hosokawa, T., Kitayama, T., Hayasaka, T., Ido, S., Uno, Y., Shibayama, A., Nakata, J., Nishimura, K., and Nakajima, M. 1989. NTT superconducting storage ring—Super-ALIS. *Rev. Sci. Instr.* **60**: 1783–1785.

Houle, F. A., Hinsberg, W. D., and Sanchez, M. I. 2002. Kinetic model for positive tone resist dissolution and roughening. *Macromolecules* **35**: 8591–8600.

Hundt, E. and Tischer, P. 1978. Influence of photoelectrons on the exposure of resists by x-rays. *J. Vac. Sci. Technol.* **15**: 1009.

Iwanenko, D. and Pomeranchuk, J. 1944. On the maximal energy attainable in a betatron. *Phys. Rev.* **65**: 343.

Kato, F., Fujinawa, S., Li, Y., and Sugiyama, S. 2007. Fabrication of high aspect ratio nano grating using SR lithography. *Microsyst. Technol.* **13**: 221–225.

Khan, M., Han, G., Maldonado, J., and Cerrina, F. 2001b. *Proc. SPIE* **4343**: 176.

Khan, M., Han, G., Tsvid, G., Kitayama, T., Maldonado, J., and Cerrina, F. 2001a. Can proximity x-ray lithography print 35 nm features? Yes. *J. Vac. Sci. Technol. B* **19**: 2423–2427.

Kim, S.-K., Lee, J.-E., Park, S.-W., and Oh, H.-K. 2006. Optical lithography simulation for the whole resist process. *Curr. Appl. Phys.* **6**: 48–53.

Kitayama, T. and Itoga, K. 2002. Progress of second generation proximity x-ray lithography (PXL-II) technology, papers of technical meeting on light application and visual science. *IEE Jpn,* **LAV-02(6–11)**: 13–16 (in Japanese).

Klauser, R., Huang, M. L., Wang, S. C., Chen, C. H., Chuang, T. J., Terfort, A., and Zharnikov, M. 2004. Lithography with a focused soft x-ray beam and a monomolecular resist. *Langmuir* **20**: 2050–2053.

Kola, R. R., Celler, G. K., Frackoviak, J., Jurgensen, C. W., and Trimble, L. E. 1991. Stable low-stress tungsten absorber technology for sub-half-micron x-ray lithography. *J. Vac. Sci. Technol. B* **6**: 3301–3305.

Kong J. R., Wilhelmi, O., and Moser, H. O. 2003. Gap optimisation for proximity x-ray lithography using the super-resolution process. *Int. J. Comp. Eng. Sci.* **4**: 585–588.

Lee, K. Y., LaBianca, N., Rishton, S. A. Zolgharnain, S., Gelorme, J. D., Shaw, J., and Chang, T. H.-P. 1995. Micromachining applications of a high resolution ultrathick photoresist. *J. Vac. Sci. Technol. B* **13**: 3012–3016.

Lenhart, J. L., Fischer, D., Sambasivan, S., Lin, E. K., Wu, W.-L., Guerrero, D. J., Wang, Y. B., and Puligadda, R. 2007. Understanding deviations in lithographic patterns near interfaces: Characterization of bottom anti-reflective coatings (BARC) and the BARC–resist interface. *Appl. Surf. Sci.* **253**: 4166–4175.

Liu, Z., Bouamrane, F., Roulliay, M., Kupka, R. K., Labèque, A., and Megtert, S. 1998. Resist dissolution rate and inclined-wall structures in deep x-ray lithography. *J. Micromech. Microeng.* **8**: 293–300.

Liu, L., Liu, G., Xiong, Y., Chen, J., Kang, C. L., Huang, X. L., and Tian, Y. C. 2008. Fabrication of Fresnel zone plates with high aspect ratio by soft x-ray lithography. *Microsyst. Technol.* **14**: 1251–1255.

Löchel, B., Schliwinski, H.-J., Huber, H.-L., Trube, J., Klages, C.-P., Lüthje, H., and Schäfer, L. 1992. Diamond membranes for x-ray masks. *Microelectron. Eng.* **17**: 175–179.

Ma, Y., Shin, J., and Cerrina, F. 2003. Line edge roughness and photoresist percolation development model. *J. Vac. Sci. Technol. B* **21**: 112–117.

Maboudian, R. and Howe, R.T. 1997. Critical review: Adhesion in surface micromechanical structures. *J. Vac. Sci. Technol. B* **15**: 1–20.

Chapter 7

Mack, C. A. 2006. *Field Guide to Optical Lithography. SPIE Field Guide Series*, Vol. FG06. Bellingham, Washington, USA: SPIE Publications.

Madou, M. 1997. *Fundamentals of Microfabrication*. Boca Raton, FL: CRC Press.

Maid, B., Ehrfeld, W., Hormes, J., Mohr, J., and Muenchmeyer, D. 1989. Anpassung der spektralen Verteilung der Synchrotronstrahlung fuer die Roentgentiefenlithographie, Report KfK 4579, Forschungszentrum Karlsruhe.

Maier, S. A. 2007. *Plasmonics: Fundamentals and Applications*. Springer.

Malueg, D. H., Taylor, J. W., Thielman, D., Quinn, L., Dhuey, S., and Cerrina, F. 2004. Modeling, fabrication, and experimental application of clear x-ray phase masks. *J. Vac. Sci. Technol. B* **22**: 3575.

Malvezzi, A. M., Cattaneo, F., Vecchi G., Falasconi, M., Guizzetti, G., Andreani, L. C., Romanato, F. et al. 2002. Second-harmonic generation in reflection and diffraction by a GaAs photonic-crystal waveguide, *J. Opt. Soc. Am. B* **19**: 2122–2128.

Marumoto, K., Yabe, H., Sunao, A., Kise, K., Ami, S., Sasaki, K., Watanabe, H., Itoga, K., and Sumitani, H. 2003. Fabrication of high resolution x-ray masks using diamond membrane for second generation x-ray lithography. *J. Vac. Sci. Technol. B* **21**: 207–213.

Mather, J. W. 1965. Formation of a high-density deuterium plasma focus. *Phys. Fluids* **8**: 366.

Matsui, S., Moriwaki, K., Hasegawa, S., Aritome, H., and Namba, S. 1979. Contrast of the x-ray mask for synchrotron radiation. *Jpn. J. Appl. Phys.* **18**: 1205–1206.

McGowan, J. W., Borwein, B., Medeiros, J. A., Beveridge, T., Brown, J. D., Spiller, E., Feder, R., Topalian, J., and Gudat, W. 1979. High resolution microchemical analysis using soft x-ray lithographic techniques. *J. Cell Biol.* **80**: 732–735.

Meyer, P., Schulz, J., and Hahn, L. 2003. DoseSim: Microsoft-Windows graphical user interface for using synchrotron x-ray exposure and subsequent development in the LIGA process. *Rev. Sci. Instrum.* **74**: 1113–1119.

MicroChem. 2001. *NANOTMSU_8, 2000 Negative Tone Photoresists Formulations* 2–15 Report.

Minkov, D., Yamada, H., Toyosugi, N., Morita, M., and Yamaguchi, T. 2007. Application of a theory for generation of soft x-ray by storage rings and its use for x-ray lithography. *Synchrotron Radiation Instrumentation: Ninth International Conference on Synchrotron Radiation Instrumentation, AIP Conference Proceedings* **879**: 268–271.

Minkov, D., Yamada, H., Toyosugi, N., Yamaguchi, T., Kadono, T., and Morita M. 2006. Theory and characteristics of transition radiation emitted by low-energy storage-ring synchrotrons for use in x-ray lithography. *J. Synchrotron Rad.* **13**: 336–342.

Moon, E. E., Lee, J., Everett, P. N., and Smith, H. I. 1998. Application of interferometric broadband imaging alignment on an experimental x-ray stepper. *J. Vac. Sci. Technol. B* **16**: 3631–3636.

Moser, H. O., Jian, L. K., Chen, H. S., Bahou, M., Kalaiselvi, S. M. P., Virasawmy, S., Maniam, S. M. et al. 2009. All-metal self-supported THz metamaterial—The meta-foil. *Opt. Express* **17**: 23914–23919.

Moser, H. O. and Lehr, H. 1989. Design of a fast electron beam scanning system for compact synchrotron light sources. *Rev. Sci. Instr.* **60**: 1771–1774.

Murata, K. 1985. Theoretical studies of the electron scattering effect on developed pattern profiles in x-ray lithography. *J. Appl. Phys.* **57**: 575.

Murphy, J. B., Bozoki, E., Galayda, J., Halama, H., Heese, R., Hsieh, H., Kalsi, S. et al. 1990. The Brookhaven Superconducting x-ray Lithography source, in *EPAC 90, Proceedings of 2nd European Particle Accelerator Conference*, Nice, France, *1990*, Editions Frontières, Gif-sur-Yvette, Vol. 2, pp. 1828.

Nakanishi, T., Ikegami, K., Kodera, I., Maruyama, A., Matsuda, T., Morita, M., Nakagawa, T. et al. 1995. Construction and beam experiment of a compact storage ring at MELCO. *Rev. Sci. Instr.* **66**: 1968–1970.

Ocola, L. E. and Cerrina, F. 1993. Parametric modeling of photoelectron effects in XRL. *J. Vac. Sci. Technol. B* **11**: 2839–2844.

Ocola, L. E. and Cerrina, F. 1994. Parametric modeling at resist-substrate interfaces. *J. Vac. Sci. Technol.* **B** 12:3986–3989.

Ogawa, T., Mochiji, K., Soda, Y., and Kimura, T. 1989. The effects of secondary electrons from a silicon substrate on SR x-ray lithography. *JJAP Series 3, Proc. 1989 Int. Symp. on MicroProcess Conf.*, Kobe, pp. 120–123.

Ohta, T., Kumar, R., Yamashita, Y., and Hoga, H. 1994. High temperature deposition of SiN films using low pressure chemical vapor deposition system for x-ray mask application. *J. Vac. Sci. Technol. B* **12**: 585–588.

Oizumi, H., Ohtani, M., Yamashita, Y., Murakami, K., Nagata, H., and Atoda, N. 1994. Resist performance in 5 nm soft x-ray projection lithography. *Jpn. J. Appl. Phys.* 33: 6919–6922.

Raccurt, O., Tardif, F., d' Avitaya, F. A., and Vareine, T. 2004. Influence of liquid surface tension on stiction of SOI MEMS. *J. Micromech. Microeng.* **14**: 1083–1090.

Raptis, I., Velessiotis, D., Vasilopoulou, M., and Argitis P. 2000. Development mechanism study by dissolution monitoring of positive methacrylate photoresists. *Microelectron. Eng.* **53**: 489–492.

Ravet, M. F. and Rousseaux F. 1996. Status of diamond as membrane material for x-ray lithography masks. *Diamond Relat. Mater.* **5**: 812–818.

Reynolds, G. W. and Taylor, J. W. 1999. Factors contributing to sidewall roughness in a positive-tone, chemically amplified resist exposed by x-ray lithography. *J. Vac. Sci. Technol. B* **17**: 334–344.

Reznikova, E., Mohr, J., Boerner, M., Nazmov, V., and Jakobs, P. 2008. Soft x-ray lithography of high aspect ratio SU8 submicron structures. *Microsyst. Technol.* **14**: 1683–1688.

Romanato, F., Di Fabrizio, E., Vaccari, L., Altissimo, M., Cojoc, D., Businaro, L., and Cabrini, S. 2001. LILIT beamline for soft and deep x-ray lithography at Elettra. *Microelectron. Eng.* **57–58**: 101–107.

Romanato, F., Businaro, L., Vaccari, L., Cabrini, S., Candeloro, P., De Vittorio, M., Passaseo, A. et al. 2003. Fabrication of 3D metallic photonic crystals by x-ray lithography. *Microelectron. Eng.* **67–68**: 479–486.

Saleh, B. E. A. and Teich, M. C. 1991. *Fundamentals of Photonics*. New York: John Wiley & Sons.

Schattenburg, M. L., Tanaka, I., and Smith, H. I. 1987. Microgap x-ray nanolithography. *Microelectron. Eng.* **6**: 273–279.

Schattenburg, M. L., Li, K., Shin, R. T., Kong, J. A., and Smith, H. I. 1991. Electromagnetic calculation of soft x-ray diffraction from 0.1 μm gold structures. *J. Vac. Sci. Technol. B* **9**: 3232–3236.

Schott, G. A. 1912. *Electromagnetic Radiation and the Mechanical Reactions Arising from It*. Cambridge: Cambridge University Press.

Schwinger, J. 1946. Electron radiation in high energy accelerators. *Phys. Rev.* **70**: 798.

Schwinger, J. 1949. On the classical radiation of accelerated electrons. *Phys. Rev.* **75**: 1912.

Scipioni, L., Sanford, C. A., Notte J., Thompson, B., and McVey, S. 2009. Understanding imaging modes in the helium ion microscope. *J. Vac. Sci. Technol. B* **27**: 3250–3255.

Shin, J., Han, G., Ma, Y., Moloni, K., and Cerrina, F. 2001. Resist line edge roughness and aerial image contrast. *J. Vac. Sci. Technol. B* **19**: 2890–2895.

Shultz, A. R., Frank, P., Griffing, B. F., and Young, A. L. 1985. Quantum yield for poly(methyl methacrylate) chain scission by 214–229 nm wavelength light. *J. Polym. Sci.: Polym. Phys. Ed.* **23**: 1749–1785.

Sidorkin, V., van Veldhoven, E., van der Drift, E., Alkemade, P., Salemink, H., and Maas, D. 2009. Sub-10-nm nanolithography with a scanning helium beam. *J. Vac. Sci. Technol. B* **27**: L18–L20.

Sijbrandij, S., Notte, J., Scipioni, L., Huynh, C., and Sanford, C. 2010. Analysis and metrology with a focused helium ion beam. *J. Vac. Sci. Technol. B* **28**: 73–77.

Smith, H. I., Carter, D. J. D., Meinhold, M., Moon, E. E., Lim, M. H., Ferrera, J., Walsh, M., Gil, D., and Menon, R. 2000. Soft x-rays for deep sub-100 nm lithography, with and without masks. *Microelectron. Eng.* **53**: 77–84.

Smith, H. I., Berggren, K. K., Mondol, M. K., and Schattenburg, M. L. 2005. *The annual Progress Report of the Research Laboratory of Electronics (RLE) at the Massachusetts Institute of Technology (MIT)* 147, p. 24–10. Available at http://www.rle.mit.edu/media/pr_no147.html.

Smith, H. I. and Schattenburg, M. L. 1994. X-ray nanolithography: Limits and applications to sub-100 nm manufacturing. In *Nanolithography: A Borderland between STM, EB, IB, and x-ray Lithographies*, eds. M. Gentili, C. Giovanella, and S. Selci, NATO ASI Series, Series E: Applied Sciences—Vol. 264, pp. 103–119. Dordrecht, Netherlands: Kluwer Academic Publishers.

Smith, H. I., Spears, D. L., and Bernacki, S. E. 1973. X-ray lithography: A complementary technique to electron beam lithography. *J. Vac. Sci. Technol.* **10**: 913–917.

Sokolov, A. A. and Ternov, I. M. 1968. *Synchrotron Radiation*. Oxford: Pergamon Press.

Spears, D. L. and Smith, H. I. 1972. High-resolution pattern replication using soft x-rays. *Electron. Lett.* **8**: 102–104.

van Spengen, W. M., Puers, R., and De Wolf, I. 2002. A physical model to predict stiction in MEMS. *J. Micromech. Microeng.* **12**: 702–713.

Spiller, E., Eastman, D., Feder, R., Grobman, W. D., Gudat, W., and Topalian, J. 1976. The application of synchrotron radiation to x-ray lithography. *J. Appl. Phys.* **47**: 5450–5459.

Sugano, K., Weiyu, S., Tsuchiya, T., and Tabata, O. 2008. Design of a soft x-ray source with periodic microstructure using resonance transition radiation for tabletop synchrotron. *IEEJ Trans. Electr. Electron. Eng.* **3**: 268–273.

Suzuki, K. and Smith, B. W. eds. 2007. *Microlithography: Science and Technology*. Boca Raton, FL: CRC Press Online, Taylor & Francis.

Takahashi, N. 1987. Compact superconducting storage ring for x-ray lithography. *Nucl. Instr. Meth. B* **24/25**: 425–428.

Takahashi, N. 1990. Compact SR light source for x-ray lithography: Aurora. *Microelectron. Eng.* **11**: 283–286.

Tanaka, T., Morinaga, M., and Atoda, N. 1993. Mechanism of resist pattern collapse during development process. *Jpn. J. Appl. Phys.* **32**: 6059.

Tanaka, T., Morigami, M., Oizumi, H., Ogawa, T., and Uchino, S. 1994b. Prevention of resist pattern collapse by flood exposure during rinsing process. *Jpn. J. Appl. Phys.* **33**: L1803.

Tanaka, T., Morigami, M., Oizumi, H., Soga, T., Ogawa, T., and Murai, F. 1994a. Prevention of resist pattern collapse by resist heating during rinsing. *J. Electrochem. Soc.* **141**: L169.

Thompson, L. F., Willson, C. G., and Bowden, M. J. eds. 1994. *Introduction to Microlithography, ACS Professional Reference Book*. Washington, DC: American Chemical Society.

Tomimasu, T. 1987. An electron undulating ring dedicated to VLSI lithography. *Jpn. J. Appl. Phys.* **26**: 741–746.

Tomimasu, T., Noguchi, T., Sugiyama, S., Yamazaki, T., and Mikado, T. 1985. *IEEE Trans. Nucl. Sci.* **NS-32**: 3403.

Torikai, A., Ohno, M., and Fueki, K. 1989. Photodegradation of poly(methyl methacrylate) by monochromatic light: Quantum yield, effect of wavelengths, and light intensity. *J. Appl. Polym. Sci.* **41**: 1023–1032.

Toyota, E., Hori, T., Khan, M., and Cerrina, F. 2001. Technique for 25 nm x-ray nanolithography. *J. Vac. Sci. Technol. B* **19**: 2428–2433.

Toyosugi, N., Yamada, H., Minkov, D., Morita, M., Yamaguchi, T., and Imai, S. 2007. Estimation of soft x-ray and EUV transition radiation power emitted from the MIRRORCLE-type tabletop synchrotron. *J. Synchrotron Rad.* **14**: 212–218.

Trinks, U., Nolden, F., and Jahnke, A. 1982. The table-top synchrotron radiation source "Klein Erna." *Nucl. Instr. Meth.* **200**: 475–479.

Utsumi, Y. and Kishimoto, T. 2005. Large area and wide dimension range x-ray lithography for lithographite, galvanoformung, and abformung process using energy variable synchrotron radiation. *J. Vac. Sci. Technol. B* **23**: 2903–2909.

Valiev, K. A. 1992. *The Physics of Submicron Lithography*. New York: Plenum Publishing Corporation.

Veigele, J. 1973. Photon cross sections from 0.1 keV to 1 MeV for elements $Z = 1$ to $Z = 94$. *Atomic Data Tables* **5**: 51.

Verma, R., Lee, P., Springham, S. V., Tan, T. L., Rawat, R. S., and Krishnan, M. 2008. Order of magnitude enhancement in x-ray yield at low pressure deuterium–krypton admixture operation in miniature plasma focus device. *Appl. Phys. Lett.* **92**: 011506.

Victoreen, J. A. 1943. Probable x-ray mass absorption coefficients for wavelengths shorter than the K critical absorption wavelength. *J. Appl. Phys.* **14**: 95.

Victoreen, J. A. 1948. The absorption of incident quanta by atoms as defined by the mass photoelectric absorption coefficient and the mass scattering coefficient. *J. Appl. Phys.* **19**: 855.

Vora, K. D., Shew, B. Y., Harvey, E. C., Hayes, J. P., and Peele, A. G. 2005. Specification of mechanical support structures to prevent SU-8 stiction in high aspect ratio structures. *J. Micromech. Microeng.* **15**: 978–983.

Wang, L., Desta, Y. M., Fettig, R. K., Goettert, J., Hein, H., Jakobs, P., and Schulz, J. 2004. High resolution x-ray mask fabrication by a 100 keV electron-beam lithography system. *J. Micromech. Microeng.* **14**: 722–726.

Ward, B. M., Notte, J., and Economou, N. P. 2006. Helium ion microscope: A new tool for nanoscale microscopy and metrology. *J. Vac. Sci. Technol. B* **24**: 2871–2874.

Watanabe, H., Marumoto, K., Sumitani, H., Yabe, H., Kise, K., Itoga, K., and Aya, S. 2002. 50 nm pattern printing by narrowband proximity x-ray lithography. *Jpn. J. Appl. Phys.* **41**: 7550–7555.

Wilson, A. D. 1993. X-ray lithography in IBM, 1980–1992, the development years. *IBM J. Res. Develop.* **37**, 299–318.

Chapter 7

Wilson, M. N., Smith, A. I. C., Kempson, V. C., Purvis, A. L., Townsend, M. C., Jorden, A. R., Anderson, R. J., and Andrews, D. E. 1990. Helios: A compact synchrotron x-ray source. *Microelectron. Eng.* **11**: 225–228.

Wilson, M. N., Smith, A. I. C., Kempson, V. C., Townsend, M. C., Schouten, J. C., Anderson, R. J., Jorden, A. R., Suller, V. P., and Poole, M. W. 1993. The Helios 1 compact superconducting storage ring x-ray source. *IBM J. Res. Develop.* **37**: 351–371.

Winston, D., Cord, B. M., Ming, B., Bell, D. C., DiNatale, W. F., Stern, L. A., Vladar, A. E. et al. 2009. Scanning-helium-ion-beam lithography with hydrogen silsesquioxane resist. *J. Vac. Sci. Technol. B* **27**: 2702–2706.

Wong, D., Patran, A., Tan, T. L., Rawat, R. S., and Lee, P. 2004. Soft x-ray optimization studies on a dense plasma focus device operated in neon and argon in repetitive mode. *IEEE Trans. Plasma Sci.* **32**: 2227–2236.

Wong, D., Tan, T. L., Lee, P., Rawat, R. S., and Patran, A. 2006. Study of x-ray lithographic conditions for SU-8 by Fourier transform infrared spectroscopy. *Microelectron. Eng.* **83**: 1912–1917.

Wood, A. 2001. *Magnetic Venture—The Story of Oxford Instruments.* Oxford: Oxford University Press.

Xia, Y. and Whitesides, G.M. 1998. Soft lithography. *Angew. Chem. Int. Ed. Engl.* **37**: 551–575.

Yamada, H. 1989. Photon storage ring, *Jpn. J. Appl. Phys.* **28**: L1665–L1668.

Yamaguchi, T., Namatsu, H., Nagase, M., Yamazaki, K., and Kurihara, K. 1997. Nanometer-scale linewidth fluctuations caused by polymer aggregates in resist films. *Appl. Phys. Lett.* **71**: 2388–2390.

Yamaguchi, T., Namatsu, H., Nagase, M., Yamazaki, K., and Kurihara, K. 2000. *J. Photopolym. Sci. Technol.* **13**: 427–433.

Yasin, S., Hasko, D. G., and Ahmed, H. 2001. Fabrication of <5 nm width lines in poly(methylmethacrylate) resist using a water:isopropyl alcohol developer and ultrasonically-assisted development. *Appl. Phys. Lett.* **78**: 2760.

Yoshimura, T., Shiraishi, H., Yamamoto, J., and Okazaki, S. 1993. Nano edge roughness in polymer resist patterns. *Appl. Phys. Lett.* **63**: 764–766.

Yue, W., Chiam, S., Ren, Y., van Kan, J. A., Osipowicz, T., Jian, L. K., Moser, H. O., and Watt, F. 2008. The fabrication of x-ray masks using proton beam writing. *J. Micromech. Microeng.* **18**: 085010.

Yue, W. S., Ren, Y. P., van Kan, J. A., Chiam, S.-Y., Jian, L. K., Moser, H. O., Osipowicz, T., and Watt, F. 2009. Proton beam writing and electroplating for the fabrication of high-aspect-ratio Au microstructures. *Nucl. Instr. Meth. B* **267**: 2376–2380.

Zaikov, G. E. ed. 1995. *Degradation and Stabilization of Polymers: Theory and Practice.* New York: Nova Science Publishers.

Zhao, M., Zhu, X., Chen, B., Xie, C. Q., Liu, M., and Cao, L. 2007. Fabrication of quantum dot array diffraction grating for soft x-ray spectroscopy. *Proc. SPIE* **6724**: 67241O.1–67241O.7.

Zhao, M., Zhu, X., Chen, B., Xie, C., Liu, M., Niu, J., Kuang, L., and Cao, L. 2008. Design, fabrication, and test of soft x-ray sinusoidal transmission grating. *Opt. Eng.* **47**: 058001.

Zweifel, H., Maier, R. D., and Schiller, M. 2009. *Plastics Additives Handbook.* Muenchen, Germany: Hanser Verlag.

8. Simulations in Micro- and Nanofabrication

Sergey Babin

Abeam Technologies, Castro Valley, California

Nanofabrication Handbook. Edited by Stefano Cabrini and Satoshi Kawata © 2012 CRC Press / Taylor & Francis Group, LLC. ISBN: 978-1-4200-9052-9

Nanofabrication is a complex and expensive process. Especially in semiconductor manufacturing, where the cost of equipment units can reach tens of millions of dollars, and every experiment in process development is costly. Thus, it is no wonder that simulations are used extensively to supplement experiments. Simulations allow engineers and researchers to understand the process, to optimize it by evaluating options, and to correct factors responsible for hindering the process. By troubleshooting process problems and determining optimum process settings, simulations help shorten time and save money by reducing the number of experiments. In addition, if simulators can predict a systematic error, then usually this error can be corrected. We focus on simulation packages that are commercially

Chapter 8

available, even though there are many good simulators developed in universities and in companies for internal use. A few areas are discussed: optical lithography, electron beam lithography (EBL), nanoimprint, dry etch, film deposition, scanning electron microscope (SEM) metrology, and optical metrology.

8.1   Optical Lithography

8.1.1   Projection Lithography

Lithography is one of the most complex and costly operations. Optical lithography is the main method used in the mass production of chips. The wavelength used in lithography is by far larger than the fabricated feature size. In 1980s, most specialists predicted the limits of optical lithography to be about 1.5–1 µm for minimum features. However, the progression of this technique is so remarkable that features of 32 nm are still printed using optical lithography and this progress is expected to continue. The industry is using all kinds of tricks to print these small features, including optical proximity correction, phase shift masks, double patterning, immersion liquids, and so on. Simulation of optical lithography is a critical step that cannot be underestimated; it is the enabler that allows lithography at such small feature sizes.[1–3]

The well-known optical lithography simulators are ProLith of KLA-Tencor, Sentaurus Lithography of Synopsys, and Tempest of Panoramic Technologies. In their simulations, optical models range from simple principles of geometrical optics to rigorous electromagnetic simulations based on Maxwell equations. They consider

- Light generation by the source, which may include configurations such as quadrupoles, dipoles, rings, sectors, and so on.
- Wave propagation through the mask. The mask can be either an approximation of the thin patterned film or a full three-dimensional (3D) representation of part of a photomask—in this case, the simulation is done using rigorous electromagnetic theory.
- Light propagation through the optical system of a stepper or scanner.
- Aerial image in the resist; the simulation includes light reflections from substrate layers, top or bottom antireflection layers, and interference of all the primary and reflected waves in the resist.
- Photokinetics and photochemistry.
- Postexposure bake time and temperature.

- Resist development model; either a simple threshold model or a complex model based on dissolution of the material according to the distribution of photoactive compounds (PACs) after the irradiation and the diffusion of PACs.

An example of optical lithography simulation is given in Figure 8.1. These results were obtained using the HyperLith(TM) lithography simulator from Panoramic Technology Inc.[4]

The simulators of optical lithography are used widely for the following tasks:

- Film stack optimization
- Source/mask optimization
- Evaluation of new resists
- Optimization of numerical aperture and partial coherence of a stepper
- Setting troubleshooting and root cause analysis
- Verification of optical proximity correction
- Improvement of the process window

The wide use of simulators in research & development (R&D) and the semiconductor industry has been a huge factor in the success of the industry to achieve successful printing of critical dimensions (CDs) down to 28 nm range and below.

8.1.2   Contact and Proximity Printing

Contact printing is often used in the fabrication of devices with feature sizes of 1 µm and above; it is difficult to make features smaller than 1 µm using contact or proximity printing. This is an inexpensive method that does not require significant spending on equipment. Simulations can be helpful here to understand the printing process, its capabilities, and tolerances for specific conditions and materials used. GenISys offers a Layout LAB simulator which models light intensity in the resist as well as resist development.[5] In the simulations presented in Figure 8.2, aerial images and resist contours are simulated for two patterns. The first simulation is compared with the experimental result. The second simulation suggests that the optimization of mask design is needed to improve the printability of the pattern.

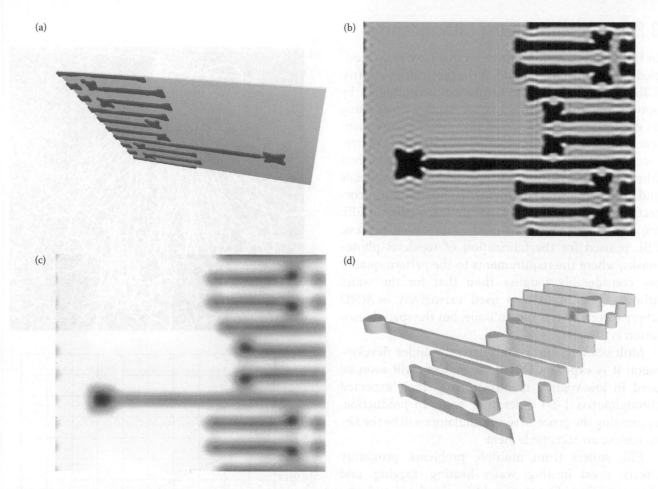

(a)

(b)

(c)

(d)

FIGURE 8.1 The scattering of light from the 3D mask geometry (a) is simulated using the finite-difference time-domain (FDTD) method. The near-fields (b) are analyzed to extract the scattered orders which are imaged by the scanner forming an image intensity pattern in the photoresist at the wafer plane (c). Finally, the photoresist exposure, postexposure bake, and development processes are simulated, resulting in the 3D resist surface (d).

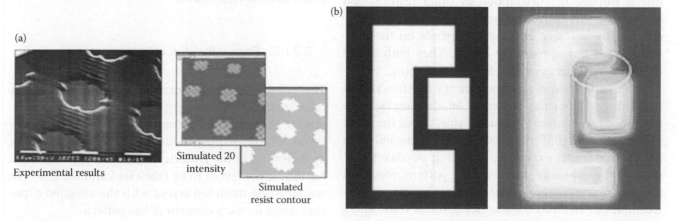

(a)

Experimental results

Simulated 20 intensity

Simulated resist contour

(b)

FIGURE 8.2 Simulator of contact and proximity printing optical lithography. (a) Experimental result, simulated aerial image, and resist image; (b) designed mask and simulated aerial image in the resist.

Chapter 8

8.2 Electron Beam Lithography

In EBL, there are a few processes that have to be modeled to optimize the quality of the fabricated patterns. EBL offers an excellent resolution unattainable by optical lithography. On the other hand, writing using an electron beam is slow and EBL systems are expensive. This is the reason why it is very desirable to predict the EBL process in order to carry out the correct fabrication on the first attempt, without many trials and consequent improvements. Simulations and corrections of effects that deteriorate the quality of EBL are an indispensable part of the fabrication process. EBL is used for the fabrication of top-level photomasks, where the requirements to the pattern quality are considerably tougher than that for the wafer lithography. EBL is also used extensively in R&D where throughput is not an issue, but the spatial resolution is the key.

Multibeam systems are currently under development; it is expected that these systems will soon be used in low-volume manufacturing. Their expected throughput is 1–50 wafers per hour. In production, optimizing the process using simulators will be the key to achieve an acceptable yield.

EBL suffers from multiple problems: proximity effects, resist heating, wafer heating, fogging, and charging. These factors should be simulated and corrected to achieve a good-quality pattern.

8.2.1 Proximity Effects and Its Correction

Electron scattering in the resist and substrate results in the redistribution of absorbed energy in the resist. The line size after development depends on whether the line is isolated or surrounded by other features and how dense the surrounding is (external proximity effect). The line size error also depends on the linewidth itself (intraproximity effect). They both are of the same nature, caused by electron scattering.

The distribution of absorbed energy as a function of the distance from the center of an electron beam is described by a sum of Gaussians, thus called the point spread function (PSF). The convolution of the PSF with the pattern under exposure, including possible variations of the exposure dose over the pattern, results in the distribution of absorbed energy in the resist. Examples of electron trajectories inside a resist and a multilayered wafer, as well as a PSF simulated using Monte Carlo software, are shown in Figure 8.3.

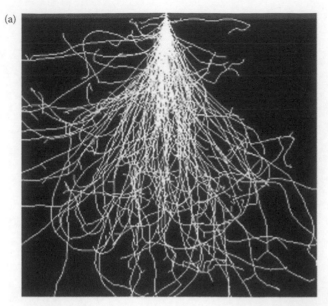

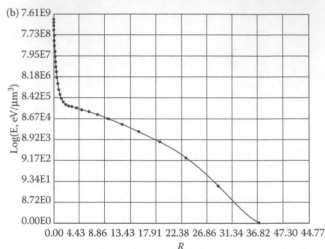

FIGURE 8.3 (a) Electron trajectories at 100 kV and (b) a PSF of absorbed energy in EBL as a function of a distance from the beam center; a multilayered substrate.

8.2.1.1 Proximity Effect Correction and Verification

The proximity effects correction (PEC) attempts to equalize the absorbed energy in the resist at the boundaries of the patterned elements. This involves a modulation of the exposure dose over the pattern, or modification of the pattern shapes, or both. The input for PEC is the layout to be fabricated and the PSF. The output is the modified layout with the assigned exposure doses for each element of the pattern.

There are commercially available packages used in EBL: Proxecco of PDF Solutions, Layout Beamer of

GenISys, and PARAPEC of Nippon Control System.[6–8] A graphical user interface of Layout Beamer is displayed in Figure 8.4a; the result of correction and corresponding experimental result are shown in Figure 8.4b.

8.2.1.1.1 Modeling of Proximity Effects and Resist Development

The verification of proximity correction is normally done by a software independent of the PEC software manufacturer. TRAVIT-EBL by aBeam

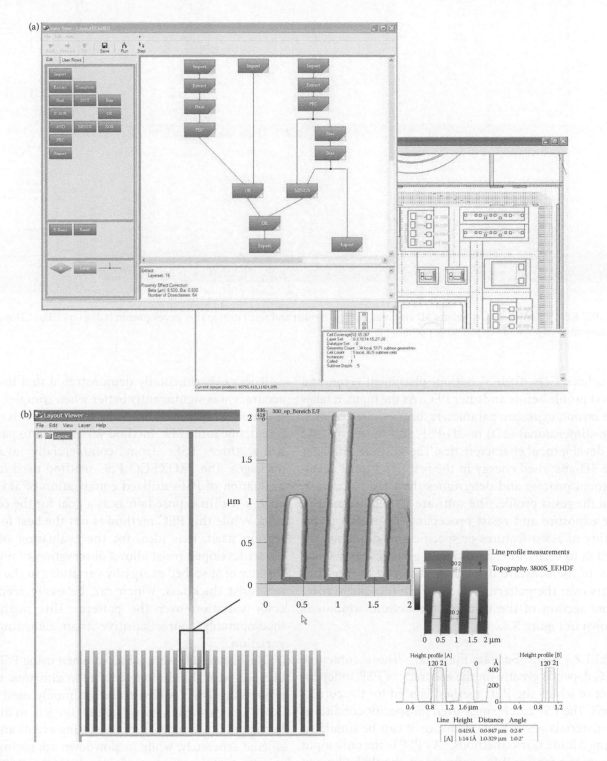

FIGURE 8.4 PEC and preceding Boolean operations in Beamer software of GenISys: (a) graphical user interface; (b) result of correction and corresponding experimental result.

Chapter 8

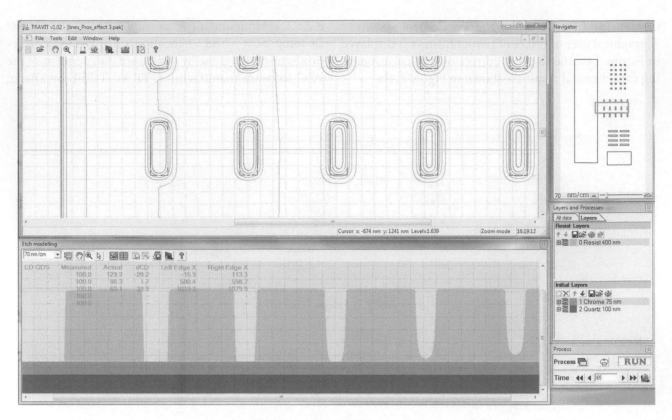

FIGURE 8.5 TRAVIT-EBL simulates 3D absorbed energy in resist and resist profile after development; it displays CDs, CD variation, placement of edges, and vertical resist profile.

simulates CDs, their variation, placement error, and resist profile before and after PEC. As the input, it takes the layout, exposure parameters, beam characteristics, two-dimensional (2D) or 3D PSF, and resist type and its development characteristics. The software simulates the 3D-absorbed energy in the resist and resist development process and determines the CDs, placement, and the resist profile. The software is used to optimize the exposure and resist processing, to predict printability of assist features or specific area of interest, as well as to verify PEC accuracy. A graphical user interface of the software displaying absorbed energy contours over the pattern in the top window and vertical cross section of the resist in the bottom window is shown in Figure 8.5.

8.2.1.1.2 Point Spread Function

The accuracy of PEC depends greatly on the accuracy of PSF, independent of which the PEC method is used for the corrections. The PSF can be measured for specific conditions of materials and EBL systems, or it can be simulated using Monte Carlo methods.[9] As PSF is the only input parameter for any PEC software, it should be known precisely.

It was experimentally demonstrated that the PEC accuracy was significantly better when complex physical models were used in Monte Carlo.[10] In this experiment, one same PEC method was applied to patterns using three PSFs from commercially available packages. The PROXECO PEC method used for the evaluation of PSFs utilized equalization of absorbed energy within exposed areas as a goal for the correction. While this PEC method is not the best for correction itself, it is ideal for the evaluation of PSF. Underdeveloped resist allows observation of nonuniformity of absorbed energy by variations in the residual resist thickness, which can be easily seen as a color variation over the pattern. This method is incomparably more sensitive than detecting CD variation.

The best results were achieved when using PSF simulations with the discrete loss approximation model rather than PSF resulted from commonly used slowdown approximation model; see Figure 8.6. In discrete loss approximation, electron scattering events are considered separately, while in slowdown approximation, results of all the effects are lumped into a single number. This difference in models results in significant

FIGURE 8.6   **(See color insert.)** Optical microscope images of resist after partial development. Various PSFs were used for the same PEC correction. Better color uniformity means better accuracy of PSF in this experiment. (Courtesy of Dr. B. Nilsson, Chalmers University.)

difference in the accuracy of simulated PSF and finally in the PEC accuracy.

8.2.2   Resist Heating

High beam current and high electron energy are used in modern EBL systems to improve spatial resolution and to decrease writing time. Both these parameters lead to significant energy density locally deposited by the electron beam. Therefore, the resist temperature is highly nonuniform during writing. This heat changes the sensitivity of the resist and leads to undesired variations of CDs. This effect is critical in mask making where the beam is of variable shape, has a high current, and the substrate is quartz with low thermal conductivity. Temperature variation up to a few hundred degree Celsius was found.[11] The resist heating is also an important factor on wafers in direct write lithography when ultrahigh resolution is targeted.[12]

Such variation typically takes place when the distance range is a few tens of microns, and was also noticeable on millimeter distances. In the example presented in Figure 8.7a time-dependent heating effect is shown. Effective absorbed dose in the resist was changed due to resist heating. The exposure goes from the bottom left corner to the right, and then backs up as a serpentine. A 50 kV variably shaped beam system, the beam current density of which was 95 A/cm^2 and the exposure dose being 19 μC/cm^2, was uniform everywhere. Subfield was 32 μm, 1 μm^2 flashes. The simulated temperature reaches over 200°C. The corre-

sponding experimental result is shown in Figure 8.7b, the optical micrograph of the resist after partial development; the white area on the top is chrome, where the resist was fully gone. The resist acts as a temperature sensor; its residual thickness is observed as color variation. The simulated result corresponds well to the experiment. The simulations were carried out by the TEMPTATION (Temperature Simulation) software produced by aBeam.

The resist heating can be predicted; optimized resist and EBL system parameters can be selected to minimize CD error due to resist heating. Furthermore, corrections of resist heating can be applied; however, there is no commercial software for the heating correction.

8.2.3   Wafer Heating

In addition to a local resist heating which leads to variation of CDs, the electron beam also heats the full wafer. The wafer expands, which results in the placement distortion. The typical global temperature elevation in direct write is a fraction of a degree. The distortion greatly depends on wafer mounting and beam settings and may take a value of 10s of nanometers to about 1 μm. Distribution of temperature over an 8 in. wafer before the exposure of the last chip in fast projection EBL is shown in Figure 8.8; the exposure dose was 5 C/cm^2, beam voltage 100 kV, and beam current density 160 mA/cm^2. Wafer cooling by helium at the bottom of the wafer and by radiation was taken into account.[13]

Chapter 8

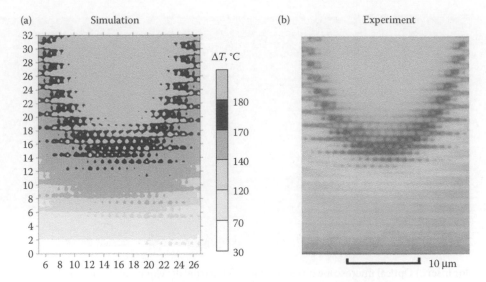

FIGURE 8.7 The effective absorbed dose in the resist was changed due to resist heating. Uniform exposure dose was applied; the exposure goes from the bottom left corner to the right and then back up as a serpentine. (a) Simulated temperature rises, the temperature goes up to over 200°C; (b) optical micrograph of resist after partial development, the white area on the top is chrome (resist was fully gone). The resist acts as a temperature sensor.

Varieties of commercial software were applied to simulate wafer heating using finite-element or finite-difference methods, such as ANSYS. In-plane displacement after the exposure of two rows of subfields is shown in Figure 8.9.[14] It was demonstrated that the software is effective in the prediction of in-plane and out-of-plane distortions on masks and wafers.

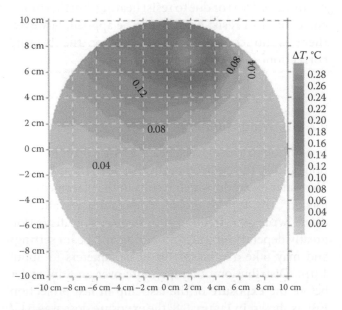

FIGURE 8.8 Distribution of a temperature over an 8 in. wafer before the exposure of the last chip using fast projection EBL. Wafer cooling by helium at the bottom of the wafer and by radiation was taken into account.

8.2.4 Shot Noise

In the early days of EBL, poly(methyl methacrylate) resist was most commonly used. With its excellent resolution, the resist required a high exposure dose which resulted in a long writing time. In addition, PMMA does not have good resistance in dry etch. Chemists put a lot of effort into the development of fast resists. The development of chemically amplified resists was a breakthrough as they require an exposure dose two to three orders less than PMMA, and correspondingly, a writing throughput becomes two orders higher than that with PMMA. It turned out that the advantage of high sensitivity has an important drawback, especially when it comes to small sizes of features.

When the resist requires a small exposure dose, the number of electrons required to do the exposure drops. When the area of the feature is also smaller, the total number of electrons per feature should still be statistically sound. Otherwise, two equal features will receive different doses due to statistical variation of current in the electron gun. This effect is called shot noise; see Figure 8.10.[15] CHARIOT-EBL software by aBeam Technologies simulates the shot noise.

8.2.5 Charging

Electron beam causes severe charging of the resist and substrate. As a result, the primary electron beam may

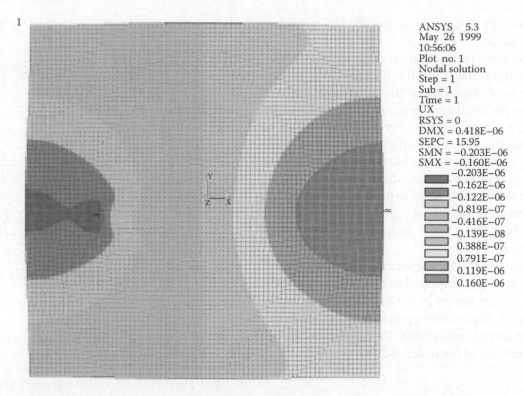

FIGURE 8.9   (See color insert.) In-plane displacement after the exposure of two rows of subfields in EBL exposure of silicon wafer. (Results by Computational Mechanics Center, University Wisconsin-Madison.)

enter the resist in an area away from the desired point, leading to placement distortion. The placement distortion due to charging is especially important in mask making where the placement error should be of the order of a couple of nanometers, while in reality the error is often one order of magnitude larger. Double patterning using complimentary masks is especially demanding for masks with low pattern distortion. The simulation and correction of placement errors due to charging were carried out by NuFlare Technology.[16] The software predicts the map of placement errors, a map that is used to predistort the pattern layout to compensate for the distortion. This is a complex task that was accomplished only partially at this time.

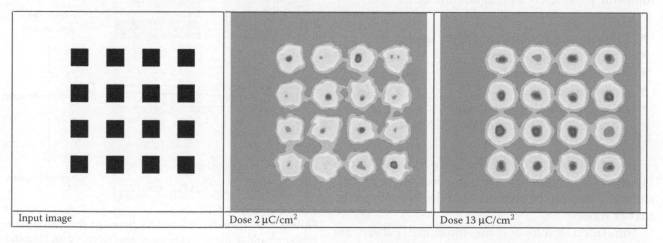

FIGURE 8.10   Effect of shot noise on contact hole cross section for different resist sensitivities. Simulation for contact dimension 45 × 45 nm. (Adapted from P. Kruit, S. Steenbrink, and M. Wieland, *J. Vac. Sci. Technol. B*, 24(6), 2006, 2931.)

Chapter 8

Another solution was offered by aBeam's software CHARIOT-Mask.

In photomask fabrication, a conductive chrome layer under the resist is used. Still, the typical placement error is of the order of 20 nm. In direct write, layers under the resist are often nonconductive, which may lead to a considerably higher placement error. In R&D, conductive layers are often used on the top or under the resist to suppress charging. In production, however, such layers cause additional defects and therefore are not used. It is expected that charging will be a severe problem in high-throughput direct write EBL when the resist is used on wafers with dielectric layers.

An example of the measurement and simulation of the distortion map is displayed in Figure 8.11. The pattern was a 40 mm square in the middle of the mask; a map of 81 × 81 displacement points is shown. The displacement pattern is well visible. The typical scale of the placement error is of the order of 20 nm.

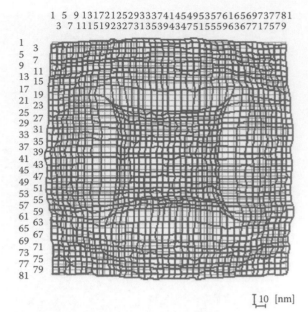

FIGURE 8.11 Distortion map due to charging in EBL writing of a mask. In this case, the pattern has an area of 40 mm in the center of the mask. (Courtesy of N. Nakayamada of NuFlare Technology). The typical placement error is of the order of 20 nm.

8.3 Nanoimprint

Nanoimprint lithography (NIL) is a process that is easy to comprehend, but it still remains a complex process with often underestimated sophistication.[16] Commercial software for modeling nanoimprinting is limited at the moment to NIL Simulation Suite (NSS) of Cognoscens.

NIL relies on a direct mechanical deformation of the resist. Optimization of the NIL process is possible when using a simulation to understand the pattern displacement and decreasing the residual layer.[17] The following parameters should be considered

- Pressure, temperature, and their gradients
- Timing of subprocesses
- Master geometry and material
- Resist properties
- Solver and solver concentration
- Ultraviolet (UV) curing (for UV NIL)

The workflow of NSS suite software is shown in Figure 8.12. It simulates the resist flow under protrusions, the buildup of pressure, and deformation in the master stamp.

Simulation results for the imprinted pattern are shown in Figure 8.13. The pattern geometry and a process can be optimized to achieve a uniform resist thickness with a homogeneous residual layer after the NIL. The addition of "dummy" features to the layout may help improve the quality of the fabricated pattern. The NSS suite helps with such optimizations.

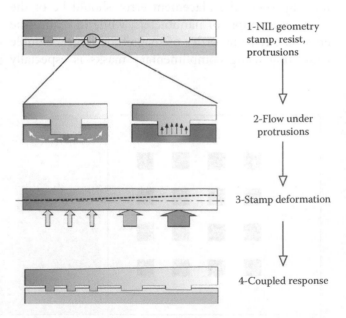

FIGURE 8.12 Workflow of the NSS: the resist flow and resulting pressure buildup under a protrusion are calculated and linked to the stamp deformation through the continuity of the pressure condition.

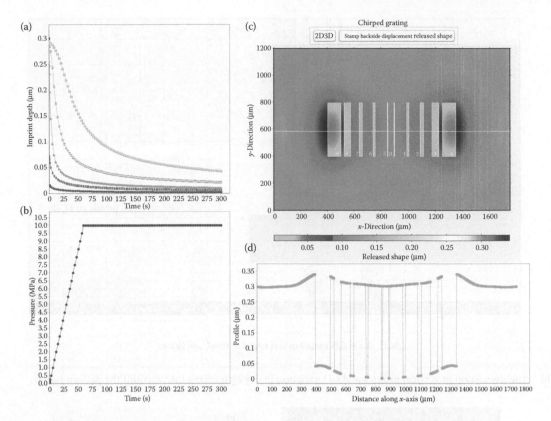

FIGURE 8.13   (See color insert.) NIL of a chirped grating: (a) imprint depth as a function of time, (b) pressure diagram, (c) top view, and (d) resulting profile along the *x*-direction.

8.4   Dry Etch and Film Deposition

The dry etch and film deposition processes are simulated using commercial software tools such as VICTORY Cell from Silvaco, TRAVIT from aBeam, as well as Sentaurus Process and T-SUPREMIV from Synopsys.

8.4.1   Dry Etch

Dry etch is an extremely complex problem. It involves plasma physics, kinetics of chemical reactions, and plasma chemistry, which are often not known in detail. When a new process is being established, the distances of at least half-a-dozen parameters should be optimized simultaneously. This task is expensive and time consuming. Any help using simulations would greatly help in understanding the process and optimizing the process conditions.

One dry etch simulation software is TRAVIT by aBeam used to simulate dry etch, film deposition, and/or electroplating. In simulating dry etch, the software takes into account the circuit layout, microloading, trenching, footing, dispersion, and so on. Complex multistep processes involving changing etch recipes

and film deposition can be simulated in a single run, such as double patterning and microelectromechanical systems (MEMS).

The output of the software are CDs- and CD-variation, vertical profile during and after the etch, and the placement error; see Figure 8.14.

8.4.2   MEMS Deep Etch

The Bosch process is often used for deep etching of silicon. The process is composed of alternating etch of silicon and passivation of walls. These two processes are repeated multiple times. While each step produces far from vertical profiles, the passivation prevents walls from etching to a high degree. The resulting profile looks nearly vertical on a macroscale.

The deep etch involves multiple effects depending on the recipe, for example, widening of the linewidth with etch depth and reduction of the size and changing the shape of scallops. Simulation software normally can track these effects, in addition to effects such as microloading described above. An example of simulation is shown in Figure 8.15.

Chapter 8

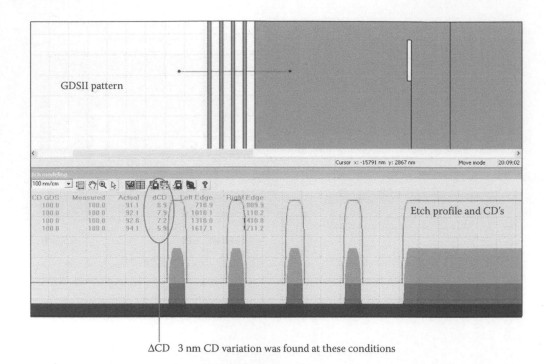

FIGURE 8.14 The simulation of vertical profiles and CD variation in dry etch taking into account the pattern layout.

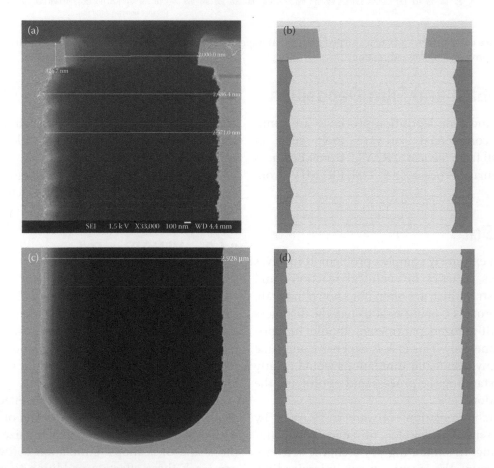

FIGURE 8.15 Experimental and simulated results of deep etch of silicon: (a, b) top part of the etched pattern and (c, d) bottom part. (Experimental image is a courtesy of R. Nagarajan of Institute of Microelectronics, Singapore, simulation is done by TRAVIT software of aBeam.)

8.4.3   Double–Patterning Technology

The combination of film deposition and dry etch allows for the increase in spatial resolution of fabricated patterns. Double-patterning technology, sometimes referred as spacer patterning, is used to improve the resolution of fabricated patterns when the process is limited by capabilities of lithography. A typical simulation flow is presented in Figure 8.16. Here, lithography is followed by resist trim, conformal deposition of sacrificial layer, anisotropic etch of the layer, resist ashing, etch of a hard mask, and, finally, etch of silicon dioxide. As a result, features with two times higher spatial density are fabricated.

There are multiple problems associated with implementation of double patterning: pattern decomposition into two mask layers, strict tolerance for the alignment of complimentary masks, as well as process-dependent CD-variation and placement error. The process development is complex because of many possible variables. The simulation is useful to predict CD-variation and placement error, and so optimize the resist trim, thickness of sacrificial layer, and the etch processes to improve the targeted placement and CDs.[19]

8.5   SEM Metrology

The measurement of CDs is the most crucial step in micro- and nanofabrication. Semiconductor companies say, "we can build if we can measure." The SEM is the main tool used in metrology.

8.5.1   Numerical Simulations of SEM Images

Monte Carlo simulations are proven to be useful tools to understand the formation of SEM images and tune SEMs for accurate metrology.[9–21] These simulations involve elastic scattering of electrons and inelastic energy loss, generation of secondary electrons. To model SEM, the energy loss model should involve discrete electron–solid interaction events, as well as charge/discharge model. In addition, the simulations need to consider electromagnetic fields over the sample as well as geometry, position, and energy transfer function of the detector.

The Monte Carlo simulation tool, CHARIOT of aBeam Technologies, simulates electron scattering in the 3D multilayered target using advanced discrete loss physical models, including electron transport at low energies.[21] Trajectories of primary and secondary electrons in graphical interface of CHARIOT are shown in Figure 8.17. Tracking the low-energy electrons and holes allows for the opportunity to find charge distribution within the target; the distribution is updated according to the discharge of the target. When the charge distribution is known, the electrical field over the sample is calculated. Primary and secondary electrons travel in this field, producing a signal on the SEM detector. The simulation involves

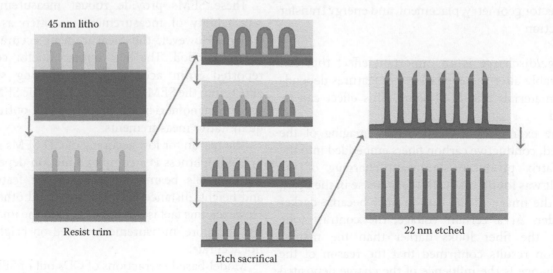

45 nm litho

Resist trim

Etch sacrifical

22 nm etched

FIGURE 8.16   Simulation and optimization of the double-patterning process. In this process, a pattern with 45 nm lines made by optical lithography resulted in 22 nm lines at half the initial pitch using combinations of dry etch and film deposition processes.

Chapter 8

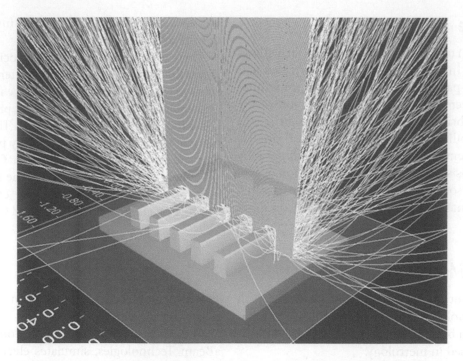

FIGURE 8.17 Monte Carlo simulation of electron scattering in a 3D target. Primary and secondary electrons over the sample are shown.

- Electron scattering inside a 3D sample
- Electrical fields formed by a potential applied to detectors and additional elements as well as local E-fields due to sample charging
- Charge and discharge of the sample
- Deflection of primary beams due to local and global fields and their landing points; trajectories of secondary electrons in the presence of electromagnetic fields until they reach a detector
- Detector geometry, placement, and energy transfer function

Charge/discharge is an important effect that can considerably alter the appearance of features depending on materials and SEM setup. This effect can be modeled.

In the example presented below, imaging of the grounded, conductive carbon fibers embedded into the SiO_2 matrix produced significant charging of the matrix. It was found that with the increase in the beam energy, the image of the carbon fiber became darker and wider. At a certain voltage, the contrast tone reverses: the fiber looks darker than the matrix. Simulation results confirmed that the reason of the image change is the influence of the charge deposited by the e-beam. The experimental and simulation results are displayed in Figures 8.18 and 8.19.[22]

8.5.2 Extraction of CDs from SEM Images

SEMs specialized for CD measurements have been developed. These tools use a well-calibrated beam deflection system, stable wafer stage, and are optimized for low distortion, low charging measurements. The image brightness is analyzed and the CDs are extracted based on a chosen brightness threshold or maximum gradient of brightness.

These SEMs provide robust measurements; the repeatability of measurements is often as good as 0.3 nm. However, the measurement accuracy is not nearly as good. The best semiconductor companies reported 5 nm accuracy, after putting significant efforts into the SEM calibration. Analytic SEMs widely used in nanofabrication are even less optimized for quantitative measurements.

The reason for low accuracy of CD SEMs is that the image brightness is a complex function depending on the materials, beam voltage, beam size, feature shape and height, distance between lines, and other factors. However, this fact is not considered in the image analysis. Therefore, measurements based on brightness are not accurate.

Model-based extractions of CDs out of SEM images provide much better results. An approach based on Monte Carlo simulations of a library of images at vari-

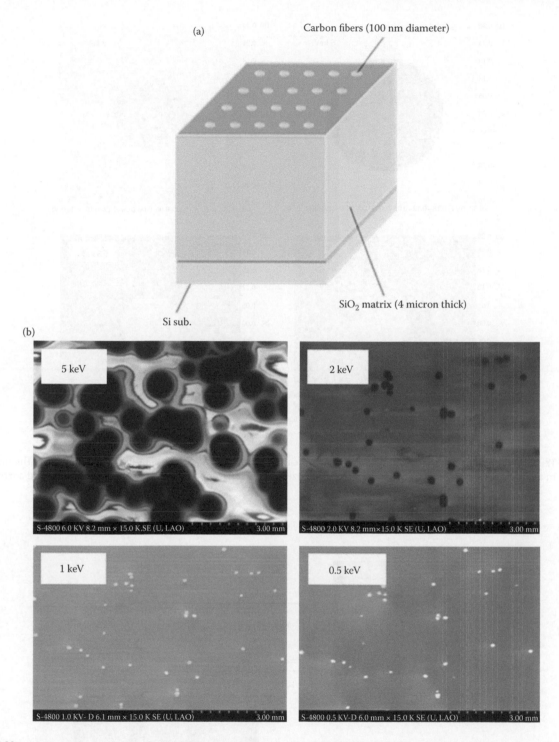

FIGURE 8.18 (a) Specimen overview. (b) SEM images of the same area at variable beam voltage. Tone reversal was found. Also, fibers of different sizes appear.

ous conditions and then matched to the SEM image was suggested.[23]

Another approach is to use an analytic model of the SEM.[24] In addition to the image, the myCD software by aBeam requires knowledge of the material and SEM setup. These are normally known to the operator. The model then finds the contours, CDs, and the shape of the feature that would produce an SEM image under analysis; see Figure 8.20. The model takes a few seconds and can be used for run-time measurements. The advantage of this method is that it does not require the simulation of a library, which may take significant time.

Chapter 8

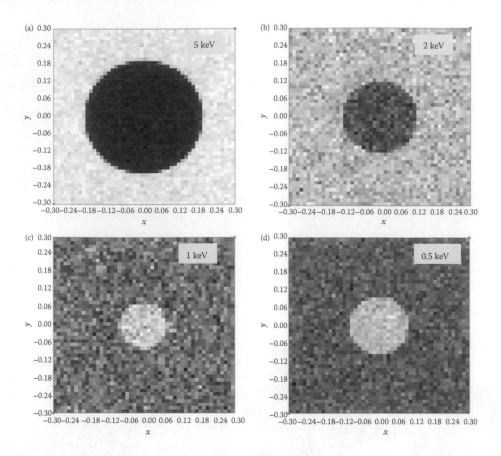

FIGURE 8.19 Simulation results: the SEM image of the single carbon fiber (a) 5 kV, (b) 2 kV, (c) 1 kV, and (d) 0.5 kV. The field of view was equal in all images. Fiber diameter was 100 nm. The fiber size appears different depending on the beam voltage and corresponding charging.

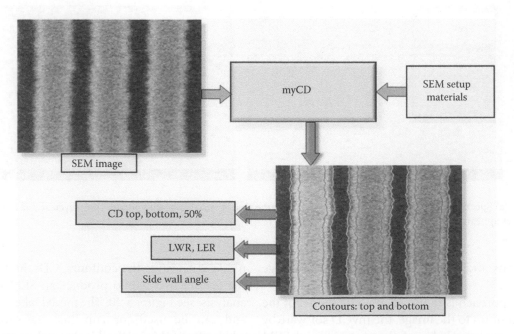

FIGURE 8.20 The myCD software analyzes SEM images based on a physical model and automatically extracts the CDs, roughness, and side wall angle of features.

8.5.3 Calibration of SEM for Metrology

In order to provide accurate measurements using SEM, the microscope should be well calibrated. The main parameters that should be calibrated are field of view, map of field distortion, and e-beam size. The field of view is measured and set using gratings or special marks fabricated with high accuracy. Measurements of the beam size are more complex.

8.6 Optical Metrology

8.6.1 Ellipsometry

Optical ellipsometry is a powerful method to measure the thicknesses of thin films and their optical properties.

Ellipsometry measures a change in polarization of the reflected or transmitted light from a layered substrate. The measured response depends on optical properties and thicknesses of films in the multilayered stack. The thin films in the substrate can be dielectrics, semiconductors, and metals. The superposition of multiple light waves introduces interference that depends on the relative phase of each light wave; see Figure 8.22. Fresnel reflection and transmission coefficients are used to calculate the response from each contributing beam. The ellipsometry technique has been well developed since the 1960s; there are many companies promoting commercial equipment and software, such as J. A. Woollam, Horiba, Scientific Computing International, Semiconsoft, to name a few.

The knife edge method is often used, which is not accurate and is operator dependent. BEAMETR is an analysis software that automatically measures the beam size. The SEM operator takes an image of the provided test sample and the software extracts beam size in the x- and y-coordinates; see Figure 8.21. The method uses an analysis of the spectral frequency of the pattern in comparison with that of the designed pattern.[25]

8.6.2 Scatterometry

Optical methods can not only deal with measuring thicknesses, but are also capable of measuring CDs and their vertical profiles.[26] These methods can only be applied to periodic patterns such as gratings. The optical setup uses polarized light diffracted from the grating; the measurement is made over a wide range of wavelengths.

A resulting measured amplitude of reflected or transmitted light as a function of the wavelength at various polarizations is then compared with a library of similar "signatures." The library is simulated based on a rigorous electromagnetic theory. These are often huge libraries; the simulation of these libraries takes from a few hours to a few days on a cluster of computers. When the library is created, a search for the best match between the measured result and the library takes only a short time. A lot of efforts have been put into optimizing the size of the required library.

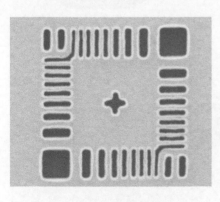

Test sample, 30 nm

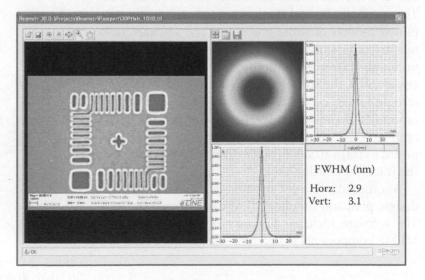

FIGURE 8.21   A test sample and software are used to automatically measure beam size in SEMs and other e-beam systems. The operator takes an image of the provided test sample and the software extracts beam size in x and y.

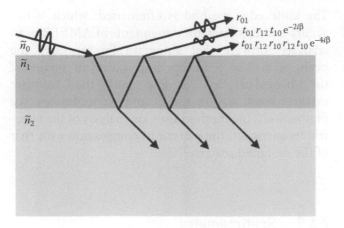

FIGURE 8.22 Reflected light from a multilayered substrate changes polarization according to thicknesses and optical properties of films. (Image courtesy of J. A. Woollam Co.)

KLA-Tencor, Nanometrix, and TEL are the major suppliers of scatterometry systems and software. In order to tune the results of a scatterometry system to the measurement results received by other methods such as CD-SEM, fitting parameters are often used by system manufacturers. This tuning takes care of the imperfections of optical elements and the method itself. Users of scatterometry systems often complain that the information about the applicability of a system to other layers or materials is unknown because neither the software code nor tuning parameters are disclosed by the system makers to users.

A software that is based on pure physics and does not involve tuning parameters has been developed by aBeam. The software solves rigorous electromagnetic equations and simulates scatterometric dependencies for multilayered 2D and 3D patterns, including patterns made of dielectrics, semiconductors, and metals. This software is used to understand the sensitivity of scatterometry to process variation, and helps select an appropriate system for specific applications.

8.7 Plotter of 3D Micropatterns

Some CAD software packages offer plotting multilayered GDSII patterns in 3D. These take design data of each layer and the layer thicknesses, and then plot the designed pattern in 3D. This is a convenient addition to the design tool, but has nothing to do with the actual fabrication process.

SEMulator-3D* is a very useful software offered by Coventor. The software takes the designed layers, adds information about overlay and vertical cross sections of layers, and builds a 3D microfabricated device from this information as shown in Figure 8.23. There are no physical models, or only extremely simplified models of some processes. The software does not directly simulate the manufacturing process, or uses only extremely simplified models of some processes. Still, the software is good to visualize the pattern and see the possible problems associated with it, especially at the design verification or process development stage. It is also great for educational purposes, to train students and engineers who are new to the microfabrication area.

Another software of Coventor, the MEMS+ is used to design, plot, and simulate MEM devices and systems.[27] The software includes a library of typical models of MEMS components that can be selected and used in a

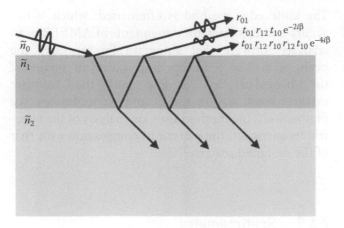

FIGURE 8.23 SEMulator displays (a) 3D microcircuit and (b) MEMS device. The software is very useful in verification of the design of devices, even though it does not simulate fabrication processes.

---

* SEMulator3D and MEMS+ are registered trademarks of Coventor, Inc.

complex MEMS system. It can also generate parametric symbol and simulation model for the Cadence Virtuoso design environment or MATLAB and then import and visualize the results of these simulations in 3D.

References

1. F. H. Dill, A. R. Neureuther, J. A. Tuttle, and E. J. Walker, Modeling projection printing of positive photoresists. *IEEE Trans. Electron Devices*, ED-22, No. 7, 1975, 456.
2. C. A. Mack, *Fundamental Principles of Optical Lithography: The Science of Microfabrication*. John Wiley & Sons, London, 2007.
3. C. A. Mack, Thirty years of lithography simulation. *Optical Microlithogr. XVII, Proc. SPIE*, 5754, 2005, 1.
4. www.panoramictech.com
5. www. genisys-gmbh.com
6. T. Waas, H. Eisenmann, O. Vollinger, and H. Hartmann, Proximity correction for high CD accuracy and process tolerance. *Microelectron. Eng.*, 27(1–4), 1995, 179.
7. N. Unal, D. Mahalu, O. Raslin, D. Ritter, C. Sambale, and U. Hofmann, Third dimension of proximity effect correction (PEC). *Microelectron. Eng.*, 87(5–8), 2010, 940.
8. M. Shoji, N. Horiuchi, T. Chikanaga, T. Niinuma, and D. Tsunoda, New proximity effect correction for under 100 nm patterns. *Proc. SPIE* 6283, 2006, 62832L.
9. D. C. Joy, *Monte Carlo Modeling for Electron Microscopy and Microanalysis*. Oxford University Press, Oxford, 1995.
10. B. Nilsson, *Proc. Int. Conf. Micro and Nano-Engineering*. 2009, Abstracts.
11. S. Babin and I. Kuzmin, Experimental verification of the TEMPTATION temperature simulation software tool. *J. Vac. Sci. Technol. B* 16(6), 1998, 3241.
12. V. Sidorkin, E. van der Drift, and H. Salemink, Influence of hydrogen silsesquioxane resist exposure temperature on ultrahigh resolution electron beam lithography. *J. Vac. Sci. Technol. B* 26(6), 2008, 2049.
13. S. Babin, I. Kuzmin, H. Yamashita, and M. Yamabe, Thermal analysis of projection electron beam lithography using complementary mask exposures. *J. Vac. Sci. Technol. B*, 21(6), 2003, 2691.
14. B. Kim, R. Engelstad, E. Lovell, S. Stanton, A. Liddle, and G. Gallatin, Finite element analysis of SCALPEL wafer heating. *J. Vac. Sci. Technol. B*, 17(6), 2003, 2883.
15. P. Kruit, S. Steenbrink, and M. Wieland, Predicted effect of shot noise on contact hole dimension in e-beam lithography. *J. Vac. Sci. Technol. B*, 24(6), 2006, 2931.
16. N. Nakayamada, S. Wake, T. Kamikubo, H. Sunaoshi, and S. Tamamushi, Modeling of charging effect and its correction by EB mask writer EBM-6000. *Proc. SPIE*, 7028, 2008, 70280C.
17. S. Y. Chou, P. R. Krauss, W. Zhang, L. Guo, and L. Zhuang, Sub-10 nm imprint lithography and applications. *J. Vac. Sci. Technol. B*, 15, 1997, 2897.
18. C. Gourgon, N. Chaix, H. Schift, M. Tormen, S. Landis, C. M. Torres, A. Kristensen. et al., Benchmarking of 50 nm features in thermal nanoimprint. *J. Vac. Sci. Technol. B*, 25(6), 2007, 2373.
19. S. Babin and K. Bay, Modeling of CD and placement error in multi-spacer patterning technology. *Proc. SPIE*, 7748, 2010, 774813.
20. J. S. Villarrubia and Z. J. Ding, Sensitivity of SEM width measurements to model assumptions. *Proc. SPIE*, 7272, 2009, 72720R.
21. S. Babin, S. Borisov, A. Ivanchikov, and I. Ruzavin, CHARIOT: Software tool for modeling SEM signal and e-beam lithography. *Phys. Procedia* 1, 2008, 305.
22. S. Babin, S. Borisov, H. Ito, A. Ivanchikov, V. Militsin, and M. Suzuki, Comprehensive simulation of SEM images taking into account local and global electromagnetic fields. *Proc. SPIE*, 7729, 2010, 77290W.
23. J. S. Villarrubia, A. E. Vladár, B. D. Bunday, and M. Bishop, Dimensional metrology of resist lines using a SEM model-based library approach. *Proc. SPIE*, 5375, 2004, 199.
24. S. Babin, K. Bay, and M. Machin, Model-based analysis of SEM images to automatically extract linewidth, edge roughness, and wall angle. *Proc. SPIE*, 7638, 2010, 76380R.
25. D. Joy, M. Gaevski, S. Babin, M. Machin, and A. Martynov, Technique to automatically measure electron-beam diameter and astigmatism: BEAMETR. *J. Vac. Sci. Technol. B*, 24(6), 2006, 2956.
26. C. J. Raymond, Overview of scatterometry applications in high volume silicon manufacturing. *Proc. AIP*, 788, 2005, 394.
27. T. Udeshi and E. Parker, MEMulator™: A fast and accurate geometric modeling, visualization and mesh generation tool for 3D MEMS design and simulation. *Nanotechnology*, 2, 2003, 480.

New Lithographic Techniques

9. Nanoimprint Lithography

Stephen Y. Chou

Department of Electrical Engineering, Princeton University, Princeton, New Jersey

Nanofabrication Handbook. Edited by Stefano Cabrini and Satoshi Kawata © 2012 CRC Press / Taylor & Francis Group, LLC. ISBN: 978-1-4200-9052-9

Chapter 9

9.1   Overview

Nanopatterning, the ability to pattern nanostructures, is essential to the advancement and industrialization of nanotechnology. For manufacturing in nanoscale, it requires the patterning technologies to produce sub-10 nm features over a large area, high-throughput and low cost. At present, the conventional patterning methods that employ radiation (photons, electrons, and ions) cannot meet these requirements, primarily limited by their intrinsic physical principles. Nanoimprint lithography (NIL), on the other hand, a mechanical patterning method that uses a mold (also called template) to mechanically deform a material into features, can meet all these requirements.

Micromolding has been used for making compact disks and other microdevices since the 1970s. In 1995, NIL was proposed and demonstrated as a technology for nanopatterning [1,2]. The work ignited a paradigm-shift in development of new nanopatterning methods from the use of radiation to mechanical deformation, and has led to a worldwide exploration and development of NIL and its applications. Today NIL has become a key nanofabrication method in research and development as well as manufacturing in a broad range of industries, from nano/microoptical devices (e.g., light-emitting diodes, cell phones, solar cells, and displays), nanoelectronics, magnetic data storage, to biotechnology and medicine, to name just a few [3].

The initial motivation to explore NIL was two: (1) meet the pressing need to find economical-viable way to manufacture the new nanodevices or nanomaterials developed in the laboratory (in early 1990s it became clear that without such manufacturing technologies, the funding to nanodevice research could dry up), and (2) provide a feasible method to manufacture "quantized magnetic disks" (also termed "bit patterned media")—a new paradigm for magnetic data storage that was proposed about one year earlier than NIL, which becomes one of the first real industrial applications of NIL [4,5].

Thermal nanoimprint lithography (T-NIL) was demonstrated first where a thermoplastic resist as used [1,6], followed by UV curable nanoimprint lithography (UV-NIL) [7], and then step-and-flash imprint lithography (SFIL) where a UV curable resist was dispensed precisely on a wafer surface in droplets [8].

Compared with radiation-based nanopatternings, NIL has three major advantages due to its fundamentally different physical principle. First, it does not have diffraction limit in resolution; second, it is intrinsically a 3D patterning; and third, it can reduce a multistep process of lithography and etching into a single step of direct imprinting of a material, hence reducing the steps, equipment, and cost of fabrication. Although NIL is far less mature than photolithography and has its drawbacks (as discussed below), it has a unique combination of high resolution, large area, high throughput, and low cost, that are unmatchable by other existing methods.

NIL has its own challenges. For example, NIL is a form of contact lithography, facing issues of defect density, 1X mold fabrication and cost, mold damage, and slow alignment speed. It is yet to be seen if all these issues can be solved, to what degree, and for what applications. Unlike conventional patterning which has been continuously developed by semiconductor industry for over 30 years with huge research expenditure, NIL is still a new technology and has received far less funding for development. In the coming years, we will see if sufficient funding, time and efforts will solve the previous issues and will allow us to fully reach the potential of NIL.

Due to its demonstrated ultrahigh patterning resolution and throughput, nanoimprint has been put on the roadmaps of many industries, including International Technology Roadmap for Semiconductors (ITRS) as a next-generation patterning method for manufacturing semiconductor-integrated circuits and the roadmap for manufacturing magnetic data storage disks.

9.2 NIL Principle and Key Attributes

9.2.1 Basic Principle

Nanoimprint patterns nanostructures by mechanical deformation of a deformable material using a mold to duplicate the nanostructures on the mold surface (Figure 9.1) [1,6]. To facilitate nanoimprint, the deformable material is often in liquid phase during the imprint and becomes solidified fully or partially before removing the mold. To achieve the intended nanoscale features, nanoimprint must use drastically different materials and processes from conventional micromolding (as discussed in Section 9.2.6).

The imprinted material can serve either as a resist for pattern transfer (as in conventional lithography and being removed later) or as a part of the devices to be built, which permanently stays on the wafer (i.e., direct imprint of functional materials). During nanoimprint, a residue layer can be formed between mold protrusions and a substrate, which often needs be removed by etching to expose the substrate.

The working principle of nanoimprint fundamentally differs from conventional lithography, which creates a pattern in the material by changing the local chemical properties of the material using radiation. Therefore, nanoimprint offers many unique advantages in patterning and manufacturing over conventional lithography, such as sub-2 nm patterning resolution, pattern transfer fidelity, 3D patterning, large area (full wafer if needed), ability of reducing total fabrication steps, high throughput, and low cost (Figures 9.2 through 9.5).

9.2.2 Sub-2 nm Patterning Solution

Since patterning in NIL is primarily a mechanical deformation process, its resolution is related to the mold surface topology, but not an optical imaging of the mold. Hence, it is not limited by the factors that limit the resolution of conventional lithography, such as wave diffraction, scattering and interference in a resist, backscattering from a substrate, and the chemistry of resist and developer. In fact, photo-curable NIL has demonstrated 6 nm half-pitch imprinted into a resist (Figure 9.2) [9,10] and thermal-NIL has demonstrated arrays of 10 nm-diameter dots separated by 40 nm (400 dots/in.^{2}) (Figure 9.3) [3]. Yet, these features are not the limits of NIL process itself, but rather the limits of our ability in making the features on the mold; NIL can achieve even smaller features if a mold of such can be made. From the faithful duplication of nanometer variations on a sidewall of a mold, it is clear in the first nanoimprint work that imprinting of sub-3 nm features is possible [1,6].

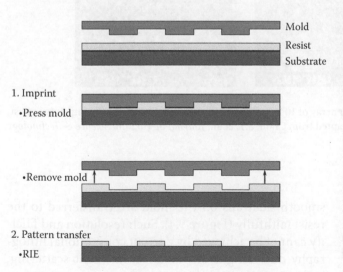

1. Imprint
•Press mold

•Remove mold

2. Pattern transfer
•RIE

FIGURE 9.1 Schematic of NIL: (a) deposit imprint material on a substrate, (b) imprinting nanostructure in the material using a mold to create a thickness contrast, and (c) pattern transfer using anisotropic etching to remove residue resist in the compressed areas to expose the substrate [1,6]. The imprint material also can be deposited in droplet form.

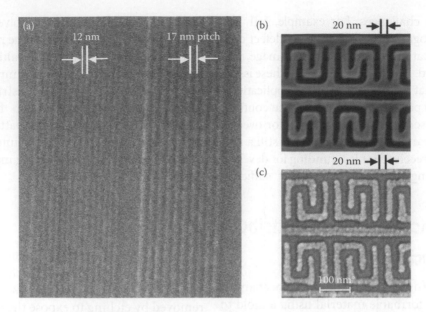

FIGURE 9.2   SEM image of (a) imprinted resist grating with a minimum of 6 nm half-pitch, (b) 20 nm half-pitch resist pattern for SRAM metal contacts fabricated by NIL, and (c) Cr lift-off of the resist in (b). (Adapted from Austin, M.D. et al., *Nanotechnology*, 2005. **16**(8): 1058–1061.)

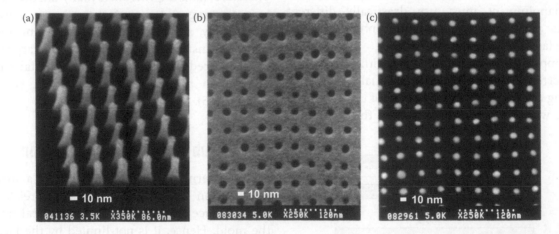

FIGURE 9.3   SEM image of (a) the nanoimprint mold of the pillar array of 10 nm diameter and 40 nm period, (b) the imprinted resist, and (c) the metal dots after lift-off using the imprinted resist. (Adapted from Chou, S.Y. et al., *Journal of Vacuum Science & Technology B*, 1997. **15**(6): 2897–2904.)

9.2.3   High Pattern Transfer Fidelity

NIL has been demonstrated to have high fidelity in pattern transfer, accurately reproducing original mold patterns and maintaining smooth vertical sidewalls in the imprint resist. For example, repeated imprinting of an SRAM metal interconnect patterns of 20 nm half-pitch has achieved a standard deviation of 1.3 nm in the variation of the imprinted feature width (Figure 9.2) [10]. High aspect ratio patterns with

smooth sidewalls on the mold are transferred to the resist faithfully (Figure 9.4). Such resolution and fidelity cannot be achieved by current conventional lithography due to light diffraction limit, light scattering, and other noise [11].

9.2.4   Three-Dimensional Patterning

The third unique feature of NIL is three-dimensional (3D) patterning, rather than the 2D patterning as in

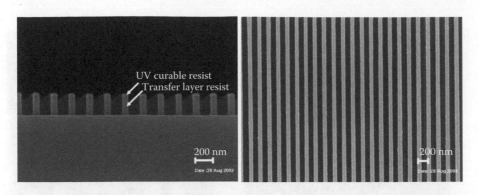

FIGURE 9.4 Imprinted resist line profile of 70 nm width 200 nm period, showing the smooth vertical sidewalls. (Adapted from Li, M.T., L. Chen, and S.Y. Chou, *Applied Physics Letters*, 2001. **78**(21): 3322–3324.)

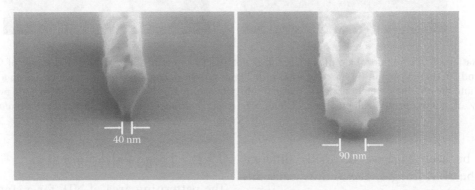

FIGURE 9.5 3D patterning. SEM micrograph of two T-gate of 40 and 90 nm foot-print, respectively, fabricated by a single NIL and a lift-off of metal. (Adapted from Li, M.T., L. Chen, and S.Y. Chou, *Applied Physics Letters*, 2001. **78**(21): 3322–3324.)

conventional lithography. 3D features are very desirable for certain applications such as microwave circuits and MEMS. For example, the T-gate for microwave transistors has a narrow foot print for high-frequency operation, but wide top for lower resistance. Fabrication of a T-gate often requires two electron beam lithography steps; one for the foot print and one for the wide top. Each electron beam exposure could take over 2 h to pattern a single 4″ wafer. With NIL, the entire 4″ wafer can be patterned in one step in less than 10 s. Figure 9.5 shows a 40 nm T-gate fabricated by a single NIL step and lift-off of metal [12]. Nanoimprint is also used to create 3D damascene oxide patterns for metal interconnect using SFIL [13].

9.2.5 Patterning Resolution Independent of Substrate Materials

Another key difference of nanoimprint from conventional lithography (based on radiation) is that its patterning resolution is independent of the substrate materials'

atomic number. For example, photolithography or electron beam lithography has high back scattering that makes it very difficult to directly achieve a high resolution in a high atomic substrate. The resolution will vary from across locations on the same substrate, depending upon the local substrate material atomic number. But for NIL, the patterning resolution is independent of the substrate atomic number [3].

Figure 9.6a shows the imprint of an array of 60 nm-deep holes with 10 nm minimum diameter and 40 nm period in 78 nm-thick poly(methylmethacrylate) (PMMA) film on a gold substrate (300 nm-thick gold on silicon) [3]. Figure 9.6b shows the nickel gratings with 20 nm linewidth and 80 nm period made by imprint on the gold substrate and nickel (13 nm thick) lift-off [3].

The above figures show two interesting facts. First, in comparison with the original molds, the imprinted features have the same dimension as the mold. This indicates that the NIL resolution was, as expected, not affected by the underlying high atomic number substrate. Second, the different gold grains on the

Chapter 9

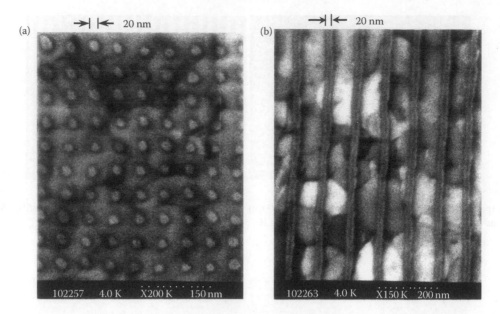

FIGURE 9.6 Variation of substrate local atomic number does not affect nanoimprint resolution. SEM image of (a) nickel dots with 20 nm diameter and 50 nm period and (b) nickel grating with 20 nm linewidth and 80 nm period, which are on a gold substrate and fabricated by NIL and lift-off. The different grains of the gold substrate, which are visible beneath the PMMA, do not affect nanometer patterning fidelity of nanoimprint. (Adapted from Chou, S.Y. et al., *Journal of Vacuum Science & Technology B*, 1997. **15**(6): 2897–2904.)

substrate can be seen to have distinctly different brightness in the SEM images, due to drastically different secondary electron emission rates from different crystalline orientations. (Note that Figure 9.6a has less brightness variation due to coverage of PMMA.) The above two SEMs images clearly show that despite a large variation in secondary electron emission rate from one gold grain to another, both the imprinted dot diameter and linewidth are uniform over the entire sample, clearly demonstrating that the resolution of NIL is independent of the substrate atomic number.

9.2.6 Direct Patterning and Fabrication Steps Reduction

When replacing an imprinting resist with functional materials that will stay on devices as a part of the device structure, nanoimprint can in fact reduce multiple fabrication steps into one (imprint). For example, for the T-gates in microwave devices or the interconnects in ICs, the conventional approaches to create a 3D dielectric structure require multiple steps of depositions, lithography, and etching, but nanoimprint needs only one step because the imprint material is already the dielectrics and the 3D shape can be made by one imprint.

9.2.7 Large Patterning Area

The patterning area of NIL for a single imprint can be much larger than the exposure field of a conventional photolithography stepper (~1 in.²). This is because NIL requires neither high-precision optics nor a well-conditioned monochromatic light source. Today, full 4 in. or 8 in. wafers are routinely imprinted in a single imprint. When air cushion press (ACP) is used (see Section 9.3.2), which presses wafers and mold by pressured air (or fluid) creating uniform pressure everywhere, excellent imprint uniformity has been achieved [14].

9.2.8 Low Cost and High Throughput

Because NIL does not use complicated and expensive optics systems and laser sources, NIL tools can be much cheaper than conventional photolithography. For a single full wafer imprint, its throughput should be higher than photolithography. However, NIL does have items that could be more expensive. For example, 1X mold for NIL is intrinsically more expensive than the 4X mask for photolithography, unless low-cost 1X mold making will be developed. Also the current step-and-repeat nanoimprint tool has about 20 wafers per hour throughput, which is only one-third of current photolithography.

9.3 Various Forms of NIL

Nanoimprint has various forms, which can be classified according to material processing methods, pressing methods, imprinting area, material dispensing methods, or imprinting objectives. A particular imprint approach can be formed by choosing one method from each of the above categories.

Some examples of the current popular approaches are given briefly in the rest of the section and their advantage and disadvantages will be discussed in detail in Section 9.4. Clearly, these examples are far from exclusive. In fact, many other nanoimprint variations are possible. For example, it has been conceived as early as 1996 that nanoimprint can be extended beyond compression molding into casting molding, vapor deposition molding, blow molding, powder molding, and deposition molding [15].

9.3.1 Different Material Processing Methods

9.3.1.1 Thermal Nanoimprint

The earliest form of NIL is thermal imprinting (Thermal-NIL), which presses the topological nanostructures of a mold into a thermoplastic thin film on a substrate [1]. During the pressing, thermoplastic film is heated above its glass transition temperature (T_g), becoming a viscous liquid (a thermoplastic is solid when below T_g) and flowing into the mold. A thermal plastic can be easily dissolved in many solvents; hence, in addition to be used as an etching mask, it is a good material for the lift-off process.

9.3.1.2 UV or Thermal Curable Nanoimprint

The UV or thermal curable nanoimprint presses a mold into a curable material, which is a liquid before the curing and a solid after the curing (the solid hardness depends on the degree of the curing) [7,8]. The curing, a process of cross-linking monomers or oligomers, can be done by using photons (often UV light) or phonons (thermal heating) or both. A curable material before a curing often has a viscosity much lower than a thermal plastic liquid phase. Once cured, the curable material cannot return to liquid by heating and is hard to be dissolved in solvents. This makes it unsuited for a lift-off (except for special dissolvable material). UV curing process allows nanoimprint and photolithography to be done at the same time with one mold (which has both topological imprint patterns and light blocking metal patterns). Such combination can use photolithography to define large patterns, hence

eliminating the need of transporting a large amount of material in a conventional NIL.

9.3.1.3 Hybrid Thermal and UV Nanoimprint

In some imprinting processes, both thermal and UV treatments are used for different purposes. In one approach, thermal heating is used first to reduce the viscosity of a UV material (hence improving the material flow) since the viscosity often drops with temperature; after the mold is pressed into the resist, the UV light is used to cure the material. In another approach, UV curing is used, during imprint, to cure the imprinted materials only partially for easy demolding; and after demolding, a thermal curing is used to solidify the imprinted materials completely. Certainly, it is also possible to do thermal curing first and then UV light curing.

9.3.1.4 Laser-Assisted Direct Imprint

One can directly imprint nanostructures into a surface of hard solid substrates, such as semiconductors (e.g., silicon), metals (e.g., Cr), or glasses, by selectively ultra-fast melting an ultra-thin surface layer while keeping the substrate solid at a low temperature, followed by imprinting the thin molten layer (Figure 9.7). The melting can be achieved using a pulsed laser with a selected wavelength (based upon the materials) that is transparent to the mold (i.e., no absorption) but is only absorbed by the first 10 nm surface layer of the substrate. An ultrafast imprint (e.g., 200 ns) of nanostructures in Si and Cr have been demonstrated. The fast imprint is due to a low viscosity. A molten Si and Cr have a viscosity lower and slightly higher than that of water, respectively [16,17]. For example, a 308 nm wavelength excimer laser pulse melts only the Si surface not a quartz mold, and sub-10 nm features was imprinted directly into Si with a quartz mold in less than 300 ns (Figure 9.8) [16].

9.3.2 Different Pressing Methods

9.3.2.1 Solid Parallel-Plate Press

Solid parallel-plate press (SPP), which uses two hard solid parallel plates to press a mold and wafer together, is the oldest press used for nanoimprint, because it has been widely available. However, SPP suffers from drawbacks, which make it a poor choice as a laboratory research tool and almost impossible for real

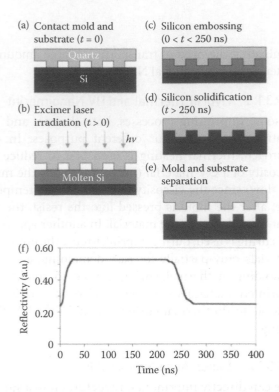

(a) Contact mold and substrate (t = 0)

Quartz
Si

(b) Excimer laser irradiation (t > 0)

hv

Molten Si

(c) Silicon embossing (0 < t < 250 ns)

(d) Silicon solidification (t > 250 ns)

(e) Mold and substrate separation

(f)

FIGURE 9.7 Schematic of laser-assisted direct imprint (LADI) of nanostructures in silicon. (a) A quartz mold is brought into contact with the silicon substrate. A force presses the mold against the substrate. (b) A single XeCl (308 nm wavelength) excimer laser pulse (20 nm pulse width) melts a thin surface layer of Si. (c) The molten silicon is imprinted while the silicon is in the liquid phase. (d) The silicon rapidly solidifies. (e) The mold and silicon substrate are separated, leaving a negative profile of the mold patterned in the silicon. (f) The reflectivity of a HeNe laser beam from the silicon surface versus the time, when the silicon surface is irradiated by a single XeCl (308 nm) laser pulse with 1.6 mJ cm 22 fluence and 20 ns pulse duration. Molten Si, becoming a metal, gives a higher reflectivity. The measured reflectivity shows the silicon in liquid state for about 220 ns. (Adapted from Chou, S.Y., C. Keimel, and J. Gu, *Nature*, 2002. **417**(6891): 835–837.)

manufacturing. The problems with SPP include: (a) severe imprint nonuniformity and hence low or nearly zero yield, which are caused by nonflat surfaces of the molds, the substrates, or the press plates; (b) easily damage the mold due to extremely high pressure at one or a few points caused by a nonuniform contact, and due to the lateral shearing forces created by the nonparalleleness in the pressing; and (c) large thermal masses, hence a long thermal imprinting cycle.

9.3.2.2 Air Cushion Press

ACP uses a gas (or fluid) to directly press the mold and substrate against each other in an enclosed chamber rather than solid parallel plates (Figure 9.9) [14,18]. ACP has five advantages over SPP: (a) ACP has much more uniform imprint pressure. The experiments show that the measured pressure distribution across a 100-mm-diameter single-imprint field (i.e., whole 4″ wafer) in ACP is nearly an order of

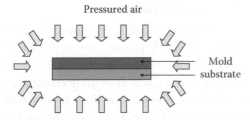

Pressured air

Mold
substrate

FIGURE 9.9 Schematic of the air cushion press (ACP) nanoimprint principle. (Adapted from Gao, H. et al., *Nano Letters*, 2006. **6**(11): 2438–2441; Chou, S.Y., US Patent 6,482,742, *Fluid Pressure Imprint Lithography*. 2002.)

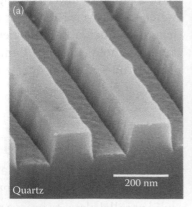

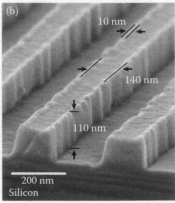

(a) Quartz
200 nm

(b) 10 nm
140 nm
110 nm
200 nm
Silicon

FIGURE 9.8 SEM image of the cross-section of samples patterned using LADI. (a) A quartz mold. (b) Imprinted patterns in silicon. The imprinted silicon grating is 140 nm wide, 110 nm deep and has a 300 nm period, an inverse of the mold. We note that the 10 nm wide and 15 nm tall silicon lines at each top corner of the silicon grating are the inverted replicas of the small notches on the mold (the notches were caused by the reactive ion etching trenching during mold fabrication). (Adapted from Chou, S.Y., C. Keimel, and J. Gu, *Nature*, 2002. **417**(6891): 835–837.)

magnitude more uniform than SPP. (b) ACP is immune to any dust and topology variations on the backside of the mold or substrate. (c) When a dust particle is between the mold and substrate, ACP reduces the damage area by orders of magnitude. (d) ACP causes much less mold damage because of uniform pressure and significantly less lateral shift between the mold and substrate. (e) ACP has much smaller thermal mass and therefore significantly faster speed for thermal imprinting. The experimen-

tal comparison of the pressure uniformity and dust effects are given in Figures 9.10 through 9.13, which show that SPP has large local and global pressure variation as much as ~10 fold from location to location [14].

9.3.2.3 Electrostatic Force Press

The electrostatic force-assisted NIL (EFAN) is a novel imprint method where the force that presses the mold into a resist is provided by an electrostatic force

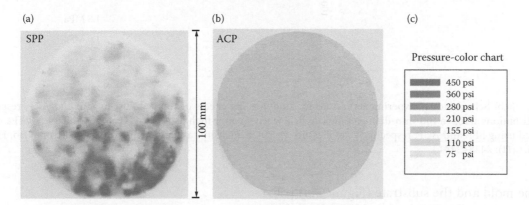

FIGURE 9.10 (a,b) Pressure distribution across a 100-mm-diameter imprint field when a 1.38 MPa nominal pressure is applied using SPP and ACP, respectively. (c) Pressure vs. color intensity calibration chart. The SPP has both local and global pressure variation, which is about 10 folds larger than ACP. (Adapted from Gao, H. et al., *Nano Letters*, 2006. **6**(11): 2438–2441.)

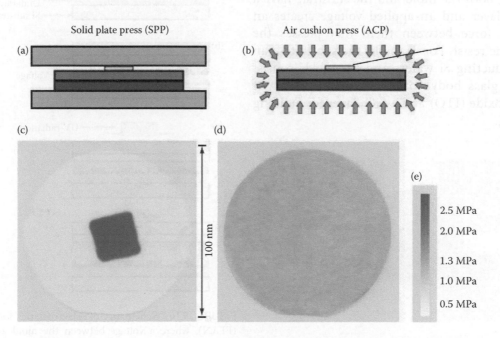

FIGURE 9.11 (a,b) Schematics of experimental setups for studying the effects of backside dust/topology in SPP and ACP, respectively. (c,d) Pressure distributions across a 100-mm-diameter imprint field when a paper piece of 2.6 × 2.6 cm² in area and 0.1 mm in height is inserted on the backside of the mold, and a 1.38 MPa nominal pressure is applied using SPP and ACP, respectively. (e) Pressure vs. color intensity calibration chart. (Adapted from Gao, H. et al., *Nano Letters*, 2006. **6**(11): 2438–2441.)

Chapter 9

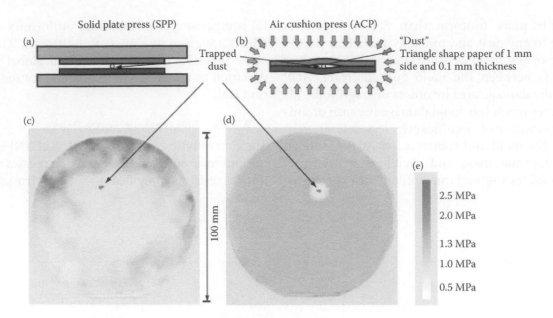

FIGURE 9.12　(a,b) Schematics of experimental setups for studying the effects of trapped dust in SPP and ACP, respectively. (c,d) Pressure distributions across a 100-mm-diameter imprint field when 0.1-mm-high paper dust is trapped and a 1.38 MPa nominal pressure is applied using SPP and ACP, respectively. (e) Pressure vs. color intensity calibration chart. (Adapted from Gao, H. et al., *Nano Letters*, 2006. **6**(11): 2438–2441.)

between the mold and the substrate (Figure 9.14) [19]. EFAN can overcome the drawbacks of SPP, offers good pressure uniformity similar to the ACP without requiring a pressure chamber, and is well suited for step-and-repeat NIL.

In EFAN, both the mold and the substrate have a conductive layer and an applied voltage creates an electrostatic force between them that presses the mold into the resist. For example, the substrates can be semiconducting Si wafers and the molds consist of a Pyrex glass body with a conducting layer of indium tin oxide (ITO) and a SiO_2 layer for molding patterns [19].

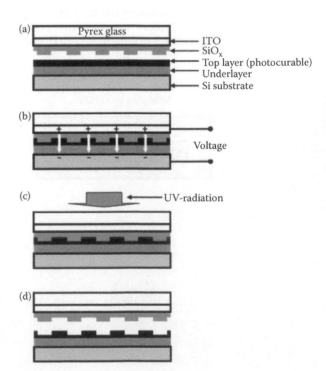

FIGURE 9.14　Schematic of electrostatic force-assisted NIL (EFAN), where a voltage between the mold and the substrate generates an electrostatic pressure. (a) Implementation of EFAN; (b) imprinting by applying a voltage; (c) UV curing of the top-layer resist; (d) separation of the mold after resist curing. (Adapted from Liang, X.G. et al., *Nano Letters*, 2005. **5**(3): 527–530.)

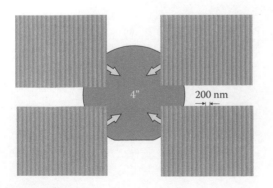

FIGURE 9.13　Top-view SEM images taken from four locations on the 100-mm-diameter sample imprinted by ACP. Uniform 200 nm period gratings are observed at all locations. (Adapted from Gao, H. et al., *Nano Letters*, 2006. **6**(11): 2438–2441.)

EFAN has demonstrated excellent imprint results, such as patterning of nanostructures in a photocurable resist spin coated on a wafer, with high fidelity and excellent uniformity over the entire the substrate, and in an ambient atmosphere without using a vacuum chamber. Furthermore, without much optimization, 100 nm half-pitch gratings with a residual layer thickness of 22 ± 5 nm were imprinted across a 100 mm-diameter wafer in about 2 s (Figure 9.15) [19].

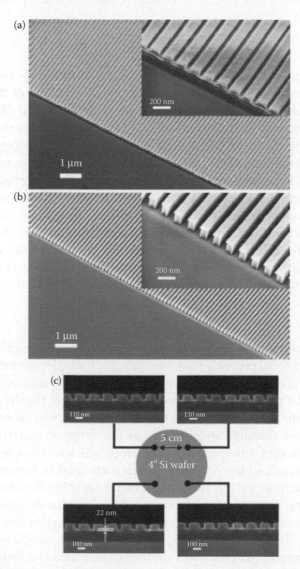

FIGURE 9.15 SEM images of (a) as-imprinted 200 nm period gratings in the top-layer (photocurable) of double-layer resist by EFAN (inset is a high-magnification image), and (b) 200 nm period gratings in the underlayer transferred from the top layer by RIE. (c) Cross-sectional SEM pictures captured from different locations on a 4-in. Si wafer showing the imprint uniformity in as-imprinted photocurable top-layer resist. (Adapted from Liang, X.G. et al., *Nano Letters*, 2005. **5**(3): 527–530.)

9.3.2.4 Roller Nanoimprint

An alternative approach to planar NIL is the roller nanoimprint lithography (RON), which was demonstrated soon after planar nanoimprint, and was motivated by even lower cost and higher throughput [20].

There are two methods for roller nanoimprint (Figure 9.16): (a) *cylinder mold method*, where the mold has a cylinder shape (e.g., bending a thin film mold around a smooth roller) rolls on a flat substrate or a sheet of film; and (b) *planar mold method*, where a flat (planar)mold is placed directly on top of a substrate and a roller of a smooth surface rolls on top of the mold, similar to a credit-card imprinting machine; and a slight bending of the flat mold under the pressure of the roller imprints the mold patterns into the resist on the substrate surface [20]. In case a mold cannot be bent slightly but a substrate can, one can put the substrate under a roller.

At the first try, it was found that roller nanoimprint not only has the same patterning resolution as a planar nanoimprint (e.g., the demonstrated 30 nm-wide resist grating with 200 nm period and 210 nm height (Figure 9.17)), but also has four additional key advantages over the planar nanoimprint that uses the solid parallel plate (SPP) [20]. The advantages are: (i) far better uniformity over large areas, (ii) much smaller total imprint force needed, (iii) almost no air bubbles even when imprinting in atmosphere, and (iv) far less effects caused by a dust. These advantages came from the fact that in roller nanoimprint, only a "line" area is in contact for a given time (rather than an entire area in a planar nanoimprint) and the movement of the roller helps push the air out and offers rather unique "smoothing" effects.

Because of these advantages, roller nanoimprint is well suited for large areas and/or high throughput and low costs, such as future large area TV screens or new thin film of metamaterials, which consists of nanostructures patterned by nanoimprint.

9.3.3 Imprint Area (Full Wafer or Step-and-Repeat or Rolling)

An entire wafer can be imprinted by a single step using a wafer size mold, or multiple steps using a much smaller mold through stepping and repeating the imprint of the small mold. *Step-and-Flash Imprint Lithography (SFIL)* is a photo-NIL process in which resist liquid droplets are dispensed and imprinted on one single die area at a time. This process is repeated as the imprint mold is "stepped" from one area (e.g., die)

Chapter 9

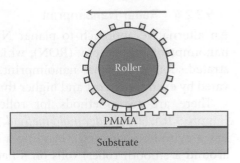

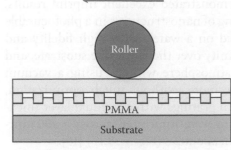

| Imprint using a cylinder mold | Imprint using a flat mold |

FIGURE 9.16 Schematics of two ways to do roller nanoimprint. (Adapted from Tan, H., A. Gilbertson, and S.Y. Chou, *Journal of Vacuum Science & Technology B*, 1998. **16**(6): 3926–3928.)

FIGURE 9.17 A 30 nm-wide resist grating (200 nm period and 210 nm height) by roller nanoimprint (RON). (Adapted from Tan, H., A. Gilbertson, and S.Y. Chou, *Journal of Vacuum Science & Technology B*, 1998. **16**(6): 3926–3928.)

to another areas (other dies) across the wafer, repeating the resist drop and imprint cycle. Two primary reasons for using a step and repeat method are: to reduce the need of making a large mold, and to achieve a better overlay between the mold and the substrate. In roller nanoimprint, one can use roll to roll to imprint an area measured by tens or hundreds of square meters or more.

9.3.4 Different Methods of Dispensing Imprinting Material

There are several ways to apply an imprint material onto a substrate, including spinning, dipping, spraying, and precision dropping. Each material dispensing method has its own advantages and disadvantages. We discuss two of the most common methods here.

9.3.4.1 Spinning

Imprint material, initially in a solution form, is randomly dropped on a substrate, then a spinning of the substrate turns the droplets into a uniform thin film with a fixed thickness which is determined by spinning speed, material solution viscosity and solvent evaporation. During the spinning, a part of the dropped material was spun way, becoming waste. Spinning has a long history, and is widely used in photolithography and is inexpensive. Clearly, the spinning method cannot be directly applied to the cases where only a selected area of a wafer needs imprint material (e.g., resist), or different areas of a wafer need different amounts of imprint materials, or the wafers are not in round shape.

9.3.4.2 Precision Dropping

Rather than spinning, an inkjet can drop an imprint material precisely on a substrate. Each droplet can have a precisely controlled volume and be placed accurately at a particular location of a substrate. After the dropping, a mold is pressed on top the substrate to cause a flow (locally) and imprint of the dropped material. Except for the solvent in the drops which will be evaporated, all the dropped materials are used in forming patterns during an imprint, hence reducing the waste and material cost. The drop volume and the dropping locations are optimized for a better final imprint result.

The precision dropping allows the depositing imprint materials only on the selected area of a wafer, while keeping the other areas from of the materials. It also allows adjustment to the amount of the materials deposited at a location, according to the pattern density, to create a uniform residue layer thickness, even for a nonuniform pattern density. (Note: In practice, such adjustment is good only to the scale of a droplet size, which is often ~100 μm in diameter.) Finally, when imprinting in atmosphere, the flow of the droplets during an imprint can be designed in a way to help repel the air bubble, and a liquid imprint material can absorb certain amount of air to further reduce the air bubble size; however, these air bubble reductions have limitations, as discussed in Ref. [21].

9.3.5   Different Nanoimprint Objectives

The imprinted material can serve as a resist for subsequent processing which will be removed afterwards, or as a functional structure of a device which will stay as a part of the device (e.g., patterning silicon oxide). In the first case, nanoimprint is lithography; and in the second case, nanoimprint is a direct patterning where one step performs several process steps: lithography, etching, and striping resist. For simplicity, we call all imprinted materials "resists" in either case.

9.3.6   Differences between Nanoimprint and Micromolding

To achieve the intended nanostructures, nanoimprint must use drastically different materials, processes and

9.4   Key Technologies in Nanoimprint

For a successful manufacturing using nanoimprint, one needs a complete solution that includes nanoimprint molds (masks), materials, machines, and processing, which are in fact intertwined. A failure in any one of the interrelated areas could lead to a failure of the entire nanoimprint manufacturing. We will subsequently discuss the key requirements and current approaches in each technology area.

9.4.1   Mold (Masks)

9.4.1.1   Requirements

A good imprinting requires an imprint mold having: (a) desired nanostructures, (b) good fidelity in pattern transfer, and (c) good facilitation in reducing defects. The requirement (a) is mainly related to the patterning of the mold, and the requirements (b) and (c) are primarily related to mold materials and structures.

9.4.1.2   Patterning of Molds

An imprint mold can be patterned using conventional patterning methods or nonconventional ones, or a hybrid of the two. Conventional methods include electron beam, ion beam, scanning laser beam, and

technologies from conventional micron or larger size molding. This includes: (i) use of an ultrathin (often mono-molecular) layer of mold release material on a mold. If the release layer is too thick, then the release material can seal off the nanoscale imprint structures on the mold. (ii) It should avoid significant distortion of the mold (caused by stresses, thermal expansion difference, etc.) during the imprint process, which can shear off imprinted structures in the resist. (iii) The protrusions on a mold should have a proper aspect ratio and sidewall slop, for easy separation of the mold and the imprinted resist and for avoiding significant shear forces that can beak imprinted resist structures. (iv) It should use proper resist thickness to avoid forming a thick residue resist layer after imprinting, since the thick residue layer can prevent a high-fidelity transfer of imprinted resist pattern to a substrate pattern. (v) It should design a resist with right properties that facilitate the mold separation and the resist mechanical strength.

photolithography. The nonconventional patterning methods include interference lithography, edge lithography, self-assembly and guided self-assembly methods, as well as combining these with conventional lithography. The nonconventional patterning is needed because many applications need a feature resolution or pattern area, or both far beyond the electron beam lithography's capability. Since NIL uses 1X masks (mold), it is far more challenging than photolithography masks.

9.4.1.3   Mold Materials and Structures

The imprint molds can be classified according to the hardness of their materials: hard mold, soft mold, and hybrid mold, which has local areas made of soft or hard materials. The reason for using a hard mold, which has high mechanical hardness, is to reduce the mold distortions both locally and globally. The reduction of local mold distortion will help individual pattern transfer fidelity, and the reduction of global mold distortion will help global overlay fidelity. The typical materials for hard molds are quartz, silicon, metals, or their combinations.

The key reasons to use a soft mold are two: (a) easy to make the soft mold surface to conformal to a substrate surface, which is not very flat and cannot have a good conformal contact with hard mold; and (b) reduce the effect of the damages by the dusts, which are unavoidable in many fabrication environment or intrinsically exist in certain processing (e.g., the growth of some compound semiconductors). Under a hard mold, a dust will prevent an imprint not only at the dust location, but also a large area surround the dust (often called "crater"). A soft mold can reduce or eliminate the crater. A very soft mold can imprint even the resist on the dust. Typical materials for soft mold are polymers. The most popular one is polydimethylsiloxane (PDMS), which has a hardness that can be adjusted by controlling its polymer components.

However, a soft mold, being soft, will be distorted easily during imprint, hence causing poor imprint fidelity of individual structures as well as poor global pattern position accuracy (namely pattern position shifted due to mold distortion). Two approaches have been used to circumvent the issues. One is to use a hard back plate with a thin soft front layer for the mold. This improves the global pattern positioning accuracy, but keeps the individual structure imprinting fidelity. Another approach is to use a hard back plate and a hard front surface with nanopatterns with a soft material in between. This sandwich mold structure will have make a mold have decent accuracy in both individual pattern imprinting and global pattern positioning, while accommodating a non-flat substrate surface and existence of dust.

9.4.1.4   Mold Release Layer

To have a good mold release property, a mold release agent is needed. There are three approaches to achieve a good mold release, which can be used individually or combined. One is to coat a release layer on the mold surface [15], the other is to add a release agent into the imprint material (called "internal release agent"), and another is to fluorinated mold materials. The release agent to be coated should have: (a) ultra-thin thickness or monolayer so that it will not fill up the nano-features, (b) good release property, and (c) good durability. The requirements for resist release are: (a) to have good release at mold but not weaken adhesion between the material and substrate, and (b) it can refurnish the mold release agent to the mold surface.

The release agents on the mold surface can be divided into two classes, fluorided polymers and diamond-like carbon. The fluorocarbon rated material has very low surface energy and a weak interaction with the imprint material. The carbon-like material also has low surface energy because of the saturated carbon bonds. The release agent often self-assembled monolayer (SAM) to reduce the distortion to the mold features. The SAM has two end functional group: one has fluorocarbon for good release and the other has chlorosilane for good bonding to the mold surface [15]. An example of the mold release agent is the fluorinated alkyl silane, such as tridecafluoro-1,1,2,2-tetrahydrooctyltrichlorosilane (CF_3-$(CF_2)_5(CH_2)_2SiCl_3$), which can be coated by vapor deposition [22].

Most internal mold release agents are also fluorinated materials. The internal mold release agent will also reduce the adhesion of the imprint materials with the substrate to cause peel off. To avoid this, an adhesion layer is often used between the substrate and the imprinted materials.

9.4.2   Imprinting Materials

9.4.2.1   Requirements

Basic requirements for a good imprinting material are: (a) it has good flow properties during the imprint, (b) good shape fidelity after imprint, (c) small shrinkage, and (d) good demolding property. Since the imprinting material can be functioned as a resist layer or a functional material that is a part of the device, this additional function of imprint material will add additional requirements to the resist. For example, when the imprint material is used as resist, the materials should have a good resistance to etching. The functional materials for devices depend upon the function to be achieved. For example, for insulating structures, oxides can be used. A broad range of materials can be used for nanoimprint. Subsequently, we briefly discuss the general types of the imprint materials, and many details can be found in Ref. [23].

9.4.2.2   Thermal Plastic Materials

Thermal plastic is a material that becomes a liquid when heated at a temperature above its glass transition temperature and becomes solid when the temperature is below the glass transition temperature. This property can be cycled indefinitely. The thermal plastic materials are linear chained polymers. The reason they can flow at high temperatures is that there is very weak bonding between the linear polymer chains, so that at high temperatures the chains can slide past each other, making them flow under pressure. The advantage of thermal plastic imprint material is that it is a very simple material. In principle, it only has one key component and therefore has less interaction between the polymer and the mold and is relatively easy to demold. Of course, to modify the flow property of thermal plastic one can add the plasticizers which lower the T_g and the mold release agent.

9.4.2.3 Curable UV or Thermal Materials

The curable must have three key components: monomers, linkers, and photo or thermal sensors. The linkers will link the polymer during the imprint and the light, photons or thermal energy will be applied by the initiators and cause the linker to work. For the linkers to get monomer form of network polymer net, the material turns from liquid to solid. The biggest advantage of curable material is that a very broad range of monomers can be used. Second, the monomer can have very low viscosity of three magnitude times lower than the thermal plastic. The disadvantage is that it has a lot of polar materials and is harder to demold compared to thermal plastic. Third is that the initiators can be somewhat sensitive to the oxygen and can be bleached and affect the cost shrinking property.

9.4.2.4 Hybrid Material

They also have hybrid materials to use the advantages of thermal or UV curable. First is to make partially soft and easy to separate, then using the hard bake to harden the material to get the strength needed. In some cases, even for a few UV curable polymers, a heated treatment imprint is used to increase the viscosity.

9.4.3 Imprint Machines

9.4.3.1 Requirement

The most fundamental requirement of an imprint machine is the imprint uniformity. The second requirement is easy demolding and the third is for some processes requiring good overlay alignment accuracy.

9.4.3.2 Different Forms of Pressing

The imprint uniformity is related to the substrate and the mold, but most importantly it is related to the pressing method. Four of the pressing methods are discussed in Section 9.3.

9.4.3.3 Alignment

The alignment methods in nanoimprint tools are very similar to that currently used in photolithography, and hence there is a very rich literature. The most accurate alignment approach is to use Moiré fringes, which have demonstrated sub-10 nm alignment accuracy. [24,25]. In a normal Moire method, the interference fringes are independent of the gap, orientation, and wavelength of the light, but when taking higher orders of diffraction, the fringes become depending on the gap between the mold and the substrate—a very useful and powerful method [26].

9.4.4 Imprinting Process

An imprint process involves many factors such as: the resist schemes (e.g., single layer or multiple layers), imprint pressure and methods of applying them, resist flow [27], curing conditions (for curable imprint), and alignments (if needed). A failure in one of them can destroy the entire process. Due to the limited space of the chapter, we discuss only the resist scheme here.

9.4.4.1 Single Layer of Resists

Single resist scheme is the most favored in manufacturing because of simple and hence low cost and higher yield. Often a monolayer adhesion layer will be used between the resist and a substrate.

9.4.4.2 Multilayer of Resists

Multilayers of resist are often used, because they (a) help the pattern transfer from resists to the substrate (e.g., many materials are good for NIL patterning resolution but are featured by poor etching resistance), (b) reduce the fabrication difficulties, and (c) planarize a nonflat wafer for better NIL (since wafer surface might not be flat but NIL needs to have conformal contact).

Here we present five basic multilayer resist schemes for NIL (Figure 9.18) [28]. All five schemes start with the same first step: the base layer of resist. The base layer can be used for surface planarization, where the resist is thicker than the step height on a wafer surface. The base layer can planarize a wafer surface by (a) a free viscous flow of the resist at an elevated baking temperature, or (b) a forced resist flow through pressing a flat mold on the resist surface while heating. For example, the samples have 100 nm height steps in SiO_2. Poly(methylmethacrylate) (PMMA) and modified Novolak resin 250 or 300 nm thick were used as the planarization layer. The free viscous flow approach works well for PMMA, when heated up to 175°C for 24 h, but not for the Novolak resin because it is a thermal-set plastic and has a high viscosity. The forced flow approach works well for both PMMA and for Novolak resin, when heated at 175°C for 20 min, with a pressure on the flat mold of 600 psi. After removing the mold, the resist over the sharp steps was planarized (Figure 9.19).

Chapter 9

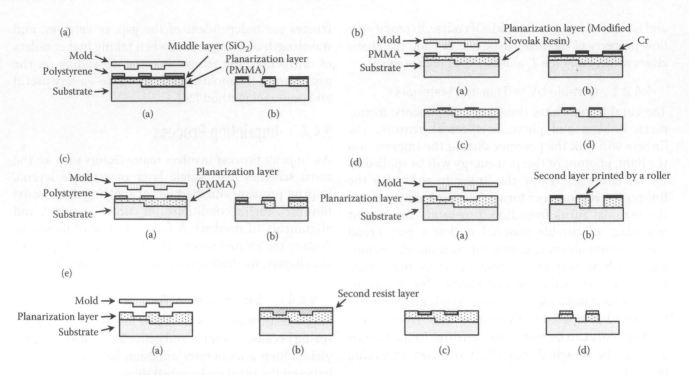

FIGURE 9.18 Five schemes for NIL on nonflat surfaces. (a) Trilayer resist scheme (positive tone), (b) imprint and lift-off scheme (negative tone), (c) imprint and RIE scheme (positive tone), (d) imprint and print scheme (positive tone), and (e) imprint and etch back scheme (negative tone). (Adapted from Sun, X.Y. et al., *Journal of Vacuum Science & Technology B*, 1998. **16**(6): 3922–3925.)

FIGURE 9.19 Resist pattern on a 100 nm SiO_2 sharp step in the imprinting and lift-off scheme (Scheme B). (Adapted from Sun, X.Y. et al., *Journal of Vacuum Science & Technology B*, 1998. **16**(6): 3922–3925.)

9.4.4.2.1 Scheme A: Trilayer Resist Scheme

The trilayer resist scheme has a "positive tone" (borrowed from photolithography), namely the resist profile after the fabrication has the area under the protrusions of the mold removed (Figure 9.18a). For example, the bottom layer is a 250 nm-thick 15 K molecular weight PMMA as the planarization layer. The middle layer is a 30 nm thick SiO_2, for the purpose of transferring patterns. The top layer is a 250 nm-thick 2 k molecular weight polystyrene, which is the patterning layer for NIL. One key point is that the glass transition temperature (T_g) of 2 k polystyrene is at about 0°C which

is much lower than the T_g of the PMMA (93°C). Therefore, the top-layer polystyrene can be imprinted at 90°C without deforming the bottom-layer PMMA.

The process began with an imprint to pattern the top layer at a temperature of 90°C and a pressure of 600 psi. Then the pattern in the top layer was transferred to the middle-layer SiO_2 by reactive ion etching (RIE), using CHF_3 chemistry. A second RIE transferred the SiO_2 pattern into PMMA using oxygen plasma, where SiO_2 has much lower etching rate than PMMA, hence serving as a good etching mask.

Figure 9.20a shows rectangular holes in the middle layer after processing with the trilayer resist scheme. The bottom layer is invisible in the SEM image. Figure 9.20b shows a resist pattern on a 100 nm SiO_2 sharp step after processing. The resist pattern consists of a 250 nm-thick PMMA layer (bottom layer), and a 30 nm-thick SiO_2 layer (middle layer). As shown, the top surface of the resist at the 100 nm step is flat, the side wall is straight, and the lateral dimension of the resist pattern is unaffected by the step.

9.4.4.2.2 Scheme B: Imprint and Lift-Off Scheme

The imprint and lift-off scheme uses two resist layers and has a "negative tone," namely the resist profile after the fabrication has the area under

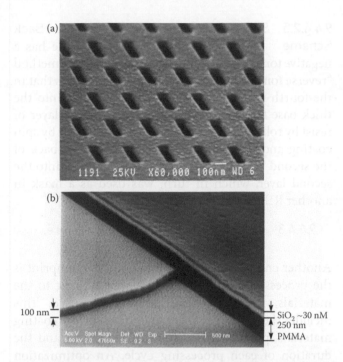

FIGURE 9.20 (a) Rectangular holes in the middle layer in the trilayer scheme (Scheme A). The underneath bottom layer is invisible in the SEM image. (b) Resist pattern on a 100 nm SiO_2 sharp step in the trilayer scheme (Scheme A). (Adapted from Sun, X.Y. et al., *Journal of Vacuum Science & Technology B*, 1998. **16**(6): 3922–3925.)

the protrusions of the mold remain (Figure 9.18b). For example, the bottom layer is a modified Novolak resin (Brewer Science XHRi 16) previously used as antireflection coating (ARC). It is a thermal-set polymer so that it hardens after a bake. It will not dissolve in acetone, and therefore it can survive a lift-off process. The top layer is 15 k molecular weight PMMA. Since the T_g of the modified Novolak resin is much higher than that of the PMMA, the imprint on the top layer will not deform the bottom layer. The process started with imprinting a pattern into the top layer with a temperature of 175°C and a pressure of 600 psi, then using lift-off to transfer it to a metal (e.g., Cr), which served as a mask for the subsequent oxygen RIE pattern transfer to the planarization layer (30 nm/min).

Figure 9.19 shows a resist pattern on a 100 nm SiO_2 sharp step after processing with the imprint and lift-off scheme. The resist pattern is a layer of 300 nm modified Novolak resin. Again the surface of the resist at the 100 nm step is smooth, the side wall is vertical, and the lateral dimension of the resist profile is not affected by the step. Figure 9.21a and b shows a 190 nm period resist grating on a 190 nm period SiO_2 grating after processing with the imprint and lift-off scheme.

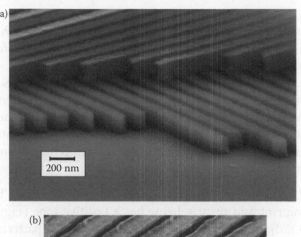

FIGURE 9.21 A 190 nm period resist grating on a 190 nm period SiO_2 grating in the imprint and lift-off (Scheme B), viewed from (a) 70° and (b) 40°. (Adapted from Sun, X.Y. et al., *Journal of Vacuum Science & Technology B*, 1998. **16**(6): 3922–3925.)

Again, the steps on the substrate have no effects on the resist linewidth and the vertical side-wall.

9.4.4.2.3 Scheme C: Imprint and RIE Scheme The imprint and RIE scheme used double-layer resists and has a positive tone (Figure 9.18c). The process started with an imprint to pattern the top layer. Then the pattern was transferred to the bottom using RIE with the top layer as the mask. For example, the bottom layer is 250 nm-thick PMMA with a molecular weight of 15 k. The top layer is 250 nm-thick polystyrene with a molecular weight of 2 k. Since polystyrene has a lower T_g than PMMA, imprinting on the top layer will not deform the bottom layer. After the imprint, oxygen RIE was used to etch PMMA, with a gas flow of 6 sccm, a pressure of

3 mTorr, and a power of 150 W. In this recipe, 2 k polystyrene has an etching rate of 25 nm/min, which is lower than that of 15 k PMMA (60 nm/min), therefore it can serve as the etching mask for PMMA.

Figure 9.20 shows a resist pattern on a 100 nm SiO_2 step after processing with the imprint and RIE scheme. Again, the surface, the side wall, and the lateral dimension of the resist are not affected by the SiO_2 step. However, due to the erosion of oxygen RIE, the polystyrene surface is rough.

9.4.4.2.4 Scheme D: Imprint and Print Scheme

The imprint and print scheme has a positive tone (Figure 9.18d). In this method, the base layer is first patterned by NIL. Then a second layer is printed only on the top surface of the imprinted pattern by a roller. The second layer will serve as the etching mask for the following RIE pattern transfer into the whole base layer. A careful selection of the material for the second layer and the RIE recipe can give a high RIE selectivity between the roller printed layer and the base layer. Thus, the roller printed layer can be used as an RIE etching mask to transfer the NIL pattern.

9.5   Key Challenges in Nanoimprint

The challenges in nanoimprint can be grouped into two areas: (i) fundamental patterning capability, and (ii) manufacturing capability. In fundamental patterning capability, the key parameters are imprinting resolution, pitch, 3D duplication fidelity (e.g., resist fluidic flow), resist mechanical properties, demolding, and in some cases, overlay accuracy. In manufacturing capability, the key parameters are imprint area, throughput, yield (hence defect density), and cost.

9.5.1   Challenges in Fundamental Patterning Capability

To address the challenges in fundamental pattering capability by nanoimprint, we need to investigate and understand the fundamental sciences in the five critical areas of nanoimprint, imprint methods, machines, molds, materials and processes. Some of these issues have been discussed in previous sections.

9.5.2   Challenges in Manufacturing Capability

In manufacturing, the mold fabrication is a high priority, since without a mold, nothing gets duplicated. The mold fabrication also strongly influences the nanoim-

9.4.4.2.5   Scheme E: Imprint and Etch Back Scheme

The imprint and etch back scheme has a negative tone (Figure 9.18e). Some later call this method "reverse tone NIL." In the first step, the same as that in the fourth method, a pattern was imprinted into the thick base layer. Instead of applying another layer of resist by rolling, a second layer of resist was cast by spin coating and was planarized. A uniform etch back of the second layer resist transferred the pattern into the second layer, which in turn, was used as a mask in another RIE etching of the bottom layer.

9.4.4.3   Different Nanoimprint Temperature, Pressure, and Processing Time

Another engineering aspect in a good nanoimprint is the process conditions, which is very specific to the materials, machines, and molds being used. This includes the suitable imprint materials, surface coating materials, press temperature, press pressure, and the duration of each processing cycle. An optimization process must be used in order to achieve a good process. There are many conditions needed to be optimized.

print area, yield, and cost. Nanoimprint throughput depends on the machine design as well imprint materials. The yield is related to the defect density. There are several kinds of defects. One kind defect is related to the imprint material filling into nanostructures on the mold. The other kinds of defects are due to air bubbles trapped. A third kind of defect is created during the demolding which sometimes breaks the nanostructures due to stresses. Several other manufacturing issues are discussed as follows.

Defect Density. The contact between an imprint mold and a wafer makes the technology susceptible to more defects than projection photolithography. However, the situation of today's nanoimprint is quite different from the problems faced by old contract lithography, which was a key reason of having migrated to projection lithography. In the old contact lithography, the mold would pick up some "dirt" at each contact, hence accumulating the "dirt," becoming worse each time, and eventually going over certain defect density limits. In contrast, in nanoimprint, the mold is coated with a thin antisticking layer which prevents "dirt" to stick on a mold, while the resist behaves more like a glue that will take a "dirt" away from a mold. Therefore, in each imprint, a "dirt" on the mold will be picked up by the resist and comes off from the mold, making a mold cleaner after each imprint. Such mold "self" cleaning in nanoimprint was long observed in its early date [29] and was further documented recently [30]. Recently, a defect density of 1.2 defects per cm² has been reported [31]. The belief is that the defect density can be further reduced.

Mold Damage is another issue in a contact printing. Currently, a variety of technologies have been devised for avoiding and reducing mold damages. For example, before imprinting a die on a wafer, the die can be previewed by a microscope and will be skipped from imprinting if a "significant" dust is observed, avoiding mold damage. A "soft" mold also can be used to reduce mold damaging.

1X Masks (Mold). In today's mold making, the 1X molds cost much more than 4X molds. However, the cost difference is getting smaller as 4X molds use more and more OPCs (for improving nanostructure patterning capabilities) which have a feature size near or the same as that in 1X molds. In parallel, some new ways of making 1X molds (nanoimprint molds) are being explored.

Wafer-Throughput for step-and-flash imprint (S-FIL) tools is ~20 wafer (8″ diameter) per hour commercially available. Although these tools can produce features sizes 5–7 times smaller, they have a throughput slower than current 65 nm photolithography tools which have a throughput of 60–80 wafers per hour. It should be clearly understood that just imprinting a pattern into a resist itself can be done in less than a microsecond. In fact, sub 200 nm imprinting in either liquid silicon or resists have been demonstrated [16,32]. The slow throughput of current S-FIL tools primarily comes from the time for dispensing resist on a die, alignment on each die, and resist curing. There is no doubt that with further development, these times can be reduced and throughput of step-and-repeat tools can be improved. For many devices, the overlay alignments may not be needed or are very coarse. In these cases, nanoimprint could use roll to roll, and have an incredible throughput similar to newspaper printing.

9.6 Applications of Nanoimprint

Because of its unmatchable advantages over the existing nanopatterning technology, nanoimprint technology, as soon as it was proposed, has been applied quickly and increasingly to a broad range of disciplines, including magnetic data storage, optics, optoelectronics, displays, biotechnology, semiconductor integrated circuits, advanced materials, chemical synthesis, to name just a few [3]. Initially, nanoimprint technology was used in laboratory demonstrations; but in recent years, nanoimprint technologies are increasingly being used by various industry sectors to become a key industrial manufacturing tool. We are just seeing a very beginning of a powerful enabling technology.

Acknowledgments

The author acknowledges valuable contributions from his former and present students and postdocs, as indicated by the relevant references, and the financial support obtained from ONR, DARPA, and ARO. This chapter is based on the author's several previously published papers as evidenced by references.

References

1. Chou, S.Y., P.R. Krauss, and P.J. Renstrom, Imprint of sub-25 Nm vias and trenches in polymers. *Applied Physics Letters*, 1995. **67**(21): 3114–3116.
2. Chou, S.Y., US Patent 5,772,905, *Nanoimprint Lithography.* 1998.
3. Chou, S.Y. et al., Sub-10 nm imprint lithography and applications. *Journal of Vacuum Science & Technology B*, 1997. **15**(6): 2897–2904.
4. Chou, S.Y. et al., Single-domain magnetic pillar Array of 35-Nm diameter and 65-Gbits/in(2) density for ultrahigh density quantum magnetic storage. *Journal of Applied Physics*, 1994. **76**(10): 6673–6675.
5. Chou, S.Y., Patterned magnetic nanostructures and quantized magnetic disks. *Proceedings of the IEEE*, 1997. **85**(4): 652–671.
6. Chou, S.Y., P.R. Krauss, and P.J. Renstrom, Imprint lithography with 25-nanometer resolution. *Science*, 1996. **272**(5258): 85–87.
7. Haisma, J. et al., Mold-assisted nanolithography: A process for reliable pattern replication. *Journal of Vacuum Science & Technology B*, 1996. **14**(6): 4124–4128.
8. Bailey, T. et al., Step and flash imprint lithography: Template surface treatment and defect analysis. *Journal of Vacuum Science & Technology B*, 2000. **18**(6): 3572–3577.
9. Austin, M.D. et al., Fabrication of 5 nm linewidth and 14 nm pitch features by nanoimprint lithography. *Applied Physics Letters*, 2004. **84**(26): 5299–5301.
10. Austin, M.D. et al., 6 nm half-pitch lines and 0.04 mu m(2) static random access memory patterns by nanoimprint lithography. *Nanotechnology*, 2005. **16**(8): 1058–1061.

Chapter 9

11. Pease, R.F. and S.Y. Chou, Lithography and other patterning techniques for future electronics. *Proceedings of the IEEE*, 2008. **96**(2): 248–270.

12. Li, M.T., L. Chen, and S.Y. Chou, Direct three-dimensional patterning using nanoimprint lithography. *Applied Physics Letters*, 2001. **78**(21): 3322–3324.

13. Willson, C.G., A decade of step and flash imprint lithography. *Journal of Photopolymer Science and Technology*, 2009. **22**(2): 147–153.

14. Gao, H. et al. Air cushion press for excellent uniformity, high yield, and fast nanoimprint across a 100 mm field. *Nano Letters*, 2006. **6**(11): 2438–2441.

15. Chou, S.Y., US Patent 6,309,580, *Release Surfaces, Particularly for Use in Nanoimprint Lithography*. 2001.

16. Chou, S.Y., C. Keimel, and J. Gu, Ultrafast and direct imprint of nanostructures in silicon. *Nature*, 2002. **417**(6891): 835–837.

17. Xia, Q.F. and S.Y. Chou, Applications of excimer laser in nanofabrication. *Applied Physics a-Materials Science & Processing*, 2010. **98**(1): 9–59.

18. Chou, S.Y., US Patent 6,482,742, *Fluid Pressure Imprint Lithography*. 2002.

19. Liang, X.G. et al., Electrostatic force-assisted nanoimprint lithography (EFAN). *Nano Letters*, 2005. **5**(3): 527–530.

20. Tan, H., A. Gilbertson, and S.Y. Chou, Roller nanoimprint lithography. *Journal of Vacuum Science & Technology B*, 1998. **16**(6): 3926–3928.

21. Liang, X., Tan, H., Fu, Z., and Chou, S.Y., Air bubble formation and dissolution in dispensing nanoimprint lithography. *Nanotechnology*, 2006. **18**: 025303.

22. Jung, G.Y. et al., Vapor-phase self-assembled monolayer for improved mold release in nanoimprint lithography. *Langmuir*, 2005. **21**(4): 1158–1161.

23. Guo, L.J., Nanoimprint lithography: Methods and material requirements. *Advanced Materials*, 2007. **19**(4): 495–513.

24. Li, N.H., W. Wu, and S.Y. Chou, Sub-20-nm alignment in nanoimprint lithography using Moire fringe. *Nano Letters*, 2006. **6**(11): 2626–2629.

25. Resnick, D.J. et al., Imprint lithography for integrated circuit fabrication. *Journal of Vacuum Science & Technology B*, 2003. **21**(6): 2624–2631.

26. Moon, E.E. et al., Interferometric-spatial-phase imaging for six-axis mask control. *Journal of Vacuum Science & Technology B*, 2003. **21**(6): 3112–3115.

27. Heyderman, L.J. et al., Flow behaviour of thin polymer films used for hot embossing lithography. *Microelectronic Engineering*, 2000. **54**(3–4): 229–245.

28. Sun, X.Y. et al., Multilayer resist methods for nanoimprint lithography on nonflat surfaces. *Journal of Vacuum Science & Technology B*, 1998. **16**(6): 3922–3925.

29. Wu, W. et al., Large area high density quantized magnetic disks fabricated using nanoimprint lithography. *Journal of Vacuum Science & Technology B*, 1998. **16**(6): 3825–3829.

30. Stewart, M.D. and C.G. Willson, Imprint materials for nanoscale devices. *MRS Bulletin*, 2005. **30**(12): 947–951.

31. Singh, L. et al., Defect reduction of high-density full-field patterns in jet and flash imprint lithography. *Journal of Micro-Nanolithography MEMS and MOEMS*, 2011. **10**(3).

32. Xia, Q.F. et al., Ultrafast patterning of nanostructures in polymers using laser assisted nanoimprint lithography. *Applied Physics Letters*, 2003. **83**(21): 4417–4419.

10. Scanning Probe Lithography

Omkar A. Nafday

Department of Biological Sciences, Florida State University, Tallahassee, Florida

10.1 Introduction

Scanning probe microscopy (SPM) is a technique that relies on surface scanning based on "feeling" rather than "seeing" as one would expect in a typical microscope. It relies on tip–surface interactions to construct three-dimensional (3D) representations of the surface. The possibility of visualizing individual atoms in real time has been a source of excitement and amazement to the scientific community. This was achieved in 1982 by Binnig and coworkers by their invention of the scanning tunneling microscope (STM) (Binnig and Rohrer 1982). During the early days of STM invention, determination of atomic surface structure was one of the most fundamental problems in surface science. In the premicroscope era, the principal tools of elucidating the atomic structure included the use of diffraction of waves. X-ray

beams were focused onto the system of interest and analysis of their postcollision angular patterns shed light on its atomic structure. The motivation behind the invention of the STM in the tunneling era was to develop a tool to image and study the electrical properties of insulating materials separated by a thin air gap. Thus, initial work focused on solving STM instrumentation problems, isolating mechanical vibrations, determining tip–sample forces, selection of piezoelectric materials, and determination of tip shape (eliminating secondary tips). An early example of a tool (similar to STM) to study the topography and the density of atomic steps on crystal surfaces was the Topographiner by Young et al. (1972). This was a type of field-emission microscope which relied on applying a high voltage to produce field-emission current rather than a tunneling current. The vertical and horizontal resolutions obtained were 30 Å and 4000 Å, respectively, although Young et al. suggested improving the horizontal resolution (200 Å) by using

Nanofabrication Handbook. Edited by Stefano Cabrini and Satoshi Kawata © 2012 CRC Press / Taylor & Francis Group, LLC. ISBN: 978-1-4200-9052-9

Chapter 10

sharper field-emission tips. Later, several criteria to be considered for good STM design were developed along with the development of STM theory and evaluation of the strengths and limitations of the STM (Tersoff and Hamman 1985, Pohl 1986, Venables et al. 1987). Today, the STM has developed into a versatile tool to measure the surface properties under a wide range of pressures (ultrahigh vacuum (UHV) to several hundred bar), temperatures (2–1000 K), and under environments of various gases (H_2, CO, O_2, NH_3). However, there was a need in the research community to be able to pattern materials at specific sites for investigation of various nanoscale phenomena, which extended beyond simple imaging. Thus, the STM was rapidly adapted for patterning materials, even though the STM was ideally suited for high-resolution imaging based on the principle of electron tunneling. Electron tunneling is an atomic-scale phenomenon, wherein electrons are transferred (tunnel) through energetically forbidden regions of space and will be covered in Section 10.1.1.

The motivation behind the invention of the atomic force microscope (AFM) (Binnig et al. 1987) was to overcome the limitation of the STM, which was incapable of imaging nonconductive substrates. More accurately, this instrument was an STM piezo drive monitoring cantilever movement by maintaining the tunneling current at a constant level with a separate piezo drive monitoring AFM sample movement. The cantilever was sandwiched between the AFM sample and the STM piezo drives thus effectively serving as the sample for the STM. They demonstrated a lateral resolution of 30 Å and a vertical resolution <1 Å when imaging at ambient conditions. The ability to deposit a controllable amount of material at specific sites under ambient conditions has captured the attention of research groups worldwide. Tools such as the AFM that facilitate controlled deposition processes, such as dip pen nanolithography (DPN), can be found in almost every research laboratory today. Since its invention, the AFM has evolved into a necessary tool for micro and nanoscale surface characterization, modification, and subsequent processing. The AFM today can be found in almost every research or industrial laboratory today; it has indeed become the method of choice for imaging, characterizing, and differentiating materials besides being the platform for creating nanoscale architectures using lithographic methods such as DPN. On the other extreme of the AFM operating conditions, UHV (Nony et al. 2004, Tanaka et al. 2006) and high-temperature studies (Ivanov et al. 2001, Weeks et al. 2002 a,b) demonstrated the versatility of the

AFM. Today, there are a multitude of AFM designs, each having its own unique set of functionality. However, all the AFM designs can be classified essentially into two groups: (A) sample moves relative to the cantilever or the other scenario and (B) the cantilever is scanned over a stationary sample. There are significant merits and demerits in both designs. For example, in design A, the optics required for laser feedback is relatively straightforward in design. However, since the heavy sample stage has to move, this design can only scan at speeds up to the resonant frequency of the stage. In design B, there is no restriction on scan speed as the moving cantilevers are relatively light, but on the other hand this design requires a more complex optical feedback setup. Applications of the AFM today involve a wide variety of areas including inorganic chemistry, biomolecules, polymers, crystallography, thin films, force–distance measurements, surface indentation and nanofabrication studies, and a host of AFM-based sensing approaches. Detailed reviews of AFM applications in biophysics, force measurements, nanofabrication, and cantilever-based sensing approaches can be found in the literature (Butt et al. 2005, Tseng et al. 2005, Gadegaard 2006).

10.1.1 Principle of STM and AFM Operation

The STM utilizes an extremely sharp tip (ideally terminating with an atom) to image the surface topography of a suitably conducting sample. The STM tip and the surface to be imaged, have to be both conducting in STM operation. When an opposite bias is applied between them, a current is registered. This current is due to electron tunneling and there is an exponential dependence between the tunneling current and the distance from the tip to the surface. Thus, even minute difference in tip–sample distance will lead to a significant change in tunnel current. The STM tip must be brought very close to the surface to register a measurable tunnel current, since a small change in the tip–substrate distance will change the tunneling current by a factor of 10. Maintaining the tip at a distance this close to the surface without crashing into the surface presents a tremendous electronic control problem. It is akin to bringing an object from the moon at rocket speed and to prevent it from crashing into the earth's water surface with <1 ft of skimming distance. However, this is possible through the use of microelectromechanical systems (MEMS) technology, where the atomically sharp tip can be fabricated and mounted on a silicon nitride cantilever. The tip and

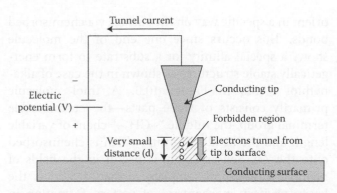

FIGURE 10.1 The electron tunneling effect in STM operation. (Reproduced from http://www.acoustics.org. With permission.)

cantilever are normally connected to a piezoelectric material actuator, usually made of lead zirconium titanate (PZT). This assembly allows the tip movement over the sample in the x-, y-, z-direction by applying a voltage across the PZT crystal faces. PZT crystals exhibit the piezoelectric effect, that is, they change shape when a voltage is applied on opposite sides of the crystal. The actuators on the cantilever are then engaged, while constantly monitoring the tunnel current, until the specified tunnel current is achieved. A feedback loop adjusts the voltage to the piezo in order to keep the tip scanning at constant tunneling current over the topography of the sample.

The tunneling phenomenon is described pictorially in Figure 10.1. The atomically sharp tip (radius ~9 Å) is moved over the sample under study, and a small voltage (tens of mV) is applied between tip and the surface. This bias between the tip and the sample causes a flow of electrons between the tip and the sample (usually metals). In metals, we can imagine that electrons in the bulk conducting solid could be described quantum mechanically as particles in a box. In the simplest approximation, the energy states of a single electron in such a box could be calculated using a finite-well potential (since there is a finite probability that the electron could leave the bulk solid). This is the quantum–mechanical representation of the valence and conduction bands in a metal. These electrons "tunnel" from filled bands in the surface to empty bands in the tip (or, if the polarity of the bias voltage is reversed, from filled bands in the tip to empty bands in the surface). In this way, using a combination of MEMS and electron tunneling, high-resolution images can be obtained with the STM where topography differences of the sample contours can be revealed to better than 1/100th of an atomic diameter (Hansma et al. 1988).

An AFM generally uses a silicon nitride (Si_3N_4) or silicon (Si) tip mounted on a "diving board" type cantilever to scan surface features. Typically, a laser is bounced off the back of the cantilever onto a photodetector divided into four quadrants, as shown in Figure 10.2. The segmented photodetector allows for both vertical and lateral displacement of the cantilever to be recorded simultaneously. The AFM tip is scanned with a piezo tube over the sample in x–y directions, in raster fashion, one line at a time as shown in the left inset (red square) of Figure 10.2. Each tip movement in the x-direction is then displaced along the y-direction and thus a two-dimensional (2D) image formed over many x–y movements collectively contains z-topography information. This $z(x, y)$ information creates a 3D image of the sample

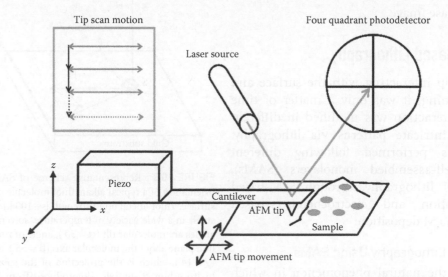

FIGURE 10.2 Schematic representation of AFM operation.

topography. As the cantilever scans the surface, minute deviations of the piezo scanner due to surface features are registered as changes in the photodetector signal intensity and a high-resolution 3D image of the surface is obtained.

The AFM can be operated in either contact mode or alternate contact mode and the image of the surface can be displayed in a variety of modes (topography, lateral force, phase, and error). Generally, if the samples are relatively soft (biomolecules), alternate contact operation is better since this mode minimizes tip–sample interactions providing a way to image without physically damaging the sample. For lithography, the AFM is generally operated in contact mode since lateral force and sample topography information is desired. The choice of which type of mode to adopt while imaging depends on the type of information desired, amount of resolution necessary, and the type of substrate to be imaged. The topography of the sample is generated by plotting the vertical correction signal applied from the feedback loop to the z-piezo. At regions of higher topography, the cantilever experiences a larger force, which is compensated by raising the cantilever up and away from the surface. This feedback mechanism is reversed for regions of low topography, effectively keeping the cantilever at the constant force setpoint. The lateral force image provides differences in friction between the patterned features, and the unpatterned substrate regions are detected as the tip scans over the features created by the inked AFM tip (in DPN). The phase image is a reflection of differences in sample stiffness, enabling distinction of the different components of a heterogeneous surface. A stiffer surface has a greater phase shift than a softer area and appears brighter in the phase image.

10.1.2 STM-Based Lithography

With the STM tip interacting with the surface and doing the patterning, it was only a matter of time before the tip interaction was modified in different ways to create intricate patterns via lithography. Lithography was performed following different routes—using self-assembled monolayers (SAMs), direct-write STM lithography, through SPM-based surface anodization and electrochemistry-based liquid-mediated STM deposition.

10.1.2.1 STM Lithography Using SAMs

Self-assembly is the natural phenomenon in which amphiphilic molecules with dissimilar end groups orient in a specific way on the surface via chemisorbed bonds. This occurs since one end of the molecule shows a special affinity for a substrate to form energetically stable structures as shown in the case of alkanethiol SAMs in Figure 10.3. A thiol molecule primarily consists of three parts—the customizable terminal group, the middle "–CH$_2$–" chain of variable length and the head "S" group which is chemisorbed onto the Au. Self-assembly is useful in the fields of nanofabrication and template formation and is the basis (primary mechanism) of pattern formation in thiol ink lithography. Alkanethiols, typically having the structure SH(CH$_2$)$_n$X, self-assemble and chemisorb either from solution or from the gas phase onto suitable surfaces and form well-ordered and oriented monolayer thin films. The alkanethiol molecules are chemisorbed to the gold surface by the formation of Au–S bonds while the terminal group ("X") can be tailored to provide specific substrate characteristics. By changing "X" it is thus possible to introduce a wide variety of functionality into the self-assembled structure. Self-assembly is not limited to thiols; a number of other systems exist including chlorosilanes on silicon

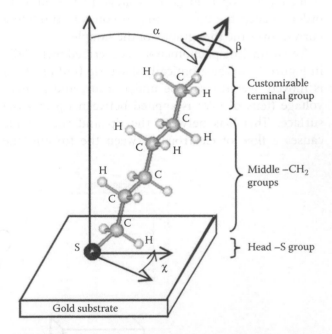

FIGURE 10.3 Representative scheme of SAM formation showing orientation of a typical alkanethiol molecule chemisorbed on a gold surface. Thiol–metal bond strength is ~100 kJ/mol, making the bond stable in a wide variety of temperature, solvents, and potentials. The angles are molecular tilt ($\alpha = 30°$), angle of rotation of the hydrocarbon plane about the molecular axis ($\beta = 55°$), and angle of precession ($\chi = 14°$), which is the projection of the substrate normal and the hydrocarbon chain axis. (Reproduced from Vericat C. et al. 2005. *Phys. Chem. Chem. Phys.* 7, 3258. With permission.)

(Si) (Haller 1978, Sagiv 1980) and carboxylic acids on metal oxides (Allara and Nuzzo 1985a,b). Organosulfur compounds on gold offer the most controllable and flexible system currently used in nanofabrication. Another advantage is their ease of preparation and analysis, particularly the self-assembly of *n*-alkanethiols. An excellent review of the applications of self-assembly can be found elsewhere (Swalen et al. 1987). Nanometer-scale patterning of monomer (stearic alcohol) and polymer films (polymethylmethacrylate (PMMA)) has been performed on different substrates (GaAs, Au, and Pt) by using the STM. The STM-generated patterns were then transferred into metal films and semiconductor substrates by chemical etching. Using an AFM, 25 nm linewidths have been imaged. UHV-STM nanolithography was demonstrated using SAMs as ultrathin resists (Kleineberg et al. 2001). Hexadecanethiol ($SH(CH_2)_{15}CH_3$), and *N*-biphenylthiol ($SH \pm (C_6H_6)_2 \pm NO_2$) monolayers were prepared on gold (Au) films and Au(111) single crystals. Organosilane SAMs such as octadecyltrichlorosilane ($SiCl_3 \pm (CH_2)_{17} \pm CH_3$) monolayers were prepared on hydroxylated Si(100) and hydroxylated chromium film surfaces. Dense line patterns were written by UHV-STM by controlling various tunneling parameters (gap voltage, tunneling current, scan speed, and orientation) and transferred onto the underlying substrate by wet-etch techniques. Best resolution was achieved without etch transfer for a 20×20 nm^2

written in hexadecanethiol/Au(111) with an edge definition of about 5 nm. Etch transfer of the STM nanopatterns in Au films resulted in 55 nm dense line patterns (15 nm deep), while 35 nm wide and 30 nm deep dense line patterns written in octadecyltrichlorosilane/Si(100) and anisotropically etched onto Si(100) surface were achieved.

Lewis and Gorman performed organothiolate SAM lithography using an STM tip. They found that on elevating the tip bias, the SAM is locally desorbed. They conducted this desorption process in the presence of a second thiol component which is then locally introduced at the desoption site of the first SAM as shown in Figure 10.4. It was found that the structure of an SAM strongly influences the process of replacement lithography. Specifically, longer and more well-ordered SAMs required a higher applied bias to undergo replacement. Conversely, shorter SAMs and SAMs with cyano, hydroxy, and carboxylic acid head groups can undergo replacement at lower applied bias. However, this latter class of SAMs underwent replacement to a lesser degree of completion and with a poorer resolution. Thus, from the standpoint of high-resolution lithography, well-packed SAMs have a benefit even if they require a slightly higher bias to perform the operation (Lewis and Gorman 2004). Sundermann et al. reported the fabrication of artificial nanostructures in ultrathin resist films of PMMA and an alkanethiol SAM (hexadecanthiol) patterned by STM-lithography in UHV.

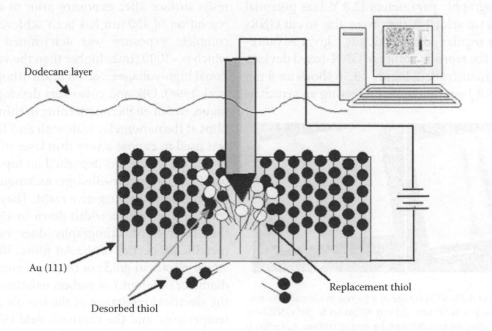

FIGURE 10.4 STM-enabled replacement lithography using alkanethiols. (Reproduced from Lewis, M.S. and C.B. Gorman. 2004. *Journal of Physical Chemistry B* 108(25, June 24): 8581. With permission.)

Chapter 10

The PMMA patterns were analyzed by an AFM, while the SAM patterns were investigated by STM. Linewidths down to 75 nm were reproducibly achieved in PMMA with bias voltages up to 10 V and tip currents of 1 nA. The PMMA pattern was also successfully transferred into the underlying Mo/Si multilayer substrate by fluorine-reactive ion etching (RIE), presenting STM lithography as an attractive alternative to conventional e-beam lithography for the fabrication of lateral nanostructures in multilayers. The SAM pattern width down to 15 nm was reproducibly achieved with a 1 V bias voltage and 1 nA tip current (Sundermann et al. 2000). In a similar study, STM lithography of very fine lines (<10 nm) in width was reported by using a resist process with the STM tip as an electron beam source. A diluted positive electron beam resist PMMA was first coated on an S-doped GaAs substrate. Using the STM with a bias voltage of 10 V and a tunneling current of 0.5 nA, lines were created using different tip scan speeds. Subsequently, they deposited titanium (Ti) layer in the sample region etched by the STM tip in the PMMA resist resulting in the formation of fine Ti lines as narrow as 50 nm (Hironaka et al. 1997).

10.1.2.2 Direct-Write STM Lithography

Tapaszto et al. developed an STM lithography-based technology that allows the engineering of graphene with true nanometer precision. Figure 10.5a shows a 10-nm-wide and 120-nm-long graphene nanoribbon (GNR) created by STM lithography. By setting the optimal lithographic parameters (2.4 V bias potential and 2.0 nm/s tip velocity), they were able to cut GNRs with suitably regular edges, which is a great advancement toward the reproducibility of GNR-based devices. As a basic demonstration, Figure 10.5b shows an 8 nm wide, 30° GNR bent junction connecting an armchair

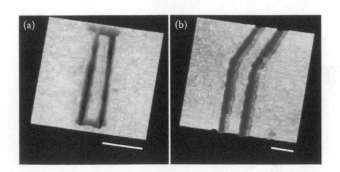

FIGURE 10.5 (a) A 3D STM image of a 10-nm-wide and 120-nm-long GNR. Scale bar is 50 nm. (b) An 8-nm-wide 30° GNR bent junction connecting an armchair and a zigzag ribbon. Scale bar is 20 nm. (Reproduced from Tapaszto, L. et al. 2008. *Nature Nanotechnology* 3(7, July): 397. With permission.)

and a zigzag ribbon, giving rise to a metal–semiconductor molecular junction (Tapaszto et al. 2008). STM lithography was used to investigate a thiol-derivated 12-mer oligonucleotide (HS-ssDNA) self-assembled Au(111) surface. Under low sample bias and high tunneling current, the repeated scanning resulted in the growth of nanostripes. The stripe orientation, the stripe width, and the spacer width between adjacent nanostripes were found to be dependent on their relative locations from dislocation points where two adjacent gold terraces overlap. Such stripes may reflect the strain distributions and the release pathway along the Au surfaces. The results also suggest that the presence of HS-ssDNA molecules enhances the lithography processes on the gold surface by acting as force transmitters (Chen et al. 2009). Weeks et al. presented a reproducible method of producing nm-size holes on a graphite surface with a high-pressure STM. Holes were only observed at pressures above 5 bar with voltage pulses in the field-emission range (±4–10 V). The depth of the hole was shown to vary with the pressure while the observed diameter was independent of the applied pressure. The results obtained in the presence of elevated pressures of dry nitrogen, oxygen, and argon gases were reported (Weeks et al. 2002). Kragler et al. used an STM operating in air to expose a 50-nm-thick electron-beam-sensitive resist with low energetic electrons. They demonstrated that it is possible to expose the resist under ambient conditions with voltages of similar to 50 V without observable modification of the resist surface after exposure prior to development. A resolution of 150 nm has been achieved. The dose for complete exposure was determined as 10 mC/cm² which is ~1000 times higher than the value for conventional high-voltage electron beam lithography (Kragler et al. 1996). Olk and coworkers developed a new technique, which enables patterning of thin evaporated Au films at the nanometer scale with an STM. The STM tip was used to expose a very thin layer of omega-tricosenoic acid, which was deposited on top of the Au films using the Langmuir–Blodgett technique and acts as an electron-sensitive, negative resist. They fabricated narrow Au lines with a width down to 15 nm and found that their STM lithography does not degrade the metallic properties of the Au films. They were able to attach electrical gold contacts to a small bundle (total diameter of 50 nm) of carbon nanotubes and measure the electrical resistance of the bundle as a function of temperature and the magnetic field (Vanhaesendonck et al. 1994). A gold STM tip was used as a miniature solid-state emission source for directly depositing nm-sized gold structures. The process was demonstrated in UHV on gold substrates, and in air on gold and platinum substrates. The emission mechanism is field evaporation of tip atoms, which is enhanced by the close proximity of the substrate (Mamin et al. 1991). Schwank et al. used a modified STM working in a controlled oxygen atmosphere to create structures on the surface of a steel sample covered with 10 nm amorphous hydrogenated carbon (a-C:H). They showed the capability of producing either μ-sized hills or holes depending on the oxygen pressure and the other parameters such as feedback settings, bias voltage, and setpoint current. These hills consisted of a locally delaminated a-C:H film induced by a dielectric breakdown between the tip and the sample and it was concluded that the process responsible for hole formation is reactive oxygen etching (Schwank et al. 1998).

10.1.2.3 SPM-Mediated Surface Oxidation

STM lithography via local oxidation was also used to create metallic patterns. Nanoscale aluminum (Al) patterns were created by Ono et al. by selective Al chemical vapor deposition on an H-terminated silicon surface using a UHV-STM (Ono et al. 1996). The H-terminated surface was created by a palladium STM tip by applying a negative sample bias. The formation of a nanoscale hexagonal pattern by light STM-enabled local oxidation of the (111) surface of a magnetite single crystal was reported by Murphy et al. using STM. This structure was found to have a periodicity of 42 ± 3Å and was comprised of three distinct regions (Murphy et al. 2005). Accurate current measurements were performed during nanolithography for the first time and were attributed to the reduction of H^+ ions at the STM tip. They reported that $\sim 60 \times 10^6$ atoms of oxygen were required to produce the volume of SiO, and that the integrated charge while writing that line was $\sim 150 \times 10^6$ C. To incorporate an oxygen atom into the surface lattice from the native water layer requires the release of two H^+ ions, which are reduced at the tip, resulting in the measured current flow and the formation of H_2. These results made it possible to improve the control of oxide growth using conducting tip SPM-based nanolithography (Ruskell et al. 1996). In a recent review, the mechanisms of STM-induced hydrogen desorption was discussed, including the postpatterning deposition of molecules and materials, and the implications for nanoscale device fabrication (Walsh and Hersam 2009). UHV-STM enabled patterning and characterization of the physical, chemical, and electronic properties of

nanostructures on surfaces with atomic precision. Specifically, on hydrogen-passivated Si(100) surfaces, selective nanopatterning with the STM probe allows the creation of atomic-scale templates of dangling bonds surrounded by a robust hydrogen resist. The STM-created patterns can be used as templates for a variety of materials to form hybrid silicon nanostructures while maintaining a pristine background resist. This nanolithography approach has led to its use on a variety of other substrates, including alternative hydrogen-passivated semiconductor surfaces, molecular resists, and native oxide resists. A new hybrid lithography method was presented which combined the key features of the AFM and STM by incorporating two independent feedback loops, one to control current (for STM) and one to control force. They demonstrated a minimum resolution of 41 nm and 100 nm resist features patterned over 180 nm of topography created by local oxidation of silicon (Wilder et al. 1997). Another example of hybrid STM lithography was demonstrated by Marrian et al. when they demonstrated e-beam lithography in the resist SAL-601 with the STM. Patterns were written by raising the tip–sample voltage above −12 V while operating the STM in the constant current mode; 50-nm-thick resist films were patterned and the pattern transferred into a GaAs substrate by RIE. The variation of feature size with applied dose and tip–sample bias voltage was also studied. The development of ultralow-voltage e-beam lithography based on STM technology was also discussed (Dobisz and Marrian 1997).

10.1.2.4 Electrochemistry-Based, Liquid-Mediated STM Lithography

Similar to the work of Ono et al., the STM was used as a tool to produce nanometer-scale modifications on hydrogen-passivated silicon(100) surfaces under positive and negative sample bias versus tip voltage. However, their experiments were performed in air and under a hydrofluoric (HF) acid solution with an electrochemical STM (ECSTM), under bipotentiostat control. In air, it was found that under both polarities the surface under the tip becomes oxidized and that the oxidation is possible although no tunneling current flows between the tip and the sample. The experiments under HF demonstrated that two different mechanisms existed, oxidation at anodic polarization and direct silicon etching under cathodic polarization. In air, it was shown that the oxidation is induced by the electrical field between the tip and the surface. In an HF acid solution, it was shown that in addition to oxide formation, silicon

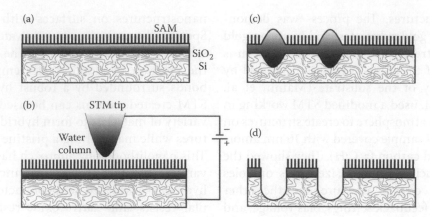

FIGURE 10.6 Schematic diagram of STM lithography. (a) Preparation of a TMS monolayer on Si substrates (TMS–Si). (b) STM patterning (c) Patterned sample. (d) Pattern transfer to the Si substrate by chemical etching. (Reproduced from Sugimura H. 2005. *Int. J. Nanotech.* 2, 314. With permission.)

dissolution also occurs by direct silicon etching (Ye et al. 1995). The ECSTM was also used to investigate the early stages of palladium electrodeposition by using a palladous chloride solution and a highly oriented pyrolytic graphite surface (Tong et al. 1995). They showed localized control of Pd nucleation and growth under the ECSTM tip. Kimizuka and Itaya investigated the underpotential deposition (UPD) of silver (Ag) on Pt(111) surfaces using ECSTM in aqueous sulfuric acid solutions. Two sets of well-defined UPD peaks in the cyclic voltammogram were observed on a well-ordered Pt(111) surface (Kimikuza and Itaya 1992). Sugimura and Nagakiri fabricated coplanar nanostructures consisting of two different types of organosilane monolayers—trimethylsilyl (TMS) and 3-aminopropyl triethoxysilane (APS) using both an STM and an AFM. The first organosilane monolayer, which had been uniformly prepared on a substrate, was shown to locally degrade through electrochemistry of adsorbed water at the junction of the monolayer and the SPM probe. This probe-scanned region chemisorbed molecules of the second organosilane, resulting in the creation of an SAM confined to the SPM-defined pattern as shown in Figure 10.6. Aldehyde-modified fluorescent latex nanoparticles and horseradish peroxidase protein molecules were shown to selectively assemble onto an APS-modified template by using glutaraldehyde as a cross-linker between the amino groups of the APS monolayer and those of the protein (Sugimura and Nagakiri 1995).

In order to realize 2D lithography at high resolution (several tens of nanometers), a new approach of the lithography of a fine pattern using a confined etchant layer technique (CELT) in an electrochemical system was presented (Tian et al. 1992). A mold plate

of conductive material with a high-resolution line was used instead of the STM tip. The etchant species is generated at the surface of the mold plate by electrochemical photochemical or photoelectrochemical methods, then diffuses away from the surface of the mold plate. The key feature of CELT is the design of a chemical reaction which rapidly destroys the etchant (within μs) following its generation. Therefore, the gradient of the concentration of etchant can be greatly enhanced and the thickness of the diffusion layer can be greatly decreased to several tens of nm. Thus, the etchant layer is confined and its outer boundary can essentially retain the fine structure of the pattern of the mold plate. Then the substrate to be corroded is adjusted by ECSTM to approach the mold plate within several tens of nm and the corroded pattern retains the fine structure.

10.1.3 AFM Lithography

The AFM has also been used creatively for creating nanostructures on different surfaces since the AFM can be operated under ambient conditions in a variety of operating modes as mentioned earlier. Another advantage of the AFM is that pattern creation can be multiplexed easily by using multiple cantilevers mounted on a suitable platform on the AFM. These different cantilevers could potentially be inked with different tips, thus enabling multiplexed pattern creation with multiple inks.

10.1.3.1 Scratching/Making Indents with AFM Cantilevers

This is the simplest lithography technique as it involves mechanically scratching, grafting, burrowing, etching, and/or removing material using the force of cantilever contact with the surface. The amount of force required is highly dependent on the substrate and this force can be modulated by varying the stiffness of the cantilever used or by operating the AFM in alternate contact mode. The cantilever can be kept immobile at a certain location to create local indents (holes) by the vertical force of the cantilever or the cantilever can be moved in a certain path to create lines or creative architectures on different surfaces including polymers, metals, inorganic compounds, and semiconductors (Jin and Unertl 1992, Cappella and Sturm 2002). One example is the mechanochemical scanning probe lithography (MC-SPL) based on the nanoscale abrasive interaction between a nanoprobe and a surface (Sung and Kim 2005). The technique consisted of two sequential processes—a mechanical scribing process of an ultrathin SAM resist film coated on the substrate by using a nanoprobe and the subsequent chemical etching process of the substrate material at regions where the resist has been removed during the mechanical scribing process. Their results showed that MC-SPL had the capability of patterning under relatively high speeds (up to 1 mm/s) without degradation of pattern integrity and patterns with widths <100 nm could be fabricated over an area of about 10,000 μm^2 of silicon and various metal surfaces with flat and nonflat geometry at the patterning speed of 1 mm/s.

Nanografting is another AFM-based chemomechanical approach which combines the displacement of selected molecules by an AFM tip and the placement of a new material at that displacement site. Nanografting allows a more precise control over the size and geometry of patterned features and their locations on surfaces. Linewidths down to 50 nm were written on hydrogen-terminated silicon and were used to explore the impact of tip radius and tip wear using Si_3N_4-coated tips (Lee et al. 2006b). Using nanografting with SAMs, the Liu group has investigated the influence of ligand local structure on biorecognition and protein immobilization (Liu et al. 2008). The patterned SAMs produced in this way open new opportunities for systematic studies of such size-dependent properties such as friction, conductivity, and spatially confined surface reactions. Nanografting has also been demonstrated in alternate contact mode using SAMs (Liang and Scoles 2007). There are several other applications for the nanografting like: creating 3D nanostructures for positive and negative pattern transfer (Liu et al. 2002), meniscus force nanografting of DNA (Schwartz 2001), placement of metalloproteins on gold (Case et al. 2003),

creating nanoscale patterns within SAMs (Xu et al. 1999), and with proteins (Wadu-Mesthrige et al. 1999).

10.1.3.2 Dip Pen Nanolithography

DPN is an inherently additive AFM-based technique that operates under ambient conditions, making it suitable to deposit a wide range of biological, organic, and inorganic materials, at (only at) specific locations as shown in Figure 10.7. DPN is generally performed in contact mode AFM operation; however, the alternate contact mode, DPN, has also been demonstrated (Agarwal et al. 2003). DPN was first demonstrated in 1995, using 1-octadecanethiol (ODT) ink on a mica substrate (Agarwal et al. 2003). They noticed that an AFM tip became coated with ODT when trying to image a gold surface under an ODT–ethanol solution. This coated tip on contact with a mica substrate resulted in irregular ODT islands which were 1.2 ± 0.3 nm in height. The Jaschke and Butt (1995) experiments were not well controlled, particularly for trying to make lines. However, the credit of developing DPN into a feasible nanopatterning technique is due to the Mirkin group at Northwestern University, Evanston, IL (Piner et al. 1999). The alkanethiol ink and the gold substrate interact chemically by formation of gold–sulfur bonds thus enabling control over the deposition process to get the type of patterns desired. Also, Mirkin and coworkers viewed the ink transport as mediated by the meniscus and controlled the humidity accordingly. They demonstrated precise control by patterning ODT dots, lines and grids with the smallest linewidth being 30 nm.

The primary mode of ink transport is attributed to the bulk water meniscus that forms between the substrate and the AFM tip. However, this mode of ink

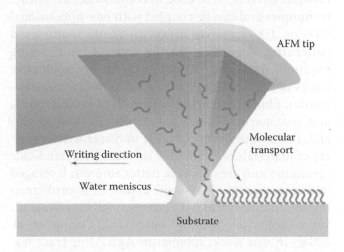

FIGURE 10.7   Schematic representation of the DPN method.

transport is immediately questionable when patterning hydrophobic inks such as ODT and when patterning under conditions of almost 0% RH. Sheehan and Whitman demonstrated facile ODT ink deposition on gold even after the tip was exposed to a 0% RH environment for over 26 h during the course of which, over 1200 ODT islands of ~630 nm radius were deposited by DPN (Sheehan and Whitman 2002). There have been many studies in the literature investigating the role of the water meniscus in ink transport and how it can affect the type of features formed (Jang et al. 2001, 2004, Peterson et al. 2004, Nafday et al. 2006b). Although the water meniscus has been attributed to be a mode of ink transport in DPN, other variants of nanolithography such as thermal DPN rely on physical melting of the material to be deposited (King et al. 2002, Nelson et al. 2006).

DPN has been demonstrated with a variety of different inks either as a patterning tool or for preparing templates for subsequent modifications. The earliest application of DPN was to study the self-assembly and growth processes of 16-mercaptohexadecanoic acid (MHA) and ODT inks under ambient conditions (Hong et al. 1999). Examples of using DPN to pattern with chemically diverse inks include iron (Gundiah et al. 2004), gold (Maynor et al. 2001), gold–silver alloy (Zhang et al. 2004b), ferroelectric copolymer nanostructures (Tang et al. 2004), polymer nanowires (Noy et al. 2002), proteins (Lee et al. 2003, 2006c, Li et al. 2005), and energetic materials (Nafday et al. 2006a). Zhang et al. (2004b) showed that thiols patterned with DPN on gold/silicon substrates can serve as etch resist layers. In this way, patterns of metal, silicon, glass, or chrome nanostructures can be generated. This approach demonstrated the complementary role of DPN, to be used in conjunction with other techniques and can be coupled with one-dimensional (1D) or 2D cantilever arrays to create multiplexed patterning, depending on the scale of multiplexing required. Low-cost, direct writing of conductive traces is highly desired for applications in nanoelectronics, photonics, circuit repair, flexible electronics, and nanoparticle-based gas detection. Toward this end, the unique ability of DPN to direct write a variety of materials onto suitable surfaces with nanoscale resolution and area-specific patterning was leveraged to demonstrate a direct-write approach toward creating conductive traces with commercially available silver nanoparticle (AgNP)-based ink (Wang et al. 2008). In this work, submicron AgNP line trace features were created on various surfaces. The example

of using DPN to create conductive silver traces between two preexisting electrode junctions is both novel and practical. Toward this end, the AFM cantilever is first plasma cleaned for 5 min to remove possible surface junk from the tip. This ensures uniform tip condition every time prior to coating the tip with the conductive silver ink. Ink, tip, and surface interaction chemistry is very critical in this DPN application as the ink needs to flow uniformly to create a continuous (and thus conductive) silver trace. In other words, the ink needs to "like" the surface more than the tip. After inking the tip, the inked tip is briefly kept in contact with the surface for a few seconds to create an inking blob on the surface. This action can be performed repeatedly until a bleeding dot of a constant size is obtained. These initial bleeding dots will be quite big (optically visible) compared to the final feature size required, but this step helps bleed the excess ink from the tip to achieve the goal of uniform ink flow from the tip. The incorporation of this bleeding step before performing the final DPN is typical of all inks which are physisorbed on the surface. This is the only feasible test of ensuring uniform ink flow for inks which are physisorbed on the surface and coated on the tip in the liquid state. It is also important to tune the conductive ink viscosity and the amount of surfactant therein and for this purpose the ink vendor can be consulted for ensuring quality control. After obtaining a uniform bleeding dot shape, the tip can be used to create conductive traces by moving the tip in a specified path using lithography algorithms or by moving the surface itself. Subsequent to creating these traces on the surface, the surface is baked to evaporate excess solvent (and or residual surfactants) to 150–300°C and solidify (cure) the conductive ink trace. This ink trace can be created between two existing electrode structures to fabricate a closed nanocircuit, the conductivity of which can be tested by using either 2- or 4-point electrical measurements. An ohmic contact between the electrodes is ensured when the voltage–current graph is a straight line with a definite slope. The slope of the graph gives the resistance of the conductive trace which can be then correlated to the resistivity value of bulk silver. The curing step has also been shown to have a significant effect on the final silver trace conductivity and there are various techniques to achieve the curing procedure effectively.

Another practical application is creating positive- and negative-type metal photomasks fabricated by DPN as shown in Figure 10.8 (Jang et al. 2009).

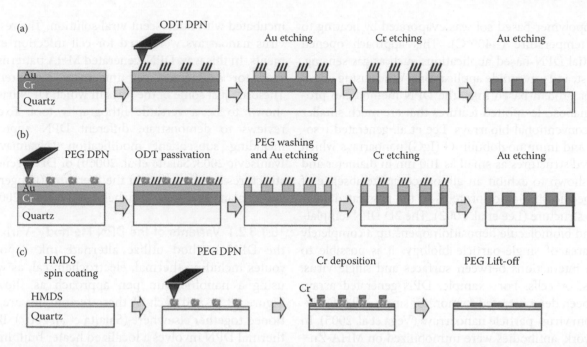

FIGURE 10.8 DPN-based photomask fabrication approaches. (a) Positive-type photomasks are generated by first using DPN to pattern ODT, which is used as a resist for subsequent Au and Cr wet-chemical etching. (b) Negative-type photomasks are generated by first using DPN to pattern polyethylene glycol (PEG). After passivation with ODT, the PEG is washed away and subsequent Au and Cr wet-chemical etching yield the resulting Cr photomask. (c) Negative-type photomasks may also be generated by a lift-off process, whereby PEG is patterned by DPN on an hexamethyldisilazane (HMDS)-coated quartz substrate. After Cr evaporation, the PEG is removed by sonication in water. (Reproduced from Jang, J.W. et al. 2009. *Small* 5(16, August 17): 1850. With permission.)

Positive-type photomasks were generated with a multiple-pen array at a rate rivaling electron beam lithography. For the negative-type photomasks, a poly(ethylene glycol) ink was used as the resist and lift-off material.

DPN-assisted templating is another area where DPN has shown promise. Wang and others created arrays of very precisely sized, positioned, and oriented single-walled carbon nanotubes by attaching them to pre-DPN-patterned MHA templates, passivated by ODT (Im et al. 2006, Wang et al. 2006). This was a powerful and elegant demonstration of the templating technique as shown by Figure 10.9, considering the traditional difficulty in precisely positioning carbon nanotubes. Integration of the nanotubes relies on the ability to control the placement, orientation, and shape of the nanotube components. This CNT templating work has been further improved with greater chemical generality (Myung et al. 2005).

Su et al. (2004) demonstrated direct patterning of aluminum and tin oxides on silicon and silicon oxide substrates based on sol–gel chemistry. This approach took advantage of the water meniscus formation as a nanowater source to hydrolyze the metal precursors. Subsequent to patterning the sol–metal inks,

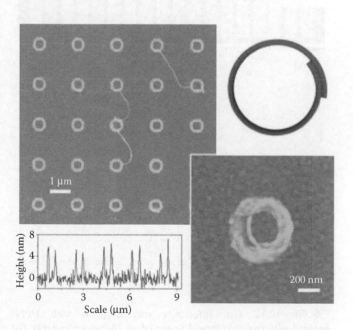

FIGURE 10.9 Example of a DPN-enabled templating approach. Single-walled carbon nanotubes assembled into rings at the MHA-ODT boundary region on a gold surface. (Reproduced from Wang, Y.H. et al. 2006. *Proceedings of the National Academy of Sciences of the United States of America* 103(7, February 14): 2026. With permission.)

Chapter 10

the copolymer-based sol was evaporated by heating to high temperature (~400°C). This approach opened potential DPN-based applications such as gas sensing, catalyst, and waveguide applications. When using oligomer or protein-based inks, the DPN method can produce nanoscale-spotted features that are much smaller than conventional bioarrays. Lee et al. generated lysozyme and immunoglobulin G (IgG) nanoarrays which featured structures as small as 100 nm in diameter and were shown to exhibit an almost complete absence of nonspecific binding of proteins to the passivated areas of the structure (Lee et al. 2002). The 2D DPN templating and biomolecule deposition opens up a completely new area of single-particle biology; it is possible to probe interactions between surfaces and single virus, spores, or cells. For example, DPN-generated arrays have been demonstrated to monitor single-cell infectivity from virus–particle nanoarrays (Vega et al. 2005). In this work, antibodies were immobilized on MHA-Zn^{2+} regions created by MHA-DPN and subsequent passivation with polyethylene glycol and immersion in an ethanol-based $Zn(NO_3)_2 \cdot 6H_2O$ solution as shown in Figure 10.10. MHA-Zn^{2+} regions were then exposed to rabbit antibody creating nanoarrays which were later

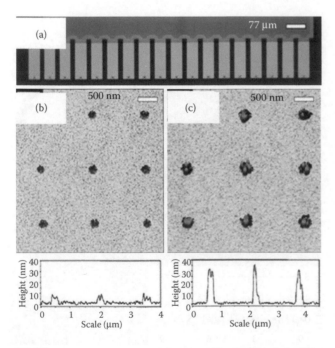

FIGURE 10.10 Cell infectivity measurements with DPN-enabled patterns. (a) Optical image of an 18-cantilever array. (b) Topography image and height profile of polyclonal rabbit immunoglobulin G immobilized on DPN-generated MHA dot patterns. (c) Topography image and height profile of Figure 10.10b, after cell-infectivity measurements, resulting in a height increase of ~22 nm. (Reproduced from Vega, R. et al. 2007. *Small*, 3, 1482. With permission.)

incubated with fluorescent viral solution. The resultant virus nanoarrays were used for cell infection experiments. In this way, DPN-generated MHA patterns were used for subsequent cell infectivity measurements. These are just some of the ways in which DPN has been shown to be a versatile lithography tool. Excellent reviews to demonstrate different DPN approaches including subsequent modification (Nyamjav and Ivanisevic 2003, Salaita et al. 2006b) of DPN-generated structures can be found in the literature (Ginger et al. 2004, Salaita et al. 2007, Haaheim and Nafday 2008).

10.1.3.2.1 Variants of the DPN Method Variants of the DPN method utilize alternate ink deposition routes including thermal, electrochemical, as well as using a nanofountain pen approach as shown in Figure 10.11 and each of these techniques are mentioned together elsewhere (Salaita et al. 2007). Briefly, thermal DPN involves a localized heater built into the cantilever to heat the AFM tip by passing current through the cantilever circuit. Details of this design can be found elsewhere and Nelson et al. have demonstrated direct deposition of indium nanostructures using thermal cantilevers (King et al. 2001, Lee and King 2007, Nelson and King 2007). Jegadesan et al. used electrochemical DPN to drive ink molecule transport by applying a voltage difference between the tip and the substrate (Jegadesan et al. 2006). A recent development is the nanofountain probe, which employs a nanopipette tip linked to ink channels housing the ink, compared to writing with an inked tip as performed in DPN (Loh et al. 2009). This probe was used in a single-cell *in vitro* transfection study where sub-100 nm resolution was demonstrated. In another study, they reported electric field-induced, sub-µm, direct delivery of proteins. Deposition was controlled by application of an electric potential of appropriate sign and magnitude between the probe reservoir and the substrate. With a view to multiplexing lipid patterning on a subcellular scale, selective adsorption of functionalized or recombinant proteins based on streptavidin or histidine-tag coupling was shown (Sekula et al. 2008). The biomimetic membrane patterns formed were then used as substrates for cell culture, as demonstrated by the selective adhesion and activation of T cells.

A variety of MEMS hardware has been developed specifically for DPN, including multicantilever arrays (1D and 2D), microfluidic ink channels, and individually actuated cantilevers as shown in Figure 10.12. These special MEMS tools developed are targeted toward

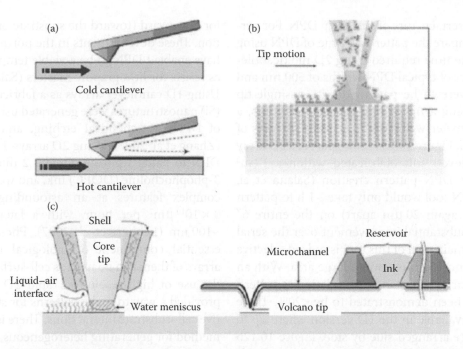

FIGURE 10.11 Variants of the DPN technique. (a) Thermal DPN, (b) electrochemical DPN, and (c) nanofountain pen-assisted DPN. (Reproduced from Haaheim, J. and O.A. Nafday. 2008. *Scanning* 30(2, April): 137. With permission.)

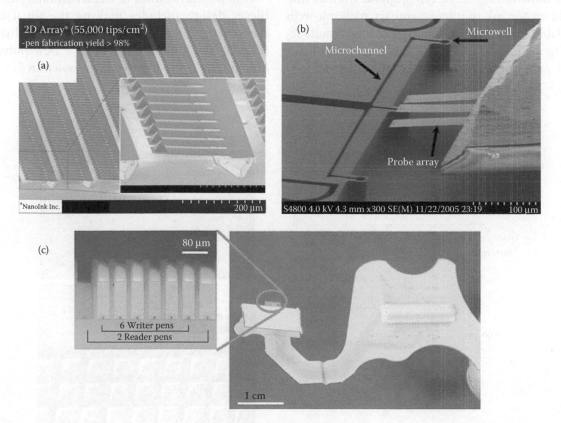

FIGURE 10.12 Advances in MEMS tools to complement DPN. (a) SEM image of a 2D cantilever array of 55,000 pens developed to overcome the serial nature of DPN. (b) SEM image of microfluidic tip coating and ink delivery channels enabling different cantilever tips to be inked with different inks. (c) Optical image of the eight individually actuated cantilever assembly enabling "writing" with one cantilever tip and imaging with another uninked cantilever. Each cantilever is wire bonded onto a circuit board platform for easy mounting and actuation through the AFM scanner. (Reproduced from Haaheim, J. and O.A. Nafday. 2008. *Scanning* 30(2, April): 137. With permission.)

Chapter 10

overcoming the serial nature of single-tip DPN. For perspective, we compare the patterning time of DPN using a single tip to the time required using 2D multicantilever arrays. If lines of typical DPN widths of 500 nm and length 10 μm were to be patterned with a single-tip 20 μm apart, using a tip writing speed of 10 μm/s, a standard 6″ diameter wafer would require ~6 years of continuous DPN. Figure 10.12a shows part of a 2D array of 55,000 cantilevers microfabricated within a 1 cm² area for parallel DPN pattern creation (Salaita et al. 2006a). This DPN tool would only take ~1 h to pattern the same lines (again 20 μm apart) on the entire 6″ wafer. This is a substantial improvement over the serial design, and the main goal of this tool is indeed selective placement of nanostructures over a large area. With an ODT-inked 2D array, the standard deviation in pattern dimensions has been demonstrated to be <20%. These arrays are also available in the 1D version where up to 52 cantilevers are arranged side by side. Figure 10.12b shows the microfluidic channels which enable individual cantilever tip inking from the microwell. The three channels shown in Figure 10.11b originate from an ink-well holding the ink of interest and six such ink-wells are available for various tip inking/holding simultaneously. Figure 10.12c shows the array of individually actuated cantilevers, using applied current (~15 mA)

for downward (toward the substrate or ink-well) actuation. These developments in the portfolio of DPN tools have enabled DPN to be a viable template creation tool as resists for lithographic masters (Salaita et al. 2006b). Using 1D cantilever arrays as a fabrication tool, silicon (Si) nanostructures were generated using a combination of DPN, wet-chemical etching, and RIE techniques (Zhang et al. 2007). Using 2D arrays, Lenhert et al. used DPN to pattern phospholipid 1,2-dioleoyl-*sn*-glycero-3-phophocholine (DOPC) ink, and were able to pattern complex features at an astounding throughput of 3×10^{10} μm² per hour, with a lateral resolution of ~100 nm (Lenhert et al. 2007). Phospholipids are an essential component of biological membranes, and arrays of them can be used as cell-surface models. Thus, the use of high-resolution rapid 2D DPN patterning provided a way to make patterns for studying cooperative cell–substrate interactions. There is no other known method for generating heterogeneous, multivalent, planar-supported lipid bilayers covering large areas with feature sizes smaller than a single cell.

A new modification to the traditional MEMS cantilever design includes making the tip itself of an elastomeric material such as polydimethylsiloxane (PDMS) and this gave birth to a new polymer-pen lithography (PPL) approach as shown in Figure 10.13

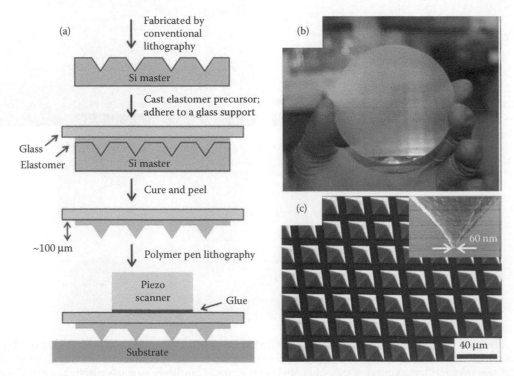

FIGURE 10.13 (a) Schematic illustration of the PPL setup. (b) A photograph of an 11-million-pen array. (c) SEM image of the polymer pen array. The average tip radius of curvature is 70 ± 10 nm (inset). (Reproduced from Huo, F.W. et al. 2008. *Science* 321(5896, 19 September): 1658. With permission.)

(Huo et al. 2008). The PPL included creation of a 11 million-pen array with each pen tip having a PDMS coating. The PDMS coating acts as an ink reservoir and readily absorbs a wide variety of DPN inks; DPN stamp tips can be used to pattern some of the common ink–substrate combinations that are currently used with the microcontact printing (µCP) technique; DPN stamp tips also can pattern liquid (solvent) inks unlike traditional DPN and DPN stamp tips can be used for AFM imaging after making DPN patterns. These novel tips essentially combine many of the advantages of DPN (high-resolution, site-specific, and simultaneous writing and imaging) and µCP methods (excellent chemical compatibility with many different ink materials). Since polyoxazoline is a hydrophilic and biocompatible polymer, these tips can easily absorb and release water-based biomolecules via their pores. Using DPN, the patterning of Human IgG and prostate specific antigen (PSA) proteins using polyoxazoline stamp tips was demonstrated by Lee et al. (2007). They found that protein nanostructures can be generated 60 times faster with this method than with a conventional silicon tip. In addition, PDMS tips have been used to generate DPN nanopatterns of MHA, ODT, dendrimers, proteins, cystamine, and inorganic salts (Zhang et al. 2004a, Choi et al. 2007).

In conclusion, these are just some of the ways in which DPN has evolved into a multifaceted, multi-ink ambient patterning technique with nanoscale registry. Future demonstrations of DPN and further development of MEMS tools to enable DPN in completely new ways will no doubt further validate the effectiveness of this method from its humble origins of uncontrollable alkanethiol patterning.

10.1.3.3 AFM Lithography Using Heated Cantilevers

AFM cantilevers having integrated heaters can be used as a source of heat at the nanometer scale. AFM cantilevers with integrated heaters were first developed for advanced data storage technology (Nonnenmacher and Wickramasinghe 1992). In thermomechanical data storage (Williams and Wickramasinghe 1986, Mamin and Rugar 1992, Nonnenmacher and Wickramasinghe 1992, Mamin 1996, Chui et al. 1998, Binnig et al. 1999, Despont et al. 2000, King et al. 2001, Raczkowska et al. 2003), a silicon AFM cantilever with an integrated solid-state heater is brought in contact with and scanned over a thin polymer film. Electrical heating pulses delivered

to the cantilever lead to a temperature rise in the cantilever and cantilever tip. The polymer in direct contact with the tip softens and is displaced by the cantilever tip under mechanical load, forming an indentation with a diameter as small as 22 nm (King et al. 2001). Figure 10.14a shows scanning electron microscope (SEM) images of heated AFM cantilevers (inset is the actual tip) fabricated in the King group at Georgia Tech (now at UIUC). Figure 10.14b also shows an infrared microscope image of one cantilever during steady-state heating. Till date, the King group is the third in the world to fabricate these cantilevers (after Stanford, Williams and Wickramasinghe 1986) and IBM (Mamin 1996, Despont et al. 2000), and the second to both fabricate and use them (after IBM workers Despont et al. 2000, King et al. 2001). The cantilevers are made from doped single-crystal silicon with an integrated differentially doped localized heater region. Detailed studies of the electrical, thermal, and mechanical characterization can be found elsewhere (Lee et al. 2006, Lee and King 2007, 2008, Park et al. 2007). The silicon cantilevers have a tip with a radius of curvature as small as 20 nm. The cantilever has a thermal time constant in the range of 1–10 µs (Williams and Wickramasinghe 1986, Chui et al. 1998, Binnig et al. 1999). The cantilever temperature can approach 1000°C. As the resistive heating element is also a temperature sensor, calibration of the cantilever temperature response is possible to within 1°C (Williams and Wickramasinghe 1986, Mamin and Rugar 1992, Chui et al. 1998, Binnig et al. 1999, King et al. 2001). An SEM image of a thermal cantilever used in this work is shown in Figure 10.15a. The ability to tailor synthetic voids in thin films of energetic material is demonstrated in Figure 10.15b which shows a simple "+" pattern written in the pentaerythritol tetranitrate (PETN) film, demonstrating the high spatial resolution and registry of the technique (King et al. 2006).

10.1.3.4 AFM Lithography/DPN with Biomolecules

Biological nanolithography on suitable surfaces is currently a very active interdisciplinary field of research at the crossroads of biotechnology and nanotechnology. Precise patterning of biomolecules on surfaces with nm resolution has great potential in many medical and biological applications ranging from molecular diagnostics to fundamental studies of molecular and cell biology. Nanolithography of biomolecules includes studies with DNA, peptides, viruses, and proteins with

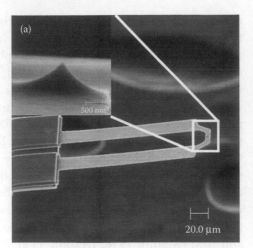
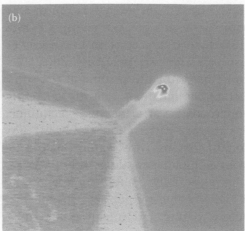

FIGURE 10.14 Cantilevers with integrated heaters fabricated by Dr. William King at UIUC. (a) SEM images of the cantilever with a close-up of the tip (left inset). The shape of the cantilever is "U" so that it can carry electrical current in a continuous circuit and be compatible with laser focusing for use in AFM imaging. This cantilever has a tip with radius of curvature <50 nm (inset). (b) Infrared microscope image of a cantilever during steady-state heating. The heating occurs primarily at the cantilever tip. (Reproduced from Lee, J. et al. 2006a. *Journal of Microelectromechanical Systems* 15(6, December): 1644. With permission.)

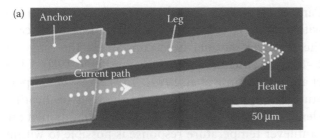

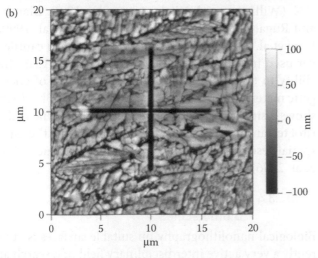

FIGURE 10.15 (a) SEM image of a typical heated AFM cantilever used for nanolithography. (b) Lithographic indents written into a thin film of an energetic material using the AFM cantilever. For each of the two lines of the "+," the cantilever was at 215°C and scanned at 0.1 Hz for 60 s. The tip load force was 7 nN. The depth of the feature was 300 nm. (Reproduced from King W.P. et al. 2006. *Nano Lett.*, 6, 2145. With permission.)

and without the presence of suitable biocompatible materials (agarose, polyethylene glycol) which aid in the deposition process. This method for the direct transfer of biomolecules encapsulated within a viscous fluid matrix was demonstrated by Senesi et al. (2009) via DPN. The method relies on the use of agarose as a carrier that is compatible with many types of biomolecules including proteins. Feature sizes as small as 50 nm were demonstrated, together with the characterization of protein and oligonucleotide structures by studying their reactivity with fluorophore-labeled antibody and complementary oligonucleotide sequences, respectively. Recently, Paxton et al. (2009) reported a way to catalyze azide–alkyne cycloadditions (CuAAC) between solvated terminal alkyne molecules and azide-terminated SAMs on silicon surfaces using heterogeneous copper-coated AFM tips. Spatially controlled surface functionalization was performed with four small molecules bearing terminal alkyne groups—propargylamine, 4-pentynoic acid, and an alkynyl-oligoethyleneoxide. Linewidths on the order of 50 nm were demonstrated. A methodology to place and stretch long DNA molecules by coupling DPN with molecular combing was also presented (Nyamjav and Ivanisevic 2003). Templates were fabricated on SiO_x surfaces using a positively charged polymer, poly(allylamine hydrochloride). Upon molecular combing, DNA molecules elongate on the surface and were localized by the presence of the charged lithographically defined structures.

10.1.3.5 Other Novel AFM Lithography Methods

Wu et al. (2006) demonstrated a successful strategy for combining the straightforward scanning probe chemical bond-breaking lithography and SAM techniques for constructing nanoscale architectural structures of nanoparticles onto modified SiO_2 surfaces. The hydroxyl-terminated surface of the sample substrate was modified by silanization with n-(2-aminoethyl)-3-aminopropyl-trimethoxysilane (AEAPTMS) molecules. Local-field-induced scanning probe bond-breaking lithography was adopted to selectively decompose the chemical bonds of AEAPTMS-SAMs on amino silane-modified SiO_2 surfaces, at a tip bias of >4.5 V between the AFM conductive tip and the SiO_2 surface under ambient conditions. After the scanning probe selective decomposition of AEAPTMS SAMs, AuNPs with negative-charged citrate surfaces were selectively anchored in the selective patterning region via Coulomb electrostatic force. Silicon oxide (SiO_x) patterns were formed on Si(100) surface by means of AFM anodization, where noncontact mode of AFM operation was used to oxidize Si wafer at the nanoscale (Hutagalung and Darsono 2009). Dot arrays with 10 nm SiO_x height and <50 nm in diameter were successfully fabricated. Kim et al. obtained scanning near-field optical microscopy images to study the excitation of surface plasmons on metallic dots fabricated using SPL. Gold nanodots were fabricated by applying electric voltages to conducting probes installed in an AFM using the mechanism of field-induced diffusion and nanooxidation plus Au-coating. In this way, they demonstrated that scanning near-field optical microscopy imaging combined with SPL was able to provide a systematic study of surface plasmon excitation on nanometallic structures (Kim et al. 2007).

Combining SPM lithography with Kelvin force microscopy (KFM) and scanning capacitance microscopy (SCM) was demonstrated by Han et al. (2007). The surface of Si substrates covered with an octadecylsilane (ODS) SAM was locally modified using the AFM tip as a minute electrode by KFM. At a sample bias >6 V, the ODS-SAM was degraded and the probe-modified regions became more frictional, while there were no lateral force microscopy (LFM) contrasts on regions probe-scanned at a lower sample bias voltage. Nevertheless, these regions showed clear KFM and SCM contrasts due to charge trapping and deposition in the ODS-SAM and SiO_2 layer. They showed the complementary use of KFM and SCM as being crucial for the electrical characterization of organic-monolayer-covered semiconductor substrates. Huo et al. showed the complementary use of nanoimprint lithography (NIL), RIE, and SPL. They showed that with closed-loop scan control of an SPM, patterned lines >100 μm and a mold pattern with a linewidth about 10 nm could be obtained. The structures on the mold are further duplicated into PMMA resists through the NIL process (Huo et al. 2004).

With all the above methods, the SPM/AFM was used in novel ways or used in complement to other methods. However, modifications to the actual AFM tip design were not performed. Wang et al. invented a new "spring-on-tip" design in which the sidewalls of the pyramidal tip AFM tip itself were modified to contain folded spring structures to reduce the overall force constant of the scanning probe (Wang et al. 2005). The spring structure was generated using the focused ion beam milling method. They demonstrated sub-100 nm lithography using a modified spring tip in the DPN writing mode (see Section 10.1.3.2). A proof-of-principle study of creating novel scanning probe tips was presented recently (Kumar 2009). The "nanobit" tips were 2–4 μm long and 120–150 nm thin flakes of Si_3N_4 or SiO_2, fabricated by electron beam lithography and standard silicon processing. Using a microgripper, they were fixed to a standard pyramidal AFM probe or alternatively inserted into a tipless cantilever equipped with a narrow slit. These modified tips were used for imaging of deep trenches, without visible deformation, wear, or dislocation after several scans. This approach allows an unprecedented freedom in adapting the shape and size of scanning probe tips to the surface topology or to the specific application.

10.1.4 Future Prospects of SPM Lithography

The two main lithography branches based on the SPM platform—using the STM and AFM and their modifications—are discussed in this chapter. Several reviews exist in the current literature using these platforms solely or in conjunction with other techniques for creating nanoscale architectures (Tseng et al. 2005, Garcia et al. 2006, Xie et al. 2006). Ultimately, the driving force for developing novel and unconventional lithography methods would be the researcher in the laboratory trying to push the envelope of new application areas. This would involve going beyond the currently established application areas of creating semiconductor devices, probing the physical and chemical properties of nanostructures, building assemblies for use in biology and biotechnology, local modification of materials and building sensors based on the SPM platform. Currently, with most research and industrial institutions embracing SPM as a standard platform for investigating nano- and microscale phenomena, the future of SPM-based lithography remains bright. Many institutions of learning are already incorporating the use of the scanning probe microscope as part of their curriculum. This would be a significant step for advancing nanotechnology right from the grass-roots level up.

The next challenge would be to incorporate SPM lithography in a manufacturing environment for rapid prototyping and multiplexed pattern creation. The issue of how small the features can be created is not an immediate concern; however, how reproducible the lithography process can be made is important to address. Besides, SPM lithography can be used in a complementary way to other techniques such as NIL, photolithography, and EBL to achieve certain goals that cannot be achieved with these techniques. Ultimately, it remains to be seen how SPM-based lithography will find a foothold on the assembly line.

Acknowledgments

I would like to thank Dr. Brandon Weeks for giving me the opportunity of sole-authorship in compiling this chapter. I have also received valuable input from my wife Renuka in proof reading this chapter and I acknowledge her support. I also thank the Editors and CRC Press for this project.

References

Agarwal, G., L.A. Sowards, R.R. Naik, and M.O. Stone. 2003. Dip-pen nanolithography in tapping mode. *Journal of the American Chemical Society* 125(2, January 15): 580–583.

Allara, D.L. and R.G. Nuzzo. 1985a. Spontaneously organised molecular assemblies. 1. Formation, dynamics and physical properties of normal-alkanoic acids adsorbed from solution on an oxidized aluminium surface. *Langmuir* 1(1): 45–52.

Allara, D.L. and R.G. Nuzzo. 1985b. Spontaneously organised molecular assemblies. 2. quantitative infrared spectroscopic determination of equilibrium structures of solution—adsorbed normal—alkanoic acids on an oxidised aluminium surface. *Langmuir* 1(1): 52–66.

Binnig, G. and H. Rohrer, 1982. Scanning tunneling microscopy. Helvetica Physica Acta, 55(6): 726–735.

Binnig, G., M. Despont, U. Drechsler, W. Haberle, P. Vettiger, H.J. Mamin, B.W. Chui, and T.W. Kenny. 1999. Ultrahigh-density atomic force microscopy data storage with erase capability. *Applied Physics Letters* 74(9, March 1): 1329–1331.

Binnig, G., C. Gerber, E. Stoll, T.R. Albrech, and C.F. Quate. 1987. Atomic resolution with atomic force microscope. *Surface Science* 189(October): 1–6.

Butt, H.J., B. Cappella, and M. Kappl. 2005. Force measurements with the atomic force microscope: Technique, interpretation and applications. *Surface Science Reports* 59(1–6, October): 1–152.

Cappella, B. and H. Sturm. 2002. Comparison between dynamic plowing lithography and nanoindentation methods. *Journal of Applied Physics* 91(1, January 1): 506–512.

Case, M.A., G.L. McLendon, Y. Hu, T.K. Vanderlick, and G. Scoles. 2003. Using nanografting to achieve directed assembly of de novo designed metalloproteins on gold. *Nano Letters* 3(4, April): 425–429.

Chen, F., A.H. Zhou, and H. Yang. 2009. The effects of strain on STM lithography on HS-ssDNA/Au(111) surface. *Applied Surface Science* 255(15, May 15): 6832–6839.

Choi, D.S., S.H. Yun, Y.C. An, M.J. Lee, D.G. Kang, S.I. Chang, H.K. Kim, K.M. Kim, and J.H. Lim. 2007. Nanopatterning proteins with a stamp tip for dip-pen nanolithography. *Biochip Journal* 1(3, September 20): 200–203.

Chui, B.W., T.D. Stowe, Y.S. Ju, K.E. Goodson, T.W. Kenny, H.J. Mamin, B.D. Terris, R.P. Ried, and D. Rugar. 1998. Low-stiffness silicon cantilevers with integrated heaters and piezoresistive sensors for high-density AFM thermomechanical data storage. *Journal of Microelectromechanical Systems* 7(1, March): 69–78.

Despont, M., J. Brugger, U. Drechsler, U. Durig, W. Haberle, M. Lutwyche, H. Rothuizen et al. 2000. VLSI-NEMS chip for parallel AFM data storage. *Sensors and Actuators A—Physical* 80(2, March 10): 100–107.

Dobisz, E.A. and C.R.K. Marrian. 1997. Control in sub-100 nm lithography in SAL-601. *Journal of Vacuum Science & Technology B* 15(6, December): 2327–2331.

Gadegaard, N. 2006. Atomic force microscopy in biology: Technology and techniques. *Biotechnic and Histochemistry* 81(2–3, June): 87–97.

Garcia, R., R.V. Martinez, and J. Martinez. 2006. Nano-chemistry and scanning probe nanolithographies. *Chemical Society Reviews* 35(1): 29–38.

Ginger, D.S., H. Zhang, and C.A. Mirkin. 2004. The evolution of dip-pen nanolithography. *Angewandte Chemie—International Edition* 43(1): 30–45.

Gundiah, G., N.S. John, P.J. Thomas, G.U. Kulkarni, C.N.R. Rao, and S. Heun. 2004. Dip-pen nanolithography with magnetic Fe_2O_3 nanocrystals. *Applied Physics Letters* 84(26, June 28): 5341–5343.

Haaheim, J. and O.A. Nafday. 2008. Dip pen nanolithography (R): A "Desktop Nanofab (TM)" approach using high-throughput flexible nanopatterning. *Scanning* 30(2, April): 137–150.

Haller, I. 1978. Covalently attached organic monolayers on semi-conductor surfaces. *Journal of the American Chemical Society* 100(26): 8050–8055.

Han, J., K.H. Lee, S. Fujii, H. Sano, Y.J. Kim, K. Murase, T. Ichii, and H. Sugimura. 2007. Scanning capacitance microscopy for alkylsilane-monolayer-covered Si substrate patterned by scanning probe lithography. *Japanese Journal of Applied Physics Part 1—Regular Papers Brief* 46(8B, August): 5621–5625.

Hansma, P.K., V.B. Elings, O. Marti, and C.E. Bracker. 1988. Scanning tunneling microscopy and atomic force microscopy—Application to biology and technology. *Science* 242(4876, October 14): 209–216.

Hironaka, K., N. Aoki, H. Hori, and S. Yamada. 1997. Nanofabrication on GaAs surface by resist process with scanning tunneling microscope lithography. *Japanese Journal of Applied Physics Part 1—Regular Papers & Short Notes* 36(6B, June): 3839–3843.

Hong, S.H., J. Zhu, and C.A. Mirkin. 1999. Multiple ink nanolithography: Toward a multiple-pen nano-plotter. *Science* 286(5439, October 15): 523–525.

Huo, F.W., Z.J. Zheng, G.F. Zheng, L.R. Giam, H. Zhang, and C.A. Mirkin. 2008. Polymer pen lithography. *Science* 321(5896, September 19): 1658–1660.

Hutagalung, S.D. and T. Darsono. 2009. On dot and out of dot electrical characteristics of silicon oxide nanodots patterned by scanning probe lithography. *Physica Status Solidi C—Current Topics in Solid State Physics* 6(4): 817–820.

Im, J., L. Huang, J. Kang, M. Lee, D.J. Lee, S.G. Rao, N.K. Lee, and S. Hong. 2006. "Sliding kinetics" of single-walled carbon nanotubes on self-assembled monolayer patterns: Beyond random adsorption. *Journal of Chemical Physics* 124(22, June 14). Article number: 224707 (6 pages).

Ivanov, D.A., Z. Amalou, and S.N. Magonov. 2001. Real-time evolution of the lamellar organization of poly(ethylene terephthalate) during crystallization from the melt: High-temperature atomic force microscopy study. *Macromolecules* 34(26, December 18): 8944–8952.

Jang, J.W., R.G. Sanedrin, A.J. Senesi, Z.J. Zheng, X.D. Chen, S. Hwang, L. Huang, and C.A. Mirkin. 2009. Generation of metal photomasks by dip-pen nanolithography. *Small* 5(16, August 17): 1850–1853.

Jang, J.Y., S.H. Hong, G.C. Schatz, and M.A. Ratner. 2001. Self-assembly of ink molecules in dip-pen nanolithography: A diffusion model. *Journal of Chemical Physics* 115(6, August 8): 2721–2729.

Jang, J.Y., G.C. Schatz, and M.A. Ratner. 2004. How narrow can a meniscus be? *Physical Review Letters* 92(8, February 27). Article number: 085504 (4 pages).

Jaschke, M. and H.J. Butt. 1995. Deposition of organic material by the tip of a scanning force microscope. *Langmuir* 11(4, April): 1061–1064.

Jegadesan, S., S. Sindhu, and S. Valiyaveettil. 2006. Easy writing of nanopatterns on a polymer film using electrostatic nanolithography. *Small* 2(4, April): 481–484.

Jin, X. and W.N. Unertl. 1992. Submicrometer modification of polymer surfaces with a surface force microscope. *Applied Physics Letters* 61(6, August 10): 657–659.

Kim, J.B., W.S. Chang, and S.J. Na. 2007. A study of the electric field intensity distribution of a modified probe for NSOM lithography. *COLA'05: 8th International Conference on Laser Ablation* 59: 674–677.

Kimikuza, N. and K. Itaya. 1992. *In-situ* scanning tunneling microscopy of underpotential deposition—Silver adlayers on Pt(111) in sulfuric acid solutions. *Faraday Discussions* 94: 117–126.

King, W.P., T.W. Kenny, K.E. Goodson, G. Cross, M. Despont, U. Durig, H. Rothuizen, G.K. Binnig, and P. Vettiger. 2001. Atomic force microscope cantilevers for combined thermomechanical data writing and reading. *Applied Physics Letters* 78(9, February 26): 1300–1302.

King, W.P., T.W. Kenny, K.E. Goodson, G.L.W. Cross, M. Despont, U.T. Durig, H. Rothuizen, G. Binnig, and P. Vettiger. 2002. Design of atomic force microscope cantilevers for combined thermomechanical writing and thermal reading in array operation. *Journal of Microelectromechanical Systems* 11(6, December): 765–774.

King, W.P., S. Saxena, B.A. Nelson, et al. 2006. Nanoscale thermal analysis of an energetic material. *Nano Letters,* 6(9): 2145–2149.

Kleineberg, U., A. Brechling, M. Sundermann, and U. Heinzmann. 2001. STM lithography in an organic self-assembled monolayer. *Advanced Functional Materials* 11(3, June): 208–212.

Kragler, K., E. Gunther, R. Leuschner, G. Falk, H. vonSeggern, and G. SaemannIschenko. 1996. Low-voltage electron-beam lithography with scanning tunneling microscopy in air: A new method for producing structures with high aspect ratios. *Journal of Vacuum Science & Technology B* 14(2, April): 1327–1330.

Lee, J., T. Beechem, T.L. Wright, B.A. Nelson, S. Graham, and W.P. King. 2006a. Electrical, thermal, and mechanical characterization of silicon microcantilever heaters. *Journal of Microelectromechanical Systems* 15(6, December): 1644–1655.

Lee, J. and W.P. King. 2007. Microcantilever hotplates: Design, fabrication, and characterization. *Sensors and Actuators A—Physical* 136(1 May 1): 291–298.

Lee, J. and W.P. King. 2008. Improved all-silicon microcantilever heaters with integrated piezoresistive sensing. *Journal of Microelectromechanical Systems* 17(2, April): 432–445.

Lee, J.-H., Y.-C. An, D.-S. Choi, M.-J. Lee, K.-M. Kim, and J.-H. Lim. 2007. Fabrication of a nano-porous polyoxazoline-coated tip for scanning probe nanolithography. *Macromolecular Symposia,* 249(1): 307–311.

Lee, K.B., J.H. Lim, and C.A. Mirkin. 2003. Protein nanostructures formed via direct-write dip-pen nanolithography. *Journal of the American Chemical Society* 125(19, May 14): 5588–5589.

Lee, K.B., S.J. Park, C.A. Mirkin, J.C. Smith, and M. Mrksich. 2002. Protein nanoarrays generated by dip-pen nanolithography. *Science* 295(5560, March 1): 1702–1705.

Lee, M.V., M.T. Hoffman, K. Barnett, J.M. Geiss, V.S. Smentkowski, M.R. Linford, and R.C. Davis. 2006b. Chemomechanical nanolithography: Nanografting on silicon and factors impacting linewidth. *Journal of Nanoscience and Nanotechnology* 6(6, June): 1639–1643.

Lee, S.W., B.K. Oh, R.G. Sanedrin, K. Salaita, T. Fujigaya, and C.A. Mirkin. 2006c. Biologically active protein nanoarrays generated using parallel dip-pen nanolithography. *Advanced Materials* 18(9, May 2): 1133–1136.

Li, B., Y. Zhang, J, Hu. et al. 2005. Fabricating protein nanopatterns on a single DNA molecule with Dip-pen nanolithography. *Ultramicroscopy* 105(1–4): 312–315.

Lenhert, S., P. Sun, Y.H. Wang, H. Fuchs, and C.A. Mirkin. 2007. Massively parallel dip-pen nanolithography of heterogeneous supported phospholipid multilayer patterns. *Small* 3(1, January): 71–75.

Lewis, M.S. and C.B. Gorman. 2004. Scanning tunneling microscope-based replacement lithography on self-assembled monolayers. Investigation of the relationship between

Chapter 10

monolayer structure and replacement bias. *Journal of Physical Chemistry B* 108(25, June 24): 8581–8583.

Liang, J. and G. Scoles. 2007. Nanografting of alkanethiols by tapping mode atomic force microscopy. *Langmuir* 23(11, May 22): 6142–6147.

Liu, J.F., S. Cruchon-Dupeyrat, J.C. Garno, J. Frommer, and G.Y. Liu. 2002. Three-dimensional nanostructure construction via nanografting: Positive and negative pattern transfer. *Nano Letters* 2(9, September): 937–940.

Liu, M., N.A. Amro, and G.Y. Liu. 2008. Nanografting for surface physical chemistry. *Annual Review of Physical Chemistry* 59: 367–386.

Loh, O., R. Lam, M. Chen, N. Moldovan, H.J. Huang, D. Ho, and H.D. Espinosa. 2009. Nanofountain-probe-based high-resolution patterning and single-cell injection of functionalized nanodiamonds. *Small* 5(14, July 17): 1667–1674.

Mamin, H.J. 1996. Thermal writing using a heated atomic force microscope tip. *Applied Physics Letters* 69(3, July 15): 433–435.

Mamin, H.J., S. Chiang, H. Birk, P.H. Guethner, and D. Rugar. 1991. Gold deposition from a scanning tunneling microscope tip. *Journal of Vacuum Science and Technology B* 9(2, April): 1398–1402.

Mamin, H.J. and D. Rugar. 1992. Thermomechanical writing with an atomic force microscope tip. *Applied Physics Letters* 61(8, August 24): 1003–1005.

Maynor, B.W., Y. Li, and J. Liu. 2001. Au "ink" for AFM "dip-pen" nanolithography. *Langmuir* 17(9, May 1): 2575–2578.

Murphy, S., A. Cazacu, N. Berdunov, I.V. Shvets, and Y.M. Mukovskii. 2005. Nanoscale pattern formation on the $Fe_3O_4(111)$ surface. *Journal of Magnetism and Magnetic Materials* 290 (April): 201–204.

Myung, S., M. Lee, G.T. Kim, J.S. Ha, and S. Hong. 2005. Large-scale "surface-programmed assembly" of pristine vanadium oxide nanowire-based devices. *Advanced Materials* 17(19, October 4): 2361–2364.

Nafday, O.A., R. Pitchimani, B.L. Weeks, and J. Haaheim. 2006a. Patterning high explosives at the nanoscale. *Propellants Explosives Pyrotechnics* 31(5, October): 376–381.

Nafday, O.A., M.W. Vaughn, and B.L. Weeks. 2006b. Evidence of meniscus interface transport in dip-pen nanolithography: An annular diffusion model. *Journal of Chemical Physics* 125(14). Article number: 144703.

Nelson, B.A. and W.P. King. 2007. Measuring material softening with nanoscale spatial resolution using heated silicon probes. *Review of Scientific Instruments* 78(2, February). http://apps.isiknowledge.com.turing.library.northwestern.edu/full_record.do?product=WOS&search_mode=GeneralSearch&qid=57&SID=2FBHj4digANpONe@nnE&page=2&doc=19. Article number: 023702 (8 pages).

Nelson, B.A., W.P. King, A.R. Laracuente, P.E. Sheehan, and L.J. Whitman. 2006. Direct deposition of continuous metal nanostructures by thermal dip-pen nanolithography. *Applied Physics Letters* 88(3, January 16). Article number: 023702 (8 pages).

Nonnenmacher, M. and H.K. Wickramasinghe. 1992. Scanning probe microscopy of thermal-conductivity and subsurface properties. *Applied Physics Letters* 61(2, July 13): 168–170.

Nony, L., R. Bennewitz, O. Pfeiffer, E. Gnecco, A. Baratoff, E. Meyer, T. Eguchi, A. Gourdon, and C. Joachim. 2004. Cu-TBPP and PTCDA molecules on insulating surfaces studied by ultrahigh-vacuum non-contact AFM. *Nanotechnology* 15(2, February): S91–S96.

Noy, A., A.E. Miller, J.E. Klare, B.L. Weeks, B.W. Woods, and J.J. DeYoreo. 2002. Fabrication of luminescent nanostructures and polymer nanowires using dip-pen nanolithography. *Nano Letters* 2(2, February): 109–112.

Nyamjav, D. and A. Ivanisevic. 2003. Alignment of long DNA molecules on templates generated via dip-pen nanolithography. *Advanced Materials* 15(21, November 4): 1805–1809.

Ono, T., H. Hamanaka, T. Kurabayashi, K. Minami, and M. Esashi. 1996. Nanoscale Al patterning on an STM-manipulated Si surface. *Thin Solid Films* 282(1–2, August 1): 640–643.

Park, K., J. Lee, Z.M.M. Zhang, and W.P. King. 2007. Frequency-dependent electrical and thermal response of heated atomic force microscope cantilevers. *Journal of Microelectromechanical Systems* 16(2, April): 213–222.

Paxton, W.F., J.M. Spruell, and J.F. Stoddart. 2009. Heterogeneous catalysis of a copper-coated atomic force microscopy tip for direct-write click chemistry. *Journal of the American Chemical Society* 131(19, May 20): 6692–6694.

Peterson, E.J., B.L. Weeks, J.J. De Yoreo, and P.V. Schwartz. 2004. Effect of environmental conditions on dip pen nanolithography of mercaptohexadecanoic acid. *Journal of Physical Chemistry B* 108(39, September 30): 15206–15210.

Piner, R.D., J. Zhu, F. Xu, S.H. Hong, and C.A. Mirkin. 1999. "Dip-pen" nanolithography. *Science* 283(5402, January 29): 661–663.

Pohl, D.W. 1986. Some design criteria in scanning tunneling microscopy. *IBM Journal of Research and Development* 30(4, July): 417–427.

Raczkowska, J., J. Rysz, A. Budkowski, J. Lekki, M. Lekka, A. Bernasik, K. Kowalski, and P. Czuba. 2003. Surface patterns in solvent-cast polymer blend films analyzed with an integral-geometry approach. *Macromolecules* 36(7 April 8): 2419–2427.

Rajendra Kumar, R.T., S.U. Hassan O.S. Sukas, et al. 2009. Nanobits: Customizable scanning probe tips. *Nanotechnology* 20(39), Article Number: 395703.

Ruskell, T.G., J.L. Pyle, R.K. Workman, X. Yao, and D. Sarid. 1996. Current-dependent silicon oxide growth during scanned probe lithography. *Electronics Letters* 32(15, July 18): 1411–1412.

Sagiv, J. 1980. Organised monolayers by adsorption. 1. formation and structure of oleophobic mixed monolayers on solid-surfaces. *Journal of the American Chemical Society* 102(1): 92–98.

Salaita, K., Y.H. Wang, J. Fragala, R.A. Vega, C. Liu, and C.A. Mirkin. 2006a. Massively parallel dip-pen nanolithography with 55000-pen two-dimensional arrays. *Angewandte Chemie—International Edition* 45(43): 7220–7223.

Salaita, K., Y.H. Wang, and C.A. Mirkin. 2007. Applications of dip-pen nanolithography. *Nature Nanotechnology* 2(3, March): 145–155.

Salaita, K.S., S.W. Lee, D.S. Ginger, and C.A. Mirkin. 2006b. DPN-generated nanostructures as positive resists for preparing lithographic masters or hole arrays. *Nano Letters* 6(11, November 8): 2493–2498.

Schwank, M., U. Muller, and E. Wintermantel. 1998. The use of a high-pressure scanning tunneling microscope as a lithography tool for modifications of amorphous hydrogenated carbon films. *Review of Scientific Instruments* 69(4, April): 1792–1799.

Schwartz, P.V. 2001. Meniscus force nanografting: Nanoscopic patterning of DNA. *Langmuir* 17(19, September 18): 5971–5977.

Sekula, S., J. Fuchs, S. Weg-Remers, P. Nagel, S. Schuppler, J. Fragala, N. Theilacker et al. 2008. Multiplexed lipid dip-pen nanolithography on subcellular scales for the templating of functional proteins and cell culture. *Small* 4(10, October): 1785–1793.

Senesi, A.J., D.I. Rozkiewicz, D.N. Reinhoudt, and C.A. Mirkin. 2009. Agarose-assisted dip-pen nanolithography of oligonucleotides and proteins. *ACS Nano* 3(8, August): 2394–2402.

Sheehan, P.E. and L.J. Whitman. 2002. Thiol diffusion and the role of humidity in "dip pen nanolithography" *Physical Review Letters* 88(15, April 15). Article number: 156104 (4 pages).

Su, C., B.Y. Hong, and C.M. Tseng. 2004. Sol-gel preparation and photocatalysis of titanium dioxide. *Catalysis Today* 96(3, October 5): 119–126.

Sugimura, H. and N. Nagakiri. 1995. Degradation of trimethylsilyl monolayer on silicon substrates induced by scanning probe anodization. *Langmuir* 11(10, October): 3623–3625.

Sugimura, H. 2005. Nanoscopic surface architecture based on molecular self-assembly and scanning probe lithography. *International Journal of Nanotechnology*: 2: 314–347.

Sundermann, M., J. Hartwich, K. Rott, D. Meyners, E. Majkova, U. Kleineberg, M. Grunze, and U. Heinzmann. 2000. Nanopatterning of Au absorber films on Mo/Si EUV multilayer mirrors by STM lithography in self-assembled monolayers. *Surface Science* 454 (May 20): 1104–1109.

Sung, I.H. and D.E. Kim. 2005. Nano-scale patterning by mechanochemical scanning probe lithography. *Applied Surface Science* 239(2, January 15): 209–221.

Swalen, J.D., D.L. Allara, J.D. Andrade, E.A. Chandross, S. Garoff, J. Israelachvili, T.J. Maccarthy et al. 1987. Molecular monolayer and films. *Langmuir* 3(6, December): 932–950.

Tanaka, S., B. Grevin, P. Rannou, H. Suzuki, and S. Mashiko. 2006. Conformational studies of self-organized regioregular poly(3-dodecylthiophene)s using non-contact atomic force microscopy in ultra high vacuum condition. *Thin Solid Films* 499(1–2 (March 21): 168–173.

Tang, Q., S.Q. Shi, H.T. Huang, and L.M. Zhou. 2004. Fabrication of highly oriented microstructures and nanostructures of ferroelectric P(VDF-TrFE) copolymer via dip-pen nanolithography. *Superlattices and Microstructures* 36(1–3, September): 21–29.

Tapaszto, L., G. Dobrik, P. Lambin, and L.P. Biro. 2008. Tailoring the atomic structure of graphene nanoribbons by scanning tunnelling microscope lithography. *Nature Nanotechnology* 3(7, July): 397–401.

Tersoff, J. and D.R. Hamman. 1985. Theory of the scanning probe microscope. *Physical Review B* 31(2): 805–813.

Tian, Z.W., Z.D. Fen, Z.Q. Tian, X.D. Zhuo, J.Q. Mu, C.Z. Li, H.S. Lin, B. Ren, Z.X. Xie, and W.L. Hu. 1992. Confined etchant layer technique for 2-D lithography at high resolution using electrochemical scanning tunneling microscopy. *Faraday Discussions* 94: 37–44.

Tong, X.Q., M. Aindow, and J.P.G. Farr. 1995. A study of the Pd highly oriented pyrolytic-graphite electrode deposition system by *in-situ* elecrochemical scanning tunneling microscopy. *Journal of Electroanalytical Chemistry* 395(1–2, October 10): 117–126.

Tseng, A.A., A. Notargiacomo, and T.P. Chen. 2005. Nanofabrication by scanning probe microscope lithography: A review. *Journal of Vacuum Science & Technology B* 23(3 June): 877–894.

Vanhaesendonck, C., L. Stockman, Y. Bruynseraede, L. Langer, V. Bayot, J.P. Issi, J.P. Heremans, and C.H. Olk. 1994. Nanolithographic patterning of metal-films with a scanning tunneling microscope. *Physica Scripta* 55: 86–89.

Vega, R.A., D. Maspoch, K. Salaita, and C.A. Mirkin. 2005. Nanoarrays of single virus particles. *Angewandte Chemie-International Edition* 44(37): 6013–6015.

Vega, R. et al. 2007. Monitoring single-cell infectivity from virus-particle nanoarrays fabricated by parallel dip-pen nanolithography. *Small* 3: 1482.

Vericat, C. et al. 2005. Self-assembled monolayers of alkanethiols on Au(111): Surface structures, defects and dynamics. *Physical Chemistry Chemical Physics* 7: 3258.

Venables, J.A., D.J. Smith, and J.M. Cowley. 1987. HREM, STEM, REM, SEM—AND STM. *Surface Science* 181(1–2, March): 235–249.

Wadu-Mesthrige, K., N.A. Amro, J.C. Garno, and G.Y. Liu. 1999. Immobilization of protein molecules on functionalized surfaces. *Scanning* 21(2, April): 75–75.

Walsh, M.A. and M.C. Hersam. 2009. Atomic-scale templates patterned by ultrahigh vacuum scanning tunneling microscopy on silicon. *Annual Review of Physical Chemistry* 60: 193–216.

Wang, H.T., O.A. Nafday, J.R. Haaheim, E. Tevaarwerk, N.A. Amro, R.G. Sanedrin, C.Y. Chang, F. Ren, and S.J. Pearton. 2008. Toward conductive traces: Dip pen nanolithography (R) of silver nanoparticle-based inks. *Applied Physics Letters* 93(14, October 6). Article number: 143105 (3 pages).

Wang, X.F., L. Vincent, D. Bullen, J. Zou, and C. Liu. 2005. Scanning probe lithography tips with spring-on-tip designs: Analysis, fabrication, and testing. *Applied Physics Letters* 87(5, August 1). Article number: 054102 (3 pages).

Wang, Y.H., D. Maspoch, S.L. Zou, G.C. Schatz, R.E. Smalley, and C.A. Mirkin. 2006. Controlling the shape, orientation, and linkage of carbon nanotube features with nano affinity templates. *Proceedings of the National Academy of Sciences of the United States of America* 103(7, February 14): 2026–2031.

Weeks, B.L., C.M. Ruddle, J.M. Zaug, and D.J. Cook. 2002a. Monitoring high-temperature solid-solid phase transition of HMX with atom microscom. *Ultramicroscopy* 93(1, October): 19–23.

Weeks, B.L., A. Vollmer, M.E. Welland, and T. Rayment. 2002b. High-pressure nanolithography using low-energy electrons from a scanning tunnelling microscope. *Nanotechnology* 13(1, February): 38–42.

Wilder, K., H.T. Soh, A. Atalar, and C.F. Quate. 1997. Hybrid atomic force scanning tunneling lithography. *Journal of Vacuum Science & Technology A—Vacuum Surfaces and Films* 15(5, October): 1811–1817.

Williams C.C. and H.K. Wickramasinghe. 1986. Scanning thermal profiler. *Applied Physics Letters*, 49(23): 1587–1589.

Wu, A.G., W.L. Cheng, Z.A. Li, J.G. Jiang, and E.K. Wang. 2006. Electrostatic-assembly metallized nanoparticles network by DNA template. *Talanta* 68(3, January 15): 693–699.

Xie, X.N., H.J. Chung, C.H. Sow, and A.T.S. Wee. 2006. Nanoscale materials patterning and engineering by atomic force microscopy nanolithography. *Materials Science & Engineering R—Reports* 54(1–2, November 1): 1–48.

Xu, S., S. Miller, P.E. Laibinis, and G.Y. Liu. 1999. Fabrication of nanometer scale patterns within self-assembled monolayers by nanografting. *Langmuir* 15(21, October 12): 7244–7251.

Ye, J.H., F. Perezmurano, N. Barniol, G. Abadal, and X. Aymerich. 1995. Nanoscale modification of H-terminated n-Si(100) surfaces in aqueous—Solutions with an *in situ* electrochemical scanning tunneling microscope. *Journal of Physical Chemistry B* 99(49, December 7): 17650–17652.

Young, R., J. Ward, and F. Scire. 1972. The Topographiner, an instrument for measuring surface microtopography. *The Review of Scientific Instruments* 43: 999–1011.

Zhang, H., N.A. Amro, S. Disawal, R. Elghanian, R. Shile, and J. Fragala. 2007. High-throughput dip-pen-nanolithography-based fabrication of Si nanostructures. *Small* 3(1, January): 81–85.

Zhang, H., R. Elghanian, N.A. Amro, S. Disawal, and R. Eby. 2004a. Dip pen nanolithography stamp tip. *Nano Letters* 4(9, September): 1649–1655.

Zhang, H., R.C. Jin, and C.A. Mirkin. 2004b. Synthesis of open-ended, cylindrical Au-Ag alloy nanostructures on a Si/SiOx surface. *Nano Letters* 4(8, August): 1493–1495.

11. Two-Photon Lithography

Satoru Shoji and Satoshi Kawata

Department of Applied Physics, Osaka University, Osaka, Japan

The two-photon absorption process makes it possible to induce photochemical reaction in a three-dimensionally restricted local spot. This feature of two-photon absorption allows us to fabricate arbitrary three-dimensional structures with high spatial resolution of the order of 100 nm. This intrinsic three-dimensional spatial resolution of two-photon lithography is a unique characteristic in contrast to other lithography techniques, and a promising tool for developing a variety of novel photonic and mechanical nanodevices. In this chapter, we will show an overview of two-photon lithography, including a historical overview, fundamental principles, optical setup, applicable materials, achievable spatial resolution, and several examples of the application of two-photon lithography to micro/nanodevices.

Nanofabrication Handbook. Edited by Stefano Cabrini and Satoshi Kawata © 2012 CRC Press / Taylor & Francis Group, LLC. ISBN: 978-1-4200-9052-9

Chapter 11

11.1   Historical Overview of Micro/Nanoprocessing Based on Two-Photon Absorption

In the 1980s, with the commercialization of ultra-short pulse lasers, two-photon absorption had started to be recognized as a powerful tool to observe, fabricate, stimulate, and manipulate materials three dimensionally with a spatial resolution beyond the diffraction limit of light. The origination of three-dimensional processing by means of two-photon absorption can be attributed to the proposal of three-dimensional data storage by Parthenopoulos and Rentzepis (1989). Bit data were recorded and erased in a photochromic dye-embedded polymer matrix through photochromic reactions, and were read through the excitation of fluorescence from the dye. They induced two-photon absorption by a tightly focused laser spot both for the writing and reading processes in order to stack the bit data in three dimensions.

After this, many types of three-dimensional optical data storage based on two-photon process were also reported, such as two-photon photorefraction in polymer (Strickler and Webb 1991), photorefractive crystals (Kawata et al. 1998), or photochromic materials (Toriumi et al. 1997, 1998; Kawata and Kawata 2000), two-photon bleaching (Gu and Day 1999) or explosion (Glezer et al. 1996), and so on. In 1990, Denk et al. (1990) proposed laser scanning fluorescence microscopy based on two-photon excited fluorescence to achieve three-dimensional spatial resolution capability. Fluorescence excited by two-photon absorption pro-vided the cross-sectional images of specimens without using a confocal system in the same manner as three-dimensional optical data storage.

The first demonstration of three-dimensional microstructures fabrication by means of two-photon absorption, that is, *"two-photon lithography,"* was reported by Maruo et al. (1997). Figure 11.1 shows one of the first reported three-dimensional structures fabricated by this group. They utilized a near-infrared femtosecond laser to induce polymerization reaction into an ultraviolet-sensitive photopolymer. As the trace of laser is focused in the volume of photopolymer, three-dimensional microstructures of solid polymer were fabricated. Since this first proposal, two-photon lithography has spread out in various science and engineering research fields from photochemistry to bioengineering, as an ideal tool for realizing arbitrary three-dimensional structures in the micrometer scale. Figure 11.2 shows just a few examples of three-dimensional micro/nanostructures achieved by two-photon lithography. So far, the spatial resolution reaches ~100 nm (Takada et al. 2005). Two-photon lithography is currently recognized to be one of the key fields of laser nanotechnologies, sometimes categorized as a discipline within *"nanophotonics."* The performance of two-photon lithography continues to improve even now with breakthroughs in many aspects of the field, such as its material science, laser technology, nonlinear optics, and micromechanics.

(a)

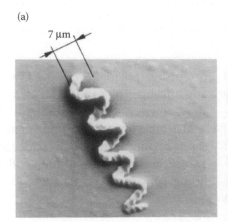

(b)

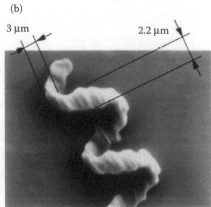

FIGURE 11.1   The first reported three-dimensional structure fabricated by two-photon lithography. (a) A 7 μm-diameter and 50 μm-length spiral coil. (b) Magnified image. (Reprinted with permission from Maruo, S., Nakamura, O., Kawata, S. 1997. Three-dimensional microfabrication with two-photon-absorbed photopolymerization. *Opt. Lett.* **22**: 132–134. Copyright 1997, Optical Society of America.)

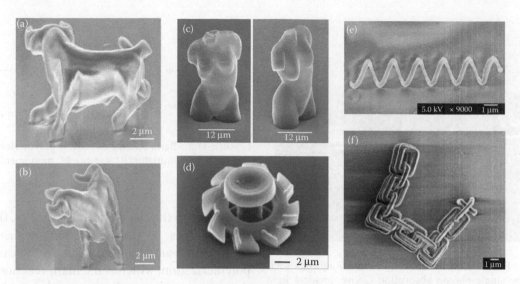

FIGURE 11.2 Variety of three-dimensional polymer microstructures made by two-photon lithography. (a,b) Microbull, (Reprinted by permission from Macmillan Publishers Ltd., *Nature*, Kawata, S., Sun, H.-B., Tanaka, T., Takada, K. 2001. Finer features for functional microdevices. **412**: 697–698. Copyright 2001.) (c) Venus statue, (Reprinted with permission from Serbin, J. et al. 2003. Femtosecond laser-induced two-photon polymerization of inorganic organic hybrid materials for applications in photonics. *Opt. Lett.* **28**: 301–303. Copyright 2003, Optical Society of America.) (d) a microgear, (Reprinted with permission from Maruo, S., Ikuta, K., Korogi, H. 2003a. Force-controllable, optically driven micromachines fabricated by single-step two-photon microstereolithography. *J. Microelectromech. Syst.* **12**: 533–549. Copyright 2003, IEEE.) (e) a microcoil, and (f) a microchain.

11.2   The Feature of Two-Photon Absorption and the Principle of Two-Photon Lithography

11.2.1   Nonlinear Property of Two-Photon Absorption

Under usual conditions, absorption of light occurs when the energy of an incident photon corresponds to the energy gap between the excited and ground state of an irradiated material (or molecule). Thus, when the energy of photon is smaller than the energy gap (i.e., the wavelength of light is longer than the absorption band), the matter is transparent and that light is transmitted without absorption. The efficiency of absorption is expressed as a material's absorption cross-section, and total number of absorbed photons is in proportion to the intensity of light. However, if the light intensity is extremely high, although the energy of a single photon is smaller than the energy gap, it is possible for the absorption of light to occur through the simultaneous absorption of two photons (or sometimes even more photons) where the total energy satisfies the energy gap to excite the material. Figure 11.3 shows the schematic diagram of single-photon and two-photon absorption. Consequently, two-photon absorption is a nonlinear process, and the number of absorbed photons is proportional to

the two-photon absorption cross-section δ and to the square of the incident intensity I (Shen 1984).

$$\frac{dI}{dz} = \delta \cdot I^2 \qquad (11.1)$$

In general, the probability of two-photon absorption is extremely low compared to single-photon absorption. In order to observe two-photon absorption experimentally, a high-power light source is usually necessary.

In practice, it is possible to create a photon density (i.e., light intensity) high enough for exciting two-photon absorption if one uses a microscope objective

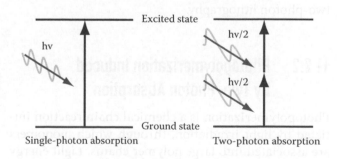

FIGURE 11.3 Energy diagram of single-photon absorption and two-photon absorption.

FIGURE 11.4 Fluorescence from rhodamine-6G aqueous solution excited by single-photon absorption (above, incident light from the left side) and two-photon absorption (below, incident light from the right side). (Image courtesy of Dr. Katsumasa Fujita, Department of Applied Physics, Osaka University.)

lens and a femtosecond-pulsed laser to focus the laser light so as to confine photons at a focus spot in space and in time. Figure 11.4 shows fluorescence from a rhodamine-6G dye solution excited by single-photon absorption and by two-photon absorption. In the case of single-photon absorption, fluorescence is excited from almost the entire light path. In contrast, in the case of two-photon absorption, fluorescence is excited only at the focus spot of the laser light. This is due to the quadratic law of two-photon absorption efficiency to light intensity. Accordingly, two-photon absorption allows us to confine photoexcitation of materials into the limited space of a focus spot. Of course, not only fluorescence, but also many other types of reactions, such as photochromism, photoisomerization, photorefraction, photoreduction, photoablation, photoionization, and photopolymerization, are also candidate subjects for two-photon absorption-assisted localization. This three-dimensional localization of photochemical reaction is the fundamental principle of two-photon lithography.

11.2.2   Photopolymerization Induced by Two–Photon Absorption

Photopolymerization is a chemical chain reaction initiated by light irradiation, through which monomers are associated into large polymer chains. Light energy produces reactive species in the material, which gives rise to start a polymerization reaction of monomers.

$$
\begin{aligned}
\text{Initiation} \quad & \left\{ \begin{aligned} I \ &\xrightarrow{\ h\nu\ }\ R^* \\ R^* + M \ &\longrightarrow\ RM^* \end{aligned} \right. \\[2ex]
\text{Propagation} \quad & RM^* + M \ \xrightarrow{\ M\ }\ RMM^* \ \text{-----} \ RMn^* \\[2ex]
\text{Termination} \quad & \left\{ \begin{aligned} RMn^* + RMm^* \ &\longrightarrow\ RMm + nR \\ RMn^* + RMm^* \ &\longrightarrow\ RMn + RMm \end{aligned} \right.
\end{aligned}
$$

FIGURE 11.5 Basic scheme of radical polymerization. Symbols denote: I, photoinitiator; R^*, radical; M, monomer; M^*, propagating radical.

According to the class of reactive species, there are two typical types of polymerization, *radical polymerization* and *ionic polymerization*. In the case of radical polymerization, which is the main reaction of focus in this chapter, free radicals are the initial reactants (Odian 1991). Figure 11.5 shows the process diagram of radical photopolymerization. In radical photopolymerization, first, the source of radicals, the so-called *photoinitiators*, absorb photons. The excited photoinitiators convert to radicals. These radicals conjugate with monomers in connection with opening unsaturated C=C bonds of monomers to form monomer radicals. The same reaction proceeds with many monomers until two monomer radicals bind each other so as to terminate the chain reaction.

11.2.3   The Principle of Two–Photon Lithography

Now, if we excite this photopolymerization reaction by using tightly focused laser light to get two-photon absorption, we can localize the source of polymerization reaction to only inside the focus spot of light, similar to the case of fluorescence emission as shown in Figure 11.6. As a result, a three-dimensionally isolated single dot of solid polymer can be formed. This solid dot is the unit volume element, so-called "voxel" in two-photon lithography. Microstructures are formed by many voxels stacked in three dimensions with sufficiently fine distance. After laser scanning, the structures are immersed in a solvent to dissolve unpolymerized resin. After drying, freestanding three-dimensional polymer structures are obtained.

Figure 11.7 shows a theoretical calculation of the spatial distribution of the polymerization rate by single- and two-photon absorption. In both cases, light is focused by an objective lens with a numerical aperture of 1.4. This is obtained from the spatial distribution of light intensity near the focus spot with the assumption

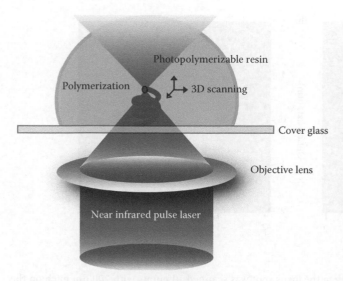

FIGURE 11.6 The principle of two-photon lithography. The polymerization occurs only at the focal point by two-photon absorption.

that the polymerization rate is in proportion to the amount of light absorption. It seems that single-photon absorption provides smaller dots at the center of the focus spot. This is because the wavelength of light for two-photon absorption is twice that for the single-photon case, and so the size of the focus spot is also doubled compared with the case of single-photon absorption. The important feature here is that in single-photon absorption, polymerization extends far from the center of focus along the optical axis. In comparison, polymerization is tightly confined at the focus spot in the case of two-photon absorption. This feature becomes more obvious when the focus spot is scanned to fabricate structures. Figure 11.8 shows the polymer-

ization rate distribution when 10 voxels were fabricated with an equal interval of 200 nm in the focal plane. In the case of single-photon absorption, not only at the focus spot but also above and below the focus, the polymerization proceeds significantly by overlapping light irradiation. In contrast, in the case of two-photon absorption, polymerized volume is still confined only in the trace of the focus spot without much blur above and below the focus. This feature is one of the most important advantages of two-photon lithography.

Another important advantage of two-photon absorption is less distortion of the laser focus during three-dimensional processing. In many photopolymerizable materials, the polymerization reaction induces the modulation of refractive index in materials, which may cause the distortion of spatial profile of the focus spot by refraction or scattering. As the wavelength of laser light is about twice as long as the absorption wavelength of materials, refraction or scattering of laser light is diminished. Consequently, the material is transparent to the operating light except at the focus spot. In this way, the laser light can penetrate deep inside the material without significant distortion.

The unique characteristics of two-photon lithography can be summarized as the following three features:

1. Three-dimensional resolution
2. High spatial resolution (subdiffraction limit resolution)
3. High penetration depth

As mentioned before, this principle is applicable for other photoinduced reactions. There are many types

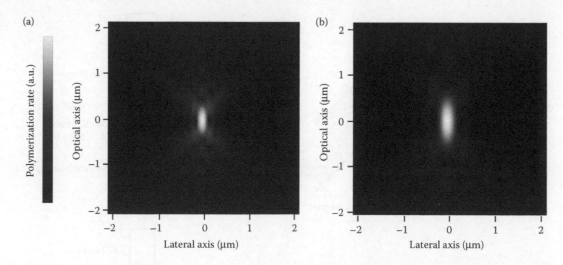

FIGURE 11.7 Polymerization distribution at the focal volume. (a) Single-photon excitation: wavelength, 390 nm; N.A., 1.4; refractive index, 1.53 and (b) two-photon excitation: wavelength, 780 nm; N.A., 1.4; refractive index, 1.53.

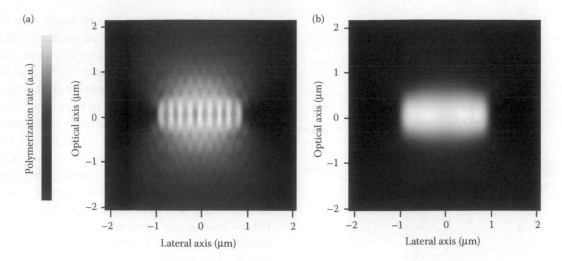

FIGURE 11.8   Polymerization distributions at the focal volume when the focus spot was scanned 10 points with 200 nm pitch on the focal plane. (a) Single-photon excitation and (b) two-photon excitation.

of microfabrication techniques based on two-photon photoreduction, two-photon isomerization, two-photon molecular orientation, two-photon ablation of glass, and so on, although we will omit the details of these cases in this chapter. In particular, two-photon reduction is a promising technique for metallic micro/nanostructure fabrication, which is described in detail in Chapter 21.

11.3   Two-Photon Lithography Apparatus

11.3.1   Optical System

The principle of two-photon lithography is to scan the focused spot of laser light in a photopolymer, by which a polymerized solid structure is formed along the trace of the laser spot through the process of two-photon absorption. Figure 11.9 shows an example of two-photon lithography apparatus. You may notice that the system is almost the same as that of laser scanning microscopy (LSM). The beam of a mode-locked Ti:Sapphire laser is expanded by a pair of lenses and focused by a high numerical aperture objective lens (N.A. 1.4, 100*, oil) into photopolymer. Typically the oscillating wavelength, pulse width, and repetition rate of the laser are 780 nm, 80 fs, and 80 MHz, respectively. Since ordinary photopolymers are ultraviolet sensitive but are transparent to near infrared light, the laser light is absorbed by the photopolymer

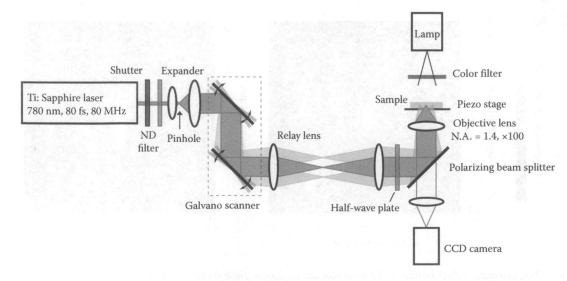

FIGURE 11.9   Schematic of the optical setup for a two-photon lithography system.

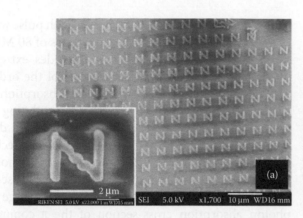 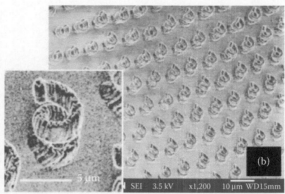

FIGURE 11.10 Mass-produced two- and three-dimensional structures array realized by multifocus spots array two-photon parallel lithography. (Reprinted with permission from Kato, J. et al. Multiple-spot parallel processing for laser micronanofabrication. *Appl. Phys. Lett.* **86**: 044102 1–3. Copyright 2005, American Institute of Physics.)

only through two-photon absorption. Laser power and duration of exposure are controlled by a neutral density (ND) filter and a mechanical shutter, respectively, to form polymer voxels with appropriate size and shape. To scan the focus spot in the material in three dimensions, typically one of two methods is used, laser scanning mode and stage scanning mode. In the laser scanning mode, a galvanometer mirror set is used. As shown in Figure 11.9, laser light reflected by the galvanomirror changes the angle of beam propagation. This light is guided into the microscope objective through a relay lens such that the light beam is always passing through the center of the aperture of the objective lens while changing the incident angle. By this mechanism, the focus spot can be scanned in the two lateral dimensions (x, y). In the vertical direction (z), a piezo stage is used to move the material relative to the laser spot. In the stage scanning mode, the focus spot is fixed and stationary, while the piezo stage moves the photopolymer in three dimensions (x, y, z) relative to the laser spot. In general, the laser scanning mode has the advantage of high scanning speed; however, scanning area is limited to the field of view of the objective lens. The stage scanning mode does not have a fundamental limit in the scanning range (only within the working distance of piezo stage), and so larger structures beyond the field of view can be fabricated. However, scanning speed is much slower than the case of laser scan mode. Choice of mode should be determined by the application. In either case, arbitrary three-dimensional structures can be fabricated within the restriction of the working distance of objective lens (~100 µm) in the z direction and the available scanning range in the x and y directions.

Rather than scanning a single focus spot, scanning of multiple spots simultaneously can increase the production yield dramatically. Batch production by means of parallel scanning of multiple focus spots has been shown by some reports (Kato et al. 2005; Dong et al. 2007). A single laser beam is divided into multiple beams by introducing a microlens array or a diffractive beam splitter, and thereby, an array of many focus spots is formed at the focal plane simultaneously. Figure 11.10 shows an example of arrays of letters "N" and microsprings fabricated by scanning hundreds of focus spots simultaneously.

11.3.2 Material Requirement in Two-Photon Lithography

Although this chapter describes mainly radical polymerization, in general any kind of photopolymerizable material can be utilized for two-photon lithography. There is a huge number of reports of two-photon lithography using various photopolymerizable materials, including acrylate resins (Maruo et al. 1997; Sun et al. 1999; Nguyen et al. 2005), epoxy resins (Witzgall et al. 1998; Teh et al. 2004; Deubel et al. 2004), optical adhesive (Galajda and Ormos 2001), organic–inorganic hybrid polymers (Serbin et al. 2003), hydrogel (Watanabe et al. 2002), biocompatible gelatine (Fujita et al. 2004), and so on. Most materials are ultraviolet sensitive because they are designed for photolithography or rapid prototyping. Figure 11.11 shows the absorption spectrum of a commercial product SCR-500 (Japan Synthetic Rubber Co., Ltd.). This compound consists mainly of urethane–acrylate monomer, oligomers, and photoinitiators (HCAP and BDMK).

Chapter 11

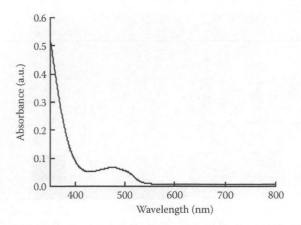

FIGURE 11.11 Absorption spectrum of the photopolymerizable resin SCR-500.

The absorption spectrum shows that the photopolymer is transparent, and so is not cured by single-photon absorption at the wavelength of 780 nm from a Ti:Sapphire laser. However, the photopolymer can be cured by the energy of two simultaneous photons, which corresponds to ultraviolet light at 390 nm. In general, in order to excite two-photon absorption dominantly and effectively, the wavelength of the laser light must be carefully chosen according to the absorption spectrum of the raw material.

11.3.3 Chromophores for Efficient Two-Photon Polymerization

In general, the two-photon absorption cross section is quite small, meaning that very high light intensity is required to generate two-photon absorption. For this purpose, femtosecond-pulsed lasers are typically used for two-photon lithography. As the laser light is tempo-rarily confined to femtosecond pulses, with pulse widths of for example, ~80 fs and repetition rates of 80 MHz, a mean laser power of a few mW provides extremely intense light pulses with a peak power of the order of kW. In order to enhance two-photon absorption efficiency, several kinds of chromophores exhibiting high two-photon absorption cross section have been developed. It was found that π-conjugation of electron-donating (D) and/or electron-accepting (A) moieties exhibits a large two-photon absorption cross section (Cumpston et al. 1999; Yan et al. 2004). The large two-photon absorption cross-section of the π-conjugated chromophores is originated from symmetric intramolecular charge transfer between the ends and the middle of the conjugated system. Accordingly, the two-photon absorption cross section of chromophores can be further enhanced by increasing the conjugation length or the donor and acceptor strength. Such chromophores exhibit more than 100 times higher efficiency of two-photon absorption compared to ordinary photopolymers, and allow us to perform two-photon lithography with much lower intensity of laser light and thus reduce thermal damage in the material. Figure 11.12 shows the examples of such two-photon chromophores developed by Cumpston et al. (1999). In addition to increasing the two-photon absorption coefficient of sensitizers, another way to enhance two-photon polymerization efficiency is to increase the concentration of sensitizers. Adronov et al. (2000) synthesized a series of dendrimers made of two-photon-absorptive chromophores, and thereby they accumulated the chromophores into dendritic macromolecules without causing molecular aggregation. It was observed that by increasing one generation of dendrimers, the two-photon absorption cross section increases by nearly twice.

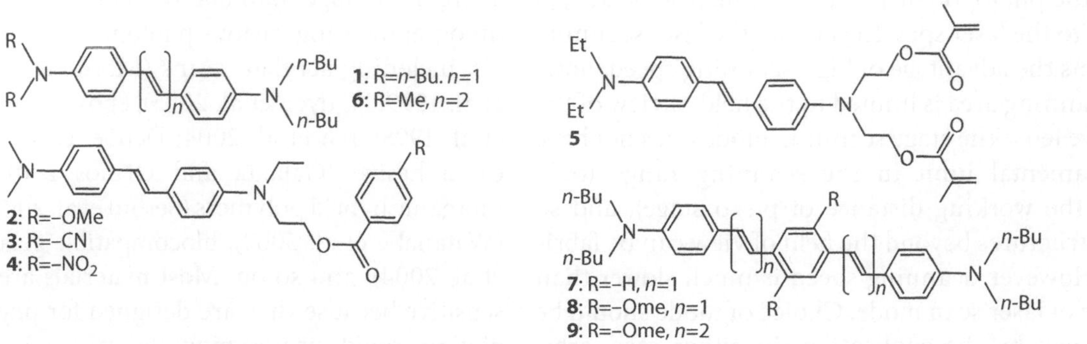

FIGURE 11.12 Molecular structures of high-efficiency two-photon chromophores. (Reprinted by permission from Macmillan Publishers Ltd., *Nature*, Cumpston, B. H. et al. 1999. Two-photon polymerization initiators for three-dimensional optical data storage and microfabrication. **398**: 51–54. Copyright 1999.)

11.4 Spatial Resolution in Two-Photon Lithography

11.4.1 Evaluation of Spatial Resolution from Voxels

The primary factor to characterize the spatial resolution and precision in two-photon lithography is a volume element, which we have called a *"voxel."* As described in the previous section, the voxel is the smallest unit of polymer volume created by a single-shot irradiation of focused laser light. Micro/nanostructures are created as aggregation of many such voxels. Indeed, the size and shape of the voxel are of crucial importance to pursue high spatial resolution and high processing accuracy. The size and shape of the voxel are affected by several factors, such as laser power, exposure time, N.A. of the objective lens, and so on. Besides these physical parameters, many chemical parameters of the constituent photopolymerizable materials, such as concentration of radical quencher or photosensitizer with high two-photon absorption cross section, temperature of the resin, and so on are important factors for voxel formation.

In order to evaluate the voxel size quantitatively, we need to observe the precise shape of freestanding voxels by scanning electron microscopy (SEM). However, if we fabricate voxels on a substrate, a part of the voxels are embedded in the substrate, and so we cannot observe the entire shape. On the other hand, if the voxels are fabricated above the substrate, the voxels will be washed away during the developing process. Sun et al. (2002) showed a solution to this problem by using an ascending scan exposure method. As shown in Figure 11.13, a series of voxels were polymerized under identical exposure parameters but with changing the height relative to the

surface of the substrate. In the beginning, voxels are formed embedded in the substrate. Where the height of the focus is about half of the voxel length, the formed voxels are seen to just contact the surface of the substrate, and often remain after developing, laying on the substrate. By observing these voxels using scanning electron microscopy, we can measure the size of the voxels in three dimensions. Figure 11.13c is an SEM image of the voxels formed by this ascending scan method. As seen in the image, the shape of voxels is actually not spherical but ellipsoidal in shape, elongated in the longitudinal direction, which reflects the distribution of light intensity at the focus spot (i.e., the point spread function of the objective lens). By changing several parameters, the finest possible size of voxels can be precisely evaluated.

11.4.2 Laser Power and Exposure Time Dependence of Voxel Size

Figure 11.14 shows the dependence of lateral voxel size on the incident laser power. In this experiment, Ti:Sapphire laser light at 780 nm and 80 fs pulse width was focused with a 1.4 N.A. objective lens into a commercial product SCR-500 photopolymerizable resin. As the incident laser power and the exposure time decrease, the voxel size becomes small. Surprisingly, as shown in the graph, at low power and short exposure time, the voxel size reaches to 120 nm, which is about eight times smaller than the wavelength of the laser light. Indeed, this is evidence of the super-resolution of two-photon lithography, which can be much higher than the optical diffraction limited focus spot of light.

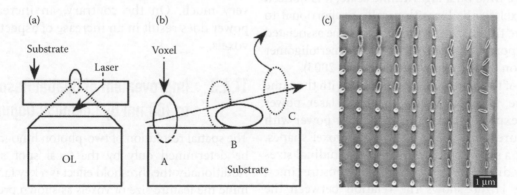

FIGURE 11.13 Schematic of the ascending scan exposure method. (a) Focusing configuration: OL, objective lens. (b) Voxels polymerized under different focusing height levels: A, an erected voxel; B, an overturned voxel, C, a floated voxel. (c) SEM image of voxels made of SCR-500 photopolymerizable resin. From certain height, voxels start to overturn on the substrate.

Chapter 11

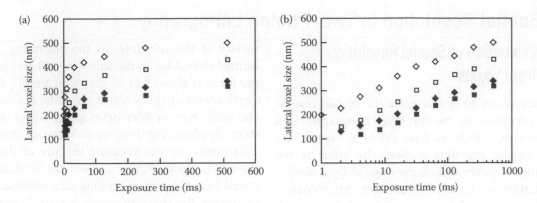

FIGURE 11.14 Voxel size as the function of the exposure time and incident laser power. (a) Linear coordinates, (b) half-logarithmic coordinates. Incident laser power: open diamonds, 12.5 mW; open squares, 10.5 mW; closed diamonds, 7.0 mW; closed squares, 5.8 mW.

In order to understand this super-resolution, one should note that neither a clear boundary of light field nor a boundary of polymerized/unpolymerized volume exists. The polymerization reaction progresses everywhere around the focus spot, but at different efficiency (or speed) depending on the local intensity of light. The resin becomes most solid at the center of the optical focus, and with increasing distance from the center the resin gradually becomes an intermediate of polymer and monomer. During the developing process, depending on the degree of polymerization progression, the under-polymerized resin below a certain level is dissolved by solvent. Here a threshold is apparent in light intensity, above which polymerized resin can withstand the developing process to remain as a solid object. In fact, this threshold defines the size of voxels. By controlling the light intensity of the focus spot near the threshold, voxels much smaller than the spot size as defined by the diffraction limit can be obtained. The schematic of this mechanism is shown in Figure 11.15. When the voxel size is plotted against the exposure time on a logarithmic scale, it is noticed that the voxel size is logarithmically proportional to the exposure time. This dependence may be associated with the exponential decay of the monomer/oligomer concentration during exposure (Sun et al. 2003).

Because of this threshold effect, even with the same energy dose, the conditions of strong laser power with short exposure time, and weak laser power with long exposure time result in different voxel shapes. Figure 11.16a shows the lateral and longitudinal sizes of voxels by different laser power and exposure times, and Figure 11.16b shows the relation between the aspect ratio of voxels and energy dose obtained from the result in Figure 11.16a. Actually, the exposure time does not affect the aspect ratio of voxels

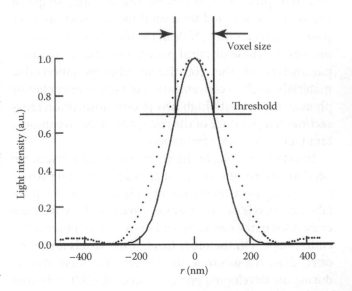

FIGURE 11.15 Light intensity distribution at the focal plane under 780 nm incident wavelength and 1.4-N.A. focusing. Light intensity (dashed line) and the square of light intensity (solid line) distribution are associated with single- and two-photon excitations, respectively.

very much. On the contrary, an increase of laser power does result in an increase of aspect ratio of the voxels.

11.4.3 Improvement of Spatial Resolution to 100 nm by Chemical Doping

The spatial resolution of two-photon lithography cannot be determined only by the focal spot size of light. Additionally, the threshold effect is a key factor to determine the feature size of voxels as shown previously, and of course this threshold is directly connected with the optical response of chemical compounds contained in photopolymerizable materials. For example, adding

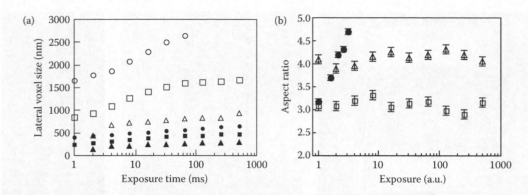

FIGURE 11.16 (a) A half-logarithmic plot of lateral and longitudinal voxel sizes as a function of the exposure time and the incident laser power. Closed symbols and open symbols indicate lateral and longitudinal sizes, respectively. Incident laser power: circles, 10.0 mW; squares, 5.0 mW; triangles, 3.2 mW. (b) Aspect ratio of voxels. Closed circles, power dependence under 64 ms exposure; open triangles, exposure time dependence under 6.8 mW exposure, open squares, exposure time dependence under 3.2 mW exposure.

chromophores with high two-photon absorption efficiency reduces the threshold intensity. By this threshold depression, it becomes possible to induce two-photon absorption-induced voxels with much smaller laser power, which leads to an enlarged dynamic range for tuning of the laser power and realize more precise control of voxel size. Kuebler et al. (2001) reported that by replacing the normal UV photoinitiator with D-pi-D chromophores, which were designed to have large two-photon absorption cross section, photopolymers with 50 times larger dynamic power range could be achieved. Also playing an important role in threshold effect are molecules of the class *"radical quenchers."* Radical quenchers are molecules that tend to combine with radicals to form a complex, which inactivate the propagation

of radical polymerization. Under common experimental conditions, oxygen molecules act as one species of such radical quenchers. (Often in order to prevent undesired quenching effects by oxygen, nitrogen purging is performed.) In the case of two-photon lithography, rather than preventing the quenching effect, adding more radical quenchers to the resin has been proposed as one of the possible methods to reduce voxel size. Figure 11.17a shows the effect of adding a two-photon initiator chromophore as well as radical quencher. The two-photon initiator and radical quencher, 1,4-bis-(2-ethylhexyloxy)-2,5-bis[2-(4-(bis-(4-bromophenyl)-amino)-phenyl)-vinyl]-benzene (EA4BPA-VB) (two-photon absorption cross section; 470×10^{-50} cm^4 s/photon at 780 nm), and *N*-alkoxy-substituted-hindered amine light stabilizer

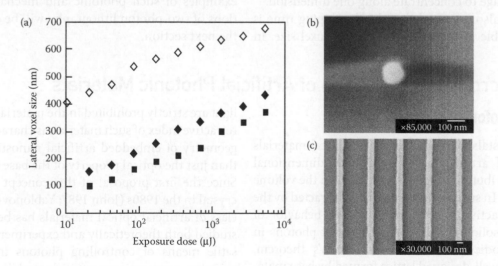

FIGURE 11.17 (a) Voxel size as a function of the exposure dose. The incident laser power was 10.5 mW. Symbols: open diamonds, SCR-500 containing 0.1 wt.% additional initiator; closed diamonds, pristine SCR-500; closed squares, SCR-500 containing 0.8 wt.% additional radical quencher. (b,c) Achievement of 100 nm spatial resolution after adding the radical quencher into SCR-500. SEM images of (b) a single voxel and (c) line structures.

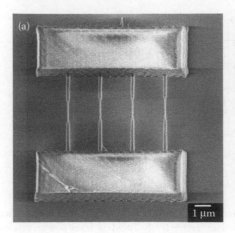 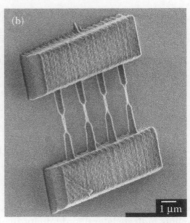 

FIGURE 11.18   SEM images of fiber arrays suspended between two anchors. The center segments of the fibers have a feature size of 65 nm. (a) Top view, (b) side view, and (c) magnified view. The lateral and longitudinal sizes of the center segments are 65 and 190 nm, respectively.

(HALS) were added in the SCR-500 resin. The use of a radical quencher leads to voxel size reduction down to 100 nm (Takada et al. 2009). Figure 11.17b and c are SEM images of single voxel and line structures with 100 nm lateral size, fabricated on a glass substrate.

Much thinner feature size is also achieved by the additional contribution of polymer shrinkage during the developing process. Figure 11.18 shows 65-nm-thick nanowires fabricated by two-photon lithography (Takada et al. 2005). Tan et al. (2007) also observed around 20-nm-thick wires fabricated by SCR-500 resin. At present such ultrahigh resolution is not always possible to achieve, but is restricted to only certain particular shapes such as suspended fibers, especially those in which the internal tension of the polymer structure can cause shrinkage to concentrate along one dimension.

It should also be noted that the developing time is not a negligible factor to determine the voxel size. In order to fabricate precise micro/nanostructures with well-controlled resolution, proper developing duration is crucial, where too short duration leaves a remainder of under-polymerized materials surrounding the structure, and too long duration may cause the dissolution of finer structural details by the solvent.

The capability for three-dimensional fabrication with a spatial resolution of 100 nm is quite promising, because this indicates that the possible feature size of the produced structures can be comparable to the wavelength of visible and infrared light, or can be comparable to the size of living cells or blood vessels. This shows us the possibility of applying two-photon lithography to optoelectronic devices and biological microelectromechanical systems (MEMS) technology. Some examples of such photonic and mechanical applications of two-photon lithography will be overviewed in the next section.

11.5   Micro/Nanofabrication of Artificial Photonic Materials

11.5.1   Photonic Crystals

Photonic crystals are a novel class of optical materials consisting of artificially produced three-dimensional periodic distribution of refractive indices in the volume of a material. In such structures, light is diffracted by the periodic refractive index. Similar to the behavior of electrons in solid crystals, the behavior of photons in such a periodic potential follows Bloch's theorem. Thereby in a well-designed lattice formed by base materials of proper refractive indices, forbidden energy bands called *"photonic bandgaps,"* appear. At the frequency of the photonic bandgaps, emission and propagation of light are strictly prohibited in the material. The effective refractive index of such materials is characterized by the geometry of embedded artificial nanostructure rather than just the optical property of the base material itself. Since the first proposal of this concept of a photonic crystal in the 1980s (John 1987; Yablonovitch 1987), this class of artificial optical materials has been intensively studied both theoretically and experimentally, as a versatile means of controlling photons including their emission, propagation, scattering, and dispersion.

Despite much effort, fabrication of photonic crystals is still challenging even at present because the refractive index distribution must be periodic in three

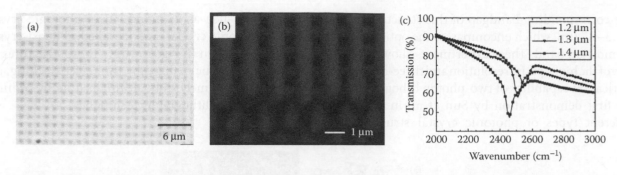

FIGURE 11.19   Layer-by-layer photonic crystal fabricated by two-photon lithography. (a) Optical microscope image. (b) Scanning electron microscope image of a cross section of the structure. (c) Transmission spectra of layer-by-layer photonic crystals with different lattice constants. The total layer number is 20. (Reprinted with permission from Sun, H.-B., Matsuo, S., Misawa, H. 1999. Three-dimensional photonic crystal structures achieved with two-photon-absorption photopolymerization of resin. *Appl. Phys. Lett.* **74**: 786–788. Copyright 1999, American Institute of Physics.)

dimensions with a period distance of around half the wavelength of the bandgap light. Currently, two-photon lithography is recognized as one of the ideal methods to fabricate such three-dimensional architectures at the micro/nanoscale. In 1999, two-photon lithography was utilized to fabricate photonic crystals for the first time (Sun et al. 1999). The first fabricated photonic crystal is shown in Figure 11.19 and was a *"wood-pile"* structure, which is known as one of the ideal structures for exhibiting a photonic bandgap. Before this report, the typical fabrication method for creating photonic crystals was a combination of semiconductor wafer-etching and wafer-bonding, in which thin two-dimensional periodic structures made of

semiconductor crystal were piled up layer-by-layer in order to build a three-dimensional lattice (Noda et al. 1996, 1999; Lin et al. 1998). In contrast to these semiconductor-processing techniques, two-photon lithography does not necessitate such a routine layer-by-layer procedure. The laser focus is simply scanned in the photopolymerizable material in three dimensions so as to directly write three-dimensional lattice structures. The first demonstration of this technique achieved a lattice constant of 1.4 μm with 20 layers, which corresponds to a photonic bandgap at around 4.0 μm. Recently, the lattice constant was further reduced down to 0.8–1.0 μm with a layer number of 40, as shown in Figure 11.20 (Deubel et al. 2004).

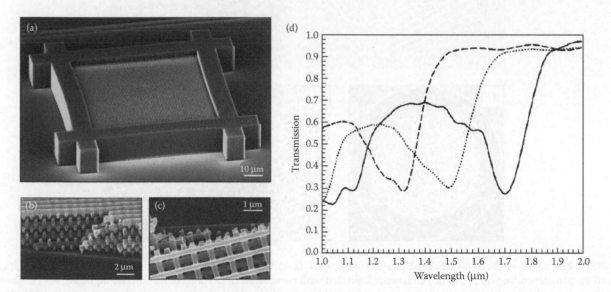

FIGURE 11.20   Wood-pile photonic crystal fabricated by two-photon lithography. A supporting frame is also fabricated surrounding the photonic crystal to prevent bending and distortion due to polymer shrink. Transmission spectra of layer-by-layer photonic crystals with different in-plane rod spacing of $a = 1.0$ μm (solid line), $a = 0.9$ μm (dotted line), and $a = 0.8$ μm (dashed line). (Reprinted by permission from Macmillan Publishers Ltd., *Nat. Mater.*, Deubel, M. et al. 2004. Direct laser writing of three-dimensional photonic-crystal templates for telecommunications. **3**: 444–447. Copyright 2004.)

Chapter 11

The corresponding bandgap of this structure is found at 1.3–1.7 μm, which encompasses the optical telecommunication band. These experiments show the advantage of the high spatial resolution and three-dimensional fabrication capability of two-photon lithography. After the first demonstration by Sun et al. in 1999, many different types of photonic crystal structures were fabricated, such as diamond lattice photonic crystal, spiral photonic crystal, and icosahedral quasi-crystal, which are shown in Figure 11.21. These structures are not possible to be realized by thin layer stacking, and hence are extremely difficult to fabricate by ordinary semiconductor lithography techniques.

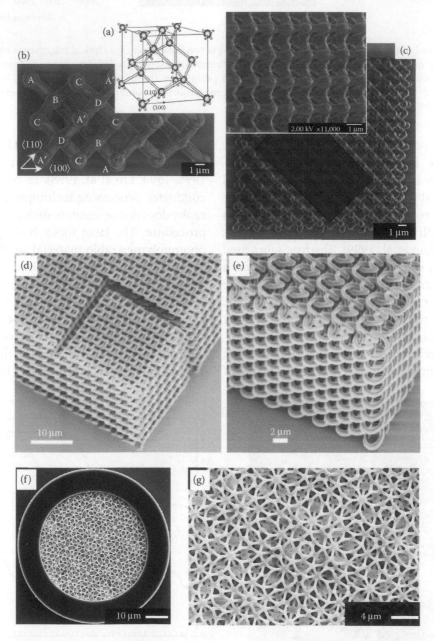

FIGURE 11.21 Series of three-dimensional photonic structures. (a–c) Diamond lattice photonic crystal. The stop band appears at around 2.5 μm, depending on the lattice constant. (Reprinted with permission from Kaneko, K., Sun, H.-B., Duan, X.-M., Kawata, S. 2003. Submicron diamond-lattice photonic crystals produced by two-photon laser nanofabrication. *Appl. Phys. Lett.* **83**: 2091–2093. Copyright 2003, American Institute of Physics.) (d,e) Spiral architecture photonic crystal fabricated by two-photon lithography. (Reprinted with permission from Seet, K. K. et al. Three-dimensional spiral-architecture photonic crystals obtained by direct laser writing. *Adv. Mater.* 2005. **17**: 541–545. Copyright 2005, Wiley-VCH Verlag.) (f,g) Three-dimensional icosahedral quasi-crystal structure. (Reprinted by permission from Macmillan Publishers Ltd., *Nat. Mater.*, Ledermann, A. et al. 2006. Three-dimensional silicon inverse photonic quasicrystals for infrared wavelengths. **5**: 942–945. Copyright 2006.)

11.5.2 Embedding Defect Cavities in Photonic Crystals

Another important advantage of two-photon lithography is that of defect structure fabrication. Photonic crystal defects are vacancy spaces in the lattice where photons can be confined, and so carry an important role in optical devices such as optical waveguides, interferometers, optical switching, laser cavities, and so on. In two-photon lithography, defect devices are designed by a preprogrammed three-dimensional model, and laser light immediately produces the photonic crystal with desired defects incorporated into it. Figure 11.22 shows an example of such a photonic crystal with defect cavities. Using two-photon localization to add defects into photonic crystals prefabricated by other methods is also possible. Artificial opal growth (Bogomolov et al. 1996) and holographic lithography techniques (Campbell et al. 2000; Shoji and Kawata 2000) are complementary methods to produce large volumes of photonic crystals. To introduce defects into prefabricated photonic crystals by these methods, at first photopolymerizable material is embedded in the interstitial space of a photonic crystal, and then two-photon lithography is performed in the photonic crystals to fill in the interstice with

polymerized material (Lee et al. 2002; Taton and Norris 2002). Refractive index matching by the infiltration of photopolymerizable material allows the laser beam to penetrate deep in the photonic crystals without significant scattering by the structure and polymerize the material inside. Figure 11.23 shows defect waveguides and cavities graved in artificial opal photonic crystals.

One crucial limitation of photonic crystals produced by two-photon lithography is the low refractive index of the base material. So far, it is considered that the refractive index of the base material must be larger than 2.0 to exhibit an absolute photonic bandgap, even with ideal lattice geometry (Ho et al. 1990). However, ordinary photopolymerizable materials can provide refractive indices of only about 1.6, much less than the requirement for complete photonic bandgaps. Doping high-refractive-index materials into photopolymerizable resins is one possible approach to enhance the refractive index of the matrix; however, the enhancement has not sufficiently been performed yet to reach an index of up to 2.0. Another promising method is replication technique using high-refractive-index materials, such as titania and silicon. Chemical vapor deposition (CVD) and sol–gel methods are potential approaches to infiltrate high-refractive-index materials

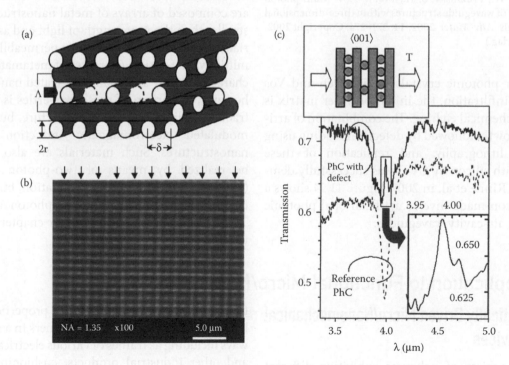

FIGURE 11.22 Wood-pile photonic crystal structure with cavities. (Left) Schematic and optical micrograph of the structure. (Right) Transmission spectra of the photonic crystal with cavities, as well as photonic crystal without defect, showing an evidence of a cavity mode in photonic bandgap. (Reprinted with permission from Sun, H.-B. et al. 2001. Microcavities in polymeric photonic crystals. *Appl. Phys. Lett.* **79**: 1–3. Copyright 2001, American Institute of Physics.)

Chapter 11

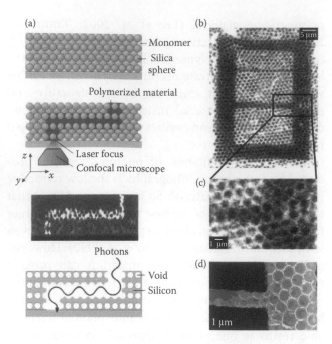

FIGURE 11.23 Defect induced inside a colloidal photonic crystal by curing immersed photopolymer. (a) Schematic of the experimental procedure. (Reprinted by permission from Macmillan Publishers Ltd., *Nature*, Taton, T. A., Norris, D. J. 2002. Device physics: Defective promise in photonics. **416**: 685–686. Copyright 2002.) (b) Fluorescence image of the photonic crystal containing photopolymerized defect structures. (c) Magnified view. (d) Scanning electron microscope image. (Reprinted with permission from Lee, W., Pruzinsky, S. A., Braun, P. V. Multi-photon polymerization of waveguide structures within three-dimensional photonic crystals. *Adv. Mater.* 2002. **14**: 271–274. Copyright 2002, Wiley-VCH Verlag.)

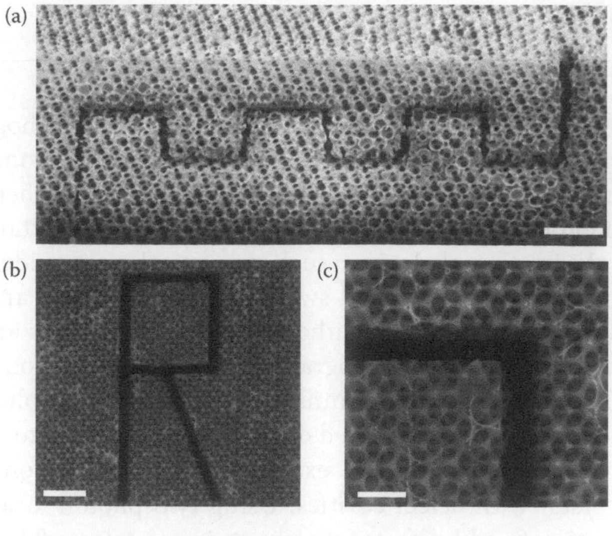

into polymer photonic crystals (Wijnhoven and Vos 1998). After infiltration, the initial polymer matrix is removed by chemical etching. The combination of artificial opal growth, followed by defect embedding using two-photon lithography, and replication of these structures with silicon by CVD was successfully demonstrated by Rinne et al. in 2007. Figure 11.24 shows a series of silicon-made inverse artificial opal photonic crystals with air-cavity waveguides.

FIGURE 11.24 Series of air-cavity waveguides embedded within silicon–air inverse opal photonic crystals observed by scanning electron microscopy. Scale bars : 3 μm for (a,b) and 1 μm for (c). (Reprinted by permission from Macmillan Publishers Ltd., *Nat. Photonics*, Rinne, S. A., Garcia-Santamaria, F., Braun, P. V. 2007. Embedded cavities and waveguides in three-dimensional silicon photonic crystals. **2**: 52–56. Copyright 2007.)

11.5.3 Metamaterials

Recently, another class of artificial optical materials has been defined, called "*metamaterials.*" Metamaterials are composed of arrays of metal nanostructures much smaller than the wavelength of light and as bulk materials; they exhibit extraordinary permeability and permittivity. The optical character of metamaterials is also characterized by artificially fabricated nanostructures; however, the origin of these properties is not the diffraction of light by periodic structure, but artificially modulated electromagnetic induction by metal nanostructures. Such materials are also possible to be realized by means of two-photon lithography (although not through polymerization, but the reduction of metal initiated by two-photon absorption). This topic is described in another chapter of this volume in detail.

11.6 Application to Functional Micro/Nanodevices

11.6.1 Optically Driven Micro/Nanomechanical Devices

There are a variety of polymers exhibiting different mechanical properties, especially with regard to their elasticity and stiffness. Currently, polymers are one of the essential materials of our daily life, due in particular to their advantageous mechanical properties and flexibility of shape. We are using polymers in a multitude of ways including as frames of various electrical appliances and other industrial products, cushioning materials, tires, plastic bags, containers, chemical textile, adhesive glue, gaskets, rubber bands, and so on. Two-photon lithography opens the door to expand the extensive use

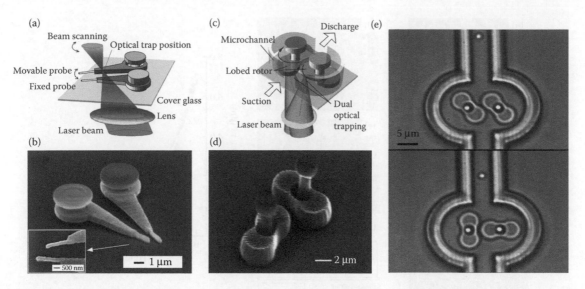

FIGURE 11.25 Microdevices driven by focus-scanned laser-trapping technique. (a,b) A micromanipulator (Reprinted with permission from Maruo, S., Ikuta, K., Korogi, H. 2003b. Submicron manipulation tools driven by light in a liquid. *Appl. Phys. Lett.* **82**: 133–135. Copyright 2003, American Institute of Physics.) (c–e) A microfluidic pump driven by a time-shared laser trapping. (Reprinted with permission from Maruo, S., Inoue, H. 2006. Optically driven micropump produced by three-dimensional two-photon microfabrication. *Appl. Phys. Lett.* **89**: 144101 1–3. Copyright 2006, American Institute of Physics.)

of polymers into micro/nanoscale applications. Figure 11.25 shows some examples of mechanical devices fabricated by two-photon lithography (Maruo et al. 2003b; Maruo and Inoue 2006). These microdevices function via laser trapping, a noncontact optical manipulation technique. A focused laser grabs parts of the microdevice by the sharp optical gradient force at the focus spot.

Scanning of the focus spot moves the position and direction of the parts so as to drive the devices. Figure 11.26 also shows other types of microdevices, which can be self-driven by stationary irradiation of light without the need of a mechanical scanning apparatus. These devices can be used as microfluidic systems or micromanipulators for bioengineering application.

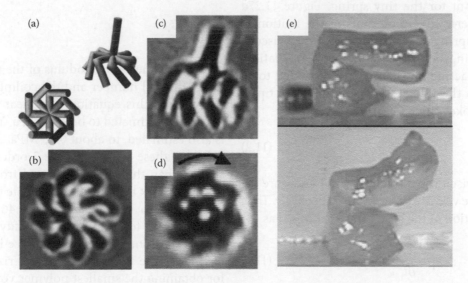

FIGURE 11.26 Self-driven microstructures by light irradiation. (a) Eight-field rotation symmetrical microrotor. (b–d) Microscope images of the rotor; (b) top view, (c) side view, and (d) top view during rotation. The torque is generated by the scattering force onto rotation symmetric propeller (Reprinted with permission from Galajda, P., Ormos, P. 2001. Complex micromachines produced and driven by light. *Appl. Phys. Lett.* **78**: 249–251. Copyright 2001, American Institute of Physics). (e) Optically movable cantilever made of photoisomerizable gelatin. (Reprinted with permission from Watanabe, T. et al. Photoreponsive hydrogel microstructure fabricated by two-photon initiated polymerization. *Adv. Funct. Mater.* 2002. **12**: 611–614. Copyright 2002, Wiley-VCH Verlag.)

Chapter 11

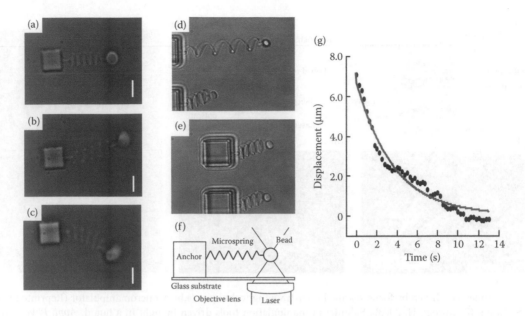

FIGURE 11.27   Microscopic images of microcoil spring fabricated by two-photon lithography. (a) Just after fabrication. Unpolymerized resin is not yet removed. (b,c) During developing, the spring flows with the convection of solvent. Scale bar; 5 μm. (d) Stretching the spring by laser trapping. (e) Recovery of the spring after the release of trapping force. (f) A schematic system for driving the microcoil spring. (g) The bead displacement versus time during the spring recovery from the stretching state.

Figure 11.27 shows another example of mechanical devices, a microcoil spring fabricated by two-photon lithography. The coil spring consists of a polymer wire with a diameter of 300 nm, and the diameter and the pitch of the coil are 4 and 2 μm, respectively. This spring shows elastic motion in aqueous solution, although the viscous drag from the surrounding solution is significant for the tiny spring. Figure 11.27d and e show a series of frame shots of the motion of the microcoil spring. Because of the strong viscous drag, the spring shows overdamped oscillation. The drag force acting on the spring is assumed to be proportional to the velocity of the bead at the tip of spring, from Stokes' law,

$$F = 6\pi\eta rv \tag{11.2}$$

where η is the viscosity of solution, and r and v are the diameter and the velocity of the bead, respectively. Then damping oscillation of the bead can be expressed as

$$m\frac{\partial^2 x}{\partial t^2} = -kx - 6\pi\eta rv\frac{\partial x}{\partial t} \tag{11.3}$$

where m and x are the mass and relative position of the bead, and k is the spring constant. Following this simple mechanical analysis, the spring constant of the microcoil can be approximately estimated to be ~1×10^{-8} N/m.

11.6.2 Mechanical Property of Two–Photon Polymerized Nanopolymer

From the obtained spring constant of the microcoil, the elasticity of the polymer formed by two-photon lithography can be derived. The spring constant of coil spring shows a relation,

$$k = \frac{G_s r^4}{4NR_c^3} \tag{11.4}$$

where G_s is the shear modulus of the material, N, and R_c are the coil number and the coiling radius, respectively. From this equation, the shear modulus of the material is estimated to be 0.25 MPa. Young's modulus is also estimated, to about 0.75 MPa. Surprisingly, the values of these coefficients are 3 orders of magnitude smaller than the values measured from a macroscale polymer block by an ordinary tensile test (shear modulus; 0.15 GPa, Young's modulus; 0.46 GPa). One possible reason for this low elasticity is insufficient polymerization. The laser irradiance is adjusted to be just above the two-photon polymerization threshold for obtaining the smallest polymer voxels and highest spatial resolution, and so the polymerization reaction may not yet be saturated. Accordingly, the spring constant was tested following the irradiance of additional ultraviolet light after the developing process, as shown in Figure 11.28 (Nakanishi et al. 2007). An increase

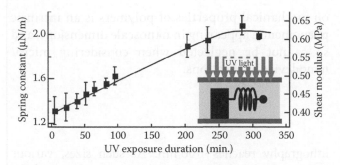

FIGURE 11.28 Change of the spring constant of a microcoil spring by the exposure of additional UV light. (Reprinted with permission from Nakanishi, S., Shoji, S., Kawata, S., Sun, H.-B. 2007. Giant elasticity of photopolymer nanowires. *Appl. Phys. Lett.* **91**: 063112 1–3. Copyright 2007, American Institute of Physics.)

was noted; however, this increase is only a few times, still not sufficient to reach the value similar to that of bulk materials.

The results of these experiments indicate the possibility that the mechanical property of such polymers dramatically changes when their size is reduced down to the micro/nanoscale. There is further evidence emphasizing this size-dependent property of the polymer. Figure 11.29 shows the temperature dependence of the stretching length of a polymer microcoil spring (Nakanishi et al. 2008). A temperature control stage was introduced into a laser-trapping microscope system, and the spring was elongated by a constant force while changing the temperature of the entire sample. By Hooke's law, the spring constant of the microcoil is obtained at varying temperatures.

$$F = -k\Delta x \tag{11.5}$$

The graph shows a sudden increase of the stretch length at a certain temperature, corresponding to a decrease in the spring constant of the microcoil, which means a decrease in the shear modulus of polymer. This result was interpreted as the phase transition of the polymer. More surprisingly, the material phase transition temperature shows remarkable size dependence. When the diameter of the polymer wire becomes <400 nm, the transition temperature starts to decrease according to the polymer diameter as shown in Figure 11.29c.

As shown earlier, the mechanical coefficients of polymer—shear modulus, elastic modulus, transition temperature, and so on—are no longer invariable but are size dependent. This size-dependent behavior is neither restricted to a specific polymer, nor only to structures formed by two-photon polymerization. The reduction of glass transition temperature is seen in different kinds of polymer thin films (Roth and Dutcher 2003; Fukao et al. 2002). It is known that the mechanical properties of polymers are governed by intramolecular and intermolecular interaction of polymer chains. For example, when polymers are heated above their transition temperature, the friction between polymer chains starts to become weak, which leads to micro-Brownian motion of the polymer chains. As a result, polymers become softer with heating. The degree of intra- and intermolecular dynamics in polymer is mainly related to the length of polymer chain, and the density of polymer cross-linking. At the surface, polymer chains are more free to move compared to the chains deep inside a bulk of polymer, since loose ends are likely to be abundant and also polymer chains are absolutely free to move in the direction perpendicular to the surface. This surface

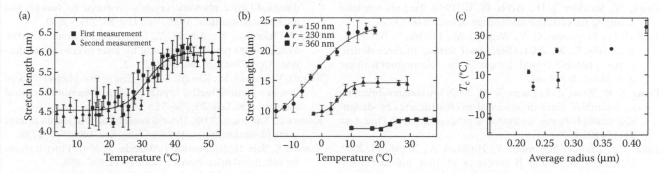

FIGURE 11.29 Temperature dependence of the elasticity of a microcoil spring. (a) Stretching length under a constant force as a function of temperature. The increase of stretching length at ~30°C indicates the phase transition of polymer. (b) Temperature-dependent stretching length of microcoil springs made of different nanowire radii. (c) Size-dependent transition temperature of polymer nanowires. (Reprinted with permission from Nakanishi, S. et al. 2008. Size dependence of transition temperature in polymer nanowires. *J. Phys. Chem. B* **112**: 3586–3589. Copyright 2008, American Chemical Society.)

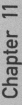

Chapter 11

effect is usually insignificant where the size of the polymer structure is large enough; however, when the size of the structure becomes small, the mechanical property of the polymer reflects a large contribution from the surface components. Indeed, the size effect on mechanical properties of polymers is an intrinsic phenomenon appearing in nanoscale dimensions, and so cannot be neglected when considering micro/nanodevice applications.

11.7 Summary

We have detailed the principles and characteristics of two-photon lithography. Two-photon lithography is characterized by the intrinsic capability for three-dimensional fabrication with a super-resolution beyond the diffraction limit of light. This unique characteristic allows us to fabricate a variety of three-dimensional structures within the scale of a few micrometers. In particular, MEMS and photonic application are utilizing two-photon lithography for producing micro/nanodevices with novel mechanical or optical functions. Consequently, due to the practical application for such novel mechanical or photonic science and technology, the performance of two-photon lithography continues to improve. So far, the spatial resolution of two-photon lithography reaches ~100 nm. At such sizes, various properties of the photopolymerization process and the resultant polymerized materials, including polymerization degree, stability, and consequent voxel size, elasticity, transition temperature, volume shrink, and so on are no longer characterized by only the profile of the laser beam at the focus spot, but are rather governed by inter- and intramolecular dynamics of monomers, radicals, solvent, and polymer chains in the nanoscale volume. Indeed the progress of two-photon lithography has made it possible to reveal the fundamental nature of polymer materials at the nanoscale. Two-photon lithography is now a key technique for both nanopolymer material science and laser micro/nanoengineering.

References

Adronov, A., Frchet, J. M. J., He, G. S., Kim, K.-S., Chung, S.-J., Swiatkiewicz, J., Prasad, P. N. 2000. Novel two-photon absorbing dendritic structures. *Chem. Mater.* **12**: 2838–2841.

Bogomolov, V. N., Gaponenko, S. V., Kapitonov, A. M., Prokofiev, A. V., Ponyavina, A. N., Silvanovich, I., Samoilovich, S. M. 1996. Photonic band gap in the visible range in a three-dimensional solid state lattice. *Appl. Phys. A* **63**: 613–616.

Campbell, M., Sharp, D. N., Harrison, M. T., Denning, R. G., Turberfield, A. J. 2000. Fabrication of photonic crystals for the visible spectrum by holographic lithography. *Nature* **404**: 53–56.

Cumpston, B. H., Ananthavel, S. P., Barlow, S., Dyer, D. L., Ehrlich, J. E., Erskine L. L., Heikal A. A., et al. 1999. Two-photon polymerization initiators for three-dimensional optical data storage and microfabrication. *Nature* **398**: 51–54.

Denk, W., Strickler, J. H., Webb, W. W. 1990. Two-photon laser scanning fluorescence microscopy. *Science* **248**: 73–76.

Deubel, M., Freymann, G. V., Wegener, M., Pereira, S., Busch, K., Soukoulis, C. M. 2004. Direct laser writing of three-dimensional photonic-crystal templates for telecommunications. *Nat. Mater.* **3**: 444–447.

Dong, X.-Z., Zhao, Z.-S., Duan, X.-M. 2007. Micronanofabrication of assembled three-dimensional microstructures by designable multiple beams multiphoton processing. *Appl. Phys. Lett.* **91**: 124103 1–3.

Fukao, K., Uno, S., Miyamoto, Y., Hoshino, A., Miyaji, H. 2002. Dynamics of α and β processes in thin polymer films: Poly(vinyl acetate) and poly(methyl methacrylate). *Phys. Rev. E* **64**: 051807 1–11.

Fujita, A., Fujita, K., Matsuda, T., Nakamura, O. 2004. Photofabrication of scaffold for cell-growth using photocurable gelatin. *Trans. Jpn. Soc. Med. Biol. Eng.* **42**: 411.

Galajda, P., Ormos, P. 2001. Complex micromachines produced and driven by light. *Appl. Phys. Lett.* **78**: 249–251.

Glezer, E. N., Molosavljevic, M., Huang, L., Finlay, R. J., Her, T.-H., Callan, J. P., Mazur, E. 1996. Three-dimensional optical storage inside transparent materials. *Opt. Lett.* **21**: 2023–2025.

Gu, M., Day, D. 1999. Use of continuous-wave illumination for two-photon three-dimensional optical bit data storage in a photobleaching polymer. *Opt. Lett.* **24**: 288–290.

Ho, K. M., Chan, C. T., Soukoulis, C. M. 1990. Existence of a photonic gap in periodic dielectric structures. *Phys. Rev. Lett.* **65**: 3152–3155.

John, S. 1987. Strong localization of photons in certain disordered dielectric superlattices. *Phys. Rev. Lett.* **58**: 2486–2489.

Kaneko, K., Sun, H.-B., Duan, X.-M., Kawata, S. 2003. Submicron diamond-lattice photonic crystals produced by two-photon laser nanofabrication. *Appl. Phys. Lett.* **83**: 2091–2093.

Kato, J., Takeyasu, N., Adachi, Y., Sun, H.-B., Kawata, S. 2005. Multiple-spot parallel processing for laser micronanofabrication. *Appl. Phys. Lett.* **86**: 044102 1–3.

Kawata, Y., Ishitobi, H., Kawata, S. 1998. Use of two-photon absorption in a photorefractive crystal for three-dimensional optical memory. *Opt. Lett.* **23**: 756–758.

Kawata, S., Kawata, Y. 2000. Three-dimensional optical data storage using photochromic materials. *Chem. Rev.* **100**: 1777–1788.

Kawata, S., Sun, H-.B., Tanaka, T., Takada, K. 2001. Finer features for functional microdevices. *Nature* **412**: 697–698.

Kuebler, S. M., Rumi, M., Watanabe, T., Braun, K., Cumpston, B. H., Heikal, A. A., Erskine, L. L., Thayumanavan, S., Barlow, S., Marder, S. R., Perry, J. W. 2001. Optimizing two-photon initiators and exposure conditions for three-dimensional lithographic microfabrication. *J. Photopolym. Sci. Tech.* **14**: 657–668.

Ledermann, A., Cademartiri, L., Hermatschweiler, M., Toninelli, C., Ozin, G. A., Wiersma, D. S., Wegener, M., Freymann, G. V. 2006. Three-dimensional silicon inverse photonic quasicrystals for infrared wavelengths. *Nat. Mater.* **5**: 942–945.

Lee, W., Pruzinsky, S. A., Braun, P. V. 2002. Multi-photon polymerization of waveguide structures within three-dimensional photonic crystals. *Adv. Mater.* **14**: 271–274.

Lin, S. Y., Fleming, J. G., Hetherington, D. L., Smith, B. K., Biswas, R., Ho, K. M., Sigalas, M. M., Zubrzycki, W., Kurtz, S. R., Bur, J. 1998. A three-dimensional photonic crystal operating at infrared wavelengths. *Nature* **394**: 251–253.

Maruo, S., Nakamura, O., Kawata, S. 1997. Three-dimensional microfabrication with two-photon-absorbed photopolymerization. *Opt. Lett.* **22**: 132–134.

Maruo, S., Ikuta, K., Korogi, H. 2003a. Force-controllable, optically driven micromachines fabricated by single-step two-photon microstereolithography. *J. Microelectromech. Syst.* **12**: 533–549.

Maruo, S., Ikuta, K., Korogi, H. 2003b. Submicron manipulation tools driven by light in a liquid. *Appl. Phys. Lett.* **82**: 133–135.

Maruo, S., Inoue, H. 2006. Optically driven micropump produced by three-dimensional two-photon microfabrication. *Appl. Phys. Lett.* **89**: 144101 1–3.

Nakanishi, S., Shoji, S., Kawata, S., Sun, H.-B. 2007. Giant elasticity of photopolymer nanowires. *Appl. Phys. Lett.* **91**: 063112 1–3.

Nakanishi, S., Shoji, S., Yoshikawa, H., Sekkat, Z., Kawata, S. 2008. Size dependence of transition temperature in polymer nanowires. *J. Phys. Chem. B* **112**: 3586–3589.

Nguyen, L. H., Straub, M., Gu, M. 2005. Acrylate-based photopolymer for two-photon microfabrication and photonic applications. *Adv. Funct. Mater.* **15**: 209–216.

Noda, S., Yamamoto, N., Sasaki, A. 1996. New realization method for three-dimensional photonic crystal in optical wavelength region. *Jpn. J. Appl. Phys.* **35**: L909–L912.

Noda, S., Yamamoto, N., Kobayashi, H., Okano, M., Tomoda, K. 1999. Optical properties of three-dimensional photonic crystals based on III–V semiconductors at infrared to near-infrared wavelengths. *Appl. Phys. Lett.* **75**: 905–907.

Odian, G. 1991. *Principle of Polymerization.* 3rd ed. New York: Wiley.

Parthenopoulos, D., Rentzepis, P. M. 1989. Three-dimensional optical storage memory. *Science* **245**: 843–845.

Rinne, S. A., Garcia Santamaria, F., Braun, P. V. 2007. Embedded cavities and waveguides in three-dimensional silicon photonic crystals. *Nat. Photonics* **2**: 52–56.

Roth, C. B., Dutcher, J. R. 2003. Glass transition temperature of freely-standing films of atactic poly(methyl methacrylate). *Eur. Phys. J. E* **12**: s01, 024.

Seet, K. K., Mizeikis, V., Matsuo, S., Juodkazis, S., Misawa, H. 2005. Three-dimensional spiral-architecture photonic crystals obtained by direct laser writing. *Adv. Mater.* **17**: 541–545.

Serbin, J., Egbert, A., Ostendorf, A., Chichkov, B. N., Houbertz, R., Domann, G., Schulz, J., Cronauer, C., Frohlich, L., Popall, M. 2003. Femtosecond laser-induced two-photon polymerization of inorganic organic hybrid materials for applications in photonics. *Opt. Lett.* **28**: 301–303.

Shen, Y. R. 1984. *The Principles of Nonlinear Optics.* New York: A Wiley-Interscience Publication.

Shoji, S., Kawata, S. 2000. Photofabrication of three-dimensional photonic crystals by multibeam laser interference into a photopolymerizable resin. *Appl. Phys. Lett.* **76**: 2668–2670.

Strickler, J. H., Webb, W. W. 1991. Three-dimensional optical data storage in refractive media by two-photon point excitation. *Opt. Lett.* **16**: 1780–1782.

Sun, H.-B., Matsuo, S., Misawa, H. 1999. Three-dimensional photonic crystal structures achieved with two-photon-absorption photopolymerization of resin. *Appl. Phys. Lett.* **74**: 786–788.

Sun, H.-B., Mizeikis, V., Xu, Y., Juodkazis, S., Ye, J.-Y., Matsuo, S., Misawa, H. 2001. Microcavities in polymeric photonic crystals. *Appl. Phys. Lett.* **79**: 1–3.

Sun, H.-B., Tanaka, T., Kawata, S. 2002. Three-dimensional focal spots related to two-photon excitation. *Appl. Phys. Lett.* **80**: 3673–3675.

Sun, H.-B., Takada, K., Kim, M.-S., Lee, K.-S., Kawata, S. 2003. Scaling laws of voxels in two-photon photopolymerization nanofabrication. *Appl. Phys. Lett.* **83**: 1104–1106.

Takada, K., Sun, H.-B., Kawata, S. 2005. Improved spatial resolution and surface roughness in photopolymerization-based laser nanowriting. *Appl. Phys. Lett.* **86**: 071122 1–3.

Takada, K., Wu, D., Chen, Q.-D., Shoji, S., Xia, H., Kawata, S., Sun, H.-B. 2009. Size-dependent behaviors of femtosecond laser-prototyped polymer micronanowires. *Opt. Lett.* **34**: 566–568.

Tan, D., Li, Y., Qi, F., Yang, H., Gong, Q., Dong, X., Duan, X. 2007. Reduction in feature size of two-photon polymerization using SCR500. *Appl. Phys. Lett.* **90**: 071106 1–3.

Taton, T. A., Norris, D. J. 2002. Device physics: Defective promise in photonics. *Nature* **416**: 685–686.

Teh, W. H., During, U., Salis, G., Harbers, R., Drechsler, U., Mahrt, R. F., Smith, C. G., Guntherodt, H.-J. 2004. SU-8 for real three-dimensional subdiffraction-limit two-photon microfabrication. *Appl. Phys. Lett.* **84**: 4095–4097.

Toriumi, A., Herrman, J. M., Kawata, S. 1997. Nondestructive read-out of a three-dimensional photochromic optical memory with a near-infrared differential phase-contrast microscope. *Opt. Lett.* **22**: 555–557.

Toriumi, A., Kawata, S., Gu, M. 1998. Reflection confocal microscope readout system for three-dimensional photochromic optical data storage. *Opt. Lett.* **23**: 1924–1926.

Watanabe, T., Akiyama, M., Totani, K., Kuebler, S. M., Stellacci, F., Wensellers, W., Braun, K., Marder, S. R., Perry, J. W. 2002. Photoreponsive hydrogel microstructure fabricated by two-photon initiated polymerization. *Adv. Func. Mater.* **12**: 611–614.

Wijnhoven, J. E. G. J., Vos, W. L. 1998. Preparation of photonic crystals made of air spheres in titania. *Science* **281**: 802–804.

Witzgall, G., Vrijen, R., Yablonovitch, E., Doan, V., Schwartz, B. J. 1998. Single-shot two-photon exposure of commercial photoresist for the production of three-dimensional structures. *Opt. Lett.* **23**: 1745–1747.

Yablonovitch, E. 1987. Inhibited spontaneous emission in solid-state physics and electronics. *Phys. Rev. Lett.* **58**: 2059–2062.

Yang, H.-K., Kim, M.-S., Kang, S.-W., Kim, K.-S., Lee, K.-S., Park, S.-H., Yang, D.-Y., Kong, H.-J., Sun H.-B., Kawata, S., Fleitz, P. 2004. Recent progress of lithographic microfabrication by the TPA-induced photopolymerization. *J. Photopolym. Sci. Tech.* **17**: 385–392.

Chapter 11

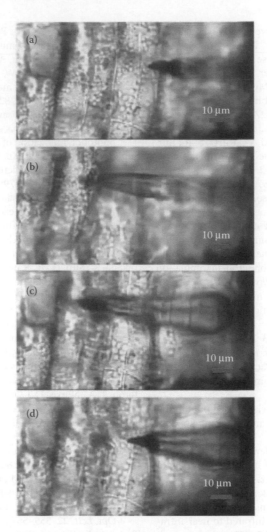

FIGURE 4.33 Optical microscope images of chloroplast capture. (a) Filtering tool approaching cell wall surface, (b) needle of filtering tool entering cell, (c) filtering tool remaining in cell for several seconds to filter organelles, and (d) filtering tool with chloroplast removed from cell.

FIGURE 8.6 Optical microscope images of resist after partial development. Various PSFs were used for the same PEC correction. Better color uniformity means better accuracy of PSF in this experiment. (Courtesy of Dr. B. Nilsson, Chalmers University.)

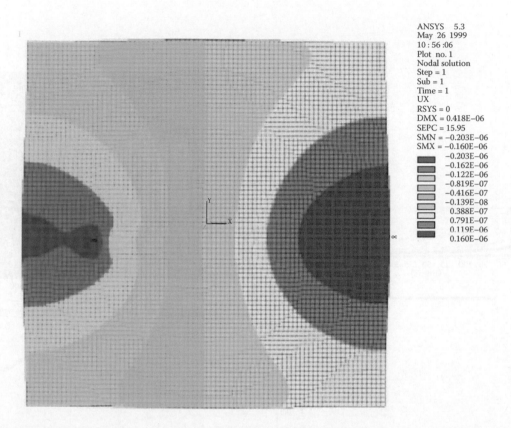

ANSYS 5.3
May 26 1999
10 : 56 :06
Plot no. 1
Nodal solution
Step = 1
Sub = 1
Time = 1
UX
RSYS = 0
DMX = 0.418E−06
SEPC = 15.95
SMN = −0.203E−06
SMX = −0.160E−06

−0.203E−06
−0.162E−06
−0.122E−06
−0.819E−07
−0.416E−07
−0.139E−08
0.388E−07
0.791E−07
0.119F−06
0.160E−06

FIGURE 8.9 In-plane displacement after the exposure of two rows of subfields in EBL exposure of silicon wafer. (Results by Computational Mechanics Center, University Wisconsin-Madison.)

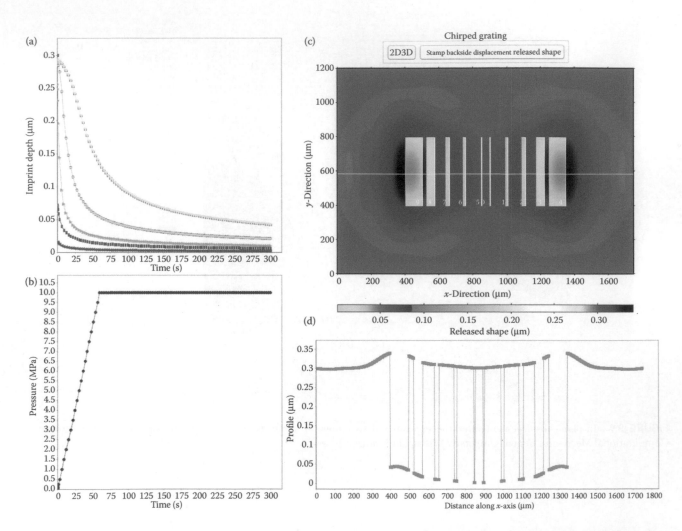

FIGURE 8.13 NIL of a chirped grating: (a) imprint depth as a function of time, (b) pressure diagram; (c) top view, and (d) resulting profile along the *x*-direction.

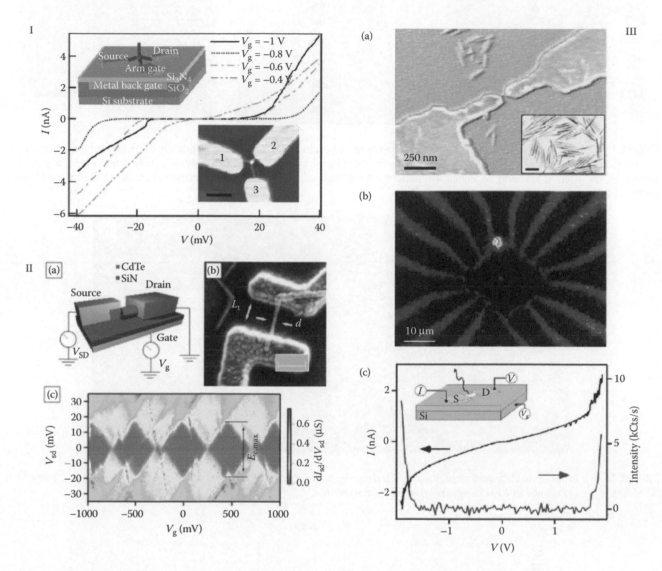

FIGURE 12.5 (I) A single CdTe tetrapod contacted by three lateral electrodes and a metal back gate. The I–V curves show source drain current of two lateral electrodes, demonstrating Coulomb blockade, for different back gate voltages. The scale bar in the inset is 100 nm. (Reprinted with permission from Cui, Y. et al., Electrical transport through a single nanoscale semiconductor branch point. *Nano Lett.* 2005, 5, 1519–1523. Copyright (2005) American Chemical Society.) (II) SET device based on a single CdTe rod shown as schematic illustration (a) and by a scanning electron microscopy image (b) with scale bar corresponding to 100 nm. The characteristic Coulomb blockade diamond structure is shown in II (c). (Reprinted with permission from Trudeau, P. E. et al., Electrical contacts to individual colloidal semiconductor nanorods. *Nano Lett.* 2008, 8, 1936–1939. Copyright (2008) American Chemical Society.) (III) Single CdSe rod that shows electroluminescence. (a) Scanning electron microscopy image of the device (false color). The inset shows transmission electron microscopy images of the nanorods, the scale bar is 50 nm. (b) Optical microscope image showing electroluminescence from a nanorod transistor. (c) Current (black) and simultaneously measured electroluminescence plotted against bias voltage. The inset illustrates the device architecture with source and drain electrodes contacting a single nanorod. (Reprinted with permission from Gudiksen, M. S. et al., Electroluminescence from a single-nanocrystal transistor. *Nano Lett.* 2005, 5, 2257–2261. Copyright (2005) American Chemical Society.)

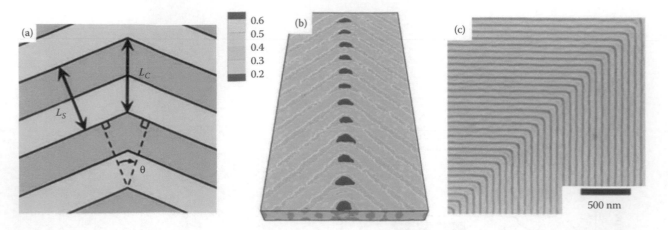

FIGURE 13.5 Directed assembly on a chemical pattern composed of chevrons. (a) Schematic showing the different local pattern periods within the single pattern. (b) SCMF simulation results showing the localization of homopolymers (red) in the corners of a pattern formed by the directed assembly of a chemical pattern having 90° bends. (c) SEM image of ternary blend directed to assemble on a chemical pattern having 90° bends.

FIGURE 14.16 Programmable assembly of Sierpinski triangle by use of computational assembly. Scale bar = 100 nm. (a) Zoomed out AFM image. (b, c) Zoomed in AFM image showing errors marked by x.

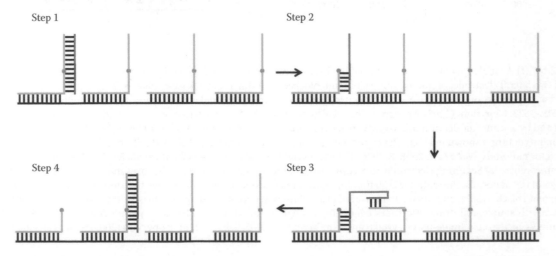

FIGURE 14.31 Mao's DNAzyme walker moving from one spot to the next on the black track with four foot holds. Red subsequence is the DNAzyme; blue dots are places where the DNAzyme can cleave.

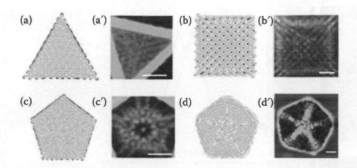

FIGURE 15.5 Simulation (a–d) and experimental results (a′–d′) using a 100-nm-thick aluminum plate with various edge/slit shapes. (a), (a′) triangle; (b), (b′) square; (c), (c′) pentagon; (d), (d′) pentagon with rounded corners. The side lengths are 2, 2, 1.5, and 1.5 μm, respectively in (a–d). The excitation light had circular polarization in all the simulations. The scale bars in all the AFM images represent 500 nm. (From Chen, W. and H. Ahmed, *Applied Physics Letters*, 1993. **62**(13): 1499–1501.)

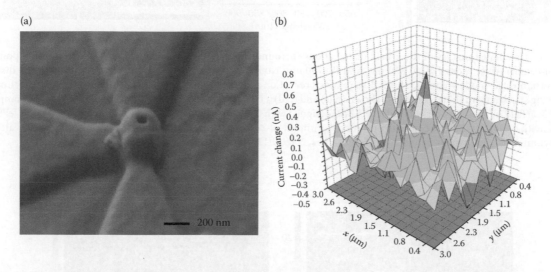

FIGURE 15.12 Scanning electron micrograph (a) and ion impact response map (b) from a FinFet with 100 nm hole in the top gate. (From Weis, C. D. et al., *Nuclear Instruments & Methods in Physics Research Section B—Beam Interactions with Materials and Atoms*, 2009. **267**(8–9): 1222–1225.)

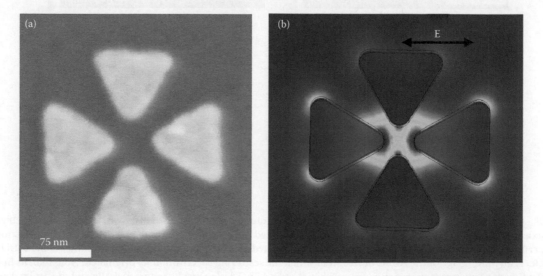

FIGURE 18.5 (a) SEM image of a symmetric bowtie "cross" nanoantenna. (b) Image of the normalized e-field amplitude distribution surrounding the nanoantenna when excited at its primary resonance of ~ 775 nm.

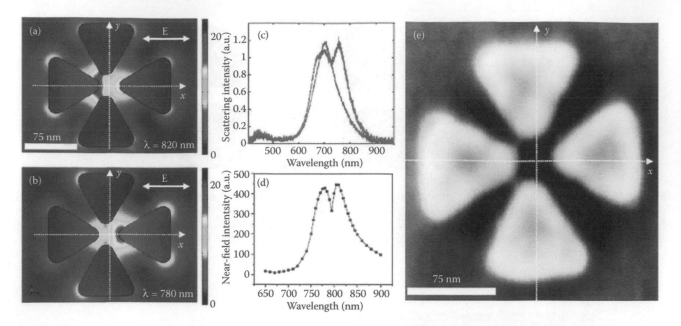

FIGURE 18.6 Calculated images of the e-field distributions surrounding an ABnC (*y*-oriented "vertical" bowtie component shifted 5 nm left-of-center) when excited at (a) 820 nm, and (b) 780 nm. Note that the two modes are spectrally and spatially distinct while maintaining nanoscale mode volumes. This ABnC retains a reflection symmetry axis oriented along the *x*-axis. (c) Experimental darkfield scattering spectra from (blue) the ABnC shown in (e) and (red) a representative symmetric cross nanoantenna for comparison. (d) Calculated nearfield intensity spectrum from the ABnC designed with the 5 nm vertical bowtie shift shown in (a) and (b). (e) SEM image of an ABnC with its vertical bowtie component shifted ~ 5 nm left-of-center. (From Zhang, Z. et al., Manipulating nanoscale light fields with the asymmetric bowtie nano-colorsorter. *Nano Letters*, 2009. 9(12): 4505–4509. With permission.)

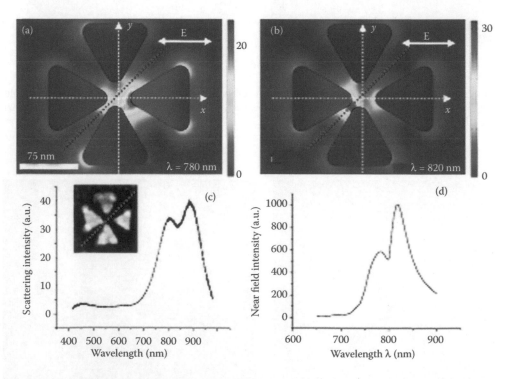

FIGURE 18.8 Calculated images of the e-field distributions surrounding an ABnC designed to have its reflection symmetry axis rotated 45° from *x*-axis, when excited at (a) 780 nm, and (b) 820 nm. The dotted red lines in (a), (b), and inset of (c) mark the reflection symmetry axis. (c) Experimental darkfield scattering spectrum and (d) calculated nearfield intensity spectrum from the ABnC with this symmetry. Inset of (c) is an SEM image of the ABnC device from which the scattering spectrum in (c) originated. Note that, as in Figures 18.1 and 18.2, retention of a single reflection symmetry axis in the cross ABnC design results in two spatially and spectrally distinct modes. (From Zhang, Z. et al., Manipulating nanoscale light fields with the asymmetric bowtie nano-colorsorter. *Nano Letters*, 2009. 9(12): 4505–4509. With permission.)

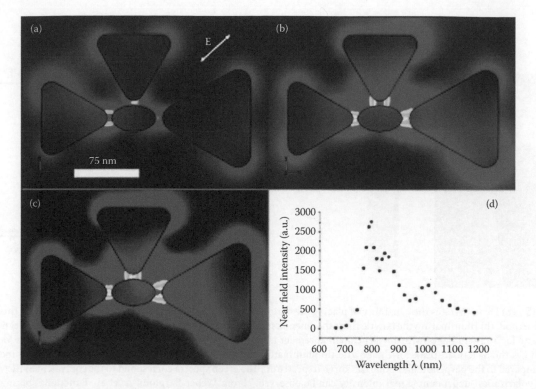

FIGURE 18.9 An example of an ABnC with an increased number of asymmetric degrees of freedom. Images of calculated field distributions showing spatially distinct field confinements due to mode hybridizations at (a) the left triangle-ellipse gap, (b) the top triangle-ellipse gap, and (c) the right triangle-ellipse gap. (d) The calculated nearfield intensity spectrum from this device, which now exhibits three spectral modes (rather than two) due to the increased asymmetry. (From Zhang, Z. et al., Manipulating nanoscale light fields with the asymmetric bowtie nano-colorsorter. *Nano Letters*, 2009. 9(12): 4505–4509. With permission.)

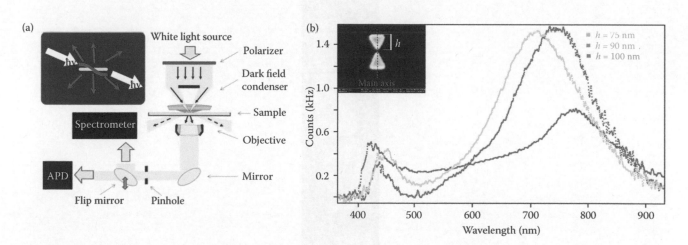

FIGURE 18.14 (a) Diagram of the darkfield spectroscopy setup used to determine the resonant frequency of the plasmonic antennae. White light is focused onto the sample under an angle greater than what the objective would collect. The resonantly scattered light of the nanoparticle is collected by the objective and sent to an avalanche photo diode (APD) of spectrometer. (b) Inset: definition of the bowties main axes and the varying height parameter of the three bowties are characterized. Main: darkfield spectra of three differently sized bowtie antennae. Increasing the height of the triangle elongates the surface plasmon cavity length and therefore redshifts the major peak position. (From Weber-Bargioni, A. et al., Functional plasmonic antenna scanning probes fabricated by induced-deposition mask lithography. *Nanotechnology*, 2010. **21**(6): 065306. With permission.)

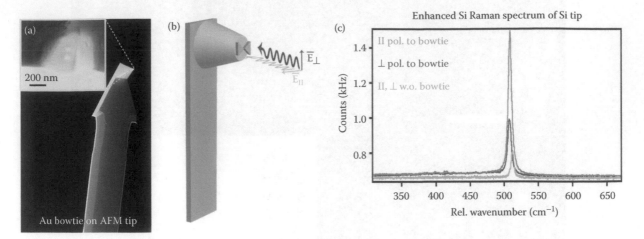

FIGURE 18.15 (a) Demonstrates the capability of placing the optical bowtie antenna well defined on a Si AFM tip with this novel nano-fabrication method. (b) Illuminating the bowtie tip at the antenna's resonance frequency with laser light polarized parallel to the bowtie's axis will create for a functional bowtie a nearfield enhancement in the bowtie gap. (c) Si Raman spectrum of the Si AFM tip. The blue spectrum shows the Si Raman peak for the bowtie tip illuminated parallel to the main axis and a 2.5-fold intensity increase can be recorded compared to the perpendicular polarization (red spectrum). The green spectra correspond to the equivalent Si tip without an Au bowtie. No polarization-dependent signal intensity can be observed. (From Weber-Bargioni, A. et al., Functional plasmonic antenna scanning probes fabricated by induced-deposition mask lithography. *Nanotechnology*, 2010. **21**(6): 065306. With permission.)

FIGURE 21.4 Transmission microscope image of the heterostructures in the catalytically grown compound semiconductor nanowires. (Courtesy of L. Samuelson. From K. A. Dick et al. 2005. *Nano Lett.* 5: 761. With permission.)

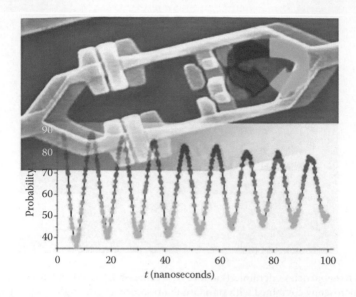

FIGURE 21.29   SEM image of the flux qubit with SQUID for readout and Rabi oscillations. (Courtesy of J. E. Mooij.)

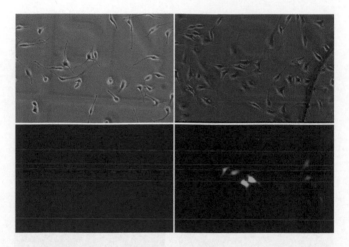

FIGURE 22.1   Hela cell GFP–plasmid transfection experiment through microinjection technique. White-light (top row) and fluorescent (bottom row) microscopy images of injected cells, respectively, soon after injection (on the left) and 24 h of incubation later (on the right). During the incubation time, only microinjected cells in a healthy state can carry out the GFP synthesis, thus becoming fluorescent green.

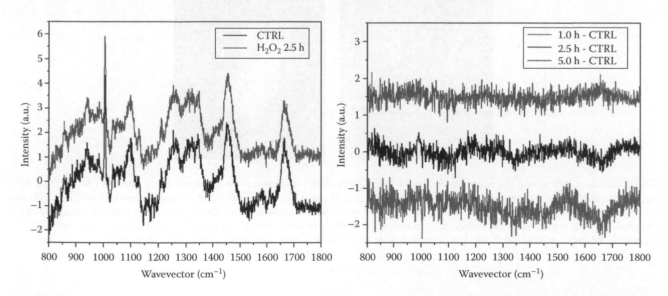

FIGURE 22.5   On the right, average Raman spectra recorded from Hela cells incubated 2.5 h with H_2O_2 9 mM (top line) and from control cell (bottom line); on the left: difference between Raman spectra from incubated cells and control cells, for Hela incubated 1.0, 2.5, and 5.0 h with 9 mM H_2O_2.

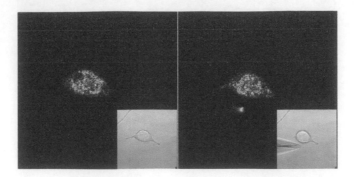

FIGURE 22.12 Nanoporous Si nanoparticles microinjection into Hela cell: just before (a) and just after (b) the microinjection. Pictures are recorded via two-photon microscopy combined with standard fluorescence. On the right, a strong rhodamine signal coming from the micropipette tip is observable.

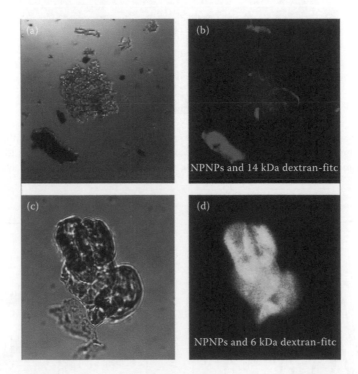

FIGURE 22.20 Optical and fluorescence images of the NPNPs after incubation with two fluorescent polymers of different MW: dextran-fitc 14 kDa (panels a and b), and dextran-fitc 6 kDa (panels c and d). The adsorption of the lighter polymer is clearly indicated by the green fluorescence emitted from the NPs (panel d). On the contrary, in the case of heavier polymer no green fluorescence can be observed (the blue one is from salt residues), confirming the molecular weight cutoff. (Reprinted from Pujia, A. et al., 2010. *J.R. Soc. Interface* 1, 79; Pujia, A. et al. 2010. *Int. J. Nanomed.* 5, 1005–1015. With permission from Dove Medical Press Ltd.)

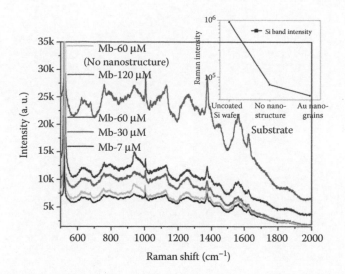

FIGURE 22.25 Raman spectra for Mb protein by varying the concentration but by keeping the laser spot centered at nanostructures. The inset graph shows the background Raman signal of the Silicon for different substrate topologies. (Reprinted from *Microelectronic Eng.*, 85(5–6), Das, G. et al., Attomole (amol) myoglobin Raman detection from plasmonic nanostructures, 1282–1285, Copyright (2008), with permission from Elsevier.)

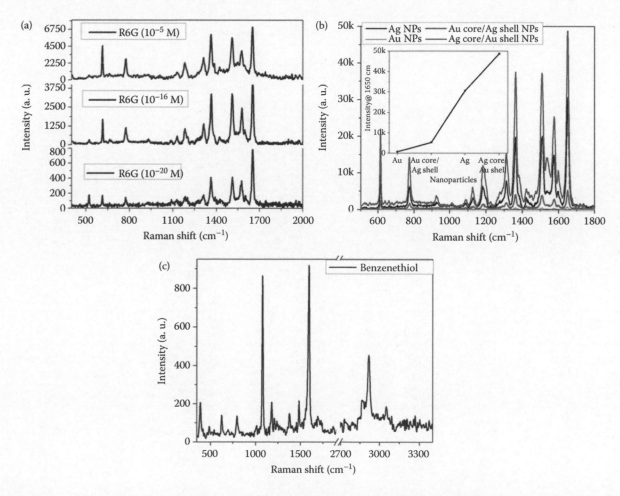

FIGURE 22.26 Raman spectra of (a) R6G at various concentrations on Ag-based *SUBS 2*; (b) R6G with a concentration of 10^{12} M, absorbed on the silver, gold, Ag/Au- and Au/Ag-based *SUBS 2*; (c) benzenethiol [10 mM] absorbed on Ag-based *SUBS 2*. (Reprinted from *Microelectronic Eng.*, 86(4–6), Coluccio, M.L. et al., Silver–based surface enhanced Raman scattering (SERS) substrate fabrication using nanolithography and site selective electroless deposition, 1085–1088, Copyright (2009), with permission from Elsevier.).

12. Colloidal Inorganic Nanocrystals
Synthesis and Controlled Assembly

Liberato Manna and Roman Krahne
Istituto Italiano di Tecnologia, Genova, Italy

12.1 Outline

The discovery of size-dependent physical properties of solids at the nanometer scale has recently triggered an intense research on the fabrication of nanostructures by a variety of approaches. Among the various nanostructures that are studied, inorganic nanocrystals are clusters composed of a number of atoms that can range from a few tens up to several tens of thousands and can be thought of as solids whose sizes along the three dimensions have been reduced down to a few nanometers or less. The properties of nanocrystals and their mutual interactions can be controlled finely by tailoring their size, composition, and surface functionalization. Nanocrystals have been investigated in a variety of

applications in many fields, among them biology and biomedicine, optics and photonics, catalysis, sensing, and high-density memory storage. Also, many strategies for assembly of nanocrystals have been proposed, with the final aims to fabricate new materials in which the single-particle properties are amplified and/or in which chemical and physical interactions among nanocrystals can give rise to novel phenomena, or to develop nanoscale devices whose behavior is dictated by the properties of individual nanocrystals. In this chapter, we will first highlight briefly the most relevant physical properties of nanocrystals, then we will describe the most popular methods for the synthesis of nanocrystals in the liquid phase, and finally we will highlight the various strategies developed so far for their assembly.

Nanofabrication Handbook. Edited by Stefano Cabrini and Satoshi Kawata © 2012 CRC Press / Taylor & Francis Group, LLC. ISBN: 978-1-4200-9052-9

Chapter 12

12.2   Size-Dependent Properties of Nanocrystals

The reasons at the base of the uniqueness of nanocrystals are: (i) the ratio of surface atoms to inner atoms is much higher than that of bulk solids; (ii) the charge carrier motion is restricted to a small material volume; (iii) the number of unit cells in a single nanocrystal is small. Owing to these reasons, the chemical and physical properties often correlate strongly with the nanocrystal size and shape.[1–3] In the following, we outline the most relevant size-dependent properties:

Melting temperature: Nanocrystals melt at much lower temperatures than those required for extended solids because of the large fraction of (more reactive) surface atoms. Such behavior can be explained by the simple capillary theory of equilibrium of finite phases. The decrease in melting temperature depends approximately on the inverse of the nanocrystals radius.[4]

Phase stability: In some cases, nanocrystals can form in phases that are unstable in the bulk under equivalent conditions.[5] An explanation is that the contribution of the surface energy to the total energy of the nanocrystal formation is not any more negligible with respect to lattice energy. Therefore, in nanocrystals the crystallization in higher energy phases can be sometimes counterbalanced by the concomitant formation of lower energy surfaces.

Phase transitions: When pressure is progressively applied to a crystal, a structural phase transition can occur and the crystal becomes more stable in a more compact phase. Real crystals have a certain number of defects (point, linear, and planar). It is documented that the structural phase transitions nucleate on one or on many of these defects and then propagate along the solid.[6] Defects therefore act as catalyst for the phase transition. This mechanism contributes to significantly lower the pressure that must be applied for the phase transition to occur. Nanocrystals are usually small enough that the probability of occurrence of defects inside them is much lower than in bulk solids. Phase transitions must then occur via different mechanisms, which usually involve higher pressures.[5]

Charging energy: In an extended solid, the addition of one extra charge, that is, an electron, does not lead to any significant change in the electronic structure. This in turn does not influence the further addition of another electron. In highly confined structures, the Coulomb repulsion caused by the addition of charges leads to a charging energy.[7–9] For spherical nanocrystals, the Coulomb energy scales as $1/r$, r being the radius of the nanocrystal.[7–9]

Optical band gap in semiconductors: In semiconductors, quantum confinement sets in when the nanocrystals size becomes smaller than a critical threshold, which typically can be related to the Bohr radius of the particle (e.g., an electron) or quasi-particle (like an exciton). The quantum confinement leads to a widening of the band gap (i.e., the energy gap between the highest occupied and the lowest unoccupied energy levels) and of the level spacing at the band edges.[5–10] Therefore, the band gap in a semiconductor nanocrystal is dependent on their size and for spherical nanocrystals it roughly scales according to the inverse of the square of the nanocrystal diameter.[5,11] The specific size dependence is strongly related to the type and the shape of semiconductor nanocrystal.

Magnetic properties:[12,13] Bulk ferro- or ferrimagnetic materials are characterized by their saturation magnetization, M_S, their remanent magnetization in the absence of an external magnetic field, M_R, and their coercive field or the external field required to suppress the spontaneous magnetization of the material, H_C. In particles with sizes around hundreds of nanometers, the inversion of the magnetic moment still occurs by the displacement of the magnetic domain walls, since they are multimagnetic domain particles. Such a process requires small amounts of energy and consequently leads to low coercive field values. At smaller sizes, the domain wall formation is no longer an energetically convenient process, and nanoparticles tend to form single magnetic domains where the switching of the magnetic dipole occurs by overcoming of the anisotropy energy barrier, $\Delta E = K_{eff} V$, where K_{eff} is the effective anisotropy constant of the particles and V is their magnetic volume. Single-domain particles show the highest value of coercivity. The higher the volume of single-domain nanoparticles and the anisotropy constant, the higher is the value of coercivity. When the size of the single-domain particles is further decreased, the anisotropy energy decreases in agreement with the previous equation, and lower values of coercivity are measured. A critical size can be found at which the thermal energy ($k_B T$) is high enough to easily overcome the ΔE value. Below this critical diameter value, the magnetic moments of the particles become independent from each other and they are spontaneously and continuously reversed, and the total magnetic moment of the particle is averaged to zero. Such phenomenon is called *superparamagnetism* and is characterized by the absence of coercivity (like paramagnetic species) while high magnetization values are maintained.

Optical properties of metal nanoparticles: Some metal nanostructures are well known to produce colors, due to their strong light absorption in the visible region of the spectrum. This property has been exploited in glass artworks like, for example, the rose window of the Notre Dame cathedral in Paris where silver and gold nanoparticles make the coloring of the glass. This effect is caused by surface plasmon resonances which consist of collective motions of the free electrons and depends also on the size and on the shape of the metal nanoparticles.[14–17] Other types of interactions of metal nanoparticles with the electromagnetic radiation are

i. Absorption processes due to electronic transitions of bound electrons from occupied to empty bands (inter-band transitions)

ii. Scattering processes due to accelerations of electrons in the nanostructures caused by the electromagnetic radiation

Electronic transitions in metal nanoparticles are rather insensitive to particle size (except in the case of sub-2 nm metal clusters, which are made of a few atoms), and are located at high energy (e.g., below 325 nm in silver nanoparticles). For nanoparticles with diameters between 10 and 30 nm, and in the simple case of spherical nanoparticles, a single plasmon mode of dipolar character is excited and its wavelength is independent of particle size, but is strongly dependent on the dielectric function of the surrounding medium, and also the peak width remains poorly correlated with size. For particles smaller than 10 nm, a size that is smaller than the mean free path of electrons, free electrons start colliding with the surface of the nanoparticles, and this leads to broadening of the plasmon resonance band. At sizes above 30 nm other effects become non-negligible, the so-called retardation effects, which again lead to broadening and loss in intensity of the plasmon band, as well as to a spectral shift in the red. At sizes above 100 nm, the optical properties are dominated by scattering effects. The shape of metal nanoparticles has perhaps the most striking influence on their optical properties. In metal nanoparticles with cylindrical (i.e., rod) shapes, even for small sizes two nondegenerate plasmon modes of dipolar nature are present. One mode corresponds to oscillations of the charge density along the diameter of the nanoparticles and is therefore a transverse mode, while the other one corresponds to oscillations along the elongation direction of the particle, and is therefore a longitudinal mode. The resonance condition for the transverse mode occurs for wavelengths close to those of a sphere with the same radius as the rod, while the longitudinal mode resonates at lower energies (i.e., longer wavelengths). The resonance condition for the longitudinal mode is dependent on depolarization factors, and for large rods is additionally dependent on retardation effects. In-depth coverage of size and shape effects on the interaction of metal nanoparticles with radiation can be found in many recent reviews.[17–20]

Redox and catalytic properties: The size-dependence of the electronic structure in nanocrystals provides a tool for tuning the redox potentials of the charge carriers and for modulating the dynamics of redox processes, while the dominance of electronic surface states in nanocrystals can lead to improved catalytic properties.[21,22] As an example, ultrasmall (1–2 nm diameter) nanocrystals of some noble metals have shown exceptionally high catalytic activity toward several reactions, possibly due to a combination of high surface area and specific electronic interactions with the substrate that are possible only for small metal clusters.[23]

12.3 Liquid-Phase Synthesis Routes to Nanocrystals

Synthesis routes to nanocrystals (Figure 12.1) can be classified based on the phase in which the reactions take place, which can either be the gas or the liquid or the solid phase. Liquid-phase approaches to nanoparticles, on which this chapter will be focused, can utilize two types of reactors:

- *A batch reactor* (Figure 12.2a) is a vessel with a stirrer, a heating setup, temperature controllers and meters, and valves/septa for injecting reactants and for collecting aliquots. A given amount of solution is processed in the reactor, after which this is emptied, cleaned, and prepared for another run.

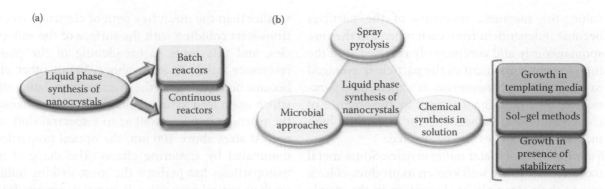

FIGURE 12.1 The various liquid-phase synthesis approaches to nanocrystals classified according to the type of reactor in which they are carried out (a) and on the type of technique, as shown in (b) (spray pyrolysis, microbial approaches, and chemical synthesis in solution).

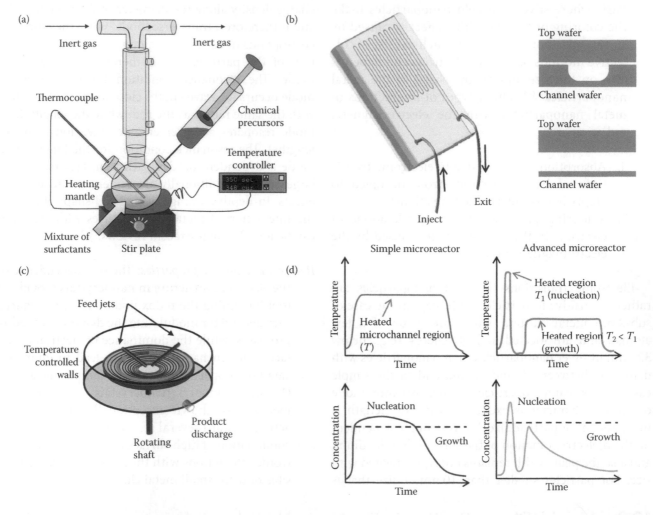

FIGURE 12.2 (a) A typical lab-scale batch reactor setup for the synthesis of nanocrystals; (b) a sketch of a simple microchannel reactor, with side view and cross-sectional view of a microchannel region; (c) a sketch of a "spinning disk processor" for the continuous synthesis of nanocrystals; (d) time dependence of the temperature of a small portion of fluid as it flows in a microchannel reactor for a simple microreactor (top left) and for an ideal microchannel reactor (top right) in which various channel regions are heated at different temperatures. The panel shows also the time dependence of the nucleation and growth stages, based on the time dependence of the concentration of active species, inside a small portion of fluid as it flows along the channel, for a simple microreactor (bottom left) and for an ideal microreactor (bottom right). (Adapted from Springer Science+Business Media: *J. Nanopart. Res.* A novel continuous microfluidic reactor design for the controlled production of high-quality semiconductor nanocrystals. *10*, 2008, 893–905, Winterton, J. D. et al.)

- A *flow reactor* (Figure 12.2b) operates instead under a flow or reactants, and yields products continuously. Syntheses in flow reactors offer potential advantages over the batch-type ones, such as higher reproducibility and the possibility to monitor and feed back the reaction conditions in real time.

The most popular techniques for the synthesis of nanoparticles in the liquid phase are

Spray pyrolysis: Spray pyrolysis is a continuous liquid-to-solid conversion process to nanocrystals: a liquid mixture of reactants containing metal salts is atomized into a furnace or a flame. Micrometer-sized droplets are formed, the solvent evaporates from them, and the chemicals in the droplet react. A powder of aggregated particles is collected.[24] The degree of polydispersity of the nanocrystals depends mainly on the distribution of the droplet sizes in the aerosol. Variations of this approach are related to how the aerosol is generated. In "thermospray," for instance, a solution of organometallic precursors is dispersed from a nebulizer in a spray of micron-sized droplets in an oven.[25] Thermospray methods yield smaller droplets (~1 μm) than common pneumatic sprays (~10 μm) and also the size distribution of the nanoparticles is narrower.[26] Nanocrystals produced by spray pyrolysis have no stabilizing agents, which can be advantageous in many applications, but they do not form stable suspensions in solution.

Microbial approaches: Various bacteria can produce and store at their interior some types of nanocrystals, and recent reports have exploited bacteria for the controlled synthesis of different types of nanomaterials.[27,28] Bacteria can be regarded as cheap and reusable bioreactors that in the future could be able to synthesize many types of nanocrystals with high selectivity. Microbial approaches are promising for the large scale synthesis of nanoparticles, with the possibility to carry out reactions at low temperatures, with little use of toxic/expensive chemicals.

Chemical synthesis in solution[11,29] includes: *Growth in templating media, sol–gel methods*, and *growth in the presence of surface stabilizing agents*. The synthesis medium can be an aqueous solution, a mixture of organic solvents and surfactants, a mixture of salts that form stable liquids (i.e., ionic liquids), or a multiphase system. All these techniques can be distinguished from each other by the way the control of size and shape of nanocrystals is achieved, by the way the particles are stabilized against aggregation, and by the type of reactor in which they are carried out. These techniques are discussed in more detail in the following.

Growth in templating media: Nanoparticles can be grown in various types of templating media. Nanowires have been fabricated, for example, by electrochemical deposition in porous membranes,[30,31] which are static templates. Micelles are instead self-organized structures that are formed for example when water and a nonpolar solvent are mixed in the presence of amphiphilic molecules. (These can be lipids, surfactants, polymers).[32,33] Various parameters regulate how the micelles are formed with sizes down to a few nanometers and with controllable shapes. Nanoparticles have been prepared by confined reactions inside micelles, and the shapes and sizes of the nanoparticles are often related to those of the micelles.[32,33]

Sol–gel methods can provide nanostructures of different materials.[34,35] A solution of chemical precursors in the form of metal alkoxides (a "sol") is hydrolyzed in the presence of water or alcohols. The alkoxides undergo condensation reactions followed by polymerization (gelation), leading to a three-dimensional network of metal-oxide bonds. Densification of the gel occurs through removal of the solvent. Depending on the composition of the sol, on its processing, on the conditions of the gelation, and on the drying process, monoliths, fibers, coatings and powders can be prepared, and different types of porous structures can be additionally fabricated.

Growth in the presence of surface stabilizing agents: This is one of the most general approaches to prepare nanocrystals. Various molecules are used as "moderators" of crystal growth, as they adhere to the surface of crystals, slowing down their growth rate. The chemical nature of these molecules depends on whether the synthesis is carried out in aqueous or in nonaqueous environments and on the nanocrystals that one wants to synthesize.[11,29] The synthesis techniques that we will discuss from now on are all variations of this approach. The energy required for the various reactions to take place is generally supplied by heating. There are, however, two general alternative approaches for supplying energy to the system, namely by sonication or by microwave irradiation. In the sonication approach, chemical reaction of metal salts or of chemical precursors is triggered by ultrasounds.[36] When ultrasounds are produced in liquids, extremely high temperatures and pressures can be produced locally, due to nucleation, growth, and collapse of cavitation bubbles. Also the use of microwave irradiation in the synthesis of nanocrystals has been increasing steadily in the last years.[37,38] The advantage of using microwave irradiation is due to the fact that one can trigger selective reactions in the system, thus many side reactions do not occur and consequently the synthesis of side products is minimized. Microwave synthesis approaches are very popular now in organic chemistry.[39]

12.4   Synthesis of Colloidal Nanocrystals in "Batch" Reactors

Hot-injection synthesis: The high-temperature thermal reaction of precursors in surfactants is a popular approach to colloidal nanoparticles in the presence of stabilizers.[11,29] Surfactants have a polar head group and one or more hydrophobic hydrocarbon chains. A mixture of surfactants or/and coordinating solvents is heated in a batch reactor under argon or nitrogen, and chemical precursors are quickly injected in it (see Figure 12.2a). The decomposition of the precursors frees reactive species, the "monomers," which induce a peak in nucleation rate, followed by the growth of these nuclei. By means of this fast injection, usually nucleation and growth are separated in time and this contributes to ensure a narrow distribution of nanocrystal sizes. The surfactants mitigate the reactivity of the monomers, helping in regulating the evolution of the nanocrystal size, and they loosely bind to the nanoparticles with their head groups, allowing for the addition/removal of chemical species to/from the nanoparticles surface. At lower temperatures instead surfactants stay bound to the nanoparticles, guaranteeing their solubility in various solvents. These can be nonpolar or moderately polar, as the surfactant-coated nanocrystals are hydrophobic. By working with specific surfactants and at high temperatures nanoparticles with narrow size distributions, with few internal defects and with a uniform coating are formed. Most hot-injection syntheses of nanocrystals originally used toxic and expensive chemicals, but recent research efforts have made this type of synthesis cheaper and less dangerous. A practical example of synthesis of colloidal nanocrystals using the hot-injection approach is reported in Box 12.1.

Using water as a solvent, the synthesis of nanocrystals in the presence of stabilizers has some limitations. In water the reaction temperatures cannot be higher than 100°C (at atmospheric pressure), which limits the degree of crystallinity, and often the size distribution is broad, plus many chemicals react with water, and so they cannot be handled in it. Water, however, is nontoxic and cheap, and nanocrystals grown in water are ready for biological applications,[41] while hydrophobic nanocrystals require treatments to become water soluble.[42,43]

A major drawback of the hot-injection reaction in a typical laboratory glass flask is that it can yield a few hundreds of milligrams of nanocrystals at best. In the case of large volumes, a sudden rise or fall of temperatures of the whole reaction mixture, or fast mixing, is hard to realize, and this limits the large-scale synthesis. These issues have been addressed in part by advances in both synthesis procedures and setups, as highlighted below and in the next section.

Slow-injection approach: In some cases, it has been found that a slow heating-up of the reaction mixture, instead of fast injection of reactants, is also capable of inducing a sharp peak in nucleation rate.[44–46] Here, a "delayed" nucleation might be due to the rise in concentration of a reactive species, which at some point reaches a critical supersaturation, triggering nucleation. As soon as nuclei are formed, further nucleation stops as the concentration of the reactive species drops below a critical threshold. This mechanism, when viable, is equally efficient in separating nucleation from growth, with the advantage of being up-scalable.

Three-phase approach: In a recently developed "liquid–solid-solution" high-yield synthesis of nanocrystals, linoleic acid, sodium linoleate, ethanol, and an aqueous solution of a metal salt are mixed in an autoclave.[47] Three phases form, namely a solid phase of sodium linoleate, a liquid phase of ethanol and linoleic acid, and a solution phase of ethanol and water containing the metal ions. At the interface between the solid sodium linoleate and the solution phase containing the metal ions, a metal

BOX 12.1   A PRACTICAL EXAMPLE OF SYNTHESIS OF COLLOIDAL NANOCRYSTALS USING THE HOT-INJECTION APPROACH: SYNTHESIS OF SPHERICAL CdSe NANOCRYSTALS WITH 3 NM DIAMETER, IN THE CUBIC SPHALERITE STRUCTURE (THE REPORTED PROCEDURE IS A MODIFICATION OF A PUBLISHED SYNTHESIS[40])

What is needed: A standard chemistry lab equipped with a properly working fume hood, inside which a Schlenk line is installed, and a nitrogen glove box (best if capable of achieving 1 ppm level of oxygen and water moisture), a temperature controller, a heating mantle, a thermocouple rod sensor, a magnetic stir plate, a centrifuge, glassware, syringes, and various chemicals.

Description of the synthesis: In a 100 mL three-neck round bottom flask, 0.48 g (0.6 mmol) of cadmium stearate (90% purity) are dissolved in 37 mL of 1-octadecene (90%). A magnetic stir bar is added to the flask. The central neck of the flask is connected via a "Liebig"-type condenser to the Schlenk line, while the other two necks are capped with rubber septa. In order to heat the flask, a heating mantle is positioned below it, and the thermocouple rod sensor is trapped between the walls of the flask and the heating mantle. Both the heating mantle and the thermocouple are connected to the temperature controller. Below the heating mantle, the magnetic stir plate is positioned and stirring is activated. The overall setup is sketched in Figure 12.2a. The flask is pumped to vacuum at 90°C for 40 min, and then the reaction mixture is cooled to room temperature. While the flask is flushed with a gentle flow of nitrogen, one of the two rubber septa is removed and 0.024 g (0.3 mmol) of Selenium powder (99.99%) are quickly added to the flask. The mixture is then degassed again under vacuum for 10 min at 50°C, after which nitrogen flow inside the flask is restored. At this point, one of the rubber septa capping one neck of the flask is punctured with the thermocouple tip up to the point that the tip region of the thermocouple is able to "fish" into the liquid mixture inside the flask. The level of immersion must be such that the thermocouple can correctly sense the temperature of the liquid while it still allows the magnetic stir bar to stir freely without hitting the thermocouple, and additionally the size of the punctured hole in the septum must be such that the septum still remains airtight after insertion of the thermocouple. After this operation, the solution is slowly heated up to 240°C under nitrogen flow. In a nitrogen glove box, a solution is prepared by mixing 0.1 mL of oleic acid (90%) with 1 mL of oleylamine (70%), and 4 mL of 1-octadecene (90%). (All solvents must have been degassed previously.) The solution is loaded into a plastic syringe equipped with an 18–20 gauge needle, and the content is added drop-wise into the reaction mixture (by puncturing with the needle the other rubber septum) 3 min after approaching 240°C, in order to stabilize the nanoparticle growth. After 1 h of reaction, the heating is stopped by removing the heating mantle, and the flask is cooled to room temperature.

Cleaning procedure: Addition of isopropanol to the solution leads to the formation of turbidity, due to the aggregation/precipitation of nanocrystals. The whole mixture is transferred to large (20 mL or bigger) vials, which are centrifuged at 3000 rpm for a few minutes. It is advisable to perform this operation by first sucking the solution using a large (30–40 mL) syringe provided with a long needle and by transferring the content of the syringe into vials having open top caps equipped with Teflon/silicone septa, under nitrogen flow. The filled vials can then be transferred in the glove box, where they can be centrifuged using a centrifuge with a brushless induction motor. By operating in such a way, and by using only anhydrous solvents, surface oxidation of the nanocrystals is prevented or at least is strongly reduced. Centrifugation accelerates the precipitation of nanocrystals and causes their sedimentation at the bottom of the vials, where they form a dark red-brown pellet, while the supernatant solution should appear colorless or pale yellow. After centrifugation, the supernatant liquid is properly discarded in a container for chemical liquid waste, and the wet precipitate (the nanocrystals) is dissolved by addition of toluene. The nanoparticles are precipitated again from this solution upon addition of methanol, the solution is centrifuged, and the supernatant is again discarded. This procedure is repeated at least twice in order to purify the nanocrystals from excess surfactants, solvents, and precursors. Finally, the "cleaned" nanocrystals can be dissolved in a nonpolar or moderately polar solvent and they can be used for experiments or for further synthesis procedures.

linoleate complex forms by ion exchange. This stays in the solid phase, while sodium ions are released in the solution phase.[47] The temperature is raised such that ethanol reduces the metal ions at the solid–liquid and solid–solution interfaces, inducing nucleation and growth of nanoparticles. These are stabilized by linoleate and are neither soluble in the liquid nor in the solution phase, and thus they precipitate out. Many types of nanocrystals can be prepared by this approach (noble metals, semiconductors), plus the synthesis conditions are mild.

12.5 Synthesis of Colloidal Nanocrystals in Continuous Reactors

We will review here two popular methods for synthesizing nanocrystals continuously, namely those based on microchannel reactors and those based on rotating disk processors. Spray pyrolysis, which is also a continuous method, has been already discussed before.

Synthesis in microchannel reactors:[48,51] The first microreactors for nanocrystal synthesis consisted of a micrometer-sized diameter channel etched inside a glass wafer, which was then thermally sealed to another glass wafer, with one hole for injection and one for extraction of the liquid, and a heating tape placed below the channel region (see Figure 12.2b). The liquid with the reactants was flown in the heated channel, inside which the nanocrystals nucleated and grew, and optical spectra were monitored on the solution exiting from the reactor.[48–51] The reasons for having a micrometer-sized channel, instead of a channel of larger section, can be understood easily: in a channel regions of liquid (hence particles) closer to the walls flow slower than regions (hence particles) closer to the center, and therefore the former spend more than the latter in the channel, so they grow bigger, contributing to spreading of the size distribution over time. In a microchannel, this is counterbalanced efficiently by particles diffusing from the regions near the walls to the central regions and back. The nanoparticle velocities can be homogenized further by entrapping liquid droplets of the solution in an inert carrier fluid, with which the solution does not mix: each droplet is now isolated from the others as it flows along the channel and the velocity profile of the fluid in each droplet follows a recirculating pattern. On average all the particles experience the same residence time in the channel, and their size distribution stays narrow.[52]

A major cause of particle size broadening inside microchannels is due to the difficulty of separating nucleation and growth events in it:[52] in a uniformly heated microchannel reactor, when the fluid enters the heated region of the channel, it reaches thermal equilibrium rapidly and stays at this temperature throughout its journey inside the channel (see Figure 12.2d, left plots). This temperature must be high enough to induce nucleation, so that the concentration of reactive species overcomes the nucleation threshold soon after the liquid enters the channel, but it remains above such threshold long enough for further nucleation events to coexist with growth of particles that had nucleated at the channel entry, causing substantial spread in nanoparticle size distribution. In an optimal microreactor (Figure 12.2, right plots), the nucleation of the particles must occur only in the initial section of the channel, which therefore must be heated at a relatively higher temperature. This section must be followed by a region where the liquid is cooled, then by other regions in which the liquid is heated at an intermediate temperature that ensures growth of the nuclei but inhibits further nucleation. Winterton et al.[52] have recently proposed designs of an optimal microreactor, and have built a simplified version of it, with encouraging results.

Synthesis in spinning disk processors:[53,54] The spinning disk processor consists of a heated spinning disk connected via a rotating shaft to a heat exchanger and a rotor, and is encased in a circular chamber with temperature-controlled walls (see Figure 12.2c). Feed jets on top of the disk deliver the solution of chemicals at the center of the disk. The disk rotation generates centrifugal forces that pull the liquid to the edge of the disk. Two zones exist on the disk: (i) a central zone where the fluid is dragged away from the center of the disk by centrifugal forces; (ii) an outer zone where the film

feels both a radial flow due to centrifugal forces and a tangential flow due to viscous drag by the spinning surface of the disk. The combination of viscous drag between the moving fluid and the disk surface and the centrifugal forces leads to turbulences contributing to efficient mixing in the fluid layer. Nucleation and growth take place while the fluid is dragged from the center of the disk to its edge. After the fluid is ejected to the walls, it percolates to the base of the chamber. The formation of a thin film of fluid guarantees a fast heat transfer from the disk to the fluid, and both spinning speed and disk temperature govern particle size and size distributions. The efficient mixing guarantees homogeneous reaction conditions, and synthesis scale-up is possible. The temperature of the walls and that of the disk can be tuned independently, and therefore the fluid can be suddenly cooled when hitting the walls. Due to the short residence time of the fluid on the disk, this can be heated at high temperatures, allowing many types of chemical reactions.

12.6   Shape Control of Colloidal Nanocrystals

In nanocrystals, the physical properties depend also on their shape.[2,55] For example: (i) in semiconductor nanocrystals the transition from a sphere to a rod involves drastic changes in the quantum confinement of carriers along the elongation direction. This strongly influences optical and electronic features of the nanocrystal, such as electronic level crossing, degree and type of polarization of the emitted light, and fluorescence quantum yield; (ii) in metal nanocrystals, the same shape transition leads to a splitting of the plasmon mode into longitudinal and transverse plasmon modes, as described earlier in this chapter; (iii) magnetic properties are strongly dependent on shape, especially if the direction of elongation coincides with that of the easy magnetization. Apart from intrinsic physical properties, the types of assemblies that one can achieve with nanocrystals, and their possible applications (e.g., in novel nanocomposite materials and devices), are strongly related to the shape of the individual nanocrystal building blocks.

Most effective strategies to synthesize nanocrystals with controlled shapes are (i) reactions in static templates (i.e., porous membranes) or in dynamic templates (i.e., micelles). This has already been discussed above; (ii) growth in the presence of chemical

additives that alter the growth rates of the various facets;[56-60] (iii) solution–liquid–solid (SLS) growth in the presence of a metal particle acting as a catalyst;[61-63] (iv) oriented attachment;[64,67] (v) seeded growth;[3,68,69] (vi) growth based on the Kirkendall effect or on a template-mediated approach for hollow-shaped nanocrystals (see also section on core–shell nanocrystals);[70-73] (vii) shape control mediated by planar defects.[74,75] We will not discuss here the approach related to the growth in templates, as this was briefly highlighted earlier. Sketches of these approaches are shown in Figure 12.3a–g.

Growth in the presence of chemical additives: The various facets in a (nano)crystal have different arrangements and densities of atoms, polarity, and number of surface-broken bonds. Chemical species present in solution, sometimes even at trace levels, can bind selectively to some of these facets, in a few cases with such a strength that the growth of these facets is almost suppressed.[56-60] The nanoparticles, in the presence of selective adsorbers, can evolve into well-faceted shapes, as they will tend to eliminate their faster growth-rate facets. Selective adhesion can yield anisotropic shapes (rods, discs, platelets) if the nanoparticles crystallize in phases with a unique

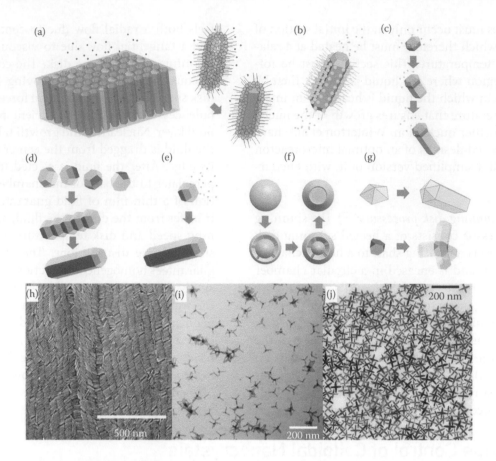

FIGURE 12.3 Various strategies for the synthesis of shape-controlled nanocrystals along with some examples of final products: (a) synthesis in templates, either static-like porous membranes (left) or dynamic-like micelles (right); (b) growth in the presence of chemical additives capable of selective adhesion; (c) solution–liquid–solid approach; (d) oriented attachment; (e) seeded growth; (f) formation of hollow particles via the Kirkendall effect; (g) shape control via formation of planar defects; (h) a scanning electron microscope (SEM) image of CdS nanorods laying flat on a substrate (GaAs), and laterally ordered on it; (i) a transmission electron microscope (TEM) image of CdTe tetrapods; (j) a TEM image of CdSe/CdS octapods. (All these materials were synthesized in our group.)

axis of symmetry. The latter will be either the fast- or the slow-growth direction, depending on the relative growth rates of the many facets involved. Examples of phases with a unique axis of symmetry are the hexagonal close-packed structure of Co,[76] the hexagonal wurtzite structure of most II–VI semiconductors,[77,78] the hexagonal hematite structure in Fe_2O_3,[79] and the tetragonal anatase structure for TiO_2.[80] Also multipods can be explained on the basis of the depression/enhancement in the growth rate of different facets due to various molecules present in the growth environment.[81]

Solution–liquid–solid growth approaches: In the vapor–liquid–solid (VLS) growth mechanism, a nanowire can be grown from a catalyst particle placed on a substrate.[82,83] This method has been extended to the SLS growth of colloidal nanorods and nanowires in the liquid phase (CdSe, InAs, InP, Si, Ge[61–63]). In most cases, the solution containing the precursors is additionally loaded with colloidal metal nanoparticles that act as catalysts. These become supersaturated with the monomers and act as a preferential site for nucleation and growth of nanoparticles. As these evolve usually only from one side of the metal particle, the growth of nanorods/nanowires is observed. When growth occurs from multiple sides, flower-like or multi-branched structures are observed.

Oriented attachment: In some cases, nanocrystals are first formed in solution, but then they spontaneously attach to each other through some specific sets of facets, either because these facets are less efficiently passivated by stabilizing molecules, or because they are high-energy facets.[64–66,84] The assembly is often induced by dipole–dipole interactions between particles. The final nanoparticles can have various shapes (wires,[84] rings,[65] rods,[66] multipods[64]). An advantage of this method is that even nanocrystals of structures with nonunique crystallographic directions (i.e., the rock-salt structure[65] or the sphalerite structure[66]) can develop anisotropic shapes.

Seeded growth: It is possible to achieve anisotropic growth of noble metal materials (Ag, Au, Pt) and semiconductors using a seed-mediated approach,[3,68,69] by which preformed nanocrystal seeds are mixed with reactants and eventually surfactants that are facet-selective absorbers. The seeds act as redox catalysts for metal ion reduction, so that their further growth is preferred over homogenous nucleation of additional particles. The process leads to nanorods, nanowires, and also branched nanostructures.[69] It was recently demonstrated that starting from cubic seeds, these could be enlarged either to isotropic cuboctahedron or to nanorods,[85] depending on the lattice mismatch between the seed and material grown on top of it (cuboctahedron in the case of small lattice mismatch, nanorods in the case of large lattice mismatch).

Growth based on the Kirkendall effect: Differences in the diffusion rate of two species across the interface between two bulk materials are known to lead to the formation of voids (the so-called Kirkendall effect).[70–72] This effect has been exploited recently in the formation of hollow nanoparticles, starting from spherical nanocrystals of a given material (for instance cobalt or iron). These were then reacted with sulfur or oxygen. Since the diffusion rate of sulfur (oxygen) inward is faster than that of Cobalt (iron, zinc) outward, CoS_2, Fe_2O_3, or ZnO hollow nanoparticles were formed. Another approach for preparing hollow nanoparticles relies on coating a sacrificial template with a desired material, followed by selective dissolution of the template.[73] For a review covering, many approaches to hollow nanoparticles (in addition to the Kirkendall effect) and their applications see for example the recent work of Hyeon.[73]

Shape control mediated by planar defects: Planar defects, such as twin planes and stacking faults, can also lead to the formation of nanoparticles with peculiar shapes. In many metal nanoparticles, penta-twinned nanocrystals are often formed

(Figure 12.3g, top sketches).[74] These are in practice mosaic structures made of five face-centered cubic (fcc) domains. Each domain shares with its two adjacent domains a 111 facet, and each interface between two domains is a twin plane. In this specific (and simple) case, this plane is a mirror plane of symmetry for the two domains that are joined through it. (We will not go into the details of how and why such twinned structures form.) These twinned particles can have shapes that range from platelets, spheres to elongated structure like rods. Many other twinned geometries have also been observed.[74] Also in these structures it is important to note that a nanocrystalline material that apparently crystallizes in a highly symmetric crystal structure (cubic fcc) can still evolve into an anisotropic shape.

Of relevant interest is also the formation of peculiar shapes arising from creation of stacking faults during the growth of nanocrystals. We mention here one of the most studied examples. For many semiconductors, both crystallization in the cubic sphalerite phase or in the hexagonal wurtzite phase is possible (one of these materials is, for instance, CdTe), and the two phases are often energetically very similar.[86] In the case of nanocrystals, it was found that by tuning the reaction kinetics in a careful way it is possible to achieve nucleation and initial growth of the nanocrystals in the sphalerite phase, while further growth occurs in the wurtzite phase, by continuation of growth on some facets of the initially formed nuclei.[75–76] In the tetrapod shape,[76] which arises from this type of "switching" during the growth (see sketches of Figure 12.3g and TEM image of Figure 12.3i), the central "branching" region corresponds to the initially formed sphalerite nucleus, while the four arms have wurtzite structure. The planes that separate the cubic region from the sphalerite region are "stacking faults." This approach has been refined over the years, and tetrapods with uniform shapes can now be prepared regularly. Another interesting shape is the octapod (see Figure 12.3j), in which eight arms evolve from a starting octahedral-shaped nucleus.[77]

12.7 Multimaterial Colloidal Nanocrystals

A further development in the liquid-phase synthesis of nanocrystals involves the fabrication of more elaborate nanocrystals comprising crystalline domains made of different materials joined together.[87,88] The main drive for increasing the degree of structural and compositional complexity in nanocrystals is to group in a single particle the chemical–physical properties that are characteristic of each material, but also to modulate

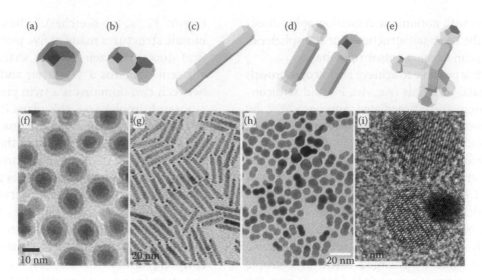

FIGURE 12.4   Top row: sketches of the various types of multimaterial nanocrystals. (a) core–shell; (b) heterodimer; (c) nanorod "barcode"; (d) dumbbells and matchsticks; (e) branched nanocrystals with decorated tips. Bottom row: TEM images of typical multimaterial nanocrystals (All syntheses were performed in our group.): (f) spherical Au(core)/Fe_3O_4 (shell nanocrystals); (g) CdS nanorods with Au domains at their tips; (h) $CoPt_3$–Au heterodimers; (i) FePt/Fe_3O_4 heterodimers.

these properties due to the mutual interactions between the proximal domains. We start by reviewing the most common geometry of multimaterial nanocrystals, which is the core–shell one (Figure 12.4).

Core–shell nanocrystals: In this configuration, a "core" of a given material is covered by a shell of another material. The materials composing the core and the shell can be semiconductors, metals, insulators, and belong to various classes of chemicals. The most common reasons for growing a core–shell or a multishell nanocrystal are

- To enhance or tune the plasmon absorption characteristics, hence the color of the colloid, if one or all the materials involved are noble metals.[89,90]
- To enhance or modulate the fluorescence emission from the nanocrystal, if the materials involved are semiconductors.[91–94] Examples of core–shell nanocrystals in which the fluorescence of the core material is greatly enhanced are CdSe(core)/ZnS(shell) and CdSe(core)/ CdS(shell), or the double-shell CdSe/CdS/ZnS nanocrystals. In all these cases, the carriers are strongly confined in the core material and forced to recombine radiatively from there. Typical examples in which the fluorescence emission of the core–shell structure is instead significantly shifted to the red (infrared) with

respect to that of both the core and the shell materials are CdSe/CdTe and ZnTe/ZnSe.[95] In these cases, the carriers are separated at the core–shell interface, due to the offsets in the relative valence and conduction bands of the two materials, and their recombination leads to emission of light at energies that are comparatively lower than the band gaps of both core and shell materials.[96]

- To modulate the magnetic properties of the nanocrystal, if one or more materials involved have a specific magnetic behavior.[97–99]
- To improve the photocatalytic and/or photoelectrochemical response of the starting material.[100–102]
- To provide a shell material to which molecules can be attached easily, or simply which protects the core material from degradation.[103]
- To synthesize a multifunctional nanoparticle, for example, a magnetic-fluorescent or a magnetic-plasmonic nanocrystal. Typical examples are magnetic Co nanocrystals coated with a fluorescent CdSe shell[99] or iron oxide (core)/ Au(shell) nanocrystals.[104]

When the core and the shell materials have the same crystal structure and their differences in lattice parameters are small, the shell growth around the core might proceed uniformly and few defects are formed at the core–shell interface, if the shell is only a few monolayers thick. The latter is found for example in core–shell nanocrystals of combinations of semiconductors like CdSe/CdS, CdSe/ZnSe, InAs/CdSe, CdS/InP, and similar other combinations. If the two materials have the same crystal structure, but differ considerably from each other in lattice parameters (i.e., above 5%), the shell still grows epitaxially on the core, but it develops many crystal defects (dislocations, for example) passed a few monolayer thickness, due to strain relaxation. Examples are CdSe/ZnS, CdS/CdTe, ZnTe/CdTe, and others. If, on the other hand, the two materials have completely different crystal structures, many scenarios are possible. In some cases, the second material can still form a shell on top of the core material, but it will be "patchy," that is, it will be a multicrystalline shell formed by various domains with different crystallographic orientations. Alternatively, the shell might not cover uniformly the core, but will grow as a separate domain, sharing with the core material only a limited set of interfaces. This case will be highlighted later on.

In core–shell nanocrystals, there is generally a concentric arrangement of the core and of the shell(s). Examples are spherical or rod-shaped cores onto which a uniform shell layer is grown, so that the thickness of the shell is more or less the same on all the facets of the core. Recent works on some systems (e.g., CdSe/CdS, ZnSe/CdS nanocrystals, and also metal/metal nanocrystals) have shown that the shell can elongate preferentially along some crystallographic directions, such that a rod-shaped shell could be grown around spherical or cubic nanocrystals.[85,105,106] A practical example of a synthesis of strongly fluorescent rod-shaped CdS nanorods, with narrow distributions or rod lengths and diameters, and encasing at their interior a spherical "core" made of CdSe, is described in Box 12.2.

The most common synthetic approaches to core–shell nanocrystals are

1. To a solution containing preformed nanocrystals, a solution of precursors of the shell materials is injected. The conditions must be carefully chosen such that the homogeneous nucleation of separate nanocrystals of this shell material is prevented, and instead the heterogeneous nucleation of the shell material on top of the core is the preferred process. In some cases the initial shell is amorphous or discontinuous, but it might be turned into a crystalline and uniform shell on thermal annealing.[99] A more elaborate approach to control the shell growth consists of alternating deposition of monolayers of each atomic species that will compose the shell material.[110]

2. In some cases, all the necessary molecular precursors required to grow the core–shell nanocrystals can be mixed at once and coreacted.[111,112] This strategy is viable whenever there is a large difference in reactivity among the various precursors, and hence some of them decompose first and form nanocrystals of a given material, upon which the shell material grows via delayed decomposition/reaction of the less reactive precursors.

3. Another strategy is to replace the outermost layer of the core with the shell material by means of a redox reaction, in which the material is oxidized and released in solution and is replaced by another material,[89] or by means of an ion exchange process.[113] In both cases, the initial core shrinks in size.

Nanocrystals made of sections of different materials: In the other possible arrangement of multimaterial nanocrystals, two or more nanoscale domains are connected to each other but no domain completely covers the other(s). The most common examples are the hetero-dimer geometry made of two roughly spherical nanodomains of two different materials sharing a small interface, or a more elaborate geometry based on multiple rod/spherical sections. The latter can involve, as most frequently developed structures, the following ones: (i) a nanorod made of linear sections of different materials;[114,115] (ii) a branched nanostructure whose branches are nanorods of different materials;[114] (iii) a dumbbell or matchstick in which a nanorod is decorated at one or at both tips with roughly spherical domains of another material;[116,117] (iv) branched nanostructures, for example, tetrapods, in which the apexes are decorated with spherical domains of another material.[118]

In analogy with the previously discussed case of core–shell nanocrystals, also in these types of nanostructures the properties inherent to each domain, such as magnetism, optical absorption, or fluorescence, can be altered substantially because of the contact junctions. The major peculiarity of these nanostructures with respect to the core–shell ones is that each domain, by being not buried completely in another material, can be accessed from the external world. A few examples

BOX 12.2   A PRACTICAL EXAMPLE OF SYNTHESIS OF STRONGLY FLUORESCENT ROD-SHAPED COLLOIDAL CORE–SHELL CdSe/CdS NANOCRYSTALS. THE REPORTED PROCEDURE HAS BEEN REPORTED BY US IN A PUBLISHED WORK.[107]

What is needed: The same setups as described in Box 12.1, but a different set of chemicals. The procedures are described more succinctly here, as it is preferable for the reader to perform first the synthesis described in Box 12.1.

Synthesis of 2.9 nm CdSe spherical nanocrystals, with wurtzite crystal structure: 3.0 g of trioctylphosphine oxide (TOPO, 99%), 0.280 g of octadecylphosphonic acid (ODPA, 99%), and 0.060 g of cadmium oxide (CdO) are mixed in a 50 mL 3-neck round-bottom flask, heated to ca. 150°C and degassed under vacuum for approximately 1 h. Then, under nitrogen, the solution is heated to 300°C and kept at this temperature for several minutes in order to dissolve CdO completely. (The reaction mixture turns optically clear and colorless.) At this point, 1.5 g of trioctylphosphine (TOP, 97%) are injected in the flask and the reaction mixture is heated up to 350°C. As soon as this temperature is stabilized (350 ± 1°C), a solution of TOP:selenium (prepared by dissolving 0.058 g of selenium in 0.360 g of TOP) is swiftly injected into the flask and the heating mantle is removed immediately. (Usually it takes a few seconds to do so.) The mixture is then left to cool down at room temperature under stirring. As-prepared CdSe nanocrystals are precipitated with addition of methanol and washed at least three times by repeated dissolution/precipitation with toluene and methanol and followed by centrifugation (3000–4000 rpm/4–5 min). Before the last precipitation, the solution of nanocrystals dissolved in toluene is filtered through a hydrophobic PTFE syringe filter with 0.2 μm pore size. Thoroughly washed and filtered CdSe nanocrystals are precipitated by addition of methanol again, centrifuged, dried under vacuum, and redispersed in 1 mL of TOP. The concentration of CdSe nanocrystals in the resulting TOP solution (referred henceforth to as "CdSe/TOP stock solution") is estimated as ≈200 μM.[108,109]

Synthesis of 6 nm diameter × 24 nm length CdSe(sphere)/CdS(rod) core–shell nanorods. 3.0 g of TOPO, 0.290 g of ODPA, 0.080 g of hexylphosphonic acid (HPA,99%), and 0.057 g of cadmium oxide (CdO, 99.5%) are mixed in a 50 mL 3-neck round-bottom flask and the mixture is heated to 350°C following the same procedure as described above for preparing the CdSe nanocrystals. The stock solution for growing core–shell nanorods is prepared by mixing a TOP:sulfur solution (prepared by dissolving 0.120 g of sulfur in 1.5 g of TOP) with 200 μL of the CdSe/TOP stock solution. As soon as the temperature of the mixture in the flask is stabilized at 350°C ± 1°C, the stock solution is swiftly injected into it. After injection, the temperature in the flask drops to 280–300°C and recovers in about 2 min. to the initial value of 350°C. The total growth time for the nanorods is 8 min after injection, after which the heating mantle is removed, and the mixture is cooled down to room temperature under stirring. The purification procedure for these nanorods is similar to that for the CdSe seeds described above, except that in this case the solution is not filtered.

highlighting the relevance and usefulness of this general arrangement are the following:

1. In semiconductor–semiconductor heterostructures, depending on the relative band gap alignment of the components, the charge carriers can be either localized preferentially in one domain or each of them is localized in a different domain, which can have important implications in optoelectronic and photovoltaic applications, since each of these carriers can be extracted (injected) from (into) these domains;

2. In metal–semiconductor heterostructures (for instance, CdSe rods with Au tips[117]), electron transfer to the metal domain can be exploited to perform spatially separated redox reactions (i.e., reduction at the metal domain, oxidation at the semiconductor domain[119,120]); on these types of heterostructures, the surface of each domain can be selectively functionalized by a specific type of molecule,[121] which can also be useful to achieve controlled assembly of the nanocrystal (see later).[122]

Again, there are various approaches that one can exploit to grow these types of nanocrystals. The most common ones are

1. Growth of a second material domain on a preformed seed. Sometimes, this second material does not form a shell at all, but nucleates only on specific locations on the starting nanocrystal. In some cases, the second material grows as an amorphous shell around the initial core, but following an annealing process this shell coalesces into one or more separate spherical grains attached at one side of the initial nanocrystal. Typical examples are FePt–CdS[123] dimers and γ-Fe$_2$O$_3$–MeS (where Me = Zn, Cd, Hg)[123,124] oligomers, Au–Fe$_3$O$_4$ dimer- and flower-like nanostructures,[125] and Au–CoPt$_3$ nanocrystals.[126]

2. *One-pot synthesis*: Simultaneous injection of all the chemical precursors can lead to the direct formation of dimer-like structures. Examples of this type of mechanism are Co–Pd and Cu–In sulfide heterodimers, which are prepared by coreaction of their respective molecular precursors.[127,128] In these cases, a selective nucleation of one material is followed by growth continuation of the second material on top of it as in the core-shell case discussed before, but now the second material does not form a shell. Another example is the aqueous synthesis of Ag–Se nanocrystals in which sequential reduction of the Ag and Se ions leads to the initial nucleation of Ag onto which the reduction of Se ions follows.[129] Depending on the reaction conditions, the nanocrystals are characterized either by a single Se domain or by more Se domains attached to each Ag nucleus.

The formation of these nanostructures, in which the interface shared by the various domains is minimal as compared to the core–shell geometry, can be driven by various mechanisms. One of the mechanisms could be just the minimization of interfacial energy, since in these nanocrystals either there is significant difference in lattice parameters between the two materials, or their crystal structures are different. Hence, the interface(s) shared by the two domains can correspond the thermodynamically most stable one(s), for instance, because the mutual orientations of the two material grains along this (these) specific interface(s) are such that there is the lowest possible lattice mismatch, or they result in a minimal number of broken/distorted bonds at the interface(s). Especially when attempting to grow a metal shell around a core of another material, this shell often coalesces to a separate domain. One indeed has to remember that in all these types of growth two energy terms play key roles, namely the interfacial energy and the cohesive energy. Suppose that the cohesive energy of the material that we wish to grow as a shell is high (like in the case of metals). Then it is very likely that this material will not grow as a shell but will rather form a separate domain.

In many cases, however, the formation of these "noncore–shell" nanostructures is dictated by kinetic reasons. For example, a material nucleates only on specific sites on the surface of a core nanocrystal because these are the most reactive ones under the given reaction conditions (i.e., there is a bigger number of unpassivated dangling bonds or they are coated less efficiently by stabilizer molecules, like the tips of wurtzite rod-shaped nanocrystals of CdSe or CdS[130]). Also, often further enlargement of these nucleated domains is preferred over nucleation of additional domains on other regions of the core surface, since the energetic barrier for growth is lower than that for nucleation. Often, a combination of the two mechanisms is operative.

Another novel approach to nanocrystals made of sections of different materials is represented by cation exchange, by which the sublattice of cations in ionic nanocrystals can be partially or completely exchanged with a new sublattice of cations, while the sublattice of anions remains basically unaltered.[131–134] In general, this method yields nanocrystals of a new material but it preserves the size/shape of the starting nanocrystals, and it was shown that novel heterostructures can be formed via partial cation exchange, for example, as a result of minimization of lattice strain. An example is represented by striped CdS–Ag$_2$S nanorods.[132]

12.8 Assembly of Colloidal Nanocrystals

Introduction: After having discussed extensively on the synthetic approaches to colloidal nanocrystals and on the different elaborate nanocrystals that are accessible nowadays, this section will describe the various strategies developed so far for assembling nanocrystals on surfaces or in device structures, both as individual nanoparticles and as nanoparticle assemblies. The purposes for doing this are disparate. In contacting individual nanoparticles to electrode structures, the main aim is to exploit the single nanocrystal properties in nanoscale electronic and optoelectronic devices (such as single electron transistors). In realizing controlled assemblies of nanocrystals on substrates, on the other hand, the aims are multiple, and among them:

- To amplify the single nanocrystal properties or to tune nanocrystal properties due to proximity effects (if the nanocrystals are arranged in a close-packed geometry)
- To provide a substrate patterned with an ordered 2D array of nanocrystals, such that molecules can be anchored preferentially to these nanocrystals
- To fabricate multilayer films that can be used in field effect transistors or in photovoltaic cells.
- To create three-dimensional superlattices of one type or of multiple types of nanocrystals, behaving as artificial solids with novel engineered properties
- To create chain- and network-like structures by head-to-tail assembly of rod-shaped or branched nanocrystals that can have controlled porosity and novel mechanical properties
- To assemble a discrete number of nanocrystals with both positional and orientational order on a substrate (for instance, tetrapod-shaped nanocrystals that touch a flat substrate with three arms and with the fourth arm pointing upward), for uses that can range from nanocrystal-functionalized atomic force microscopy (AFM) tips to field emitters

We will start this section by discussing approaches for the controlled positioning of individual nanocrystals, and then we will target nanocrystal assemblies.

Assembly of individual nanocrystals for single-crystal functional optoelectronic devices: Individual metal and semiconductor nanocrystals can function as single-electron transistors (SETs)[7,135,136] or as light-emitting elements in electronic circuits.[137] For SET functionality, the nanocrystal needs to be connected to source and drain electrodes by tunnel junctions and to a third gate electrode via a thin isolating layer. The Coulomb charging energy required to add an extra charge on the nanocrystal leads to discrete energy levels that can be tuned with respect to the potentials of the source and drain electrodes via the gate electrode. The main challenge in this respect is to fabricate electrodes of compatible size and to control the electrical contacts to the functional nanoparticle. Here the type of contact between semiconductor nanoparticles and metal electrodes is of particular interest. Fabrication of an all-inorganic interface can lead to either an ohmic or a Schottky type of contact, and the challenges are on how to achieve control over this process.[138–142] Other contact schemes comprise linking the nanoparticles via organic molecules to the metal electrodes, or positioning them via external electric fields into the desired positions. In both cases, the presence of organic molecules in the contact leads to a rather large tunnel resistance (typically larger than 10 MΩ).[143,144] In the following, we will discuss some of the most relevant applications of nanocrystals as functional elements and we will describe the related schemes for electrode fabrication with junctions having gaps smaller than 50 nm.

EBL-based devices with electrode pattern alignment: The most straightforward method to fabricate electrode patterns with nanoscale resolution is by electron beam lithography (EBL). For details on this technology, please see the corresponding chapter of this handbook. By direct EBL patterning of the electrodes, junctions with gaps down to 20 nm can be achieved. This approach is, however, very costly, work intensive, and has a low throughput because it is based on a sequential fabrication technique. Furthermore, if the goal is to contact a specific particle on a given substrate, it additionally poses the difficulty of precise alignment of the electrode pattern with respect to the nanoparticle position and orientation (especially for anisotropic-shaped nanocrystals). Some examples of how these challenges have been tackled are briefly reviewed below (Figure 12.5).

Alivisatos et al. reported the successful fabrication of three terminal contacts to CdTe tetrapods (arm lengths around 100 nm) and demonstrated single-electron transistor functionality controlled either via a planar back gate, or in some cases via one of the tetrapod arms that showed a higher contact resistance.[145] The group was also able to show transistor action of single CdSe and CdTe nanorods.[139] Although the tetrapods used in those

experiments were made of semiconducting materials, the zero-current voltage range could be attributed to Coulomb blockade, and therefore to charging of a small conductive island.[146] The commonly accepted explanation for this finding is that the band alignment between the metal electrodes and the semiconductor nanocrystal is such that the Fermi level gets pinned within the valence band of the semiconductor material, whose dense-level

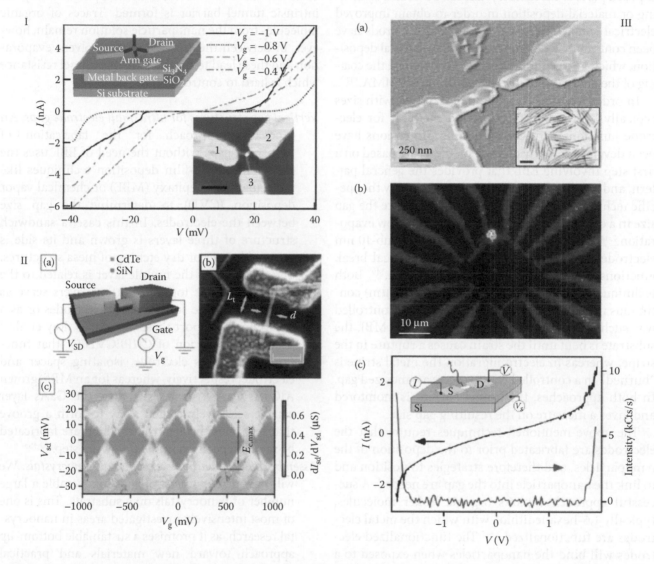

FIGURE 12.5 (See color insert.) (I) A single CdTe tetrapod contacted by three lateral electrodes and a metal back gate. The I–V curves show source drain current of two lateral electrodes, demonstrating Coulomb blockade, for different back gate voltages. The scale bar in the inset is 100 nm. (Reprinted with permission from Cui, Y. et al., Electrical transport through a single nanoscale semiconductor branch point. *Nano Lett.* 2005, 5, 1519–1523. Copyright (2005) American Chemical Society.) (II) SET device based on a single CdTe rod shown as schematic illustration (a) and by a scanning electron microscopy image (b) with scale bar corresponding to 100 nm. The characteristic Coulomb blockade diamond structure is shown in II (c). (Reprinted with permission from Trudeau, P. E. et al., Electrical contacts to individual colloidal semiconductor nanorods. *Nano Lett.* 2008, 8, 1936–1939. Copyright (2008) American Chemical Society.) (III) Single CdSe rod that shows electroluminescence. (a) Scanning electron microscopy image of the device (false color). The inset shows transmission electron microscopy images of the nanorods, the scale bar is 50 nm. (b) Optical microscope image showing electroluminescence from a nanorod transistor. (c) Current (black) and simultaneously measured electroluminescence plotted against bias voltage. The inset illustrates the device architecture with source and drain electrodes contacting a single nanorod. (Reprinted with permission from Gudiksen, M. S. et al., Electroluminescence from a single-nanocrystal transistor. *Nano Lett.* 2005, 5, 2257–2261. Copyright (2005) American Chemical Society.)

structure is not resolved.[147] Park et al. found that single CdSe nanorods contacted by EBL and metal evaporation showed electroluminescence when the bias voltage overcame the band gap energy.[137] One advantage of the technique involving EBL and subsequent metal evaporation after the nanocrystals have been deposited on the substrate is that it allows for modifications of the metal semiconductor contact area. For example, after EBL, the exposed parts of the nanocrystals can be treated by etching or material deposition in order to obtain improved electrical contact. Recently, single CdSe nanorods have been contacted by electron-beam-induced metal deposition, which is a technique that does not require the coating of the nanocrystal-covered substrate by PMMA.[142]

In order to contact smaller nanocrystals, with sizes typically below 10 nm, fabrication methods for electrode junctions with corresponding dimensions have been developed. Many of these methods are based on a first step involving EBL that provides the general pattern, and are distinguished from each other by the specific technique that is then exploited to reduce the gap size in a controlled way, for example, by shadow evaporation.[7, 147] Other fabrication approaches for sub-10 nm electrode gaps are based either on mechanical break junctions (MBJ)[148] or on electromigration.[149] Both techniques start from a narrow (typically <1 μm) continuous metal stripe which is damaged in a controlled way such that a nanoscale gap is created. In MBJ, the substrate is bent until the strain causes a rupture in the stripe, whereas in electromigration the metal stripe is "burned" in a controlled way to create a nanosized gap. In both approaches, the tunnel current is monitored and gives a measure on the resulting gap size.

The above-mentioned techniques require that the electrodes are fabricated prior to the deposition of the nanoparticles, and therefore strategies to position and to link the nanoparticle into the gap are needed. A successful approach exploits organic linker molecules, typically 1,6-hexanedithiol, with which the metal electrodes are functionalized.[7,147] The functionalized electrodes will bind the nanoparticles when exposed to a solution containing the nanoparticles in suspension. Most nanoparticles will be bound on the surface of the metal electrodes, but some will bridge the electrode gap. The critical parameters involved in the control of the number of particles in the gap are gap size, thiol coverage, and nanoparticle concentration. The electrical contact is formed via the thiol molecules, which act as a tunnel barrier. The tunnel resistance can be tuned in a certain range by varying the length of the hydrocarbon chain of the dithiol molecules[143] (Figure 12.6).

An alternative method for positioning nanoparticles in between electrode junctions is electrostatic trapping or dielectrophoresis.[150–153] In this case, an external electric field is applied while the electrodes are exposed to the nanoparticles suspended in solution. Via dipole–dipole interaction, the nanoparticles get polarized and are attracted toward the region of strongest electric field, that is, the electrode gap. Since no linker molecules are used in electrostatic trapping, ideally no intrinsic tunnel barrier is formed. Traces of organic molecules from the nanoparticle solution remain, however, in between the electrodes after solvent evaporation and these lead to the creation of a tunnel resistance which is hard to control (Figure 12.7).

Vertical gap structures for ultrasmall electrode gaps: An innovative approach for the fabrication of ultrasmall gaps without the need of EBL uses the precision of thin-film deposition techniques like molecular beam epitaxy (MBE) or chemical vapor deposition (CVD) to determine the gap size between the electrodes. In this case, a sandwich structure of three layers is grown and its side is exposed by wet or dry etching of mesa structures. The thickness of the middle layer is related to the gap size and the top and bottom layers serve as substrates for the source–drain electrodes or as a substrate to support these electrodes. Ray et al.[154] used a combination of Cr/PECVD/Cr that functions directly as electrode, isolating spacer and electrode, respectively, whereas for an MBE-grown AlGaAs/GaAs/AlGaAs structure the GaAs layer can be selectively etched away to form a groove that separates the electrodes which are fabricated by metal deposition in the following step.[152,155]

Assembly of large numbers of colloidal nanocrystals: We will now discuss the strategies to assemble a large number of nanocrystals on a substrate. This is one of most intensively investigated areas in nanocrystal research, as it promises a sustainable bottom-up approach toward new materials and practical device applications in many fields. Especially for the case of spherical nanocrystals, substantial progress has been achieved regarding their controlled assembly into ordered superstructures over large areas. Examples include the preparation of long-range-ordered superlattices of nearly mono-disperse spherical nanocrystals, and the self-assembly of combinations of spherical nanocrystals of different sizes and materials in binary or ternary superlattices.[156,157] The topic of self-assembly of mainly

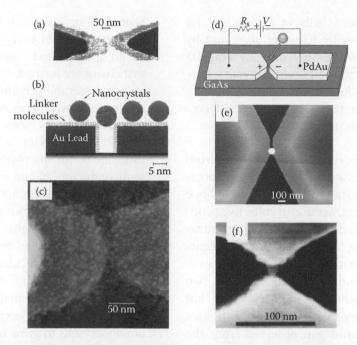

FIGURE 12.6   (a) The electrode gap size defined by EBL is reduced by shadow evaporation of the gold from +/– 15° angles of the normal. (b) Schematic illustration of the single nanocrystal device in which the nanocrystal is bound by linker molecules. (Reprinted with permission from Klein, D. L. et al., An approach to electrical studies of single nanocrystals. *Appl. Phys. Lett.* 1996, 68, 2574–2576. Copyright (1996) American Institute of Physics.) (c) SEM image of a nanojunction bridged by nanocrystals. (Reprinted with permission from Klein, D. L. et al., *Nature* 1997, 389, 699–701. Copyright *Nature*.) (d) Illustration of the method of electrostatic trapping. (e–f) Single-metal nanoparticles (Au, Pd) positioned into a nanojunction by electrostatic trapping. ((e) Reprinted from *Phys. E 17*, Krahne, R. et al., Nanoparticles and nanogaps: Controlled positioning and fabrication. 498–502. Copyright (2003), with permission from Elsevier and (f) Reprinted with permission from Bezryadin, A., Dekker, C., Schmid, G., Electrostatic trapping of single conducting nanoparticles between nanoelectrodes. *Appl. Phys. Lett.* 1997, 71, 1273–1275. Copyright (1997) American Institute of Physics.)

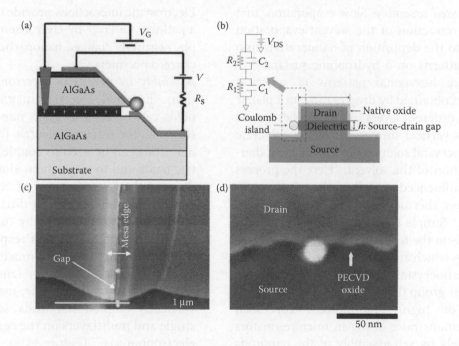

FIGURE 12.7   Illustrations of the vertical gap structure obtained by MBE and selective etching at the side of a predefined mesa structure (a) and by CVD and subsequent reactive ion etching (b). (c–d) SEM images of single gold nanoparticles positioned in the vertical gap structures. (Reprinted with permission from Krahne, R. et al., Fabrication of nanoscale gaps in integrated circuits. *Appl. Phys. Lett.* 2002, 81, 730–732. Copyright (2002) American Institute of Physics, (a,c) and Adapted from Ray, V. et al., Cmos-compatible fabrication of room-temperature single-electron devices. *Nat. Nanotechnol.* 2008, 3, 603–608, copyright *Nature* (b,d).)

Chapter 12

spherical nanoparticles is subject of detailed and comprehensive reviews by Kinge et al.,[158] and by Rogach et al.[159] (self-assembly into 2D and 3D arrays by slow evaporation of monodisperse spherical nanocrystals, including bimodal superstructures). The role of interparticle and external forces for self-assembly is discussed by Min et al.[160]

Ordered arrays of nanocrystals are interesting from the fundamental point of view as they mimic the organization of atoms into crystals, and there is great hope that it will be possible to extract useful collective properties arising from ordered superstructure organization. Scanning tunneling spectroscopy experiments have demonstrated, for instance, signatures of superlattice effects in the conductive properties of two dimensional arrays of spherical nanocrystals,[161] but also of rod-shaped nanocrystals,[162] which in both cases manifest in a reduced band gap observed from the superlattice with respect to individual nanocrystals. In the present context, however, we will not limit ourselves to the assembly of nanoparticles into ordered 2D and 3D arrays, but we will extend the discussion also to "disordered" assemblies.

The various possible strategies for the assembly of nanoparticles are highlighted below:

1. *Drying-mediated assembly:* Slow evaporation, that is, the slow retraction of the solvent evaporation front, leads to the deposition of nanocrystals into organized patterns on a hydrophilic surface.[163,164] Examples are hexagonal patterns of spherical nanoparticles obtained by drop casting on a planar substrate,[165] ordered layers of nanospheres and nanorods by vertical immersion of the substrate into the nanocrystal solution, and subsequent thermal evaporation of the solvent. (Here the process can be also influenced by pulling the substrate out of the solution, thermal heating, and retraction of the solution.) Simple drop casting of nanocrystal solution leads to the formation of well defined coffee stain rings which are composed of dense ordered layers of nanocrystal material.[166] Recently, we showed in our group that the coffee stain-mediated deposition of highly luminescent core–shell nanorods demonstrates lasing in microresonators obtained solely by self-assembly of the nanorods via the microfluidic dynamics involved.[167]

2. *Chemically assisted assembly:* Self-assembled monolayers of organic molecules can be employed to anchor the nanocrystals, for example, silanization of Si or glass surfaces where the nanocrystals attach via the functional SH end groups (thiols). Functional SH groups can also be used to link nanocrystals such that chains are formed.[117,168] Hydrogen bonding can serve for assembling nanocrystals with biomolecules and to take advantage of the specific recognition properties of many biomolecules (ssDNA, streptavidin/biotin, etc.). Nanorod chains were fabricated by using the biotin–streptavidin recognition to link the semiconductor nanorods tip-to-tip,[169] or metal nanorods side-to-side into stripes.[170] Another road to assembly of nanorods into chain-like structure is via exploiting the higher reactivity of the nanorod end tips, which allows them to be selectively functionalized with molecules that mediate end-to-end nanorod organization.[171–174] An approach that avoids the use of organic material for the end-to-end assembly of nanorods exploits the shape anisotropy of nanocrystals to grow small metallic Au nanoparticles on selected locations of their surface. These Au domains can be used to weld the nanocrystals into an all-inorganic network by chemical reactions of the Au with iodine.[175] The interest in end-to-end assembly of nanorods is that this might lead to nanocomposite fibers with controllable and combined properties such as tunable mechanical, electric-, thermoelectric, or photo conductive behavior. Electrostatic interactions provide on the other hand a pathway for layer-by-layer assembly of, for example, negatively charged nanoparticles and positively charged polymers.[176]

3. *Assembly by means of electromagnetic interactions:* External electric, magnetic, or optical fields can be used to align nanocrystals. As an example, the intrinsic electric dipole moment of nanorods can be used to couple to external electric fields and to align them along the field lines during the evaporation of the solvent.[177,178] Alignment of nanocrystals during solvent evaporation is always affected by turbulences of the evaporating liquid. In this respect, assembly in solution should lead to much more uniform results especially on larger length scales. Ryan and coworkers, for example, managed to obtain vertically aligned nanorods self-assembled in single and multilayers on the centimeter scale by electrophoresis[179] (Figure 12.8).

4. *Assembly at liquid–liquid interfaces:* Liquid–liquid interfaces provide a suitable means for assemblies over large length scales. For example, the water–solvent interface can be employed for the assembly

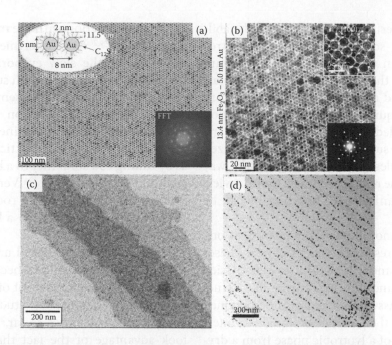

FIGURE 12.8 Self-assembly of isotropic nanocrystals: (a) Hexagonal lattice of gold nanocrystal obtained by drop casting. (Reprinted by permission from Macmillan Publishers Ltd. *Nat. Mater.* Bigioni, T. P. et al., Kinetically driven self assembly of highly ordered nanoparticle monolayers. 2006, 5, 265–270. Copyright (2006).) (b) Binary superlattice consisting of Fe_2O_3 and gold nanoparticles fabricated by slow solvent evaporation in a low-pressure chamber. (Reprinted with permission from Shevchenko, E. V. et al., Structural diversity in binary nanoparticle superlattices. *Nature* 2006, 439, 55–59. Copyright (2006) American Chemical Society.) (c) Ordered Au nanocrystal layers self-assembled on a water surface. (Reprinted with permission from Santhanam, V. et al., Self-assembly of uniform monolayer arrays of nanoparticles. *Langmuir* 2003, 19, 7881–7887. Copyright (2003) American Chemical Society.) (d) DNA template self-assembly of nanocrystals. (Reprinted with permission from Le, J. D. et al., DNA-templated self-assembly of metallic nanocomponent arrays on a surface. *Nano Lett.* 2004, 4, 2343. Copyright (2004) American Chemical Society.)

of nanocrystals into ordered films.[58,183,184] All these strategies are based on the well-known evidence that nanoparticles tend to self-segregate at the interface between water and the solvent in which they are dispersed (usually toluene)[185] as this seems to reduce the interfacial energy between the two liquids.

5. *Template-assisted assembly:* Nanowires, carbon nanotubes, nanoholes, or trenches as well as biological templates such as DNA strands, peptides, proteins, and viruses are possible candidates that have been successfully used for nanocrystal assembly. For example, protein-encapsulated fluorescent nanoparticles can be fabricated by self-assembly based on a genetically modified chaperonin protein from the hyperthermophilic archaeon *Sulfolobus shibatae*.[186,187] The combination of diblock copolymer micelle nanolithography and hydroxylamine[188] can be employed to achieve highly ordered metallic nanoparticle arrays with tunable particle sizes and interparticle spacing.[189]

Assembly of anisotropic particles: Although in the previous section we have already discussed several cases of assembly of nanorods when listing the various strategies for assembly, we would like to highlight in a separate section some important concepts concerning the assembly of anisotropic nanoparticles "in general." First of all, it is important to stress that assembly of anisotropic particles is much more difficult to achieve than for spherical nanoparticles since both positional and orientational order of the individual objects is required. At the same time, the more complex shape can lead to improved properties and performance. To cite an example, ordered arrays of nanorods must necessarily show coherent orientation of the anisotropic nanocrystals along a given direction, as opposed to ordered multilayers of spherical nanocrystals, in which the orientation of each individual nanocrystal is practically undefined. A defined geometrical arrangement, coupled to distinctive physical properties of individual nanorods (e.g., linearly polarized absorption and emission), could be translated into a unique and predictable macroscopic property of the ensemble.

A recent perspective work has indicated indeed that many promises from assembly of nanoparticles, for

Chapter 12

instance, in meta-materials, will be based on the ability to control the large-scale arrangement of anisotropic nanoparticles.[190] In this direction, an important role will be played by the multimaterial nanocrystals that we have described earlier in this chapter, which will come already equipped with the information required for a controlled assembly (e.g., because various regions on their surface will be functionalized with different molecules, which will dictate the way they will assemble). The reader can find a useful review on the assembly of anisotropic building blocks by Glotzer et al.[191]

Drop casting of nanorod solutions has led to oriented assemblies that resemble smectic crystal phases (for gold and silver nanorods[193]), and to side-by-side alignment of CdS nanorods into stripe-like structures.[194–196] Nanorod assemblies can also be obtained by solvent fluidics and the presence of liquid–air interface in the deposition of a lyotropic phase from a drying solution,[197] by a Langmuir–Blodgett approach,[198] or by unidirectional alignment through attachment of nanorods to the surface of a single-cleaved semiconductor monolayer (Figure 12.9).[199]

The self-assembly of nanorods crystallizing in the wurtzite crystal phase into laterally or vertically aligned arrays can be mediated by external electric fields via interaction with the intrinsic dipole of these nanorods.[177,178,200–202] Other methods for the vertical assembly are the slow evaporation of the nanorod solution in between a smooth substrate and a block of HOPG,[203] and the direct assembly in solution mediated by depletion attraction forces.[107,204] Nanorod assembly into liquid crystalline phase-like arrays can be achieved also by interparticle interactions as well as by tuning rod solubility in a binary solvent/nonsolvent liquid mixture.[58,183,184] Vertical assembly can be promoted by controlling the rod interfacial energy, by slow solvent evaporation on a liquid–solid–air interface[205] (Figure 12.10).

The assembly of branched nanostructures, like tetrapods, is much less studied because it is more difficult to realize superstructures out of them. In this respect, there have been only a few studies on the assembly of tetrapods on substrates so far.[206–208] All these studies took advantage of the fact that tetrapods self-align when deposited on a planar surface, with three arms touching the surface and the fourth arm pointing vertically upward. The degree of order can be enhanced by specifically patterned substrate surfaces. As an example, Cui et al.[206] fabricated nanoscale trenches in a polymer film on Au-coated Si substrates, after which they immersed the patterned substrates vertically into

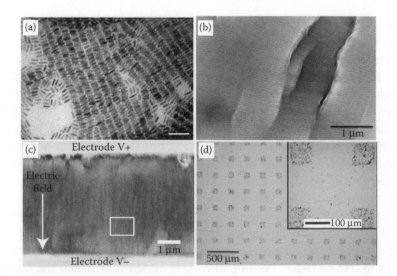

FIGURE 12.9 Lateral nanorod assemblies: (a) Lateral alignment of nanorods achieved by drop casting the nanorod solution of a TEM grid. (b) Dense, micron-sized multilayers of laterally aligned nanorods that were formed at the border of an evaporating drop of nanorod solution via coffee stain fluid dynamics. (Reprinted with permission from Nobile, C. et al., Self-assembly of highly fluorescent semiconductor nanorods into large scale smectic liquid crystal structures by coffee stain evaporation dynamics. *Journal of Physics-Condensed Matter* 2009, *21*, 264013. Copyright, Institute of Physics.) (c) Nanorod alignment induced by an external electric field. (Reprinted with permission from Carbone, L. et al., Synthesis and micrometer-scale assembly of colloidal CdSe/CdS nanorods prepared by a seeded growth approach. *Nano Lett.* 2007, *7*, 2942–2950. Copyright (2007), American Chemical Society.) (d) Nanorod patterns obtained on a prepatterned substrate with regions with different wettability. (Liu, S. H., Tok, J. B. H., Locklin, J., Bao, Z. N., Assembly and alignment of metallic nanorods on surfaces with patterned wettability. *Small* 2006, *2*, 1448–1453. Copyright Wiley-VCH Verlag GmbH & Co. KGaA. Reprinted with permission.)

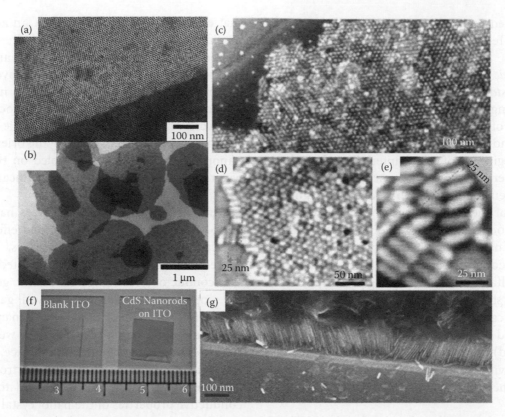

FIGURE 12.10 Vertical nanorod assembly: (a) Vertical nanorod alignment at a water–solvent interface and fished on a TEM grid. (b) Bundles of vertically aligned nanorods obtained by external electric fields applied during solvent evaporation. (Reprinted with permission from Carbone, L. et al., Synthesis and micrometer-scale assembly of colloidal CdSe/CdS nanorods prepared by a seeded growth approach. *Nano Lett.* 2007, *7*, 2942–2950. Copyright (2007) American Chemical Society.) (c–e) Assembly of CdS nanorods into perpendicular superlattices by controlled evaporation of a nanorod solution trapped between a smooth substrate and a block of highly oriented pyrolytic graphite. (Reprinted with permission from Ahmed, S., Ryan, K. M., Self-assembly of vertically aligned nanorod supercrystals using highly oriented pyrolytic graphite. *Nano Lett.* 2007, *7*, 2480–2485. Copyright (2007) American Chemical Society.) (f–g) Centimeter-scale layers of vertically aligned nanorods obtained by electrophoresis. (Ahmed, S., Ryan, K. M., Centimeter scale assembly of vertically aligned and close packed semiconductor nanorods from solution. *Chem. Commun.* 2009, 6421–6423, Reproduced by permission of The Royal Society of Chemistry.)

a solvent solution containing CdTe tetrapods, and found that the capillary forces during solvent evaporation lead to oriented assemblies of the tetrapods inside the trenches. Tetrapods have been assembled also via electrostatic trapping.[208] By this method, the tetrapods are forced toward the region of strongest electric field, for example, onto the extremity of a metalized AFM tip, or in between electrode pairs. This approach can be used to position single tetrapods in between electrodes with gaps of few 10s of nanometers.

Assembly of tetrapods in 3D structures has been proposed recently by our group. II–VI semiconductor tetrapods were organized into network structures using gold domains as linkers, which resulted in an end-to-end connection in between the arms of different tetrapods.[175] This approach exploits the shape anisotropy of nanocrystals to grow small metallic Au nanoparticles on selected locations of their surface, basically at their

tips. (An approach that was reported for the first time by Banin et al.[117]). Small amounts of molecular iodine are used to destabilize the Au domains grown on the arm tips . and to induce the coalescence of Au domains belonging to different nanocrystals, thus forming larger Au particles, each of them bridging two or more tetrapods through their tips. This strategy introduces an inorganic and robust junction between nanocrystals and hence avoids the use of molecular organic spacers for the assembly.[169] It works also in connecting nanorods in chain-like structures, as described earlier.

Site-selective decoration of one of the tetrapod tips with Au nanoparticles was also achieved by spin coating a polymer onto a substrates covered with tetrapods such that the tetrapods were partially protected.[209] The Au nanoparticles were attached to the uncovered tips of the vertical arms via dithiol linkers. The authors also demonstrated that it is possible to break off the uncovered,

Chapter 12

gold-decorated vertical arms and in this way they obtained CdTe rods with Au particles on only one end.

Optoelectronic devices based on self-assembled layers of nanocrystals: Colloidal nanocrystals represent interesting materials for optoelectronic devices because of the cost-effective way they can be produced and their appealing properties (as discussed at the beginning of this chapter). The photoelectrical properties of oriented assemblies of nanorods have been studied which revealed a significant impact of the nanorod order and orientation on the photocurrent response.[210,211] The fabrication of layers of nanocrystals by spin coating nanocrystals suspended in solution have been employed for photovoltaic and light-emitting prototype devices. Nanocrystal–polymer blends have been studied for applications in solar cells.[212–214] In this respect, elongated nanocrystals (e.g., rod-shaped or hyperbranched) are especially appealing because these can provide improved percolation pathways to collect the photogenerated charges.[215–217]

Nanocrystals have been widely exploited also in light-emitting diodes.[218–220] Light-emitting devices that emit in the blue consisting of an active layer of magic sites CdSe/ZnSe core–shell nanocrystals blended with diphenylcarbazole and an evaporated electron transporting/hole blocking layer have been demonstrated.[221] Also, LED devices in which the nanocrystals were transferred by microcontact printing were realized,[222] and recently this method has been further extended by employing oriented layers of nanorods that were self-assembled on the surface of water to achieve LEDs that emit linearly polarized light.[223]

Semiconductor nanocrystal layers have also been employed as active elements in field effect transistor devices.[197,224,225] In these approaches, micrometer-spaced source and drain were fabricated by standard semiconductor technology in planar geometry and the doped Si/SiO$_2$ substrate-served gate electrode. Thin films of nanocrystals (PbTe nanocrystals in ref.[224] and ZnO nanorods in ref.[197]) were deposited from solution onto the substrate surface, covering the area between the source and drain electrodes. Thermal and chemical treatment can be used to modify the conductive properties of the nanocrystal film. Carrier mobilities obtained from such devices are in the range of 0.1–0.7 cm^2 V^{-1} s^{-1}.

12.9 Conclusions and Perspectives

In this chapter, we have highlighted the most developed approaches toward the synthesis and the assembly of colloidal inorganic nanoparticles. As the number of proposed applications of nanocrystals widens, more stringent requirements are casted on the way these nanocrystals need to be fabricated, handled, and assembled. Industrial applications of nanoparticles that are being mainly targeted at present and that will be intensively developed in the coming decade are still related to the fabrication of bulk nanomaterials. For such applications, the manipulation of individual nanoparticles (i.e., the ability to locate, address, manipulate, and program a single and specific nanocrystal) is yet not strongly required. These involve, for example, the field of cosmetics, of biomedical applications of nanocrystals, and of nanocomposite materials in general, for energy-related materials/devices and for materials with enhanced mechanical/surface properties. In all these applications, a disordered assembly of nanocrystals or perhaps only a partial ordering on large scales is already sufficient, while critical parameters are clearly the uniformity in sample size, shape, surface functionalization, stability of the nanocrystals, the economic and environmental viability in their fabrication (which in addition must target large quantities of material), and not least their reduced toxicity. In this respect, the field of nanocrystal synthesis is already well advanced, with new sets of procedures for preparing nanocrystals of increasing complexity and homogeneity being reported day after day. In this scenario, in addition to the various reactor schemes described in this chapter, a remarkable engineering advance for what concerns the synthesis is represented by the small-scale automated "batch type" reactor developed by Chan, Milliron, and coworkers at the Molecular Foundry in Berkeley.[226] This is capable of performing in an automated way a large number of syntheses with a reproducibility, speed, and precision that is superior to that of a human operator; hence, it is capable of scanning the parameter space much faster and more precisely than done manually by any researcher. Additionally, it does not rely on the complex and not yet popular microfluidic type of reactor, but rather on the traditional "glass flask" scheme still adopted in the vast majority of research labs around the world. Therefore, is it expected to increase the pace at which new syntheses of colloidal nanocrystals are discovered and/or refined in research labs, as soon as an increasing number of labs start conforming to this new standard.

The fabrication of ordered assemblies from nanocrystal building blocks over large areas and volumes, the exact and reproducible positioning of individual nanocrystals, or of small ensembles of nanocrystals, at well-defined locations, and the understanding and mastering of interactions of these nanocrystals with contacts/substrates is perhaps the key to the engineering of future devices. In the spatial organization of nanoscale objects, many types of interactions can play a role,[160,227,228] and a deep understanding and handling of these interactions is far from being achieved. In spite of recent successes in self-organization of nearly monodisperse spherical colloidal nanoparticles,[180] and of some key advances in the self-assembly of shape-controlled nanocrystals,[191] this field is still in its infancy. As an example, long-range assembly of anisotropic nanocrystals is still difficult, and this is unfortunate, since most of the promises from assembly of nanoparticles, for example, in meta-materials, will rely heavily on the ability to control the large-scale spatial arrangement of anisotropic nanoparticles.[190] Success in this direction will likely come from the fabrication of anisotropic building blocks which already have the information for programmed assembly embedded, for example, topological control in the chemical composition and surface functionalization.[191] Ideally, at one point it will be possible to functionalize the individual nanocrystals with various types of molecules, such that specific types of molecules will be located on specific regions on the nanocrystal surface and thereby allow for a complete control of its spatial and orientation positioning with neighboring building blocks.

References

1. Hens, Z., Vanmaekelbergh, D., Stoffels, E., van Kempen, H., Effects of crystal shape on the energy levels of zero-dimensional Pbs quantum dots. *Phys. Rev. Lett.* 2002, *88*, art. no. 236803.
2. Burda, C., Chen, X. B., Narayanan, R., El-Sayed, M. A., Chemistry and properties of nanocrystals of different shapes. *Chem. Rev.* 2005, *105*, 1025–1102.
3. Murphy, C. J., Sau, T. K., Gole, A. M., Orendorff, C. J., Gao, J., Gou, L., Hunyadi, S. E., Li, T., Anisotropic metal nanoparticles: Synthesis, assembly, and optical applications. *J. Phys. Chem. B* 2005, *109*, 13857–13870.
4. Goldstein, A. N., Echer, C. M., Alivisatos, A. P., Melting in semiconductor nanocrystals. *Science* 1992, *256*, 1425–1427.
5. Alivisatos, A. P., Perspectives on the physical chemistry of semiconductor nanocrystals. *J. Phys. Chem.* 1996, *100*, 13226–13239.
6. Alivisatos, A. P., Scaling law for structural metastability in semiconductor nanocrystals. *Ber. Bunsen-Ges. Phys. Chem.* 1997, *101*, 1573–1577.
7. Klein, D. L., McEuen, P. L., Katari, J. E. B., Roth, R., Alivisatos, A. P., An approach to electrical studies of single nanocrystals. *Appl. Phys. Lett.* 1996, *68*, 2574–2576.
8. Kim, S. H., Markovich, G., Rezvani, S., Choi, S. H., Wang, K. L., Heath, J. R., Tunnel diodes fabricated from CdSe nanocrystal monolayers. *Appl. Phys. Lett.* 1999, *74*, 317–319.
9. Davydov, D. N., Haruyama, J., Routkevitch, D., Statt, B. W., Ellis, D., Moskovits, M., Xu, J. M., Nonlithographic nanowire-array tunnel device: Fabrication, aero-bias anomalies, and coulomb blockade. *Phys. Rev. B, Condens. Matter (USA)* 1998, *57*, 13550–13553.
10. Heath, J. R., Shiang, J. J., Covalency in semiconductor quantum dots. *Chem. Soc. Rev.* 1998, *27*, 65–71.
11. Schmidt, G., *Nanoparticles: From Theory to Applications—Fourth Edition*. Wiley: Weinheim, 2006.
12. Battle, X., Lebarta, A., Finite-size effects in fine particles: Magnetic and transport properties. *J. Phys. D* 2002, *35*, R15.
13. Fiorani, D., *Surface Effects in Magnetic Nanoparticles*. Springer: New York, 2005.
14. Klimov, V., *Semiconductor and Metal Nanocrystals*. Marcel Dekker: New York, 2004.
15. Noguez, C., Optical properties of isolated and supported metal nanoparticles. *Optic. Mater.* 2005, *27*, 1204–1211.
16. Link, S., El-Sayed, M. A., Shape and size dependence of radiative, non-radiative and photothermal properties of gold nanocrystals. *Int. Rev. Phys. Chem.* 2000, *19*, 409–453.
17. Perez-Juste, J., Pastoriza-Santos, I., Liz-Marzan, L. M., Mulvaney, P., Gold nanorods: Synthesis, characterization and applications. *Coord. Chem. Rev.* 2005, *249*, 1870–1901.
18. Eustis, S., El-Sayed, M. A., Why gold nanoparticles are more precious Than pretty gold: Noble metal surface plasmon resonance and its enhancement of the radiative and nonradiative properties of nanocrystals of different shapes. *Chem. Soc. Rev.* 2006, *35*, 209–217.
19. Myroshnychenko, V., Rodriguez-Fernandez, J., Pastoriza-Santos, I., Funston, A. M., Novo, C., Mulvaney, P., Liz-Marzan, L. M., de Abajo, F. J. G., Modelling the optical response of gold nanoparticles. *Chem. Soc. Rev.* 2008, *37*, 1792–1805.
20. Zhang, J. Z., Noguez, C., Plasmonic optical properties and applications of metal nanostructures. *Plasmonics* 2008, *3*, 127–150.
21. Narayanan, R., El-Sayed, M., Catalysis with transition metal nanoparticles in colloidal solution: Nanoparticle shape dependence and stability. *J. Phys. Chem. B* 2005, *109*, 12663–12676.
22. Chen, M. S., Goodman, D. W. Active structure of supported Au catalysts. *Catal. Today* 2006, *111*, 22–33.
23. Herzing, A. A., Kiely, C. J., Carley, A. F., Landon, P., Hutchings, G. J., Identification of active gold nanoclusters on iron oxide supports for Co oxidation. *Science* 2008, *321*, 1331–1335.
24. Messing, G. L., Zhang, S. C., Jayanthi, G. V., Ceramic powder synthesis by spray-pyrolysis. *J. Am. Ceram. Soc.* 1993, *76*, 2707–2726.

Chapter 12

25. Amirav, L., Lifshitz, E., Thermospray: A method for producing high quality semiconductor nanocrystals. *J. Phys. Chem. C* 2008, *112*, 13105–13113.

26. Koropchak, J. A., Veber, M., Thermospray sample introduction to atomic spectrometry. *Crit. Rev. Anal. Chem.* 1992, *23*, 113–141.

27. Roh, Y., Vali, H., Phelps, T. J., Moon, J. W., Extracellular synthesis of magnetite and metal-substituted magnetite nanoparticles. *J. Nanosci. Nanotechnol.* 2006, *6*, 3517–3520.

28. Gericke, M., Pinches, A., Microbial production of gold nanoparticles. *Gold Bull.* 2006, *39*, 22–28.

29. Ozin, G. A., Arsenault, A. C., Cademartiri, L., *Nanochemistry: A Chemical Approach to Nanomaterials.* Royal Society of Chemistry: London, 2009.

30. Reiss, B. D., Freeman, R. G., Walton, I. D., Norton, S. M., Smith, P. C., Stonas, W. G., Keating, C. D., Natan, M. J., Electrochemical synthesis and optical readout of striped metal rods with submicron features. *J. Electroanal. Chem.* 2002, *522*, 95–103.

31. Piraux, L., Encinas, A., Vila, L., Matefi-Tempfli, S., Matefi-Tempfli, M., Darques, M., Elhoussine, F., Michotte, S., Magnetic and superconducting nanowires. *J. Nanosci. Nanotechnol.* 2005, *5*, 372–389.

32. Lisiecki, I., Size, shape, and structural control of metallic nanocrystals. *J. Phys. Chem. B* 2005, *109*, 12231–12244.

33. Pileni, M. P., Reverse micelles used as templates: A new understanding in nanocrystal growth. *J. Exp. Nanosci.* 2006, *1*, 13–27.

34. Niederberger, M., Nonaqueous sol–gel routes to metal oxide nanoparticles. *Acc. Chem. Res.* 2007, *40*, 793–800.

35. Mackenzie, J. D., Bescher, E. P., Chemical routes in the synthesis of nanomaterials using the sol-gel process. *Acc. Chem. Res.* 2007, *40*, 810–818.

36. Suslick, K. S., Price, G. J., Applications of ultrasound to materials chemistry. *Ann. Rev. Mater. Sci.* 1999, *29*, 295–326.

37. Li, L., Qian, H. F., Ren, J. C., Rapid synthesis of highly luminescent CdTe nanocrystals in the aqueous phase by microwave irradiation with controllable temperature. *Chem. Comm.* 2005, *4*, 528–530.

38. Hu, X., Li, G. B., Yu, J. C., Design, fabrication, and modification of nanostructured semiconductor materials for environmental and energy applications. *Langmuir* 2009, DOI: 10.1021/la902142b.

39. Kappe, C. O., Dallinger, D., Murphree, S., *Practical Microwave Synthesis for Organic Chemists: Strategies, Instruments, and Protocols.* Wiley: Weinheim, 2008.

40. Talapin, D. V., Nelson, J. H., Shevchenko, E. V., Aloni, S., Sadtler, B., Alivisatos, A. P., Seeded growth of highly luminescent CdSe/CdS nanoheterostructures with rod and tetrapod morphologies. *Nano Lett.* 2007, *7*, 2951–2959.

41. Medintz, I. L., Mattoussi, H., Quantum dot-based resonance energy transfer and its growing application in biology. *Phys. Chem. Chem. Phys.* 2009, *11*, 17–45.

42. Pellegrino, T., Kudera, S., Liedl, T., Javier, A. M., Manna, L., Parak, W. J., On the development of colloidal nanoparticles towards multifunctional structures and their possible use for biological applications. *Small* 2005, *1*, 48–63.

43. Yu, W. W., Semiconductor quantum dots: Synthesis and water-solubilization for biomedical applications. *Expert Opin. On Bio. Ther.* 2008, *8*, 1571–1581.

44. Casula, M. F., Jun, Y. W., Zaziski, D. J., Chan, E. M., Corrias, A., Alivisatos, A. P., The concept of delayed nucleation in nanocrystal growth demonstrated for the case of iron oxide nanodisks. *J. Am. Chem. Soc.* 2006, *128*, 1675–1682.

45. Kwon, S. G., Piao, Y., Park, J., Angappane, S., Jo, Y., Hwang, N. M., Park, J. G., Hyeon, T., Kinetics of monodisperse iron oxide nanocrystal formation by "Heating-Up" process. *J. Am. Chem. Soc.* 2007, *129*, 12571–12584.

46. Hyeon, T., Chemical synthesis of magnetic nanoparticles. *Chem. Comm.* 2003, 927–934, doi: 10.1039/B207789B.

47. Wang, X., Zhuang, J., Peng, Q., Li, Y. D., A general strategy for nanocrystal synthesis. *Nature* 2005, *437*, 121–124.

48. Edel, J. B., Fortt, R., deMello, J. C., deMello, A. J., Microfluidic routes to the controlled production of nanoparticles. *Chem. Comm.* 2002, 1136–1137, 10.1039/B202998G.

49. Nakamura, H., Yamaguchi, Y., Miyazaki, M., Uehara, M., Maeda, H., Mulvaney, P., Continuous preparation of CdSe nanocrystals by a microreactor. *Chem. Lett.* 2002, 31, 1072–1073.

50. Yen, B. K. H., Stott, N. E., Jensen, K. F., Bawendi, M. G., A continuous-flow microcapillary reactor for the preparation of a size Series of CdSe nanocrystals. *Adv. Mater.* 2003, *15*, 1858–1862.

51. Chan, E. M., Mathies, R. A., Alivisatos, A. P., Size-controlled growth of CdSe nanocrystals in microfluidic reactors. *Nano Lett.* 2003, *3*, 199–201.

52. Winterton, J. D., Myers, D. R., Lippmann, J. M., Pisano, A. P., Doyle, F. M., A novel continuous microfluidic reactor design for the controlled production of high-quality semiconductor nanocrystals. *J. Nanopart. Res.* 2008, *10*, 893–905.

53. Hartlieb, K. J., Raston, C. L., Saunders, M., Controlled scalable synthesis of Zno nanoparticles. *Chem. Mater.* 2007, *19*, 5453–5459.

54. Chin, S. F., Iyer, K. S., Raston, C. L., Saunders, M., Size selective synthesis of superparamagnetic nanoparticles in thin fluids under continuous flow conditions. *Adv. Funct. Mater.* 2008, *18*, 922–927.

55. Kumar, S., Nann, T., Shape control of Ii-Vi semiconductor nanomateriats. *Small* 2006, *2*, 316–329.

56. Zhao, N., Qi, L. M., Low-temperature synthesis of star-shaped Pbs nanocrystals in aqueous solutions of mixed cationic/anionic surfactants. *Adv. Mater.* 2006, *18*, 359–362.

57. Sun, Y. G., Xia, Y. N., Shape-controlled synthesis of gold and silver nanoparticles. *Science* 2002, *298*, 2176–2179.

58. Dumestre, F., Chaudret, B., Amiens, C., Respaud, M., Fejes, P., Renaud, P., Zurcher, P., Unprecedented crystalline super-lattices of monodisperse cobalt nanorods. *Angew. Chem. Int. Ed.* 2003, *42*, 5213–5216.

59. Yin, Y., Alivisatos, A. P., Colloidal nanocrystal synthesis and the organic-inorganic interface. *Nature* 2005, *437*, 664–670.

60. Tang, K. B., Qian, Y. T., Zeng, J. H., Yang, X. G., Solvothermal route to semicouductor nanowires. *Adv. Mater.* 2003, *15*, 448–450.

61. Ahrenkiel, S. P., Micic, O. I., Miedaner, A., Curtis, C. J., Nedeljkovic, J. M., Nozik, A. J., Synthesis and characterization of colloidal Inp quantum rods. *Nano Lett.* 2003, *3*, 833–837.

62. Grebinski, J. W., Hull, K. L., Zhang, J., Kosel, T. H., Kuno, M., Solution-based straight and branched CdSe nanowires. *Chem. Mater.* 2004, *16*, 5260–5272.

63. Kan, S. H., Aharoni, A., Mokari, T., Banin, U., Shape control of Iii-V semiconductor nanocrystals: Synthesis and properties of Inas quantum rods. *Faraday Discus.* 2004, *125*, 23–38.

64. Zitoun, D., Pinna, N., Frolet, N., Belin, C., Single crystal manganese oxide multipods by oriented attachment. *J. Am. Chem. Soc.* 2005, *127*, 15034–15035.

65. Cho, K. S., Talapin, D. V., Gaschler, W., Murray, C. B., Designing PbSe nanowires and nanorings through oriented attachment of nanoparticles *J. Am. Chem. Soc.* 2005, *127*, 7140–7147.

66. Yu, J. H., Joo, J., Park, H. M., Baik, S. I., Kim, Y. W., Kim, S. C., Hyeon, T., Synthesis of quantum-sized cubic Zns nanorods by the oriented attachment mechanism *J. Am. Chem. Soc.* 2005, *127*, 5662–5670.

67. Lee, E. J. H., Ribeiro, C., Longo, E., Leite, E. R., Oriented attachment: An effective mechanism in the formation of anisotropic nanocrystals. *J. Phys. Chem. B* 2005, *109*, 20842–20846.

68. Gou, L. F., Murphy, C. J., Fine-tuning the shape of gold nanorods. *Chem. Mater.* 2005, *17*, 3668–3672.

69. Kuo, C. H., Huang, M. H., Synthesis of branched gold nanocrystals by a seeding growth approach. *Langmuir* 2005, *21*, 2012–2016.

70. Yin, Y. D., Rioux, R. M., Erdonmez, C. K., Hughes, S., Somorjai, G. A., Alivisatos, A. P., Formation of hollow nanocrystals through the nanoscale Kirkendall effect. *Science* 2004, *304*, 711–714.

71. Cabot, A., Smith, R. K., Yin, Y. D., Zheng, H. M., Reinhard, B. M., Liu, H. T., Alivisatos, A. P., Sulfidation of cadmium at the nanoscale. *Acs Nano* 2008, *2*, 1452–1458.

72. Qiu, Y. F., Yang, S. H., Kirkendall approach to the fabrication of ultra-thin Zno nanotubes with high resistive sensitivity to humidity. *Nanotechnology* 2008, *19*, art. n. 265606.

73. An, K., Hyeon, T., Synthesis and biomedical applications of hollow nanostructures. *Nano Today* 2009, *4*, 359–373.

74. Elechiguerra, J. L., Reyes-Gasga, J., Yacaman, M. J., The role of twinning in shape evolution of anisotropic noble metal nanostructures. *J. Mater. Chem.* 2006, *16*, 3906–3919.

75. Manna, L., Milliron, D. J., Meisel, A., Scher, E. C., Alivisatos, A. P., Controlled growth of tetrapod-branched inorganic nanocrystals. *Nat. Mater.* 2003, *2*, 382–385.

76. Puntes, V. F., Krishnan, K., Alivisatos, A. P., Synthesis of colloidal cobalt nanoparticles with controlled size and shapes. *Top. Catal.* 2002, *19*, 145–148.

77. Deka, S., Mistza, K., Dorfs, D., Genovese, A., Bertoni, G., Manna, L., Octapod-shaped colloidal nanocrystals of cadmium chalcogenides via "one-pot" cation exchange and seeded growth. *Nano Lett.* 2010, 10, 3770–3776.

78. Cozzoli, P. D., Manna, L., Curri, M. L., Kudera, S., Giannini, C., Striccoli, M., Agostiano, A., Shape and phase control of colloidal Znse nanocrystals. *Chem. Mater.* 2005, *17*, 1296–1306.

79. Sugimoto, T., Itoh, H., Mochida, T., Shape control of monodisperse hematite particles by organic additives in the Gel-Sol system. *J. Colloid Interface Sci.* 1998, *205*, 42–52.

80. Cozzoli, P. D., Kornowski, A., Weller, H., Low-temperature synthesis of soluble and processable organic- capped anatase Tio$_2$ nanorods. *J. Am. Chem. Soc.* 2003, *125*, 14539–14548.

81. Jun, Y. W., Choi, J. S., Cheon, J., Shape control of semiconductor and metal oxide nanocrystals through nonhydrolytic colloidal routes. *Angew. Chem. Int. Ed.* 2006, *45*, 3414–3439.

82. Yang, P. D., Yan, H. Q., Mao, S., Russo, R., Johnson, J., Saykally, R., Morris, N., Pham, J., He, R. R., Choi, H. J., Controlled growth of Zno nanowires and their optical properties. *Adv. Funct. Mater.* 2002, *12*, 323–331.

83. Bjork, M. T., Ohlsson, B. J., Sass, T., Persson, A. I., Thelander, C., Magnusson, M. H., Deppert, K., Wallenberg, L. R., Samuelson, L., One-dimensional steeplechase for electrons realized. *Nano Lett.* 2002, *2*, 87–89.

84. Tang, Z. Y., Kotov, N. A., Giersig, M., Spontaneous organization of single CdTe nanoparticles into luminescent nanowires. *Science* 2002, *297*, 237–240.

85. Habas, S. E., Lee, H., Radmilovic, V., Somorjai, G. A., Yang, P., Shaping binary metal nanocrystals through epitaxial seeded growth. *Nat. Mater.* 2007, *6*, 692–697.

86. Yeh, C. Y., Lu, Z. W., Froyen, S., Zunger, A., Zinc-Blende-Wurtzite polytypism in semiconductors. *Phys. Rev. B* 1992, *46*, 10086–10097.

87. Cozzoli, P. D., Pellegrino, T., Manna, L., Synthesis, properties and perspectives of hybrid nanocrystal structures. *Chem. Soc. Rev.* 2006, *35*, 1195–1208.

88. Salgueirino-Maceira, V., Correa-Duarte, M. A., Lopez-Quintela, M. A., Rivas, J., Advanced hybrid nanoparticles. *J. Nanosci. Nanotechnol.* 2009, *9*, 3684–3688.

89. Yang, J., Lee, J. Y., Too, H. P., Core-shell Ag-Au nanoparticles from replacement reaction in organic medium. *J. Phys. Chem. B* 2005, *109*, 19208–19212.

90. Rodriguez-Gonzalez, B., Burrows, A., Watanabe, M., Kiely, C. J., Marzan, L. M. L., Multishell bimetallic AuAg nanoparticles: Synthesis, structure and optical properties. *J. Mater. Chem.* 2005, *15*, 1755–1759.

91. Manna, L., Scher, E. C., Li, L. S., Alivisatos, A. P., Epitaxial growth and photochemical annealing of graded CdS/ZnS shells on colloidal CdSe nanorods. *J. Am. Chem. Soc.* 2002, *124*, 7136–7145.

92. Kim, S., Fisher, B., Eisler, H. J., Bawendi, M., Type-Ii quantum Dots: CdTe/CdSe(Core–Shell) and CdSe/ZnTe(Core–Shell) heterostructures. *J. Am. Chem. Soc.* 2003, *125*, 11466–11467.

93. Yu, K., Zaman, B., Romanova, S., Wang, D. S., Ripmeester, J. A., Sequential synthesis of Type Ii colloidal CdTe/CdSe core-shell nanocrystals. *Small* 2005, *1*, 332–338.

94. Eychmuller, A., Mews, A., Weller, H., A Quantum dot quantum well - CdS/Hgs/CdS. *Chem. Phys. Lett.* 1993, *208*, 59–62.

95. Chen, C. Y., Cheng, C. T., Lai, C. W., Hu, Y. H., Chou, P. T., Chou, Y. H., Chiu, H. T., Type-Ii CdSe/CdTe/ZnTe (Core-Shell-Shell) quantum dots with cascade band edges: The Separation of electron (at CdSe) and hole (at ZnTe) by the CdTe layer. *Small* 2005, *1*, 1215–1220.

96. Reiss, P., Protiere, M., Li, L., Core–shell semiconductor nanocrystats. *Small* 2009, *5*, 154–168.

97. Ban, Z. H., Barnakov, Y. A., Golub, V. O., O'Connor, C. J., The synthesis of Core-Shell Iron@Gold nanoparticles and their characterization. *J. Mater. Chem.* 2005, *15*, 4660–4662.

98. Wang, H., Brandl, D. W., Le, F., Nordlander, P., Halas, N. J., Nanorice: A hybrid plasmonic nanostructure. *Nano Lett.* 2006, *6*, 827–832.

99. Kim, H., Achermann, M., Balet, L. P., Hollingsworth, J. A., Klimov, V. I., Synthesis and characterization of Co/CdSe Core–shell nanocomposites: Bifunctional magnetic-optical nanocrystals. *J. Am. Chem. Soc.* 2005, *127*, 544–546.

100. Liz-Marzan, L. M., Mulvaney, P., The assembly of coated nanocrystal. *J. Phys. Chem. B* 2003, *107*, 7312–7326.

101. Hirakawa, T., Kamat, P. V., Charge separation and catalytic activity of Ag@Tio$_2$ core-shell composite clusters under UV-irradiation. *J. Am. Chem. Soc.* 2005, *127*, 3928–3934.

102. Mulvaney, P., Liz-Marzan, L. M., Giersig, M., Ung, T., Silica encapsulation of quantum dots and metal clusters. *J. Mater. Chem.* 2000, *10*, 1259–1270.

103. Park, H. Y., Schadt, M. J., Wang, L., Lim, I. I. S., Njoki, P. N., Kim, S. H., Jang, M. Y., Luo, J., Zhong, C. J., Fabrication of magnetic core @ Shell Fe Oxide @ Au nanoparticles for interfacial bioactivity and bio-separation. *Langmuir* 2007, *23*, 9050–9056.

104. Levin, C. S., Hofmann, C., Ali, T. A., Kelly, A. T., Morosan, E., Nordlander, P., Whitmire, K. H., Halas, N. J., Magnetic-plasmonic core-shell nanoparticles. *Acs Nano* 2009, *3*, 1379–1388.

Chapter 12

105. Talapin, D. V., Koeppe, R., Gotzinger, S., Kornowski, A., Lupton, J. M., Rogach, A. L., Benson, O., Feldmann, J., Weller, H., Highly emissive colloidal CdSe/CdS heterostructures of mixed dimensionality. *Nano Lett.* 2003, *3*, 1677–1681.

106. Carbone, L., Nobile, C., De Giorgi, M., Della Sala, F., Morello, G., Pompa, P. P., Hytch, M. J. et al., Synthesis and micrometer-scale assembly of colloidal CdSe/CdS nanorods prepared by a seeded growth approach. *Nano Lett.* 2007, *7*, 2942–2950.

107. Baranov, D., Fiore, A., Van Huis, M., Giannini, C., Falqui, A., Lafont, U., Zandbergen, H., Zanella, M., Cingolani, R., Manna, L., Assembly of colloidal semiconductor nanorods in solution by depletion attraction. *Nano Lett.* 2010, *10*, 743–749.

108. Yu, W. W., Qu, L. H., Guo, W. Z., Peng, X. G., Experimental determination of the extinction coefficient of CdTe, CdSe, and CdS nanocrystals. *Chem. Mater.* 2003, *15*, 2854–2860.

109. Yu, W. W., Qu, L. H., Guo, W. Z., Peng, X. G., Experimental determination of the extinction coefficient of CdTe, CdSe and CdS nanocrystals (Vol 15, Pg 2854, 2003). *Chem. Mater.* 2004, *16*, 560–560.

110. Li, J. J., Wang, Y. A., Guo, W. Z., Keay, J. C., Mishima, T. D., Johnson, M. B., Peng, X. G., Large-scale synthesis of nearly monodisperse CdSe/CdS Core–shell nanocrystals using air-stable reagents via successive ion layer adsorption and reaction. *J. Am. Chem. Soc.* 2003, *125*, 12567–12575.

111. Pastoriza-Santos, I., Koktysh, D. S., Mamedov, A. A., Giersig, M., Kotov, N. A., Liz-Marzan, L. M., One-pot synthesis of Ag@ TiO₂ core-shell nanoparticles and their layer-by-layer assembly. *Langmuir* 2000, *16*, 2731–2735.

112. Zeng, H., Li, J., Wang, Z. L., Liu, J. P., Sun, S. H., Bimagnetic core–shell FePt/Fe₃O₄ nanoparticles. *Nano Lett.* 2004, *4*, 187–190.

113. Kershaw, S. V., Burt, M., Harrison, M., Rogach, A., Weller, H., Eychmuller, A., Colloidal CdTe/HgTe quantum dots with high photoluminescence quantum efficiency at room temperature. *Appl. Phys. Lett.* 1999, *75*, 1694–1696.

114. Milliron, D. J., Hughes, S. M., Cui, Y., Manna, L., Li, J. B., Wang, L. W., Alivisatos, A. P., Colloidal nanocrystal heterostructures with linear and branched topology. *Nature* 2004, *430*, 190–195.

115. Shieh, F., Saunders, A. E., Korgel, B. A., General shape control of colloidal CdS, CdSe, CdTe quantum rods and quantum rod heterostructures. *J. Phys. Chem. B* 2005, *109*, 8538–8542.

116. Kudera, S., Carbone, L., Casula, M. F., Cingolani, R., Falqui, A., Snoeck, E., Parak, W. J., Manna, L., Selective growth of PbSe on one or both tips of colloidal semiconductor nanorods. *Nano Lett.* 2005, *5*, 445–449.

117. Mokari, T., Rothenberg, E., Popov, I., Costi, R., Banin, U., Selective growth of metal tips onto semiconductor quantum rods and tetrapods. *Science* 2004, *304*, 1787–1790.

118. Mokari, T., Sztrum, C. G., Salant, A., Rabani, E., Banin, U., Formation of asymmetric one-sided metal-tipped semiconductor nanocrystal dots and rods. *Nat. Mater.* 2005, *4*, 855–863.

119. Elmaletn, E., Saunders, A. E., Costi, R., Salant, A., Banin, U., Growth of photocatalytic CdSe-Pt nanorods and nanonets. *Adv. Mater.* 2008, *20*, 4312–4317.

120. Costi, R., Saunders, A. E., Elmalem, E., Salant, A., Banin, U., Visible light-induced charge retention and photocatalysis with hybrid CdSe-Au nanodumbbells. *Nano Lett.* 2008, *8*, 637–641.

121. Choi, J. S., Jun, Y. W., Yeon, S. I., Kim, H. C., Shin, J. S., Cheon, J., Biocompatible heterostructured nanoparticles for multimodal biological detection. *J. Am. Chem. Soc.* 2006.

122. Salant, A., Amitay-Sadovsky, A., Banin, U., Directed self assembly of gold-tipped semiconductor nanorods. *J. Am. Chem. Soc.* 2006, *128*, 10006–10007.

123. Gu, H. W., Zheng, R. K., Zhang, X. X., Xu, B., Facile one-pot synthesis of bifunctional heterodimers of nanoparticles: A Conjugate of quantum dot and magnetic nanoparticles. *J. Am. Chem. Soc.* 2004, *126*, 5664–5665.

124. Kwon, K. W., Shim, M., Gamma-Fe₂O₃/Ii-Vi sulfide nanocrystal heterojunctions. *J. Am. Chem. Soc.* 2005, *127*, 10269–10275.

125. Yu, H., Chen, M., Rice, P. M., Wang, S. X., White, R. L., Sun, S. H., Dumbbell-like bifunctional Au-Fe₃O₄ nanoparticles. *Nano Lett.* 2005, *5*, 379–382.

126. Pellegrino, T., Fiore, A., Carlino, E., Giannini, C., Cozzoli, P. D., Ciccarella, G., Respaud, M., Palmirotta, L., Cingolani, R., Manna, L., Heterodimers based on CoPt₃-Au nanocrystals with tunable domain size. *J. Am. Chem. Soc.* 2006, *128*, 6690–6698.

127. Teranishi, T., Inoue, Y., Nakaya, M., Oumi, Y., Sano, T., Nanoacorns: Anisotropically phase-segregated CoPd sulfide nanoparticles. *J. Am. Chem. Soc.* 2004, *126*, 9914–9915.

128. Choi, S. H., Kim, E. G., Hyeon, T., One-pot synthesis of copper-indium sulfide nanocrystal heterostructures with acorn, bottle, and larva shapes. *J. Am. Chem. Soc.* 2006, *128*, 2520–2521.

129. Gao, X. Y., Yu, L. T., MacCuspie, R. I., Matsui, H., Controlled growth of Se nanoparticles on Ag nanoparticles in different ratios. *Adv. Mater.* 2005, *17*, 426ff.

130. Carbone, L., Kudera, S., Giannini, C., Ciccarella, G., Cingolani, R., Cozzoli, P. D., Manna, L., Selective reactions on the tips of colloidal semiconductor nanorods. *J. Mater. Chem.* 2006, *16*, 3952.

131. Son, D. H., Hughes, S. M., Yin, Y. D., Alivisatos, A. P., Cation exchange reactions-in ionic nanocrystals. *Science* 2004, *306*, 1009–1012.

132. Robinson, R. D., Sadtler, B., Demchenko, D. O., Erdonmez, C. K., Wang, L. W., Alivisatos, A. P., Spontaneous superlattice formation in nanorods through partial cation exchange. *Science* 2007, *317*, 355–358.

133. Jain, P. K., Amirav, L., Aloni, S., Alivisatos, A. P., Nanoheterostructure cation exchange: Anionic framework conservation. *J. Am. Chem. Soc.* 2010, *132*, 9997–9999.

134. Li, J. S., Zhang, T. R., Ge, H. P., Yin, Y. D., Zhong, W. W., Fluorescence signal amplification by cation exchange in ionic nanocrystals. *Angew. Chem. Int. Ed.* 2009, *48*, 1588–1591.

135. Fulton, T. A., Dolan, G. J., Observation of single-electron charging effects in small tunnel junctions. *Phys. Rev. Lett.* 1987, *59*, 109.

136. Ralph, D. C., Black, C. T., Tinkham, M., Spectroscopic measurements of discrete electronic states in single metal particles. *Phys. Rev. Lett.* 1995, *74*, 3241–3244.

137. Gudiksen, M. S., Maher, K. N., Ouyang, L., Park, H., Electroluminescence from a single-nanocrystal transistor. *Nano Lett.* 2005, *5*, 2257–2261.

138. Steiner, D., Mokari, T., Banin, U., Millo, O., Electronic structure of metal-semiconductor nanojunctions in gold CdSe nanodumbbells. *Phys. Rev. Lett.* 2005, *95*, 056805 (4 pages).

139. Trudeau, P. E., Sheldon, M., Altoe, V., Alivisatos, A. P., Electrical contacts to individual colloidal semiconductor nanorods. *Nano Lett.* 2008, *8*, 1936–1939.

140. Leonard, F., Talin, A. A., Size-dependent effects on electrical contacts to nanotubes and nanowires. *Phys. Rev. Lett.* 2006, *97*, 026804 (4 pages).

141. Demchenko, D. O., Wang, L. W., Localized electron states near a metal/semiconductor nanocontact. *Nano Lett.* 2007, *7*, 3219–3222.

142. Steinberg, H., Lilach, Y., Salant, A., Wolf, O., Faust, A., Millo, O., Banin, U., Anomalous temperature dependent transport through single colloidal nanorods strongly coupled to metallic leads. *Nano Lett.* 2009, *9*, 3671–3675.

143. Rampi, M. A., Whitesides, G. M., A Versatile experimental approach for understanding electron transport through organic materials. *Chem. Phys.* 2002, *281*, 373–391.

144. Weiss, E. A., Kriebel, J. K., Rampi, M. A., Whitesides, G. M., The study of charge transport through organic thin films: Mechanism, tools and applications. *Philos. Trans. R. Soc. London, A* 2007, *365*, 1509–1537.

145. Cui, Y., Banin, U., Bjork, M. T., Alivisatos, A. P., Electrical transport through a single nanoscale semiconductor branch point. *Nano Lett.* 2005, *5*, 1519–1523.

146. Devoret, M. H., Grabert, H., Introduction to single charge tunneling. *Single Charge Tunnel.* 1992, *294*, 1–19.

147. Klein, D. L., Roth, R., Lim, A. K. L., Alivisatos, A. P., McEuen, P. L., A Single-electron transistor made from a cadmium selenide nanocrystal. *Nature* 1997, *389*, 699–701.

148. Reed, M. A., Zhou, C., Muller, C. J., Burgin, T. P., Tour, J. M., Conductance of a molecular junction. *Science* 1997, *278*, 252–254.

149. Park, H., Lim, A. K. L., Alivisatos, A. P., Park, J., McEuen, P. L., Fabrication of metallic electrodes with nanometer separation by electromigration. *Appl. Phys. Lett.* 1999, *75*, 301–303.

150. Krahne, R., Dadosh, T., Gordin, Y., Yacoby, A., Shtrikman, H., Mahalu, D., Sperling, J., Bar Joseph, I., Nanoparticles and nanogaps: Controlled positioning and fabrication. *Phys. E* 2003, *17*, 498–502.

151. Bezryadin, A., Dekker, C., Schmid, G., Electrostatic trapping of single conducting nanoparticles between nanoelectrodes. *Appl. Phys. Lett.* 1997, *71*, 1273–1275.

152. Krahne, R., Yacoby, A., Shtrikman, H., Bar-Joseph, I., Dadosh, T., Sperling, J., Fabrication of nanoscale gaps in integrated circuits. *Appl. Phys. Lett.* 2002, *81*, 730–732.

153. Barsotti, R. J., Vahey, M. D., Wartena, R., Chiang, Y. M., Voldman, J., Stellacci, F., Assembly of metal nanoparticles into nanogaps. *Small* 2007, *3*, 488–499.

154. Ray, V., Subramanian, R., Bhadrachalam, P., Ma, L. C., Kim, C. U., Koh, S. J., Cmos-compatible fabrication of room-temperature single-electron devices. *Nat. Nanotechnol.* 2008, *3*, 603–608.

155. Luber, S. M., Strobel, S., Tranitz, H. P., Wegscheider, W., Schuh, D., Tornow, M., Nanometre spaced electrodes on a cleaved algaas surface. *Nanotechnology.* 2005, *16*, 1182–1185.

156. Shevchenko, E. V., Kortright, J. B., Talapin, D. V., Aloni, S., Alivisatos, A. P., Quasi-ternary nanoparticle superlattices through nanoparticle design. *Adv. Mater.* 2007, *19*, 4183–4188.

157. Shevchenko, E. V., Ringler, M., Schwemer, A., Talapin, D. V., Klar, T. A., Rogach, A. L., Feldmann, J., Alivisatos, A. P., Self-assembled binary superlattices of CdSe and Au nanocrystals and their fluorescence properties. *J. Am. Chem. Soc.* 2008, *130*, 3274ff.

158. Kinge, S., Crego-Calama, M., Reinhoudt, D. N., Self-assembling nanoparticles at surfaces and interfaces. *ChemPhysChem* 2008, *9*, 20–42.

159. Rogach, A. L., Talapin, D. V., Shevchenko, E. V., Kornowski, A., Haase, M., Weller, H., Organization of matter on different size scales: Monodisperse nanocrystals and their superstructures. *Adv. Funct. Mater.* 2002, *12*, 653–664.

160. Min, Y. J., Akbulut, M., Kristiansen, K., Golan, Y., Israelachvili, J., The role of interparticle and external forces in nanoparticle assembly. *Nat. Mater.* 2008, *7*, 527–538.

161. Steiner, D., Aharoni, A., Banin, U., Millo, O., Level structure of inas quantum dots in two-dimensional assemblies. *Nano Lett.* 2006, *6*, 2201–2205.

162. Steiner, D., Azulay, D., Aharoni, A., Salant, A., Banin, U., Millo, O., Electronic structure and self-assembly of cross-linked semiconductor nanocrystal arrays. *Nanotechnol.* 2008, *19 doi: 10.1088/0957-4484/19/6/065201.*

163. Rabani, E., Reichman, D. R., Geissler, P. L., Brus, L. E., Drying-mediated self-assembly of nanoparticles. *Nature* 2003, *426*, 271–274.

164. Kletenik-Edelman, O., Ploshnik, E., Salant, A., Shenhar, R., Banin, U., Rabani, E., Drying-mediated hierarchical self-assembly of nanoparticles: A dynamical coarse-grained approach. *J. Phys. Chem. C* 2008, *112*, 4498–4506.

165. Talapin, D. V., Shevchenko, E. V., Murray, C. B., Titov, A. V., Kral, P., Dipole-dipole interactions in nanoparticle superlattices. *Nano Lett.* 2007, *7*, 1213–1219.

166. Bigioni, T. P., Lin, X. M., Nguyen, T. T., Corwin, E. I., Witten, T. A., Jaeger, H. M., Kinetically driven self assembly of highly ordered nanoparticle monolayers. *Nat. Mater.* 2006, *5*, 265–270.

167. Zavelani-Rossi, M., Lupo, M. G., Krahne, R., Manna, L., Lanzani, G., Lasing in self-assembled microcavities of CdSe/CdS core–shell colloidal quantum rods. *Nanoscale* 2010, *DOI: 10.1039/b9nr00434c.*

168. Joseph, S. T. S., Ipe, B. I., Pramod, P., Thomas, K. G., Gold nanorods to nanochains: Mechanistic investigations on their longitudinal assembly using alpha,omega-alkanedithiols and interplasmon coupling. *J. Phys. Chem. B* 2006, *110*, 150–157.

169. Salant, A., Amitay-Sadovsky, E., Banin, U., Directed self-assembly of gold-tipped CdSe nanorods. *J. Am. Chem. Soc.* 2006, *128*, 10006–10007.

170. Gole, A., Murphy, C. J., Biotin-streptavidin-induced aggregation of gold nanorods: Tuning rod-rod orientation. *Langmuir* 2005, *21*, 10756–10762.

171. Caswell, K. K., Wilson, J. N., Bunz, U. H. F., Murphy, C. J., Preferential end-to-end assembly of gold nanorods by biotin-streptavidin Connectors. *J. Am. Chem. Soc.* 2003, *125*, 13914–13915.

172. Thomas, K. G., Barazzouk, S., Ipe, B. I., Joseph, S. T. S., Kamat, P. V., Uniaxial plasmon coupling through longitudinal self-assembly of gold nanorods. *J. Phys. Chem. B* 2004, *108*, 13066–13068.

173. Pan, B. F., Ao, L. M., Gao, F., Tian, H. Y., He, R., Cui, D. X., End-to-end self-assembly and colorimetric characterization of gold nanorods and nanospheres via oligonucleotide hybridization. *Nanotechnol.* 2005, *16*, 1776–1780.

174. Salem, A. K., Chen, M., Hayden, J., Leong, K. W., Searson, P. C., Directed assembly of multisegment Au/Pt/Au nanowires. *Nano Lett.* 2004, *4*, 1163–1165.

175. Figuerola, A., Franchini, I. R., Fiore, A., Mastria, R., Falqui, A., Bertoni, G., Bals, S., Van Tendeloo, G., Kudera, S., Cingolani, R., Manna, A., End-to-end assembly of shape-controlled nanocrystals via a nanowelding approach mediated by gold domains. *Adv. Mater.* 2009, *21*, 550–554.

176. Kotov, N. A., Dekany, I., Fendler, J. H., Layer-by-layer self-assembly of polyelectrolyte-semiconductor nanoparticle composite films. *J. Phys. Chem.* 1995, *99*, 13065–13069.

177. Ryan, K. M., Mastroianni, A., Stancil, K. A., Liu, H. T., Alivisatos, A. P., Electric-field-assisted assembly of perpendicularly oriented nanorod superlattices. *Nano Lett.* 2006, *6*, 1479–1482.

Chapter 12

178. Carbone, L., Nobile, C., De Giorgi, M., Sala, F. D., Morello, G., Pompa, P., Hytch, M. et al., Synthesis and micrometer-scale assembly of colloidal CdSe/CdS nanorods prepared by a seeded growth approach. *Nano Lett.* 2007, *7*, 2942–2950.

179. Ahmed, S., Ryan, K. M., Centimetre scale assembly of vertically aligned and close packed semiconductor nanorods from solution. *Chem. Commun.* 2009, 6421–6423, doi: 10.1039/B914478A.

180. Shevchenko, E. V., Talapin, D. V., Kotov, N. A., O'Brien, S., Murray, C. B., Structural diversity in binary nanoparticle superlattices. *Nature* 2006, *439*, 55–59.

181. Santhanam, V., Liu, J., Agarwal, R., Andres, R. P., Self-assembly of uniform monolayer arrays of nanoparticles. *Langmuir* 2003, *19*, 7881–7887.

182. Le, J. D., Pinto, Y., Seeman, N. C., Musier-Forsyth, K., Taton, T. A., Kiehl, R. A., DNA-templated self-assembly of metallic nanocomponent arrays on a surface. *Nano Lett.* 2004, *4*, 2343.

183. Li, L. S., Walda, J., Manna, L., Alivisatos, A. P., Semiconductor nanorod liquid crystals. *Nano Lett.* 2002, *2*, 557–560.

184. Talapin, D. V., Shevchenko, E. V., Murray, C. B., Kornowski, A., Forster, S., Weller, H., CdSe and CdSe/CdS nanorod solids. *J. Am. Chem. Soc.* 2004, *126*, 12984–12988.

185. Lin, Y., Skaff, H., Emrick, T., Dinsmore, A. D., Russell, T. P., Nanoparticle assembly and transport at liquid-liquid interfaces. *Science* 2003, *299*, 226–229.

186. McMillan, R. A., Paavola, C. D., Howard, J., Chan, S. L., Zaluzec, N. J., Trent, J. D., Ordered nanoparticle arrays formed on engineered chaperonin protein templates. *Nat. Mater.* 2002, *1*, 247–252.

187. Xie, H., Li, Y. F., Kagawa, H. K., Trent, J. D., Mudalige, K., Cotlet, M., Swanson, B. I., An intrinsically fluorescent recognition ligand scaffold based on chaperonin protein and semiconductor quantum-dot conjugates. *Small* 2009, *5*, 1036–1042.

188. Spatz, J. P., Mossmer, S., Hartmann, C., Moller, M., Herzog, T., Krieger, M., Boyen, H. G., Ziemann, P., Kabius, B., Ordered deposition of inorganic clusters from micellar block copolymer films. *Langmuir* 2000, *16*, 407–415.

189. Lohmueller, T., Bock, E., Spatz, J. P., Synthesis of quasi-hexagonal ordered arrays of metallic nanoparticles with tuneable particle size. *Adv. Mater.* 2008, *20*, 2297ff.

190. Stebe, K. J., Lewandowski, E., Ghosh, M., Oriented assembly of metamaterials. *Science* 2009, *325*, 159–160.

191. Glotzer, S. C., Solomon, M. J., Anisotropy of building blocks and their assembly into complex structures. *Nat. Mater.* 2007, *6*, 557–562.

192. Liu, S. H., Tok, J. B. H., Locklin, J., Bao, Z. N., Assembly and alignment of metallic nanorods on surfaces with patterned wettability. *Small* 2006, *2*, 1448–1453.

193. Jana, N. R., Shape effect in nanoparticle self-assembly. *Angew. Chem. Int. Ed.* 2004, *43*, 1536–1540.

194. Ghezelbash, A., Koo, B., Korgel, B. A., Self-assembled stripe patterns of CdS nanorods. *Nano Lett.* 2006, *6*, 1832–1836.

195. Querner, C., Fischbein, M. D., Heiney, P. A., Drndic, M., Millimeter-scale assembly of CdSe nanorods into smectic superstructures by solvent drying kinetics. *Adv. Mater.* 2008, *20*, 2308ff.

196. Nobile, C., Carbone, L., Fiore, A., Cingolani, R., Manna, L., Krahne, R., Self-assembly of highly fluorescent semiconductor nanorods into large scale smectic liquid crystal structures by coffee stain evaporation dynamics. *J. Phys. Condens. Matter* 2009, *21*, 264013.

197. Sun, B., Sirringhaus, H., Surface tension and fluid flow driven self-assembly of ordered ZnO nanorod films for high-performance field effect transistors. *J. Am. Chem. Soc.* 2006, *128*, 16231–16237.

198. Kim, F., Kwan, S., Akana, J., Yang, P. D., Langmuir-blodgett nanorod assembly. *J. Am. Chem. Soc.* 2001, *123*, 4360–4361.

199. Artemyev, M., Moller, B., Woggon, U., Unidirectional alignment of CdSe nanorods. *Nano Lett.* 2003, *3*, 509–512.

200. Nobile, C., Fonoberov, V. A., Kudera, S., Della Torre, A., Ruffino, A., Chilla, G., Kipp, T. et al., Confined optical phonon modes in aligned nanorod arrays detected by resonant inelastic light scattering. *Nano Lett.* 2007, *7*, 476–479.

201. Hu, Z., Fischbein, M. D., Querner, C., Drndić, M., Electric field driven accumulation and alignement of CdSe and CdTe nanorods in nanoscale devices. *Nano Lett.* 2006, *6*, 2585–2591.

202. Harnack, O., Pacholski, C., Weller, H., Yasuda, A., Wessels, J. M., Rectifying behavior of electrically aligned ZnO nanorods. *Nano Lett.* 2003, *3*, 1097–1101.

203. Ahmed, S., Ryan, K. M., Self-assembly of vertically aligned nanorod supercrystals using highly oriented pyrolytic graphite. *Nano Lett.* 2007, *7*, 2480–2485.

204. Zanella, M., Bertoni, G., Franchini, I. R., Brescia, R., Baranov, D., Manna, L., Assembly of shape-controlled nanocrystals by depletion attraction. *Chem. Commun.* 2011, *47*, 203–205.

205. Li, L. S., Alivisatos, A. P., Semiconductor nanorod liquid crystals and their assembly on a substrate. *Adv. Mater.* 2003, *15*, 408ff.

206. Cui, Y., Bjork, M. T., Liddle, J. A., Sonnichsen, C., Boussert, B., Alivisatos, A. P., Integration of colloidal nanocrystals into lithographically patterned devices. *Nano Lett.* 2004, *4*, 1093–1098.

207. Fang, L., Park, J. Y., Cui, Y., Alivisatos, A. P., Shcrier, J., Lee, B., Wang, L.-W., Salmeron, M., Mechanical and electrical properties of CdTe tetrapods studied by atomic force microscopy. *J. Chem. Phys.* 2007, *127*, 184704.

208. Nobile, C., Ashby, P. D., Schuck, P. J., Fiore, A., Mastria, R., Cingolani, R., Manna, L., Krahne, R., Probe tips functionalized with colloidal nanocrystal tetrapods for high-resolution atomic force microscopy imaging. *Small* 2008, *4*, 2123–2126.

209. Liu, H. T., Alivisatos, A. P., Preparation of asymmetric nanostructures through site selective modification of tetrapods. *Nano Lett.* 2004, *4*, 2397–2401.

210. Steiner, D., Azulay, D., Aharoni, A., Salant, A., Banin, U., Millo, O., Photoconductivity in aligned CdSe nanorod arrays. *Phys. Rev. B* 2009, *80*, art. no. 195308.

211. Persano, A., De Giorgi, M., Fiore, A., Cingolani, R., Manna, L., Cola, A., Krahne, R., Photoconduction properties in aligned assemblies of colloidal CdSe/CdS nanorods. *Acs Nano* 2010, *4*, 1646–1652.

212. Huynh, W. U., Dittmer, J. J., Teclemariam, N., Milliron, D. J., Alivisatos, A. P., Barnham, K. W. J., Charge transport in hybrid nanorod-polymer composite photovoltaic cells. *Phys. Rev. B* 2003, *67*, art. no.-115326.

213. Huynh, W. U., Peng, X. G., Alivisatos, A. P., CdSe nanocrystal Rods/Poly(3-Hexylthiophene) composite photovoltaic devices. *Adv. Mater.* 1999, *11*, 923–927.

214. Huynh, W. U., Dittmer, J. J., Alivisatos, A. P., Hybrid nanorod-polymer solar cells. *Science* 2002, *295*, 2425–2427.

215. Wu, Y., Wadia, C., Ma, W. L., Sadtler, B., Alivisatos, A. P., Synthesis and photovoltaic application of Copper(I) sulfide nanocrystals. *Nano Lett.* 2008, *8*, 2551–2555.

216. Gur, I., Fromer, N. A., Geier, M. L., Alivisatos, A. P., Air-stable all-inorganic nanocrystal solar cells processed from solution. *Science* 2005, *310*, 462–465.

217. Gur, I., Fromer, N. A., Chen, C. P., Kanaras, A. G., Alivisatos, A. P., Hybrid solar cells with prescribed nanoscale morphologies based on hyperbranched semiconductor nanocrystals. *Nano Lett.* 2007, *7*, 409–414.

218. Medvedev, V., Kazes, M., Kan, S., Banin, U., Talmon, Y., Tessler, N., Near infrared polymer nanocrystal leds. *Synthetic Metals* 2003, *137*, 1047–1048.

219. Schlamp, M. C., Peng, X. G., Alivisatos, A. P., Improved efficiencies in light emitting diodes made with CdSe(CdS) core-shell type nanocrystals and a semiconducting polymer. *J. Appl. Phys.* 1997, *82*, 5837–5842.

220. Hikmet, R. A. M., Chin, P. T. K., Talapin, D. V., Weller, H., Polarized-light-emitting quantum-rod diodes. *Adv. Mater.* 2005, *17*, 1436–1439.

221. Rizzo, A., Li, Y. Q., Kudera, S., Della Sala, F., Zanella, M., Parak, W. J., Cingolani, R., Manna, L., Gigli, G., Blue light emitting diodes based on fluorescent CdSe/ZnS nanocrystals. *Appl. Phys. Lett.* 2007, *90*, art. no. 051106, doi:10.1063/1.2426899.

222. Rizzo, A., Mazzeo, M., Palumbo, M., Lerario, G., D'Amone, S., Cingolani, R., Gigli, G., Hybrid light-emitting diodes from microcontact-printing double-transfer of colloidal semiconductor CdSe/ZnS quantum dots onto organic layers. *Adv. Mater.* 2008, *20*, 1886–1891.

223. Rizzo, A., Nobile, C., Mazzeo, M., De Giorgi, M., Fiore, A., Carbone, L., Cingolani, R., Manna, L., Gigli, G., Polarized light emitting diode by long-range nanorod self-assembling on a water surface. *Acs Nano* 2009, *3*, 1506–1512.

224. Talapin, D. V., Murray, C. B., PbSe Nanocrystal Solids for N- and P-Channel Thin Film Field-Effect Transistors. *Science* 2005, *310*, 86–89.

225. Sun, B., Sirringhaus, H., Solution-processed Zinc Oxide field-effect transistors based on self-assembly of Colloidal Nanorods. *Nano Lett.* 2005, *5*, 2408–2413.

226. Chan, E., Xu, C. X., Mao, A., Han, G., Cohen, B., Milliron, D., Reproducible, high-throughput synthesis of colloidal nanocrystals for optimization in multidimensional parameter space. *Nano Lett.* 2010, *10*, 1874–1885.

227. Anderson, V. J., Lekkerkerker, H. N. W., Insights into phase transition kinetics from colloid science. *Nature* 2002, *416*, 811–815.

228. Bishop, K. J. M., Wilmer, C. E., Soh, S., Grzybowski, B. A., Nanoscale forces and their uses in self-assembly. *Small* 2009, *5*, 1600–1630.

Chapter 12

13. Directed Self–Assembly

Gordon S. W. Craig

Nanoscale Science and Engineering Center, University of Wisconsin–Madison, Madison, Wisconsin

Paul F. Nealey

Department of Chemical and Biological Engineering, University of Wisconsin–Madison, Madison, Wisconsin

13.1 Introduction

The advent of microelectronics has been at the heart of technological development over the past four decades, and advances in lithography have been at the heart of the development of microelectronics. Indeed, the submicron length scales of the features in today's integrated circuits (ICs) make the term "microelectronics" something of a misnomer. Continuing the decrease in feature size, or the increase in circuit element density, requires overcoming the inherent physical challenges of current optical lithographic systems, while at the same time developing next-generation lithographic techniques to be used when the limits of optical lithography become comparatively cost prohibitive.[1] In the case of magnetic storage devices, the demand for increasing storage density has led to a need to reduce feature size that exceeds even that of ICs.[2–4]

One potential method to address the lithographic challenge of ever-decreasing pattern features is to make use of the assembled domains of microphase-separated block copolymers.[5] Block copolymers consist of two or more chemically different polymer chains joined by covalent bonds. In the case of a diblock copolymer such as polystyrene-*block*-poly(methyl methacrylate) (PS-*b*-PMMA), the PS and PMMA chains are incompatible. In a blend of PS and PMMA homopolymers, these two polymers would phase separate. However, because of the covalent bond connecting the two chains, PS-*b*-PMMA spontaneously self-assembles into a microphase-separated morphology. The size and shape of microphase separated domains of an A–B block copolymer in the bulk depend on the length of the A and B blocks, as well as the A–B segment–segment interaction parameter, χ, and the temperature. Typically, microphase-separated block copolymers assemble into either lamellar, cylindrical, or spherical domains, with length scales ranging from 5 to 100 nm. With appropriate orientation in a

Nanofabrication Handbook. Edited by Stefano Cabrini and Satoshi Kawata © 2012 CRC Press / Taylor & Francis Group, LLC. ISBN: 978-1-4200-9052-9

Chapter 13

thin film, these domains have shapes, such as dots and lines that are at the core of IC pattern design. Moreover, they naturally form with length scales that are suitable for use in lithography, as was originally suggested by Park et al.[6] In their seminal work, Park et al. demonstrated that both lamellar and spherical domains in thin films of either polystyrene-*block*-polyisoprene (PS-*b*-PI) or polystyrene-*block*-polybutadiene (PS-*b*-PB) could serve as a template for etching an underlying layer of silicon nitride on a silicon wafer, resulting in etched features with a pattern period of approximately 30 nm.

While the self-assembly of block copolymer domains in thin films provides useful shapes with the appropriate length scale for lithographic applications, for the broadest technological application, it is necessary that the assembled domains provide a pattern with a high degree of perfection as well as registration to the underlying substrate. Additionally, it is necessary for the domains to have the appropriate orientation with respect to the substrate to make the desired pattern. For example, if a pattern consisting of a series of parallel lines were desired, such a pattern could be achieved with lamellae, but only if the lamellae were oriented perpendicular to the substrate; lamellae oriented parallel to the substrate would provide a featureless pattern. To achieve control of the domain orientation and registration, as well as the perfection of the overall assembly, researchers have turned to directing the assembly of the domains in a variety of ways.[7,8]

Topographically or chemically patterned substrates provide two of the most commonly studied methods of directing the assembly of block copolymer domains in thin films. A schematic of these two types of directed assembly is presented in Figure 13.1. The schematic begins with a patterned substrate, which can be generated in several ways. Topographically patterned substrates have been formed by a variety of methods, such as Si[9] or Au[10] deposition on patterned resist and lift-off, or patterning followed by chemical etching.[11] In some cases, such as the topographically directed assembly of lamellae, it is also necessary to apply a surface treatment to the topographical pattern to force the orientation and the alignment of the domains.[10] Chemical patterns typically have been formed by the deposition of a polymer brush followed by lithography and oxygen plasma treatment.[12] Chemical patterns are typically characterized by the pattern period L_s and the width W of the pattern feature. As will be discussed in more detail, one important factor in directed assembly is how closely L_s matches the bulk period of the block

copolymer, L_o. Once the topographical or chemical pattern has been formed, a thin film of block copolymer is spin coated onto the chemical pattern. Initially, the spin-coated film of block copolymer lacks any long-range structure. Annealing the block copolymer film enables the domains to assemble on the patterned substrate, resulting in oriented, ordered domains, as shown in the last row in Figure 13.1.

In the case of topographically patterned substrates directing the assembly, often referred to as "graphoepitaxy," the block copolymer domains are forced to organize within ridges or channels previously patterned in the substrate. One inherent aspect of topographically directed assembly is that it has the ability to increase pattern resolution; the feature size of the pattern created by the assembled block copolymer is typically significantly smaller than that of the initial topographical pattern. One primary drawback of topographically directed assembly is that it is more challenging to achieve registration with underlying features on the substrate. Topographically directed assembly can provide uniform, well-ordered arrays of spheres and cylinders,[7,11] but assembly of lamellar domains has proven to be more difficult.[9,10,13]

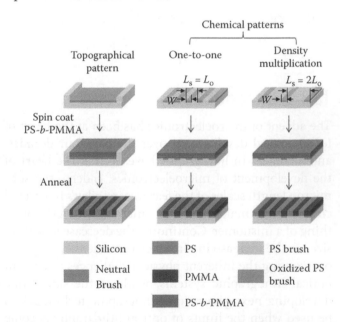

FIGURE 13.1 Methods of directing the assembly of lamellar block copolymer domains in thin films, starting with a patterned substrate. The topographical pattern schematic shows PS-*b*-PMMA lamellae assembling within a trench. Two examples of directed assembly on a chemical pattern are shown: One-to-one directed assembly, and directed assembly with density multiplication. Both patterns have the same pattern line width W, but the one-to-one chemical pattern has a pattern period L_s equal to the bulk lamellar period L_o of the PS-*b*-PMMA, whereas in the pattern designed for density multiplication, $L_s = nL_o$ ($n = 2$ in this schematic).

Whereas topographically directed assembly relies on physical confinement of the block copolymer to drive the organization of the domains, assembly on a chemical pattern is directed by the different affinity that each block of the copolymer has for neighboring regions of the chemical pattern.[14,15] As shown in the schematic of one-to-one directed assembly on a chemical pattern in Figure 13.1, the blue block prefers to wet the blue portion of the chemical pattern, and the red block prefers to wet the pink portion of the chemical pattern. As each domain is assembled over a region of the chemical pattern, one-to-one directed assembly offers no opportunity for resolution enhancement, in contrast to topographically directed assembly. The trade-off for the loss of resolution enhancement, when compared with topographically directed assembly, is that directed assembly on a chemical pattern inherently provides registration to underlying features on the substrate. Additionally, the assembled patterns have the potential to improve line edge roughness (LER) and critical dimension (CD) control in the patterning process,[16,17] which is viewed by the International Technology Roadmap for Semiconductors (ITRS) as one of the most difficult challenges facing lithography.[1] Previous research has shown that chemical patterns can direct the assembly of spheres, cylinders, and lamellae to create a variety of technologically useful patterns.[7,10,12,18–23]

Recent research has demonstrated the ability of a chemical pattern to direct the assembly of a block copolymer in such a way that the pattern generated by the assembled block copolymer has increased resolution compared to the underlying chemical pattern.[4,24,25] As shown in Figure 13.1, by forming the chemical pattern with a pattern period L_s that is an integral multiple of the bulk pattern period of the block copolymer, L_o, the chemical pattern can induce the block copolymer domains to form a well-ordered assembly, with some of the domains located over regions of the chemical pattern that have less affinity for the domains. In this type of directed assembly on a chemical pattern, known as directed assembly with density multiplication, many of the benefits of one-to-one directed assembly, such as registration, pattern perfection, and improved LER and CD control, are still possible, but resolution enhancement can also be achieved.

For the remainder of this chapter, we will focus on directed assembly on chemical patterns, in both the one-to-one and density multiplied methods. We will first review the directed assembly process, and point out the key factors that control domain formation in thin films of block copolymers on chemically patterned substrates. We will then discuss applications of directed assembly for specific patterning needs. While we understand that there are a variety of methods for directing the assembly of block copolymer domains, for the rest of this chapter we will use the term "directed assembly" to refer to directed assembly on a chemical pattern.

13.2 Controlling Factors in the Directed Assembly Process

The directed assembly process, shown in Figure 13.1, starts with the creation of a chemical pattern on the surface of a substrate. Because the interfacial interaction between the distinct regions of the chemical pattern and the blocks of the block copolymer are the driving force for the assembly of the block copolymer film, the nature of the chemical pattern deserves special attention. The first investigation of directed assembly used a chemical pattern made out of stripes of gold on native oxide.[26] Shortly thereafter, lithographically patterned self-assembled monolayers of alkylsiloxanes showed an improved ability to direct the assembly of block copolymers.[15,27,28] However, the optimal results obtained to date for one-to-one directed assembly of PS-*b*-PMMA are typically achieved with a chemical pattern that is based on a low-molecular-weight polystyrene (PS) brush.[22] After covalently bonding the PS brush to the substrate, it can be lithographically patterned. After lithography, some regions of the PS brush will be covered by the lithographic photoresist, while other regions will be exposed. The exposed regions can be etched briefly with oxygen plasma. Subsequent stripping of the photoresist leaves a chemical pattern consisting of regions of PS brush, which prefer to be wet by the PS block of PS-*b*-PMMA, and oxidized PS brush or native oxide, which prefer to be wet by the PMMA block.

Once the chemical pattern is formed, the block copolymer is spin coated onto the patterned substrate. The spin-coated block copolymer is disordered, but annealing the block copolymer film above its T_g enables it to equilibrate in the presence of the chemical pattern. Under the appropriate conditions of pattern–copolymer interfacial energy, $L_s - L_o$ commensurability, surface energy of the free surface, film thickness t, and χN, where N is the total overall degree of polymerization of the block copolymer, the block copolymer will assemble to yield well-ordered domains. The role that each of the aforementioned factors play in affecting the directed assembly process will be detailed in the pages to come. Once the desired assembly is achieved, one of the blocks can be selectively removed, leaving the remaining block to serve as a template for technological applications.

The most fundamental aspect of directed assembly is that it is a thermodynamically driven process. This thermodynamic control of directed assembly contrasts with the predominant chemically amplified resists, which are kinetically controlled,[29] and therefore more susceptible to minor process variations, especially as the CDs of patterns become more and more fine. In contrast to standard, kinetically controlled photoresists, directed assembly should be more robust because the final assembly is at thermodynamic equilibrium, and therefore should be dependent on predetermined intrinsic and extrinsic factors. Many of these factors relate to properties of the block copolymer itself, but it is important to realize that the equilibrium achieved in directed assembly is an equilibrium of the block copolymer film in the presence of the chemical pattern, and therefore the final assembled morphology also depends on attributes of the chemical pattern as well as the material at the free surface of the film. In the following sections, each of these contributions to the thermodynamic equilibrium of the entire directed assembly system will be reviewed.

Of course, the intrinsic properties of the block copolymer play a very important role in directed assembly. For directed assembly to work, the block copolymer must microphase separate in the bulk. The thermodynamics of microphase separation of an A–B block copolymer in the bulk are driven by two competing factors: an enthalpic free-energy term, F_{A-B}, that favors a decreased number of A–B domain interfaces, balanced by an entropic free-energy term, F_{ent}, caused by the limited extensibility of each chain of the block copolymer.[5] The balance of these two terms leads to different morphologies[30] and domain spacings that are determined primarily by N, the volume fraction of component A f, and χ. For a lamellar morphology of a symmetric block copolymer ($f = 0.5$), the free energy per copolymer chain of the microphase-separated block copolymer in the bulk in the strong segregation limit, F_{bulk}, can be expressed as

$$F_{bulk} = F_{ent} + F_{A-B} = kT \left(\frac{3L^2}{8l^2 N} + \frac{2lN}{L} \sqrt{\frac{\chi}{6}} \right), \quad (13.1)$$

where k is the Boltzmann constant, T is absolute temperature, l is the Kuhn segment length, and L is the period of the lamellar morphology.[31,32] F_{bulk} is minimized

when $L = L_o$. The equation for F_{bulk} assumes that the role of the surfaces of the block copolymer is negligible. While Equation 13.1 was developed for a symmetric diblock copolymer, the same general concept of the competition of the enthalpic and entropic free energy terms determining the equilibrium structure also applies to multiblock copolymers as well as block copolymer–homopolymer blends.

In thin (<300-nm thick) block copolymer films, the thickness of the film also plays a role in determining the equilibrium morphology of the block copolymer. The thinner the film is, the more significant its role is. Knoll et al. demonstrated this effect in thin films of cylinder-forming polystyrene-*block*-polybutadiene-*block*-polystyrene (SBS) triblock copolymer, as shown in the tapping mode scanning force microscopy (TM-SFM) height images and the simulation results presented in Figure 13.2.[33] With both experiments and simulations, they showed that as the block copolymer thickness increases from a very thin (~10 nm) layer to thicknesses that are several times L_o of the SBS triblock copolymer, an array of morphologies are observed, including perforated lamellae, cylinders ordered either perpendicular or parallel to the substrate, isolated domains, and ellipsoidal cylinders.

It is especially important to understand the impact of film thickness in instances when a particular domain orientation or pattern is desired. For example, useful patterns can be generated by orienting lamellae or cylinders perpendicular to the substrate. One approach for doing this is to bond to the substrate a random copolymer brush made with the same monomers as in the block copolymer.[34] Self-assembly of the block copolymer on an unpatterned random copolymer brush will cause the domains to orient perpendicular to substrate, but whether the perpendicular orientation of the lamellae or cylinders traverses the thickness of the film, resulting in patterns of lines or dots, respectively, at the free surface of the film, depends on the difference in surface energy of the two blocks, as described later, as well as the thickness of the film.[35] Beyond just orientation, Guarini et al. found that thickness also affected the uniformity and long-range order of self-assembled cylindrical domains on a random copolymer brush, and reported an optimal thickness of ~40 nm for cylindrical domains with a diameter of ~20 nm.[36] The dependence of the domain orientation and the pattern quality on film thickness also applies in the case of block copolymers assembled on chemical nanopatterns. As shown in the scanning electron microscope (SEM) images of Park et al. (Figure 13.3), well-ordered

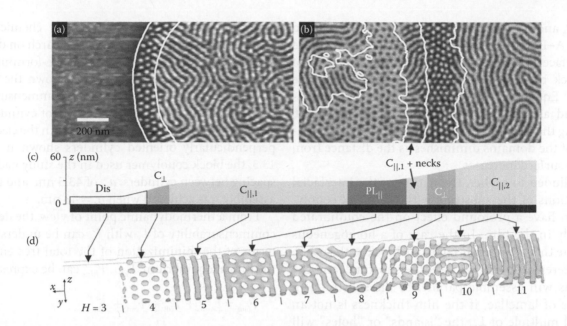

FIGURE 13.2 (a,b) TM-SFM phase images of thin SBS films on Si substrates after annealing in chloroform vapor. The surface is everywhere covered with a ~10-nm-thick PB layer. Bright (dark) corresponds to PS (PB) microdomains below this top PB layer. Contour lines calculated from the corresponding height images are superimposed. (c) Schematic height profile of the phase images shown in (a,b). (d) Simulation results of a thin film of triblock copolymer. (Reprinted with permission from Knoll, A. et al., *Phys. Rev. Lett.* **89**: 035501, 2002. Copyright (2002) by the American Physical Society.)

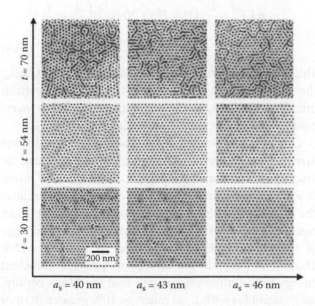

FIGURE 13.3 Dependence of film thickness on the pattern quality of cylindrical domains directed to assemble on a chemical nanopattern comprised of a hexagonal array of spots with various spot spacings, a_s. The block copolymer used in the study had a bulk cylinder spacing, a_o, of 43.5 nm. (Reprinted with permission from Park, S.M. et al., *Macromolecules*, **41**: 9118–9123, 2008. Copyright (2008) by the American Physical Society.)

arrays of vertical cylinders could be assembled, but the quality of the assembly depended on the thickness of the film.[20,37] Figure 13.3 shows cylinders oriented perpendicular to the substrate, but the general effect of increasing thickness also applies to assemblies of perpendicular lamellae,[38] parallel cylinders,[18] and spheres.[19]

One method for controlling the domain orientation at the top surface of the film is to place another material in contact with the film at the top surface. Just as the nature of the substrate can cause the domains near the substrate to be either parallel or perpendicular,[39,40] a material at the top surface of the film can have the same effect. Typically, the material at the free surface of the film when it anneals is either air or vacuum, but the block copolymer film can be sandwiched between two chemically treated substrates. When that chemical treatment is nonpreferential to the two blocks of the copolymer, as can be the case with a random copolymer brush, the domains are directed to orient themselves perpendicular to the substrate and the film surface.[41,42]

The effect of the free surface can be put in terms of the free energy of the surface, F_{surf}. In the absence of stronger driving forces, F_{surf} will lead the block with the lowest surface energy to populate the surface layer. F_{surf} per chain in the film can be expressed as

$$F_{surf} = (f_{AS}\gamma_A + (1-f_{AS})\gamma_B)\left(\frac{M_n}{\rho Nt}\right), \quad (13.2)$$

Chapter 13

where γ_A and γ_B are the surface energies of blocks A and B of an A–B block copolymer, f_{AS} is the fraction of the free surface covered by the A block, ρ is the density of the block copolymer, and t is the thickness of the film.[14,43] Equation 13.2 applies to the entire film system, and as such, has an inverse dependence on t, implying that the effect of the free surface on the orientation of the domains diminishes as the distance from the free surface increases.

As alluded to earlier, the nature of the interfacial interactions of the substrate with the block copolymer can have a profound effect on the equilibrated assembly in the film. In the case of a homogeneous substrate that is preferential to one block or the other, the preferential block will wet the substrate, and the domains will assemble parallel to the substrate. In the case of lamellae, if the film thickness is not an integral multiple of L_o, the "islands" or "holes" will form in the film to maintain the parallel orientation of the domains.[40] In the case of nonpreferential substrates, as mentioned earlier, domains will orient perpendicular to the substrate. Finally, in the case of a patterned substrate in which one region of the pattern prefers to be wet by the A block of an A–B block copolymer, and the neighboring region prefers to be wet by the B block, the ability of the chemical pattern to direct the assembly depends on the interfacial energy contrast between neighboring regions of the chemical pattern with the block copolymer. For a chemical pattern having equal areas of the different regions of the chemical pattern, the interfacial energy per chain in the film, F_{int}, between the block copolymer and the chemically patterned interface can be expressed as

$$F_{int} = (f_{A1}\gamma_{A1} + (1-f_{A1})\gamma_{B1}$$
$$+ f_{A2}\gamma_{A2} + (1-f_{A2})\gamma_{B2})\left(\frac{M_n}{2\rho Nt}\right), \quad (13.3)$$

where f_{A1} and f_{A2} are the fractions of the regions 1 and 2 of the chemical pattern, respectively, that are covered by the A block, and γ_{A1}, γ_{A2}, γ_{B1}, and γ_{B2} are the interfacial energies of the A and B blocks with each region of the chemical pattern.

Even when F_{int} is significant, the best examples of directed assembly in terms of the perfection of the assembly, and the film thickness through which the assembled domains persist, occur when L_o and L_s are commensurate. This result has been observed consistently throughout the experimental development of

one-to-one directed assembly on a chemical pattern.[10,14,15,22,26,44,45] The majority of research on directed assembly has been done with lamellae-forming block copolymers, but research has also shown the importance of pattern/block copolymer commensurability for the one-to-one directed assembly of cylinder- and sphere-forming block copolymers.[18,19] In the case of the perpendicularly oriented cylinders shown in Figure 13.3, the block copolymer used in the study had a bulk spacing between cylinders, a_o, of 43.5 nm, and the best assembly was achieved when $a_s = 43$ nm.

From a thermodynamic point of view, the desire for commensurability of L_s with L_o can be understood in terms of the minimization of the total free energy of the film at equilibrium, F_{film}. F_{film} can be expressed as[14]

$$F_{film} = F_{A-B} + F_{ent} + F_{surf} + F_{int}. \quad (13.4)$$

In the case of F_{bulk} $(=F_{A-B} + F_{ent})$, for any block copolymer directed to assemble on a chemical pattern with period L_s, F_{bulk} will be minimized when $L = L_s = L_o$. Edwards et al. estimated the free-energy penalty, ΔF_{bulk}, associated with assembling a film of a symmetric block copolymer on a substrate when $L_s \neq L_o$ as[14]

$$\Delta F_{bulk} = \Delta F_{ent} + \Delta F_{A-B}$$
$$= kT\left\{\frac{3}{8l^2N}(L_s^2 - L_o^2) + 2lN\sqrt{\frac{\chi}{6}}\left(\frac{1}{L_s} - \frac{1}{L_o}\right)\right\}, \quad (13.5)$$

Thus, for successful directed assembly when $L_s \neq L_o$, the free-energy associated with the chemical pattern interface with the block copolymer, F_{int}, must overcome the energetic penalty associated with ΔF_{bulk}.

One of the impressive aspects of one-to-one directed assembly is that block copolymers can be assembled with a high degree of perfection when $L_s \neq L_o$. Such assemblies are possible precisely because the final morphology of the block copolymer is an equilibrium structure that minimizes the free energy of the entire system, as defined in Equation 13.4. Thus, it is possible for an F_{int} to provide sufficient energy to the film system such that the block copolymer assembles with L_s as much as 10% greater than or less than L_o. Edwards et al. adroitly demonstrated the competing effects of F_{int} and ΔF_{bulk} by forming an array of chemical patterns in which both L_s and the chemical composition of the brush layer in the chemical pattern were varied.[14] The results of directed assembly of a PS-*b*-PMMA block copolymer

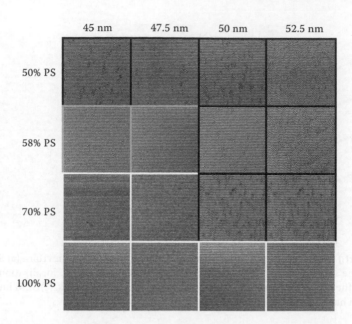

FIGURE 13.4 Top-down SEM images of a lamellae-forming PS-*b*-PMMA block copolymer (L_o = 48.0 nm) directed to assemble on chemical patterns with varying L_s (columns) and PS-*r*-PMMA brush composition (rows, expressed in terms of % PS). White, gray, and black frames denote assemblies with very few, few, and many pattern defects, respectively. Each micrograph shows a 2 × 2 μm area.

(L_o = 48 nm) on these different chemical patterns is shown in Figure 13.4, in which the chemical pattern made with a pure PS brush was able to direct the assembly of the block copolymer over a range of L_s values. When the chemical pattern was made with a nonpreferential random copolymer brush (PS content = 58%[34]), the best assemblies only occurred when L_s and L_o were commensurate.

Along with the commensurability of length scales affecting the quality of the assembly, the commensurability of pattern geometry is also important. For example, Park et al. showed the challenges of assembling block copolymer cylinders, which normally self-assemble into hexagonal arrays, on a chemical pattern consisting of a square array of spots.[37] Even though the pattern spot diameter and spacing matched the bulk values of the block copolymer, it was still difficult to achieve assemblies with a high degree of perfection. In contrast, the assembly of lamellae into different shapes was possible, as long as the localized pattern period of the underlying chemical pattern did not differ significantly from L_o. For example, Wilmes et al. were able to assemble lamellae into concentric rings,[23] and Park et al. were able to replicate fingerprint patterns.[46]

It is when a chemical pattern has multiple length scales that it becomes more difficult to direct the assembly of a block copolymer on the chemical

pattern. For example, in the case of a pattern consisting of a series of right angle bends laid out as chevrons, the period between lines along the line of vertices, $L_c = \sqrt{2}L_s$ as shown in Figure 13.5a. Stoykovich et al. found that the addition of PS and PMMA homopolymer to the PS-*b*-PMMA facilitated the assembly of the angled structures. Single chain in mean field (SCMF) simulations of the directed assembly of PS-*b*-PMMA/PS/PMMA ternary blends on chemical patterns consisting of 90° bends suggests that the homopolymer in the blend collects at the vertex of each bend, as shown in Figure 13.5b.[22] The result of the blend equilibrating on the chemical pattern is a series of right angle bends, as shown in the SEM image in Figure 13.5c.

With an understanding of the basic concepts that affect one-to-one directed assembly, a wide range of useful structures has been fabricated. Initial research focused on patterns formed from naturally occurring structures, such as arrays of dots and parallel lines. With an improved understanding of the factors that affect directed assembly, a wide range of patterns have been made, including many of the fundamental elements of semiconductor fabric architecture, such as jogs, t-junctions, terminations, and isolated lines and spots.[12,21] In the following section, we will outline how to apply directed assembly to achieve specific patterning requirements.

Chapter 13

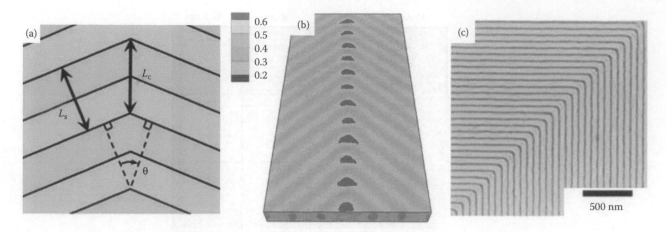

FIGURE 13.5　**(See color insert.)** Directed assembly on a chemical pattern composed of chevrons. (a) Schematic showing the different local pattern periods within the single pattern. (b) SCMF simulation results showing the localization of homopolymers (red) in the corners of a pattern formed by the directed assembly of a chemical pattern having 90° bends. (c) SEM image of ternary blend directed to assemble on a chemical pattern having 90° bends.

13.3　Application of One-to-One Directed Assembly for Specific Patterning Needs

13.3.1　Arrays of Spots

As mentioned above, initial work on block copolymer lithography focused on a pattern consisting of an array of spots. Such arrays have potential applications ranging from nanoparticle arrays[47] to bit patterned media[2,48] to IC design.[1,21] The first approach that many consider for generating arrays of spots with block copolymers is to use sphere-forming block copolymers, as was the case in the seminal work of Park et al.[6] Used as an etch template, the three-dimensional spherical domains naturally yield two-dimensional-spotted patterns. However, in the case of Park et al., the spheres were self-assembled on an unpatterned substrate, resulting in a lack of registration and relatively poor order compared to later work with directed assembly.

Directing the assembly of spherical domains is more difficult than directing the assembly of lamellae or cylinders because in the case of cylinders and lamellae, the domains can propagate continuously away from the chemical pattern, but in the case of spheres, discrete sphere domains must rely on neighbor–neighbor interactions to drive the alignment of domains. As a result, well-ordered arrays of spherical domains can be formed, both with topographical[7,49] and chemical[19] patterns, but precise registration of the spheres to underlying features is a challenge. Another approach to achieve an ordered assembly of spheres is to start with the directed assembly of a cylinder-forming block copolymer such that the cylinders lie parallel to the substrate,

and then alter the chemistry of the minority block such that it loses volumes and converts to spherical domains at equilibrium.[50] While this approach can register the domains in the direction of the lines that directed the assembly of the initial cylinders, it cannot precisely define the location of the spherical domains along those lines. Another approach is to generate a spotted pattern to direct the assembly of very thin ($t \approx L_0/2$) films of lamellae-forming block copolymers on a spotted chemical pattern.[51] However, such a spotted pattern would only exist in the half of the film closest to the substrate, and due to the thinness of the film as well as other morphological features such as "bridges" that rise to the top surface, such a morphology would not be likely to serve as an effective template.

Fortunately, by directing the assembly of a cylinder-forming block copolymer such that the domains are oriented perpendicular to the substrate, a spotted pattern on the free surface can be formed. Hexagonal arrays of spots that are registered to the underlying substrate can be readily formed via directed assembly.[18,20] However, because cylindrical domains naturally occur in hexagonal arrays in the bulk, there is an energetic penalty associated with directing the domains to assemble in square arrays. As a result, directing the assembly of cylinders of a diblock copolymer is more challenging, and prone to defect formation, especially in thicker films.[37]

Recent work has shown two routes toward the self-assembly of block copolymer films into square arrays.

Tang et al. found that by incorporating hydrogen bonding between the chains of two block copolymers, the resulting block copolymer blend would self-assemble into square arrays of cylinders.[52] They modified the polystyrene in PS-*b*-PMMA and polystyrene-*block*-poly (ethylene oxide) (PS-*b*-PEO) with small amounts of randomly incorporated 4-vinylpyridene and 4-hydroxy-styrene units, respectively. After solvent annealing, square arrays of PEO and PMMA cylinders formed in a PS matrix. More recently, Chuang et al. achieved similar results with a polyisoprene-*block*-polystyrene-*block*-polyferrocenylsilane (PI-*b*-PS-*b*-PFS) triblock terpolymer, which on self-assembly yielded square arrays of PI and PFS cylinders in a PS matrix.[53] Chuang et al. demonstrated that they could direct the assembly of the cylindrical domains with topographical patterns. It is reasonable to assume for both cases that the appropriate chemical pattern could direct the assembly of the domains, adding registration to the highly ordered square arrays.

13.3.2   Arrays of Lines

Lines and line segments are key components of virtually all IC patterns. Linear patterns can be readily formed by the self-assembly of lamellae-forming block copolymers on neutral substrates,[35,46] or of cylinder-forming block copolymers on preferential substrates.[13] However, as

mentioned above, self-assembly results in a random, "fingerprint" pattern that may be scientifically interesting, but rarely technologically useful. Appropriate choice of an underlying chemical pattern can direct the morphology to assemble in an ordered pattern.

Directed assembly of cylindrical morphologies on linear chemical patterns can yield ordered, linear patterns on the surface of the block copolymer film, as shown in Figure 13.6,[18] but the utility of these films will most likely be limited to films with thickness $\leq L_o$ for two reasons. First, if the film is to be used as an etch template, the film thickness should be less than the radius of the cylindrical domains. Otherwise, the matrix block surrounding the cylindrical domains would block the formation of the template. Second, as shown in the schematic in Figure 13.6, for films with thicknesses greater than the radius of the domains, the cylindrical domains at the free surface of the film are aligned by the cylindrical domains at the chemical pattern interface without having any direct, physical contact with the domains at the interface. Despite the lack of direct contact with the chemical pattern at the interface, a high degree of pattern perfection can be achieved, as shown in the SEM of PS-*b*-PMMA having $L_o = 44.8$ nm and $L_s = 45.0$ nm (Figure 13.6). Also, a high degree of order could be achieved in films in which L_s differed from L_o by ~5%. However, previous work has shown that the degree of ordering that can be achieved by cylindrical

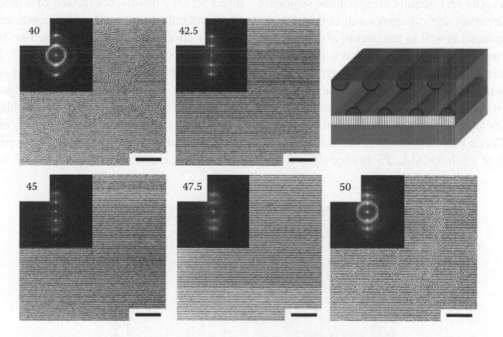

FIGURE 13.6   Scanning electron micrographs of self-assembled cylindrical domains in a 39-nm-thick film of PS-*b*-PMMA with $L_o = 44.8$ nm on chemically nanopatterned polystyrene brushes after annealing for 72 h at 190°C. L_s values of the chemically nanopatterned substrates are shown (in nm) in the upper left-hand corner of each SEM. The black scale bars represent 500 nm. The schematic provides an example of block copolymer domain behavior for this thickness of film.

domains that are not in contact with the underlying is very sensitive to both film thickness and variations in the width (*W*) of the chemical pattern lines.[18]

A more customary route to achieve linear patterns is to direct the assembly of lamellar domains so that they orient perpendicularly to the substrate surface, as shown in Figure 13.1.[15] Directed assembly can yield uniform, ordered domains that span the thickness of the film, even in 200-nm-thick films. As mentioned earlier and shown in Figure 13.4, by using a chemical pattern made from a PS brush, with a large interfacial energy contrast between the two blocks of the copolymer, ordered assemblies could be achieved with L_s as much as 10% greater than or less than L_o.[14] Thus, one batch of a block copolymer, with a given L_o value, could satisfy a range of pattern periods. Additional research has shown that the block copolymer can correct for minor variations in *W* of the lines in the chemical pattern.[16] This self-correcting capability means that a block copolymer has the potential to improve CD control and LER created in the chemical pattern by the lithography process, which would be an important gain for current lithographic techniques.[1,12,17]

For device fabrication, it is necessary to control the location and the dimensions of the interpolated domains, such that the domains create a density multiplied pattern only where such a pattern is desired. For example, multigate field effect transistors (FET), such as finFET, nanowire FET, and tri-gate FET, require arrays of line segments in which the location and dimensions of both the individual line segments as well as the overall array must be precisely controlled.[54–57] In cases where it is desirable to prohibit the formation of any domain pattern in certain regions on the substrate, it is necessary to use a chemical pattern with elements that are highly preferential toward each of the blocks. For example, the use of a chemical pattern based on a PS brush enabled the assembly of a ternary blend of PS-*b*-PMMA, PS homopolymer, and

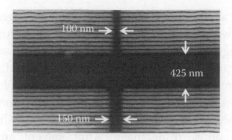

FIGURE 13.7 Four arrays of line patterns are packed together with spacings of 100 nm and 150 nm between left and right columns, and a spacing of 450 nm between top and bottom rows. Ternary blends were assembled into periodic lines only in the desired areas and there were no pattern features outside the desired areas.

PMMA homopolymer, into discrete arrays of line segments, with well-defined gaps between arrays, as shown in Figure 13.7.[58] In the gap regions, the strong preference of the PS brush for the PS domains of the blend forced the domains of the blend to orient parallel to the substrate, resulting in the gap observed in the domain pattern at the free surface of the films. If a nonpreferential brush, such as a PS-*r*-PMMA random copolymer, were used instead of the PS brush, linear patterns could still be formed in the patterned region of the substrate, but the gap regions would be filled with lamellar domains oriented perpendicularly to the substrate, resulting in the presence of fingerprint patterns in the gap.

Recent research has also demonstrated the ability to direct ternary blends, comprised of a block copolymer and homopolymers of each of its substituent blocks, into complex checkerboard shapes, as well as "trimming patterns" that combine spots and lines.[59] For example, through careful selection of the amount of each homopolymer in a PS-*b*-PMMA/PS/PMMA blend, as well as the ratio of PS to PMMA in the overall blend, the blend could be directed to assemble into a rectangular trimming pattern consisting of alternating parallel dashed and solid lines, as shown in Figure 13.8. The ternary

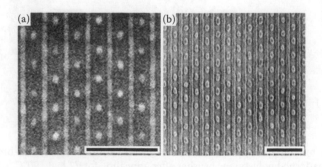

FIGURE 13.8 Top-down SEM images of rectangular trimming pattern of resist (a) and PS-*b*-PMMA/PS/PMMA ternary blend thin films assembled on a similar rectangular trimming pattern (b). The scale bar is 500 nm.

blend used in Figure 13.8 had a homopolymer fraction of 0.29 and a styrene fraction of 0.43. Figure 13.8 demonstrates one facet of the range of structures that can be achieved with directed assembly.

13.3.3   Nonregular, Device-Oriented Structures

One of the most impressive features of directed assembly is that it allows for the formation of complex geometries beyond ordered assemblies of naturally occurring patterns, such as the arrays of spots or lines discussed in the preceding section. While arrays of lines and spots definitely have technical applications, the ability to form nonregular patterns opens the door for an array of potential lithographic applications for ICs. The Semiconductor Industry Association has defined an essential set of pattern geometries to include long lines, short segments, 90° bends, jogs, T-junctions, and arrays of spots.[60]

The first study of the use of directed assembly to generate nonregular, device-oriented structures focused on the formation of bends, such as those shown in Figure 13.5.[22] As mentioned earlier, a PS-*b*-PMMA/PS/PMMA ternary blend was used to facilitate the assembly of patterns with multiple length scales in one pattern. Subsequent work demonstrated that chemical patterns made from lithographically patterned PS brush could direct ternary blends to assemble in jogs, T-junctions, and terminations, as shown in Figure 13.9.[21]

While the preponderance of research on directed assembly has focused on generating a repeating pattern of one type or another, it is often necessary to pattern an individual line or spot.[60] Stoykovich et al. demonstrated one method of making isolated line

segments that is based on the assembly of lamellar domains on a chemical pattern that has regions where $L_s \gg L_o$, and other regions where $L_s = L_o$, resulting in a combination of parallel and perpendicular domain orientations. As shown in Figure 13.4, block copolymers assembled on chemical patterns with $L_s = L_o$ yield domains that are oriented perpendicularly to the substrate. In contrast, block copolymers form domains oriented parallel to the substrate on chemically patterned surfaces with $L_s \gg L_o$.[61,62] Island and hole morphologies may be generated in parallel lamellae when the copolymer film has a thickness greater than L_o.[63] Stoykovich et al. used films with thicknesses less than L_o, which have been shown to maintain uniform film thicknesses and do not form island or hole morphologies.[63,64]

A schematic of the process used to form isolated segments is presented in Figure 13.10a. The top figure represents the chemically patterned surface, which has a region with $L_s = L_o$ between two regions with $L_s \gg L_o$. The next figure demonstrates the orientation of the block copolymer domains assembled on these surfaces. In this schematic, red and blue represent the PS and PMMA domains, respectively. The third figure in the schematic shows the results of the selective elimination of the PMMA domains to provide a suitable mask for etching or depositing material on the substrate. The presence of the parallel-oriented lamellae is important because they protect the substrate from etching, as there exists a continuous layer of resistant material (PS in this case). The figure at the bottom of the schematic provides an example of the isolated structures that could be reactive-ion-etched into the substrate using the self-assembled layer as a mask. In this case, the removal of the PMMA lines led to the transfer of two lines to the substrate, but alternatively, the PS domains

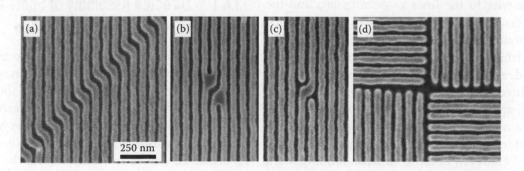

FIGURE 13.9   Top-down SEM images of the PS-*b*-PMMA/PS/PMMA ternary blend directed to assemble into (a) nested arrays of jogs, (b) isolated PMMA jogs, (c) isolated PS jogs, and (d) arrays of T-junctions. In all the SEM images, the PS domains are displayed in light gray, while the PMMA domains are dark gray or black. (Reprinted with permission from Stoykovich, M.P. et al. 2007. *ACS Nano*, 1: 168–175. Copyright (2007) by the American Chemical Society.)

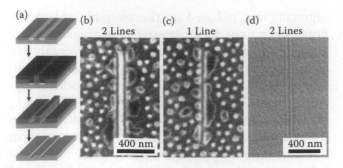

FIGURE 13.10 (a) Schematic for directing the assembly of block copolymers into isolated features using chemically patterned surfaces. The top figure represents the chemically patterned surface (PS-preferential with lighter gray PMMA-preferential stripes on a substrate). The second figure demonstrates the orientation of the block copolymer domains (lighter gray, PS domains; darker gray, PMMA domains) assembled on these surfaces. The perpendicular domains in panels b and c can be selectively removed (in this case, the blue domains) to provide a suitable mask. The bottom row demonstrates the isolated structures that could be reactive-ion-etched into the substrate using the self-assembled layer as a mask. (b) and (c) Top-down SEM images of the double- and single-line PS structures formed after selective removal of the PMMA domains from the block copolymer blend films. (d) Top-down SEM images of defect-free isolated segment structures etched into a SiO$_2$ substrate.

could be removed, leading to the transfer of a single, isolated line to the substrate. Additionally, while the schematic shows the fabrication of one PS line (and two PMMA lines), any number of lines can be generated with a periodicity of L_o.[21]

The efficacy of this method to form isolated lines is shown in the top-down SEM images of isolated PS structures in Figure 13.10b. The SEMs show PS structures formed after directed assembly on a chemical pattern such as that shown in Figure 13.10a, followed by selective removal of PMMA domains. The bright spots in the image are nodules of the PS domains that rose to the surface of the film; they do not affect the ability of the parallel sections of the PS domains to serve as an etch mask for the underlying substrate. The PS domains were used as an etch mask, as shown in the SEM image of the underlying SiO$_2$ substrate after HF wet etching (Figure 13.10c). The etch depth in the patterned segments was 5–10 nm, and the root-mean-square roughness outside of the segments was <0.4 nm, comparable to that of the as-deposited SiO$_2$ layer.[21]

13.4 Application of Directed Assembly for Improved Resolution

Recent research has shown the ability of directed assembly on chemical patterns to form patterns with a greater feature density than can be obtained with lithographic tools.[24,25,58] The general approach to achieve pattern density multiplication (Figure 13.1) is similar to that of one-to-one directed assembly, except that the spacing between lines or spots in the chemical pattern is increased. In the case of directing the assembly of lamellae with density multiplication, the chemical pattern will have lines with W equal to the width of the domains, but with gaps between the lines such that $L_s = nL_o$, where n, the density multiplication factor, is an integer greater than 1. We refer to the lines as *guiding lines* and the space between the lines as the *background region*. One challenge of directed assembly with density multiplication is that while one block of the copolymer will wet the background region, the other block will not, and yet that block is directed to assemble over the background region. One approach used to address this concern was to employ a nonpreferential random copolymer in the background region, which led to a fourfold density multiplication.[24] However, because lamellae-forming block copolymers self-assemble into "fingerprint" structures on nonpreferential substrates, the use of a non-preferential material in the background region will lead to defect formation in areas in which no pattern is desired. The

background region can be made preferential to one of the blocks, for example, by using a PS brush, which then forces the domains to be parallel to the substrate in regions where no patterning is desired, such as the arrays of line segments shown in Figure 13.7. The trade-off in using a preferential material in the background region is that the maximum value of n that will yield defect free assemblies will be lower. The highest n value reported to date for a system with a preferential background layer was 3.[58] Keeping these trade-offs in mind, various geometries can be assembled.

13.4.1 Directed Assembly of Spots and Lines with Density Multiplication

Dense, spotted patterns have been formed via the directed assembly of cylindrical domains on chemical patterns comprised of a sparse array of spots.[25] In that work, PS-b-PMMA was directed to assemble on a chemical pattern composed of spots of oxygen-plasma-treated PS brush and a background region of untreated PS brush. As shown in Figure 13.11b, the assembly of PS-b-PMMA on the chemical pattern led to a fourfold increase in feature density. Importantly, pattern rectification was also achieved, as shown in the spot size distribution of the e-beam patterned resist used to

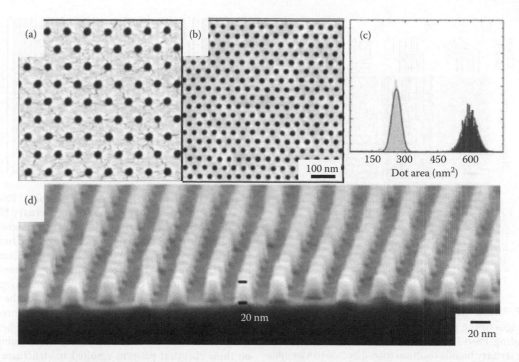

FIGURE 13.11 SEM images of (a) developed e-beam resist used to form a chemical pattern with $L_s = 78$ nm and (b) PS-b-PMMA with $L_0 = 39$ nm directed to assemble on the chemical pattern. (c) Dot size distribution of the e-beam (dark gray) and guided block copolymer patterns (light gray), revealing the improved uniformity of the assembled block copolymer film. (d) Pattern transfer using the pattern in (a) to form 20-nm-tall Si pillars.

define the chemical pattern, and the assembled cylindrical domains (Figure 13.11c).

Despite the fact that some cylindrical PMMA domains had to assemble above the background region of PS brush, and therefore likely had a thin region of PS between the domain and the substrate, these patterns were used to form a template for etching the underlying silicon substrate. The PMMA domains were selectively removed from the film, and then 7-nm-thick spots of Cr were deposited using a lift-off technique. The Cr was used as an etch mask for a CF_4 reactive ion etch, resulting in the formation of 20 nm Si pillars, as shown in Figure 13.11d, that were uniform over the entire sample (3 mm long).[25]

An alternative approach to make spotted patterns with increased resolution involves directing the assembly of spherical domains on a spotted pattern similar to that shown in Figure 13.11a. Xiao et al. assembled a sphere-forming polystyrene-*block*-polydimethylsiloxane (PS-b-PDMS) block copolymer on a chemical pattern formed in a PS brush.[4] The spots in the chemical pattern served as pinning sites for some of the spherical domains, with the other spherical domains assembling themselves between the spherical domains. PS-b-PDMS was used in part because of its large χ value, which enables it to form a

microphase-separated morphology at relatively low M_n and correspondingly small domain sizes. Although the use of these morphologies as an etch template were not reported, based on the seminal work of Park et al. on block copolymer lithography, which used spherical domains of PS-b-PI and PS-b-PB as an etch template,[6] it is likely that a method could be found to use the spherical PS-b-PDMS domains as an etch template, as well.

Directed assembly with density multiplication has also been demonstrated to form patterns of parallel lines. Cheng et al. demonstrated the ability to achieve a fourfold increase in feature density by assembling films of PS-b-PMMA on striped chemical patterns composed of guiding lines of hydrogen silsesquioxane (HSQ), which is preferentially wet by PMMA, and background regions composed of a film of cross-linked poly(styrene-r-epoxydicyclopentadiene methacrylate), which provides a nonpreferential surface for PS and PMMA.[24] In addition to achieving directed assembly with density multiplication, Cheng et al. demonstrated pattern rectification with their system. They intentionally formed a chemical pattern that contained alternating dashed and solid lines of HSQ as a representation of a poorly defined lithographic pattern. When a PS-b-PMMA film was annealed on top of this chemical

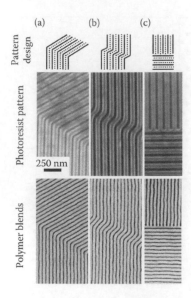

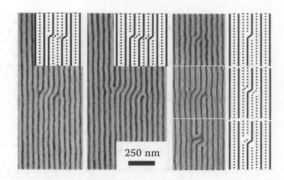

FIGURE 13.12 The schematic chemical pattern design (top row), SEM images of the photoresist pattern (middle row), and the corresponding PS-*b*-PMMA/PS/PMMA block copolymer–homopolymer ternary blends (bottom row) directed to assemble into device-oriented structures, such as (a) 120 bends, (b) jogs, (c) T-junctions and terminations on the chemical patterns, with 2× density multiplication. The solid and dotted lines in the design layouts denote patterned lines and the lines to be interpolated, respectively. The PS and PMMA domains are light and dark, respectively, in the images. The scale bar applies to all SEM images. (Reprinted with permission from Liu, G. et al. 2010. *Adv. Funct. Mater.*, **20**: 1251–1257. Copyright (2010). Wiley-VCH.)

FIGURE 13.13 SEM images of assembled PS-*b*-PMMA/PS/PMMA block copolymer–homopolymer ternary blends on complicated chemical patterns, with a schematic showing the corresponding chemical pattern for each image. The two SEMs on the left demonstrate successful interpolation of the lamellar domains in jogs that are nested between Y-junctions. To ensure defects did not form, one-to-one chemical patterns were designed where the jog diverted from the parallel lines, except for the three patterns with complete interpolation on the right. The three SEM images on the right show assembled ternary blends on chemical patterns designed without any one-to-one features. Assembly of the blends on these chemical patterns resulted in structures having either zero PS jogs, two PS jogs, or defects, from top to bottom in the figure. The solid and dotted lines in the schematic denote chemically patterned lines and the location of interpolated domains, respectively. The scale bar applies to all SEM images.

pattern, uniform PMMA lines were formed, despite the gaps in some of the underlying HSQ lines.

13.4.2 Density Multiplication of Device-Oriented Structures

While the initial research on directed assembly with density multiplication understandably focused on the formation of naturally occurring patterns, such as arrays of lines or spots, the next step in the investigation of density multiplication was to investigate which of the device-oriented geometries could be generated. Liu et al. examined directed assembly with density multiplication on a number of semiconductor pattern elements, including jogs, terminations, angles, as well as combinations of these elements.[58] They were able to achieve density multiplication of PS-*b*-PMMA on chemical patterns containing T-junctions, jogs, bends, and terminations, as shown in Figure 13.12. However, when a jog was placed between parallel lines without any one-to-one pattern features to direct the lamellar domains where the jog diverts from the parallel lines, a perfect pattern could not be formed with density multiplication (Figure 13.13).

13.5 Conclusion

Directed assembly of block copolymers on chemical patterns offers the possibility of enhancing state-of-the-art lithographic processes at a time when these processes are reaching fundamental limits. As directed assembly is a thermodynamically limited process that depends on free-energy minimization of the entire film system, including the free surface and the interface with the chemical pattern, it can reliably attain desired length scales and structures, free of the variability that accompanies kinetically limited processes. Moreover, because the equilibrium is determined by the entire system, the length scales and shapes of the final pattern can be affected not just by the properties of the block copolymer, but also by the size and shape of the underlying chemical pattern. Therefore, the chemical pattern can create patterns in the block copolymer that have length scales of up to 10% different from what the block copolymer would naturally have. Correspondingly, the block copolymer can rectify defects in the lithographically defined chemical pattern. With an understanding of the factors that control directed assembly, it is possible to achieve a collection of desired features, including registration, LER improvement, resolution enhance- ment, and the ability to form a variety of naturally occurring as well as device-oriented patterns. This combination of capabilities should benefit a number of technologies, from bit-patterned media to IC fabrication.

References

1. International Technology Roadmap for Semiconductors—Lithography. 2007(ed.). *International SEMATECH*: Austin, TX.
2. Service, R.F. 2006. Data storage—is the terabit within reach? *Science.* **314**: 1868–1870.
3. Terris, B.D., and Thomson, T. 2005. Nanofabricated and self-assembled magnetic structures as data storage media. *J. Phys. D-Appl. Phys.* **38**: R199–R222.
4. Xiao, S., Yang, X., Park, S. et al. 2009. A novel approach to addressable 4 teradot/in.^{2} patterned media. *Adv. Mater.* **21**: 2516–2519.
5. Bates, F.S., and Fredrickson, G.H. 1990. Block copolymer thermodynamics—Theory and experiment. *Annu. Rev. Phys. Chem.* **41**: 525–557.
6. Park, M., Harrison, C., Chaikin, P.M. et al. 1997. Block copolymer lithography: Periodic arrays of similar to 10(11) holes in 1 square centimeter. *Science.* **276**: 1401–1404.
7. Cheng, J.Y., Ross, C.A., Smith, H.I. et al. 2006. Templated self-assembly of block copolymers: Top-down helps bottom-up. *Adv. Mater.* **18**: 2505–2521.
8. Hawker, C.J., and Russell, T.P. 2005. Block copolymer lithography: Merging "bottom-up" with "top-down" processes. *MRS Bull.* **30**: 952–966.
9. Ruiz, R., Ruiz, N., Zhang, Y. et al. 2007. Local defectivity control of 2D self-assembled block copolymer patterns. *Adv. Mater.* **19**: 2157–2162.
10. Park, S.M., Stoykovich, M.P., Ruiz, R. et al. 2007. Directed assembly of lamellae-forming block copolymers by using chemically and topographically patterned substrates. *Adv. Mater.* **19**: 607–611.
11. Segalman, R.A., Yokoyama, H., and Kramer, E.J. 2001. Graphoepitaxy of spherical domain block copolymer films. *Adv. Mater.* **13**: 1152–1155.
12. Stoykovich, M.P., and Nealey, P.F. 2006. Block copolymers and conventional lithography. *Mater. Today.* **9**: 20–29.
13. Ruiz, R., Sandstrom, R.L., and Black, C.T. 2007. Induced orientational order in symmetric diblock copolymer thin films. *Adv. Mater.* **19**: 587–591.
14. Edwards, E.W., Montague, M.F., Solak, H.H. et al. 2004. Precise control over molecular dimensions of block-copolymer domains using the interfacial energy of chemically nanopatterned substrates. *Adv. Mater.* **16**: 1315–1319.
15. Kim, S.O., Solak, H.H., Stoykovich, M.P. et al. 2003. Epitaxial self-assembly of block copolymers on lithographically defined nanopatterned substrates. *Nature.* **424**: 411–414.
16. Edwards, E.W., Muller, M., Stoykovich, M.P. et al. 2007. Dimensions and shapes of block copolymer domains assembled on lithographically defined chemically patterned substrates. *Macromolecules.* **40**: 90–96.
17. Daoulas, K.C., Muller, M., and Stoykovich, M.P. 2008. Directed copolymer assembly on chemical substrate patterns: A phenomenological and single-chain-in-mean-field simulations study of the influence of roughness in the substrate pattern. *Langmuir.* **24**: 1284–1295.
18. Edwards, E.W., Stoykovich, M.P., Solak, H.H. et al. 2006. Long-range order and orientation of cylinder-forming block copolymers on chemically nanopatterned striped surfaces. *Macromolecules.* **39**: 3598–3607.
19. Park, S.M., Craig, G.S.W., La, Y.H. et al. 2008. Morphological reconstruction and ordering in films of sphere-forming block copolymers on striped chemically patterned surfaces. *Macromolecules.* **41**: 9124–9129.
20. Park, S.M., Craig, G.S.W., Liu, C.C. et al. 2008. Characterization of cylinder-forming block copolymers directed to assemble on spotted chemical patterns. *Macromolecules.* **41**: 9118–9123.
21. Stoykovich, M.P., Kang, H., Daoulas, K.C. et al. 2007. Directed self-assembly of block copolymers for nanolithography: Fabrication of isolated features and essential integrated circuit geometries. *ACS Nano.* **1**: 168–175.
22. Stoykovich, M.P., Muller, M., Kim, S.O. et al. 2005. Directed assembly of block copolymer blends into nonregular device-oriented structures. *Science.* **308**: 1442–1446.
23. Wilmes, G.M., Durkee, D.A., Balsara, N.P. et al. 2006. Bending soft block copolymer nanostructures by lithographically directed assembly. *Macromolecules.* **39**: 2435–2437.
24. Cheng, J.Y., Rettner, C.T., Sanders, D.P. et al. 2008. Dense self-assembly on sparse chemical patterns: Rectifying and multiplying lithographic patterns using block copolymers. *Adv. Mater.* **20**: 3155–3158.
25. Ruiz, R., Kang, H., Detcheverry, F.A. et al. 2008. Density multiplication and improved lithography by directed block copolymer assembly. *Science.* **321**: 936–939.
26. Rockford, L., Liu, Y., Mansky, P. et al. 1999. Polymers on nanoperiodic, heterogeneous surfaces. *Phys. Rev. Lett.* **82**: 2602–2605.
27. Peters, R.D., Yang, X.M., Wang, Q. et al. 2000. Combining advanced lithographic techniques and self-assembly of thin films of diblock copolymers to produce templates for nanofabrication. *J. Vac. Sci. Technol. B.* **18**: 3530–3534.
28. Yang, X.M., Peters, R.D., Nealey, P.F. et al. 2000. Guided self-assembly of symmetric diblock copolymer films on chemically nanopatterned substrates. *Macromolecules.* **33**: 9575–9582.
29. Ito, H. 2003. Chemical amplification resists: Inception, implementation in device manufacture, and new developments. *J. Polym. Sci., Part A: Polym. Chem.* **41**: 3863–3870.
30. Bates, F.S., and Fredrickson, G.H. 1999. Block copolymers—designer soft materials. *Phys. Today.* **52**: 32–38.
31. Matsen, M.W., and Bates, F.S. 1997. Block copolymer microstructures in the intermediate-segregation regime. *J. Chem. Phys.* **106**: 2436–2448.
32. Wang, Q., Nath, S.K., Graham, M.D. et al. 2000. Symmetric diblock copolymer thin films confined between homogeneous

and patterned surfaces: Simulations and theory. *J. Chem. Phys.* **112**: 9996–10010.

33. Knoll, A., Horvat, A., Lyakhova, K.S. et al. 2002. Phase behavior in thin films of cylinder-forming block copolymers. *Phys. Rev. Lett.* **89**: 035501.

34. Mansky, P., Liu, Y., Huang, E. et al. 1997. Controlling polymer-surface interactions with random copolymer brushes. *Science.* **275**: 1458–1460.

35. Huang, E., Rockford, L., Russell, T.P. et al. 1998. Nanodomain control in copolymer thin films. *Nature.* **395**: 757–758.

36. Guarini, K.W., Black, C.T., and Yeung, S.H.I. 2002. Optimization of diblock copolymer thin film self assembly. *Adv. Mater.* **14**: 1290–1294.

37. Park, S.M., Craig, G.S.W., La, Y.H. et al. 2007. Square arrays of vertical cylinders of PS-*b*-PMMA on chemically nanopatterned surfaces. *Macromolecules.* **40**: 5084–5094.

38. Welander, A.M. 2009. *Directed Self-Assembly of Diblock and Triblock Copolymer Films on Chemically Nanopatterned Surfaces.* University of Wisconsin—Madison, Madison, WI.

39. Huang, E., Russell, T.P., Harrison, C. et al. 1998. Using surface active random copolymers to control the domain orientation in diblock copolymer thin films. *Macromolecules.* **31**: 7641–7650.

40. Mansky, P., Russell, T.P., Hawker, C.J. et al. 1997. Ordered diblock copolymer films on random copolymer brushes. *Macromolecules.* **30**: 6810–6813.

41. Kellogg, G.J., Walton, D.G., Mayes, A.M. et al. 1996. Observed surface energy effects in confined diblock copolymers. *Phys. Rev. Lett.* **76**: 2503–2506.

42. Ji, S., Liu, C.C., Son, J.G. et al. 2008. Generalization of the use of random copolymers to control the wetting behavior of block copolymer films. *Macromolecules.* **41**: 9098–9103.

43. Helfand, E., and Wasserman, Z.R. 1976. Block copolymer theory 4. Narrow interphase approximation. *Macromolecules.* **9**: 879–888.

44. Stoykovich, M.P., Edwards, E.W., Solak, H.H. et al. 2006. Phase behavior of symmetric ternary block copolymer-homopolymer blends in thin films and on chemically patterned surfaces. *Phys. Rev. Lett.* **97**: 147802.

45. Welander, A.M., Kang, H.M., Stuen, K.O. et al. 2008. Rapid directed assembly of block copolymer films at elevated temperatures. *Macromolecules.* **41**: 2759–2761.

46. Park, S.M., La, Y.H., Ravindran, P. et al. 2007. Combinatorial generation and replicating directed assembly of complex and varied geometries with thin films of diblock copolymers. *Langmuir.* **23**: 9037–9045.

47. Tang, Z.Y., and Kotov, N.A. 2005. One-dimensional assemblies of nanoparticles: Preparation, properties, and promise. *Adv. Mater.* **17**: 951–962.

48. Yang, X., Xiao, S., Wu, W. et al. 2007. Challenges in 1 Teradot/in.(2) dot patterning using electron beam lithography for bit-patterned media. *J. Vac. Sci. Technol. B.* **25**: 2202–2209.

49. Cheng, J.Y., Mayes, A.M., and Ross, C.A. 2004. Nanostructure engineering by templated self-assembly of block copolymers. *Nat. Mater.* **3**: 823–828.

50. La, Y.H., Edwards, E.W., Park, S.M. et al. 2005. Directed assembly of cylinder-forming block copolymer films and thermochemically induced cylinder to sphere transition: A hierarchical route to linear arrays of nanodots. *Nano Lett.* **5**: 1379–1384.

51. Daoulas, K.C., Muller, M., Stoykovich, M.P. et al. 2006. Fabrication of complex three-dimensional nanostructures from self-assembling block copolymer materials on two-dimensional chemically patterned templates with mismatched symmetry. *Phys. Rev. Lett.* **96**: 036104/1–036104/4.

52. Tang, C.B., Lennon, E.M., Fredrickson, G.H. et al. 2008. Evolution of block copolymer lithography to highly ordered square arrays. *Science.* **322**: 429–432.

53. Chuang, V.P., Gwyther, J., Mickiewicz, R.A. et al. 2009. Templated self-assembly of square symmetry arrays from an ABC triblock terpolymer. *Nano Lett.* **9**: 4364–4369.

54. Black, C.T. 2005. Self-aligned self assembly of multi-nanowire silicon field effect transistors. *Appl. Phys. Lett.* **87**: 163116.

55. Hisamoto, D., Lee, W.C., Kedzierski, J. et al. 2000. FinFET—A self-aligned double-gate MOSFET scalable to 20 nm. *IEEE Trans. Electron Devices.* **47**: 2320–2325.

56. Huang, Y., Duan, X.F., Cui, Y. et al. 2001. Logic gates and computation from assembled nanowire building blocks. *Science.* **294**: 1313–1317.

57. Choi, Y.K., King, T.J., and Hu, C.M. 2002. A spacer patterning technology for nanoscale CMOS. *IEEE Trans. Electron Devices.* **49**: 436–441.

58. Liu, G., Thomas, C.S., Craig, G.S.W. et al. 2010. Integration of density multiplication in the formation of device-oriented structures by directed assembly of block copolymer-homopolymer blends. *Adv. Funct. Mater.* **20**: 1251–1257.

59. Kang, H., Craig, G.S.W., and Nealey, P.F. 2008. Directed assembly of asymmetric ternary block copolymer-homopolymer blends using symmetric block copolymer into checkerboard trimming chemical pattern. *J. Vac. Sci. Technol. B.* **26**: 2495–2499.

60. Herr, D.J.C. 2006. Update on the extensability of optical patterning via directed self-assembly. In *Future Fab International*, Dustrud, B., ed. Montgomery Research Incorporated: San Francisco; Vol. 20, pp. 82–86.

61. Russell, T.P., Coulon, G., Deline, V.R. et al. 1989. Characteristics of the surface-induced orientation for symmetric diblock Ps/Pmma copolymers. *Macromolecules.* **22**: 4600–4606.

62. Coulon, G., Russell, T.P., Deline, V.R. et al. 1989. Surface-induced orientation of symmetric diblock copolymers—A secondary ion mass-spectrometry study. *Macromolecules.* **22**: 2581–2589.

63. Coulon, G., Collin, B., Ausserre, D. et al. 1990. Islands and holes on the free-surface of thin diblock copolymer films 0.1. Characteristics of formation and growth. *J. Phys.* **51**: 2801–2811.

64. Fasolka, M.J., Banerjee, P., Mayes, A.M. et al. 2000. Morphology of ultrathin supported diblock copolymer films: Theory and experiment. *Macromolecules.* **33**: 5702–5712.

14. Self-Assembled DNA Nanostructures and DNA Devices

John H. Reif, Harish Chandran, and Nikhil Gopalkrishnan
Department of Computer Science, Duke University, Durham, North Carolina
Faculty of Computing and Information Technology, King Abdulaziz University, Jeddah, Saudi Arabia

Thomas LaBean
Department of Computer Science and Department of Chemistry, Duke University, Durham, North Carolina

Nanofabrication Handbook. Edited by Stefano Cabrini and Satoshi Kawata © 2012 CRC Press / Taylor & Francis Group, LLC. ISBN: 978-1-4200-9052-9

Chapter 14

14.1 Introduction

14.1.1 Some Unique Advantages of DNA Nanostructures

The particular molecular-scale constructs that are the topic of this chapter are known as DNA nanostructures.

As will be explained, DNA nanostructures have some unique advantages among nanostructures: they are relatively easy to design, fairly predictable in their geometric structures, and have been experimentally implemented in a growing number of labs around the world. They are constructed primarily of synthetic DNA.

14.1.2 Use of Bottom–Up Self-Assembly

Construction of molecular-scale structures and devices is one of the key challenges facing science and technology in the twenty-first century. This challenge is at the core of an emerging discipline of nanoscience. A key challenge is the need for robust, error-free methods for self-assembly of complex devices out of a large number of molecular components. This requires novel approaches. For example, the microelectronics industry is now reaching the limit of miniaturization made possible by top-down lithographic fabrication techniques. New bottom-up methods are needed for self-assembling complex, aperiodic structures for nanofabrication of molecular electronic circuits that are significantly smaller than conventional electronics.

A key principle in the study of DNA nanostructures is the use of self-assembly processes to actuate the molecular assembly. Since self-assembly operates naturally at the molecular scale, it does not suffer from the limitation in scale reduction that restricts lithography or other more conventional top-down manufacturing techniques.

In attempting to understand the modern development of DNA self-assembly, it is interesting to recall that mechanical methods for computation date back to the very onset of computer science, for example, to the cog-based mechanical computing machine of Babbage. Lovelace stated in 1843 that Babbage's "Analytical Engine weaves algebraic patterns just as the Jacquard-loom weaves flowers and leaves." In some of the recently demonstrated methods for biomolecular computation described here, computational patterns were essentially woven into molecular fabric (DNA lattices) via carefully controlled and designed self-assembly processes.

14.1.3 The Dual Role of Theory and Experimental Practice

In many cases, self-assembly processes are programmable in ways analogous to more conventional computational processes. We will overview theoretical principles and techniques (such as tiling assemblies and molecular transducers) developed for a number of DNA self-assembly processes that have their roots in computer science theory (e.g., abstract tiling models

and finite-state transducers). However, the area of DNA self-assembled nanostructures and molecular robotics is by no means simply a theoretical topic—many dramatic experimental demonstrations have already been made and a number of these will be discussed.

14.1.4 The Interdisciplinary Nature of the Field

DNA self-assembly is highly interdisciplinary and uses techniques from multiple disciplines such as biochemistry, physics, chemistry, material science, computer science, and mathematics. While this makes the topic quite intellectually exciting, it also makes it challenging for a typical reader.

14.1.5 The Rapid Progress of Complexity of DNA Nanostructures

The complexity of experimental demonstrations of DNA nanostructures has increased at an impressive rate (even in comparison to the rate of improvement of silicon-based technologies). This chapter discusses the accelerating scale of complexity of DNA nanostructures (such as the number of addressable pixels of 2D patterned DNA nanostructures) and provides some predictions for the future. Other surveys are given by Seeman (2004), Deng et al. (2006), and Amin et al. (2009).

14.1.6 Programmable DNA Nanostructures and Devices

We particularly emphasize molecular assemblies that are *autonomous*, executing steps with no exterior mediation after starting, and *programmable*, the tasks executed can be modified without entirely redesigning the nanostructure. In many cases, the self-assembly processes are programmable in ways analogous to more conventional computational processes. Computer-based design and simulation are also essential to the development of many complex DNA self-assembled nanostructures and systems. Error-correction techniques for correct assembly and repair of DNA self-assemblies are also discussed.

14.1.7 Applications of DNA Nanostructures

Molecular-scale devices using DNA nanostructures have been engineered to have various capabilities, ranging from (i) execution of molecular-scale computation,

Chapter 14

(ii) use as scaffolds or templates for the further assembly of other materials (such as scaffolds for various hybrid molecular electronic architectures or perhaps high-efficiency solar-cells), (iii) robotic movement and molecular transport (akin to artificial, programmable versions of cellular transport mechanisms), (iv) exquisitely sensitive molecular detection and amplification of single molecular events, (v) transduction of molecular sensing to provide drug delivery, (vi) vehicles for drug delivery inside cells, and (vii) protein structure determination. Error-correction techniques for correct assembly and repair of DNA self-assemblies have also been recently developed. Computer-based design and simulation are also essential to the development of many complex DNA self-assembled nanostructures and systems.

14.1.8   Organization

Section 14.2 gives a brief introduction to DNA, some known enzymes used for manipulation of DNA nanostructures, and some reasons why DNA is uniquely suited for assembly of molecular-scale devices. Section 14.3 narrates the first experimental demonstration of autonomous biomolecular computation and its shortcomings. Section 14.4 describes common DNA motifs, DNA tiles, DNA lattices composed of assemblies of these tiles, and software for tile design. Section 14.5 describes autonomous finite-state computations using linear DNA nanostructures. Section 14.6 discusses various methods for assembling patterned and addressable 2D DNA nanostructures and algorithmic self-assembly. Section 14.7 overviews methods for error correction and self-repair of DNA tiling assemblies. Section 14.8 covers 3D DNA nanostructures, including wireframe polyhedra, 3D DNA lattices, and 3D DNA origami. Section 14.9 reviews protocols for detection of molecular targets (DNA, RNA) and its application to autonomous molecular computation. Section 14.10 describes autonomous molecular transport devices self-assembled from DNA. Section 14.11 makes concluding remarks and sets out future challenges for the field.

14.2   DNA, Its Structure, and Its Manipulation

14.2.1   Introducing DNA

DNA self-assembly research is highly interdisciplinary and uses techniques from biochemistry, physics, chemistry, material science, computer science, and mathematics. A reader having no training in biochemistry must obtain a coherent understanding of the topic from a short chapter. This section is written with the expectation that the reader has little background knowledge of chemistry or biochemistry. On the other hand, a reader with a basic knowledge of DNA, its structure, and its enzymes can skip this section and proceed to the next.

14.2.2   DNA and Its Structure

Single-stranded DNA (ssDNA) is a long polymer made from repeating units called *nucleotides*. The nucleotide repeats contain both the segment of the *backbone* of the molecule, which holds the chain together, and a *base*. A base linked to a sugar is called a *nucleoside* and a base linked to a sugar and one or more phosphate groups is called a *nucleotide*. The backbone of the DNA strand is made from alternating phosphate and sugar residues. The sugar in DNA is 2-deoxyribose, which is a pentose (five-carbon) sugar. The sugars are joined together by phosphate groups that form phosphodiester bonds between the third and fifth carbon atoms of adjacent sugar rings. These asymmetric bonds mean a strand of DNA has a direction. The asymmetric ends of DNA strands are called the *5-prime* and *3-prime* ends, with the 5-prime end having a terminal phosphate group and the 3-prime end a terminal hydroxyl group. The four bases found in DNA are adenine (abbreviated A), cytosine (C), guanine (G), and thymine (T). These bases form the alphabet of DNA; the specific sequence comprises DNA's information content. Each base is attached to a sugar/phosphate to form a complete nucleotide. These bases are classified into two types; adenine and guanine are fused five- and six-membered heterocyclic compounds called *purines*, while cytosine and thymine are six-membered rings called *pyrimidines*. Each type of base on one strand overwhelmingly prefers a bond with just one type of base on the other strand. This is called *complementary base pairing*. Here, purines form hydrogen bonds to pyrimidines, with A bonding preferentially to T, and C bonding preferentially to G. This arrangement of two nucleotides binding together across the double helix is called a *base pair*. In living organisms, DNA does not usually exist as a single molecule, but instead as a pair of molecules called *double-stranded DNA* (dsDNA) (see

FIGURE 14.1 Structure of a DNA double helix. (Image by Michael Ströck and released under the GNU Free Documentation License.)

Figure 14.1) that are held tightly together via a reaction known as *DNA hybridization*. These two long strands entwine like vines, in the shape of a double helix. DNA hybridization occurs in a physiologic-like buffer solution with appropriate temperature, pH, and salinity.

In a double helix, the direction of the nucleotides in one strand is opposite to their direction in the other strand: the strands are *antiparallel*. The DNA double helix is stabilized by hydrogen bonds between the bases attached to the two strands and stacking between contiguous base pairs. As hydrogen bonds are not covalent, they can be broken and rejoined relatively easily. The two strands of DNA in a double helix can therefore be pulled apart like a zipper, either by a mechanical force or high temperature. The two types of base pairs form different number of hydrogen bonds, AT forming two hydrogen bonds, and GC forming three hydrogen bonds. The association strength of hybridization depends on the sequence of complementary bases, stability increasing with length of DNA sequence and GC content. This association strength can be approximated by software packages. The *melting temperature* of a DNA helix is the temperature at which half of all the molecules are fully hybridized as double helix, while the

other half are single stranded. The kinetics of the DNA hybridization process is quite well understood; it often occurs in a (random) zipper-like manner, similar to a biased one-dimensional random walk. Single-stranded DNA is flexible and has a small persistence length when compared to double-stranded DNA of comparable length. Single-stranded DNA is sometimes thought of as a freely jointed chain while double-stranded DNA is more like a worm-like chain. The exact geometry (angles and positions) of each segment of a double helix depends slightly on the component bases of its strands and can be determined from known tables. There are about 10.5 bases per full rotation on the helical axis. The width of the DNA double helix is 2.2–2.6 nm and the helical pitch is about 3.4 nm. A *DNA nanostructure* is a multimolecular complex consisting of a number of ssDNA that have partially hybridized along their subsegments.

14.2.3 Manipulation of DNA

Here, we list some techniques and known enzymes used for manipulation of DNA nanostructures. *Strand displacement* is the displacement of a single strand of DNA from a double helix by an incoming strand with a longer complementary region to the template strand. The incoming strand has a *toehold*, an empty single-stranded region on the template strand complementary to a subsequence of the incoming strand, to which it binds initially. It eventually displaces the outgoing strand via a kinetic process modeled as a one-dimensional random walk. Strand displacement is a key process in many of the DNA protocols for running DNA autonomous devices. Figure 14.2 illustrates DNA strand displacement via branch migration.

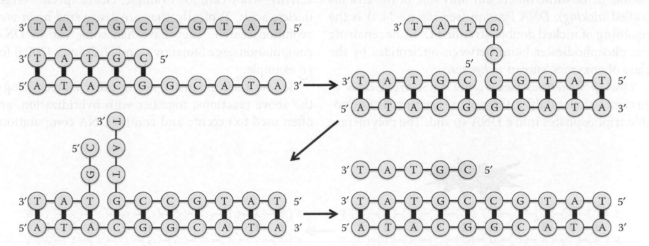

FIGURE 14.2 Strand displacement of dsDNA via a branch migration hybridization reaction. The figure illustrates DNA strand displacement of a DNA strand induced by the hybridization of a longer strand, allowing the structure to reach a lower energy state.

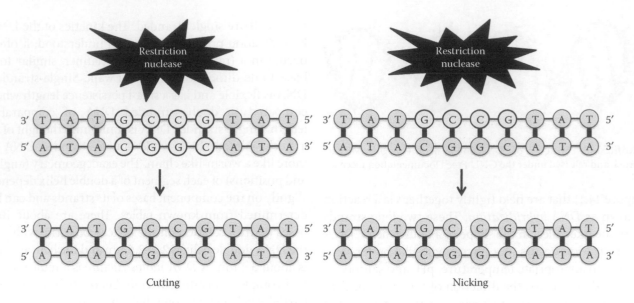

FIGURE 14.3   Example of restriction enzyme cuts of a single-stranded DNA sequence. The subsequence recognized by the nuclease is unshaded.

In addition to the hybridization reaction described above, there are a wide variety of known enzymes and other proteins used for manipulation of DNA nanostructures that have predictable effects. Interestingly, these proteins were discovered in natural bacterial cells and tailored for laboratory use.

DNA restriction (see Figure 14.3) is the cleaving of phosphodiester bonds between the nucleotide subunits at specific locations determined by short (4–8 base) sequences by a class of enzymes called *nucleases*. *Endonucleases* cleave the phosphodiester bond within a polynucleotide chain while *exonucleases* cleave the phosphodiester bond at the end of a polynucleotide chain. Some nucleases have both these abilities. Some restriction enzymes cut both the strands of a DNA double helix while others cut only one of the strands (called *nicking*). *DNA ligation* (see Figure 14.4) is the rejoining of nicked double-stranded DNA by repairing the phosphodiester bond between nucleotides by the class of enzymes known as *ligases*.

DNA polymerases (see Figure 14.5) are a class of enzymes that catalyze the polymerization of nucleoside triphosphates into a DNA strand. The polymerase "reads" an intact DNA strand as a template and uses it to synthesize the new strand. The newly polymerized molecule is complementary to the template strand. DNA polymerases can only add a nucleotide onto a preexisting 3-prime hydroxyl group. Therefore, it needs a *primer*, a DNA strand attached to the template strand, to which it can add the first nucleotide. Certain polymerase enzymes (e.g., phi-29) can, as a side effect of their polymerization reaction, efficiently displace previously hybridized strands. Isothermal denaturation (breaking of base pairings) can also be achieved by *helicases*, which are motor proteins that move directionally along a DNA backbone, denaturing the double helix. In addition, *deoxyribozymes (DNAzymes)* are a class of nucleic acid molecules that possess enzymatic activity—they can, for example, cleave specific target nucleic acids. Typically, they are discovered by *in vivo* evolution search. They have had some use in DNA computations; see Stojanovic and Stefanovic (2003) for an example.

Besides their extensive use in other biotechnology, the above reactions, together with hybridization, are often used to execute and control DNA computations

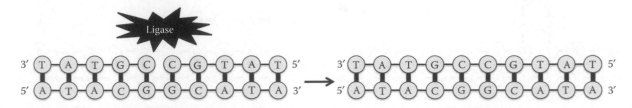

FIGURE 14.4   Ligase healing a single-stranded nick. Note that the two parts are bound to the same template.

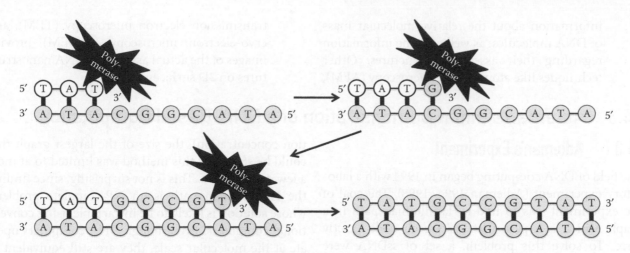

FIGURE 14.5 Extension of primer strand (unshaded) bound to the template by DNA polymerase.

and DNA robotic operations. The restriction enzyme reactions are programmable in the sense that they are site specific, only executed as determined by the appropriate DNA base sequence. Ligation and polymerization require the expenditure of energy via consumption of ATP molecules, and thus can be controlled by ATP concentration.

14.2.4 Why Use DNA to Assemble Molecular-Scale Devices?

There are many advantages of DNA as a material for building things at the molecular scale. Below, we list some reasons why DNA is uniquely suited for assembly of molecular-scale devices.

1. From the perspective of design, the advantages are
   - A variety of geometries can be achieved by carefully programming DNA sequences to interact among themselves in a predictable manner. The shape of the DNA nanostructure is controlled by its component DNA strands and this gives us an ability to program a myriad of nanostructures.
   - The structure of most complex DNA nanostructures can be reduced to determining the structure of short segments of dsDNA. The basic geometric and thermodynamic properties of dsDNA are well understood and can be predicted by available software systems from key relevant parameters like sequence composition, temperature, and buffer conditions.
   - Design of DNA nanostructures can be assisted by software. To design a DNA nanostructure or device, one needs to design a library of

ssDNA strands with specific segments that hybridize to (and only to) specific complementary segments on other ssDNA. There are a number of software systems (developed at NYU, Caltech, Arizona State, and Duke University) for design of the DNA sequences composing DNA tiles and for optimizing their stability, which employ heuristic optimization procedures for this combinatorial sequence design task (see Section 14.4.4 for more details).

2. From the perspective of experiments, the advantages are
   - The solid-phase chemical synthesis of custom ssDNA is now routine and inexpensive; a test tube of ssDNA consisting of any specified short sequence of bases (<150) can be obtained from commercial sources for modest cost (about half a U.S. dollar per base at this time); it will contain a very large number (typically at least 10^{12}) of identical ssDNA molecules. The synthesized ssDNA can have errors (premature termination of the synthesis is the most frequent error), but can be easily purified by well-known techniques (e.g., electrophoresis as mentioned below).
   - The assembly of DNA nanostructures is a very simple experimental process: in many cases, one simply combines the various component ssDNA into a single test tube with an appropriate buffer solution at an initial temperature above the melting temperature, and then slowly cools the test tube below the melting temperature.
   - The assembled DNA nanostructures can be characterized by a variety of techniques. One such technique is electrophoresis. It can provide

Chapter 14

information about the relative molecular mass of DNA molecules, as well as some information regarding their assembled structures. Other techniques like atomic force microscopy (AFM), transmission electron microscopy (TEM), and cryo-electron microscopy (cyroEM) provide images of the actual assembled DNA nanostructures on 2D surfaces and in 3D.

14.3 Adleman's Initial Demonstration of a DNA-Based Computation

14.3.1 Adleman's Experiment

The field of DNA computing began in 1994 with a laboratory experiment (Adleman 1994, 1998). The goal of the experiment was to find a Hamiltonian path in a graph, which is a path that visits each node exactly once. To solve this problem, a set of ssDNA were designed based on the set of edges of the graph. When combined in a test tube and annealed, they self-assembled into dsDNA. Each of these DNA nanostructures was a linear DNA double helix that corresponded to a path in the graph. If the graph had a Hamiltonian path, then one (or a subset) of these DNA nanostructures encoded the Hamiltonian path. By conventional biochemical extraction methods, Adleman was able to isolate only DNA nanostructures encoding Hamiltonian paths, and by determining their sequence, the explicit Hamiltonian path. It should be mentioned that this landmark experiment was designed and experimentally demonstrated by Adleman alone, a computer scientist with limited training in biochemistry.

14.3.2 The Nonscalability of Adleman's Experiment

While this experiment founded the field of DNA computing, it was not scalable in practice, since the number of different DNA strands needed increased exponentially with the number of nodes of the graph. Although there can be an enormous number of DNA strands in a test tube (10^{15} or more, depending on solution concentration), the size of the largest graph that could be solved by this method was limited to at most a few dozen nodes. This is not surprising, since finding the Hamiltonian path is an NP-complete problem, whose solution is likely to be intractable using conventional computers. Even though DNA computers operate at the molecular scale, they are still equivalent to conventional computers (e.g., deterministic Turing machines) in computational power. This experiment taught a healthy lesson to the DNA computing community (which is now well recognized): to carefully examine scalability issues and to judge any proposed experimental methodology by its scalability.

14.3.3 Autonomous Biomolecular Computation

Shortly following Adleman's experiment, there was a burst of further experiments in DNA computing, many of which were quite ingenious. However, almost none of these DNA computing methods were autonomous, and instead required many tedious laboratory steps to execute. In retrospect, one of the most notable aspects of Adleman's experiment was that the self-assembly phase of the experiment was completely autonomous—it required no external mediation. This autonomous property makes an experimental laboratory demonstration much more feasible as the scale increases. The remaining part of this chapter mostly discusses autonomous devices for biomolecular computation based on self-assembly.

14.4 Self-Assembled DNA Tiles and Lattices

14.4.1 DNA Nanostructures

Recall that a DNA nanostructure is a multimolecular complex consisting of a number of ssDNA that have partially hybridized along their subsegments. The field of DNA nanostructures was pioneered by Nadrian Seeman (Robinson and Seeman 1987). Particularly useful types of motifs often found in DNA nanostructures include

- *Stem loop* (also called hairpins) (see Figure 14.6a): an ssDNA that loops back to hybridize on itself, that is, one segment of the ssDNA (near the 5-prime end) hybridizes with another segment further along (nearer the 3-prime end) on the same ssDNA strand. The stem loop in Figure 14.6a has an unpaired region (with sequence TTTT), which is typical for this motif. Stem loops are often used to form patterning on DNA nanostructures.

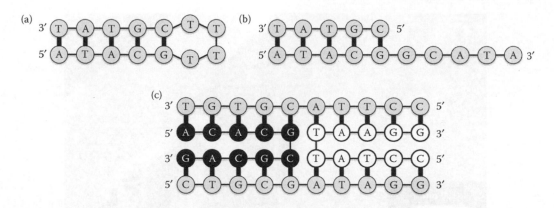

FIGURE 14.6   Common DNA motifs. (a) DNA hairpin. (b) Sticky end. (c) Holliday junction.

- *Sticky end* (see Figure 14.6b): an unhybridized ssDNA that protrudes from the end of a double helix. The sticky end shown (GCATA) protrudes from dsDNA (ATACG on the bottom strand). Sticky ends are often used to combine two DNA nanostructures together via hybridization of their complementary ssDNA.

- *Holliday junction* (see Figure 14.6c): two parallel DNA helices form a junction with one strand of each DNA helix crossing over to the other DNA helix. Holliday junctions are often used to hold together various parts of a DNA nanostructure.

14.4.2   Computation by Self-Assembly

The most basic way by which computer science ideas have impacted DNA nanostructure design is via the pioneering work by theoretical computer scientists on a formal model of 2D tiling due to Wang (1961), which culminated in a proof by Berger (1966), and later Robinson (1971), that universal computation could be done via tiling assemblies. Winfree (1995) was the first to propose applying the concepts of computational tiling assemblies to DNA molecular constructs. His core idea was to use tiles composed of DNA to perform computations during their self-assembly process. To understand this idea, we will need an overview of DNA nanostructures, as presented in Section 14.4.3.

14.4.3   DNA Tiles and Lattices

A *DNA tile* is a DNA nanostructure that has a number of sticky ends on its sides, which are termed *pads*. A DNA lattice is a DNA nanostructure composed of a group of DNA tiles that are assembled together via hybridization of their pads. Generally the strands composing the DNA tiles are designed to have a melting temperature above those of the pads, ensuring that when the component DNA molecules are combined together in solution, the DNA tiles assemble first, and only then, as the solution is further cooled, do the tiles bind together via hybridization of their pads.

Figure 14.7 illustrates some principal DNA tiles. Also see LaBean et al. (2007). Winfree et al. (1996) developed a family of DNA tiles known collectively as DX tiles (see Figure 14.7) that consisted of two parallel DNA helices linked by immobile Holliday junctions. They demonstrated that these tiles formed large 2D lattices, as viewed by AFM (see Figure 14.8a).

Subsequently, other DNA tiles were developed by LaBean et al. (2000) to provide for more complex strand topology and interconnections, including a family of DNA tiles known as *TX tiles* (see Figure 14.7) composed of three DNA helices. Both the DX tiles and the TX tiles are rectangular in shape, where two opposing

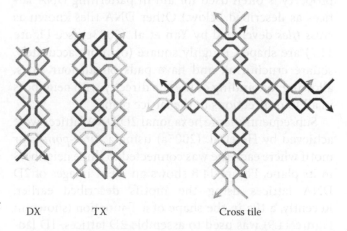

DX   TX   Cross tile

FIGURE 14.7   DNA tiles: DX, TX, and the cross tile.

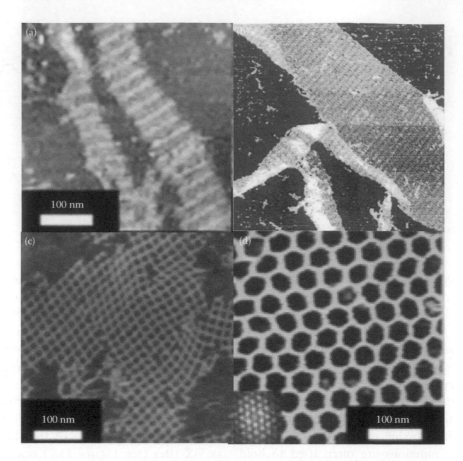

FIGURE 14.8 DNA lattices. (a) DX lattice. (Reprinted by permission from Macmillan Publishers Ltd: *Nature*. Winfree, E. et al., Design and self- assembly of two-dimensional DNA crystals, 394, 539–544, © 1998.) (b) TX ribbons. (Reprinted with permission from LaBean, T. et al., Construction, analysis, ligation, and self-assembly of DNA triple crossover complexes, *122*(9), 1848–1860. © 2000 American Chemical Society.) (c) Cross tile lattice. (From Yan, H. et al., DNA-templated self-assembly of protein arrays and highly conductive nanowires, 2003c, *Science 301*(5641), 1882–1884. Reprinted with permission from AAAS.) (d) Three-point star hexagonal lattice. (Reprinted with permission from He, Y. et al., Self-assembly of hexagonal DNA two-dimensional (2D) arrays, *127*(35), 12202–12203. © 2005a American Chemical Society.)

edges of the tile have pads consisting of ssDNA sticky ends of the component strands. In addition, TX tiles have topological properties that allow for strands to propagate in useful ways through tile lattices. (This property is often used for aid in patterning DNA lattices as described below.) Other DNA tiles known as *cross tiles* developed by Yan et al. (2003c) (see Figure 14.7) are shaped roughly square (or more accurately, square cruciform), and have pads on all four sides, allowing for binding of the tile directly with neighbors in all four directions in the lattice plane.

Subsequently, large hexagonal 2D DNA lattices were achieved by He et al. (2005a) using a *three-point-star* motif where each tile was connected to three neighbors in its plane. Figure 14.8 shows an AFM images of 2D DNA lattices using the motifs described earlier. Recently, a tile in the shape of a T-junction (shown in Figure 14.9) was used to assemble 2D lattices, 1D ladders, and rings (Hamada and Murata 2009). These tiles

are different from the tiles described earlier as they do not use Holliday junction.

The tiles described above are designed to be planar. But in reality they possess a small curvature, thus preventing large planar lattices. To counter this, a strategy called *corrugation* developed by Yan et al. (2003c) was introduced in which neighboring tiles are flipped with respect to each other, thus canceling out their curvature. Another technique to minimize defects due to curvature and obtain large assemblies was *sequence symmetry* introduced by He et al. (2005b) in which geometrically symmetric parts of the tile are given the same sequence, thus ensuring symmetric curvature.

To program a tiling assembly, the DNA sequence of the pads are designed so that tiles assemble together as intended. Proper designs ensure that only the adjacent pads (two pairs of sticky ends in the case of cross tiles) of neighboring tiles are complementary, so only those pads hybridize together.

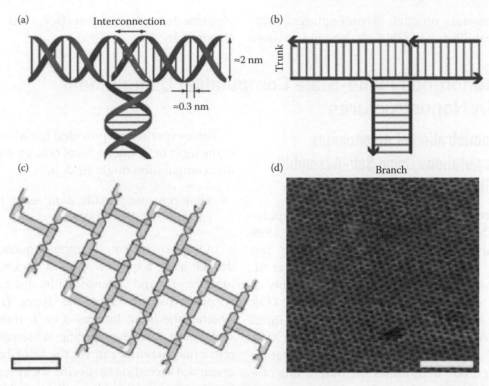

FIGURE 14.9 T-junction tiling. (a,b) Design of the T-junction. (c) Lattice schematics. (d) AFM image of the lattice. (From Hamada, S. and Murata, S. Substrate-assisted assembly of interconnected single-duplex DNA nanostructures. *Angewandte Chemie International Edition.* 2009. *48*(37), 6820–6823. Copyright Wiley-VCH Verlag GmbH & Co. KGaA. Reproduced with permission.)

14.4.4 Software for Design of DNA Tiles

A number of prototype computer software systems have been developed for the design of the DNA sequences composing DNA tiles, and for optimizing their stability. Figure 14.10 gives a screen shot of a software system known as TileSoft, developed jointly by Duke and Caltech, which provides a graphically interfaced sequence optimization system for designing DNA secondary structures (Yin et al. 2004a). A more

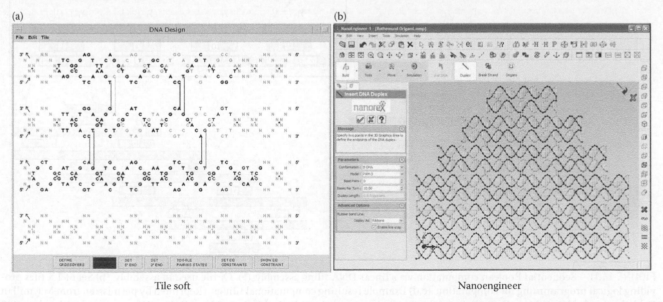

FIGURE 14.10 (a) TileSoft: sequence optimization software for designing DNA secondary structures. (b) Nanoengineer.

Chapter 14

recent commercial product, NanoEngineer, with enhanced capabilities for DNA design and a more sophisticated graphic interface, was developed by Nanorex, Inc.

14.5 Autonomous Finite-State Computation Using Linear DNA Nanostructures

14.5.1 Demonstration of Autonomous Computations Using Self-Assembly of DNA Nanostructures

The first experimental demonstrations of computation using DNA tile assembly were done in 1999 (LaBean et al. 1999, 2000; Mao et al. 2000; Yan et al. 2003a). Among the experiments, Mao et al. (2000) demonstrated a two-layer, linear assembly of TX tiles that executed a bit-wise cumulative XOR computation. In this computation, n bits are input and n bits are output, where the ith output is the XOR of the first i input bits. This is the computation occurring when one determines the output bits of a full-carry binary adder circuit found in most computer processors. This experiment is illustrated in Figure 14.11.

These experiments provided initial answers to some of the most basic questions of how autonomous molecular computation might be done:

- *How can one provide data input to a molecular computation using DNA tiles?*

In this experiment, the input sequence of n bits was defined using a specific series of "input" tiles with the input bits (1s and 0s) encoded by distinct short subsequences. Two different tile types (Depending on whether the input bit was 0 or 1, these had specific sticky ends and also specific subsequences at which restriction enzymes can cut the DNA backbone.) were assembled according to specific sticky end associations, forming the blue input layer illustrated in Figure 14.11.

Figure 14.11a shows a unit TX tile and the sets of input and Figure 14.11b output tiles with geometric shapes

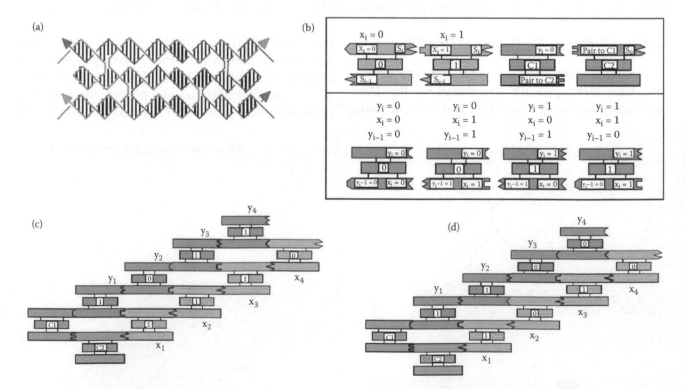

FIGURE 14.11 Sequential Boolean computation via a linear DNA tiling assembly. (a) TX tile used in assembly. (b) Set of TX tiles providing logical programming for computation. (c,d) Example resulting computational tilings. (Reprinted by permission from Macmillan Publishers Ltd: *Nature* Mao, C. et al., Logical computation using algorithmic self-assembly of DNA triple-crossover molecules, *407*, 49–496. © 2000.)

conveying sticky end complementary matching. The tiles of (b) execute binary computations depending on their pads, as indicated by the table in (b). The (blue) input layer and (green) corner condition tiles were designed to assemble first (see example computational assemblies (c) and (d)). The (red) output layer then assembles specifically starting from the bottom left using the inputs from the blue layer. See Mao et al. (2000) for more details of this molecular computation. The tiles were designed such that an output reporter strand ran through all the *n* tiles of the assembly by bridges across the adjoining pads in input, corner, and output tiles. This reporter strand was pasted together from the short ssDNA sequences within the tiles using ligation enzyme mentioned previously. When the solution was warmed, this output strand was isolated and identified. The output data was read by experimentally determining the sequence of cut sites (see below). In principle, the output could be used for subsequent computations.

The next question of concern is

- *How can one execute a step of computation using DNA tiles?*

To execute steps of computation, the TX tiles were designed to have pads at one end that encoded the cumulative XOR value. Also, since the reporter strand segments ran through each such tile, the appropriate input bit was also provided within its structure. These two values implied the opposing pad on the other side of the tile would be the XOR of these two bits.

A final question of concern is

- *How can one determine and/or display the output values of a DNA tiling computation?*

The output in this case was read by determining which of the two possible cut sites (endonuclease cleavage sites) were present at each position in the tile assembly. This was executed by first isolating the reporter strand, then digesting separate aliquots with each endonuclease separately and the two together, and finally these samples were examined by gel electrophoresis and the output values were displayed as banding patterns on the gel. Another method for output (presented below) is the use of AFM observable patterning. The patterning was made by designing the tiles computing a bit 1 to have a stem loop protruding from the top of the tile. This molecular patterning was clearly observable under appropriate AFM imaging conditions.

Although they are only very simple computations, these experiments did demonstrate for the first time methods for autonomous execution of a sequence of finite-state operations via algorithmic self-assembly, as well as for providing inputs and for outputting the results. Further DNA tile assembly computations will be presented below in Section 14.5.2.

14.5.2 Autonomous Finite-State Computations via Disassembly of DNA Nanostructures

An alternative method for autonomous execution of a sequence of finite-state transitions was subsequently developed by Shapiro and Benenson (2006). Their technique essentially operated in the reverse of the assembly methods described above, and instead can be thought of as disassembly. They began with a linear double-stranded DNA nanostructure whose sequence encoded the inputs, and then they executed series of steps that digested the DNA nanostructure from one end (see Figure 14.12). On each step, a sticky end at one end of the nanostructure encoded the current state, and the finite transition was determined by hybridization of the current sticky end with a small "rule" nanostructure encoding the finite-state transition rule. Then a restriction enzyme, which recognized the sequence encoding the current input as well

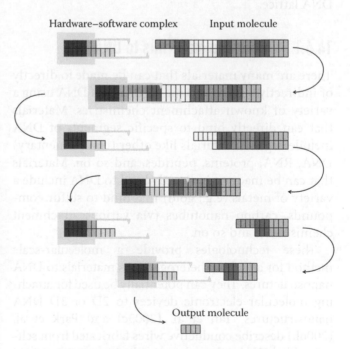

FIGURE 14.12 Autonomous finite-state computations via disassembly of a double-stranded DNA nanostructure.

Hardware–software complex   Input molecule

Output molecule

as the current state, cut the appended end of the linear DNA nanostructure to expose a new sticky end encoding the next state.

The hardware–software complex for this molecular device is composed of dsDNA with an ssDNA overhang (shown at top left ready to bind with the input molecule) and a protein restriction enzyme (shown as gray pinchers).

14.6 Assembling Patterned and Addressable 2D DNA Lattices

One of the most appealing applications of tiling computations is their use to form patterned nanostructures to which other materials can be selectively bound.

An *addressable* 2D DNA lattice is one that has a number of sites with distinct ssDNA. This provides a superstructure for selectively attaching other molecules at addressable locations. Examples of addressable 2D DNA lattices will be given in Section 14.6.2.

As discussed below, there are many types of molecules that we can attach to DNA. Known attachment chemistry allows them to be tagged with a given sequence of ssDNA. Each of these DNA-tagged molecules can then be assembled by hybridization of their DNA tags to a complementary sequence of ssDNA located within an addressable 2D DNA lattice. In this way, we can program the assembly of each DNA-tagged molecule onto a particular site of the addressable 2D DNA lattice.

14.6.1 Attaching Materials to DNA

There are many materials that can be made to directly or indirectly bind to specific segments of DNA using a variety of known attachment chemistries. Materials that can directly bind to specific segments of DNA include organic materials like other (complementary) DNA, RNA, proteins, peptides, and so on. Materials that can be made to indirectly bind to DNA include a variety of metals (e.g., gold) that bind to sulfur compounds, carbon nanotubes (via various attachment chemistries), and so on.

These technologies provide a molecular-scale method for attaching heterogeneous materials to DNA nanostructures. They can potentially be used for attaching molecular electronic devices to 2D or 3D DNA nanostructures. Yan et al. (2003c) and Park et al. (2006b) describe conductive wires fabricated from self-assembled DNA tubes plated with silver, as illustrated in Figure 14.13.

This ingenious design is an excellent demonstration that there is often more than one way to do any task at the molecular scale. Adar et al. (2004) demonstrated in the test tube a potential application of such a finite-state computing device to medical diagnosis and therapeutics. See the conclusion in Section 14.11 for further discussion.

14.6.2 Methods for Programmable Assembly of Patterned 2D DNA Lattices

The first experimental demonstration of 2D DNA lattices by Winfree et al. (1998) provided very simple patterning by repeated stripes determined by a stem loop projecting from every DNA tile on an odd column. This limited sort of patterning needed to be extended to large classes of patterns.

In particular, the key capability needed is a programmable method for forming distinct patterns on 2D DNA lattices, without having to completely redesign the lattice to achieve any given pattern. There are at least three methods for assembling patterned 2D DNA lattices that have been experimentally demonstrated, as described in the next few sections.

14.6.2.1 Programmable Assembly of Patterned 2D DNA Lattices by Use of Scaffold Strands

A *scaffold strand* is a long ssDNA around which shorter ssDNA assemble to form structures larger than individual tiles. Scaffold strands were used to demonstrate programmable patterning of 2D DNA lattices by propagating 1D information from the scaffold into a second dimension to create AFM observable patterns (Yan et al. 2003b). The scaffold strand weaves through the resulting DNA lattice to form the desired distinct sequence of 2D barcode patterns (Figure 14.14a). In this demonstration, identical scaffold strands ran through each row of the 2D lattices, using short stem loops extending above the lattice to form pixels. This determined a bar code sequence of stripes over the 2D lattice that was viewed by AFM. In principle, this method may be extended to allow for each row's patterning to be determined by a distinct scaffold strand, defining an arbitrary 2D pixel image.

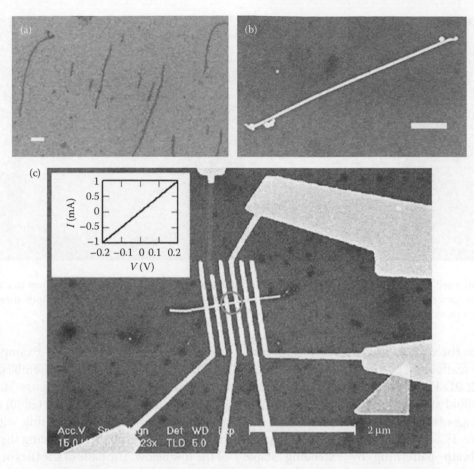

FIGURE 14.13 Conductive wires fabricated from self-assembled DNA tubes plated with silver. (a) DNA tubes prior to plating. (b) DNA tubes after silver plating. (c) SEM image of conductivity test on silicon oxide substrate. (From Yan, H. et al., DNA-templated self-assembly of protein arrays and highly conductive nanowires, 2003c, *Science*, *301*(5641), 1882–1884. Reprinted with permission from AAAS.)

A spectacular experimental demonstration of patterning via scaffold strand is also known as *DNA origami* (Rothemund 2006). This approach makes use of a long strand of "scaffold" ssDNA (such as from the genome of a viral phage) that has only weak secondary structure and few long repeated or self-complementary subsequences. To this is added a large number of relatively short "staple" ssDNA sequences, with subsequences complementary to certain subsequences of the scaffold ssDNA. These staple sequences are chosen so

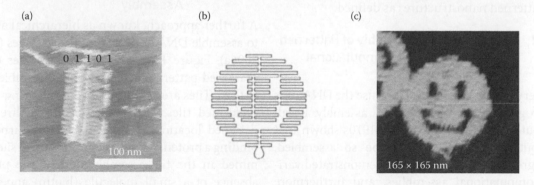

FIGURE 14.14 Methods for programmable assembly of patterned 2D DNA lattices by use of scaffold strands. (a) Barcode patterning. (From Yan, H. et al. Directed nucleation assembly of DNA tile complexes for barcode-patterned lattices, *Proceedings of the National Academy of Sciences of the United States of America* 100(14), 8103–8108. © 2003b National Academy of Sciences, U.S.A; Reprinted by permission from Macmillan Publishers Ltd: *Nature* Mao, C. et al. Logical computation using algorithmic self-assembly of DNA triple-crossover molecule, *407*, 493–496. © 2000.) (b) DNA origami design. (c) AFM imaging of DNA origami. (From Rothemund, P. Folding DNA to create nanoscale shapes and patterns, *Nature*, © 2006.)

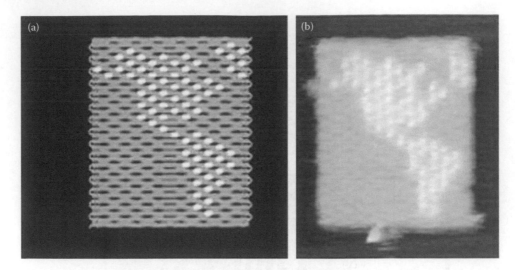

FIGURE 14.15 Patterned origami. Bright dots are staples extended into a stem-loop structure, causing them to stick out of the plan. (a) Schematics of the patterning. (b) Atomic force microscopy image of the patterned DNA origami. (From Rothemund, P. Folding DNA to create nanoscale shapes and patterns, *Nature 440*, 297–302 © 2006.)

that they bind to the scaffold ssDNA by hybridization, and induce the scaffold ssDNA to fold together into a fully addressable 2D DNA nanostructure. A schematic trace of the scaffold strand is shown in Figure 14.14b, and an AFM image of the resulting assembled origami is shown in Figure 14.14c. This method can be slightly modified to obtain patterning by extending staple strands at the end into a stem loop structure. These stem loops will stick out of the plane of the nanostructure and will appear as a bright dot on an AFM image (see Figure 14.15). This landmark work of Rothemund (2006) very substantially increases the scale of 2D patterned assemblies to hundreds of molecular pixels (i.e., stem loops viewable via AFM) within square area <100 nm on a side. In principle, this "molecular origami" method with staple strands can be used to form arbitrary complex 2D patterned nanostructures as defined.

14.6.2.2 Programmable Assembly of Patterned 2D DNA Lattices by Computational Assembly

Another very promising method is to use the DNA tile's pads to program a 2D computational assembly. Recall that computer scientists have in the 1970s shown that any computable 2D pattern can be so assembled. Winfree's group has experimentally demonstrated various 2D computational assemblies, and furthermore provided AFM images of the resulting nanostructures (Barish et al. 2005; Fujibayashi et al. 2008). Figure 14.16 gives a modulo-2 version of Pascal's triangle (known as the Sierpinski triangle), where each tile determines and outputs to neighborhood pads the XOR of two of the tile

pads (Rothemund et al. 2004). Example AFM images (scale bars = 100 nm) of the assembled structures are shown in the three panels of Figure 14.16. Figure 14.17 gives Rothemund and Winfree's (2000) design for a self-assembled binary counter, starting with 0 at the first row, and on each further row being the increment by 1 of the row below. The pads of the tiles of each row of this computational lattice were designed in a similar way to that of the linear XOR lattice assemblies described in the previous section. The resulting 2D counting lattice is found in MUX designs for address memory, and so this patterning may have major applications to patterning molecular electronic circuits.

14.6.2.3 Programmable Assembly of Patterned 2D DNA Lattices by Hierarchical Assembly

A further approach, known as hierarchical assembly, is to assemble DNA lattices in multiple stages (Park et al. 2006a). Figure 14.18 gives three examples of preprogrammed patterns displayed on addressable DNA tile lattices. Tiles are assembled prior to mixing with other preformed tiles. Unique ssDNA pads direct tiles to designed locations. White pixels are "turned on" by binding a protein (avidin) at programmed sites as determined in the tile assembly step by the presence or absence of a small molecule (biotin) appended to a DNA strand within the tile. Addressable, hierarchical assembly has been demonstrated for lattices of size 8 × 8 (140 × 140 nm) to date by Pistol and Dwyer (2007) and has considerable potential particularly in conjunction with the above methods for patterned assembly.

FIGURE 14.16 **(See color insert.)** Programmable assembly of Sierpinski triangle by use of computational assembly. Scale bar = 100 nm. (a) Zoomed out AFM image. (b,c) Zoomed in AFM image showing errors marked by x.

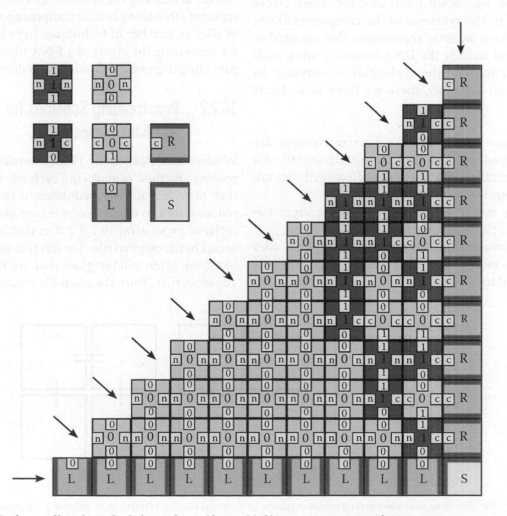

FIGURE 14.17 Rothemund's and Winfree's design for a self-assembled binary counter using tilings.

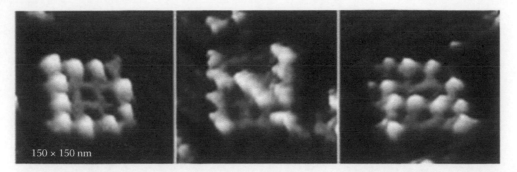

FIGURE 14.18 2D Patterns by hierarchical assembly. AFM images of characters D, N, and A. (From Park, S. H. et al., Finite-size, fully addressable DNA tile lattices formed by hierarchical assembly procedures. *Angewandte Chemie International Edition.* 2006a. *45*(5), 735–739. Copyright Wiley-VCH Verlag GmbH & Co. KGaA. Reproduced with permission.)

14.7 Error Correction and Self-Repair at the Molecular Scale

14.7.1 The Need for Error Correction at the Molecular Scale

In many of the self-assembled devices described here, there can be significant levels of error. These errors occur both in the synthesis of the component DNA, and in the basic molecular processes that are used to assemble and modify the DNA nanostructures, such as hybridization and the application of enzymes. In tile-based self-assembly, there are three main kinds of errors:

- Nucleation error: Tile-based nanostructures are grown from a special tile known as the *seed tile*. All nanostructures that grow out of nonseed tiles are erroneous assemblies.
- Growth error: Attachment of an incorrect tile instead of a better matched tile.
- Facet (roughening) error: Attachment of tiles along a facet (boundary) where no growth was intended to occur.

There are various purification and optimization procedures developed in biochemistry for minimization of many of these types of errors. However, there remains a need for development of algorithmic methods for decreasing the errors of assembly and for self-repair of DNA tiling lattices comprising a large number of tiles. A number of techniques have been proposed for decreasing the errors of a DNA tiling assembly, by providing increased redundancy, as described below.

14.7.2 Proofreading Schemes for Error-Resilient Tilings

Winfree and Bekbolatov (2003) developed a "proofreading" method of replacing each tile with four tiles that provide sufficient redundancy to quadratically reduce errors, as illustrated in Figure 14.19. Each tile is replaced by an array of 2×2 tiles that logically correspond to the original tile. The internal sides of the new block are given unique glues that are not present on any other tiles. Thus, the assembly proceeds like for the

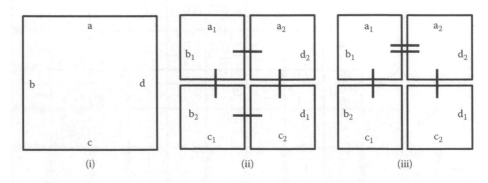

FIGURE 14.19 Proofreading schemes for error-resilient tilings. (i) Original tile. (ii) Winfree et al. general 2×2 proofreading scheme. (iii) Chen et al. general 2×2 snaked proofreading scheme. The lines represent pad strengths.

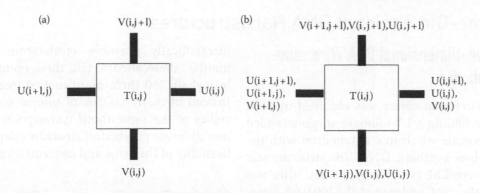

FIGURE 14.20   A compact scheme for error-resilient tilings. (a) Original tile. (b) Error resilient tile.

original tile set but scaled up by a factor of 4 in area. When a mismatched tile is incorporated in this new tiling at some position, further assembly cannot proceed at that position without making an additional error. This gives mismatched tiles time to dissociate and thus the tiling is resilient to growth errors. Reif et al. (2004) proposed a more compact method for decreasing assembly errors, as illustrated in Figure 14.20. This method modifies the pads of each tile, so that essentially each tile executes the original computation required at that location, as well as the computation of a particular neighbor, providing a quadratic reduction of errors without increasing the assembly size. Chen and Goel (2004) proposed *snaked proofreading* (see Figure 14.19) to correct facet errors in addition to growth errors. Both these techniques were experimentally tested by Chen et al. (2007). Nucleation errors were handled in Schulman and Winfree (2009) by constructing tile sets that introduce arbitrarily large barriers to incorrect nucleation.

By combining all the aforementioned techniques, it might be possible to design robust tile sets to perform tiling-based computations. The experimental testing of these and related error-reduction methods is ongoing. It seems possible that other error-correction techniques (such as error-correcting codes) developed in computer science may also be utilized.

14.7.3   Activatable Tiles for Reducing Errors

The uncontrolled assembly of tiling assemblies in reverse directions is potentially a major source of errors in computational tiling assemblies, and a roadblock in the development of applications of large patterned computational DNA lattices. Methods for controlled directional assembly of tiling assemblies would eliminate these errors. Majumder et al. (2007) have recently developed novel designs for an enhanced class of error-resilient DNA tiles (known as *activatable tiles*) for controlled directional assembly of tiles. While conventional DNA tiles store no state, the activatable tiling systems make use of a powerful DNA polymerase enzyme that allows the tiles to transition between active (allowing assembly) and inactive states (Figure 14.21). A *protection–deprotection* process strictly enforces the direction of tiling assembly growth so that the assembly process is robust against entire classes of growth errors. Initially, prior to binding with other tiles, some pads of the tile will be in an inactive state, where the tile is protected from unwanted binding with other tiles and thus preventing lattice growth in the (unwanted) reverse direction. After appropriate bindings and subsequent deprotections, the tile transitions to an active state, allowing further growth.

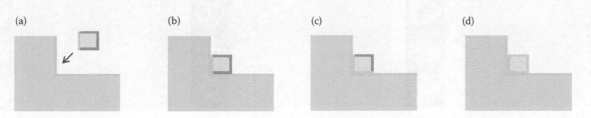

FIGURE 14.21   Activatable tiles. (a) Partially formed assembly with two activated boundaries and a protected tile. (b) Protected tile binds to the boundary. (c) The other input pad is activated. (d) When both the inputs pads bind, the output pads are activated.

14.8 Three-Dimensional DNA Nanostructures

14.8.1 Three-Dimensional DNA Wireframe Polyhedra

The first 3D wireframe object was obtained by Shih et al. (2004) by folding a 1.7-kilobase single-stranded DNA into nanoscale wireframe octahedron with the help of five 40 base synthetic DNA. The structure was imaged using cyroEM (see Figure 14.22). This was followed soon after by Goodman et al. (2004) who constructed a wireframe DNA regular tetrahedron from four 55 base ssDNA (see Figure 14.22) in a single synthesis step. The structure was experimentally demonstrated to be structurally robust and the fabrication process was quick and simple. Another approach toward wireframe structures was demonstrated by He et al. (2008) when they used a three-point-star motif to

hierarchically assembly tetrahedrons (4 three-point motifs), dodecahedra (20 three-point motifs), and buckyballs (60 three-point motifs) (see Figure 14.23). Instead of many ssDNA of unique sequences, many copies of the same motif (three-point-star) assemble into different polyhedral structures depending on the flexibility of the arms and concentration of the motif.

14.8.2 Three-Dimensional DNA Lattices

Most of the DNA lattices described in this chapter have been limited to 2D sheets. It appears to be much more challenging to assemble 3D DNA lattices of high regularity. There are some very important applications to nanoelectronics and biology if this can be done, as described below.

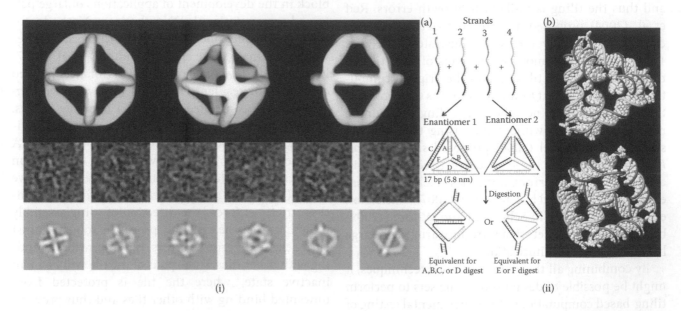

FIGURE 14.22 Wireframe polyhedra. (i) Truncated octahedron. (Reprinted by permission from Macmillan Publishers Ltd: *Nature* Shih, W., Quispe, J., and Joyce, G. A 1.7-kilobase single-stranded DNA that folds into a nanoscale octahedron, 427(6975), 618–621. © 2004.) (ii) Tetrahedron. (From Goodman, R., Berry, R., and Turberfield, A. 2004. The single-step synthesis of a DNA tetrahedron. *Chemical Communications*, 1372–1373. Reproduced by permission of The Royal Society of Chemistry.)

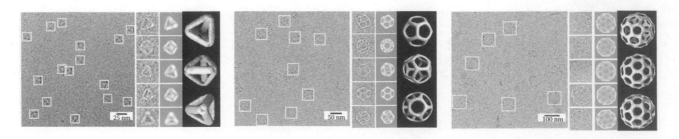

FIGURE 14.23 Creating various polyhedra using the three-point motifs. (Reprinted by permission from Macmillan Publishers Ltd: *Nature*, He, Y. et al., Hierarchical self-assembly of DNA into symmetric supramolecular polyhedra, 452(7184), 198–201. © 2008.)

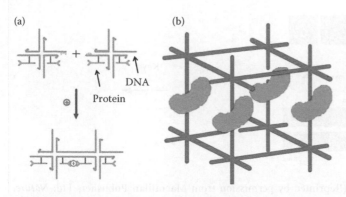

FIGURE 14.24 Scaffolding of (a) 3D nanoelectronic architectures and (b) proteins into regular 3D arrays.

The density of conventional nanoelectronics is limited by lithographic techniques to only a small number of layers. The assembly of even quite simple 3D nanoelectronic devices such as memory would provide much improvement in density. Figure 14.24a shows DNA and protein organizing functional electronic structures.

It has been estimated that at least one-half of all natural proteins cannot be readily crystallized, and have unknown structure, and determining these structures would have a major impact in the biological sciences. Suppose a 3D DNA lattice can be assembled with sufficient regularity and with regular interstices (say within each DNA tile comprising the lattice). Then a given protein might be captured within each of the lattice's interstices, allowing it to be in a fixed orientation at each of its regularly spaced locations in 3D (see Figure 14.24b). This would allow the protein to be arranged in 3D in a regular way to allow for x-ray crystallography studies of its structure. This visionary idea is due to Seeman. So far, there has been only limited success in assembling

3D DNA lattices, and they do not yet have the degree of regularity (down to 2 or 3 Å) required for the envisioned x-ray crystallography studies. The best effort thus far has been achieved by Zheng et al. (2009) through the tensegrity triangle, which is a rigid DNA motif with three helical arms oriented along three linearly independent axes (see Figure 14.25). Rhombohedral crystals of 4 Å resolution were obtained.

14.8.3 Three-Dimensional DNA Origami

Rothemund's origami demonstrated arbitrary flat 2D nanostructures. Andersen et al. (2009) extended this technique to construct hollow containers (box) with walls of flat 2D origami. A cube-like hollow box with a hinged lid that can be opened and closed by a DNA strand as a key was constructed and imaged (see Figure 14.26).

DNA origami was extended to simple 3D cylindrical filaments that were used to partially orient membrane proteins in solution for structural studies employing NMR (Douglas et al. 2007). In a new approach, Douglas et al. (2009) created stunning 3D origami by carving out 3D shapes from a honeycomb-like solid 3D structure (see Figure 14.27). In addition, they provided design automation software, caDNAno (www.cadnano.org), that enables rapid prototyping of arbitrary 3D nanostructure with about 6 nm resolutions. Dietz et al. (2009) demonstrated the ability to bend and twist the honeycomb lattice by underwinding or overwinding the DNA double helix (see Figure 14.27). Honeycomb lattice-based nanostructures have higher charge density and hence require longer annealing times than 2D DNA origami and carefully controlled salt concentrations, and usually had lower yields than flat 2D DNA origami.

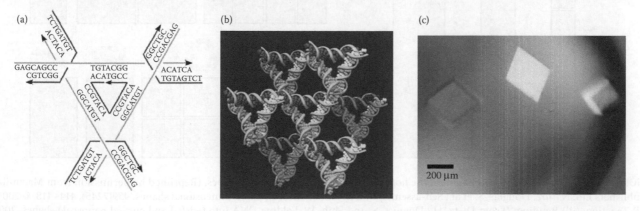

FIGURE 14.25 (a) Schematics of the tensegrity tile. (b) Lattice structure. (c) Optical image of the 3D lattice. (Reprinted by permission from Macmillan Publishers Ltd: *Nature,* Zheng, J. et al., From molecular to macroscopic via the rational design of a self-assembled 3D DNA crystal, *461*(7260), 74–78. © 2009.)

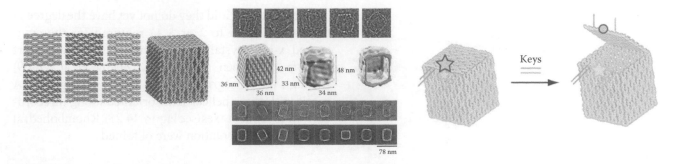

FIGURE 14.26 DNA box made by folding up planar origami. (Reprinted by permission from Macmillan Publishers Ltd: *Nature*, Andersen, E. et al., Self-assembly of a nanoscale DNA box with a controllable lid, *459*(7243), 73–76. © 2009.)

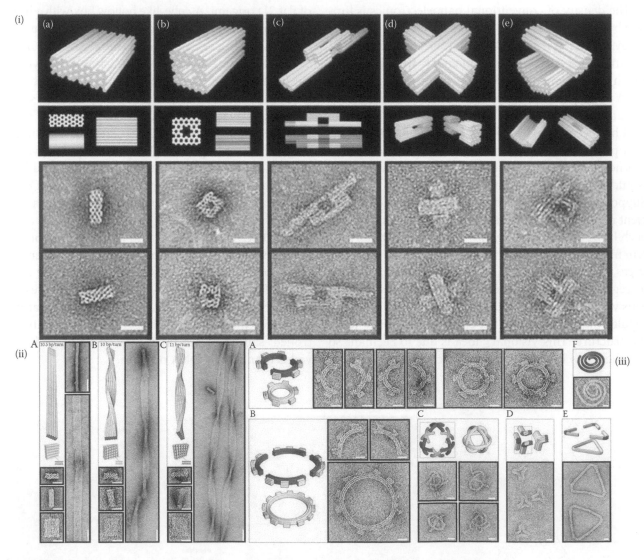

FIGURE 14.27 3D DNA origami based on the honeycomb lattice. (i) Various 3D shapes. (Reprinted by permission from Macmillan Publishers Ltd: *Nature*, Douglas, S. et al., Self-assembly of DNA into nanoscale three-dimensional shapes, *459*(7245), 414–418. © 2009.) (ii) Twisting. (iii) Bending. (From Dietz, H., Douglas, S. and Shih, W. Folding DNA into twisted and curved nanoscale shapes, 2009, *Science, 325*(5941), 725–730. Reprinted with permission of AAAS.)

14.9   From Nucleic Detection Protocols to Autonomous Computation

14.9.1   The Detection Problem

A fundamental task of many biochemical protocols is to sense a particular molecule and then amplify the response. In particular, the detection of specific strands of RNA or DNA is an important problem for medicine. Typically, a protocol for nucleic detection is specialized to a subsequence of single-stranded nucleic acid (DNA or RNA oligonucleotide) to be detected. Given a sample containing a very small number of the nucleic strand molecules to be detected, a detection protocol must amplify this to a much larger signal. Ideally, the detection protocol is exquisitely sensitive, providing a response from the presence of only a few of the target molecules.

There are a number of novel methods for doing DNA computation that can be viewed as being derived from protocols for detection of DNA. Therefore, understanding the variety of detection protocols can provide insight into these methods for DNA computation.

14.9.2   Methods for Autonomous Molecular Computation Derived from PCR

14.9.2.1   The Polymerase Chain Reaction

The original and still the most frequently used method for DNA detection is the *polymerase chain reaction (PCR)*, which makes use of DNA polymerase to amplify a strand of DNA by repeated replication, using rounds of thermal cycling (Saiki et al. 1985). (Recall that given an initial "primer" DNA strand hybridized onto a segment of a template DNA strand, the polymerase enzyme can extend the primer strand by appending free DNA nucleotides complementary to the template's nucleotides.) In addition to DNA polymerase, the protocol requires a pair of "primer" DNA strands, which are extended by the DNA polymerase, each followed by heating and cooling, to allow displacement of the product strands.

14.9.2.2   Whiplash PCR: A Method for Local Molecular Computation

A method for DNA computation, known as *whiplash PCR*, introduced by Sakamoto et al. (1999) (see Figure 14.28), makes use of a strand of DNA that essentially encodes a "program" describing state transition rules of a finite-state computing machine; the strand is comprised of a sequence of "rule" subsequences (each encoding a state transition rule), and each separated by stopper sequences (which can stop the action of DNA polymerase). On each step of the computation, the 3-prime end of the DNA strand has a final sequence encoding a state of the computation. A computation step is executed when this 3-prime end hybridizes to a portion of a "rule" subsequence, and the action of DNA polymerase extends the 3-prime end to a further subsequence encoding a new state.

Whiplash PCR is interesting, since it executes a local molecular computation (recall that a molecular computation is local if the computation within a single molecule, possibly in parallel with other molecular computing devices). In contrast, most methods for autonomous molecular computation (such as those based on the self-assembly of tiles) provide only a capability for distributed parallel molecular computation since to execute a computation they require multiple distinct molecules that interact to execute steps of each computation.

14.9.3   Isothermal and Autonomous PCR Detection and Whiplash PCR Computation Protocols

Neither the original PCR protocol nor the whiplash PCR executes autonomously—they require thermal cycling for each step of their protocols. Walker et al. (1992a,b) developed isothermal (requiring no thermal cycling) methods for PCR known as strand

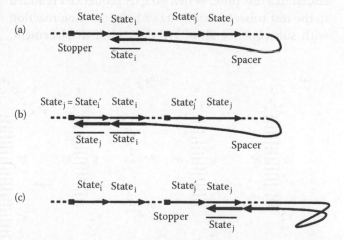

FIGURE 14.28   Whiplash PCR state transitions. The current state is annealed onto the transition table by forming a hairpin structure (a). The current state is then extended by polymerase and the next state is copied from the transition table (b). After denaturation, the new current state is annealed to another part of the transition table to enable the next transition (c). (Reprinted from *Biosystems*, Sakamoto, K. et al., State transitions by molecules, 81–91, © 1999, with permission from Elsevier.)

Chapter 14

displacement amplification (SDA) in which strands displaced from DNA polymerase are used for the further stages of the amplification reaction. Reif and Majumder (2008) recently developed an autonomously executing version of whiplash PCR (known as isothermal reactivating whiplash PCR) that makes use of a strand-displacing polymerization enzyme (recall however that certain polymerase enzymes such as phi-29 can, as a side effect of their polymerization reaction, displace previously hybridized strands) with techniques to allow the reaction to proceed isothermally. In summary, an isothermal variant (strand displacement PCR) of the basic PCR detection protocol provided insight on how to design an autonomous method for DNA computation. Like whiplash PCR, this new isothermal reactivating whiplash PCR provides for local molecular computation.

14.9.4 Autonomous Molecular Cascades for DNA Detection

Dirks and Pierce (2004) demonstrated an isothermal, enzyme-free (most known detection protocols require the use of protein enzymes) method for highly sensitive detection of a particular DNA strand. This protocol makes a triggered amplification by hybridization chain reaction briefly illustrated in Figure 14.29.

The protocol made use of multiple copies of two distinct DNA hairpins H1 and H2 that are initially added to a test tube. When ssDNA sequence I is added to the test tube, I initially has a hybridization reaction with subsequence ab of H1 via strand displacement,

thus exposing c that had been previously hidden within the stem loop of H1. Next, cb* has a hybridization reaction with the subsequence c*b of H2, thus exposing a second copy of a* that had been previously hidden within the stem loop of H2. That other copy of a*b* then repeats the process with other similar (but so far unaltered) copies of H1 and H2, allowing a cascade effect to occur completely autonomously. Such autonomous molecular cascade devices have applications to a variety of medical applications, where a larger response (e.g., a cascade response) is required in response to one of multiple molecular detection events. Note that the response is linear in the concentration of strand I.

14.9.5 Hybridization Reactions for Autonomous DNA Computation

Zhang et al. (2007) developed a general methodology for designing systems of DNA molecules by the use of catalytic reactions that are driven by entropy. In particular, it demonstrates a general, powerful scheme for executing any Boolean circuit computation via cascades of DNA hybridization reactions. The unique common property of the above detection protocol of Dirks and Pierce (2004) and the molecular computations of Zhang et al. (2007) are their use only of hybridization, making no use of restriction enzyme or any other protein enzymes.

Following on this work, Yin et al. (2008) developed an elegant and highly descriptive labeled scheme (with nodes indicating inputs, products, etc.) for illustrating

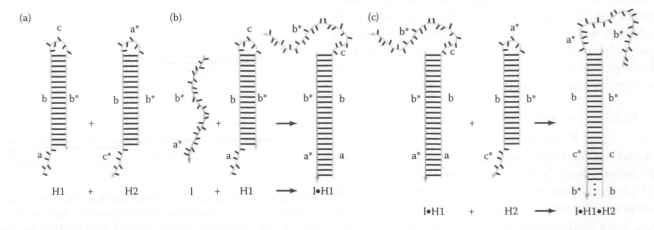

FIGURE 14.29 Autonomous molecular cascade for signal amplification. (a) Hairpins H1 and H2. (b) Step 1 of the reaction: I + H1 -> I-H1. (c) Step 2 of the reaction: I-H1 + H2 -> I-H1- H2. (From Dirks, R. and Pierce, N. Triggered amplification by hybridization chain reaction, *Proceedings of the National Academy of Sciences of the United States of America 101*(43), 15275–15278. © 2004 National Academy of Sciences, U.S.A.)

the programming of biomolecular self-assembly and reaction pathways.

14.9.6   Autonomous Detection Protocols and Molecular Computations Using DNAzyme

In addition, Tian et al. (2006) demonstrated a novel method for DNA detection that involves amplification of the target strand via rolling circle amplification followed by the use of a dual set of DNAzyme (recall a DNAzyme is a DNA molecule that possess enzymatic activity, in particular, cutting particular single-stranded DNA) that provided for colorimetric DNA detection at a limit of 1 pM. This led to the DNAzyme-based autonomous DNA walker Tian et al. (2005) described in Section 14.10.4.2.

14.10   Autonomous Molecular Transport Devices Self-Assembled from DNA

14.10.1   Molecular Transport

Many molecular-scale tasks may require the transport of molecules and there are a number of other tasks that can be done at the molecular scale that would be considerably aided by an ability to transport within and/or along nanostructures. For example of the importance of molecular transport in nanoscale systems, consider the cell, which uses protein motors fueled by ATP to do this.

14.10.2   Nonautonomous DNA Motor Devices

In the early 2000s, a number of researchers developed and demonstrated motors composed of DNA nanostructures; for example, Yurke et al. (2000) demonstrated a DNA actuator powered by DNA hybridization (complementary pairing between DNA strands). However, all these DNA motor devices required some sort of externally mediated changes (such as temperature-cycling, addition or elimination of a reagent, etc.) per work-cycle of the device, and so did not operate autonomously.

14.10.3   The Need for Autonomous Molecular Transport

Almost all the conventionally scaled motors used by mankind run without external mediation, and almost all natural systems for molecular motors are also autonomous (e.g., the cell's protein motors are all autonomous). The practical applications of molecular devices requiring externally mediated changes per work-cycle are quite limited. So it is essential to develop autonomous DNA devices that do not require external mediation while executing the movements.

14.10.4   Autonomous DNA Walkers

Reif (2003) first described the challenge of autonomous molecular transport devices, which he called "DNA walkers" that traversed DNA nanostructures, and proposed two designs that gave bidirectional movement. Sherman and Seeman (2004) demonstrated a DNA walker, but it was nonautonomous since it required external mediation for every step it made.

14.10.4.1   Restriction Enzyme-Based Autonomous DNA Walkers

The first autonomous DNA walker was experimentally demonstrated by Yin et al. (2004c). It employed restriction enzymes and ligase; see Yin et al. (2004b) for its detailed general design.

The device is described in Figure 14.30.

- Initially, a linear DNA nanostructure (the "road") with a series of attached ssDNA strands (the "steps") is self-assembled. Also, a fixed-length segment of DNA helix (the "walker") with short sticky ends (its "feet") hybridized to the first two steps of the road.
- Then the walker proceeds to make a sequential movement along the road, where at the start of each step, the feet of the walker are hybridized to two further consecutive two steps of the road.
- Then a restriction enzyme cuts the DNA helix where the backward foot is attached, exposing a new sticky end forming a new replacement foot

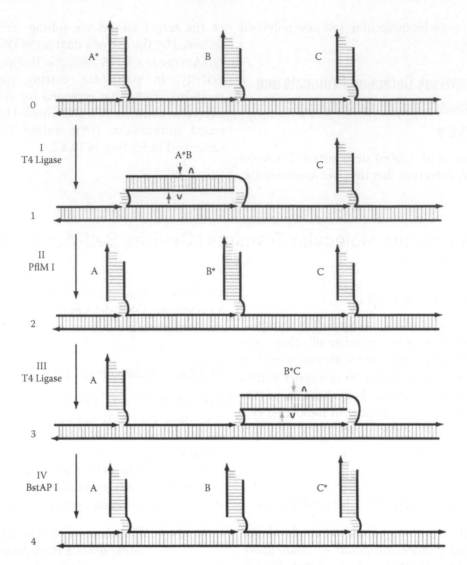

FIGURE 14.30 Autonomous molecular transport devices self-assembled from DNA. (From Yin, P. et al., A unidirectional DNA walker that moves autonomously along a linear track. *Angewandte Chemie International Edition.* 2004c. *116*(37), 4906–4911. Copyright Wiley-VCH Verlag GmbH & Co. KGaA. Reproduced with permission.)

that can hybridize to the next step that is free, which can be the step just after the step where the other foot is currently attached. A somewhat complex combinatorial design for the sequences composing the steps and the walker ensures that there is unidirectional motion forward along the road.

14.10.4.2 DNAzyme-Based Autonomous DNA Walkers

Subsequently, Tian et al. (2005) demonstrated an autonomous DNA walker (Figure 14.31) that made use of a DNAzyme motor, designed by Chen et al. (2004), which used the cuts provided by the enzymatic activity of DNAzyme to progress along a DNA nanostructure.

Bath and Turberfield (2007) also give an extensive survey of these and further recent DNA motor and walker devices.

14.10.5 Programmable Autonomous DNA Devices: Nanobots

There are some important applications of these autonomous DNA walkers, including transport of molecules within large self-assembled DNA nanostructures. However, the potential applications may be vastly increased if they can be made to execute computations while moving along a DNA nanostructure. This would allow them, for example, to make programmable changes to their state and to make movements

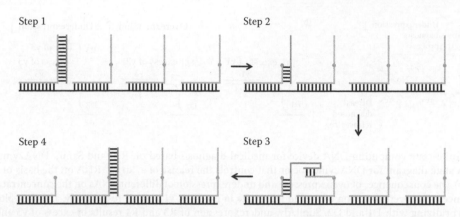

FIGURE 14.31 (**See color insert.**) Mao's DNAzyme walker moving from one spot to the next on the black track with four foot holds. Red subsequence is the DNAzyme; blue dots are places where the DNAzyme can cleave.

programmable (Figure 14.32). We will call such programmable autonomous DNA walker devices "programmable DNA nanobots." Yin et al. (2005) describe an extension of the design of the restriction-enzyme-based autonomous DNA walker of Yin et al. (2004b), described in Section 14.10.4.3, to allow programmed computation while moving along a DNA nanostructure.

Another DNA nanobot design (see Figures 14.33 and 14.34) for programmed computation while moving along a DNA nanostructure was developed by Reif and Sahu (2007) using in this case an extension of the design of the DNAzyme-based autonomous DNA walker of Tian et al. (2005) also described above. It remains a challenge to experimentally demonstrate these.

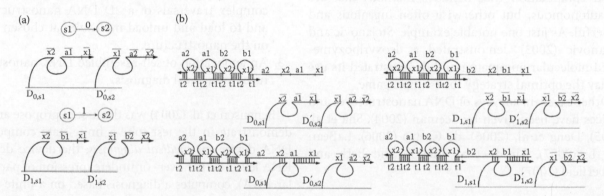

FIGURE 14.32 Reif and Sahu's DNA nanobot. (a) The implementation of a state transition through DNAzymes. (b) $D_{0,s1}$ in the transition machinery for state transition at 0 combines with input nanostructure when active input symbol encoded by the sticky end is 0. When the active input symbol encoded by the sticky end is 1, $D_{1,s1}$ in the transition machinery for state transition at 1 combines with the input nanostructure.

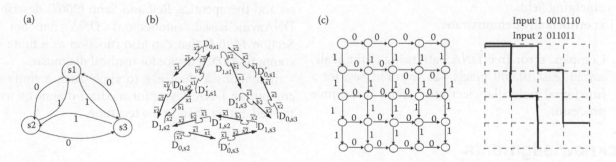

FIGURE 14.33 Programmed traversal of a grid DNA nanostructure. (a) Transition diagram of a finite-state machine. (b) The DNAzyme implementation of the finite-state machine shown in (a). (c) Illustration of programmable routing in 2D.

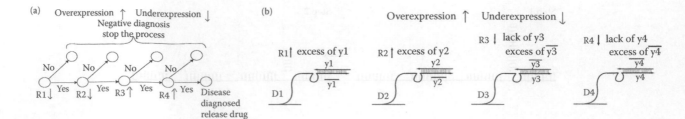

FIGURE 14.34 A finite-state computing DNA device for medical diagnosis based on Reif and Sahu's DNAzyme-based autonomous DNA nanobot. (a) A state diagram for DNAzyme doctor that controls the release of a "drug" RNA on the basis of the RNA expression tests for a disease. (b) The consequences of overexpression and underexpression of different RNAs on the concentrations of the respective characteristic sequences. The overexpression of R1 and R2 results in excess of y1 and y2, respectively, and they block the path of input nanostructure by hybridizing with D1 and D2. Similarly, underexpression of R3 and R4 results in excess of y3 and y4, respectively, to block the path of input nanostructure.

14.11 Conclusions and Challenges

14.11.1 What Was Covered and What Was Missed: Further Reading

This chapter has covered most of the major known techniques and results for autonomous methods for DNA-based computation and transport.

However, there is a much larger literature of DNA-based computation that includes methods that are nonautonomous, but otherwise often ingenious and powerful. As just one notable example, Stojanovic and Stefanovic (2003) demonstrated a deoxyribozyme-based molecular automaton and demonstrated its use to play the optimal strategy for a simple game.

Other excellent surveys of DNA nanostructures and devices have been given by Seeman (2004), Sha et al. (2005), Deng et al. (2006), de Castro (2006), LaBean and Li (2007), Lund et al. (2006), and Bath and Turberfield (2007).

14.11.2 Future Challenges for Self-Assembled DNA Nanostructures

There are a number of key challenges still confronting this emerging field:

Experimentally demonstrate:

1. Complex, error-free DNA patterning to the scale, say, at least 10,000 pixels—as required, say, for a functional molecular electronic circuit for a simple processor.

Note: This would probably entail the use of a DNA tiling error-correction method as well as a significant improvement over existing DNA patterning techniques.

2. A programmable DNA nanobot autonomously executing a task critical to nanoassembly.

Note: The first stage might be a DNA walker that can be programmed to execute various distinct, complex traversals of a 2D DNA nanostructure, and to load and unload molecules at chosen sites on the nanostructure.

3. An application of self-assembled DNA nanostructures to medical diagnosis.

Benenson et al. (2004) was the first to propose and to demonstrate in the test tube a finite-state computing DNA device for *medical diagnosis*: the device detects RNA levels (either over- or underexpression of particular RNA), computes a diagnosis based on a finite-state computation, and then provides an appropriate response (e.g., the controlled release of a single-stranded DNA that either promotes or interfere with expression). They demonstrated in the test tube a potential application of such a finite-state computing device to medical diagnosis and therapeutics. Reif and Sahu (2007) described a DNAzyme-based autonomous DNA nanobot (see Section 14.10.4) that can also function as a finite-state computing DNA device for medical diagnosis.

It remains a challenge to apply such a finite-state computing DNA device for medical diagnosis within the cell, rather than in the test tube.

Acknowledgments

We would like to thank Sudheer Sahu and Urmi Majumder for their valuable suggestion on a preliminary version of this chapter. This work was supported by NSF Grants CCF-0829797 and CCF-0829798.

References

Adar, R., Benenson, Y., Linshiz, G., Rosner, A., Tishby, N., and Shapiro, E. 2004, Stochastic computing with biomolecular automata, *Proceedings of the National Academy of Sciences of the United States of America 101*(27), 9960–9965.

Adleman, L. 1994, Molecular computation of solutions to combinatorial problems, *Science 266*(5178), 1021–1024.

Adleman, L. 1998, Computing with DNA, *Scientific American 279*(52), 54–61.

Amin, R., Kim, S., Park, S. H., and LaBean, T. 2009, Artificially designed DNA nanostructures, *NANO: Brief Reports and Reviews 4*(3), 119–139.

Andersen, E., Dong, M., Nielsen, M., Jahn, K., Subramani, R., Mamdouh, W., Golas, M. et al. 2009, Self-assembly of a nanoscale DNA box with a controllable lid, *Nature 459*(7243), 73–76.

Barish, R., Rothemund, P., and Winfree, E. 2005, Two computational primitives for algorithmic self-assembly: Copying and counting, *Nano Letters 5*, 2586–2592.

Bath, J. and Turberfield, A. 2007, DNA nanomachines, *Nature Nanotechnology 2*, 275–284.

Benenson, Y., Gil, B., Ben-Dor, U., Adar, R., and Shapiro, E. 2004, An autonomous molecular computer for logical control of gene expression, *Nature 429*(6990), 423–429.

Berger, R. 1966, The undecidability of the domino problem, *Memoirs of American Mathematical Society 66*, 1–72.

Chen, H.-L. and Goel, A. 2004, Error free self-assembly using error prone tiles, *DNA Computing*, 62–75.

Chen, H.-L., Schulman, R., Goel, A., and Winfree, E. 2007, Reducing facet nucleation during algorithmic self-assembly, *Nano Letters 7*, 2913–2919.

Chen, Y., Wang, M., and Mao, C. 2004, An autonomous DNA nanomotor powered by a DNA enzyme, *Angewandte Chemie International Edition 43*(27), 3554–3557.

de Castro, L. 2006, *Fundamentals of Natural Computing: Basic Concepts, Algorithms, and Applications*, Boca Raton, FL: Chapman and Hall.

Deng, Z., Chen, Y., Tian, Y., and Mao, C. 2006, A fresh look at DNA nanotechnology, *Nanotechnology: Science and Computation* 23–34.

Dietz, H., Douglas, S., and Shih, W. 2009, Folding DNA into twisted and curved nanoscale shapes, *Science 325*(5941), 725–730.

Dirks, R. and Pierce, N. 2004, Triggered amplification by hybridization chain reaction, *Proceedings of the National Academy of Sciences of the United States of America 101*(43), 15275–15278.

Douglas, S., Chou, J., and Shih, W. 2007, DNA-nanotube-induced alignment of membrane proteins for NMR structure determination, *Proceedings of the National Academy of Sciences of the United States of America 104*(16), 6644–6648.

Douglas, S., Dietz, H., Liedl, T., Hogberg, B., Graf, F., and Shih, W. 2009, Self-assembly of DNA into nanoscale three-dimensional shapes, *Nature 459*(7245), 414–418.

Fujibayashi, K., Hariadi, R., Park, S. H., Winfree, E., and Murata, S. 2008, Toward reliable algorithmic self-assembly of DNA tiles: A fixed-width cellular automaton pattern, *Nano Letters 8*(7), 1791–1797.

Goodman, R., Berry, R., and Turberfield, A. 2004, The single-step synthesis of a DNA tetrahedron, *Chemical Communications* 1372–1373.

Hamada, S. and Murata, S. 2009, Substrate-assisted assembly of interconnected single-duplex DNA nanostructures, *Angewandte Chemie International Edition 48*(37), 6820–6823.

He, Y., Chen, Y., Liu, H., Ribbe, A., and Mao, C. 2005a, Self-assembly of hexagonal DNA two-dimensional (2D) arrays, *Journal of the American Chemical Society 127*(35), 12202–12203.

He, Y., Tian, Y., Chen, Y., Deng, Z., Ribbe, A., and Mao, C. 2005b, Sequence symmetry as a tool for designing DNA nanostructures, *Angewandte Chemie International Edition 44*(41), 6694–6696.

He, Y., Ye, T., Su, M., Zhang, C., Ribbe, A., Jiang, W., and Mao, C. 2008, Hierarchical self-assembly of DNA into symmetric supramolecular polyhedra, *Nature 452*(7184), 198–201.

LaBean, T., Gothelf, K., and Reif, J. 2007, Self-assembling DNA nanostructures for patterned molecular assembly, *Nanobiotechnology II* 79–97.

LaBean, T. and Li, H. 2007, Constructing novel materials with DNA, *NanoToday 2*(2), 26–35.

LaBean, T., Winfree, E., and Reif, J. 1999, Experimental progress in computation by self-assembly of DNA tilings, *DNA Based Computers V, DIMACS 54*, 123–140.

LaBean, T., Yan, H., Kopatsch, J., Liu, F., Winfree, E., Reif, J., and Seeman, N. 2000, Construction, analysis, ligation, and self-assembly of DNA triple crossover complexes, *Journal of the American Chemical Society 122*(9), 1848–1860.

Lund, K., Williams, B., Ke, Y., Liu, Y., and Yan, H. 2006, DNA nanotechnology: A rapidly evolving field, *Current Nanoscience 2*, 113–122.

Majumder, U., LaBean, T., and Reif, J. 2007, Activatable tiles: Compact, robust programmable assembly and other applications, *DNA Computing*, 15–25.

Mao, C., Labean, T., Reif, J., and Seeman, N. 2000, Logical computation using algorithmic self-assembly of DNA triple-crossover molecules, *Nature 407*, 493–496.

Park, S. H., Pistol, C., Ahn, S. J., Reif, J., Lebeck, A., and LaBean, C. D. T. 2006a, Finite-size, fully addressable DNA tile lattices formed by hierarchical assembly procedures, *Angewandte Chemie International Edition 45*(5), 735–739.

Park, S. H., Prior, M., LaBean, T., and Finkelstein, G. 2006b, Optimized fabrication and electrical analysis of silver nanowires templated on DNA molecules, *Applied Physics Letters 89*(3), 033901 (3 pages), doi: 10.1063/1.2234282.

Pistol, C. and Dwyer, C. 2007, Scalable, low-cost, hierarchical assembly of programmable DNA nanostructures, *Nanotechnology 18*(12), 125305–125309.

Reif, J. 2003, The design of autonomous DNA nano-mechanical devices: Walking and rolling DNA, *DNA 8*, 439–461.

Reif, J. and Majumder, U. 2008, Isothermal reactivating whiplash PCR for locally programmable molecular computation, *DNA Computing* 41–56.

Reif, J. and Sahu, S. 2007, Autonomous programmable nanorobotic devices using DNAzymes, *DNA Computing* 66–78.

Reif, J., Sahu, S., and Yin, P. 2004, Compact error-resilient computational DNA tiling assemblies, *DNA Computing* 293–307.

Robinson, B. and Seeman, N. 1987, The design of a biochip: A self-assembling molecular-scale memory device, *Protein Engineering 1*(4), 295–300.

Robinson, R. 1971, Undecidability and nonperiodicity for tilings of the plane, *Inventiones Mathematicae 12*, 177–209.

Rothemund, P. 2006, Folding DNA to create nanoscale shapes and patterns, *Nature 440*, 297–302.

Rothemund, PWK., Papadakis, N. and Winfree, E. 2004, Algorithmic self-assembly of DNA sierpinski triangles. PLoS Biology 2(12): e424, doi: 10.1371/journal.pbio.0020424.

Rothemund, P. and Winfree, E. 2000, The program-size complexity of self-assembled squares, *Symposium on Theory of Computation* 459–468.

Saiki, R., Scharf, S., Faloona, F., Mullis, K., Horn, G., Erlich, H., and Arnheim, N. 1985, Enzymatic amplification of beta-globin genomic sequences and restriction site analysis for diagnosis of sickle cell anemia, *Science 230*(4732), 1350–1354.

Sakamoto, K., Kiga, D., Momiya, K., Gouzu, H., Yokoyama, S., Ikeda, S., Sugiyama, H., and Hagiya, M. 1999, State transitions by molecules, *Biosystems* 81–91.

Schulman, R. and Winfree, E. 2009, Programmable control of nucleation for algorithmic self-assembly, *SIAM Journal on Computing 39*(4), 1581–1616.

Seeman, N. 2004, Nanotechnology and the double helix, *Scientific American 290*(6), 64–75.

Sha, R., Zhang, X., Liao, S., Constantinou, P., Ding, B., Wang, T., Garibotti, A. et al. 2005, Structural DNA nanotechnology: Molecular construction and computation, *Unconventional Computing* 20–31.

Shapiro, E. and Benenson, Y. 2006, Bringing DNA computers to life, *Scientific American 17*(3), 4047.

Sherman, W. and Seeman, N. 2004, A precisely controlled DNA biped walking device, *Nano Letters 4*, 1203–1207.

Shih, W., Quispe, J., and Joyce, G. 2004, A 1.7-kilobase single-stranded DNA that folds into a nanoscale octahedron, *Nature 427*(6975), 618–621.

Stojanovic, M. and Stefanovic, D. 2003, A deoxyribozyme-based molecular automaton, *Nature Biotechnology 21*(9), 1069–1074.

Tian, Y., He, Y., Chen, Y., Yin, P., and Mao, C. 2005, A DNAzyme that walks processively and autonomously along a one-dimensional track, *Angewandte Chemie International Edition 44*(28), 4355–4358.

Tian, Y., He, Y., and Mao, C. 2006, Cascade signal amplification for DNA detection, *ChemBioChem 7*(12), 1882–1864.

Walker, T., Fraiser, M., Schram, J., Little, M., Nadeau, J., and Malinowski, D. 1992a, Strand displacement amplification—An isothermal, *in vitro* DNA amplification technique, *Nucleic Acid Research 20*(7), 1691–1696.

Walker, T., Little, M., Nadeau, J., and Shank, D. 1992b, Isothermal *in vitro* amplification of DNA by a restriction enzyme/DNA polymerase system, *Proceedings of the National Academy of Sciences of the United States of America 89*(1), 392–396.

Wang, H. 1961, Proving theorems by pattern recognition II, *Bell Systems Technical Journal 40*, 1–41.

Winfree, E. 1995, On the computational power of DNA annealing and ligation, *DNA Based Computers, Proceeding of DIMACS Workshop. American Mathematical Society*, pp. 199–221.

Winfree, E. and Bekbolatov, R. 2003, Proofreading tile sets: Error correction for algorithmic self-assembly, *DNA Computing* 126–144.

Winfree, E., Liu, F., Wenzler, L., and Seeman, N. 1998, Design and self-assembly of two-dimensional DNA crystals, *Nature 394*, 539–544.

Winfree, E., Yang, X., and Seeman, N. 1996, Universal computation via self-assembly of DNA: Some theory and experiments, *DNA Based Computers II, DIMACS 44*, 191–213.

Yan, H., Feng, L., Labean, T., and Reif, J. 2003a, Parallel molecular computations of pairwise exclusive-or (XOR) using DNA string tile self-assembly, *Nature 125*(47), 14246–14247.

Yan, H., LaBean, T., Feng, L., and Reif, J. 2003b, Directed nucleation assembly of DNA tile complexes for barcode-patterned lattices, *Proceedings of the National Academy of Sciences of the United States of America 100*(14), 8103–8108.

Yan, H., Park, S. H., Finkelstein, G., Reif, J., and LaBean, T. 2003c, DNA-templated self-assembly of protein arrays and highly conductive nanowires, *Science 301*(5641), 1882–1884.

Yin, P., Choi, H., Calvert, C., and Pierce, N. 2008, Programming biomolecular self-assembly pathways, *Nature 451*(7176), 318–322.

Yin, P., Guo, B., Belmore, C., Palmeri, W., Winfree, E., LaBean, T., and Reif, J. 2004a, TileSoft: Sequence Optimization Software for Designing DNA Secondary Structures, Technical report, Duke and Caltech.

Yin, P., Turberfield, A., Sahu, S., and Reif, J. 2004b, Designs for autonomous unidirectional walking DNA devices, *DNA Computing* 410–425.

Yin, P., Yan, H., Daniell, X., Turberfield, A., and Reif, J. 2004c, A unidirectional DNA walker that moves autonomously along a linear track, *Angewandte Chemie International Edition 116*(37), 4906–4911.

Yin, P., Sahu, S., Turberfield, A., and Reif, J. 2005, Design of autonomous DNA cellular automata, *DNA Computing* 376–387.

Yurke, B., Turberfield, A., Mills, A., Simmel, F., and Neumann, J. 2000, A DNA-fuelled molecular machine made of DNA, *Nature 406*(6796), 605–608.

Zhang, D., Turberfield, A., Yurke, B., and Winfree, E. 2007, Engineering entropy-driven reactions and networks catalyzed by DNA, *Science 318*, 1121–1125.

Zheng, J., Birktoft, J., Chen, Y., Wang, T., Sha, R., Constantinou, P., Ginell, S., Mao, C., and Nadrian 2009, From molecular to macroscopic via the rational design of a self-assembled 3D DNA crystal, *Nature 461*(7260), 74–78.

15. New Emerging Techniques

Thomas Schenkel

Ion Beam Technology Group, Accelerator and Fusion Research Division, Lawrence Berkeley National Laboratory, Berkeley, California

T.-C. Shen

Department of Physics, Utah State University, Logan, Utah

Yuan Wang

Nanoscale Science and Engineering Center, University of California, Berkeley, California

Xiang Zhang

Mechanical Engineering Department, University of California, Berkeley, California

Nanofabrication Handbook. Edited by Stefano Cabrini and Satoshi Kawata © 2012 CRC Press / Taylor & Francis Group, LLC. ISBN: 978-1-4200-9052-9

Chapter 15

In the previous chapters, we discussed several techniques that are used to fabricate and to manipulate nanostructures and nanomaterials. The main purpose of nanofabrication is to find the most robust and reproducible way to control materials and their geometries in nanometer range. The list of techniques has not been completed in past chapters and there are always new intriguing "tricks" played by scientists to achieve their goal. It is very important to control the "phenomena" during the fabrication process and to control carefully those "tricks" into real new nanofabrication techniques. In this chapter, we discuss three interesting examples of new approaches that were established to solve specific problems and that are becoming solid and promising techniques. These will open up new possibilities in the nanofabrication field.

15.1 Low-Energy e-Beam Nanolithography on Silicon Hydride Surfaces

15.1.1 Introduction

Scanning-probe lithography can arguably achieve the highest resolution in nanoscale patterning. However, the condition on sample surfaces is equally demanding. In the following sections, we focus on some technical issues of practicing low-energy e-beam lithography on hydrogen (H)-terminated silicon surfaces, which could create atom-scale patterns and be compatible with conventional silicon processing. To achieve the desired linewidths in device patterning, one has to consider both the interaction between the lithographic source and the resist and the compatibility with the rest of the processes. In conventional e-beam or photolithography, the organic resist is quite robust and needs to be developed *ex situ* after the electron or photon exposure. Hydrogen resist, however, does not need postlithography development. Patterning is achieved by selective removal of hydrogen by ~6 eV electrons and photons from silicon surfaces in ultrahigh vacuum (UHV). The H-depleted region can then be reacted with various gas molecules via chemical vapor deposition (CVD). Thus, if the overall processing can accommodate a UHV digression—from H-deposition, through e-beam lithography, to subsequent CVD—then the e-beam lithography on H-terminated silicon surfaces could be considered for ultrafine patterning.

15.1.2 Preparation for Pristine Silicon Surfaces

A pristine silicon surface is preferred before H-termination. An excellent review of the surface structures of silicon can be found in ref. [1]. The lowest energy surface on Si(100) at room temperature is the 2×1 reconstructed surface which consists of rows of Si-dimers, 7.7 Å apart as depicted in Figure 15.1a. This surface is quite reactive because each Si in a dimer has an unsaturated dangling bond and the dangling bond density is $6.8 \times 10^{14}/cm^2$ on this surface. The lowest energy surface at room temperature on Si(111) is 7×7 reconstructed as shown in Figure 15.1b. The 7×7 reconstruction reduces the dangling bond density to $3.0 \times 10^{14}/cm^2$; hence, it is less prone to contamination than the Si(100)-2×1. Depending on the quenching rate after high-temperature annealing, other $(2n + 1) \times (2n + 1)$ regions may be present if the adatoms have little time to diffuse to the step edge [2].

The following procedures are commonly adopted to prepare pristine Si surfaces in UHV [3] and the rational will become clear in the following sections:

1. Outgas the sample holder and sample below the oxide desorption temperature (~780°C), until the pressure stays in the 10^{-10} Torr range. This may take several hours.

FIGURE 15.1 (a) Filled-state STM image of Si(100)-2 × 1 surface. The inset shows an empty-state image where a unit cell (7.7 Å × 3.8 Å) is marked. Note that the dark line is through the center of the dimer. (b) Filled-state image of Si(111)-7 × 7 surface. A unit cell (26.9 Å each side) is marked. The inset is an empty-state image. Scale bar, 5 nm.

2. Raise the sample temperature rapidly to 1250°C and hold for 1–2 min. The pressure must stay in the 10^{-9} Torr range or less for this to be successful. It is important to go through the temperature range from 950°C to 1250°C quickly the first time the oxide is removed.

3. Cooling rates may lead to different surface reconstructions.

4. Once a sample is cleaned in this manner, it can be cleaned repeatedly.

Before introducing the silicon sample into a UHV chamber, it can be cleaned simply by rinsing with ethanol [4], acetone [5], or by more elaborate chemical etching, for example, the Shiraki method [6] or the RCA method [7]. If organic carbon contamination is a concern, then ultraviolet (UV) ozone cleaning [8] or ozone-injected water [9] could be included. If the sample has an oxide layer, high-temperature flashing should be able to lift off most of the contamination on top of the oxide, but one has to consider indiffusion of certain contaminants during flashing, which may reappear on the surface later at room temperature (e.g., nickel). For highly doped samples, it is most convenient to pass a direct current (dc) to heat the sample to 1250°C, but for lightly doped samples one can use radiative heating or electron bombardment [10]. After high-temperature flashing, a high-quality 2 × 1 surface with <1% defects can usually be obtained. For samples with prefabricated structures, high-temperature flashing may not be acceptable. Ion sputtering by Ne^+ [11], Ar^+ [12], or Xe^+ [13] followed by annealing in the range of 920–1200°C can also clean the silicon surfaces. Flashing briefly to 700°C after 300 eV Ar^+ sputtering creates a relatively flat surface for nanolithography, but with substantially more defects than by high-temperature anneals [14].

15.1.2.1 Step Bunching Induced by DC Heating

DC current heating could cause step bunching on Si(111) [15] and Si(100) [16] with and without step pinning by the contamination particles (e.g., SiC). If suitable, one can even take advantage of the step bunching effect to create large step-free regions for nanofabrication. On Si(100) surfaces by first creating grating structures of ~5 μm in lateral dimensions and ~0.5 μm deep by photolithography followed by 0.5–12 h annealing at ~1100°C results in 5–10 μm periodic step-free regions [17], as opposed to the usual <1-μm-wide terraces found in commercial wafers with 1/4° miscut. On the Si(111) surface, annealing the samples with prefabricated square craters with sides of 150 μm and a depth of ~1 μm to 1200°C for 20 min can create a single terrace 60-μm wide [18]. Elaborated networks of steps were also possible after Si epitaxy [19].

Besides steps and terrace sizes, impurities on the surface may also inhibit nanofabrication. Since the development of scanning tunneling microscope (STM), real-space surface structures down to atom scale can be routinely achieved. We thus have a means to identify potential contamination sources during surface preparation. Reactions of Si with numerous elemental and molecular species in UHV have been

Chapter 15

intensively studied in the past few decades. For reviews, see refs. [1,20,21]. However, it is not straightforward to identify contaminants based on surface structures alone. Auger electron spectroscopy in principle can identify surface chemical species with sufficient concentration, but it is hard to quantify the surface coverage of the contaminants. Besides, the 3–5 keV e-beams can also induce unintended surface reaction. In the next sections, a brief discussion on a few common contaminants on the Si surfaces is given.

15.1.2.2 Carbon-Induced Surface Reconstruction after Annealing

An earlier study by Henderson et al. reports that a carbon-free surface which has been exposed to a vacuum $<5 \times 10^{-10}$ Torr for 20 h at room temperature produces carbide on its surface after it was heated to 800°C [22]. The culprit is attributed to the residue C-containing molecules in the UHV chamber and solvent from chemical cleaning. In a stainless-steel UHV chamber, the major residual gases, roughly in the order of their partial pressures, are H_2, CO, H_2O, CO_2, CH_4, and other hydrocarbons. H_2, H_2O, CO, Ar, and CO_2 could be originated from outgassing of the stainless-steel chamber itself [23] and the rest of the hydrocarbon species could be formed by the hot filaments of vacuum gauges and high voltages in the ion pump [24]. It has been observed that when the ion pump is gated off, the partial pressures of CO, H_2O, and CH_4 drop. A liquid nitrogen shroud could reduce the partial pressure of most of the aforementioned gases by two to three orders of magnitude, but is only marginally effective for H_2O reducing by about a factor of 2. Although the room temperature sticking coefficient for H_2 and O_2 on Si surfaces are of the order of 10^{-10} and 10^{-1}, respectively [25,26], hot filaments in the UHV chamber can dissociate the molecules into neutral and charged atoms that are ready to react with Si.

The carbon-containing molecules can dissociate and react with Si when the surface is heated between 400°C and 600°C. It has been reported that by annealing a clean Si surface at 600°C extensively, the 2×1 surfaces changes into c(4×4) [27,28]. To verify the source of this transformation, Jemander et al. deposited 0.07 ML (monolayer) carbon on Si(100) followed by a 600°C anneal for <3 min and a nearly complete c(4×4) surface reconstruction appeared [29]. Clearly, the concentration of the C atoms is not enough to be a regular part of the reconstruction, but they can create sufficient strains to induce a new surface reconstruction on the silicon surfaces.

Continuous annealing at 750°C results in SiC islands and defected domains [29]. When heated above 1100°C, the carbide will dissolve into the silicon crystal resulting in a clean surface [22,30]. These studies suggest that one should be careful about the adsorption of carbon-containing molecules onto silicon surfaces from stainless-steel UHV chambers, pump oil in the load-lock, and wet-cleaning solutions. Prolonged annealing at 600–800°C will create SiC crystallites which will roughen the surfaces rendering them unsuitable for nanofabrication.

15.1.2.3 Other Contaminants: Water, Nickel, and Dopant

Water is usually the second or third abundant species in a UHV chamber. At room temperature, H_2O adsorbs dissociatively on Si surfaces to produce Si–H and Si–OH surface species with an initial sticking coefficient near unity on Si(100)-2×1 [31]. The sticking coefficient is less, ~0.23, on Si(111)-7×7 due to its much larger distance, 4.6 Å, between neighboring dangling bonds [32]. Under STM, the water-reacted sites on Si(100)-2×1 appear as depressions in the filled-state images and protrusions in the empty-state images [33]. These defects increase linearly with exposure to the residual gas in a UHV environment [34]. On heating the samples, the adsorbed hydrogen, both from the Si–H and Si–OH bonds desorb recombinatively at ~530°C leaving a residual surface oxide. At temperatures >700°C, the oxide desorbs as SiO resulting in surface voids [35].

Nickel is one of the common sources of metal contamination on Si surfaces. Contact with stainless-steel tweezers, Ni-containing thermocouples, and stainless machine screws during annealing can easily introduce Ni onto the Si surface [36]. Less than 1% of Ni on the Si(100) surface can induce aligned dimer vacancies perpendicular to the dimer rows after 1250°C flashes [37,38]. Repeated flashing cannot remove Ni from the surface, because Ni diffuses into the bulk at high temperature but segregates in the surface during cooling [39]. Nickel on the Si(111) surface forms $\sqrt{19}$ rings (diameter ~11.5 Å) [40]. One can find $\sqrt{19}$ rings embedded between the 7×7 domains after high-temperature flashes even with a trace amount of Ni contamination [41].

Dopant segregation during high-temperature annealing is another source for surface "novel" features. Boron is easy to segregate in the subsurface via outdiffusion from the bulk by annealing. On B-doped Si(100) samples, lower surface B coverage can create paired protrusions [42] while heavy B buildup leads to significant changes in the surface morphology [43]. Annealing heavily B-doped Si(111) samples could create adatom-rich 1×1 regions among the B-rich $\sqrt{3} \times \sqrt{3}$ [44] and the 7×7 domains [45].

The two common n-type dopants, As and P, do not leave detectable surface structures on the Si surface after high-temperature flashing. Arsenic begins to desorb from Si(100) surfaces at ~530°C and peak around 750°C [46]. Similar As desorption temperatures were found on Si(111) [47]. Phosphorus desorbs from Si surfaces from 600°C based on the experiments on PH_3-covered surfaces [48,49]. The high evaporation rate of As and P leaves the surface clean after 1250°C flashes.

15.1.3 Hydrogen–Terminated Surfaces

The surface science of H-terminated silicon surfaces and their interactions with various metals have been reviewed in refs. [50,51]. Here only a few technical issues on sample preparation are discussed.

15.1.3.1 UHV Preparation

In contrast to molecular hydrogen, atomic hydrogen is readily to react with Si surfaces. Atomic hydrogen flux can be generated by a hot filament a few centimeters in front of the sample in an H_2-backfilled UHV chamber (~10^{-6} Torr). However, hydrogen can also break the Si–Si back bonds to form volatile silane. Therefore, an equilibrium surface structure is a balance between many reactions dictated by sample temperatures [52,53]. At elevated sample temperatures (~380°C), saturation H-exposure results in monohydride Si(100)- 2×1 surfaces. This surface resembles the clean Si-2×1 surface but can be distinguished by the location of the dimer rows in the empty-state STM images (Figure 15.2a). As the sample temperature is lowered during H-exposure, more regions of monohydride and dihydride alternating rows, the 3×1 reconstruction, appear on the surface. At sample temperature at ~100°C, a complete coverage of the 3×1 reconstruction will be attained (Figure 15.2b). Further lowering the sample temperature during H-exposure leads to mostly dihydride 1×1 surfaces with significant surface etching [54]. It is much more difficult to produce a smooth H-terminated Si(111) surface in UHV, because the adatoms of the 7×7 structures tend to be etched off to form three-dimensional (3D) islands. Relatively smooth Si(111)-1×1 monohydride surfaces amid H-terminated Si islands can be achieved at 380°C [55,56].

It is also interesting to note that monohydride Si(100) surfaces can also be obtained without explicitly using atomic hydrogen but by exposing a clean Si surface to molecular hydrogen at a partial pressure of 0.1 Torr at ~1000°C [57].

15.1.3.2 Wet Chemical Preparation

High-temperature annealing could cause undesirable impurity diffusion changing the designed impurity concentration profile within the Si substrate. In addition, crystal defects, such as dislocations and stacking faults, tend to increase and slip lines are often generated across the Si substrate during high T treatment. Therefore, it is desirable to prepare the H-terminated Si surfaces by wet-chemical processes. It was pointed out first by Yablonovitch et al. that the standard hydrofluoric acid (HF) etch to remove silicon oxide

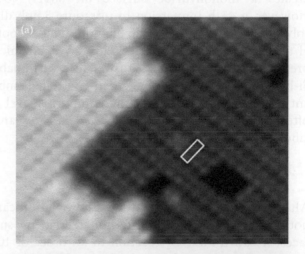

FIGURE 15.2 (a) Empty-state STM image of H/Si(100)-2×1 monohydride surface. A unit cell (7.7 Å × 3.8 Å) is marked. Note that the dark lines lie between dimers. (b) Empty-state STM image of H/Si(100)-3×1 surface. A unit cell (11.6 × 3.8 Å) is marked. The monohydride and dihydride units form bright and dark rows, respectively.

leaves H-terminated Si surfaces not F-terminated ones [58]. An ideal, at least based on the infrared (IR) spectra, monohydride Si(111)-1 × 1 surface can be achieved by pH-buffered HF solution [59]. However, real-space STM images reveal that the surface morphology is sensitive to pH values and the dissolved oxygen in the solutions [60,61]. When applying the same recipe, 40% aqueous NH_4F dip, to Si(100) surfaces, however, the surface is far from smooth—it fills with hillocks [62]. Diluted HF etching with more careful procedures, including ozonized ultrapure water, results in atom resolving images at small regions, but it is not clear about surface morphologies at large scales [63]. The role that water plays in wet-chemical preparation has been studied by Watanabe et al. They found that based on IR spectra, boiling deionized water immersion after a diluted HF dip creates a monohydride surface on Si(111) [64]. However, a 45 h immersion in ultrapure 5 ppb dissolved oxygen after an HF dip did not lead to smooth H-terminated Si(100) surfaces [65]. Wet-chemical preparation is certainly an answer to a sustainable Si-based nanotechnology, but a well-controlled ambient environment at the required impurity level is often beyond the capability of academic research laboratories.

15.1.3.3 Robustness of H-Terminated Si Surfaces

Although a simple dip in a diluted HF solution cannot prepare a defect-free H-terminated silicon surfaces, it can inhibit O_2 reaction to an order of 10^{12} compared to a clean Si(111)-7 × 7 [66] and Si(100)-2 × 1 [67]. Monohydride Si(100)-2 × 1 surfaces can still react with water, albeit with a much lower rate [68,69]. In a UHV chamber at a pressure of 10^{-10} Torr, the silicon hydride surface can be maintained for more than 10 h, if not days. However, no quantitative sticking coefficient measurements are available for water on various silicon hydride surfaces. To further explore the compatibility of silicon hydride surfaces with subsequent processes, UHV-prepared monohydride Si(100) surfaces were found degraded when exposed to ambient air and water, but in relatively good conditions if exposed to dry N_2, dissolved oxygen-free water, and some organic solvents [70].

The inertness of the H-terminated Si surfaces is the basis for patterning—only H-removed area can react with desired molecules, such as oxygen [71], phosphine [72,73], dimethyl-aluminum hydride [74,75], titanium tetrachloride [76], and iron carbonyl [77].

15.1.4 e-Beam Lithography on Hydrogen Resist

It is easy to pattern H-terminated surfaces by removing hydrogen from the selected areas either by electrons [78] or photons [79]. The key factors are the beam size, which dictates the minimal dimensions of the pattern, and the current density, which determines the time and quality of the patterned area. For the Si–H bond, it takes only ~6.5 eV (depends on doping [80]) to remove H by electronic excitation and even less by vibrational excitation [81]. It is also possible to achieve hydrogen desorption from a hole resonance by a negative sample bias [82]. Recently, a commercial low-energy e-gun (<15 eV) has been used to measure the hydrogen desorption yield, but the effectiveness for large pattern generation is not clear [83]. High-energy e-beam (10–40 keV) irradiation on H-terminated Si surfaces leads to oxidation after removing hydrogen in a standard scanning electron microscope chamber [84]. Even in UHV, 25 keV e-beams on H-terminated Si(100) create contamination but it was deemed feasible to pattern large δ-doped contact pads [85]. The surface defects created by a beam of 2 keV electrons in UHV at a density of 9.6×10^{18} cm^{-2} are clearly noticeable by STM [86]. Without a low-energy, high-current e-gun, scanning probes are still the most economical choice for research [87].

15.1.4.1 Scanning Probe Lithography in UHV

In the field emission regime, STM can easily generate a current density of 10^3 A/cm^2. Given the yield of H-atoms/electrons ~2.5×10^{-6} [81], if the emission current is 0.1 nA, it will take <70 ms to remove all the hydrogen in a 4 nm square. In the tunneling regime (tip–sample bias is less than the work function of the tip), the entire tunneling current can go through one atom on the surface creating atom-scale patterns [88,89]. Figure 15.3 illustrates that even an imperfect H/Si(111)-1 × 1 monohydride surface prepared by NH_4F etch can be patterned at 3 nm pitch by a 6-nA tunneling current in UHV. At low temperatures, 0.3 eV STM e-beams at a current of 22.5 nA can even dissociate a single O_2 molecule on Pt(111) surfaces [90]. Contrary to thermal desorption, lower substrate temperatures often increase the H-yield by electron-stimulated vibrational excitation [91] and hole resonance desorption [92]. A detailed study by Adams et al. reveals that the H-yield is nearly a constant from 10 to 30 eV and the dimension of the desorption area is a result of tip shape, bias, and emission current [93]. More details about UHV STM lithography can be found in a recent review [94].

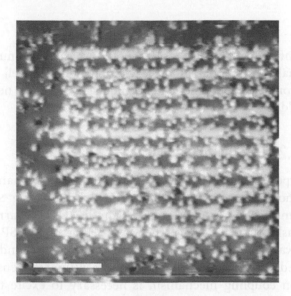

FIGURE 15.3 Silicon dangling-bond lines written 3 nm apart on an NH$_4$F-etched H/Si(111)-1 × 1 monohydride surface by an STM in UHV. Writing current is 6 nA at a sample bias of 4 V. The dangling bonds appear brighter than the H-terminated area. Scale bar, 10 nm.

We are indebted to Binnig and coworkers [95,96] for the invention of the simple, but powerful, scanning-probe microscopes (SPMs) as a tool for both imaging and manipulation at atom scale. But to contact the nanometer patterns created by SPMs with microscale contacts is still a daunting task. One way is to create prefabricated contacts (e.g., by dopant implant), then use low-energy Ar$^+$ sputtering and low-temperature annealing to create an H-terminated surface in UHV followed by SPM lithography. Simple two-terminal δ-layer wires at minimal width of 50 nm were fabricated by STM lithography, phosphine deposition, and Si epilayer growth and their transport was characterized at 0.3 K [97]. The contact between the As-implanted contact and P δ-layers is ohmic. A more involved scheme using an *in situ* STM–SEM system to create registration markers was developed in Simmons group [98]. Using this contact scheme, the group has successfully fabricated many interesting phosphorus δ-layer devices, including wires [99], quantum dots [100], tunneling gaps [101], and Aharanov–Bohm rings [102].

Similar to the standard e-beam lithography, STM e-beam lithography is a sequential process. It would be much widely accepted if an array of tips can be controlled individually to create large patterns. A 1D and a two-dimensional (2D) array of 50 and 100 individual actuation cantilevers, respectively, have been developed [103,104] but no UHV application has been reported.

15.1.4.2 Scanning Probe Lithography in Ambient

Because of the ease to prepare a relatively good H-terminated Si surface by wet chemical etch, there have been many reports using SPM in ambient to create oxide patterns which in turn can be used as a mask for subsequent nanofabrication [105,106]. In particular, because SPM can create very small structures, it is the easiest way to fabricate lateral tunnel junctions. The same oxidation principle can be applied to Si/SiGe heterolayers to create isolated quantum dots which show Coulomb blockade oscillation at 0.5 K [107]. Anodic oxidation can also be applied to metals. Room-temperature single-electron transistors were fabricated by SPM with a titanium oxide tunnel gap of 5–30 nm [108]; a metal-insulator–tunnel transistor was demonstrated with 58.5 nm TiO$_x$ gap [109]; superconducting single photon detectors defined by SPM oxidation of Nb was fabricated [110]. Although all these research devices are usually one of a kind, the reproducibility issue could be addressed by better control of the humidity, tip curvature, tip–sample distance, and surface contamination. Together with the development of individual actuation tip arrays, it is plausible that SPM e-beam lithography could be part of an industrial process in nanoscale device fabrication.

15.1.5 Outlook

Thanks to a few decades of rigorous research in surface science, a significant amount of knowledge on the physical and chemical interactions occurring on a large variety of surfaces and a rich selection of analytic and imaging tools to control and characterize these phenomena have been accumulated. SPM e-beam lithography on H-terminated Si surfaces is just one example of using proximity to focus e-beam at a distance <1 nm to stimulate surface interaction. This kind of precision can only be achieved in a very controlled environment, such as a UHV chamber. Since the same tip is used for both imaging and lithography, it is difficult to address both high-resolution and large area patterning simultaneously. The geometry of the tip up to the last atom is also uncontrollable, which dictates the resolution of SPM in both imaging and lithography. Recent SPM images of pentacene by attaching a CO molecule to a Cu tip is one example of scientific innovation [111]. A new tool is in order to establish an industry-scale nanofabrication technology based on the atom-scale science obtained by the current SPM.

15.2   Plasmonic Nanolithography

This chapter reviews the recent development of plasmonic nanolithography techniques for exceeding the diffraction limit in high-resolution photolithographic nanometer-scale device fabrication. Maskless nanolithography, such as electron-beam and scanning-probe lithography, offers the desired resolution and flexibility. The applications are limited, however, especially for the commercialization of nanoscale devices, due to the low throughput and high systematic cost. Surface plasmons (SPs) (electron oscillations at a metal surface) can offer access to much shorter wavelength compared to the excitation light wavelength. In addition, at the resonant state, the local field intensity of SPs can be boosted by several orders of magnitude compared to the excitation light. A variety of approaches to utilize these attractive features of SPs are covered in this chapter. These approaches promise a new route toward next-generation nanomanufacturing.

15.2.1   Introduction

Developing a highly efficient lithography process has become more and more crucial because it determines the throughput and the cost of electronic industry. Optical lithography is the major technology used in manufacturing electronic devices over the past several decades because of its high yield. Continued improvement of optical lithography by pushing the optical resolution limit allows rapid pattern replication with high density and accuracy. However, the minimum printable size or the optical resolution of an optical lithography system is limited by the intrinsic diffraction property of light. Advanced photolithography techniques utilizing short optical wavelengths in the deep UV [112], the extreme UV [113], or the x-ray region [114] can offer sub-100 nm resolution. The finest feature size and spacing between patterns are still limited by the diffraction of light. And it is difficult and complex to develop the new associated light source, photoresist, and optics. The emerging nanoscale device commercialization calls for breakthroughs in high-throughput nanofabrication technologies that are flexible for frequent design changes.

SPs are collective oscillations of electrons coupled to electromagnetic waves and are excited at the interface of a metal and dielectric [115]. SPs exhibit a maximum intensity at the surface with an exponentially decaying amplitude perpendicular to the interface. The surface confinement of SPs results in the wavelength much smaller than that of the excitation light, as well as strong electromagnetic field enhancement in the near field.

15.2.2   Plasmonic Masks

Experimental evidence of high-resolution UV nanolithography was first demonstrated by utilizing SPs transmitted through subwavelength 2D hole array masks [116,117]. Since the wave vector of the SP is always larger than that of the light in the surrounded media at the same frequency, a momentum compensation coupling mechanism is necessary to excite the SPs. Rough surface texturing, grating coupler, or attenuated total reflection (ATR) coupler have been routinely applied to convert light into SPs and vice versa. For example, a 2D square array of holes can be treated as a 2D grating coupler. By properly selecting the array periodicity and the dielectric constant of surrounding medium, SPs can be resonantly excited and coupled through the holes to the other side of the metal mask. This selective resonance nature of SPs has a potential to pattern nanoscale features using conventional near-UV light sources.

The configuration of the SP optical lithography is illustrated in Figure 15.4 [117]. A plasmonic mask designed for lithography in the UV range is composed of an 80-nm-thick aluminum layer perforated with 2D periodic hole arrays by focused ion beam (FIB) milling and two surrounding dielectric layers on each side. Aluminum is chosen since it can excite the SPs in the UV range. A negative near-UV photoresist (SU-8) is spun on the top of the spacer layer and polymerized on the mask in the lithography process. After the development, the topography of exposure features is characterized by atomic force microscopy (AFM). The result shows an exposure result obtained from the 170 nm period array. Features as small as 90 nm (equivalent to $\lambda/4$, where λ is the exposure light wavelength) have been achieved, far beyond the diffraction limit of farfield lithography. It should be noted that smaller features can be achieved by reducing the spacer layer thickness, which opens up a possibility of high-resolution and density nanolithography with high transmittance using a conventional light source without the complicated setup and vacuum requirement such as in the extreme UV lithography.

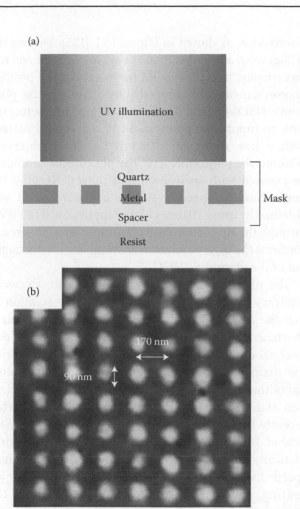

FIGURE 15.4 (a) Schematic drawing of the SP optical lithography. FIB is used to fabricate 2D hole arrays on aluminum substrates. PMMA is chosen as the spacer layer to match the dielectric constant of the quartz substrate. (b) AFM image of a pattern with 90 nm features on a 170 nm period obtained by SP lithography. The mask is an array of 40 nm holes on a 170 nm period. The spacer layer thickness is 30 nm, and the exposure dose is 72 mJ/cm². (From Liu, Z.W. et al., *Nano Letters*, 2005. **5**(9): 1726–1729.)

15.2.3   Surface Plasmonic Interference

Grating is considered as a very efficient SP coupler; however, it can only provide a discrete wave vector, which limits the application for complex pattern lithography. A single sharp edge (or a slit) can also be used to excite SPs, where the light diffracted from the corner gains very broad band of wave vectors, and the interface will automatically select the components with matched wave vector to SPs and support their propagation [118]. While the nonresonant nature of this process will reduce the coupling efficiency, these simple structures will drastically reduce the complexity of sample fabrication.

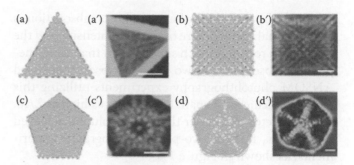

FIGURE 15.5 (**See color insert.**) Simulation (a–d) and experimental results (a′–d′) using a 100-nm-thick aluminum plate with various edge/slit shapes. (a), (a′) triangle; (b), (b′) square; (c), (c′) pentagon; (d), (d′) pentagon with rounded corners. The side lengths are 2, 2, 1.5, and 1.5 μm, respectively in (a–d). The excitation light had circular polarization in all the simulations. The scale bars in all the AFM images represent 500 nm. (From Chen, W. and H. Ahmed, *Applied Physics Letters*, 1993. **62**(13): 1499–1501.)

Since SPs are essentially two dimensionally confined on metal–dielectric interface, their interference patterns can be controlled by arranging different 2D geometries of the slits/edges [119]. Figure 15.5a–d shows simulated SP interference patterns with four exampled geometries. Obviously, not only periodic but also quasi-periodic and even more complicated 2D patterns can be realized. The lithography experimental results (see Figure 15.5a′–d′) clearly show the formation of different interference patterns predicted by simulations.

15.2.4   Near-Field Plasmonic Nanolithography

Several maskless lithography methods have been developed to expose a resist layer as the system is scanning the substrate. By utilizing focused particle beams such as electrons or ion beams [120,121] or involving a local sharp tip such as an AFM probe or a near-field scanning optical microscope (NSOM) probe [122,123], these techniques have been demonstrated with resolution in the range of a few tens of nanometers or even smaller.

The combination of apertureless near-field enhancement and nonlinear absorption techniques has been applied in photolithography [124]. Apertureless near-field scanning optical microscopy (ANSOM) uses sharp metallic tips instead of fiber apertures to achieve nanometer-scale resolution. In ANSOM, light polarized along the sharp axis of a metallic tip induces a high concentration of SPs resulting in strong enhancement of the electromagnetic field in the local vicinity of the tip. When high-intensity light shines on a material, the probability for two-photon absorption is proportional to the square of the field intensity, and the absorption region is much smaller than that using one-photon absorption. Two-photon absorption-based ANSOM photolithography experiments utilizing this idea have achieved a spatial resolution as high as $\lambda/10$, nearly a factor of 2 better than the resolution achieved in previous farfield two-photon lithography experiments, as shown in Figure 15.6.

An alternative approach to improve the scanning-type nanolithography has been presented by integrating a conic plasmonic lens on the tapered tip of an NSOM probe to enhance the transmission and to generate an intense light spot with subwavelength

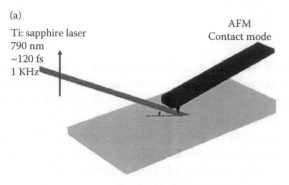

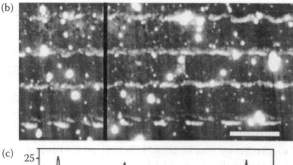

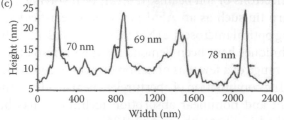

FIGURE 15.6 (a) Schematic diagram of the experimental setup for two-photon apertureless near-field photolithography. (b) AFM images of two-photon produced line structures in SU-8 exposed using the field enhancement of the ANSOM tip. (c) The height profile along the dark vertical line in (b), suggesting that two-photon apertureless near-field lithography can produce ~72 ± 10 nm features using 790 nm light. The scale bars in (b) are 1 μm. (From Wang, Y. et al., *Nano Letters*, 2008. **8**(9): 3041–3045.)

dimensions, as shown in Figure 15.7 [125]. During the lithography, a laser beam at the wavelength of 365 nm was coupled into the NSOM tip to expose the positive photoresist on a Si substrate. By scanning the plasmonic NSOM probe on the top surface of photoresist with an input laser power of 8 μW, a recorded pattern with a line width of ~100 nm has been observed. Without the plasmonic lens, the photoresist can never be exposed with such low input power. Keeping the same scanning speed, a similar exposure result was obtained by using 10 times higher input power (80 μW) through the tip without the plasmonic lens. This result confirms the high-resolution and high optical throughput of the plasmonic lens.

The maskless lithography methods directly write arbitrary patterns with sub-100 nm high resolution so that the cycle time for nanoscale device validation is shortened. However, the low throughput, which is due to serial scanning natures of these methods, remains the main bottleneck. A high-throughput plasmonic nanolithography to circumvent the critical parallelization and slow scanning challenges has been reported recently by flying a plasmonic lens array in the near field at 10 m/s above a rotating substrate, which dramatically increases the throughput [126]. The high-speed flying plasmonic lens arrays utilize an air bearing, similar to a hovercraft hovering above the surface of a lake, as shown in Figure 15.8. The unique air bearing head, named plasmonic flying head, inspired by the magnetic recording head in the hard-disk drive, was designed to fly the plasmonic lens arrays at the height of tens of nanometers above the disk surface at speeds of meters per second. A spindle was used to rotate the disk at 2000 rpm which is equivalent to the linear speed of ~10 m/s at the outer radius, which generates an aerodynamic lift force balanced with the spring force supplied by the top suspension arm to precisely retain a nanoscale gap between the plasmonic lens arrays and the rotating substrate. A UV laser was focused down to a several micrometer spot on top of the plasmonic lens array which further focuses the light beam to sub-100 nm spot on the rotating substrate for patterns writing on a thermal photoresist deposited on the substrate. The result demonstrated high-speed patterning with 80 nm resolution at 10 m/s.

15.2.5 Concluding Remarks

The goals of plasmonic nanolithography are to develop a manufacturing tool with nanometer-scale spatial

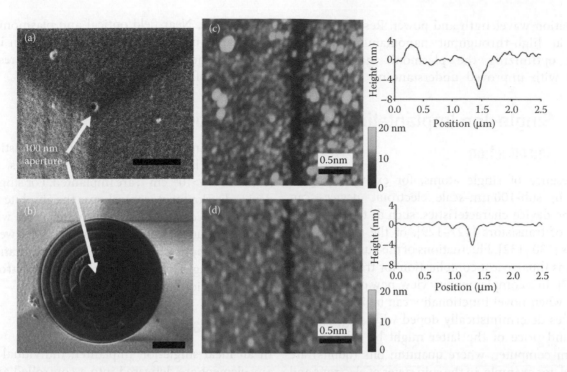

FIGURE 15.7 NSOM probe consists of nanostructured plasmonic lens being fabricated on the end of an optical fiber. A length of 100 nm aperture on the NSOM tip (a) before and (b) after the fabrication of the plasmonic lens. Scale bars in (a) and (b) are 1 μm. (c) AFM image of the photoresist after a near-field scanning exposure of 8 μW using the plasmonic lens. (d) Control experiment by single-aperture NSOM probe with an input laser power of 80 μW. The width and depth of patterned lines in (c) and (d) are 124, 6.8, 97, and 4.2 nm, respectively. (From Srituravanich, W. et al., *Nature Nanotechnology*, 2008. **3**(12): 733–737.)

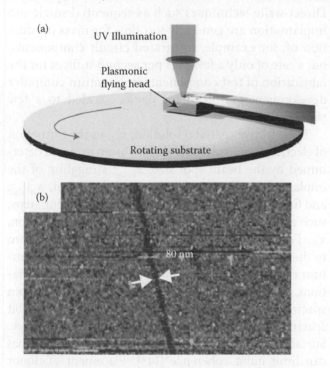

FIGURE 15.8 (a) Schematic of a high-throughput maskless nanolithography using plasmonic lens arrays flying at a near field on a rotating disks. (b) Maskless lithography by flying plasmonic lenses at the near field. AFM image of pattern with 80 nm linewidth on the TeO_x-based thermal photoresist. (From Keyes, R.W., *Reports on Progress in Physics*, 2005. **68**(12): 2701–2746.)

resolution that can dramatically increase the productivity and reduce the cost in the electronics industry. Innovations in the fabrication processes which provide a potential to replicate a pattern rapidly with high density and accuracy are critical for the improvement of the speed and performance of integrated circuits and their packaging. For conventional optical lithography techniques, the abilities of generating arbitrary features with sub-100 nm feature size are extremely challenging, not only because of the diffraction limit, but also because of the cost and complexity for the associated light source, photoresist, and optics.

Featuring a wavelength much smaller than that of the excitation light, as well as strong electromagnetic field enhancement in the near field, SPs have been demonstrated to achieve tens of nanometer resolution with conventional light source. The plasmonic technique presents additional advantages besides high-resolution lithography. The large field intensity at the focal point of plasmonic structures such as a plasmonic lens has the potential for high-speed lithography, which has been demonstrated with a plasmonic flying head with a linear scanning speed of ~10 m/s. This property is essential to explore new avenues in nanoscale fabrication with regular

Chapter 15

illumination wavelength and power. Research efforts aimed at high-throughput nanomanufactory will involve optimizing the plasmonic nanostructure designs with improved understanding of the near-field physics. Near-field optical and plasmonic-based lithography may lead to exciting revolution in semiconductor industry as well as fundamental researches in the field of nanoscience.

15.3   Single-Ion Implantation and Deterministic Doping

15.3.1   Introduction

The presence of single atoms, for example, dopant atoms, in sub-100 nm-scale electronic devices can affect the device characteristics, such as the threshold voltage of transistors [127–129], or the subthreshold currents [130–132]. Fluctuations of the number of dopant atoms thus pose a complication for transistor scaling [127]. In a complementary view, new opportunities emerge when novel functionality can be implemented in devices deterministically doped with single atoms. The grand price of the latter might be a large-scale quantum computer, where quantum bits (qubits) are encoded, for example, in the spin states of electrons and nuclei of single-dopant atoms in silicon [133], or in color centers in diamond [134,135]. Both the possible detrimental effects of dopant fluctuations and single-atom device ideas motivate the development of reliable single-atom doping techniques which are the subject of this chapter.

Single-atom doping can be approached with top-down and bottom-up techniques. Top-down refers to the placement of dopant atoms into a more or less structured matrix environment, like a transistor in silicon [136,137]. Bottom-up refers to approaches to introduce single-dopant atoms during the growth of the host matrix, for example, by directed self-assembly and scanning-probe-assisted lithography [138,139]. Bottom-up approaches are discussed in Section 15.1.

Since the late 1960s, ion implantation has been a widely used technique to introduce dopant atoms into silicon and other materials to modify their electronic properties [140]. It works particularly well in silicon since the damage to the crystal lattice that is induced by ion implantation can be repaired by thermal annealing. In addition, the introduced dopant atoms can be incorporated with high efficiency into lattice positions in the silicon host crystal which makes them electrically active. This is not the case for, for example, diamond, which makes ion implantation doping to engineer the electrical properties of diamond, especially for *n*-type doping much harder than for silicon [141,142].

Ion implantation is usually a highly statistical process, where high fluences of energetic ions, ranging from ~10^9 to >10^{16} cm^{-2}, are implanted. For single-atom device development, control over the absolute number of ions is needed and ions have to be placed with high spatial resolution. In the following sections, we discuss a series of approaches to single-ion implantation with regard to single-ion impact sensing and control of single-ion positioning.

15.3.2   Placement of Single Ions

In an ideal single-ion implanter, individual ions of any element are delivered into a controlled area on a wafer at a reasonable rate, each ion impact is registered, and the ion beam is turned off fast enough to prevent the impact of the next ion before the sample has been moved to the next implant position. Direct-write techniques such as sequential single-ion implantation are generally too slow for mass production of, for example, integrated circuit components, but a rate of only a few ions per second suffices for the fabrication of test components in quantum computer development and even for devices scaled to a few thousand qubits.

The effective spatial resolution, x_{eff}, in the formation of electrically active single-dopant atom arrays is determined by the beam spot size, x_{beam}, straggling of the implanted ion during slowdown in the target, $x_{straggl}$, and finally by diffusion in consecutive processing steps, such as annealing and gate oxide growth or deposition, x_{diff}. For a donor spin qubit spacing d, with $d = 10$–20 nm in the original Kane proposal for a silicon-based quantum computer [133] and $d = $ ~100 nm in several variations, the effective resolution in phosphorus atom spacing should be a faction of the qubit spacing. Qubit spacing is one critical metric and another is coherence. Surfaces and interfaces are often a source of noise that can limit qubit coherence [143]. Placement of donor qubits at greater depth thus protects them from this noise source. But there is a trade-off with the achievable placement precision since epitaxial overgrowth or implantation at higher energies is needed to achieve greater depth and these can degrade the effective

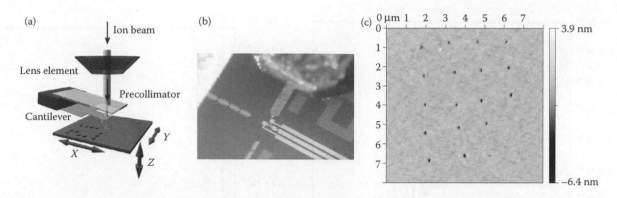

FIGURE 15.9 Schematic (left) and photograph (right) of setup for ion implantation with scanning-probe alignment and example of pattern formed by ion implantation with scanning-probe alignment. (From MoberlyChan, W.J. et al., *Mrs Bulletin*, 2007. **32**(5): 424–432; Persaud, A. et al. *Nano Letters*, 2005. **5**(6): 1087–1091.)

placement precision due to increased diffusion or range straggling, respectively.

We now briefly discuss strategies for control of these three placement resolution limiting factors. In a well-optimized single-ion placement experiment, all contributions have to be addressed in parallel and use, for example, of a beam focused to <5 nm spot size is ineffective if dopants have a much larger range straggling or diffuse significantly during annealing.

15.3.2.1 Beam Spot Size

Control of the ion position is addressed in the ion optical column of the implanter. There are two approaches to achieving small, that is, <10 nm, spot sizes. One is to focus ions to a tight spot using a high-brightness ion source, the other is to use a broad beam of ions and define the effective beam spot with a nanometer-scale stencil mask.

Commercial FIB systems can deliver pA currents of ions from liquid metal ion guns—mostly Ga⁺—with kinetic energies of ~30 keV into beam spots with diameters of about 5–10 nm [144,145]. Shinada et al. have reported focusing of a 60 keV phosphorus ion beam with an aiming accuracy of 60 nm [129]. Here, the phosphorus ions were formed from a liquid metal ion source with a phosphorus-containing nickel alloy. Greater variability in the ion species for FIBs is highly desired and is being addressed, for example, in the recent development of ion-trap-based ion sources [146]. Very low ion temperatures in the laser-cooled ion traps and very small virtual source sizes result in ultra-high-brightness ion beams that could enable highly precise ion implantation.

Alternatively, beams of low-energy dopant ions can be collimated in nanostencils that can be integrated with a scanning probe as a dynamic shadow mask [147].

Figure 15.9 shows a schematic of a setup that integrates ion beams with a scanning force microscope. The scanning probe provides imaging and alignment functions, and the nanostencil limits the effective beam spot size. Nanostencils with diameters as small as 5 nm have been formed in silicon-based cantilevers using a combination of Ga-FIB drilling and local thin-film deposition [148].

15.3.2.2 Range Straggling

Range straggling quantifies the lateral and longitudinal spread of the distribution of implanted ions [140]. Range straggling results from statistical energy loss processes during the slowdown of impinging ions. Ion ranges and range straggling can be estimated with the widely used SRIM code (www.srim.org) [149] and quantified using secondary ion mass spectrometry (SIMS). For phosphorus in silicon, straggling amounts to ~35 nm for a 60 keV implant with a 70 nm range. For an implantation energy of 10 keV, the range is about 15 nm with a longitudinal straggle of 8 nm, and for 1 keV, both range and straggling are only a couple of nm. Straggling thus sets a limit to the kinetic energy at which an effective implant resolution can be achieved. A consequence of reducing the impact energy is that single-ion registration through detection of secondary electrons becomes impractical in a regime of kinetic electron emission because of the decrease of secondary electron yields [150]. Use of highly charged projectiles avoids this limitation as electron emission following the impact of low-energy, high-charge-state ions results from deposition of potential energy, not kinetic energy [151,152]. For implantation into a given depth, straggling is lower for heavier ions than for lighter ions (see Figure 15.10). For example, for implantation of donors into silicon at a peak depth of about 20 nm, the

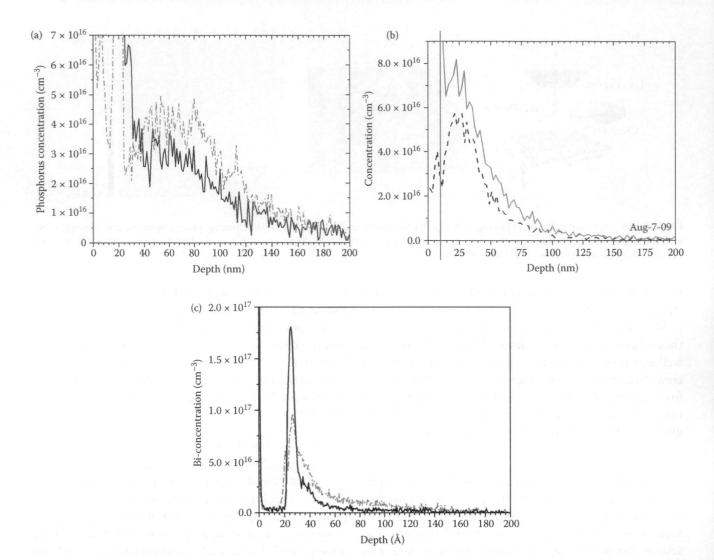

FIGURE 15.10 SIMS depth profiles of as-implanted and annealed silicon samples. (a): ^{31}P (annealed for 10 s at 950°C), (b): ^{121}Sb, annealed for 10 s at 850°C. The vertical line at 10 nm indicates the interface between the top SiO$_2$ layer and silicon. (c): ^{209}Bi, implant before and after annealing (1000°C, 10 s). The implant energy was 60 keV and the implantation fluence was 2 × 10^{11} cm^{-2} for all three ion species. (From Schenkel, T. et al., *Nuclear Instruments & Methods in Physics Research Section B—Beam Interactions with Materials and Atoms*, 2009. **267**(16): 2563–2566; Weis, C.D. et al., *Nuclear Instruments & Methods in Physics Research Section B—Beam Interactions with Materials and Atoms*, 2009. **267**(8–9): 1222–1225.)

straggling for 13 keV ^{31}P ions is ~10 nm, while for ^{121}Sb at 25 keV, straggling is only about 6 nm.

While useful for quick estimates, SRIM does not include channeling effects, nor effects of accumulated damage on the range of ions.

15.3.3 Diffusion during Annealing

Ion implantation damages the host material because ions transfer momentum to target atoms during the collision cascade as they slow down. In silicon, damage above the amorphization threshold can be repaired by annealing above 550°C [140,153]. In diamond, damage above a threshold results in graphitization on

annealing, and there is no epitaxial regrowth as in silicon, making ion implantation doping of diamond much more challenging than for silicon [141]. Besides damage repair, annealing is needed to electrically activate dopants, that is, to incorporate them into the host lattice. Dopants diffuse through coupling to defects, interstitials, and vacancies, which are present as a result of the implant process or which are generated at an equilibrium rate during annealing at a given temperature. Further, defects can also be injected from interfaces or during annealing in reactive environments [154]. For example, annealing of silicon under oxidizing conditions results in injection of interstitials. Phosphorus is an interstitial diffuser and oxidation-enhanced diffusion can lead to dopant segregation to the SiO_2–Si interface and to dopant loss. Both effects are detrimental to single-atom placement, where the position of single-dopant atoms should be determined by the spot size of the focused beam or the collimating aperture and where dopant movement has to be minimized while 100% efficient electrical activation is required for efficient single-atom device integration. Values for intrinsic and bulk diffusivities, for example, $D_0 = 10^{-14}$ cm^2/s for P in Si at 1000°C predict a minimal broadening of the implant profile during a few-seconds-long annealing step, t, $x_{diff} = 2\sqrt{D_0 t}$, of only a few nanometers. But dopant movement for shallow implants is often dominated by defect injection from the interface. Antimony is a vacancy diffuser and Sb movement is suppressed during annealing in the presence of an oxide interface in silicon [143]. For dopant atoms in silicon, very high electrical activation efficiencies are routinely achieved [140,154]. For color center formation in diamond, process development is much less mature and more complicated process sequences are required to achieve highly efficient color center formation, for example, involving cold implantation of nitrogen ions followed by coimplantation of other ions (e.g., carbon or noble gases) at controlled sample temperatures (hot or cold) to increase the vacancy density followed by rapid thermal annealing for NV-center formation [141,142,155] (Table 15.1).

15.3.3.1 Examples of Characterization of Range Straggling and Diffusion

One atom in a transistor channel volume of $10 \times 10 \times 10$ nm^3 is equivalent to a bulk concentration of 10^{18} $atoms/cm^3$, and many material analysis techniques that are typically used for analysis of impurities at higher

Table 15.1 Summary of Factors Limiting the Placement of Single-Dopant Atoms by Ion Implantation

Placement Limiting Factor	Comment	References
Ion beam spot size	FIBs or dynamic nanostencil	[146–147,166]
Range straggling	Increases with implant energy, decreases with mass of implanted ion	[140,149]
Diffusion during annealing	Dopant-specific diffusion mechanisms, surface and interface effects	[140,153,154]

concentrations are also very useful when the goal is to master the placement and integration of single atoms.

Dynamic SIMS is a widely used metrology tool for characterization of depth profiles of implanted ions [156], while spreading resistance analysis (SRA) is widely used for depth profiling of the electrical resistivity of samples, often after ion implantation and annealing [157]. SIMS is sensitive to trace concentrations of elements down to the ppb level (parts per billion, where 1 ppb = 5×10^{13} $atoms/cm^3$ in silicon). SRA provides no elemental sensitivity, only the sign of the carrier concentration, and is sensitive to carrier concentrations as low as a ~10^{11} cm^{-3}. Both SIMS and SRA can provide depth profiles with a few-nm-depth resolution, with SIMS providing higher-depth resolution than SRA for high-enough concentrations of impurities. Often there are trade-offs between achievable depth resolution and sensitivity. Figure 15.10a shows SIMS depth profiles of P atoms in silicon before and after annealing. The SIMS spectra are relatively noisy, even though the P atom peak concentration is above 10^{16} $atoms/cm^3$. This is due to possible mass interferences, which necessitates running of the SIMS analysis at very high mass resolution, which reduces signal levels at a given depth resolution [158]. The profile taken after rapid thermal annealing in a nitrogen ambient and in the presence of an oxide showed an effect of segregation of ^{31}P atoms to the SiO_2–Si interface. Figure 15.10b shows SIMS depth profiles for similar implant and annealing conditions but for ^{121}Sb ions. The higher ion mass leads to a shallower implant range and lower-range straggling, and no dopant segregation is observed. Figure 15.10c further illustrates the effect of higher ion mass in reducing straggling for a given implant energy with a 60 keV bismuth (Bi) implant (^{209}Bi) before and after annealing. The SIMS profile also indicates that a small degree of redistribution of Bi atoms during annealing (1000°C, 10 s). This is unexpected on the basis of low Bi diffusivity values and might be a result of the intense lattice damage induced by the heavy Bi ions.

15.3.4 Detection of Single Ions

The ease or difficulty of single-ion detection depends on the ions kinetic energy and charge state. For precise single-ion placement, kinetic energies have to be low, so that the positioning uncertainty from range straggling remains smaller than a characteristic device scale, such as the nearest-neighbor qubit coupling distance. When ions impinge on solids, secondary electrons are emitted, electron–hole pairs are generated inside the solid when the ion transfers its kinetic energy in elastic and inelastic collisions, target material is sputtered off the surface into vacuum, and the surface topography can be modified. In some materials, light is emitted in radiative relaxation processes following electronic excitation of target atoms and molecules. Also, x-rays can be emitted following inner-shell ionization of target atoms. For some materials (e.g., graphite, mica, and diamond), the problem of single-ion detection has been addressed by imaging of topological modifications, that is, extended defects, generated on surfaces by single-ion impacts (Table 15.2).

Figure 15.11 shows an example of single-ion impact detection through measurements of current changes in

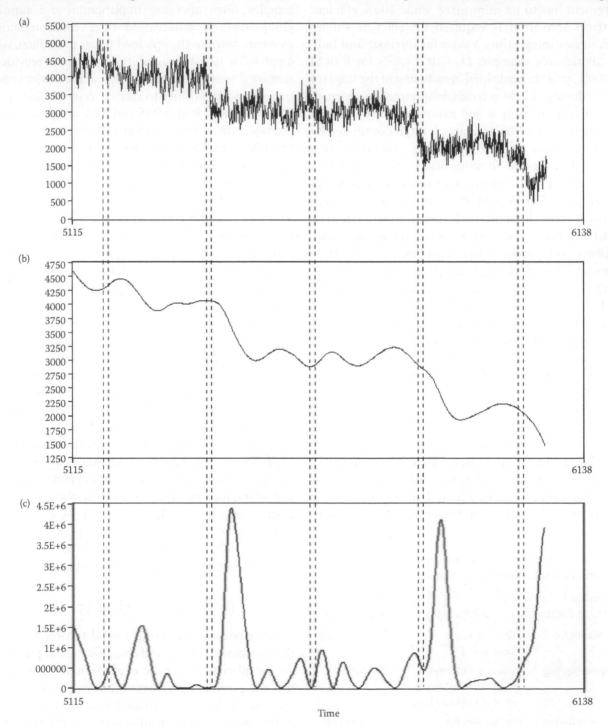

FIGURE 15.11   Source–drain current as a function of time during pulsed exposure of a transistor to Xe^{6+} ions, $E_{kin} = 48$ keV, (a) raw data, (b) smoothed data, and (c) derivative of (b), with current steps induced by single-ion impacts at room temperature. (From Weis, C.D. et al., *Journal of Vacuum Science & Technology B*, 2008. **26**(6): 2596–2600.)

Table 15.2 Summary of Methods for Single-Ion Impact Detection

Method for Single Ion Impact Detection	Comment	References
Secondary electrons	Secondary electron yields increase with increasing ion kinetic energy and charge state, and with decreasing surface work functions	[129,137, 152,159]
Electron–hole pairs	Requires integration with diode structures, signal increases with kinetic energy of ions	[136]
Current changes in transport channels	Requires integration with transport channels (resistors or transistors)	[160,161]
Topology modification	Requires high-resolution *in situ* imaging and flat sample surfaces	[162–165]

transistor channels [166]. Transistors were processed with tungsten metallization for postimplant annealing and holes in the transistor channels were opened using a combination of focused ion drilling and reactive gas etching in a dual-beam FIB [160]. Scanning the

dynamic shadow mask over the transistor results in a response map, similar to ion-beam-induced charge mapping (which is usually done with MeV ion beams [167] with about µm-scale imaging resolution limits). An example of an ion impact response map is shown in Figure 15.12. This method can be applied for studies of the response of scaled device components to ionizing radiation at a spatial resolution limited by the opening diameter of the nanostencil (80 nm in the example of Figure 15.12). Transistors like these have been used for electrical detection of spin resonance [168]. Thus, the same device structure is used for single-ion impact detection and, after annealing, for electrical detection of spin resonance. Scaling to single nuclear spin-state readout is subject of ongoing research [169,170].

15.3.5   Outlook

As lithographic access to sub-25 nm-scale features becomes more and more routine in many laboratories, effects of single atoms on process variability and function in device structures will become more and more common. This poses challenges and opportunities. Techniques for deterministic doping of nanoscale structures with single-dopant atoms enable paths for understanding of single-atom effects both where they are undesired (e.g., random dopant fluctuations in scaled transistors) and where they offer tantalizing new opportunities (e.g., in single-atom-based quantum bits and quantum computer development).

(a)   (b)

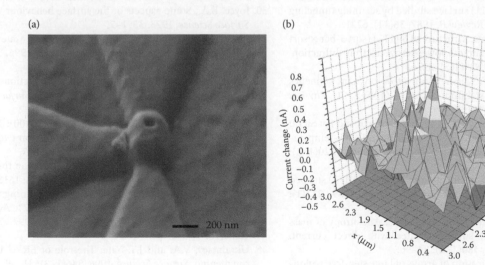

FIGURE 15.12 (**See color insert.**) Scanning electron micrograph (a) and ion impact response map (b) from a FinFet with 100 nm hole in the top gate. (From Weis, C.D. et al., *Nuclear Instruments & Methods in Physics Research Section B—Beam Interactions with Materials and Atoms*, 2009. **267**(8–9): 1222–1225.)

Chapter 15

Acknowledgments

This work was supported by NSA under contract number MOD 713106A, and by the Director, Office of Science, of the Department of Energy under Contract No. DE-AC02-05CH11231.

References

1. Neergaard Waltenburg, H. and J.T. Yates, Surface chemistry of silicon. *Chemical Reviews*, 1995. **95**(5): 1589–1673.
2. Yang, Y.N. and E.D. Williams, High atom density in the "1 × 1" phase and origin of the metastable reconstructions on Si(111). *Physical Review Letters*, 1994. **72**(12): 1862.
3. Swartzentruber, B.S. et al., Scanning tunneling microscopy studies of structural disorder and steps on Si surfaces. *Journal of Vacuum Science & Technology A: Vacuum, Surfaces, and Films*, 1989. **7**(4): 2901–2905.
4. Tromp, R.M., R.J. Hamers, and J.E. Demuth, *Si(001)* Dimer structure observed with scanning tunneling microscopy. *Physical Review Letters*, 1985. **55**(12): 1303.
5. Ichimiya, A., Y. Tanaka, and K. Ishiyama, Quantitative measurements of thermal relaxation of isolated silicon hillocks and craters on the Si(111)-(7 × 7) surface by scanning tunneling microscopy. *Physical Review Letters*, 1996. **76**(25): 4721.
6. Ishizaka, A. and Y. Shiraki, Low temperature surface cleaning of silicon and its application to silicon MBE. *Journal of the Electrochemical Society*, 1986. **133**(4): 666–671.
7. Kern, W., ed. *Handbook of Semiconductor Wafer Cleaning Technology*, 1993, Park Ridge, NJ: Noyes Publications, pp. 3–67.
8. Krusor, B.S. et al., Ultraviolet–ozone cleaning of silicon surfaces studied by Auger spectroscopy. *Journal of Vacuum Science & Technology B: Microelectronics and Nanometer Structures*, 1989. **7**(1): 129–130.
9. Ohmi, T. et al., Native oxide growth and organic impurity removal on Si surface with ozone-injected ultrapure water. *Journal of the Electrochemical Society*, 1993. **140**(3): 804–810.
10. Tromp, R.M. and M.C. Reuter, Wavy steps on Si(001). *Physical Review Letters*, 1992. **68**(6): 820.
11. Nogami, J., S.-i. Park, and C.F. Quate, Indium-induced reconstructions of the Si(111) surface studied by scanning tunneling microscopy. *Physical Review B*, 1987. **36**(11): 6221.
12. Huang, H. et al., Atomic structure of Si(111) (sqrt 3-bar × sqrt 3-bar)R30°-B by dynamical low-energy electron diffraction. *Physical Review B*, 1990. **41**(5): 3276.
13. Bedrossian, P. and T. Klitsner, Anisotropic vacancy kinetics and single-domain stabilization on Si(100)-2 × 1. *Physical Review Letters*, 1992. **68**(5): 646.
14. Kim, J.C. et al., Preparation of atomically clean and flat Si(1 0 0) surfaces by low-energy ion sputtering and low-temperature annealing. *Applied Surface Science*, 2003. **220**(1–4): 293–297.
15. Latyshev, A.V. et al., Transformations on clean Si(111) stepped surface during sublimation. *Surface Science*, 1989. **213**(1): 157–169.
16. Nielsen, J.F., M.S. Pettersen, and J.P. Pelz, Anisotropy of mass transport on Si(001) surfaces heated with direct current. *Surface Science*, 2001. **480**(1–2): 84–96.
17. Tanaka, S. et al., Fabrication of arrays of large step-free regions on Si(001). *Applied Physics Letters*, 1996. **69**(9): 1235–1237.
18. Homma, Y. et al., Sublimation of the Si(111) surface in ultrahigh vacuum. *Physical Review B*, 1997. **55**(16): R10237.
19. Ogino, T., H. Hibino, and Y. Homma, Kinetics and thermodynamics of surface steps on semiconductors. *Critical Reviews in Solid State and Materials Sciences*, 1999. **24**(3): 227–263.
20. Lifshits, A.A.S.a.A.V.Z.L., Viktor Grigorevic *Surface Phases on Silicon Preparation, Structures and Properties*, 1994, Chichester: John Wiley & Sons.
21. Hamers, R.J. and Y. Wang, Atomically-resolved studies of the chemistry and bonding at silicon surfaces. *Chemical Reviews*, 1996. **96**(4): 1261–1290.
22. Henderson, R.C., R.B. Marcus, and W.J. Polito, Carbide contamination of silicon surfaces. *Journal of Applied Physics*, 1971. **42**(3): 1208–1215.
23. O'Hanlon, J.F., *A User's Guide to Vacuum Technology* (Second Edition), 1989, New York: John Wiley & Sons.
24. Zikovsky, J. et al., Reaction of a hydrogen-terminated Si(100) surface in UHV with ion-pump generated radicals. *Journal of Vacuum Science & Technology A*, 2009. **27**(2): 248–252.
25. Dürr, M. and U. Höfer, Dissociative adsorption of molecular hydrogen on silicon surfaces. *Surface Science Reports*, 2006. **61**(12): 465–526.
26. Shklyaev, A.A. and T. Suzuki, Initial reactive sticking coefficient of O_2 on Si(111)-7 × 7 at elevated temperatures. *Surface Science*, 1996. **351**(1–3): 64–74.
27. Miki, K., K. Sakamoto, and T. Sakamoto, Is the c(4 × 4) reconstruction of Si(001) associated with the presence of carbon? *Applied Physics Letters*, 1997. **71**(22): 3266–3268.
28. Nörenberg, H. and G.A.D. Briggs, The Si(001) c(4 × 4) surface reconstruction: A comprehensive experimental study. *Surface Science*, 1999. **430**(1–3): 154–164.
29. Jemander, S.T. et al., STM study of the C-induced Si(100)-c(4 × 4) reconstruction. *Physical Review B*, 2002. **65**(11): 115321.
30. Joyce, B.A., Some aspects of the surface behaviour of silicon. *Surface Science*, 1973. **35**: 1–7.
31. Ranke, W., Precursor kinetics of dissociative water adsorption on the Si(001) surface. *Surface Science*, 1996. **369**(1–3): 137–145.
32. Flowers, M.C. et al., The adsorption and reactions of water on Si(100)-2 × 1 and Si(111)-7 × 7 surfaces. *Surface Science*, 1996. **351**(1–3): 87–102.
33. Hossain, M.Z. et al., Model for C defect on Si(100): The dissociative adsorption of a single water molecule on two adjacent dimers. *Physical Review B*, 2003. **67**(15): 153307.
34. Nishizawa, M. et al., Origin of type-C defects on the Si(100)-(2 × 1) surface. *Physical Review B*, 2002. **65**(16): 161302.
35. Self, K.W., C. Yan, and W.H. Weinberg, A scanning tunneling microscopy study of water on Si(111)-(7 × 7): Adsorption and oxide desorption. *Surface Science*, 1997. **380**(2–3): 408–416.
36. Ukraintsev, V.A. and J.T. Yate, The role of nickel in Si(001) roughening. *Surface Science*, 1996. **346**(1–3): 31–39.
37. Niehus, H. et al., A real space investigation of the dimer defect structure of si(001)-(2 × 8). *Journal of Microscopy—Oxford*, 1988. **152**: 735–742.

38. Koo, J.-Y. et al., Dimer-vacancy defects on the Si(001)-2 × 1 and the Ni-contaminated Si(001)-2 × n surfaces. *Physical Review B*, 1995. **52**(24): 17269.

39. Kato, K. et al., Si(100)2 × n structures induced by Ni contamination. *Surface Science*, 1988. **194**(1–2): L87–L94.

40. Wilson, R.J. and S. Chiang, Surface modifications induced by adsorbates at low coverage: A scanning tunneling microscopy study of the Ni/Si(111) sqrt 19 surface. *Physical Review Letters*, 1987. **58**(24): 2575.

41. Parikh, S.A., M.Y. Lee, and P.A. Bennett, *Transition Metal Induced Ring—Cluster Structures on Si(111)*, 1995, Denver, CO: AVS.

42. Zhang, Z. et al., Epitaxial growth of ultrathin Si caps on Si(100):B surface studied by scanning tunneling microscopy. *Applied Physics Letters*, 1996. **69**(4): 494–496.

43. Jones, D.E. et al., Striped phase and temperature dependent step shape transition on highly B-doped Si(001)-(2 × 1) surfaces. *Physical Review Letters*, 1996. **77**(2): 330.

44. Bedrossian, P. et al., Surface doping and stabilization of Si(111) with boron. *Physical Review Letters*, 1989. **63**(12): 1257.

45. Shen, T.C. et al., STM study of surface reconstructions of Si(111): B. *Physical Review B*, 1994. **50**(11): 7453.

46. Alstrin, A.L., P.G. Strupp, and S.R. Leone, Direct detection of atomic arsenic desorption from Si(100). *Applied Physics Letters*, 1993. **63**(6): 815–817.

47. Uhrberg, R.I.G. et al., Electronic structure, atomic structure, and the passivated nature of the arsenic-terminated Si(111) surface. *Physical Review B*, 1987. **35**(8): 3945.

48. Kipp, L. et al., Phosphine adsorption and decomposition on Si(100) 2 × 1 studied by STM. *Physical Review B*, 1995. **52**(8): 5843.

49. Wallace, R.M. et al., PH[sub 3] surface chemistry on Si(111)-(7 × 7): A study by Auger spectroscopy and electron stimulated desorption methods. *Journal of Applied Physics*, 1990. **68**(7): 3669–3678.

50. Boland, J.J., Scanning tunnelling microscopy of the interaction of hydrogen with silicon surfaces. *Advances in Physics*, 1993. **42**(2): 129–171.

51. Oura, K. et al., Hydrogen interaction with clean and modified silicon surfaces. *Surface Science Reports*, 1999. **35**(1–2): 1–69.

52. Kubo, A., Y. Ishii, and M. Kitajima, Abstraction and desorption kinetics in the reaction of H+ D/Si(100) and the relation to surface structure. *The Journal of Chemical Physics*, 2002. **117**(24): 11336–11346.

53. Dinger, A., C. Lutterloh, and J. Kuppers, Interaction of hydrogen atoms with Si(111) surfaces: Adsorption, abstraction, and etching. *The Journal of Chemical Physics*, 2001. **114**(12): 5338–5350.

54. Wei, Y., L. Li, and I.S.T. Tsong, Etching of Si(111)-(7 × 7) and Si(100)-(2 × 1) surfaces by atomic hydrogen. *Applied Physics Letters*, 1995. **66**(14): 1818–1820.

55. Owman, F. and P. Mårtensson, STM study of Si(111)1 × 1-H surfaces prepared by *in situ* hydrogen exposure. *Surface Science*, 1994. **303**(3): L367–L372.

56. Kraus, A. et al., Strain relief of Si(111)7 × 7 by hydrogen adsorption. *Applied Surface Science*, 2001. **177**(4): 292–297.

57. Komeda, T. and Y. Kumagai, Si(001) surface variation with annealing in ambient H_{2}. *Physical Review B*, 1998. **58**(3): 1385.

58. Yablonovitch, E. et al., Unusually low surface-recombination velocity on silicon and germanium surfaces. *Physical Review Letters*, 1986. **57**(2): 249.

59. Higashi, G.S. et al., Ideal hydrogen termination of the Si (111) surface. *Applied Physics Letters*, 1990. **56**(7): 656–658.

60. Wade, C.P. and C.E.D. Chidsey, Etch-pit initiation by dissolved oxygen on terraces of H-Si(111). *Applied Physics Letters*, 1997. **71**(12): 1679–1681.

61. Hines, M.A., In search of perfection: Understanding the highly defect-selective chemistry of anisotropic etching. *Annual Review of Physical Chemistry*, 2003. **54**(1): 29–56.

62. Neuwald, U. et al., Wet chemical etching of Si(100) surfaces in concentrated NH₄F solution: Formation of (2 × 1)H reconstructed Si(100) terraces versus (111) facetting. *Surface Science*, 1993. **296**(1): L8–L14.

63. Endo, K. et al., Atomic image of hydrogen-terminated Si(001) surfaces after wet cleaning and its first-principles study. *Journal of Applied Physics*, 2002. **91**(7): 4065–4072.

64. Watanabe, S., N. Nakayama, and T. Ito, Homogeneous hydrogen-terminated Si(111) surface formed using aqueous HF solution and water. *Applied Physics Letters*, 1991. **59**(12): 1458–1460.

65. Kanaya, H., K. Usuda, and K. Yamada, Examination of Si(100) surfaces treated by ultrapure water with 5 ppb dissolved oxygen concentration. *Applied Physics Letters*, 1995. **67**(5): 682–684.

66. Stockhausen, A., T.U. Kampen, and W. Mönch, Oxidation of clean and H-passivated Si(111) surfaces. *Applied Surface Science*, 1992. **56–58**(Part 2): 795–801.

67. Westermann, J., H. Nienhaus, and W. Mönch, Oxidation stages of clean and H-terminated Si(001) surfaces at room temperature. *Surface Science*, 1994. **311**(1–2): 101–106.

68. Wang, Z.-H., H. Noda, Y. Nonogaki, N. Yabumoto, and T. Urisu, Hydrogen diffusion and chemical reactivity with water on nearly ideally H-terminated Si(100) surface. *Japan Journal of Applied Physics*, 2002. **41**: 4275–4278.

69. Ranga Rao, G. et al., A comparative infrared study of H₂O reactivity on Si(100)-(2 × 1), (2 × 1)-H, (1 × 1)-H and (3 × 1)-H surfaces. *Surface Science*, 2004. **570**(3): 178–188.

70. Baluch, A.S. et al., Atomic-level robustness of the Si(100)-2 × 1:H surface following liquid phase chemical treatments in atmospheric pressure environments. *Journal of Vacuum Science & Technology A: Vacuum, Surfaces, and Films*, 2004. **22**(3): L1–L5.

71. Shen, T.C. et al., Nanoscale oxide patterns on Si(100) surfaces. *Applied Physics Letters*, 1995. **66**(8): 976–978.

72. Shen, T.C. et al., *Nanoscale Electronics Based on Two-Dimensional Dopant Patterns in Silicon*, 2004, San Diego, CA: AVS.

73. Hallam, T. et al., Effective removal of hydrogen resists used to pattern devices in silicon using scanning tunneling microscopy. *Applied Physics Letters*, 2005. **86**(14): 143116.

74. Shen, T.C., C. Wang, and J.R. Tucker, The initial stage of nucleation and growth of Al on H/Si(100)-1 × 1 by dimethylaluminum hydride vapor deposition. *Applied Surface Science*, 1999. **141**(3–4): 228–236.

75. Mitsui, T., E. Hill, and E. Ganz, Nanolithography by selective chemical vapor deposition with an atomic hydrogen resist. *Journal of Applied Physics*, 1999. **85**(1): 522–524.

76. Mitsui, T., R. Curtis, and E. Ganz, Selective nanoscale growth of titanium on the Si(001) surface using an atomic hydrogen resist. *Journal of Applied Physics*, 1999. **86**(3): 1676–1679.

77. Adams, D.P., T.M. Mayer, and B.S. Swartzentruber, Selective area growth of metal nanostructures. *Applied Physics Letters*, 1996. **68**(16): 2210–2212.

Chapter 15

78. Becker, R.S. et al., Atomic-scale conversion of clean Si(111):H-1 × 1 to Si(111)-2 × 1 by electron-stimulated desorption. *Physical Review Letters*, 1990. **65**(15): 1917.

79. Riedel, D., A.J. Mayne, and G. Dujardin, Atomic-scale analysis of hydrogen bond breaking from Si(100):H induced by optical electronic excitation. *Physical Review B*, 2005. **72**(23): 233304.

80. Shen, T.C. and P. Avouris, Electron stimulated desorption induced by the scanning tunneling microscope. *Surface Science*, 1997. **390**(1–3): 35–44.

81. Shen, T.-C. et al., Atomic-scale desorption through electronic and vibrational excitation mechanisms. *Science*, 1995. **268**(5217): 1590–1592.

82. Stokbro, K. et al., STM-induced hydrogen desorption via a hole resonance. *Physical Review Letters*, 1998. **80**(12): 2618.

83. Kanasaki, J.I. and K. Tanimura, Scanning tunnelling microscopy study on hydrogen removal from Si(0 0 1)-(2 × 1):H surface excited with low-energy electron beams. *Surface Science*, 2008. **602**(7): 1322.

84. Wei, Y.Y. and G. Eres, Self-limiting behavior of scanning-electron-beam-induced local oxidation of hydrogen-passivated silicon surfaces. *Applied Physics Letters*, 2000. **76**(2): 194–196.

85. Hallam, T. et al., Use of a scanning electron microscope to pattern large areas of a hydrogen resist for electrical contacts. *Journal of Applied Physics*, 2007. **102**(3): 034308–034305.

86. Nakayama, K. and J.H. Weaver, Electron-stimulated modification of Si surfaces. *Physical Review Letters*, 1999. **82**(5): 980.

87. Shen, T.-C., Role of scanning probes in nanoelectronics: A critical review. *Surface Review and Letters (SRL)*, 2000. **7**(5–6): 683–688.

88. Schofield, S.R. et al., Atomically precise placement of single Dopants in Si. *Physical Review Letters*, 2003. **91**(13): 136104.

89. Soukiassian, L. et al., Atomic wire fabrication by STM induced hydrogen desorption. *Surface Science*, 2003. **528**(1–3): 121–126.

90. Stipe, B.C. et al., Single-molecule dissociation by tunneling electrons. *Physical Review Letters*, 1997. **78**(23): 4410.

91. Foley, E.T. et al., Cryogenic UHV-STM study of hydrogen and deuterium desorption from Si(100). *Physical Review Letters*, 1998. **80**(6): 1336.

92. Thirstrup, C. et al., Temperature suppression of STM-induced desorption of hydrogen on Si(100) surfaces. *Surface Science*, 1999. **424**(2–3): L329–L334.

93. Adams, D.P., T.M. Mayer, and B.S. Swartzentruber, Nanometer-scale lithography on Si(001) using adsorbed H as an atomic layer resist. *Journal of Vacuum Science & Technology B: Microelectronics and Nanometer Structures*, 1996. **14**(3): 1642–1649.

94. Walsh, M.A. and M.C. Hersam, Atomic-scale templates patterned by ultrahigh vacuum scanning tunneling microscopy on silicon. *Annual Review of Physical Chemistry*, 2009. **60**(1): 193–216.

95. Binnig, G. et al., Surface studies by scanning tunneling microscopy. *Physical Review Letters*, 1982. **49**(1): 57.

96. Binnig, G., C.F. Quate, and C. Gerber, Atomic force microscope. *Physical Review Letters*, 1986. **56**(9): 930.

97. Robinson, S.J. et al., Electron transport in laterally confined phosphorus delta layers in silicon. *Physical Review B*, 2006. **74**(15): 153311.

98. Rueß, F.J. et al., The use of etched registration markers to make four-terminal electrical contacts to STM-patterned nanostructures. *Nanotechnology*, 2005. **16**(10): 2446.

99. Rueß, F.J. et al., One-dimensional conduction properties of highly phosphorus-doped planar nanowires patterned by scanning probe microscopy. *Physical Review B*, 2007. **76**(8): 085403.

100. B,F.J. Fuhrer, A. et al., Atomic-scale, all epitaxial in-plane gated donor quantum dot in silicon. *Nano Letters*, 2009. **9**(2): 707–710.

101. Rueß, F.J. et al., Electronic properties of atomically abrupt tunnel junctions in silicon. *Physical Review B*, 2007. **75**(12): 121303.

102. Reusch, T.C.G. et al., Aharonov–Bohm oscillations in a nanoscale dopant ring in silicon. *Applied Physics Letters*, 2009. **95**(3): 032110–032113.

103. Minne, S.C. et al., Automated parallel high-speed atomic force microscopy. *Applied Physics Letters*, 1998. **72**(18): 2340–2342.

104. Akiyama, T. et al., Concept and demonstration of individual probe actuation in two-dimensional parallel atomic force microscope system. *Japanese Journal of Applied Physics*, 2007. **46**(9B): 6458–6462.

105. Stiévenard, D. and B. Legrand, Silicon surface nano-oxidation using scanning probe microscopy. *Progress in Surface Science*, 2006. **81**(2–3): 112–140.

106. Snow, E.S. et al., Ultrathin PtSi layers patterned by scanned probe lithography. *Applied Physics Letters*, 2001. **79**(8): 1109–1111.

107. Bo, X.-Z. et al., Nanopatterning of Si/SiGe electrical devices by atomic force microscopy oxidation. *Applied Physics Letters*, 2002. **81**(17): 3263–3265.

108. Gotoh, Y. et al., *Experimental and Theoretical Results of Room-Temperature Single-Electron Transistor Formed by the Atomic Force Microscope Nano-Oxidation Process*, 2000, Seattle, Washington, DC: AVS.

109. Chiu, F.-C. et al., Electrical characterization of tunnel insulator in metal/insulator tunnel transistors fabricated by atomic force microscope. *Applied Physics Letters*, 2005. **87**(24): 243506–243513.

110. Delacour, C. et al., Superconducting single photon detectors made by local oxidation with an atomic force microscope. *Applied Physics Letters*, 2007. **90**(19): 191116–191123.

111. Gross, L. et al., The chemical structure of a molecule resolved by atomic force microscopy. *Science*, 2009. **325**(5944): 1110–1114.

112. Rothschild, M. et al., 157 nm: Deepest deep-ultraviolet yet. *Journal of Vacuum Science & Technology B*, 1999. **17**(6): 3262–3266.

113. Gwyn, C.W. et al., Extreme ultraviolet lithography. *Journal of Vacuum Science & Technology B*, 1998. **16**(6): 3142–3149.

114. Silverman, J.P., Challenges and progress in x-ray lithography. *Journal of Vacuum Science & Technology B*, 1998. **16**(6): 3137–3141.

115. Raether, H., *Surface Plasmons*, 1988, Berlin: Springer.

116. Srituravanich, W. et al., Sub-100 nm lithography using ultra-short wavelength of surface plasmons. *Journal of Vacuum Science & Technology B*, 2004. **22**(6): 3475–3478.

117. Srituravanich, W. et al., Plasmonic nanolithography. *Nano Letters*, 2004. **4**(6): 1085–1088.

118. Liu, Z.W. et al., Focusing surface plasmons with a plasmonic lens. *Nano Letters*, 2005. **5**(9): 1726–1729.

119. Liu, Z.W. et al., Broad band two-dimensional manipulation of surface plasmons. *Nano Letters*, 2009. **9**(1): 462–466.

120. Chen, W. and H. Ahmed, Fabrication of 5–7 Nm wide etched lines in silicon using 100 keV electron-beam lithography and polymethylmethacrylate resist. *Applied Physics Letters*, 1993. **62**(13): 1499–1501.

121. Matsui, S., Y. Kojima, and Y. Ochiai, High-resolution focused ion-beam lithography. *Applied Physics Letters*, 1988. **53**(10): 868–870.

122. Betzig, E. and J.K. Trautman, Near-field optics—Microscopy, spectroscopy, and surface modification beyond the diffraction limit. *Science*, 1992. **257**(5067): 189–195.

123. Majumdar, A. et al., Nanometer-scale lithography using the atomic force microscope. *Applied Physics Letters*, 1992. **61**(19): 2293–2295.

124. Yin, X.B. et al., Near-field two-photon nanolithography using an apertureless optical probe. *Applied Physics Letters*, 2002. **81**(19): 3663–3665.

125. Wang, Y. et al., Plasmonic nearfield scanning probe with high transmission. *Nano Letters*, 2008. **8**(9): 3041–3045.

126. Srituravanich, W. et al., Flying plasmonic lens in the near field for high-speed nanolithography. *Nature Nanotechnology*, 2008. **3**(12): 733–737.

127. Keyes, R.W., Physical limits of silicon transistors and circuits. *Reports on Progress in Physics*, 2005. **68**(12): 2701–2746.

128. Keyes, R.W., Information, computing technology, and quantum computing. *Journal of Physics—Condensed Matter*, 2006. **18**(21): S703–S719.

129. Shinada, T. et al., Enhancing semiconductor device performance using ordered dopant arrays. *Nature*, 2005. **437**(7062): 1128–1131.

130. Sellier, H. et al., Transport spectroscopy of a single dopant in a gated silicon nanowire. *Physical Review Letters*, 2006. **97**(20): 206805(1)–206805(4).

131. Ono, Y. et al., Conductance modulation by individual acceptors in Si nanoscale field-effect transistors. *Applied Physics Letters*, 2007. **90**(10): 102106-1–102106-3.

132. Calvet, L.E., R.G. Wheeler, and M.A. Reed, Observation of the linear Stark effect in a single acceptor in Si. *Physical Review Letters*, 2007. **98**(9): 096805-1–096805-4.

133. Kane, B.E., A silicon-based nuclear spin quantum computer. *Nature*, 1998. **393**(6681): 133–137.

134. Jelezko, F. et al., Observation of coherent oscillation of a single nuclear spin and realization of a two-qubit conditional quantum gate. *Physical Review Letters*, 2004. **93**(13): 130501-1–130501-4.

135. Fuchs, G.D. et al., Excited-state spectroscopy using single spin manipulation in diamond. *Physical Review Letters*, 2008. **101**(11): 117601-1–117601-4.

136. Jamieson, D.N. et al., Controlled shallow single-ion implantation in silicon using an active substrate for sub-20-keV ions. *Applied Physics Letters*, 2005. **86**(20): 202101-1–202101-3.

137. Schenkel, T. et al., Solid state quantum computer development in silicon with single ion implantation. *Journal of Applied Physics*, 2003. **94**(11): 7017–7024.

138. Shen, T.C. et al., Nanoscale electronics based on two-dimensional dopant patterns in silicon. *Journal of Vacuum Science & Technology B*, 2004. **22**(6): 3182–3185.

139. Ruess, F.J. et al., One-dimensional conduction properties of highly phosphorus-doped planar nanowires patterned by scanning probe microscopy. *Physical Review B*, 2007. **76**(8): 085403-1–085403-5.

140. Nastasi, M.A. and J.W. Mayer, *Ion Implantation and Synthesis of Materials*, 2006, New York: Springer.

141. Prins, J.F., Ion implantation of diamond for electronic applications. *Semiconductor Science and Technology*, 2003. **18**(3): S27–S33.

142. Kalish, R., *Doping Diamond by Ion-Implantation, in Thin-Film Diamond I*, 2003, San Diego, CA: Academic Press Inc. pp. 145–181.

143. Schenkel, T. et al., Electrical activation and electron spin coherence of ultralow dose antimony implants in silicon. *Applied Physics Letters*, 2006. **88**(11):3.

144. Orloff, J. et al., *High Resolution Focused Ion Beams: FIB and Its Applications*, 2003, New York: Kluwer.

145. MoberlyChan, W.J. et al., Fundamentals of focused ion beam nanostructural processing: Below, at, and above the surface. *Mrs Bulletin*, 2007. **32**(5): 424–432.

146. Schnitzler, W. et al., Deterministic Ultracold ion source targeting the Heisenberg limit. *Physical Review Letters*, 2009. **102**(7): 4.

147. Persaud, A. et al., Integration of scanning probes and ion beams. *Nano Letters*, 2005. **5**(6): 1087–1091.

148. Schenkel, T. et al., Formation of a few nanometer wide holes in membranes with a dual beam focused ion beam system. *Journal of Vacuum Science & Technology B*, 2003. **21**(6): 2720–2723.

149. Ziegler, J.F., SRIM-2003. *Nuclear Instruments & Methods in Physics Research Section B-Beam Interactions with Materials and Atoms*, 2004. **219**: 1027–1036.

150. Baragiola, R.A., Principles and mechansims of ion-induced electron-emission. *Nuclear Instruments & Methods in Physics Research Section B—Beam Interactions with Materials and Atoms*, 1993. **78**(1–4): 223–238.

151. Arnau, A. et al., Interaction of slow multicharged ions with solid surfaces. *Surface Science Reports*, 1997. **27**(4–6): 117–239.

152. Park, S.J. et al., Processing issues in top-down approaches to quantum computer development in silicon. *Microelectronic Engineering*, 2004. **73–74**: 695–700.

153. Borisenko, V.E. and P.J. Hesketh, *Rapid Thermal Processing of Semiconductors*, 1997, New York: Plenum Press.

154. Ural, A., P.B. Griffin, and J.D. Plummer, Fractional contributions of microscopic diffusion mechanisms for common dopants and self-diffusion in silicon. *Journal of Applied Physics*, 1999. **85**(9): 6440–6446.

155. Weis, C.D. et al., Single atom doping for quantum device development in diamond and silicon. *Journal of Vacuum Science & Technology B*, 2008. **26**(6): 2596–2600.

156. Brundle, C.R., C.A. Evans, and S. Wilson, *Encyclopedia of Materials Characterization: Surfaces, Interfaces, Thin Films*, 1992, Boston; Greenwich, CT: Butterworth-Heinemann, Manning.

157. Dickey, D.H., Developments in ultrashallow spreading resistance analysis. *Journal of Vacuum Science & Technology B*, 2002. **20**(1): 467–470.

158. Schenkel, T. et al., Critical issues in the formation of quantum computer test structures by ion implantation. *Nuclear Instruments & Methods in Physics Research Section B—Beam Interactions with Materials and Atoms*, 2009. **267**(16): 2563–2566.

159. Persaud, A. et al., Ion implantation with scanning probe alignment. *Journal of Vacuum Science & Technology B*, 2005. **23**(6): 2798–2800.

160. Batra, A. et al., Detection of low energy single ion impacts in micron scale transistors at room temperature. *Applied Physics Letters*, 2007. **91**(19): 193502-1–193502-3.

161. Shinada, T. et al., A reliable method for the counting and control of single ions for single-dopant controlled devices. *Nanotechnology*, 2008. **19**(34): 345202-1–345202-4.

162. Sideras-Haddad, E. et al., Electron emission and defect formation in the interaction of slow, highly charged ions with diamond surfaces. *Nuclear Instruments & Methods in Physics Research Section B—Beam Interactions with Materials and Atoms*, 2007. **256**(1): 464–467.

163. Sideras-Haddad, E. et al., Possible diamond-like nanoscale structures induced by slow highly-charged ions on graphite (HOPG). *Nuclear Instruments & Methods in Physics Research*

Chapter 15

Section B—Beam Interactions with Materials and Atoms, 2009. **267**(16): 2774–2777.

164. Schneider, D. et al., Atomic displacement due to the electrostatic potential-energy of very highly charged ions at solid-surfaces, surface. *Science*, 1993. **294**(3): 403–408.

165. Schenkel, T. et al., Electronic desorption of alkyl monolayers from silicon by very highly charged ions. *Journal of Vacuum Science & Technology B*, 1998. **16**(6): 3298–3300.

166. Weis, C.D. et al., Mapping of ion beam induced current changes in FinFETs. *Nuclear Instruments & Methods in Physics Research Section B—Beam Interactions with Materials and Atoms*, 2009. **267**(8–9): 1222–1225.

167. Breese, M.B.H. et al., A review of ion beam induced charge microscopy. *Nuclear Instruments & Methods in Physics Research Section B—Beam Interactions with Materials and Atoms*, 2007. **264**(2): 345–360.

168. Lo, C.C. et al., Spin-dependent scattering off neutral antimony donors in Si-28 field-effect transistors. *Applied Physics Letters*, 2007. **91**(24): 242106-1–242106-3.

169. Sarovar, M. et al., Quantum nondemolition measurements of single donor spins in semiconductors. *Physical Review B*, 2008. **78**(24): 245302-1–245302-8.

170. Morello, A. et al., Single-shot readout of an electron spin in silicon. *Nature*, **467**(7316): 687–691.

Nanofabrication
Applications

16. Nanoelectromechanical Systems

Hiroyuki Fujita
CIRMM, IIS, The University of Tokyo, Tokyo, Japan

Yoshio Mita
School of Engineering, The University of Tokyo, Tokyo, Japan

16.1 Introduction

Recently, we have experienced astonishing expansion in nanotechnology research. The field covers a wide range including materials, electronics, surface science, and life science. Basic knowledge on physical, chemical, and biological properties of nanoscopic objects is necessary for the development of nanotechnology. In order to obtain such data, tools that can handle, measure, and modify nanoobjects are highly required.

Micromachining technology has created a great variety of micro/nanoelectromechanical systems (MEMS/NEMS) through the integration of moving mechanisms, sensors, and electronics in a chip-size system. One of the promising applications of NEMS is the tool for the investigation of nano- and bioworlds. NEMS serves as the interface between the nanoscopic world and our macroscopic world. We can reveal material properties specific to the nanoscopic region by evaluating nanomatters with NEMS tools (Fujita 2003). The fabrication technique is called micromachining. Micromachining processes are based on silicon-integrated circuits technology and used to build three-dimensional structures and movable parts by the combination of lithography, etching, film deposition, and wafer bonding.

Nanofabrication Handbook. Edited by Stefano Cabrini and Satoshi Kawata © 2012 CRC Press / Taylor & Francis Group, LLC. ISBN: 978-1-4200-9052-9

Chapter 16

16.2 High-Aspect-Ratio Nanostructures

The increasing demand for silicon high-aspect-ratio microstructure (HARMS) has accelerated the development of deep-reactive ion etching (Deep RIE or DRIE) technology since 1990s. Typical feature size of HARMS etching goes down to 5 µm in lithography pattern, and deeper than 10 µm in etching depth. Since 2000, a couple of research institutes were successful in pushing the DRIE technological limit down to submicron era. The structure features submicron in opening and around 10 µm in etching depth (Marty et al. 2005). In short, pattern resolution became finer than ordinal MEMS in one order of magnitude while keeping MEMS-compatible etching depth. This kind of fine and deep structures can be called as high-aspect-ratio nanostructures (HARNS).

HARNS is expected to introduce discontinuous evolution to open new research fields. Applications in optoelectromechanical devices and fluidics are reported. Critical merits of using DRIE HARNS are summarized in Figure 16.1:

1. Surface ratio to a given volume becomes larger.
2. Devices can be made orthogonal to the substrate surface; this allows serial arrangement of devices and structures in-plane.
3. Geometry controlled nanostructures show nanoscale-specific physical phenomena.
4. HARNS can be integrated with silicon large scale integration (LSI) circuits.

16.2.1 High-Aspect-Ratio Nanoetching

A high-aspect-ratio nanoetching discussed in this section refers to a deep etching of narrow opening pattern as shown in Figure 16.2. Nanoholes and trenches are

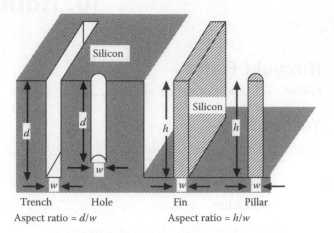

FIGURE 16.2 High-aspect nanostructure examples.

the resulting structures. Nanostructures such as sparsely placed pillars and/or fins, obtained by large-area etching, are out of this chapter's scope because large-area etching is essentially the same to ordinal DRIE and is relatively easy to obtain.

Figure 16.3 shows an etching result of a standard Bosch's process with submicron trench opening mask. Bosch's process is the de facto standard etching process in silicon DRIE, composed with alternating sidewall passivation and trench bottom etching. Table 16.1 summarizes the etching condition of an "ordinal" Bosch's process that is widely used in conventional MEMS technology. With this condition, a 5 µm opening and 100-µm-deep trenches can be easily obtained. Etching speed is around 3–4 µm/min, depending on the opening width (slower with narrower opening due to reactive ion etching (RIE) lag). It is clearly shown that an

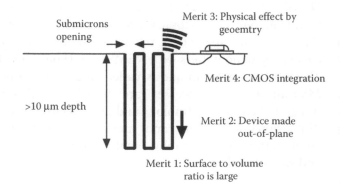

FIGURE 16.1 Merits of HARNS DRIE.

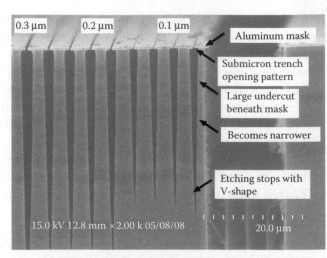

FIGURE 16.3 Submicron mask with standard Bosch's process.

Table 16.1 A "Standard" Bosch's Process Example

Apparatus: Alcatel MS-100

ICP source: 13.56 MHz, 1800 W

Substrate bias: Around 250 kHz, 100 W, 10% pulsed modulation

Deposition: C4F8 150 sccm, 2.8 Pa, 2 s

Etching: SF6 300 sccm, 3.8 Pa, 5 s

ordinal Bosch's process cannot etch out submicron openings. By analyzing the cross-sectional view, it can be said that

1. Etching plasma could not enter deeply in a narrow trench.
2. Sidewall setback (called erosion) occurred: passivation was insufficient.

Therefore for HARNS etching, it is necessary to either search for optimized etching condition or develop an alternate etching process (Blauw et al. 2001). The principles are clear:

1. Etching plasma should be accelerated toward the trench bottom.
2. More sidewall passivation is necessary.
   One can find an optimal condition by parameter tuning such as:
   1-1. Increase effective substrate bias to accelerate ions more. This is attainable by increasing substrate bias power and/or pulse on period.
   1-2. Increasing source power results in severer ion bombardment so that etching masks should be thicker. The authors are using either thick electron beam (EB) resist (Tokyo Oka Co. OEBR CAP-112) or hard mask such as with aluminum.
   1-3. Decrease the plasma pressure to increase the mean free path.
   2-1. Decrease the cycle time for smaller undercut due to isotropic etching of SF6.
   2-2. Increase the sidewall passivation deposition portion to one cycle to avoid sidewall erosion.
3. Rebalance passivation and etching cycle time to obtain a vertical cross section.

Figure 16.3 shows an etching result of a "submicron feature-optimized" Bosch's process. This first example (Figure 16.4) shows a 360 nm opening—15 μm depth (aspect ratio 1:41). Thick EB resist (OEBR-CAP112 PM by Tokyo Oka) was used for direct 4-in. EB exposure with high-throughput EB writer (F5112 + VD01

FIGURE 16.4 A 360 nm opening 190 nm remaining Si trenches A.R. = 1:41.

ADVANTEST). Designed line and space dimension on computer-aided design data was a 300 nm trench for a 550 nm pitch (250 nm for remaining silicon). A 30 nm of undercuts resulted in a 360 nm opening and 190 nm silicon fin structures. A 100 nm opening trench is tried as well. An 80-nm-thick aluminum (Al) mask was patterned with lift-off on a negative EB resist (OEBR-CAN028T2 PF). Because of 35 nm of undercut on both trench sides, trench opening width resulted in 170 nm (Figure 16.5). Average etching depth was 11.5 μm (aspect ratio 1:68). Etching time was 20 min (10 min + 2 min of pause + 10 min).

Two interesting phenomena are found in the examples. First, the etching depth decreases depending on the opening width (the narrower the opening is, the

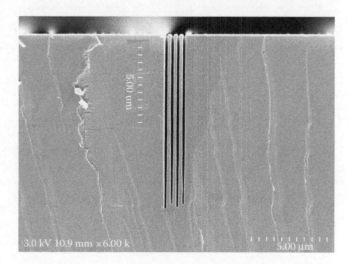

FIGURE 16.5 A 170 nm-opening 400 nm remaining Si trenches A.R. = 1:68 (Marty et al. 2005).

shallower the trench becomes for the same etching time). This phenomenon is called RIE lag (Jansen et al. 1997), local loading effect, and or aspect-ratio-dependent etching. However, the depth decrease ratio is smaller than that of the trench width. Therefore, the aspect ratio that is calculated by (depth)/(width) increases as the width decreases. The same order of bowing is found independent of trench depth (around 30 nm) that would eventually limit the "top data" of aspect ratio in narrow trenches; bowing disappears when the aspect ratio exceeds a certain value. Second, an interesting phenomenon is "aspect ratio-dependent scalloping attenuation" (Mita et al. 2006b) (Figure 16.6). Scalloping refers to the periodic peak-and-valley stripes that are found on every wall of Bosch's process. Scalloping occurs due to the alternation of isotropic etching and passivation. The interesting finding is that scalloping decreases in proportion to the etched aspect ratio; typically, no scalloping is observable on the trench walls at the depth that gives an aspect ratio larger than 1:8.

The important principle in HARNS DRIE is the same as ordinal DRIE: how to protect the sidewall and how to etch the bottom by removing the protection layer. Therefore, an engineer can anticipate many alternative methods with different process steps and gas combinations. A couple of process steps optimization examples are reported already.

1. Cryogenic process: Etching with a mixture of SF6 and O_2 with a cooled substrate below $-100°C$. A silicon oxide (SiO_2) protection layer is formed by plasma O_2. Due to the simultaneous etching and passivation, scalloping does not appear. Practical limit can mask compatibility (sometimes resist mask cracks in low temperature), etching selectivity when using resist mask, and larger dependency of wall verticality on pattern geometry.

2. Intentional O_2 irradiation during Bosch's process: Plasma O_2 irradiation step is inserted after some cycles of Bosch's process. This intentional irradiation removes excess polymer as well as forms thin (in the order of 10 nm) SiO_2 layer on the sidewall. A grass-free (black silicon free) surface has been obtained for large and small opening as a side-effect of the O_2 irradiation. An aspect ratio over 1:60 for 1 μm of opening is reported (Ohara et al. 2008).

16.2.2 Application Devices in Optical and Fluidic Devices

The submicron feature geometry made the HARNS applicable to microoptics and microfluidics. A couple of device application reports are already found in the literature. In optics, the wavelength of visible light is from 400 (violet) to 600 nm (red), and that for infrared communication is 1.3–1.5 μm. HARNS feature size is in the same order of magnitude so that HARNS can show optical effects such as interference. Reported application devices are, for example, an HARNS-integrated MEMS in-plane optical interferometer and polarization-sensitive optoelectrical devices. Key breakthrough is the capability to pattern finer than the wavelength in two dimensions and longer than the wavelength in one dimension. In channel-based fluidics, physical phenomena on the HARNS surface become important due to the increase in surface-to-volume ratio. Chromatography is the application directly benefited by HARNS.

1. MEMS in-plane optical interferometer
Figure 16.7 shows a one-mask-etched MEMS tunable wavelength division multiplexing (WDM) filtering device. The device is composed of three parts: a distributed Bragg's reflector (DBR) mirror-type optical interferometer by HARNS vertical trenches, guiding trenches for fiber insertion, and MEMS microactuators. All parts are made with a one-mask DRIE (Saadany et al. 2006; Mita et al. 2006). Guiding trenches enable in-plane passive alignment of optical fibers and DBRs, which has not been possible before HARNS technology.

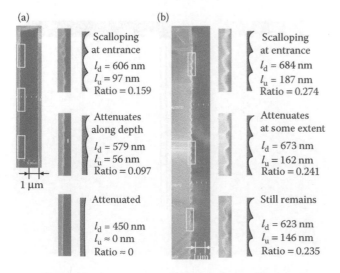

(a)   (b)

Scalloping at entrance
$l_d = 606$ nm
$l_u = 97$ nm
Ratio = 0.159

Scalloping at entrance
$l_d = 684$ nm
$l_u = 187$ nm
Ratio = 0.274

Attenuates along depth
$l_d = 579$ nm
$l_u = 56$ nm
Ratio = 0.097

Attenuates at some extent
$l_d = 673$ nm
$l_u = 162$ nm
Ratio = 0.241

1 μm

Attenuated
$l_d = 450$ nm
$l_u ≈ 0$ nm
Ratio ≈ 0

Still remains
$l_d = 623$ nm
$l_u = 146$ nm
Ratio = 0.235

FIGURE 16.6 Scalloping attenuates in high-aspect-ratio bottom. (a) 1.82 μm small opening. (b) 201 μm large opening (Mita et al. 2006b).

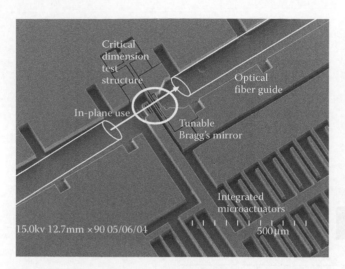

FIGURE 16.7 One-mask tunable WDM filter.

The light whose electrical field is perpendicular to the long axis of the silicon fin passes through the structure, whereas that parallel to the long axis is reflected. An extinction ratio of 1:4 is measured for 675 nm laser light. A p–n junction is made on the silicon structure so that the device can recover the "lost" light electrically. Potential application is a polarizer for liquid crystal display.

The device is made with single HARNS DRIE as well; the unique feature of the process is dry lift-off. Subsequent to HARNS trench etching, an intentionally long etching is performed. The long isotropic etching makes eaves so that the device can easily be peeled off with tweezers. Dry release is particularly interesting for this device to avoid sticking of fin structures.

2. Polarization-transmissive photovoltaic cells

A thin (5 μm) silicon film composed of submicron-width fins and trenches is made from bulk silicon wafer (Figure 16.8) (Hirose et al. 2008). The periodic silicon fins structure acts as a polarization filter so that transmitted light shows dependency on polarization.

3. Polarization-sensitive photodetector

A p–n junction that conformally covers over submicron trench fin structure can also act as a photodetector with polarization sensitivity (Figure 16.9). To dope over such high-aspect-ratio structure, a classic vapor-phase boron doping method using a solid

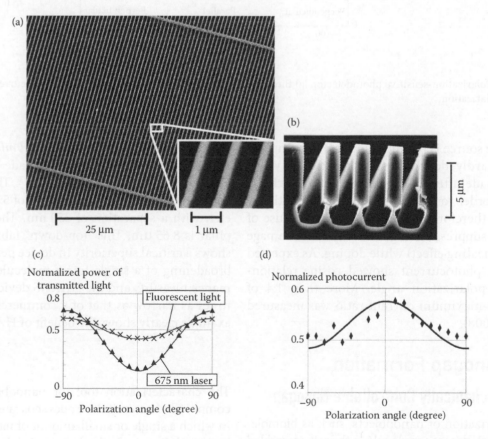

FIGURE 16.8 Polarization–transmissive photovoltaic cells. (a) Bird's eye view, (b) cross-sectional view, (c) transmitted power polarization dependence, and (d) photocurrent to polarization dependence.

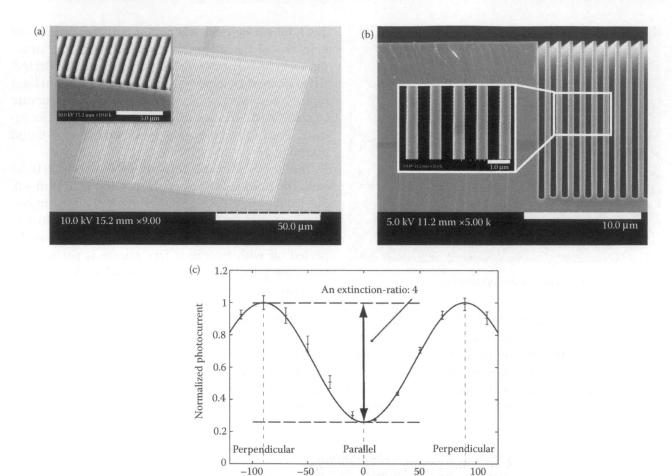

FIGURE 16.9 Polarization-sensitive photodetector. (a) Bird's eye view, (b) cross-sectional view, and (c) photocurrent dependency on 675 nm laser polarization.

boron nitride source is the most suitable. Ion implantation can hardly dope over vertical trench walls. Doping is made after DRIE fins structure etching. The process order forms a p–n junction beneath the surface, and therefore is believed to be the cause of leak current suppression together with RIE damage recovery (annealing effect) while doping. As expected theoretically, photocurrent showed cosine relationship to the polarization angle. More than 1:4 of minimum-to-maximum current ratio was measured (Imai et al. 2008).

16.3 Nanogap Formation

16.3.1 Mechanically Controllable Nanogap

The characterization of nanoobjects, such as biomolecules, nanoparticles, or a molecule in the self-assembled monolayer, is the important part of nanotechnology.

4. Chromatography device with submicron pillars
A chromatography column was made with silicon by HARNS DRIE (Tezcan et al. 2007). The column was composed of densely placed pillars of 550 nm in diameter, with a separation of 350 nm. The height of the pillar is 8.65 μm. This "top-down" fabricated column shows a critical superiority in device performance. The broadening of a single tracer molecule is the performance measure, and the HARNS device showed eight times as narrow as that of a commercial device. This example clearly shows the benefit of HARNS.

The characterization tool for nanoobjects is mainly composed of a pair of electrodes separated by a nanogap, in which a single or small amount of nanoobject(s) can be placed. Currently, the nanogap is fabricated either by electromigration (Park et al. 1999) or mechanically breaking the junction (Reed et al. 1997) on a substrate applied with external stress. By using electromigration technique, electrodes separated by a distance of a few nanometers can be obtained. However, due to the stochastic nature of this process, obtaining controllable results remain as an issue. On the other hand, the gaps created by mechanically controlled junctions have the advantage of adjustability of the gap size according to the size of the nanomaterial under investigation, which can range from subnanometer to 10s of nanometers. Typically a substrate surface is strained on which electrodes are fabricated by pushing substrate against two countersupports (Moreland and Ekin 1985). This requires using an external actuation mechanism such as a piezoelement which has to be precisely aligned. Three-point bending configuration and alignment issues bring complexity to the experiment setup.

On the other hand, the integration of electrodes and microactuators can open up a new possibility; precise and stable control of the gap between electrodes in the range of sub-nm can be achieved by using the integrated device. A mechanically controlled quantum contact was fabricated with on-chip electrostatic MEMS actuators and characterized (Gel et al. 2007). Fabricating the electrodes and actuation mechanism on the same chip brings extreme simplicity to the experimental setup and allows the experiments to be easily performed in different environments, even inside of a transmission electron microscope (TEM), while observing the motion of the tips and the gap between.

The mechanically controllable junction with an on-chip MEMS actuator makes use of microcantilevers that are bent by electrostatic force. The working principle of this mechanism is illustrated in Figure 16.10. Two cantilevers shown with black-filled area located in the middle and two electrodes for electrostatic actuation are located at the sides of cantilevers. When voltage is applied to the electrode on the left, the cantilevers bend toward the electrode by electrostatic force. The deflection of the beams allows the edges of the cantilevers to approach each other. The separation gap between tips located at the cantilever edge decreases as shown in the close view schematic. When the voltage is applied to right electrode, gap size between tips increases and retraction is obtained. High-resolution displacements at nanometer level can be obtained by this approach/retract mechanism, thanks to the intrinsic reduction factor. Typically, when the cantilevers deflect about 100 nm toward the actuation electrode, the separation gap between tips shows a change of ~20 nm. By using this mechanism, it is possible to approach and contact two electrodes separated by a distance of 10s of nanometers. Figure 16.11a shows the scanning electron microscope (SEM) picture of the fabricated device by using this concept. Two cantilevers in the middle can be seen. A close view of the tips that are located at the edge of the cantilevers is shown on the right. The initial gap is of the order of 30 nm. High-voltage electrode on the left is used for tip approach, while the one on the right is used for retraction. In order to prevent cantilevers from contacting with high-voltage electrodes, four stopper structures denoted by S can be seen at the center. Cantilevers have a length of 50 μm, while the width and the thickness are 2 μm. Figure 16.11b shows a device that is fabricated for operation in TEM.

Figure 16.12 shows the fabrication process flow for a regular device as well as a device for operation in TEM. Silicon-on-insulator (SOI) wafer is used as the starting material. Silicon thickness is 2 m and buried oxide thickness is 1 μm. After cleaning the wafer, the device

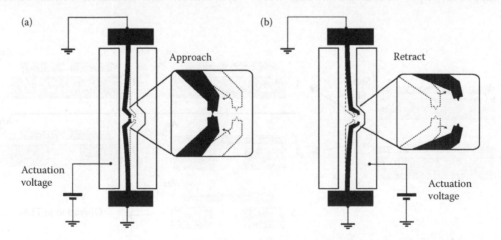

FIGURE 16.10 Working principle of the approach/retract mechanism based on deflection of two cantilevers. (a) Approach by applying voltage to electrodes on the left. (b) Retraction by applying voltage to electrode on the right.

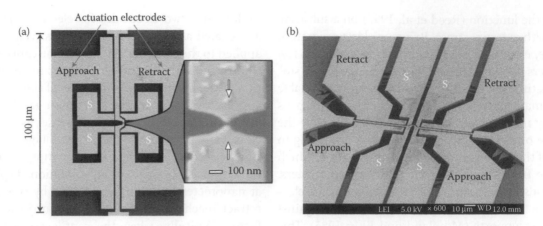

FIGURE 16.11 SEM image of the fabricated device. Parts denoted by s are stopper structures to avoid contact between cantilever and actuation electrode. (a) Device fabricated for regular measurments under probe station, a close view of the gold tips separated by a ~30 nm gap. (b) Similar device designed for operation in TEM with a through-hole in the handle wafer.

shape is defined by the EB lithography. A 50 nm Au layer with a 5 nm Cr adhesion layer is patterned by a lift-off technique. Then, 2 μm silicon is etched by anisotropic dry etching. Finally, vapor phase HF oxide etching is performed to release cantilevers on the chip. Fabrication of devices for TEM observation involves deep etching of a hole in the handle wafer from the backside after the lift-off process. After deep etching, the front-side silicon is etched by using the lift-off gold layer as a mask. Finally, wet etching of oxide with HF for 1 min is performed. The fabrication process has high yield, and reproducibility mainly depends on the development step in EB lithography, which defines the initial gap between the tips.

Devices fabricated for operation in TEM are used to measure initial gap (as fabricated) and the change in the gap size when voltage is applied to actuation electrodes to approach tips. The displacements were measured from the video image, which is calibrated by the magnification factor.

Figure 16.13a shows the experimentally measured change in gap size with black-filled circles. Error bars are showing the uncertainty in displacement measurement which is mainly due to the edge roughness of the tips that are of the order of 5 nm. Simulation results with assumed values of undercut are also shown in the same figure. Figure 16.13b shows a side-view SEM image of the tips and instances from TEM video images from which displacement measurements are performed. Edge roughness of the tips can be seen in high-magnification view.

Experimentation with gold tips coated with self-assembled monolayer (SAM) is performed. After the last step of fabrication process, the chips are dipped into 1 mM of 1,4-benzene dithiol solution in tetrahydrofuran and dried in argon atmosphere 3-1-4.

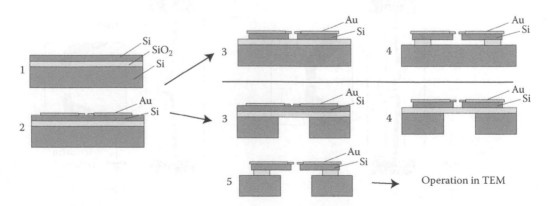

FIGURE 16.12 Device fabrication process for a regular test device and for a device operated in TEM. Both processes are the same until step 2.

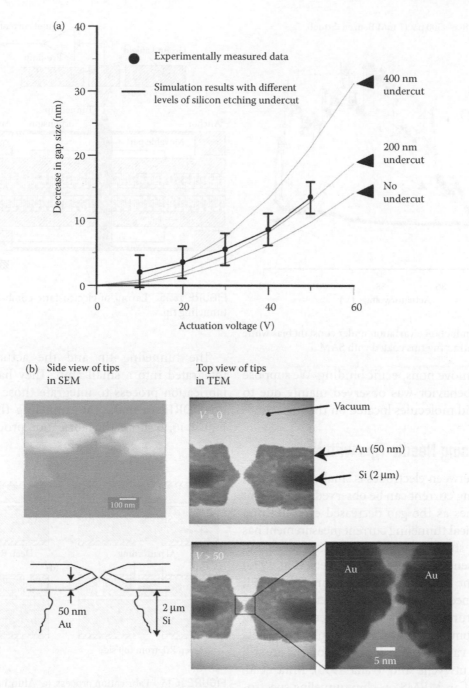

FIGURE 16.13 (a) Comparison of experimental (filled circles with error bars) and simulation results for decrease in gap size when actuation voltage is applied to electrodes. Simulation is done by assuming three different levels of undercut. (b) Side view of the SEM image and top view of the TEM images during the gap size measurements.

Measurements are made by applying a constant bias voltage of 500 mV while increasing the voltage at the actuation electrode to approach tips until current is >1 nA. After that point, the tips are allowed to separate from each other by slowly decreasing the actuation voltage. Figure 16.14 shows the measured data during the experiment. The hysteresis in the case of an SAM-coated chip is different from the case of uncoated tips. Since the measured conductance value is well smaller than the value of direct metal-to-metal contact, we tentatively attributed the reason of this hysteresis behavior as the stretching of an organic bridge formed between two metal surfaces. After the coating of SAM, the measurement was conducted without any rinsing

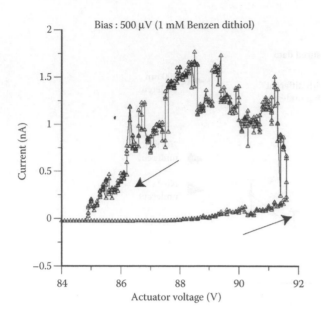

FIGURE 16.14 Conductance variation under constant bias while approaching and retracting tips coated with SAM.

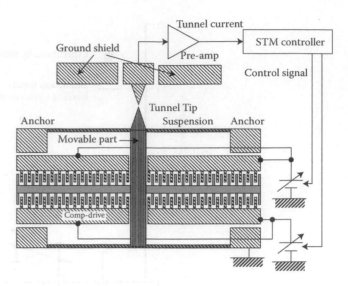

FIGURE 16.15 Layout of electrostatic comb-drive actuator and tunneling tip.

of the chip to remove nonspecific binding. We suppose that hysteresis behavior was observed mainly due to the loosely bound molecules located on the surface.

16.3.2 Opposing Needle Tips with Nanogap

When the gap between electrodes is smaller than 1 nm, vacuum tunneling current can be observed. The current increases 10 times as the gap decreased every 0.1 nm. Therefore, electrical tunneling current measurement has been intensively studied as a highly sensitive detection mechanism for sensing acceleration (Kubena et al. 1996) and displacement (Yao et al. 1992; Kenny et al. 1991, 1994). Highly concentrated electrical current and electrical field of the tunneling gap are also very interesting environments from electro- and mechanophysical points of view in which various mesoscopic phenomena are still undiscovered (Lutwyche and Wada 1995). Mita et al. (2005) developed an MEMS on-chip tunneling spectroscope made by silicon bulk micromachining technology. A very sharp tunneling tip with a small tunneling gap is controlled by a microactuator. They have successfully developed tunneling tips integrated with an electrostatic micropositioning actuator made by DRIE of an SOI wafer. The tunneling current has been observed in the air and in the vacuum TEM chamber.

The microtunneling unit (Figure 16.15) consists of two parts: an electrostatic microactuator of high aspect ratio and a nanowire tunneling tip. A sharp tunneling tip is made with the same etching process used to make the positioning mechanism.

The tunneling tip and the actuator should be integrated into a small chip. They have developed a fabrication process to integrate those elements. They used DRIE to make the tunneling tip and actuator. Figure 16.16 shows the fabrication process.

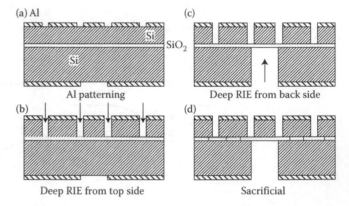

FIGURE 16.16 Fabrication process. (a) Aluminum masking layers are prepared on both sides of a 50-2-500 µm SOI wafer. The front side aluminum has been patterned into the shape of a comb drive with a tunneling tip and anchoring pads. The backside aluminum is used to define isolation trenches around the chip and for making a through-hole for TEM observation. (b) The front side silicon layer is patterned by DRIE to define the actuator. At the same time, a sharp tip is automatically formed, thanks to the slanted sidewalls meeting at the tip, where the lower part of silicon has been removed to leave a thin silicon bridge on the top (Toshiyoshi et al. 1999a, b; Milanovic et al. 2000). (c) A through-hole and isolation grooves are formed by the second DRIE from the backside. (d) Finally, the structure is sacrificial released by removing the buried oxide in HF. During this step, each piece of chip is released from the silicon wafer. After drying, the chips are ready to be metallized by vacuum evaporation of titanium, platinum, and gold.

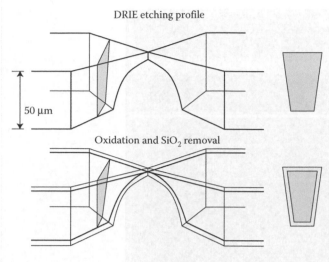

DRIE etching profile

50 µm

Oxidation and SiO₂ removal

FIGURE 16.17 Tip fabrication process.

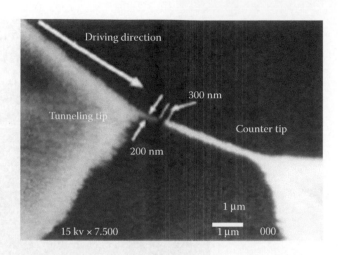

Driving direction

300 nm

Tunneling tip

Counter tip

200 nm

1 µm

15 kv × 7.500   1 µm   000

FIGURE 16.19 Close-up SEM view of the tunneling tips.

The detailed view of the tunneling tip fabrication process is shown in Figure 16.17. The profile of Deep RIE is not completely vertical. Therefore, the slanted sidewalls make a bridge at the neck of the pattern to make a pair of tunnel tips.

Figure 16.18 shows an SEM view of a fabricated tunneling device. The chip size is 2.5 × 2.5 mm min. area and 0.5 mm in thickness. The suspension width is 8 µm, and its length is 800 µm. The thickness of the structure is 50 µm. Figure 16.19 shows a close-up view of the tunneling tip on the left-hand side, which is driven by the actuator toward the counter tip seen on the right-hand side. The radius of the tip has been measured to be 200 nm, and the length to be 8 µm.

The result of controlling the tunneling gap is shown in Figure 16.20. The controller adjusted the

actuation voltage to obtain typical tunneling current of 0.7 nA under a bias voltage of 0.5 V. In this experiment, our device controlled the tunneling gap at a constant level by monitoring the tunneling current. The fluctuation of the control signal was 20 mV. They estimate the fluctuation of displacement from this voltage and mechanical stiffness of the moving part. The fluctuation of displacement is estimated to be only 0.2 nm.

We measured the tunneling current and controlled the tunneling gap in the TEM by using a sample with counter tip. Figure 16.21 shows a TEM view of the tunneling gap under control. The tunneling current flows between a small tip on the upper electrode and the lower electrode. The current actuator does not control the tip height but lateral displacement only. Therefore, one of the tips is seen over or under the other, which would not enable us to directly visualize the tunneling spot.

16.3.3 Nanogap Made by Sacrificial Etching

A nanogap over large electrodes is useful for a capacitive motion sensor when one of the electrodes displaces in a sub-nm range. A typical case is the vibration detection of a high-frequency MEMS resonator. Several electromechanical resonators based on a capacitive transduction scheme have already been demonstrated. Up to now, the maximum frequency attained with such transduction mechanism is about 1.51 GHz (Wang et al. 2004); this corresponds to the second mode frequency of the resonator, where the transmission level at resonance is about −86 dBm in vacuum.

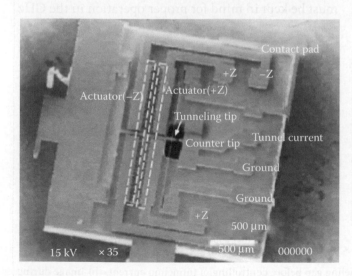

Contact pad

+Z   −Z

Actuator(−Z)   Actuator(+Z)

Tunneling tip

Counter tip   Tunnel current

Ground

Ground

+Z

500 µm

15 kV   × 35   500 µm   000000

FIGURE 16.18 SEM view of the tunneling unit chip.

Chapter 16

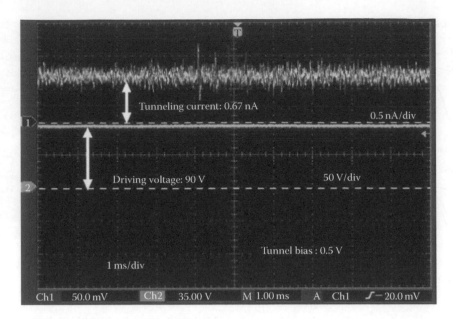

FIGURE 16.20   Tunnel current controlling result.

Agache et al. (2005) proposed a nanoelectromechanical resonator based on blade geometry, depending on capacitive transduction for both excitation and detection schemes. The theoretical resonance frequency modes for this geometry, in the case of lateral flexion solicitation, depend only on the structure sharpness. This feature allows self-alignment of the resonator with its lateral electrodes and is able to reach radio frequency band ranges (>1 GHz).

As shown in Figure 16.22, the device architecture comprises a blade geometry-based mechanical resonator, and is made of silicon. Coupling to the resonator is done capacitively using two lateral transducers. To bias and excite the device, a dc bias voltage (V_{dc}) is applied to the blade resonator, while an ac excitation voltage V_{in}

with the frequency fin is applied to one of the lateral electrodes. This combination leads to a laterally acting force inducing lateral vibrations of the resonator at the frequency fin. The oscillation amplitude is maximum, at constant V_{dc} and V_{in} excitation voltage, as soon as the frequency fin corresponds to one resonance mode of the resonator. Then, the mechanical motion of the structure modulates the output transducer capacitance in the time domain. A motional current is then detected at the output electrode and can be further processed by subsequent transceiver electronics. In summary, the input electrical signal is restituted in the output port, and filtered by the mechanical response of the resonator.

Concerning the device design, two major key points must be kept in mind for proper operation in the GHz

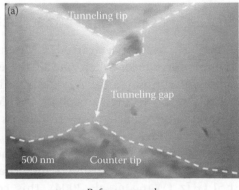

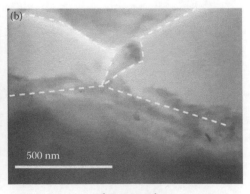

FIGURE 16.21   Tunnel current controlling in the TEM. (a) Tunneling gap before controlling of tunneling current. (b) Image during controlling tunnel current of the tunneling gap.

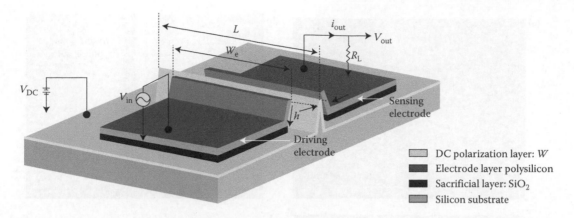

FIGURE 16.22 Perspective view of the two-port capacitively transduced silicon blade resonator and details depicting the bias and excitation scheme. The typical length is comprised between 10 mm, L, 30 mm.

frequency range. First, the resonator material must exhibit large acoustic velocities. This is the case for single-crystal silicon, which also exhibits good advantageous properties in terms of structural quality (minimization of the intrinsic loss dissipation mechanisms). Second, extremely small (sub-100 nm) electrode gaps $d0$ are needed in order to achieve low motional resistances and low insertion losses.

The fabrication process shown in Figure 16.23 requires only two lithography steps. A thermal oxidation step is performed in order to sharpen the apex of a silicon trapezoidal wall (Figure 16.24a). After the removal of the thermal SiO_2, a thin tungsten metal layer is sputtered over the structure in order to avoid the free carrier depletion phenomenon, which

would limit the transducer coupling efficiency. Then a 30-nm-thick SiO_2 layer is deposited, followed by a 300-nm-thick *in situ*-doped polysilicon layer providing the material for lateral transducers. The polysilicon layer covering the blade apex is removed by RIE after planarization based on a thick photoresist layer deposition. Then, the blade apex is released by 45-min wet etching of SiO_2 layer (Figure 16.24b). Finally, photolithography and RIE etching are used to pattern lateral electrodes (Figure 16.24c). Blade resonators, the geometry of which predicts a first resonance frequency centered at 1.05 GHz, and the second mode at 3 GHz (from the Euler–Bernoulli theory) have been fabricated, with a 20 nm lateral gap $d0$ as shown in (Figure 16.24c).

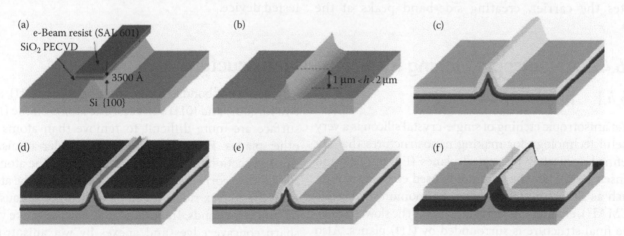

FIGURE 16.23 Fabrication process flow blade resonator self-aligned to its lateral electrodes. (a) RIE formation of silicon trapezoidal line using SiO_2 mask patterned after e-beam lithography. (b) Thermal oxidation (dry 1100°C) + SiO_2 wet etching (BE 7:1) => blade structure sharpending. (c) W sputtering deposition (35 nm) + SiO_2 PECVD deposition (30 nm) + LPCVD deposition of *in situ* doped polysilicon (300 nm). (d) Planarization by spinning and partial O_2 plasma etching of thick photoresist AZ4562 + RIE etching of polysilicon ($SF_6/N_2/O_2$) => lateral electrodes separation. (e) Resist removal + time-controlled wet etching (BE 7:1) of PECVD silicon oxide. (f) Photolithography + RIE ($SF_6/CHF_3/O_2$) etching of polysilicon (electrodes patterning).

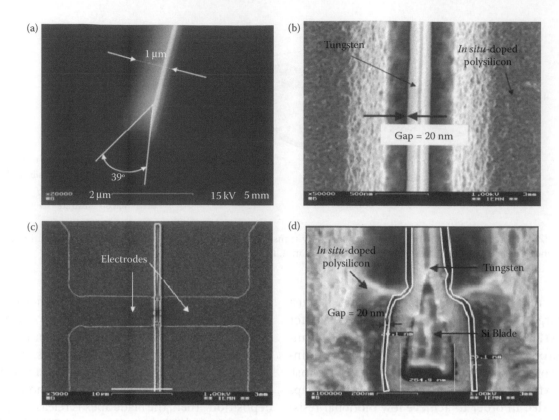

FIGURE 16.24 SEM micrographs at different stages of the process. (a) Silicon blade after dry thermal oxidation assisted shapering step. (b) Top view of Tungsten covered silicon blade after partial wet etching of SiO_2. (c) Top view of the blade resonator and its lateral electrodes. (d) View of the extremity of the blade resonator after electrodes patterning.

A carrier signal nc with a frequency $f_c = 50$ MHz is applied to the resonator, added to the V_{dc}, while the driving excitation voltage is still applied to the input electrode. The resonator displacement modulates the carrier, creating side-band peaks at the frequencies $(f_{in} + f_c)$ and $(f_{in} - f_c)$, the amplitude of which is maximum when the driving signal frequency is around 1.104 GHz (Figure 16.25), namely the fundamental resonance frequency of the tested device.

16.4 Anisotropic Etching for Silicon Nanostructure

16.4.1 Nanowires

Wet anisotropic etching of single-crystal silicon is a very useful technology for making nanostructures that are defined accurately by crystal planes (Elwenspoek and Jansen 1998). For the commonly used etching solution such as KOH and tetra-methyl ammonium hydroxide (TMAH), etching rate of (111) planes is the slowest; thus, the final structure is surrounded by (111) planes. Also SiO_2 and nitride films serve as the etching stop layer.

The slow etching rate for (111) planes can be understood from the number of atomic bonds. If a silicon atom is in the bulk crystal, it has four bonds with adjacent atoms. If the atom is exposed on the surface, it has less number of bonds, that is, three bonds for (111), and two for (001) and (011) surfaces. The atoms on the (111) surface are more difficult to remove than atoms on other planes. Furthermore, let us consider atoms on the intersection line between (111) planes. The atom at the concave corner is stable against etching. If the atom is on the convex ridge, it can be etched fast because it has only two bonds. In other words, you can make very sharp concave edges and apexes by wet anisotropic etching of silicon but no convex edges and apexes. This is inconvenient. A new process to solve the problem is explained in the following text.

Recently, a new type of silicon substrate, the so-called SOI substrate, becomes popular for both

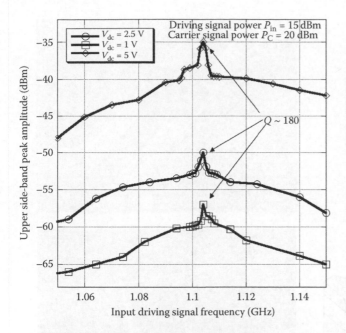

FIGURE 16.25 Amplitude (dBm) of the upper side-band peak in reflection versus the V_{dc} bias voltage and the driving signal frequency f_{in}, given a 50-MHz carrier signal frequency f_C, a 20-dBm carrier signal power P_C, and a 15-dBm driving signal power P_{in}.

microelectronics and MEMS. The SOI substrate is composed of three layers: a thin single-crystal silicon layer (active layer), a SiO_2 dielectric layer (buried oxide layer), and a thick single crystal silicon layer (handling layer). The new process depends on the SOI substrate. Suppose you have an etching mask, say a silicon nitride layer, on the active layer having (001) surface. The mask opening is aligned with [100] direction as shown in

Figure 16.26a. The wet anisotropic etching of the exposed area makes a slanted (111) plane along the [100] mask opening. The intersection between the (111) plane and the buried oxide layer remains stable because both stable planes meet at the concave corner as shown in Figure 16.26b. If you use HF etching, you can have a sharp convex edge of silicon by selectively etching the SiO_2 layer.

If you continue the process, you can have a triangular silicon wire (Hashiguchi and Mimura 1994). After the process step shown in Figure 16.26b, the wafer is put in the thermal oxidation furnace. The oxidation occurs only on the exposed (111) plane, because the silicon nitride layer does not allow the penetration of oxygen; this is called the local oxidation of silicon (LOCOS) process. We have a SiO_2 layer on the slanted (111) plane as shown in Figure 16.26c. Then the silicon nitride mask is selectively removed, for example, by wet etching using hot H_3PO_4:HNO_3 solution. Second, wet anisotropic etching makes the triangular wire defined by two (111) planes and the bottom (100) plane protected by the buried oxide layer as shown in Figure 16.26d. Finally, the silicon wire is obtained by selective removal of SiO_2 with HF.

The features of the process are as follows: (1) all edges of the wire are sharp; (2) the dimension of the wire is defined by the initial thickness of the active layer. If the thickness is in 10s of nanometers, the dimensions of the wire are also in 10s of nanometers; this means you can use conventional photolithography to make nanowires; (3) the wire has very flat surfaces

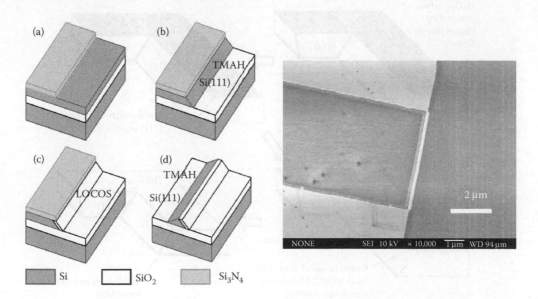

FIGURE 16.26 Fabrication process and results of a silicon nanowire on SOI wafer by KOH/TMAH etching combined with LOCOS. Please note that the wire dimension depends only on the thickness of the upper silicon layer.

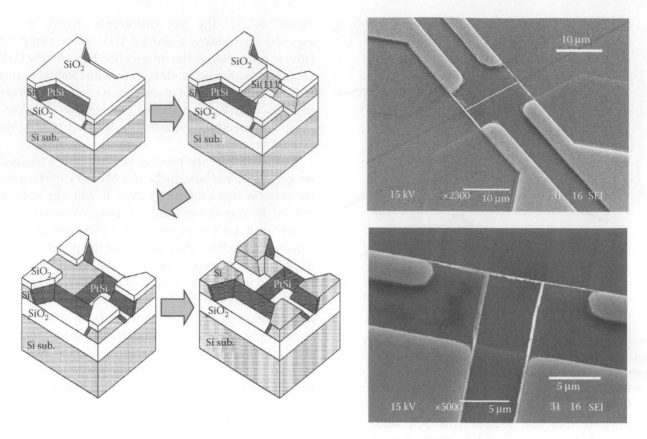

FIGURE 16.27 Fabrication process and results of H-shape and Π-shape wires.

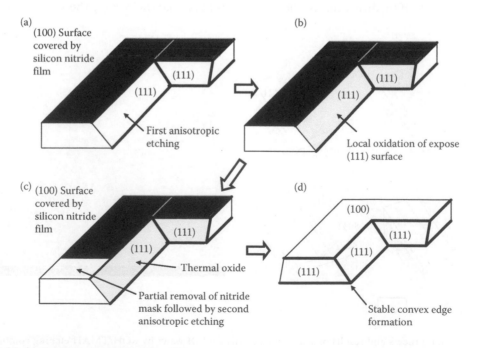

FIGURE 16.28 Stable fabrication process of convex edge.

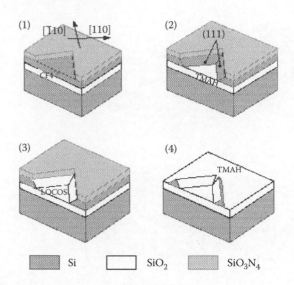

FIGURE 16.29 Fabrication process for twin probes. (1) Patterning Si_3N_4 by RIE. (2) First anisotropic TMAH etching. (3) LOCOS. (4) Second anisotropic TMAH etching.

composed of crystal facets; (4) all critical etching steps are terminated with etching stop layers; the process is very stable and uniform over the wafer area; and (5) more complicated networks of nanowires are available. Fabrication process and resulting structures are shown in Figure 16.27 (Kakushima et al. 2001).

The two-step anisotropic etching process combined with LOCOS sidewall protection can be applied to make a convex ridge as shown in Figure 16.28. Furthermore, the silicon nanowire process was also extended to realize nanoprobes that meet each other at an angle of 90°. The length and width of the probe are around 100 nm, while the separation between the tips is 100–150 nm, which can be further reduced by

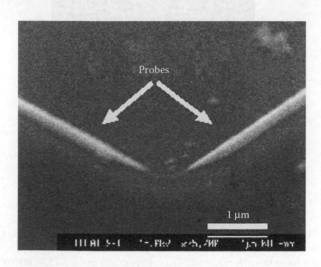

FIGURE 16.30 SEM view of twin probes.

microactuators. The fabrication process is shown in Figure 16.29. The major change with respect to the nanowire process is the shape of the mask pattern used to etch Si_3N_4 in the first process step (Figure 16.29). The second TMAH etching (Figure 16.29d) is critical for precise adjustment of the tip spacing. The result is shown in Figure 16.30.

16.4.2 Tips with Nanoradius

Sharp tips with around 10 nm radius are useful for the probes of atomic force microscopy (AFM), field-enhanced electron emitters, and biological tools for piercing in cells or soft tissues. Fabrication processes for such sharp tips include self-terminated deposition through small holes for metallic field emitters (Spindt et al. 1976), isotropic etching of silicon (Mita et al. 2005), deep reactive etching scalloped sidewalls (Miranovic et al. 2001), replication of KOH-etched pits (Hashiguchi et al. 1996), and carbon nanotubes grown or attached at the tip apex (Guillorn et al. 2003). However, there are limitations and shortcomings in those processes. For example, the self-terminated deposition is only applicable to make arrays of field emitters. The isotropic etching is easy but difficult to control the uniformity over the substrate surface. As the replicated tips are embedded in the substrate, the substrate should be completely removed to expose the tips. Batch fabrication of carbon nanotube tips in a large number is extremely difficult. In order to solve the problem, we have developed a well-controlled process to make sharp tips.

Opposing nanotips of 10–30 nm tip radius were fabricated by combining silicon anisotropic etching and LOCOS. A starting SOI substrate has a 2 μm single crystalline (001) silicon layer and a 1 μm buried oxide layer on a 500 μm handling wafer. The thin silicon layer was doped with phosphorus to have good conductivity. The fabrication steps of opposing nanotips are shown in Figure 16.31.

1. Low-pressure chemical vapor deposition (LPCVD) deposition of a silicon nitride film and patterning by RIE. The pattern edge is aligned with [100] direction. This nitride film serves as a mask for LOCOS and anisotropic wet etching.
2. Isotropic etching of the top silicon layer by RIE.
3. The exposed silicon surface at the edge is selectively oxidized (LOCOS process). The oxide film serves as a mask for anisotropic wet etching.
4. Second patterning of the silicon nitride film along the [010] direction.

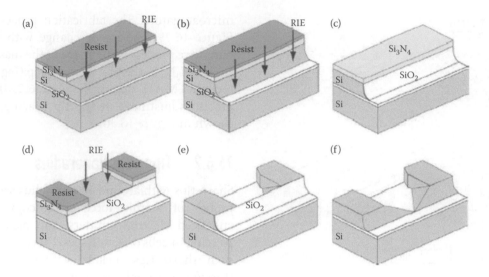

FIGURE 16.31 Fabrication steps of the opposing nanotips. (a) Patterning of silicon nitride. (b) Isotropic etching of silicon. (c) Local oxidation of silicon. (d) Patterning of silicon nitride. (e) Anisotropic etching of silicon. (f) Removing silicon dioxide.

5. TMAH anisotropic wet etching of the unprotected silicon part. (111) crystallographic planes appear at the side of tips.
6. Removal of the oxide film to release tips.

Please note that the tip apex is defined by a (111) etching-stop plane and two silicon surfaces protected by LOCOS and buried oxide layers, respectively. Therefore, very sharp tip defined at the intersecting point of three planes can be made uniformly over the substrate. Another advantage of the process is that it does not require any nanolithography to obtain nanoscopic tips.

After making nanotips, we opened a through-hole from the backside of the wafer for allowing TEM observation of tips (Nozawa et al. 2007). Scanning and transmission electron micrographs of the device are shown in Figure 16.32. Two tips with electrical connections overhang from the edge of the through-hole. The spacing between tips was 200 nm. The tip radius is typically 5–10 nm after tip sharpening process (Ravi et al. 1991); the process consists of thermal oxidation of silicon at 1223 K for 1 h followed by quick oxide layer removal with HF.

The device was characterized as the field emitter while tips were observed by TEM (Nozawa et al. 2007). Initial tip radii of both anode and cathode were ~5 nm. The gap spacing was 500 nm. The measured *I–V* characteristic of the device is shown in Figure 16.33 together with TEM images at different applied voltages. The field emission current started flowing at 86 V. It increased rapidly up to 12 μA at

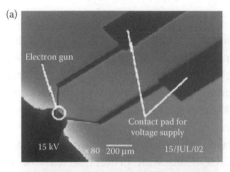

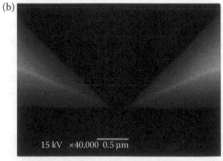

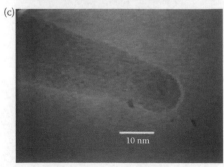

FIGURE 16.32 Electron micrographs of opposing nanotips. (a) SEM micrographs of the device. (b) SEM micrographs of opposing nano tips. The gap spacing is approx. 200 nm. (c) TEM micrographs of nano tip of the approximately 5 nm in radius.

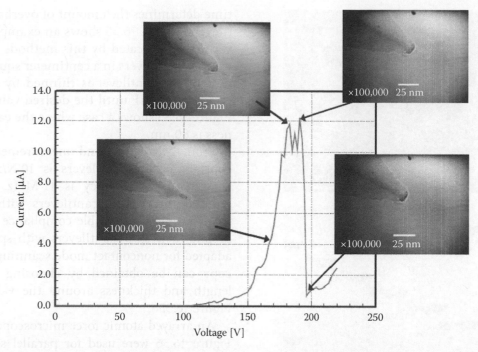

FIGURE 16.33   Simultaneous observation of the cathode tip with $I–V$ measurement of a nano-field-emitter.

180 V and saturated. It decreased suddenly at 194 V down to 1 μA. The current increased but remained low between 195 and 220 V.

The tip shape of the anode remained very sharp after the experiment. On the contrary, the tip of the cathode changed distinctively. Our TEM observation revealed that the cathode tip was still sharp up to 180 V while the current increased rapidly. At around 180 V, the tip started to lose its sharpness. The end of the tip became less dark; this suggests that some part of the tip disappeared. While the current saturated and fluctuated, the tip shape remained the same. The thin portion of the tip disappeared suddenly at 194 V; this was associated with steep current decrease. The current stayed at a low level because the tip was rounded.

This process can be extended for making sharp tips directing out-of-plane on the substrate. Even large numbers of sharp tips attached on microcantilevers can be realized (Kawakatsu et al. 2002). Figure 16.34 depicts the fabrication process. The process resembles the method introduced above, but the etching of the SOI layer is stopped ~100 nm before the etching surface reaches the silicon dioxide layer. The triangular cantilever shape is also formed by anisotropic etching of silicon by KOH. This enables exact positioning of each tip on the edge of the triangular cantilever. The silicon dioxide layer is etched by buffered hydro fluoric acid (BHF). The etching

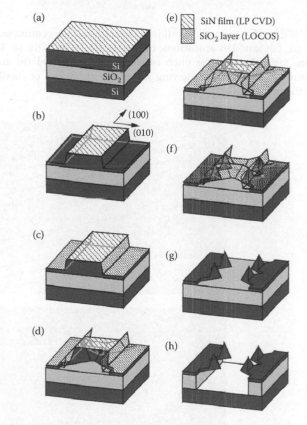

FIGURE 16.34   Process for fabricating millions of cantilevers on a centimeter square chip. (a) Deposition of SiN on SOI. (b) First anisotropic etching by KOH. (c) Local oxidation of Si (LOCOS). (d) Second anisotropic etching by KOH. (e) Local oxidation of Si (LOCOS). (f) Third anisotropic etching by KOH. (g) Removal of LOCOS SiO$_2$ film. (h) Etching of SiO$_2$ layer.

Chapter 16

(a)

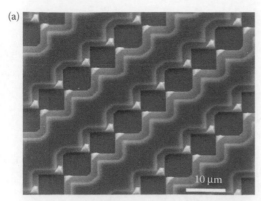

(b)

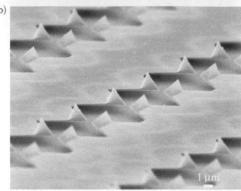

FIGURE 16.35 Example of millions of cantilevers per centimeter square, fabricated by anisotropic etching of silicon by KOH. (a) A plane view showing the pitch between probe tips, and (b) an expanded oblique view showing sharp tips at the end of small cantilevers.

time determines the amount of overhang of the cantilevers. Figure 16.35 shows an example of a cantilever array fabricated by this method. There are ~1.5 million cantilevers in a centimeter square. The thickness of the cantilever is thinned by oxidation and removal by BHF until the desired value is obtained. Figure 16.35 shows a case where the cantilever thickness is 80 nm.

From calculation and measurement, the spring constant of the cantilevers is 10 N/m. The calculated natural frequency is 11 MHz. Due to their reduced thicknesses, cantilevers with high natural frequency and reasonable compliance were obtained despite their size. Cantilevers with spring constants adapted for noncontact mode scanning force microscopy can be obtained by choosing the cantilever length and thickness around the values shown in Figure 16.35b.

An arrayed atomic force microscope cantilevers in Figure 16.36 were used for parallel scanning probe lithography (Kakushima et al. 2004). SAM films of *n*-octadecyltrimethoxysilane were attached on the silicon surface. Tips were scanned over the film while 14 V bias voltage was applied. Lines of a few 10s of nanometers were successfully drawn in parallel by anodic oxidation of the film.

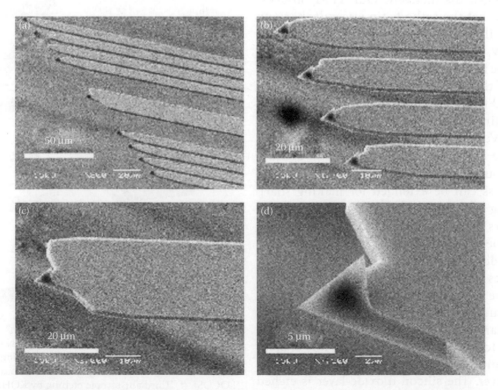

FIGURE 16.36 (a) An SEM image of a fabricated AFM probe array. (b) Four probes for lithography. (c–d) Reference probe.

16.5 Needle Array

16.5.1 Microcapillary Array

We developed an industrially compatible bulk micromachining process for microcapillaries based on the idea of sidewall protection by LOCOS (Hashiguchi and Mimura 1994; Chun et al. 1999; Ohigashi et al. 2001). The key steps are (1) making an array of holes through a silicon wafer covered with a Si_3N_4 layer by deep RIE etching, (2) local oxidation of the inner wall of etched holes for protecting the surface in the following etching (LOCOS), and (3) formation of capillaries by etching Si_3N_4 and Si with selective RIE as shown in Figure 16.37. This process is highly compatible with industrial production, because it is insensitive to etching speed distribution. The process flow is as follows:

1. Deposition of 0.1 µm silicon nitride layer for LOCOS on the wafer by LPCVD.
2. Sputtering of 0.2 µm aluminum mask layer against deep etching on front and bottom sides of the wafer.

3. Photolithography and patterning of the aluminum layer at the top side. Removal of photoresist (Figure 16.37a).
4. Etching through the wafer by inductively coupled plasma (ICP)-RIE Bosch's process.
5. Removal of aluminum (Figure 16.37b).
6. LOCOS. Only the inner walls of through holes are oxidized because both surfaces are protected by Si_3N_4 layers (Figure 16.37c).
7. Formation of SiO_2 capillaries by selectively etching Si_3N_4 and Si using ICP-RIE (Figure 16.37d).
8. Deposition of 4–5-µm-thick SiO_2 and 70-nm-thick fluorocarbon films on the wall of capillaries by LPCVD and plasma-enhanced chemical vapor deposition (PECVD), respectively (Figure 16.37e).

Figures 16.38 and 16.39 show SEM images of the microcapillary array right after the second etching (step 7 of Figure 16.37), and after thick-SiO_2 deposition (step 8 of Figure 16.37), respectively. A 5-mm-long linear array of capillaries was fabricated at the center of a chip of 17×17 mm. We tested several configurations of different dimensions; those range from 10 to 100 µm in diameter, 20 to 500 µm for spacing, and 70 to 140 µm for the height of capillaries before deposition of a thick SiO_2-layer. Height of capillaries can be controlled by the etching time in the 7th step. The capillary wall was SiO_2 film of ~500 nm in thickness. To make the capillaries stronger, we reinforced them by depositing ~4-µm-thick SiO_2 film by LPCVD in the eighth step. The difference in capillary thickness is clearly seen between Figure 16.38 (before deposition) and Figure 16.39 (after

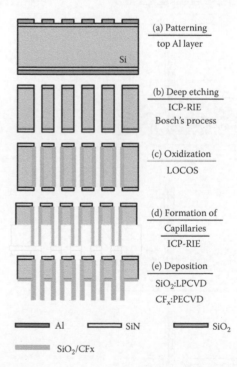

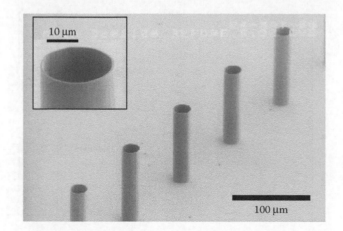

FIGURE 16.37 Fabrication process of a microcapillary array. (a) Depositing 0.1 µm SiN and 0.2 µm Al on Si surface by LPCVD and sputtering, respectively. The Al layer is a mask for deep etching. (b) Etching by ICP-RIE. (c) Oxidation of inner walls of etched holes. (d) Formation of capillaries by selective Si etching. (e) Depositing SiO_2 and CF_x films by LPCVD and PECVD, respectively. CF_x film acts as a hydrophobic layer.

FIGURE 16.38 SEM image of a microcapillary array before SiO_2 deposition. Left-upper photo is a magnification of the top part of a capillary. The diameter of capillaries was 23 µm, height 100 µm, and spacing 100 µm.

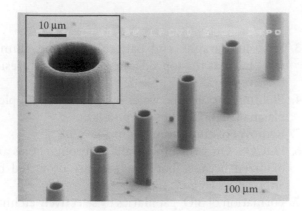

FIGURE 16.39 SEM image of a microcapillary array after SiO$_2$ deposition by LPCVD. Diameter of the inside was 20 μm and that of the outside was 30 μm. Left-upper photo is a magnification of the top part of a capillary. Thickness of the capillary wall was ~5 μm. The wall becomes 10 times as thick as that before SiO$_2$ deposition.

deposition). The breaking strengths measured by a shear tester (Dage PC2400-T/TX) were 0.16 g/capillary before SiO$_2$ deposition and 0.68 g/capillary after deposition. Hence, reinforced capillaries were nearly four times as strong as those before thick-SiO$_2$ deposition.

16.5.2 Nanoneedle Growth Process

Self-organized growth of nanoneedles is a convenient way to produce arrays of nanoneedles in large quantity. Among many processes for such nanoneedle growth, of particular interest is the vapor–liquid–solid (VLS) process (Wagner and Ellis 1964). Recently, the VLS growth technique has been extended to nanofabrication by utilizing nm-sized Au particles. Nanoneedles of a variety of semiconductor materials including GaAs (Hiruma et al. 1995; Plante and Lapierre 2008), ZnO

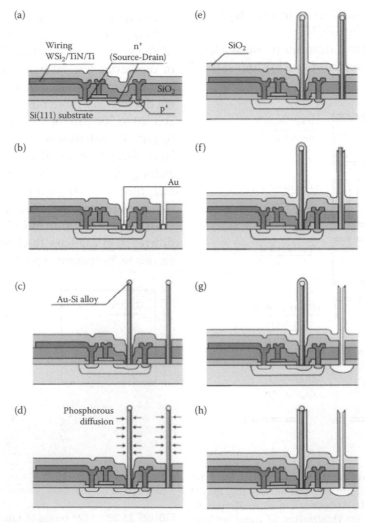

FIGURE 16.40 Process sequence of probe, tube with on-chip MOSFETs. (a) nMOSPET process. (b) Au evaporation for probe and tube region. (c) VLS growth. (d) Thermal phosphorous diffusion at 900°C. (e) SiO$_2$ deposition. (f) Etching SiO2, Au-Si alloy. (g) Etching Si by XeF$_2$ gas. (h) SiO$_2$ etching for probe region.

(Zhao et al. 2003) and silicon nanoneedles (Cui et al. 2001) have been grown. In the case of the silicon nanoneedle, it grows from the Au particles on the (111) silicon surface. The particle melts and forms an Au–Si eutectic droplet at elevated temperature. The gas mixture of $SiCl_4$ and H_2 is introduced to the droplet. It absorbs the silicon atoms over saturation concentration because of its large surface-to-volume ratio. Then, excess silicon atoms solidify where the droplet contacts with the substrate and a silicon nanoneedle grows. The needle diameter is as small as the droplet size.

Ishida et al. grew silicon nanoneedles on the top of an MOS transistor for the probe to detect electrical signals associated with neural activity (Takei et al. 2008). They also made SiO_2 capillaries by oxidizing the sidewall of the nanoneedle and removing the inner silicon by XeF_2 gas etching; this is the similar process as the one described in Section 16.5.1. After the MOS transistor process (Figure 16.40a), Au with a diameter of 4 µm was evaporated selectively on the drain region and the tube fabrication region (Figure 16.40b). Then VLS growth was carried out by a gas-source molecular beam epitaxy shown in Figure 16.40c. Phosphorus was diffused into the probes by a thermal furnace at 900°C to make the resistance of probe lower (Figure 16.40d). For metal wiring of the circuits, WSi2/TiN/Ti materials were used to withstand for the high-temperature process.

SiO_2 by plasma-enhanced CVD (PECVD) was deposited over the probe as an insulator for probe electrodes and the wall of tubes after VLS growth (Figure 16.40e). Furthermore, thick photoresist (PMER P-LA900PM) was applied by spin coating. The thickness was about 40 µm. Then photolithography process was carried out to expose the tips of probes, which would be used for the fabricated tubes. Then SiO_2 and Au–Si alloy from the tips of probes were etched out by buffered HF and aqua regia, respectively (Figure 16.40f). In addition, Si was removed by XeF_2 gas (Figure 16.40g). Hence, tube was formed completely. Then SiO_2 were etched by buffered HF to expose the tips of remaining probes, which would be used for penetrating electrodes (Figure 16.40h). Finally, hydrogen annealing was applied for terminating dangling bonds and decreasing interface states for metal–oxide–semiconductor field-effect transistors (MOSFETs).

Figure 16.41a shows the photograph of chip overview after whole processes. The chip was integrated with microtubes, probe electrode on the drain region of MOSFETs, and array selector. It has 5 × 5 array-alternated Si micro probes and SiO_2 microtubes. SEM images of the detailed structures are shown in Figure 16.41b–d.

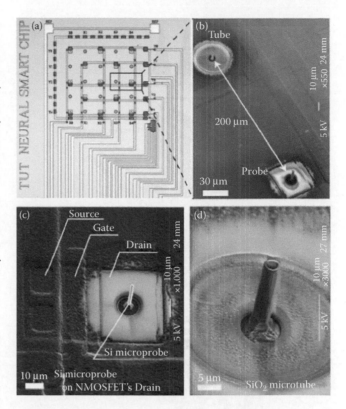

FIGURE 16.41 Fabricated Si microprobe and SiO_2 microtube array with on-chip MOSFETs; (a) a photograph shows the chip overview, (b) an SEM image of Si probe, SiO_2, and MOSFET, (c) an SEM view of the Si probe electrodes grown on the drain region, (d) an SEM image of the SiO_2 tube.

In this experiment, diameter and length for Si microprobe are 2 and 30 µm, respectively. Also SiO_2 tube is 2.7 µm in inner diameter, 3.6 µm in outer diameter, and 29 µm in length.

16.5.3 Lithography Process for a Nanoneedle

NEMS devices are composed of three-dimensional nano- and microstructures. Those structures can be fabricated with lithography process as well as etching and growth processes. Resist layers as thick as 10s or 100s of micrometers are patterned by ultraviolet (UV) light or x-ray illumination followed by the development step. SU-8 is the major thick resist for UV exposure. Polymethylmethacrylate (PMMA) is commonly used for x-ray lithography. Typical structures made by lithography have vertical sidewalls against the substrate surface, because the mask patterns are transferred by the parallel light/x-ray beam. This results in the structure that represents the translational image of the mask pattern. Therefore, such structures are sometimes called 2.5-dimension structures.

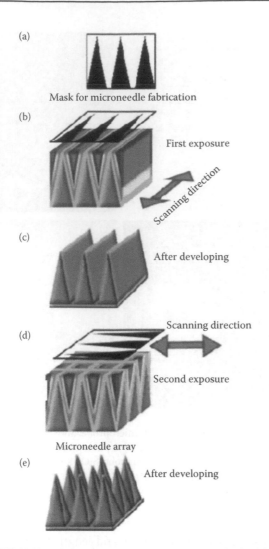

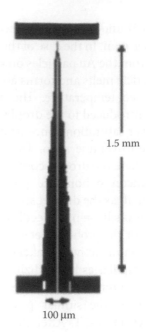

FIGURE 16.44 Shape and dimensions of an actual x-ray mask for fabrication of a quadruplet microneedle array.

FIGURE 16.42 Plain-pattern to cross-section transfer (PCT) steps for shaping a microneedle array (a) Mask for microneedle, (b) first exposure by x-ray with linear scanning, (c) first development, (d) second exposure with linear scanning in orthogonal direction to the first exposure, and (e) final development to gain microneedle array.

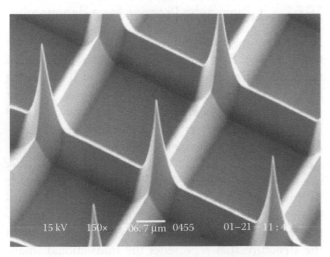

FIGURE 16.43 A three-dimensional structure fabricated by the PCT technique: an array of single-tip microneedles.

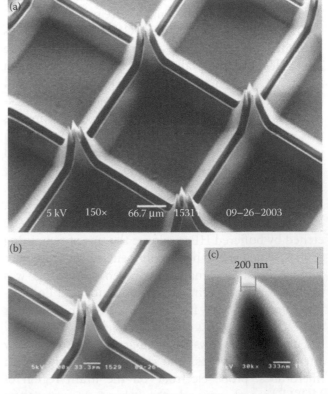

FIGURE 16.45 (a) An SEM photo of quadruplet microneedles, (b) close-up image of a quadruplet microneedle, and (c) a photo of the nanoscale tip.

Making truly three-dimensional structures require sloped sidewalls and curved surfaces. Extension of UV or x-ray lithography to make such structures has been proposed, for example, moving mask lithography (Tabata et al. 1999) and multi-angel exposure (Ehrfeld et al. 1999). Sugiyama et al. applied moving x-ray lithography twice on a resist layer and obtained a sharp needle array (Sugiyama et al. 2004). Figure 16.42 represents the principle of the process.

1. The mask has triangle patterns.
2. The substrate with a PMMA resist layer is translated in the direction pointed by the triangles while the x-ray from a synchrotron radiation source is illuminated through the mask. The PMMA resist is positive; this means that the resist material is removed proportionally to the x-ray dosage in the exposure step. Because of the translational motion, the triangular mask causes different average exposure time.
3. The resist is patterned into triangular ridges by development.

4. The same moving lithography as the step (b) is applied in the vertical direction to the first exposure.
5. An array of sharp needles, defined by the three-dimensional intersection of orthogonal ridges, appears after development.

Figure 16.43 is the scanning electron micrograph of the needle array. The height of the needle is of the order of 100 μm with the tip sharpness of 870 nm. One of the applications of the array is blood extraction through the skin. In order to fabricate reservoir for blood, they made quadruplet needles. The isosceles triangles in the original mask were separated by a thin line (Figure 16.44). They successfully fabricated an array of quadruplet needles as shown in Figure 16.45. The tip was as sharp as 200 nm. The cost of x-ray lithography is very expensive. Therefore, the resist pattern should be replicated to reduce the cost of final devices. They propose to use electroplating for obtaining a negative replication, which serves as the mold in plastic hot embossing to make many polymer needle arrays with cheap cost.

References

Agache, V., Legrand, B., Collard, D. et al. 2005. Fabrication and characterization of 1.1 GHz blade nanoelectromechanical resonator. *Appl. Phys. Lett.* **86**: 213104-1-3.

Blauw, M. A., Zijlstra, T., and van der Drift, E. 2001. Balancing the etching and passivation in time-multiplexed deep dry etching of silicon. *J. Vac. Sci. Technol. B* **19**: 2930–2934.

Chun, K., Hashiguchi, G., Toshiyoshi, H. et al. 1999. Fabrication of array of hollow microcapillaries used for injection of genetic materials into animal/plant cells. *Jpn. J. Appl. Phys., Part 2 (Lett.)* **38**: L279–L281.

Cui, Y., Lauhom, L. J., Gudiksen, M. S. et al. 2001. Diameter controlled synthesis of single-crystal silicon nanowires. *Appl. Phys. Lett.* **78**: 2214–2216.

Ehrfeld, W., Hessel, V., Lowe, H., Schulz, Ch., and Weber, L. 1999. Materials of LIGA technology. *Microsyst. Technol.* **5**: 105–112.

Elwenspoek, M. and Jansen, H. V. 1998. *Silicon Micromachining.* Cambridge: Cambridge University Press.

Fujita, H. 2003. *Micromachines as Tools for Nanotechnology.* Berlin: Springer.

Gel, M., Ishida, T., Akasaka, T. et al. 2007. Mechanically controlled quantum contact with on-chip MEMS actuator. *J. Microelectromech. Systems* **16**: 1–6.

Guillorn, M. A., Hale, M. D., Merkulov, V. I. et al. 2003. Integrally gated carbon nanotube field emission cathodes produced by standard microfabrication techniques. *J. Vac. Sci. Technol. B* **21** 957–959.

Hashiguchi, G. and Mimura, H. 1994. Fabrication of silicon quantum wire using separation by implanted oxygen wafer. *Jpn. J. Appl. Phys., Part 2 (Lett.)* **33**: 1649–1650.

Hashiguchi, G., Mimura, H., and Fujita, H. 1996. Monolithic fabrication and electrical characteristics of polycrystalline silicon field emitters and thin film transistor. *Jpn. J. Appl. Phys., Part 2 (Lett.)* **35**: L84–L86.

Hirose, K., Mita, Y., Imai, Y. et al. 2008. Polarization transmissive photovoltaic film device consisting of Si photodiode wire-grid. *J. Opt. A: Pure Appl. Opt.* **10**: 044014-1–9.

Hiruma, K., Yazawa, M., and Katsuyama, T. et al. 1995. Growth and optical properties of nanometer-scale GaAs and InAs whiskers. *J. Appl. Phys.* **77**: 447–462.

Imai, Y., Mita, Y., Hirose, K. et al. 2008. Surface corrugated P–N junction on deep submicron trenches for polarization detection with improved efficiency. *Asia-Pacific Conference of Transducers (APCOT)*, Taiwan 3B1–3.

Jansen, H., de Boer, M., Wiegerink, R. et al. 1997. RIE lag in high aspect ratio trench etching of silicon. *Microelectron. Eng.* **35**: 45–50.

Kakushima, K., Mita, M., Mita, Y. et al. 2001. Fabrication of various shapes nano structures using Si anisotropic etching and silicidation. *TRANSDUCERS '01. EUROSENSORS XV. 11th International Conference on Solid-State Sensors and Actuators*, Digest of Technical Papers **2**, Munich, pp. 1090–1091.

Kakushima, K., Watanabe, T., Shimamoto, K. et al. 2004. Arrayed AFM probes for parallel lithography. *IEEE Trans. Sens. Micromach.* **124**: 248–254.

Kawakatsu, H., Saya, D., Kato, A. et al. 2002. Millions of cantilevers for atomic force microscopy. *Rev. Sci. Instrum.* **73**: 1188–1192.

Kubena, R. L., Atkinson, G. M., Robinson, W. P. et al. 1996. A new miniaturized surface micromachined tunneling accelerometer. *J. Vac. Sci. Technol. B* **14**: 306–308.

Kenny, T. W., Kaiser, W. J., Rockstad, H. K. et al. 1994. Wide-bandwidth electromechanical actuators for tunneling displacement transducers. *J. Microelectromech. Systems* **3**: 97–104.

Chapter 16

Kenny, T. Y., Waltman, S. B., Reynolds, J. K. et al. 1991. Micromachined silicon tunnel sensor for motion detection. *Appl. Phys. Lett.* **58**: 100–102.

Lutwyche, M. I. and Wada, Y. 1995. Observation of a vacuum tunnel gap in a transmission electron microscope using a micromechanical tunneling microscope. *Appl. Phys. Lett.* **66**: 2807–2809.

Marty, F., Rousseau, L., Saadany, B. et al. 2005. Advanced etching of silicon based on deep reactive ion etching for silicon high aspect ratio micro structures and three-dimensional micro- and nano-structures. *Microelectron. J. Circuits Systems Sect.* **36**: 673–677.

Moreland, J. and Ekin, J. M. 1985. Electron tunneling experiments using Nb–Sn 'break' junctions. *J. Appl. Phys.* **58**: 3888–3895.

Mita, M., Kawara, H., Toshiyoshi, H. et al. 2005. Bulk micromachined tunneling tips integrated with positioning actuators. *J. Microelectromech. Systems* **14**: 23–28.

Mita, Y., Kubota, M., Harada, T. et al. 2006a. Contour lithography methods for DRIE fabrication of nanometre-millimetre-scale coexisting microsystems. *J. Micromech. Microeng.* **16**: 135–141.

Mita, Y., Kubota, M., Sugiyama, M. et al. 2006b. Aspect ratio dependent scalloping attenuation in DRIE and an application to low-loss fiber-optical switch. In *Proceedings of IEEE International Conference on Microelectromechanical Systems* (MEMS 2006), Istanbul, Turkey, pp. 114–117.

Miranovic, V., Doherty, L., Teasdale, D. et al. 2001. Micromachining technology for lateral field emission devices. *IEEE Trans. Electron Device* **48**: 166–173.

Nozawa, N., Kakushima, K., Hashiguchi, G. et al. 2007. *In situ* visualization of degradation of silicon field emitter tips. *IEEJ Trans. Electric. Electron. Eng.* **20**: 284–288.

Ohara, J., Takeuchi, Y., and Sato, K. 2008. Development of Si DRIE process allowing simultaneous etching from narrow and wide mask openings. *IEEJ Trans. Sens. Micromach.* **128-E**: 442–446.

Ohigashi, R., Tsuchiya, K., Mita, Y. et al. 2001. Micro capillaries array head for direct drawing of fine patterns. In *14th IEEE International Conference on Micro Electro Mechanical Systems*, Technical Digest, MEMS 2001 (Cat. No.01CH37090), Interlaken, pp. 389–392.

Park, H., Lim, A. K. L., Park, J. et al. 1999. Fabrication of metallic electrodes with nanometer separation by electromigration. *Appl. Phys. Lett.* **75**: 301–303.

Plante, M. C. and LaPierre, R. R. 2008. Au-assisted growth of GaAs nanowires by gas source molecular beam epitaxy: Tapering, sidewall faceting and crystal structure. *J. Cryst. Growth* **310**: 356–363.

Ravi, T. S., Marcus, R. B., and Liu, D. 1991. Oxidation sharpening of silicon tips. *J. Vac. Sci. Technol. B* **9**: 2733–2737.

Reed, M. A., Zhou C., Muller, C. J. et al. 1997. Conduction of a molecular junction. *Science* **278**: 252–254.

Saadany, B., Malak, M., Kubota, M. et al. 2006. Free-space tunable and drop optical filters using vertical Bragg mirros on silicon. *IEEE J. Sel. Top. Quantum Electron.* **12**: 1480–1488.

Spindt, C. A., Brodie, I., Humphrey, L. et al. 1976. Physical properties of thin-film field emission cathodes with molybdenum cones. *J. Appl. Phys.* **47**: 5248–5263.

Sugiyama, S., Khumpuang, S., and Kawaguchi, G. 2004. Plain-pattern to cross-section transfer (PCT) technique for deep x-ray lithography and applications Department of Robotics. *J. Micromech. Microeng.* **14**: 1399–1404.

Tabata, O., Agawa, T. K. et al. 1999. Moving mask LIGA (M2 LIGA) process for control of side wall inclination. *12th IEEE International Conference on Micro Electro Mechanical Systems*, (Cat. No. 99CH36291), Orland, FL, pp. 252–256.

Takei, K., Kawashima, T., Kawano, T. H. et al. 2008. Integration of out-of-plane silicon dioxide microtubes, silicon microprobes and on-chip NMOSFETs by selective vapor–liquid–solid growth. *J. Micromech. Microeng.* **18**: 035033-1-9.

Tezcan, D. S., Verbist, A., De Malsche, W. et al. 2007. Improved liquid phase chromatography separation using sub-micron micromachining technology. *IEEE IEDM* (Washington D.C.), 839–842.

Toshiyoshi, H., Goto, M., Mita, M. et al. 1999a. Fabrication of micromechanical tunneling probes and actuators on a silicon chip (in japanese). *Jpn. J. Appl. Phys., Part 1: Regul. Pap. Short Notes Rev. Pap.* **38B**: 7185–7189.

Toshiyoshi, H., Mita, M., Goto, M. et al. 1999b. Micromechanical tunneling probes and actuators on a silicon chip. In *Microprocesses and Nanotechnology '99 International Microprocesses and Nanotechnology Conference*, Digest of Papers, Yokohama, pp. 180–181.

Wagner, R. S. and Ellis W. C. 1964. A vapor–liquid–solid mechanism for growing 3C-SiC single-domain layers on 6H-SiC(0001). *Adv. Funct. Mater.* **16**: 975–979.

Wang, J., Butler, J. E., Feygelson, T. et al. 2004. 1.51-GHz nanocrystalline diamond micromechanical disk resonator with material-mismatched isolating support. In *17th IEEE International Conference on Micro Electro Mechanical Systems*, Maastricht MEMS 2004, Technical Digest (IEEE Cat. No.04CH37517), Maastricht, pp. 641–644.

Yao, J. J., Arney, S. C., and MacDonald, N.C. 1992. Fabrication of high frequency two-dimensional nanoactuators for scanned probe devices. *J. Microelectromech. Systems* **1**: 14–22.

Zhao, Q. X., Willander, M., Morjan, R. E. et al. 2003. Optical recombination of ZnO nanowires grown on sapphire and Si substrates. *Appl. Phys. Lett.* **83**: 165–167.

17. Micro- and Nanofluidics

Arata Aota

Institute of Microchemical Technology, Kanagawa, Japan

Takehiko Kitamori

Institute of Microchemical Technology, Kanagawa, Japan
The University of Tokyo, Tokyo, Japan

17.1   Introduction

Integrated microchemical systems are essential tools for high-speed, functional, and compact instruments used for analysis, synthesis, biological sciences, and technologies. In the 1990s, most microchip-based systems were used for gene or protein analysis and employed electrophoretic separation with laser-induced fluorescence (LIF) detection (Reyes et al. 2002; Auroux et al. 2002). However, other analytical and synthesis methods were needed for more general analytical, combinatorial, physical, and biochemical applications involving complicated chemical processes, organic solvents, neutral species, and nonfluorescent molecule detection. For these applications, general microintegration methods on microchips became very important.

General methods for microintegration of chemical systems that are similar to systems used in electronics have been developed (Figure 17.1). Instead of the resistor,

Nanofabrication Handbook. Edited by Stefano Cabrini and Satoshi Kawata © 2012 CRC Press / Taylor & Francis Group, LLC. ISBN: 978-1-4200-9052-9

Chapter 17

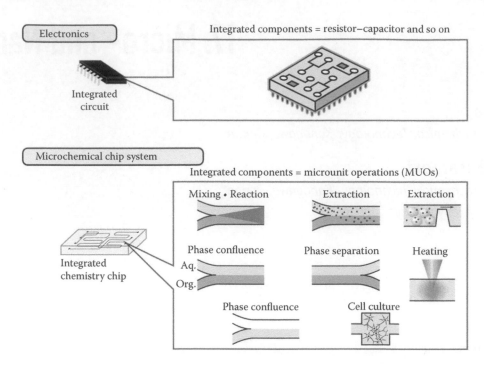

FIGURE 17.1   A microchemical chip system compared with an electronic system.

capacitor, and diode seen in electronic systems, mixing, extraction, phase separation, and other unit operations of chemical processes are integrated as components. These unit operations are known as microunit operations (MUOs), similar to the components of electric circuits. They can be combined with one another in parallel or in series by continuous-flow chemical processing (CFCP). The function of the microchemical chip is analogous to that of a chemical central processing unit. In order to realize these basic concepts, fluidic control methods are very important. Parallel multiphase microflows were realized in previous studies by partial surface modification and channel structures, which enabled various MUOs and flexible integration of the MUOs connected by CFCP. This technique has demonstrated superior performance of chemical processes in shorter processing times (from days or hours to minutes or seconds), smaller sample and reagent volumes (down to a single drop of blood), easier operation (from professional to personal), and smaller system sizes (from 10 m chemical plants to desktop plants to mobile systems) compared to conventional analysis, diagnosis, and chemical synthesis systems. Practical prototype systems have also been realized in environmental analysis, clinical diagnosis, cell analysis, gas analysis, medicine synthesis, microparticle synthesis, etc. For example, a portable micro-ELISA (enzyme-linked immunosorbent assay) system was recently constructed wherein the analysis time was several minutes, rather than several hours typical of the conventional technique. The sample volume (5 µL) was also considerably smaller than the volume (several milliliters) required for conventional diagnosis. In addition, the operation itself was much easier, and its correlation with conventional methods was confirmed with real blood samples. These features are promising for point-of-care (POC) clinical diagnosis.

The technology for microchemical processes is currently moving along two directions. The first is toward the practical applications. Although some practical prototype systems have been realized, attaining long-term stability in parallel liquid/liquid and liquid/gas microflows can sometimes be problematic, and the interfaces are distorted. For robust fluidic control, methods for stabilizing or recovering parallel multiphase microflows are essential. For microfluidic devices, the ability to process small (nano- to microliter volumes) samples and to interface with microchips is important. In addition, smart fluidic devices (valves or pumps) for flexible and reliable liquid handling on the microchips are required for more complex and high-throughput chemical processing.

The second direction involves new science in nanoscale (10–1000 nm) space, termed the extended nanospace. The extended nanospace is an important region in bridging the gap between single molecules and the condensed phase. Recently, general methodologies in the microspace were applied to the extended nanospace (see Section 17.4), demonstrating that the basic methodologies in microspace are applicable to the extended nanospace. Water displays remarkably different properties in an extended nanospace channel. A model related fundamentally to the surface chemistry of a nanochannel has been proposed. This model was applied to protein analysis of a countable number of molecules, and the same type of integrated approach

(extended nanospace channels embedded in microfluidic systems) was applied to control and analyze cell immobilization and culture.

Here, we introduce these basic methodologies and applications. First, the general concepts (MUO and CFCP) for microintegration are discussed in Section 17.2, and fluid control methods to support the MUO and CFCP are discussed in Section 17.3. Finally, new science fields in the extended nanospace are discussed in Section 17.4.

17.2 Microunit Operations and Continuous-Flow Chemical Processing

17.2.1 Design and Construction Methodology for Integration of Complicated Microchemical Systems

Integrated microchemical systems are anticipated to be applied in various fields. In order to realize general analytical, combinatorial, physical, and biological applications, general microintegration methods on microchips are important. Conventional macroscale chemical plants or analytical systems are constructed combining unit operations such as mixers, reactors, and separators. A similar methodology can be applied to the

microchemical systems. By combining MUOs with different functions in series and in parallel, various chemical processes can be integrated into microchips through the use of a multiphase microflow networks such as that shown in Figure 17.2. This methodology is termed CFCP (Tokeshi et al. 2002). Miniaturizing conventional unit operations is often not effective and sometimes does not work at all because many physical properties (e.g., heat and mass transfer efficiency, specific interfacial area, and vanishingly small gravity force) are significantly different in the microspace. Therefore, novel MUOs taking these issues into account are needed. Kitamori and colleagues have developed MUOs for

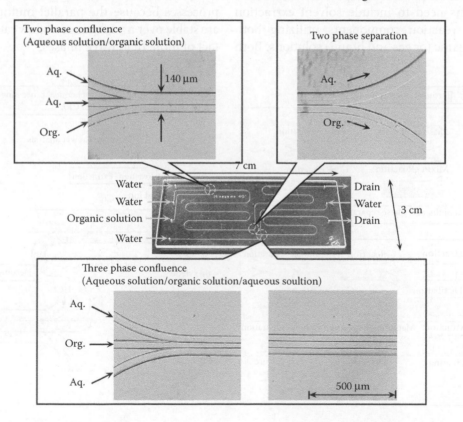

FIGURE 17.2 Multiphase microflow network.

various purposes, such as mixing and reaction (Sato et al. 1999; Sorouraddin et al. 2000, 2001), phase confluence and separation (Tokeshi et al. 2000a,b, 2001, 2002; Surmeian et al. 2001, 2002; Hisamoto et al. 2001a,b, 2003; Hibara et al. 2002a, 2003; Sato et al. 2000a; Minagawa et al. 2001; Aota et al. 2007a,b; Kikutani et al. 2004; Miyaguchi et al. 2006; Smirnova et al. 2006, 2007; Hotokezaka et al. 2005), solvent extraction (Tokeshi et al. 2000a,b, 2002; Hisamoto et al. 2001a,b; Surmeian et al. 2001, 2002; Sato et al. 2000a; Minagawa et al. 2001; Aota et al. 2007c; Kikutani et al. 2004; Miyaguchi et al. 2006; Smirnova et al. 2006, 2007; Hotokezaka et al. 2005; Hibara et al. 2001, 2003), gas–liquid extraction (Hachiya et al. 2004; Aota et al. 2009a), solid-phase extraction and reaction on surfaces (Sato et al. 2000b, 2001, 2002, 2004; Kakuta et al. 2006; Ohashi et al. 2006), heating (Tanaka et al. 2000; Slyadnev et al. 2001; Goto et al. 2005), cell culture (Goto et al. 2005, 2008; Tamaki et al. 2002; Tanaka et al. 2004, 2006), and ultrasensitive detection (Tamaki et al. 2002, 2003, 2005; Tokeshi et al. 2001, 2005; Proskurnin et al. 2003; Mawatari et al. 2006; Yamauchi et al. 2006; Hiki et al. 2006) (Figure 17.3).

17.2.2 Multiphase Microflow Network

For many chemical processing applications, microchemical systems need to include solvent extraction and interfacial reaction components utilizing both aqueous and organic (or gas and liquid) solutions. Both solutions must be controlled to realize general chemistry in a microchip. In the 1990s, electroosmotic flow was used in microchip electrophoresis (Reyes et al. 2002; Auroux et al. 2002); however, the electroosmotic flow is restricted to the flow control of only one type of solution (an aqueous buffer). Therefore, electroosmotic flow is not suitable for a flow-control method to integrate various chemical processes that require other types of solvents. Simple pressure-driven flow has been used by many researchers as a nonelectrical pumping method (Kopp et al. 1998; Brody and Yager 1997; Weigl and Yager 1999).

Recently, microsegmented flows of immiscible solutions were used for microchemical processes (Burns and Ramshaw 2001; Song et al. 2003; Kralj et al. 2005; Sahoo et al. 2007). Although the microsegmented flows are advantageous in mixing and rapid molecular transport, phase separation and contact of more than three phases are difficult. Furthermore, controlling the microsegmented flows over a wide range of flow rates is difficult because the size of the segments changes along with the flow rate ratio. Kitamori et al. have developed parallel multiphase microflows, which flow side by side along the microchannels (Hibara et al. 2001). Parallel multiphase microflows in the laminar flow regime allow for better design and control of microchemical processes because the parallel multiphase microflows are stable over a wide range of flow rates and allow contact of more than three phases.

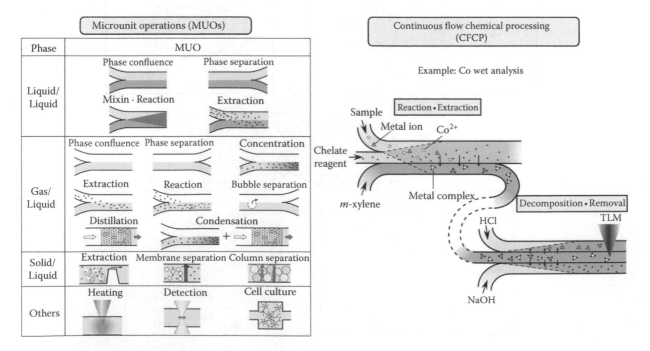

FIGURE 17.3   Microunit operations (MUOs) and continuous-flow chemical processing (CFCP). TLM, thermal lens microscope.

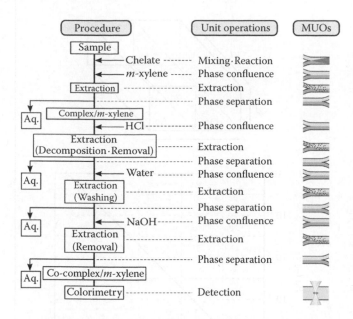

FIGURE 17.4 Designing of Co wet analysis systems based on CFCP. MUO, microunit operation.

17.2.3 Example of CFCP

Micro cobalt wet analysis is an example of CFCP that has been demonstrated (Figure 17.4) (Tokeshi et al. 2002). Conventional procedures for such analysis consist of a chelating reaction, a solvent extraction of the complex, and a decomposition and removal of the coexisting metal complex. These procedures are analogous to the conventional unit operations, including mixing and reaction, phase confluence, solvent extraction, phase separation, and solvent extraction. These unit operations similarly are analogous to the MUOs. CFCP is designed by combining MUOs in series and in parallel, as shown in Figure 17.3. The analysis time in this system is only 50 s versus 6 h for conventional devices. Microchemical processing through use of CFCP has been demonstrated on numerous occasions (Table 17.1).

The CFCP approach can be applied to develop a more complicated processing system. More rapid

Table 17.1 Examples of Microchemical Processing

Method	Analyte	LOD	Analysis Time	Reference
Solvent Extraction	Iron complex	7.7 zmol	60 s	Tokeshi et al. (2000a)
	Cobalt complex	0.13 zmol; 0.72 zmol; 0.072 zmol	60 s; 50 s; 10 min	Tokeshi et al. (2000b), Tokeshi et al. (2002), Minagawa et al. (2001)
	Nickel complex	NA	5 min	Sato et al. (2000a)
	K^+ and Na^+	4.5 zmol	1 s	Hisamoto et al. (2001a, 2000b)
	Dye molecules	NA; NA	6 s; 4 s	Surmeian et al. (2001, 2002)
	Amphetamines	$0.5\ \mu g\ mL^{-1}$	15 min	Miyaguchi et al. (2006)
	Carbaryl	0.5 zmol; 0.079 zmol	4.5 min; 5 min	Smirnova et al. (2006, 2007)
	Uran (VI)	0.86 amol	1 s	Hotokezaka et al. (2005)
Gas extraction	Formaldehyde	8.9 ppb	30 min	Hachiya et al. (2004)
	Ammonia	1.4 ppb	16 min	Aota et al. (2009a)
Immunoassay	S-IgA	$<1\ \mu g\ mL^{-1}$	<1 h	Sato et al. (2000b)
	Carcinoembryonic antigen	$30\ pg\ mL^{-1}$	35 min	Sato et al. (2001)
	Interferon-γ	$10\ pg\ mL^{-1}$	50 min (four samples)	Sato et al. (2002)
	B-type natriuretic peptide	$0.1\ pg\ mL^{-1}$	35 min	Sato et al. (2004)
	IgE	$2\ ng\ mL^{-1}$	12 min	Kakuta et al. (2006)
	Prion	$50\ pg\ mL^{-1}$	15 min	—
	C-reactive protein	$20\ ng\ mL^{-1}$	8 min	—
	Amphetamines	$0.1\ ng\ mL^{-1}$	10 min	—
Flow injection	Fe^{2+}	6 zmol	150 s	Sato et al. (1999)
	Ascorbic acid	1 zmol	30 s	Sorouraddin et al. (2000)
	Catecholamines	2 zmol	15 s	Sorouraddin et al. (2001)
Enzymatic assay	H_2O_2	NA; 50 zmol; NA	250 s; NA; NA	Hisamoto et al. (2003), Tanaka et al. (2000, 2004)

Chapter 17

analysis, bioassays, and immunoassays, as well as more efficient reaction and extraction, can be achieved through CFCP systems as compared to conventional devices. However, the development of complicated microchemical processing systems is time consuming because of the need to determine appropriate conditions for each reaction and separation process at the miniature scale, and also to design and optimize the microchannel structure. In semiconductor circuit device design, computer-aided design (CAD) is typically used to perform circuit analysis, integrated device analysis, layout design, and logical simulation. If CAD can be developed specifically for microchemical processing, including microfluid simulation, reaction time analysis, extraction time analysis, and microchannel structure design, the time for the development of these systems would be drastically shortened.

17.3 Microfluidics and Control

17.3.1 Fundamental Physical Properties of Multiphase Microflows

The control of multiphase microflows is an important basic technology for the integration of MUOs. Therefore, control methods for the stable phase separation of the multiphase microflows over a wide range of flow rates are required. However, the methods used for conventional-sized devices cannot be applied because the physical properties in the microspace are different from those in macrospace. In conventional devices, aqueous and organic phases are separated by gravity. In the microspace, however, the fluid is greatly influenced by the liquid–solid, liquid–gas, and liquid–liquid interfaces because of the large specific interfacial area. To clarify the main physical forces in the microchannels, which include the viscous force, the interfacial parameters can be analyzed using the dimensionless Reynolds (Re) and Bond (Bo) numbers, defined as the inertia-to-viscous force ratio and a gravity-to-tension ratio, respectively. Figure 17.5a shows a cross section of a model microchannel with dimensions included in these parameters.

Bo is defined as

$$Bo = \frac{(\Delta\rho) g d_h^2}{\gamma},$$ (17.1)

where $\Delta\rho$, γ, and d_h are the density difference, the interfacial tension between two phases, and the equivalent

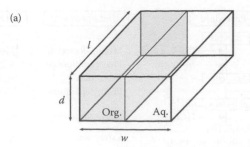

(a)

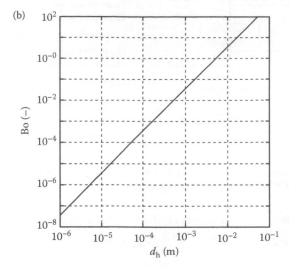

(b)

FIGURE 17.5 (a) The liquid–liquid interface between the aqueous and organic phases (Aq. and Org.) in a model microchannel. The width, depth, and length along the microchannel are represented as w, d, and l, respectively. (b) Bond number (Bo) dependence on the hydrodynamic diameter for water–toluene microflows.

diameter, respectively, and g is the gravitational acceleration (9.8 m s^{-2}). The variable d_h corresponds to the mean hydraulic diameter defined as

$$d_h = \frac{4S}{l_p}$$ (17.2)

where S is the cross section and l_p is the perimeter of a section of the microchannel. Figure 17.5b shows Bo of water–toluene two-phase flows as a function of d_h ($\Delta\rho = 0.132 \times 10^3$ kg m^{-3}, $\gamma = 36.3$ mN m^{-1}). In microspace with d_h of 100 μm, the interfacial tension exceeds gravity by three orders of magnitude.

Re is defined as

$$\begin{aligned} Re &= \frac{\rho u d_h}{\mu} \\ &= \frac{2\rho u d w}{\mu(w+2d)}, \end{aligned}$$ (17.3)

where ρ, u, and μ are the density, mean velocity, and viscosity, respectively. When water flows in the microchannel at a flow rate of 1 μL min^{-1}, Re becomes only 0.05 in microspace with a d_h of 100 μm. Therefore, multiphase microflows are considered to be laminar flow.

Multiphase microflows are dominated by pressures (Aota et al. 2007a, 2009b). One important parameter needed to describe the multiphase microflows is the pressure that drives the fluids. The pressure decreases in the downstream part of the flow because of the viscosity of the fluids. When two fluids in contact with one another have different viscosities, the pressure difference (ΔP_{Flow}) between the two phases is a function of the contact length and the flow velocity. Another important parameter is the Laplace pressure ($\Delta P_{Laplace}$) caused by the interfacial tension between two phases. The position of the interface is fixed within a point in the microchannel by the balance established between $\Delta P_{Laplace}$ and ΔP_{Flow}.

Figure 17.6a illustrates the pressure balance at the liquid–liquid interface of the two-phase microflows. The liquid–liquid interface curves toward the organic phase in a glass surface because of the hydrophilicity of

the glass. $\Delta P_{Laplace}$ is generated at the curved liquid–liquid interface. On the basis of the Young–Laplace equation, $\Delta P_{Laplace}$ is estimated as

$$\Delta P_{Laplace} = \frac{\gamma}{R} = \frac{2\gamma \sin(\theta - 90°)}{d}, \tag{17.4}$$

where R is the curvature radius of the liquid–liquid interface and θ is the contact angle. The contact angle is restricted to the values between the advancing contact angle of the aqueous phase θ_{aq} and that of the organic phase θ_{org}. Therefore, $\Delta P_{Laplace}$ is restricted as follows:

$$\frac{2\gamma \sin(\theta_{aq} - 90°)}{d} < \Delta P_{Laplace} < \frac{2\gamma \sin(\theta_{org} - 90°)}{d}. \tag{17.5}$$

When ΔP_{Flow} exceeds the maximum $\Delta P_{Laplace}$, the organic phase flows toward the aqueous phase (Figure 17.6b). When ΔP_{Flow} is lower than the minimum $\Delta P_{Laplace}$, the aqueous phase flows toward the organic phase (Figure 17.6c). When the flow rate ratio is changed, the pressure balance is maintained by changing the position of the liquid–liquid interface. This model indicates that the important parameters for microfluid control are the interfacial tension, the dynamic contact angle, and the depth of the microchannel. This model can also be applied to gas–liquid microflows.

Considering the interfacial pressure model, the flow rates should be between the higher and lower limits. ΔP_{Flow} can be evaluated by considering the pressure loss in the microchannels. Assuming that the pressure at the opened outlet is atmospheric pressure, P_{atm}, then the pressure P of each phase from the pressure loss ΔP is expressed as

$$P = P_{atm} + \Delta P_{tube} + \Delta P_{channel}$$
$$= P_{atm} + \frac{2f\rho v_{tube}^2 L_{tube}}{d_{h, tube}} + \frac{2f\rho v_{channel}^2 L_{channel}}{d_{h, channel}}, \tag{17.6}$$

where f, ρ, v, and L are the friction factor, the density, the mean velocity of the fluid, and the length of the tube and channel. The subscripts of tube and channel correspond to the parts of the outlet tube and the microchannel. When the gas–liquid and liquid–liquid microflows are considered to be laminar flow, f is expressed as

$$f = \frac{16}{Re} = \frac{16\mu}{\rho v d_h}. \tag{17.7}$$

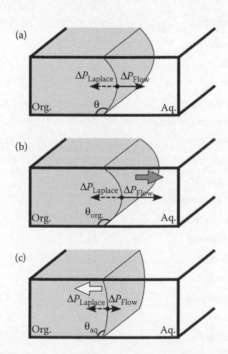

FIGURE 17.6 (a) Pressure balance at the liquid–liquid interface. Pressure difference is balanced with Laplace pressure when the two phases are separated. (b) When the pressure difference between the two phases is larger than the maximum Laplace pressure, the organic phase moves toward the aqueous phase. (c) When the pressure difference is lower than the minimum Laplace pressure, the aqueous phase moves toward the organic phase. Aq., aqueous phase; Org., organic phase.

Therefore, ΔP_{Flow} can be expressed as

$$\Delta P_{\text{Flow}} = P_{\text{org}} - P_{\text{aq}}$$

$$= \frac{32\mu_{\text{org}}v_{\text{channel,org}}L_{\text{channel,org}}}{d_{\text{h, channel,org}}^2} + \frac{32\mu_{\text{org}}v_{\text{tube,org}}L_{\text{tube,org}}}{d_{\text{h, tube,org}}^2}$$

$$- \frac{32\mu_{\text{aq}}v_{\text{channel,aq}}L_{\text{channel,aq}}}{d_{\text{h, channel,aq}}^2} - \frac{32\mu_{\text{org}}v_{\text{tube,aq}}L_{\text{tube,aq}}}{d_{\text{h, tube,aq}}^2},$$

(17.8)

where the subscripts org and aq correspond to the organic and aqueous phases, respectively. By utilizing Equation 17.8, the higher and lower limits of ΔP_{Flow}

Table 17.2 Summary of Viscosity of Solvent at 20°C

Solvent	Viscosity (cP)
Acetone	0.315[a]
Aniline	4.40[b]
1-Butanol	2.948[b]
Butylacetate	0.732[b]
Carbon tetrachloride	0.969[b]
Chloroform	0.58[b]
Cyclohexane (at 17°C)	1.02[b]
1-Decanol	12.96[c]
Dichloromethane	0.445[d]
Diethyl ether (at 25°C)	0.224[e]
1,4-Dioxane (at 25°C)	1.177[e]
Dodecane (at 25°C)	1.383[e]
Ethanol	1.20[b]
Ethylacetate	0.455[b]
Hexane	0.326[b]
Methanol	0.597[b]
Nitrobenzene	2.03[b]
1-Octanol (at 25°C)	7.288[e]
Pentane	0.240[b]
1-Propanol	2.256[b]
Toluene	0.590[b]
Water	1.002[b]
m-Xylene	0.620[b]

[a] Howard and Mcallister (1958).
[b] Weast (1985).
[c] Magallanes et al. (1989).
[d] Mueller and Ignatowski (1960).
[e] Lide (2001).

were evaluated. Table 17.2 summarizes the literature values of the viscosities of various solvents. $\Delta P_{\text{Laplace}}$ can be evaluated by utilizing Equation 17.4. Table 17.3 summarizes the interfacial tension of various solvents. The advancing and receding contact angles of water on a glass plate in toluene are 10.2 ± 4.9 and 64.5 ± 4.3 degrees, respectively. These values were also used for the contact angles in other organic solvents.

The phase separation conditions of liquid–liquid microflows for various organic solvents in the double-Y-type microchannel with a width of 215 μm, a depth of 34 μm, and a contact length of 20 mm are shown in Figure 17.7. The flow rate of the aqueous phase was fixed to be 10 μL min⁻¹. The experimental results agreed well with the theoretical values. The experimental results also agreed with simulated values

Table 17.3 Summary of Interfacial Tension between Air–Solvent and Aqueous–Organic Interfaces at 20°C

Solvent	For Air (mN m⁻¹)	For Water (mN m⁻¹)
Acetone	25.3	
Aniline		6.1
1-Butanol		1.6
Butylacetate		13.5
Carbon tetrachloride	27.6	
Chloroform	28.7	30.8
Cyclohexane		47.7
1-Decanol	28.6	
Dichloromethane	31.1	28.4
Diethyl ether	19.1	
1,4-Dioxane	34.2	
Dodecane	25.3	
Ethanol	22.7	
Ethylacetate	24.3	6.3
Hexane	19.4	50.0
Methanol	23.5	
Nitrobenzene	43.3	25.2
1-Octanol	27.7	
Pentane	18.5	
1-Propanol	23.9	
Toluene	28.8	35.3
Water	72.0	
m-Xylene	28.7	36.6

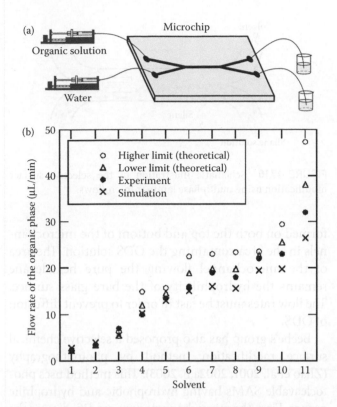

(a) Organic solution / Water / Microchip

(b)

FIGURE 17.7   (a) Illustration of the microchip having a width of 215 µm, a depth of 35 µm, and a contact length of 20 mm. (b) Phase separation conditions of the liquid–liquid microflows. Solvent: *1* aniline, *2* 1-butanol, *3* nitrobenzene, *4* cyclohexane, *5* butylacetate, *6* m-xylene, *7* chloroform, *8* toluene, *9* ethylacetate, *10* dichloromethane, *11* hexane. The *opened circles* show the theoretical higher limit, the *opened triangles* the theoretical lower limit, the *solid circles* experimental results, and the *crosses* the results of the simulation.

obtained by the three-dimensional simulation using the volume of fluid method. Controlling ΔP_{Flow} and $\Delta P_{Laplace}$ permits phase separation to be achieved in the microchannel. With the help of this, the researchers can design various microfluidic chemical processes using liquid–liquid microflows.

The phase separation conditions of air–liquid microflows for various solvents in the double-Y-type microchannel with a width of 100 µm, a depth of 45 µm, and a contact length of 20 mm are shown in Figure 17.8. The flow rate of the air phase was fixed to be 1 mL min^{-1}. The results, however, did not agree with the theoretical values. Assuming only the pressure loss and the Laplace pressure, the phase separation should be achieved within the narrow range of the flow rates. However, phase separation was achieved over a wide range of flow rates. This disparity may be explained by considering the compression of the gas phase, evaporation of the liquid phase, and wetting at the outlet port of the liquid phase. Therefore, when gas–liquid microflows form in microchannels, the compressivity, vapor pressure, humidity, and pressure of the outlet should be carefully considered.

17.3.2   Methods of Stabilization of Multiphase Microflows

Disturbances in the flow rates of pumps can destabilize multiphase microflows. However, the control method

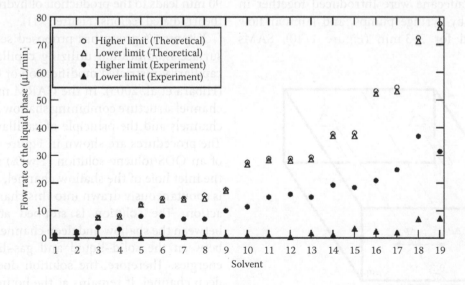

FIGURE 17.8   Phase separation conditions of the liquid–liquid microflows. Solvent: *1* 1-decanol, *2* 1-octanol, *3* 1-propanol, *4* nitrobenzene, *5* dodecane, *6* 1,4-dioxane, *7* ethanol, *8* water, *9* carbon tetrachloride, *10* m-xylene, *11* hexane, *12* toluene, *13* chloroform, *14* ethylacetate, *15* dichloromethane, *16* hexane, *17* acetone, *18* pentane, *19* diethyl ether. The *open circles* show the theoretical higher limit, the *open triangles* the theoretical lower limit, the *solid circles* the experimental results of the higher limit, and the *solid triangles* the experimental results of the lower limit.

utilizing the microchannel structure and surface energy is effective for maintaining robust control over microfluidics. Researchers have suggested that multiphase microflows can be stabilized by altering the structure or adding features to the microchannels, specifically by including a guide structure (Tokeshi et al. 2002) or a pillar structure (Maruyama et al. 2004) along the microchannel, which permit a wider range of contact angle for the fluid interface than a flat surface.

Other groups have proposed selective chemical surface modification for stabilization of the multiphase microflows (Hibara et al. 2002a, 2003, 2005, 2008; Aota et al. 2007a,b; Zhao et al. 2001, 2002a,b, 2003; van der Linden et al. 2006). Figure 17.9 illustrates the shape of the liquid–liquid interface in a microchannel with chemically patterned surfaces. The contact angle of water on a hydrophobic surface can be larger than 90°. Therefore, multiphase microflows in microchannels with patterned surfaces can form under a wider range of conditions compared to microchannels with a guide structure or pillar structure.

Beebe's group has proposed a selective chemical surface modification method by combining multiphase laminar microflow and self-assembled monolayer (SAM) chemistry (Zhao et al. 2001, 2002a,b, 2003). The flow in the microchannels is laminar flow as described in Section 17.3.1. Therefore, multiphase microflows of miscible solutions can flow side by side without turbulent mixing. A stream of pure hexadecane and a stream of octadecyltrichlorosilane (ODS) solution in hexadecane were introduced together in microchannels by syringe pumps, and laminar flow was maintained for 2–3 min (Figure 17.10). SAMs

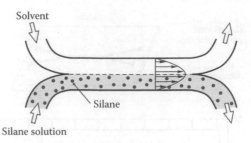

FIGURE 17.10 Schematic illustration of the selective surface modification using multiphase laminar microflows.

formed on both the top and bottom of the microchannels in the area containing the ODS solution. The area of the microchannel flowing the pure hexadecane remains the hydrophilicity of the bare glass surface. The flow rates must be fast in order to prevent diffusion of ODS.

Beebe's group has also proposed a selective chemical surface modification method by photolithography (Zhao et al. 2001, 2002a,b, 2003). This method uses photocleavable SAMs having hydrophobic and hydrophilic groups. First, the microchannels were modified using the photocleavable SAMs. The microchannels were cleaned by sequentially flushing with hexane and methanol after monolayer deposition from a 0.5 w/w% solution of the corresponding trichlorosilane in hexadecane, and then dried with nitrogen. A photomask is placed on top of the SAM-modified microchannels filled with NaOH solution. Ultraviolet (UV) irradiation through a mask for 90 min leads to the production of hydrophilic groups in the irradiated regions (Figure 17.11).

Kitamori's group has proposed selective chemical surface modification utilizing capillarity (called the capillarity-restricted modification or CARM method) (Hibara et al. 2005). In the CARM method, a microchannel structure combining shallow and deep microchannels and the principle of capillarity are utilized. The procedures are shown in Figure 17.12. A portion of an ODS/toluene solution (1 wt%) is dropped onto the inlet hole of the shallow channel, and the solution is spontaneously drawn into this channel by capillary action. The solution is stopped at the boundary between the shallow and deep channels by the balance between the solid–liquid and gas–liquid interfacial energies. Therefore, the solution does not enter the deep channel. It remains at the boundary for several minutes and is then pushed from the deep channel side by air pressure.

Since $\Delta P_{Laplace}$ depends on the depth of the microchannels, more effective stabilization of multiphase

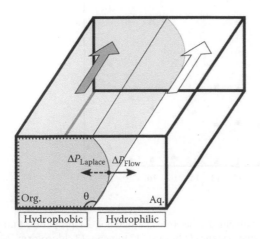

FIGURE 17.9 Shape of the liquid–liquid interface, whose contact line is pinned at the boundary between the hydrophilic and the hydrophobic surfaces.

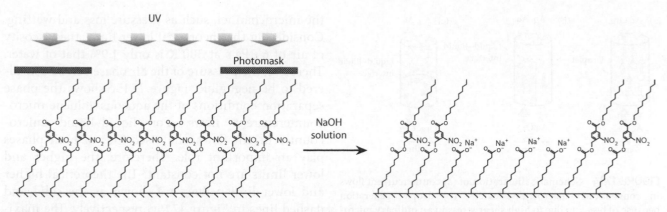

FIGURE 17.11 Ultraviolet photopatterning method. The molecular structure of a photocleavable SAM formed on glass surfaces. Ultraviolet (UV) irradiation through masks placed on top of SAM-modified microchannels leads to the production of hydrophilic carboxylate groups in the irradiated regions.

microflows can be attained by utilizing shallow microchannels. For example, a microstructure combining shallow and deep microchannels with hydrophobic and hydrophilic surfaces, respectively, can yield large $\Delta P_{Laplace}$. In a shallow 10-μm-deep microchannel, $\Delta P_{Laplace}$ of the gas–water interface is 6 kPa. Microchannels with an asymmetric cross section with the patterned surfaces mentioned above are effective not only for the stability of the multiphase microflows, but also for a fail-safe system required for practical systems. Air bubbles contamination can disturb the stability of systems, and liquid–liquid

two-phase microflows sometimes are unstable due to some disturbances. By using microchannels having asymmetric cross sections along with the patterned surface, one can convert plug flow into two-phase microflows (Figure 17.13) (Hibara et al. 2005, 2008).

In chemically patterned microchannels, microcountercurrent flows can form (Aota et al. 2007a–c). In conventional macroscale devices, countercurrent flow is attained by gravitational segregation involving droplets (Figure 17.14a). However, parallel countercurrent flow in the laminar flow regime cannot be easily realized. In an ordinary microchannel, countercurrent

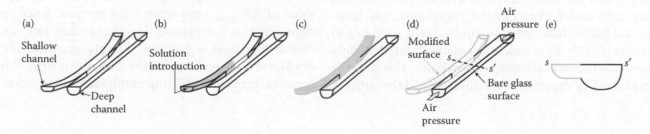

FIGURE 17.12 Modification procedures by CARM method. (a) The shallow and deep microchannels have separate inlet holes and contact points in the microchip. (b) A solution containing modification compounds is introduced from the inlet of the shallow microchannel by capillarity. (c) The solution does not leak to the deep microchannel and only the shallow microchannel is modified. (d) The solution is pushed away with air pressure from the deep microchannel. (e) A sectional illustration along the s–s′ dashed line in (d).

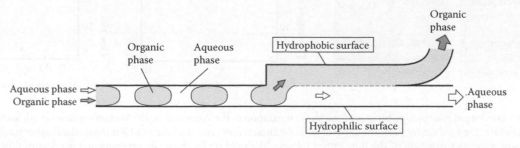

FIGURE 17.13 Conversion of plug flow into parallel two-phase microflows.

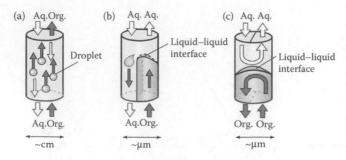

FIGURE 17.14 Schematic illustration of (a) countercurrent flows in conventional macroscale devices, (b) droplet generation because of breakup due to high shear stress in an ordinary microchannel, and (c) collision of two phases in an ordinary microchannel. Aq., aqueous phase; Org., organic phase.

flow cannot form because high shear stress at the interface causes breakup (Figure 17.14b) and the two phases collide (Figure 17.14c). To form parallel microcountercurrent flows, the aqueous solution must flow along one side of the channel and the organic solution must flow along the other side without breakup. Considering pressure balance at the interface, microcountercurrent flows can be formed in a microchannel with patterned surfaces. The phase separation conditions for gas–liquid and liquid–liquid microcountercurrent flows in the microchannel with an asymmetric cross section are shown in Figure 17.15. Since the viscosity of air can be thought of as being negligible, the maximum flow rate of water becomes constant under experimental conditions. The theoretical higher limit value is shown as the solid line in Figure 17.15b. Experimental results differed slightly from the theoretical value, a discrepancy that could be explained by experimental issues around the outlet of

the microchannel, such as pressure loss and wetting. Considering the theoretical lower limit, the viscosity of air of 6 μPa s at 300 K is only 1.9% that of water. Therefore, the pressure of the air phase can be considered to be negligible. Figure 17.15c shows the phase separation conditions of the aqueous–toluene microcountercurrent flows. For liquid–liquid microcountercurrent flows, the viscosities of both phases play an important role. Therefore, the higher and lower limits are not constant. The theoretical higher and lower limit values are shown as the solid and dashed lines in Figure 17.15c, respectively. The maximum flow rate in the upstream part of the aqueous phase agreed well with the theoretical value. The minimum flow rate however did not agree with the theoretical values. This disparity could be explained in terms of the microchannel geometry. When water leaks onto the hydrophobic surface, the contact angle on the upper wall is the same as that on the lower wall because the hydrophobic channel is flat. However, when toluene leaks onto the hydrophilic surface, the contact angle on the upper wall is different from that on the lower wall because the hydrophilic channels have a semicircular cross section and the boundary between the hydrophilic and hydrophobic surfaces has an edge. Since the range of contact angles of the liquid–liquid interface at the edge becomes wider compared to that on a flat surface, the liquid–liquid interface at the edge structure can produce a larger value of $\Delta P_{Laplace}$ than from a flat surface. Since the edge structure can expand the range of flow rate conditions for phase separation, this disagreement may not be a serious problem for the design of microfluidic chemical processes utilizing multiphase microflows.

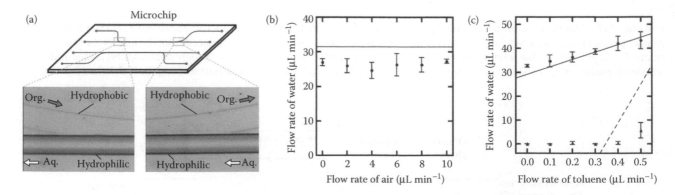

FIGURE 17.15 (a) Optical microscope images of the phase separation at the confluences. (b) Maximum flow rate of water as a function of the flow rate of air. The *solid circles* show the experimental maximum flow rates, and the *solid line* theoretical higher limit. (c) Maximum and flow rates of water as a function of the flow rate of toluene. The *solid circles* show the experimental maximum flow rates, the *solid triangles* the experimental minimum flow rates, the *solid line* theoretical higher limit, and the *dashed line* the theoretical lower limit.

Gas–liquid and liquid–liquid microflows are effective for highly efficient extraction processes. When samples comprise adsorbent molecules at a fluid–liquid interface, molecular adsorption changes the interfacial tension of the fluid–liquid interface. Since the specific interfacial area is very large in a microchip, the change in interfacial tension due to molecular adsorption can be an important factor affecting phase separation of the gas–liquid and liquid–liquid microflows. The molecular concentration is higher in the downstream portion of the microflows. Thus, when designing a chemical process that includes molecular transport through the interface, researchers need to modify the proposed model by considering molecular adsorption at the interface. For microextraction design, the *in situ* interfacial tension measurement is especially important. The microscopic quasielastic laser scattering method could be a powerful tool for this purpose (Hibara et al. 2003).

Superhydrophilic and superhydrophobic surfaces are more effective at stabilizing two-phase microflows. These surfaces can be obtained by creating roughness utilizing titanium nanoparticles. Titanium modification of a microchannel yields nanometer-scale surface roughness, and subsequent hydrophobic treatment creates a superhydrophobic surface. Photocatalytic decomposition of the coated hydrophobic molecules was used to pattern the surface wettability, which was tuned from superhydrophobic to superhydrophilic under controlled photoirradiation (Takei et al. 2007a). This method can also be applied to the conversion of plug flow into two-phase microflows (Takei et al. 2007b).

17.3.3 Wettability-Based Microvalve

Numerous miniaturized mechanical valves fabricated by micromachining technologies have been developed (Reyes et al. 2002; Auroux et al. 2002; Dittrich et al. 2006). Pneumatic-controlled valves made of soft material have also been reported (Unger et al. 2000). All these valves require both fabrication and construction of mechanical moving parts or the pressure of external control parts in the microchannels. Also, there have been reports of nonmechanical methods utilizing phase transition solution (Gui and Liu 2004), a thermoresponsive polymer (Yu et al. 2003), or hydrogel (Beebe et al. 2000). Additionally, patterned hydrophobic surfaces have been demonstrated as valves for microfluid control in a variety of applications. The patterned titanium surfaces described previously can be used very effectively as valves in a microchip due to their superhydrophobic nature (Takei et al. 2007a). Batch operation in the microchannels with patterned surfaces has been demonstrated as shown in Figure 17.16. First, fluorescent solution was introduced at a pressure below the maximum Laplace pressure (Figure 17.16a). Second, air was introduced at the same pressure to push out excess solution and retain the plug of picoliter volume (Figure 17.16b). Third, air was introduced at a pressure higher than the maximum Laplace pressure to expel the plug (Figure 17.16c and d). This microvalve utilizing chemical surface modification has no dead volume and can supply a constant volume of liquid for reaction or analysis. Moreover, the repeatability of this operation allows it to be applied to titration.

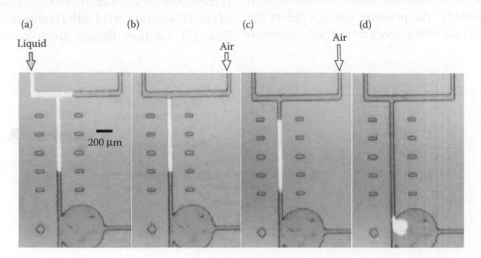

FIGURE 17.16 Fluorescence images of the liquid motion during batch operation in a microchannel. The liquid is (a) being introduced, (b) measured, (c) transferred, and (d) dispensed to the other channel.

Chapter 17

17.4 Nanofluidics

17.4.1 Control Method of Nanofluid

The 10–1000-nm-scale space (called the extended nanospace), which is a technologically unexplored region between 1-nm-scale materials and microchip, is expected to be a promising experimental area for implementing highly integrated micro/nanofluidic devices and understanding molecular physical chemistry. Kitamori's group has developed two-dimensional (2-D) extended nanospaces using top-down-type micro/nanofabrication technologies (Goto et al. 2008; Hibara et al. 2002b; Tamaki et al. 2006; Tsukahara et al. 2008a). From the viewpoint of integration, in microfluidic devices, diffusion time is very short and the specific interfacial area is very high, and effective chemical processes have been realized by utilizing the these characteristics as mentioned above. Further integration of chemical systems will enable us to utilize the advantages of microfluidic devices even more effectively, which will lead to nanofluidic chemical processing. From the viewpoint of molecular physical chemistry, since the extended nanospace is an area of transition in molecular behavior from that of individual molecules to that of the bulk condensed phase, this space can be applied to characterize the collective properties of the liquid-phase molecular clusters of 10–1000 nm size. In order to realize nanofluidic chemical processing and applications to experimental tools for investigations on collective properties of the liquid-phase molecular clusters in the extended nanospaces, it is essential to control the nanofluidics precisely.

Both a high pressure (of MPa) and a low flow rate (of pL min^{-1}) are required to develop nanofluidic control systems because of the relatively large pressure drop in extended nanospaces. The pressure balance before the Y-intersection is one of the most important factors to consider when attempting to realize stable flow in a Y-shaped extended nanospace channel. An air-pressure-based nanofluidic control system has been developed (Tsukahara et al. 2008a). Figure 17.17 shows a schematic illustration of the air-pressure-based nanofluidic control system, including a Y-shaped extended nanospace channel. The sample solution (A) is poured into a vial, which had two gateways. The outlet is connected through a capillary tube to one of the U-shaped microchannels on the chip. Air is pressurized into a sample solution (A) from the inlet by using the air-pressure controller, and the sample solution (A) in the vial is delivered to one side of the U-shaped microchannel from the outlet of the vial. After the valve that is connected to the other side of this U-shaped microchannel is closed, the sample solution (A) inside the microchannel can be introduced into the Y-shaped extended nanospace channel based on the pressure-driven flow. In this system, the pressures in the U-shaped microchannel and Y-shaped extended nanospace channel are kept constant. When the sample solution (B) is similarly introduced, the two different solutions (A) and (B) can be diffusion-mixed in the Y-shaped extended nanospace channel. The molecular diffusion distance in the nanochannel is quite short, and solution mixing can be achieved in milliseconds.

The performance of the nanofluidic control system has been evaluated by examining the variation in fluorescence intensities generated by mixing two different solutions: an aqueous solution of the fluorescein derivative Tokyo Green (TG, 5.0×10^{-5} M) and 4-(2-hydroxyethyl)-1-piperazineethanesulfonic acid (HEPES) buffer solution (pH 8.0). A fluorescence image of the TG solution at 0.1 MPa is shown in Figure 17.18b. The TG solution flowed from nanochannel A to

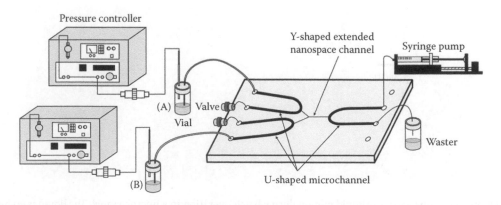

FIGURE 17.17 Schematic illustration of the air-pressure-based nanofluidic control system.

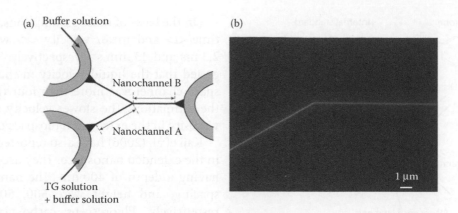

FIGURE 17.18 (a) Schematic illustration of the complex structure with microchannels and an extended nanospace channel on a chip. (b) Fluorescent images of nanochannel during the mixing.

nanochannel B without backward flow into the other side of nanochannel A or plug-like flow. When the fluorescence intensities of the TG solutions in nanochannels A and B were compared with each other, nanochannel B was found to have an intensity that was about 0.6 times lower than that in nanochannel A. When the introduction of TG solution and HEPES buffer solution into nanochannel A was switched with each other, this fluorescence intensity ratio was not changed. The phenomenon in which the intensity ratio of nanochannel A to nanochannel B was not 1.0:0.5 but 1.0:0.6 could be caused by differences in position resolution in the observed fluorescence images. The fluorescence intensities of TG solutions were measured with various known pH values. The intensity ratio of 1.0:0.6 was almost the same over the whole pH range. These results indicated that mixing equal quantities of TG solutions could be realized in the Y-shaped extended nanospace channel. The fluorescence intensities of the TG solutions were measured for pressures of 0.001–0.5 MPa. The TG solutions flowed stably in the nanochannel without the appearance of backward and plug-like flows in the pressure range of 0.0003–0.4 MPa. Since the pressure obtained is related to the pressure drop (ΔP) in the nanochannel, the ΔP value is theoretically expressed by Hagen–Poiseuille's law as

$$\Delta P = \frac{128\mu Vl}{\pi d_{\mathrm{h}}^4},\tag{17.9}$$

where l and V are the channel length and the volumetric flow rate. As the ΔP value was increased from 0.003 to 0.4 MPa at a viscosity of 1 cP, the flow rate increased from 0.16 to 21.2 pL min^{-1} in nanochannel B. Therefore, this method allows a high pressure of MPa and a low flow rate of pL min^{-1} for stable control nanofluidics.

17.4.2 Viscosity of Nanofluid

The physicochemical and reaction properties of liquids in microspace are not different from those in bulk. However, the extended nanospace is an area of transition in molecular behavior from that of individual molecules to that of the bulk condensed phase. Therefore, the physical parameters may be changed in the extended nanospace. In order to realize nanofluidic chemical processes, the physical parameters must be determined. In this section, the viscosity of nanofluid as one of the physical parameters is discussed from the measurements of the flow rates and capillary action.

The flow rates have been measured in the extended nanospace channels, which had lengths of 190 μm, widths of 380 nm, and depths of 240 nm, illustrated in Figure 17.19a (Tamaki et al. 2006). A probe was added to the liquid introduced in the left-side microchannel and an internal standard was added to the liquid in the right-side channel. By determining the ratio of the probe and internal standard molecules in the acceptor flow, the flow rate in the nanochannel can be measured. Aqueous solutions of resorcinol (1.0×10^{-2} M) and methylresorcinol (1.0×10^{-5} M) were used as the probe and internal standard, respectively. Figure 17.19b shows the flow rate dependency on the applied pressure. The flow rate depends linearly on the pressure, which implies the Hagen–Poiseuille's law can be applied to the nanochannel flow. In Figure 17.19b, the dashed line shows the theoretical line, where the viscosity of 1.0 cP was used for the aqueous solution. The experimental flow rates were systematically lower than those calculated from the equation, and the viscosity in the nanochannels could be estimated as 2.5 cP based on Equation 17.6. This result implied a size-dependent viscosity change in the extended nanospace channel.

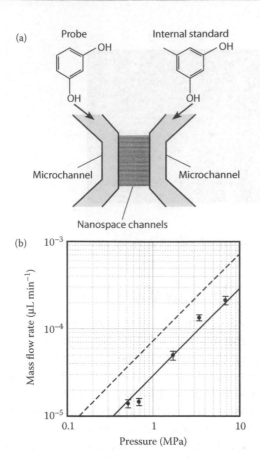

FIGURE 17.19 (a) Experimental scheme for evaluation of the flow system. (b) Plots of mass flow rate in the nanochannels versus applied pressure. The *dashed line* shows Hagen–Poiseuille's law when the viscosity was 1.0 cP. The *solid line* shows Hagen–Poiseuille's law when the viscosity was 2.5 cP.

Other liquid introduction method utilizes capillary action (Hibara et al. 2002b). If the viscosity changes in the extended nanospace channel, liquid introduction behavior should change. The introduction time has been measured in the extended nanospace channel having lengths of 100 µm, widths of 330 nm, and depths of 220 nm. The linear flow velocity was estimated as 10 mm s^{-1}. The driving force was $\Delta P_{Laplace}$. In order to estimate theoretical introduction time, $<t>$, and mean velocity, $<u>$, $\Delta P_{Laplace} = \Delta P$ was assumed. In the case of channel length l', $<t>$ can be formulized as

$$
\begin{aligned}
\langle t \rangle &= \int_0^r \frac{dl}{u} \\
&= \int_0^r \frac{32\mu l}{d_h^2 \Delta P} dl \\
&= \frac{16\mu l'^2}{d_h^2 \Delta P},
\end{aligned}
\tag{17.10}
$$

On the basis of these assumptions, the introduction time $<t>$ and mean velocity $<v>$ were estimated as 2.3 ms and 43 mm s^{-1}, respectively. The analysis suggested that the liquid velocity in the extended nanospace channel was more than four times slower than the estimation. The slower velocity suggested higher viscosity in the extended nanospace channel.

Kaji et al. (2006) have also reported higher viscosity in the extended nanospace. They used nanopillar chip having a depth of 400 nm. The nanopillar diameter, spacing, and height were 500, 500, and 400 nm, respectively. Fluorescent carboxylated polystyrene nanoparticles having a diameter of 50 nm were introduced in the nanopillar chip, and the diffusion coefficient of the nanoparticles was evaluated from the measurement of the trajectory of the nanoparticle. The evaluated diffusion coefficient was one-third of the theoretical value derived from the Stokes–Einstein equation. This result gives indirect proof that water viscosity in the extended nanospace is higher than that in a bulk solution.

By the air pressure control method, the miscible solutions have been mixed. In order to realize various nanochemical processing, the fluid control of the immiscible solutions must be realized. Therefore, the hydrophilic and hydrophobic pattern on nanoscale is essential for stable multiphase nanoflows. Hagen–Poiseuille's law can be applied to the nanofluid. However, the increase of the water viscosity in the extended nanospace channel should be carefully considered. In order to clarify the reasons for the property changes, the investigations from the viewpoint of the molecular physical chemistry should be essential.

Tsukahara et al. (2007) have reported size-confinement effects of the molecular structure, motions, protonic mobility, and localization of proton-charge distribution of water and water-surface proton exchange in 295–5000 nm extended nanospaces by NMR spectroscopy measurements. Their results indicated that the water molecules confined in extended nanospaces kept the four-coordinated hydrogen bond structures without changing O–O distance between water molecules seen for ordinary liquid water, the molecular motions of water in extended nanospaces are inhibited compared to those of bulk water, and water molecules in extended nanospace are loosely coupled in a direction perpendicular to the surfaces. The proton transfer rate of water in 120 nm space was about 20 times faster than that in bulk space.

It has been expected that unique properties of water in extended nanospaces would affect reaction properties, and that a nanochemical reaction would be quite different from that in a microspace. Tsukahara et al. (2008a) has reported an enzyme reaction, in which the fluorogenic substrate was hydrolyzed to fluorescein derivative by the enzyme acting as a catalyst in a Y-shaped extended nanospace channel. The first-order rate constant of the enzyme reaction, k_{cat}, in an extended nanospace was determined and compared with those in bulk and microspace. Comparison showed that the enzyme reaction rate in the Y-shaped extended nanospace channel increased by a factor of about 2 compared with rates in the bulk and microspaces. Furthermore, Tsukahara et al. (2008b) studied size dependence of keto-enol tautomeric equilibrium of acetylacetone (Hacac) in water confined in 202–5000 nm extended nanospaces based on 1H-NMR spectroscopy results. They found that keto tautomers of Hacac increased with decrease in the sizes below 500 nm because proton chemical exchange between Hacac and water could be enhanced by charged surfaces. Similar size-confinement phenomena have been established from experimental results of electrophoretic transports of ionic species and ion conductivities of fluidics. Enhancement of proton conductivities and ion enrichment appeared in 1-D extended nanospaces because the electric double layer overlap between charged surfaces could be induced by size confinement, and counterions in an extended nanospace were repelled from the charged surfaces (Liu et al. 2005; Pu et al. 2004; Stein et al. 2004; Plecis et al. 2005). It has been reported that DNA fragments (1–100 kbp) in 1-D extended nanospaces or nanopillars could also be separated according to the variation of space sizes. The size-dependent separations were attributed to the electrostatic interactions between DNA molecules and charged surfaces (Kaji et al. 2004; Abgrall and Nguyen 2008; Cross et al. 2007).

17.5 Conclusions

The basic methodologies and applications for microfluidics and nanofluidics were introduced. In Section 17.2, the general concepts (MUO and CFCP) for microintegration were discussed. The CFCP approach can be applied to develop a more complicated chemical processing system. If CAD was used for microchemical processing, which includes microfluid simulation, reaction time analysis, extraction time analysis, and microchannel structure design, the time of development for microchemical systems would be drastically

17.4.3 Expected Analytical Applications

As illustrated in the previous sections, extended nanospace is an important region in bridging the gap between single molecules and the condensed phase, and basic sciences such as fluidics, chemistry, and biochemistry in the extended nanospace still remain unsolved. From an application point of view, since the volume of the extended nanospace is typically smaller than fL (=1 μm^3), it is very suitable for single-cell analysis because cell volumes are typically in the range of 10 pL. Since the analytical scale is much smaller than cell volumes, future medical and bioanalysis advances may rely on the application of extended nanospace to implement single-cell proteomics or metabolomics. In addition to the quite small volume, the surface-to-volume ratio is several orders larger than that in microspace; therefore, precise trap and reaction at single-molecule level can be expected. Integration of immunoassay into the extended nanospace will become one of the key future technologies. However, basic methods for the integration were not developed. So far, the basic methods were reported for the integration, which was nano-in-microchannel fused silica chip, pressure-driven flow by a pressure controller. It was thus demonstrated, for the first time, that immunoassay could be successfully performed, and detected with alpha-fetoprotein (AFP), in an extended nanospace channel (Kojima et al. 2009). Currently, sensitivity enhancement utilizing ELISA format and thermal lens microscope (TLM) detection are in development, and ~20 molecules could be successfully detected. In addition, selective single-cell patterning method was also realized (Jang et al. 2009). By coupling single-cell patterning on microchannels and ELISA on extended nanospace channels, these new technologies become very powerful tools for single-cell biology or single-molecule analysis.

shortened. In Section 17.3, fluid control methods were discussed. The viscous force and the interfacial tension are effective forces in microspaces. In order to control these forces, surface chemistry and microchannel structures play important roles. In Section 17.4, new science fields in extended nanospace were discussed. The 10–1000-nm-scale space is expected to be a promising experimental area for implementing highly integrated micro/nanofluidic devices and understanding molecular physical chemistry. When the solutions are introduced into the extended nanospace channel, the rapid mixing in milliseconds can be realized. The water properties in the extended nanospace are changed compared with those in bulk, for example, higher viscosity of water, inhibited motion of water molecules, and faster proton transfer rate of water.

Since the basic methodologies have already been developed, development of microfluidic systems for practical applications is required in the next stage. For nanofluidics, nanounit operations should be developed and fundamental physical and chemical properties should be investigated.

References

Abgrall, P. and Nguyen, N. T. 2008. Nanofluidic devices and their applications. *Anal. Chem.* **80**: 2326–41.

Aota, A., Hibara, A., and Kitamori, T. 2007a. Pressure balance at the liquid-liquid interface of micro counter-current flows in microchips. *Anal. Chem.* **79**: 3919–24.

Aota, A., Hibara, A., Shinohara, K. et al. 2007b. Flow velocity profile of micro counter-current flows. *Anal. Sci.* **23**: 131–33.

Aota, A., Mawatari, K., Kihira, Y. et al. 2009a. Micro continuous gas analysis system of ammonia in cleanroom. In *Proceedings Micro Total Analysis Systems 2009*, eds. Kim, T., Lee, Y. S., Chung, T. D., Jeon, N. L., Lee, S. H., Suh, K. Y., Choo, J., Kim, Y. K., 609–11. Jeju, Korea: Chemical and Biological Microsystems Society.

Aota, A., Mawatari, K., Takahashi, S. et al. 2009b. Phase separation of gas-liquid and liquid-liquid microflows in microchips. *Microchim. Acta* **164**: 249–55.

Aota, A., Nonaka, M., Hibara, A. et al. 2007c. Countercurrent laminar microflow for highly efficient solvent extraction. *Angew. Chem. Int. Ed.* **46**: 878–80.

Auroux, P. A., Lossifidis, D., Reyes, D. R. et al. 2002. Micro total analysis systems. 2. Analytical standard operations and applications. *Anal. Chem.* **74**: 2637–52.

Beebe, D. J., Moore, J. S., Bauer, J. M. et al. 2000. Functional structures for autonomous flow control inside micro fluidic channels. *Nature* **404**: 588–90.

Brody, J. P. and Yager, P. 1997. Diffusion-based extraction in a microfabricated device. *Sens. Actuators A* **58**:13–18.

Burns, J. R. and Ramshaw, C. 2001. The intensification of rapid reactions in multiphase systems using slug flow in capillaries. *Lab Chip* **1**: 10–15.

Cross, J. D., Strychalski, E. A., and Craighead, H. G. 2007. Size-dependent DNA mobility in nanochannels. *J. Appl. Phys.* **102**: 024701-1-5.

Dittrich, P. S., Tachiakwa, K., and Manz, A. 2006. Micro total analysis systems. Latest advancements and trends. *Anal. Chem.* **78**: 3887–908.

Goto, M., Sato, K., Murakami, A. et al. 2005. Development of a microchip-based bioassay system using cultured cells. *Anal. Chem.* **77**: 2125–31.

Goto, M., Tsukahara, T., Sato, K. et al. 2008. Micro- and nanometer-scale patterned surface in a microchannel for cell culture in microfluidic devices. *Anal. Bioanal. Chem.* **390**: 817–23.

Gui, L. and Liu, J. 2004. Ice valve for a mini/micro flow channel. *J. Micromech. Microeng.* **14**: 242–46.

Hachiya, H., Matsumoto, T., Kanda, K. et al. 2004. Micro environmental gas analysis system by using gas-liquid two phase flow. In *Proceedings of Micro Total Analysis Systems 2004*, eds. Laurell, T., Nillson, J., Jensen, K. F., Harrison, D. J., Kutter, J. P, 91–101. Molmö, Sweden: Royal Society of Chemistry.

Hibara, A., Iwayama, S., Matsuoka, S. et al. 2005. Surface modification method of microchannels for gas-liquid two-phase flow in microchips. *Anal. Chem.* **77**: 943–47.

Hibara, A., Kasai, K., Miyaguchi, H. et al. 2008. Novel two-phase flow control concept and multi-step extraction microchip. In *Proceedings of Micro Total Analysis Systems 2008*, eds. Locascio, L. E., Gaitan, M., Paegel, B. M., Ross, D. J., Vreeland, W. N, 1326–28. San Diego, CA: Chemical and Biological Microsystems Society.

Hibara, A., Nonaka, M., Hisamoto, H. et al. 2002a. Stabilization of liquid interface and control of two-phase confluence and separation in glass microchips by utilizing octadecylsilane modification of microchannels. *Anal. Chem.* **74**: 1724–28.

Hibara, A., Nonaka, M., Tokeshi, M. et al. 2003. Spectroscopic analysis of liquid/liquid interfaces in multiphase microflows. *J. Am. Chem. Soc.* **125**: 14954–55.

Hibara, A., Saito, T., Kim, H. B. et al. 2002b. Nanochannels on a fused-silica microchip and liquid properties investigation by time-resolved fluorescence measurements. *Anal. Chem.* **74**: 6170–76.

Hibara, A., Tokeshi, M., Uchiyama, K. et al. 2001. Integrated multi-layer flow system on a microchip. *Anal. Sci.* **17**: 89–93.

Hiki, S., Mawatari, K., Hibara, A. et al. 2006. UV-excitation thermal lens microscope for non-labeled and ultrasensitive detection of non-fluorescent molecules. *Anal. Chem.* **78**: 2859–63.

Hisamoto, H., Horiuchi, T., Tokeshi, M. et al. 2001a. On-chip integration of neutral ionophore-based ion pair extraction reaction. *Anal. Chem.* **73**: 1382–86.

Hisamoto, H., Horiuchi, T., Uchiyama, K. et al. 2001b. On-chip integration of sequential ion sensing system based on intermittent reagent pumping and formation of two-layer flow. *Anal. Chem.* **73**: 5551–56.

Hisamoto, H., Shimizu, Y., and Uchiyama, K. et al. 2003. Chemico-functional membrane for integrated chemical processes on a microchip. *Anal. Chem.* **75**: 350–54.

Hotokezaka, H., Tokeshi, M., Harada, M. et al. 2005. Development of the innovative nuclide separation system for high-level radioactive waste using microchannel chip-extraction behavior of metal ions from aqueous phase to organic phase in microchannel. *Prog. Nucl. Energy* **47**: 439–47.

Howard, K. S. and Mcallister, R. A. 1958. The viscosity of acetone-water solutions up to their normal boiling points. *AIChE J.* **4**: 362–66.

Jang, K., Sato, K., Mawatari, K. et al. 2009. An efficient surface modification for cell micropatterning using photochemical reaction. *Biomaterials* **30**: 1413–20.

Kaji, N., Ogawa, R., Oki, A. et al. 2006. Study of water properties in nanospace. *Anal. Bioanal. Chem.* **386**: 759–64.

Kaji, N., Tezuka, Y., Takamura, Y. et al. 2004. Separation of long DNA molecules by quartz nanopillar chips under a direct current electric field. *Anal. Chem.* **76**: 15–22.

Kakuta, M., Takahashi, H., Kazuno, S. et al. 2006. Development of the microchip-based repeatable immunoassay system for clinical diagnosis. *Meas. Sci. Technol.* **17**: 3189–94.

Kikutani, Y., Hisamoto, H., Tokeshi, M. et al. 2004. Micro wet analysis system using multi-phase laminar flows in three-dimensional microchannel network. *Lab Chip* 4: 328–32.

Kojima, R., Mawatari, K., Tsukahara, T. et al. 2009. Integration of immunoassay into extended nanospace. *Microchim. Acta* 164: 307–10.

Kopp, M. U., de Mello, A. J., and Manz, A. 1998. Chemical amplification: Continuous-flow PCR on a chip. *Science* 280: 1046–48.

Kralj, J. G., Schmidt, M. A., and Jensen, K. F. 2005. Surfactant-enhanced liquid–liquid extraction in microfluidic channels with inline electric-field enhanced coalescence. *Lab Chip* 5: 531–35.

Lide, D. R. 2001. *CRC Handbook of Chemistry and Physics.* 82nd ed. Boca Raton, FL: CRC Press Inc.

Liu, S., Pu, Q., Gao, L. et al. 2005. From nanochannel-induced proton conduction enhancement to a nanochannel-based fuel cell. *Nano Lett.* 5: 1389–1393.

Magallanes, C., Catenaccio, A., and Mechetti, H. 1989. Relaxation time and viscosity of several *n*-alcohol/heptane systems. *J. Mol. Liq.* 40: 53–63.

Maruyama, T., Kaji, T., Ohkawa, T. et al. 2004. Intermittent partition walls promote solvent extraction of metal ions in a microfluidic device. *Analyst* 129: 1008–13.

Mawatari, K., Tokeshi, M., and Kitamori, T. 2006. Quantitative detection and fixation of single and multiple gold nanoparticles on a microfluidic chip by thermal lens microscope. *Anal. Sci.* 22: 781–84.

Minagawa, T., Tokeshi, M., and Kitamori, T. 2001. Integration of a wet analysis system on a glass chip: Determination of Co(II) as 2-nitroso-1-naphtol chelates by solvent extraction and thermal lens microscope. *Lab Chip* 1: 72–75.

Miyaguchi, H., Tokeshi, M., Kikutani, Y. et al. 2006. Microchip-based liquid-liquid extraction for gas-chromatography analysis of amphetamine-type stimulants in urine. *J. Chromatogr. A* 1129: 105–10.

Mueller, C. R. and Ignatowski, A. J. 1960. Equilibrium and transport properties of the carbon tetrachloride-methylene chloride system. *J. Chem. Phys.* 32: 1430–34.

Ohashi, T., Matsuoka, Y., Mawatari, K. et al. 2006. Automated micro-ELISA system for allergy checker: A prototype and clinical test. In *Proceedings of Micro Total Analysis Systems 2006,* eds. Kitamori, T., Fujita, H., Hasebe, S., 858–60. Japan: Japan Academic Association Inc.

Plecis, A., Schoch, R. B., and Renaud, P. 2005. Ionic transport phenomena in nanofluidics: Experimental and theoretical study of the exclusion-enrichment effect on a chip. *Nano Lett.* 5: 1147–55.

Proskurnin, M. A., Slyadnev, M. N., Tokeshi, M. et al. 2003. Optimization of thermal lens microscopic measurements in a microchip. *Anal. Chim. Acta* 480: 79–95.

Pu, Q., Yun, J., and Temkin, H. et al. 2004. Ion-enrichment and ion-depletion effect of nanochannel structures. *Nano Lett.* 4: 1099–103.

Reyes, D. R., Lossifidis, D., Auroux, P. A. et al. 2002. Micro total analysis systems. 1. Introduction, theory, and technology. *Anal. Chem.* 74: 2623–36.

Sahoo, H. R., Kralj, J. G., and Jensen, K. F. 2007. Multistep continuous-flow microchemical synthesis involving multiple reactions and separations. *Angew. Chem. Int. Ed.* 46: 5704–08.

Sato, K., Tokeshi, M., Kimura, H. et al. 2001. Determination of carcinoembryonic antigen in human sera by integrated bead-bed immunoassay in a microchip for cancer diagnosis. *Anal. Chem.* 73: 1213–18.

Sato, K., Tokeshi, M., Kitamori, T. et al. 1999. Integration of flow injection analysis and zeptomole-level detection of the Fe(II)-o-phenanthroline complex. *Anal. Sci.* 15: 641–45.

Sato, K., Tokeshi, M., Odake, T. et al. 2000b. Integration of an immunosorbent assay system: Analysis of secretory human immunoglobulin A on polystyrene beads in a microchip. *Anal. Chem.* 72: 1144–47.

Sato, K., Tokeshi, M., Sawada, T. et al. 2000a. Molecular transport between two phases in a microchannel. *Anal. Sci.* 16: 455–56.

Sato, K., Yamanaka, M., and Hagino, T. et al. 2004. Microchip-based enzyme-linked immunosorbent assay (microELISA) system with thermal lens detection. *Lab Chip* 4: 570–75.

Sato, K., Yamanaka, M., Takahashi, H. et al. 2002. Microchip-based immunoassay system with branching multichannels for simultaneous determination of interferon gamma. *Electrophoresis* 23: 734–39.

Slyadnev, M. N., Tanaka, Y., Tokeshi, M. et al. 2001. Photothermal temperature control of a chemical reaction on a microchip using an infrared diode laser. *Anal. Chem.* 73: 4037–44.

Smirnova, A., Mawatari, K., Hibara, A. et al. 2006. Micro-multiphase laminar flows for the extraction and detection of carbaryl derivative. *Anal. Chim. Acta* 558: 69–74.

Smirnova, A., Shimura, K., Hibara, A. et al. 2007. Application of a micro multiphase laminar flow on a microchip for extraction and determination of derivatized carbamate pesticides. *Anal. Sci.* 23: 103–07.

Song, H., Tice, J. D., and Ismgailov, R. F. 2003. A microfluidic system for controlling reaction networks in time. *Angew. Chem. Int. Ed.* 42: 767–72.

Sorouraddin, H. M., Hibara, A., and Kitamori, T. 2001. Use of a thermal lens microscope in integrated catecholamine determination on a microchip. *Fresenius' J. Anal. Chem.* 371: 91–96.

Sorouraddin, H. M., Hibara, A., Proskrunin, M. A. et al. 2000. Integrated FIA for the detection of ascorbic acid and dehydroascorbic acid in microfabricated glass-channel by thermal-lens microscopy. *Anal. Sci.* 16: 1033–37.

Stein, D., Kruithof, M., and Dekker, C. 2004. Surface-charge-governed ion transport in nanofluidic channels. *Phys. Rev. Lett.* 93: 03590-1–035901-4.

Surmeian, M., Hibara, A., Slyadnev, M. et al. 2001. Distribution of methyl red on water-organic liquid interface in microchannel. *Anal. Lett.* 34: 1421–29.

Surmeian, M., Sladnev, M. N., Hisamoto, H. et al. 2002. Three-layer flow membrane system on a microchip for investigation of molecular transport. *Anal. Chem.* 74: 2014–20.

Takei, G., Aota, A., Hibara, A. et al. 2007b. Phase separation of segmented flow by the photocatalytic wettability patterning and tuning of microchannel surface. In *Proceedings of Micro Total Analysis Systems 2007,* eds. Viovy, J. L., Tabeling, P., Descroix, S., and Malaquin, L., 1213–15. San Diego: Chemical and Biological Microsystems Society.

Takei, G., Nonogi, M., Hibara, A. et al. 2007a. Tuning microchannel wettability and fabrication of multiple-step Laplace valves. *Lab Chip* 7: 596–602.

Tamaki, E., Hibara, A., Kim, H. B. et al. 2006. Pressure-driven flow control system for nanofluidic chemical process. *J. Chromatogr. A* 1137: 256–62.

Tamaki, E., Hibara, A., Tokeshi, M. et al. 2003. Microchannel-assisted thermal-lens spectrometry for microchip analysis. *J. Chromatogr. A* 987: 197–204.

Tamaki, E., Hibara, A., Tokeshi, M. et al. 2005. Tunable thermal lens spectrometry utilizing microchannel-assisted thermal lens spectrometry. *Lab Chip* 5: 129–31.

Chapter 17

Tamaki, E., Sato, K., Tokeshi, M. et al. 2002. Single cell analysis by a scanning thermal lens microscope with a microchip: Direct monitoring of cytochrome-c distribution during apoptosis process. *Anal. Chem.* **74**: 1560–64.

Tanaka, Y., Sato, K., Yamato, M. et al. 2004. Drug response assay system in a microchip using human hepatoma cells. *Anal. Sci.* **20**: 411–23.

Tanaka, Y., Sato, K., Yamato, M. et al. 2006. Cell culture and life support system for microbio reactor and bioassay. *J. Chromatogr. A* **1111**: 233–37.

Tanaka, Y., Slyadnev, M. N., Hibara, A. et al. 2000. Non-contact photothermal control of enzyme reaction on a microchip by using a compact diode laser. *J. Chromatogr. A* **894**: 45–51.

Tokeshi, M., Minagawa, T., and Kitamori, T. 2000a. Integration of a microextraction system on a glass chip: Ion-pair solvent extraction on Fe(II) with 4,7-diphenyl-1,10-phenanthroline-disulfonic acid and tri-*n*-octylmethylammonium chloride. *Anal. Chem.* **72**: 1711–14.

Tokeshi, M., Minagawa, T., and Kitamori, T. 2000b. Integration of a microextraction system: Solvent extraction of Co-2-nitroso-5-dimethylaminophenol complex on a microchip. *J. Chromatogr. A* **894**: 19–23.

Tokeshi, M., Minagawa, T., Uchiyama, K. et al. 2002. Continuous-flow chemical processing on a microchip by combining microunit operations and a multiphase flow network. *Anal. Chem.* **74**: 1565–71.

Tokeshi, M., Uchida, M., Hibara, A. et al. 2001. Determination of sub-yoctomole amounts of non-fluorescent molecules using a thermal lens microscope: Sub-single molecule determination. *Anal. Chem.* **73**: 2112–16.

Tokeshi, M., Yamaguchi, J., Hattori, A., et al. 2005. Micro thermal lens optical systems. *Anal. Chem.* **77**: 626–30.

Tsukahara, T., Hibara, A., Ikeda, Y. et al. 2007. NMR study of water molecules confined in extended nanospaces. *Angew. Chem. Int. Ed.* **46**: 1180–83.

Tsukahara, T., Mawatari, K., Hibara, A. et al. 2008a. Development of a pressure-driven nanofluidic control system and its application to an enzymatic reaction. *Anal. Bioanal. Chem.* **391**: 2745–52.

Tsukahara, T., Nagaoka, K., and Kitamori, T. 2008b. Deuterium-substitution and solvent effects on tautomeric reaction dynamics in extended-nano spaces on a chip. In *Proceedings of Micro Total Analysis Systems 2008*, eds. Locascio, L. E., Gaitan, M., Paegel, B. M., Ross, D. J., and Vreeland, W. N., 1326–28, 91–101. San Diego, CA: Chemical and Biological Microsystems Society.

Unger, M. A., Chou, H. P., Thorsen, T. et al. 2000. Monolithic microfabricated valves and pumps by multilayer soft lithography. *Science* **288**: 113–16.

van der Linden, H. J., Jellema, L. C., Holwerda, M. et al. 2006. Stabilization of two-phase octanol/water flows inside poly (dimethylsiloxane) microchannels using polymer coatings. *Anal. Bioanal. Chem.* **385**: 1376–83.

Weast, R. C. 1985. *CRC Handbook of Chemistry and Physics*. 66th ed. Boca Raton, FL: CRC Press Inc.

Weigl, B. H. and Yager, P. 1999. Microfluidics—Microfluidic diffusion-based separation and detection. *Science* **283**: 346–47.

Yamauchi, M., Mawatari, K., Hibara, A. et al. 2006. Circular dichroism thermal lens microscope for sensitive chiral analysis on microchip. *Anal. Chem.* **78**: 2646–50.

Yu, C., Mutlu, S., Selvaganapathy, P., Mastrangelo, C. H. et al. 2003. Flow control valves for analytical microfluidic chips without mechanical parts based on thermally responsive monolithic polymers. *Anal. Chem.* **75**: 1958–61.

Zhao, B., Moore, J. S., and Beebe, D. J. 2001. Surface-directed liquid flow inside microchannels. *Science* **291**: 1023–26.

Zhao, B., Moore, J. S., and Beebe, D. J. 2002a. Principles of surface-directed liquid flow in microfluidic channels. *Anal. Chem.* **74**: 4259–68.

Zhao, B., Moore, J. S., and Beebe, D. J. 2003. Pressure-sensitive microfluidic gates fabricated by patterning surface free energies inside microchannels. *Langmuir* **19**: 1873–79.

Zhao, B., Viernes, N. O. L., Moore, J. S. et al. 2002b. Control and applications of immiscible liquids in microchannels. *J. Am. Chem. Soc.* **124**: 5284–85.

18. Fabrication of Nanophotonic Structures

Stefano Cabrini, P. James Schuck, and Alexander Weber-Bargioni

Molecular Foundry, Lawrence Berkeley National Laboratory, Berkeley, California

Allan Chang

Lawrence Livermore National Laboratory, Livermore, California

Manipulating light requires fabrication at multiple length scales. At nanometer length scales, nanofabrication capabilities enable the realization of small features required for the most advanced nanophotonic structures, and also provide the precise shape and roughness control needed to create high-efficiency photonic devices with larger features. Indeed, nanofabrication science and technology sits at the foundation of the nanophotonics field [1].

This chapter discusses three cutting-edge applications that highlight the essential role of nanofabrication in the achievement of important results. In the first case, conventional fabrication techniques were used to obtain photonic crystal (PhC) structures demonstrating negative refractive index behavior, Here, high-precision, well-controlled lithographic, and dry-etching processes were key in creating the final device.

The second case focuses on an application of very-high-resolution electron beam lithography, which reproducibly fabricated plasmonic structures with sub-10-nm (the so-called "single-digit-nano") resolution. This enabled the systematic study of the nanophotonic

Nanofabrication Handbook. Edited by Stefano Cabrini and Satoshi Kawata © 2012 CRC Press / Taylor & Francis Group, LLC. ISBN: 978-1-4200-9052-9

Chapter 18

properties of these devices, yielding important insights into the plasmonic behavior of subwavelength optical antennae.

The third case describes a particularly demanding investigation, which required the fabrication of high-resolution optical antenna structures on an unusual substrate. In this case, all the standard techniques were unable to reproducibly place the plasmonic nanodevices at the apex of an atomic force microscope (AFM) tip. To successfully accomplish such a hard task, a completely new method was invented.

This led to the creation of a practical macroscale device built to facilitate the use of the nanodevice's unique properties in the real world. In fact, with these new tip-based structures, we now have the opportunity to realize a next-generation scanning microscope capable of providing optical information with single-digit-nanospatial resolution.

In short, nanofabrication affords enormous possibilities for expanding the field of nanophotonics through controlling the shapes, materials, geometries, and compositions of matter at the nanoscale.

18.1   PhC Metamaterials

Metamaterials are artificial composite materials consisting of structural features much smaller than the incident electromagnetic radiation wavelength, such that they display electromagnetic properties not found in natural materials. With the correct design, metamaterials can be engineered to exhibit an effective index of refraction that appears negative to the incident light. The advent of negatively refracting metamaterials (NRMs) [2] opens up numerous intriguing possibilities [3–6] for controlling light waves that are unavailable in natural positive refractive index materials and it has thus generated enormous scientific interests. In particular, negative refracting dielectric PhCs [7–9] represents a viable way to realize NRMs for controlling long-range light propagation due to their low absorption loss. Subdiffraction imaging in the near-infrared was experimentally demonstrated using negative refracting PhCs, verifying the amplification of evanescent waves by these metamaterial [10]. The use of negative refracting PhCs for collimation can thus be very attractive because, contrary to their positive counterpart, they can have all-angle acceptance and ability to capture evanescent components, meaning subwavelength beam spot size and transfer of nearfield information are conceivable. Inspired by the principle of complementary media [11–13], it was demonstrated that a metamaterial comprising alternating stripe layers of negative (2D PhC) and positive (air) refracting materials [14,15], precisely a quasi-zero-average-index metamaterial [16,17], leads to strong self-collimation of a near-infrared beam over a large distance of 2 mm [18]. This distance represents more than 1000 times the input wavelength $\mu = 1.55$ μm while the beam spot size is fully preserved throughout the entire sample. Key to the achievement of this result is the ability to implement delicate high-aspect-ratio photonic nanostructure through high-precision nanofabrication.

Figure 18.1 shows the scanning electron microscope (SEM) images of the actual device made by a high-precision nanofabrication process using high-voltage electron beam lithography and a gas-chopping inductively coupled plasma (ICP) etching process [19]. The device is made on an silicon on insulator (SOI) substrate and the total area is 2×2 mm. The SEM images show holes with 360 nm diameter and periodicity of 470 nm in every PhC stripe. The depth of the patterns is 1.5 μm.

The fabrication starts with the electron-beam (e-beam) evaporation deposition of a thin layer (40 nm) of chromium onto an SOI wafer. This chromium layer serves as both a charge dissipation layer and etching mask for the later steps of e-beam lithography and silicon etching, respectively. Its thickness is chosen so that it is thick enough to act as an etching mask and yet thin enough for high-fidelity pattern transfer from e-beam resists. Next, the patterns are

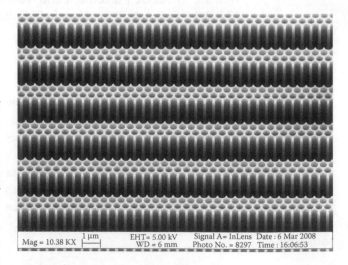

FIGURE 18.1   Scanning electron micrograph of a device made by alternating positive refractive index material (air) and negative one (PhC). The right proportion of the two sections made a "quasi-zero-average-index metamaterial."

created using 100 kV e-beam lithography (Vistec VB300) in a 300 nm layer of spun-on positive e-beam resist (ZEP 520 A). The dosage is carefully chosen to obtain the desired final holes diameter. After oxygen plasma descum, the pattern is transferred into the underlying chromium layer using a chlorine/oxygen plasma reactive ion etching (RIE) process (Oxford system 100).

The pattern is then transferred into the top silicon layer of the SOI wafer in gas-chopping ICP process cycles that alternates between SF_6/Ar and CHF_3/CH_4 plasmas. A hard mask is required in this case for the pattern to go down the full 1.5 μm in silicon. ICP systems can sustain high-density plasma at low pressure, resulting in faster etch rates and/or anisotropic etching at lower substrate dc bias than conventional RIE. However, a single-step ICP process such as HBr plasma is inadequate for the required silicon etching depth of 1.5 μm due to insufficient selectivity with the mask. In a gas-chopping plasma process, a first short etching step is performed that etches the silicon at fast rate but with low anisotropicity. An undercut is thus developed in this step. It is therefore necessary to follow up by a short deposition step to deposit a thin layer of protective polymer film to passivate the sidewall. This two-step cycle is then repeated again to achieve the desired total etched depth. This process involves a delicate balance between the etching and deposition steps, and it gives a typical scallop effect of the sidewall. The scallops can be of the order of 10s to 100s of nanometers. If the etching dominates, then unacceptably large undercut and poor feature fidelity will develop. On the other hand, if the deposition dominates, then etching will stop eventually before the desired etching thickness is reached. For micron-scale features on which this type of process is commonly used, the scallop effect is less limiting than the nanoscale features. To achieve nanoscale features, the key is to be able to decrease each step time to a minimum with fast-switching gas valves. By controlling the chopping timing and the pressure in the chamber accurately, it is possible to reduce this effect to a dimension much smaller than the propagating wavelength.

For the etching step, flow rates of 4 sccm for SF_6 and 7 sccm for Ar are used. The chamber pressure is 6 mTorr. The forward power is 30 W while the ICP power is 1000 W. The temperature of the substrate holder is controllably cooled by liquid nitrogen to 5°C. The time of a single etching step is 5 s. For the deposition step, we used 7 sccm of CHF_3 and 4 sccm of CH_4.

The chamber pressure is 25 mTorr. The forward power is 30 W while the ICP power is 1000 W. The substrate holder temperature is 5°C. The time of a single deposition step is 7 s.

It should be noted that because of the different scales between the air stripes (micron-scale) and the air holes (nanoscale), there is a lagging effect in the etching. The etch rate of the air holes is up to 50% slower than the etch rate of the air stripes. This is because it takes more time for the same amount of ions to enter the holes to react with the silicon and more time for the etching residuals to be removed from inside the holes. This effect becomes more prominent as the holes develop deeper during the process. As a result, overetching is necessary to bring the holes to the desired thickness. A process of 55 cycles of the gas-chopping steps is used to achieve an etching depth of 1.5 μm inside the holes. This is confirmed by focused ion beam (FIB) cross section (Figure 18.2).

Apart from the lagging, a sidewall slope may also develop as the process proceeds. To straighten the sidewall in etching a high-aspect-ratio nanostructure using gas-chopping process, the step time can be varied as the process proceeds to compensate for this effect. Essentially, the ratio of the deposition and etching is changed as the hole develops, such that the net sidewall etching is increased to reduce the sidewall slope. If a balance is achieved, an overall straight sidewall can be obtained. By using a variable step-time process (other process parameters being the same as above) that has 20 cycles of 6 s deposition/7 s etching, then 20 cycles of 5 s deposition/7 s etching, then 20 cycles of 2 s dep/7 s

FIGURE 18.2 Close-up scanning electron micrograph of a cross section made by focused ion beam in the PhC section showing the etch depth through the bottom of the holes.

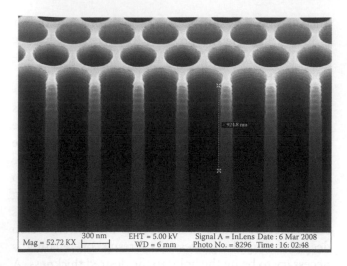

FIGURE 18.3 Close-up scanning electron micrograph of the PhC section showing vertical sidewalls.

etching, followed by 60 cycles of 5 s deposition/9 s etching, a straightened sidewall is obtained as shown in Figure 18.3.

This precise nanofabrication process enables the implementation of the first demonstration of self-collimation of light $\lambda = 1.55$ µm over a distance of 2 mm, as shown in Figure 18.4. For the optical characterization, a 1.55 µm continuous wave (CW) laser is connected to a lensed input fiber that produces an incident full-width-half-maximum beam spot size FWHM = 2.9 µm as experimentally measured using a knife edge integrated on the head of an optical fiber, made by FIB. The input fiber is fixed on a three-axis nanopositioning stage with a resolution of 20 nm.

The light propagation inside the heterostructure metamaterial is directly observed from the top using an infrared camera Xenics Xeva 185 and a high numerical aperture (NA = 0.42) objective lens with long working distance. For the implementation of this setup, the input lensed fiber is moved slightly in y-direction which produces a slight misalignment between the input fiber and silicon guiding layer, resulting in a small vertical component for the incident wave that couples with propagating wavevectors in air. As shown in Figure 18.4, the beam spot size of around 3 µm is preserved throughout the entire metamaterial. The result also shows that loss is low as the beam travels through the metamaterial.

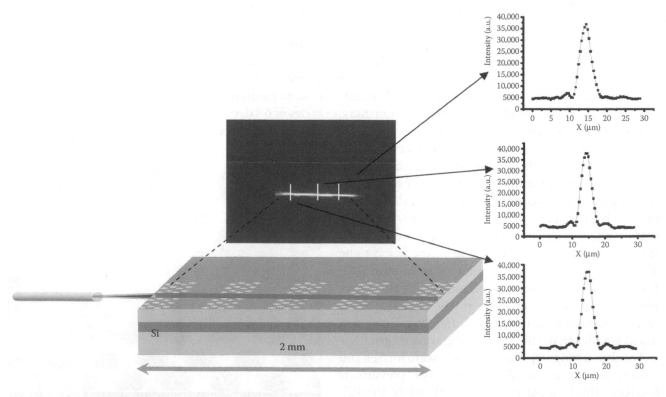

FIGURE 18.4 The schematic of experimental measurements with the image of the scattered radiation along the vertical plane. It reveals a well-collimated beam along the whole 2 mm length of the sample. On the left side is shown the measured beam profile of the scattered radiation in the first part, in the middle, and in the final part of the quasi-zero-average metamaterial. (From Mocella, V. et al., Self-collimation of light over millimeter-scale distance in a quasi-zero-average-index metamaterial. *Physical Review Letters*, 2009. 102(13): 133902. With permission.)

18.2   Applications of Nanofabrication: Plasmonics

"Plasmonics," which can be considered a subset of the more general metal optics field, deals with the unique optical properties of metallic nanostructures. Because metals no longer behave as "ideal" conductors at optical frequencies, plasmonic devices exploit the plasmon–polariton waves in metals so as to achieve optimum interaction with optical fields. In a metal, plasmons, defined as collective, coherent oscillations of charge, may oscillate at optical frequencies, but can have nanoscale wavelengths (i.e., very large momenta, or k-vectors), allowing one to manipulate optical signals at length scales far below the conventional diffraction limit. Driven by recent advances in sample characterization technologies, theoretical simulation tools, and, perhaps most importantly, nanofabrication capabilities, plasmonics has been one of the most exciting and fast-moving areas of photonics (see, e.g., [20–23]). Innovation has been fueled by a multitude of potential applications, including (bio)molecular sensing, cancer therapy, solar power generation, and deeply subwavelength light concentration, and guiding. Of particular interest is the concept of using surface plasmon–polaritons (SPPs) as information carriers in future highly integrated, ultrafast nanophotonic devices. Indeed, plasmonic technologies are poised to seed the next "nanoelectronics revolution" by combining the nanoscale component dimensions of current electronic technologies with the speed and bandwidth of photonics technologies.

Of course, completely controlling optical signals at the nanoscale requires the fabrication of plasmonic devices with "single-digit-nano" precision and resolution. Below, as an example of the critical role played by state-of-the-art nanofabrication in plasmonics, we first describe a class of plasmonic nanostructures, termed as Plasmonic Color nano-Sorters (PCoNs), whose local properties intimately depend on nm-scale changes in the orientation of the constituent metallic pieces—and particularly on small changes to the degree of asymmetry present in the device. This will be followed by a discussion on a novel all-optical characterization concept that enables the visualization of nanoscale-controlled shifts of zeptoliter mode volumes within plasmonic nanostructures. As a proof of concept, we show that it is possible to image the local energy-dependent changes in nearfield distributions within individual gold asymmetric bowtie nanocolorsorters (ABnCs), a specific type of PCoN. In the final section of this chapter, we reveal how small structural variations often

significantly impact on plasmonic properties, particularly for devices with ~zeptoliter mode volumes.

18.2.1   Manipulation of Nano-Optical Fields Enabled by "Single-Digit-Nano" Fabrication

18.2.1.1   A Device for Capturing, Concentrating, and Sorting Light: The ABnCs

A central goal of plasmonics is complete control over optical signals at deeply subwavelength scales. The recent invention of optical nanoantennae has led to a number of device designs that provide confinement of optical fields at nanometer length scales [24–36]. For photonic applications, however, the effectiveness of these structures would be *significantly improved* by the added ability to spatially sort the optical signals based on a physically accessible parameter such as energy/color [37–41]. Here, we present our experimental and theoretical study of a class of devices, termed as ABnCs, which demonstrate both the ability to efficiently capture and strongly confine broadband optical fields, as well as to spectrally filter and steer them while maintaining nanoscale field distributions [42]. The latter property is important because it allows for manipulation while preserving the physical match, created by the optical antenna, between the localized field distribution and important physical factors such as semiconductor carrier diffusion lengths and zeptoliter volumes occupied by individual nano- and quantum-objects. Because of these capabilities, ABnCs are expected to have a profound impact on a wide range of optoelectronic and plasmonic applications, including ultrafast color-sensitive photodetection, solar power light harvesting, super-resolution imaging, and multiplexed chemical sensing.

Our proof-of-principle ABnC devices are based on *asymmetric* variations of double-bowtie nanoantennae oriented in a "cross" geometry. As an essential experimental complement to our theoretical modeling, an initial step of our investigation was to demonstrate the feasibility of fabrication of these devices. The nanoantennae are fabricated using e-beam lithography (Vistec VB300, 100 keV beam energy) and lift-off on indium–tin–oxide (ITO)-coated fused silica substrates (ITO thickness = 50 nm), and consist of approximately 17-nm-thick Au on top of a 3 nm Ti adhesion layer. Each constituent Au triangle in the cross nanoantenna is designed to be equilateral in shape with a perpendicular bisector length of 75 nm. Experimentally, the nanoantenna resonances are measured by collecting darkfield scattering spectra from individual structures in a transmission confocal modality: white light is focused on the back of the transparent sample with a high numerical aperture (NA) oil condenser (NA = 1.43–1.2) and scattered light is collected with a 100×0.95 NA air objective, focused through a 150-μm-diameter pinhole, then directed into 0.3 m spectrometer (PI-Acton) and dispersed onto a liquid-nitrogen-cooled charge-coupled device (CCD) camera.

Calculations of the fields surrounding our devices were done using finite-element method (FEM) software from COMSOL. Our analyzed volume consisted of: a 500 (x-axis) $\times$ 500 (y-axis) $\times$ 250 nm (z-axis) top layer of air (index of refraction, $n = 1$), a $500 \times 500 \times 50$ nm middle layer of ITO ($n_{ITO} = 1.91$), and a $500 \times 500 \times 250$ nm bottom layer of glass ($n_{glass} = 1.5$). Perfectly matched layers (PMLs) surrounded the volume (100 nm thick for the sides, 250 nm thick at the bottom). Au nanoantennae of 20 nm thick and ABnC devices were placed directly on the ITO layer, and all triangles had 10 nm radii of curvature at the corners. These parameters are very close to the specifications of the fabricated devices [43]. The wavelength-dependent dielectric constant used for the Au was taken from Palik et al. [44]. Plane wave continuous excitation was incident from the top (air side) with electric field component amplitudes of $e_x = 1$, $e_y = 0$, $e_z = 0$. All calculated field distribution images are plots of normalized electric field (e-field) amplitude

in a plane that cuts through the center of the devices' thickness (i.e., a plane with z set to 10 nm above the Au–ITO interface). Calculated nearfield intensity spectra are plots of the maximum normalized e-field amplitude within this plane at each wavelength. We note that for experimental applications (e.g., nanoscale photodetectors [45] and nanoantenna-based nearfield scanning probes [46–48]), the relevant fields are those located directly in, and just outside, the nanoantenna gap region [49,50], which is why the nearfield spectral and spatial distribution information are calculated and presented here.

To help lend insight into the ABnC, we start by briefly describing the main resonance properties of the *symmetric* "cross" nanoantenna structure (see Figure 18.5; the symmetric cross antenna has C_{4v} symmetry. More detailed discussion on symmetric "cross" nanoantennae can be found in [51]). An experimental darkfield scattering spectrum from an individual symmetric cross nanoantenna is also shown in Figure 18.6 (red curve), with the incident white light linearly polarized in the x-direction. The main mode primarily resembles the dipolar mode of a single bowtie that is excited with e-field polarization-oriented parallel to the bowtie axis [43,52]. A weak shoulder on the blue side of the peak is also present, which is due to the x-polarized light interacting with the y-oriented "vertical" bowtie component of the cross (see [43,52]).

When x-polarized incident light is energetically resonant with the primary dipole nanoantenna mode (at the wavelength corresponding to the red curve peak in Figure 18.6c), the calculated e-field distribution (normalized e-field amplitude) surrounding the

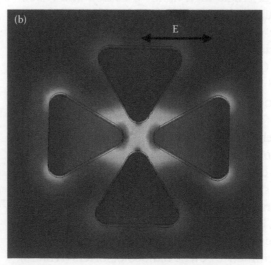

FIGURE 18.5 (**See color insert**.) (a) SEM image of a symmetric bowtie "cross" nanoantenna. (b) Image of the normalized e-field amplitude distribution surrounding the nanoantenna when excited at its primary resonance of ~775 nm.

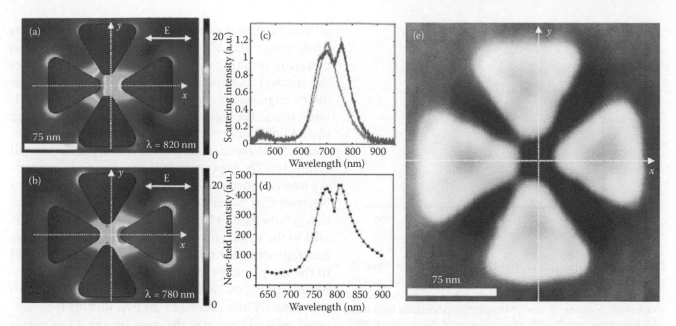

FIGURE 18.6 (See color insert.) Calculated images of the e-field distributions surrounding an ABnC (y-oriented "vertical" bowtie component shifted 5 nm left-of-center) when excited at (a) 820 nm, and (b) 780 nm. Note that the two modes are spectrally and spatially distinct while maintaining nanoscale mode volumes. This ABnC retains a reflection symmetry axis oriented along the x-axis. (c) Experimental darkfield scattering spectra from (blue) the ABnC shown in (e) and (red) a representative symmetric cross nanoantenna for comparison. (d) Calculated nearfield intensity spectrum from the ABnC designed with the 5 nm vertical bowtie shift shown in (a) and (b). (e) SEM image of an ABnC with its vertical bowtie component shifted ~5 nm left-of-center. (From Zhang, Z. et al., Manipulating nanoscale light fields with the asymmetric bowtie nano-colorsorter. *Nano Letters*, 2009. 9(12): 4505–4509. With permission.)

symmetric cross is symmetric as well. As expected, the most intense fields at the mid-height of the nanoantenna are located near the tips of the two triangles comprising the horizontal bowtie [51]. This is consistent with the nearfield distribution associated with simple bowtie nanoantennae with similar gap sizes [53], though the local intensity near the center of the structure falls off more quickly in the y-direction due to added confinement by, and interaction with, the vertical bowtie.

Next, we use an asymmetric bowtie "cross" nanoantenna to demonstrate the basic principles of an ABnC (Figure 18.6). In this case, the asymmetry has been created by moving the y-oriented "vertical" bowtie component of the cross left-of-center by 5 nm, thereby reducing the cross symmetry from C_{4v} to C_s (i.e., the ABnC now has only one symmetry axis). Note the precision required for reproducibly fabricating such structures. The effect of this symmetry breaking can be seen in the field distributions and the scattering spectrum shown in Figure 18.6. By shifting the "vertical" bowtie to the left only a few nm, the degeneracy of the nanoantenna's primary (polarization-aligned dipole) plasmon resonance mode breaks into two, which is observed as a doublet in the scattering spectrum (Figure 18.6c, blue curve) and calculated nearfield

intensity spectrum (Figure 18.6d). The small differences in experimental versus theoretical peak wavelengths are due primarily to small discrepancies in (1) the ITO dielectric constant and (2) metal particle sizes between the real values and those used in our FEM modeling.

The normalized e-field distributions for each of these two resonant peaks are shown in Figure 18.6a and b. It is clear that each resonance not only possesses a spatially distinct field distribution, but also retains a nanoscale mode volume. Each of these zeptoliter volumes can be *individually addressed* simply by adjusting the incident wavelength.

It is straightforward to tune the spectral shift between the two modes; it is controlled by changing how far the vertical bowtie is translated (in the left or right direction for horizontally polarized light) from center (i.e., by increasing the asymmetry and controlling the gap sizes between the constituent parts of the nanoantenna). The effects of shifting the vertical bowtie by 5, 10, and 15 nm, respectively, on the nearfield intensity spectral response of the ABnC are illustrated in Figure 18.7. The peak wavelengths of both modes change as a function of vertical bowtie shift. When the vertical bowtie is moved left-of-center by 15 nm the high-energy mode blueshifts by 13.4 nm, while

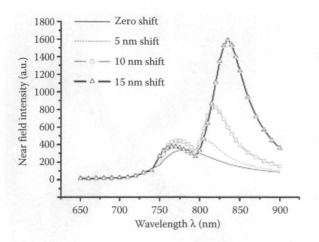

FIGURE 18.7 Calculated nearfield intensity spectra for ABnCs in which the vertical bowtie component has been shifted horizontally by 0 nm (i.e., symmetric cross; black curve), 5 nm, 10 nm, and 15 nm. (From Zhang, Z. et al., Manipulating nanoscale light fields with the asymmetric bowtie nano-colorsorter. *Nano Letters*, 2009. 9(12): 4505–4509. With permission.)

the low-energy mode displays a marked redshift of 54 nm, demonstrating clear coupling between ABnC components.

Both modes result from the interactions between all constituents of the ABnC, though symmetry-breaking affects the modes in different ways. These behaviors can be understood by examining the spatial field distribution associated with each mode (Figure 18.6a,b). As the vertical bowtie shifts away from the right-hand triangle, the coherent coupling between the right-hand triangle and the other ABnC components decreases, leading to the relatively weak spectral shift as a function of increasing asymmetry (i.e., weaker interactions result in less sensitivity to component arrangements). This also explains why the mode associated with shorter-wavelength peak in the doublet bears similarity to the dipole resonance mode of an isolated triangle. On the other hand, the longer-wavelength mode is more hybridized in nature [54]. In this case, the increase in coherent coupling energy—resulting from the decreasing (and small) gap sizes—between the three triangles on the left in Figure 18.6a dominates the mode properties. In the context of the plasmon hybridization model, the increased coupling energy manifests itself as an increased redshift in the lowest energy mode (dipolar-like in this case, which is why it is clearly observed in the scattering spectra). Note that the electric field enhancement for this mode also increases with decreasing gap size (increased asymmetry), as expected.

It is important to emphasize, however, that each mode originates from the complex coupling between

all constituents comprising an ABnC. In fact, this is what makes the concept of the nanocolorsorter so powerful, resulting in the high degree of control over the spectral and spatial properties through the symmetry engineering enabled by high-resolution nanofabrication. The significance of the interplay between each ABnC part is verified by changing the arrangement of the components, while retaining the overall symmetry. The breaking of the dipolar mode degeneracy into two spectrally and spatially distinct modes is a behavior that occurs not only when the single reflection symmetry axis lies along the x-direction (and parallel to the incident e-field orientation), but also more generally when the reflection symmetry axis is rotated in the x–y plane (i.e., whenever C_s symmetry is maintained). In Figure 18.8, we have created an ABnC with its symmetry axis rotated 45° by first shifting the y-oriented vertical bowtie of the cross left (by 5 nm in this case), then up (by 5 nm). The experimental scattering spectrum (Figure 18.8c) and calculated nearfield enhancement spectrum (Figure 18.8d) both still exhibit the signature doublet peaks. (Excitation is x-polarized.) Clearly, it is the overall arrangement/symmetry of the ABnC that determines the spectral mode structure, confirming the importance of all interacting components in defining the properties of the device. Field distributions associated with each of the two peaks remain localized and separated spatially from one another, though the exact spatial nature of the modes differs from that of the ABnC in Figure 18.4 due to the changed geometry.

The ABnC, with its multiple metallic nanostructures separated by nanoscale gaps, is also an example of a *plasmonic molecule*. A number of recent publications have discussed simple asymmetric plasmonic molecule geometries, showing that Fano resonances, often created by broken-symmetry-permitted optical interactions of bright and dark modes, result in intricate (and tunable) spectral behavior [55–60]. This is certainly also the case in the context of the ABnC, where Fano-like resonances are observed in the spectral response. One particular advantage of the cross geometry is that it allows for strong interactions with all polarization states [51], as well as for efficient cross-coupling of an x-polarized incident field into modes with significant y-oriented components (and vice versa).

We note that the device described above is a two-color ABnC. However, one can easily extend the sorting/multiplexing functionality to a greater number of wavelengths by designing the device to possess more

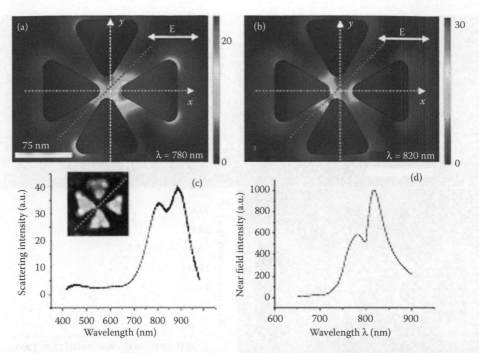

FIGURE 18.8 **(See color insert.)** Calculated images of the e-field distributions surrounding an ABnC designed to have its reflection symmetry axis rotated 45° from x-axis, when excited at (a) 780 nm, and (b) 820 nm. The dotted red lines in (a), (b), and inset of (c) mark the reflection symmetry axis. (c) Experimental darkfield scattering spectrum and (d) calculated nearfield intensity spectrum from the ABnC with this symmetry. Inset of (c) is an SEM image of the ABnC device from which the scattering spectrum in (c) originated. Note that, as in Figures 18.1 and 18.2, retention of a single reflection symmetry axis in the cross ABnC design results in two spatially and spectrally distinct modes. (From Zhang, Z. et al., Manipulating nanoscale light fields with the asymmetric bowtie nano-colorsorter. *Nano Letters*, 2009. 9(12): 4505–4509. With permission.)

asymmetric degrees of freedom. Based on the plasmon interactions that dictated the properties of the two-color ABnC discussed above, we have simulated the properties of a more asymmetric device designed to have three spatially and spectrally distinct modes (Figure 18.9). Here, we exchanged the bottom triangle of the cross for an ellipse (major axis = 50 nm, minor axis = 25 nm), moved it between the tips of the remaining triangles, and enlarged the right-most triangle (perpendicular bisector = 100 nm). In addition, the incident radiation polarization is rotated 45° from the x-axis in the calculations. This device design allowed us to engineer hybridized modes between (a) the long axis of the ellipse and a 75 nm triangle, (b) the short axis of the ellipse and a 75 nm triangle, and (c) the long axis of the ellipse and a 100 nm triangle. Because of the complexity of this device, combinations of these modes and higher-order modes are also present (see supplementary movie). The calculated plasmon mode nearfield spectra for the individual constituents of this device (75 nm triangle, 100 nm triangle, and ellipse) are given in the supplementary information. The nearfield enhancement spectrum calculated for this device is presented in Figure 18.9d. Three peaks

are observed in the spectrum, clearly demonstrating the extension of the basic two-color ABnC to multiple wavelengths.

The fact that spatially distinct, nanoscale mode volumes can be individually addressed by tuning wavelength is of considerable importance to technological applications such as multicolor photon detection and sorting (e.g., red–green–blue (RGB) photodetectors). There are significant advantages to scaling photodetectors down to the nanoscale in all three dimensions, including efficiency and modulation speed [45,61]. When implemented in conjunction with an ABnC, these advantages are combined with near-unity collection efficiency afforded by the optical nanonantenna (and subwavelength device footprint), necessitating the design of high-density, ultrafast multicolor detector arrays.

Thus, advanced nanofabrication techniques have led to the creation of the ABnC category of devices, which now make it possible to collect, concentrate, and manipulate optical fields all while maintaining deeply subwavelength field distributions. This is accomplished by engineering the internal symmetries of nanoantenna structures such that their constituent metallic

Chapter 18

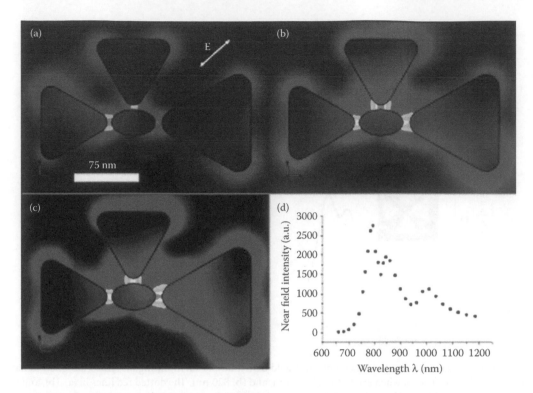

FIGURE 18.9 **(See color insert.)** An example of an ABnC with an increased number of asymmetric degrees of freedom. Images of calculated field distributions showing spatially distinct field confinements due to mode hybridizations at (a) the left triangle-ellipse gap, (b) the top triangle-ellipse gap, and (c) the right triangle-ellipse gap. (d) The calculated nearfield intensity spectrum from this device, which now exhibits three spectral modes (rather than two) due to the increased asymmetry. (From Zhang, Z. et al., Manipulating nanoscale light fields with the asymmetric bowtie nano-colorsorter. *Nano Letters*, 2009. 9(12): 4505–4509. With permission.)

elements create hybridized modes at selected positions and wavelengths. Because of the tunability and controllability of the localized optical fields, we expect these specifically designed plasmonic molecules to be useful for a wide range of multicolor photodetection and optical/plasmonic filtering/sorting applications.

18.2.1.2 Elucidating Effects of Nanoscale Structural Variations on Local Plasmonic Modes

As described above, plasmonic devices are primarily based on metallic (nano)structures with nanoscale feature sizes, and many are designed to enhance and concentrate fields at ~10 nm length scales. This leads to an enhanced sensitivity of local plasmonic modes to locally varying material properties, device substructure, and environment, which must be taken into account for each specific application. Advancement of plasmonics technologies depends on reliable fabrication of these structures, often repeatably over large areas. However, the resolution and reproducibility limits of current nanofabrication techniques, such as e-beam lithography, FIB milling, and e-beam- and FIB-induced deposition, tend also to be ~10 nm, often

struggling to break the so-called "single-digit-nano" barrier. In some cases, "bottom-up" chemically synthesized nanoplasmonic structures can have less structural variation, but these require specific arrangements and orientations for most devices, and specific binding events and placement accuracies are again limited to ~10 nm or greater [62]. Therefore, it is expected that the properties of plasmon modes with localized fields will be influenced by these unavoidable structural variations.

Below, we discuss how the recently demonstrated nonperturbative nanoscale probing technique of two-photon photoluminescence (TPPL)-based photon localization microscopy [63] has been combined with electromagnetic simulations to reveal how small structural variations often significantly impact on plasmonic properties, particularly for devices with ~zeptoliter mode volumes. Specifically, it was shown that for the studied devices, local plasmonic behavior is primarily influenced by two classes of fabrication-related variations: (1) random, "incoherent" localized defects; and (2) small changes in structure size, which are directly related to the cavity length of the plasmon resonator. For this second case, it was shown that relatively modest changes in device length of about 10% can nearly double the spatial separation between modes.

To investigate the effects of fabrication-related imperfections, we chose as our test samples the ABnC optical antennae described in the previous section. These devices are excellent test structures for this study because of their localized field distributions. Here, relative spatial movements of the two modes are correlated with the structural variations we observed via SEM characterization.

To image the local modes, we collect the TPPL signal [24,64–66] from the ABnCs when excited by a pulsed titanium sapphire laser tuned to a mode resonance (in a sample-scanning confocal modality). The TPPL originates from the interaction of the intrinsic local field with the Au nanostructure itself and therefore involves no external perturbation of the device or its local environment. TPPL has previously been used to measure plasmonic field properties such as enhancement and nearfield resonance [24,67,68], as well as mode distributions with diffraction-limited spatial resolution [69–71].

Recently, photon localization microscopy has gained significant attention within the molecular and cellular imaging communities [72–79] due to its ability to determine the position of a single emitter with nanoscale accuracy. A more detailed discussion can be found in [80] and references therein. By applying photon localization microscopy centroid analysis [81,82] to TPPL data from plasmonic systems, we can determine the center position of a local field mode in the same way, separately localizing nanoscale plasmon modes within the same diffraction-limited focal volume. For our sample-scanning confocal TPPL images of the ABnCs, the position accuracy is limited primarily by the number of TPPL photons we collect, and we demonstrate here a 95% confidence interval accuracy of ±2.5 nm. Along with being a novel technique for scientific investigation, one can think of this method as a potential alternative to SEM characterization for fabricated photonic/plasmonic device metrology, with the additional benefit of yielding information about light–matter interaction physics at the nanoscale [63].

The relative spatial separation of the two modes in our ABnCs is dependent on the amount of this asymmetric offset within the structure [42]. This dependence is plotted in Figure 18.6, where it is clear that as the amount of asymmetry increases, so too does the relative separation of the mode centroids. To more carefully compare devices with structure and gap size variations, the relative centroid positions are plotted as a function of a normalized asymmetry parameter

defined as the vertical bowtie offset divided by the measured gap size of the horizontal bowtie component. What is also clear, however, is that, within experimental uncertainty, the experimentally determined points do not all overlap with the curve predicted by finite-difference time-domain (FDTD) simulation. Because of the nanoscale accuracy of the TPPL photon localization technique, we are sensitive to the inevitable structural imperfections that arise during top-down fabrication.

To correlate optical function with structure, each ABnC was imaged in the SEM to determine actual device geometries and morphologies. There are a number of geometrical factors that are known from simulation and experiment to influence local field properties (e.g., radii of curvature and triangle aspect ratio). Here, we found the experimental scatter in relative centroid positions of our ABnCs to correlate primarily with two types of structural variations. The first are small changes of the size of the triangles in the horizontally oriented bowtie component of the ABnC, particularly the altitude or length (perpendicular bisector) of the triangle. Mode positions are most sensitive to length variations within the horizontal bowtie component (vs. the vertical bowtie component) since this is the bowtie aligned with the incident polarization. Our structures were designed to have nominal triangle lengths of 95 nm. But as is seen in Figure 18.11, FDTD simulations show that just a 10% increase in length is expected to result in a 50% increase in the separation of mode positions (meanwhile there is no significant change in resonance wavelengths; only ~3% for the 10 nm altitude increase). Comparing the optical data with the SEM images, we find that all experimental points located above the FDTD curve in Figure 18.10 have horizontal bowtie triangle lengths >95 nm. Additionally, the larger the discrepancy between simulation and experiment, the larger the length deviation from 95 nm (data not shown), demonstrating that fabrication-based variations in the length parameter are the most significant factor in causing increases (often major) in localized centroid mode separations in the ABnC devices studied here.

Figure 18.11 also shows that a decrease in triangle length will lead to a decrease in relative mode positions, which could in principle explain the deviation between experiment and theory for points below the FDTD curve in Figure 18.10. However, this is *not* the case; it is observed that triangle lengths are also 95 nm and greater for ABnCs with smaller-than-expected relative mode separations. Instead, the ABnCs that exhibit smaller-than-expected mode separations have at least one anomalous localized defect, such as a nanoscale

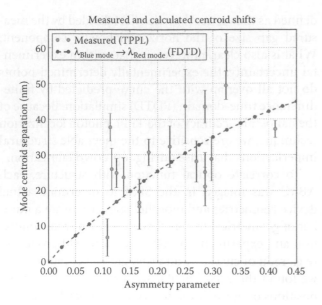

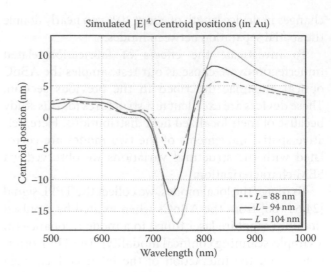

FIGURE 18.10 TPPL image mode centroid separations plotted as a function of normalized offset of the vertical bowtie component in the +*x*-direction. Experimentally measured values are marked by red circles and FDTD-calculated values are shown by the blue line and circles. Error bars correspond to the one-sigma accuracy of the experimentally determined shifts along the *x*-axis. (From McLeod, A. et al., Nonperturbative visualization of nanoscale plasmonic field distributions via photon localization microscopy. *Physical Review Letters*, 2011. 106(3): 037402. With permission.)

FIGURE 18.11 Small variations in triangle size greatly effect E4 centroid separations, leading to deviations from the theory line in Figure 18.1c. Centroid separations are determined by subtracting the minimum centroid position (near 780 nm) from the maximum position (near 850 nm). Here, triangle altitudes, *L*, for the horizontal-oriented bowtie are varied (triangles remain equilateral) while holding the gap size constant.

edge divot or a small excess Au particle, located randomly on the structure. Using FDTD simulation, we tested a number of different types of localized defects, and found that they all have the same general effect on mode centroid positions: namely, the defects tend to strongly localize fields across the entire spectral range of interest. In the ABnC devices, this has the effect of damping the measured mode centroid shift as wavelength is tuned from 780 to 840 nm since now all mode centroids are weighted toward the defect(s). It is defects such as these that are the primary cause of data points lying below the theoretical curve in Figure 18.10, and in fact should significantly affect local mode behavior for any nanoplasmonic device containing ultraconfined mode volumes (such as optical antennae).

18.2.2 Conclusion

We therefore see that nanoscale structural variations are found to significantly impact on local mode properties of nanoplasmonic devices. In particular, there are two specific classes of variations that primarily influence local plasmon properties in nanoantenna-based color sorting devices—localized anomalous defects and antenna length variations. In addition, these results demonstrate that the nonperturbative TPPL-based localization microscopy technique is appealing not only for basic investigations in research labs but also for characterization and metrology in more industrial/development settings. In fact, they highlight the need for next-generation device fabrication methods that combine the advantages of top-down and bottom-up techniques, such as assisted self-assembly.

18.3 Novel Approach to Place Well-Defined and Functional Optical Antennae on Scanning Probe Tips Based on Induced Deposition Mask Lithography

18.3.1 Introduction

Nanophotonics and Nanoplasmonics are emerging fields that attract large interest since they attempt to direct and control light at the nanometer scale. Landmark applications include nonlinear optical enhancement for possible optical transistors [83,84], plasmonic waveguides [85], exploiting novel optical properties of matter [86,87], or bypassing the diffraction limit of light [25,88] via plasmonic (optical) antennae. Optical antennae couple photons efficiently to an SPP of metal nanoparticles and allow circumventing efficiently the diffraction limit by an order of magnitude. Well-defined optical antenna will impact on nearfield optics substantially due to their distinct resonance and high nearfield localization [53]. Optical antennae placed on scanning probe tips enable imaging of single proteins [89,90] and will eventually lead to perform reliably tip-enhanced Raman spectroscopy on individual molecules [91]. However, the performance of these antennae depends, among other factors, sensitively on the shape of the metal nanostructure employed [92]. This novel concept of optical antennae demands reproducible and flexible nanofabrication methods. This highly fascinating and fast-expanding field of nanoplasmonics has been enabled only recently due to the most current developments in nanofabrication. The fabrication of plasmonic devices requires state-of-the-art nanofabrication or nanosynthesis, very high resolution (<10 nm), and high reproducibility. Still today, the performance and the implementation into actual devices are often limited by the available fabrication methods. In this chapter, we introduce a novel fabrication method—induced deposition mask lithography (IDML) [93]—developed for prototyping, allowing high resolution, high reproducibility, as well as flexibility to develop plasmonic antennae, plasmonic waveguides, or plasmonic circuit elements on a large variety of materials and topologies.

So far optical antennae have been fabricated by a number of nanofabrication methods. State-of-the-art e-beam lithography enables high-resolution fabrication [52] but requires the employment of resist, flat, and conductive surfaces, which decreases this technique's flexibility. Nanosynthesis [94] can create single nanocrystals, enabling high control over the shape; however, the size variation is broad and the concerted manipulation, as well as attachment to functional devices, is so far very difficult. Shadow evaporation [95] is restricted to flat surfaces and the variation of possible shapes is limited. FIB milling is very flexible [27] but the Ga contamination of the resulting metal structure can compromise the performance of the optical antenna [96]. Finally, focused electron beam-induced deposition (FEBID) [97] is a novel technique that enables the direct writing of high-resolution structures and precise placement, independent of the local topography. Yet, the variety and purity of material that can be deposited is so far very limited. FEBID [97] employs the secondary elec-

trons (SEs) of a focused electron beam to crack precursor molecules, which are bound to the substrate that is exposed by the electron beam. The resolution of the deposited target material is of the order of the electron beam resolution and is topographically very flexible, as long as the substrate is conductive. However, for plasmonic antennae in the visible regime, Au and Ag are used, for which there are currently hardly any viable precursors and the few that exist are instable and the purity of the deposited Au or Ag is low [98]. Since contaminations represent scattering centers for surface plasmons, it is important for the reproducible performance of optical antennae to employ pure materials.

To compensate for the limitations by the described fabrication techniques that constrain specifically the progress in nanoplasmonics, IDML was developed, which will be introduced in this chapter [93]. IDML is based on FEBID and allows the precise, topography-independent placement of uncontaminated, reproducible, and well-defined optical antennae. The exact process is described in the following section, followed by an example were Au bowtie antennae are fabricated on a glass coverslip, illustrating resolution and reproducibility. To show the technique's flexibility, a plasmonic antenna is fabricated on the apex of a scanning probe tip. With a basic optical characterization of the fabricated antennae, the functional reproducibility is demonstrated.

18.3.2   Novel Nanofabrication of Optical Antennae

Figure 18.12 depicts the IDML process, which consists of three steps: (1) deposition of the desired antenna material, (2) precise placement of the etch mask via FEBID, and (3) Ar ion milling, exploiting the energy-dependent relative etchrates for the antenna and mask materials, to transfer the mask into the material of interest.

Here, the capability and resolution of IDML is demonstrated by fabricating bowtie-like optical antennae, formed by two opposing triangles with a reproducible 12 nm gap.

A modified Zeiss XB150 crossbeam was employed enabling *in situ* FEBID and Ar+ milling. This instrumentation includes a commercial Zeiss gas injection system, a Xenos pattern generator that steers either e-beam or ion-beam, and an NTI argon ion gun for *in situ* Ar+ milling. The Ar+ milling incident angle, the ion energy, and current density can be varied.

Chapter 18

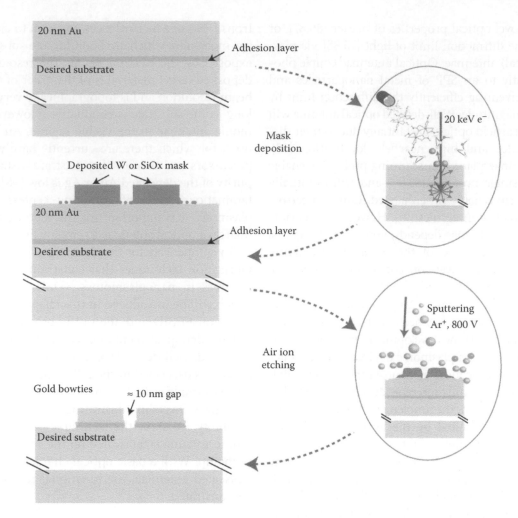

FIGURE 18.12   Schematic illustration of the novel nanofabrication process. First the antenna material is evaporated on the substrate of interest. Electron beam-induced deposition is used to place etch masks precisely at the desired place. Subsequent Ar⁺ milling transfers the mask shape into the underlying material. (From Weber-Bargioni, A. et al., Functional plasmonic antenna scanning probes fabricated by induced-deposition mask lithography. *Nanotechnology*, 2010. **21**(6): 065306. With permission.)

18.3.2.1 Material Deposition

The material to be patterned—in this case Au—is deposited on the substrate of choice. The quality (purity, grain size) of the deposited material influences both the performance of the optical antennae as well as the reproducibility and resolution of the consecutive etching process. For this work, 20-nm-thick Au films were used. Initially, the Au film was deposited on commercial Si wafers or commercial silicon oxide coverslips with 1 nm of Cr or Ti as the adhesion layer. Three evaporation schemes were utilized: (1) thermal, (2) electron beam, and (3) sputtering. Thermal evaporation produced the highest-purity Au film but large grain size (≈30 nm) and surface roughness (RMS ≈ 5 nm, measured via AFM). Electron beam evaporation results in the smallest grain size (≈10 nm) and a surface roughness of ≈ 3 nm. Sputter coating yields 30 nm grain size

but the smoothest surfaces (RMS < 1 nm). Use of the latter turns out to provide the most reproducible results in terms of sputter etching. These numbers are specific to the evaporators and their specific parameters used for this work.

18.3.2.2 Mask Deposition via FEBID

The masks are directly deposited on the Au substrate via high-resolution FEBID, which permits the placement of the mask (and subsequently the optical antenna) with high accuracy to the desired location. In FEBID, the precursor gas (organic matrix containing the material to be deposited) is introduced by a gas injection system in close proximity to the substrate [97]. The primary high-energy electrons of an electron beam pass the molecules mainly unhindered and create in the substrate a plume of SEs. The SEs (2–3 eV) that emerge at the surface in close proximity to where the primary e-beam enters the substrate have a high cross section for interacting and cracking the precursor molecules, releasing the mask material while most of the organic, volatile residues vaporizes off into the vacuum chamber [99]. This technique has been employed by Mitsuishi et al. to fabricate 3.5 nm structures [100].

Here this technique is used to directly place the etch masks on the material of interest by carefully steering the electron beam of an SEM during the deposition. The final resolution of the masks is a sensitive function of (1) the type of precursor used, (2) the primary electron energy, (3) the dwell time per written pixel, and (4) the number of written passages. Based on the precursors available, the parameter space to achieve the highest resolution for the mask was systematically investigated, examples of which are depicted in Figure 18.13a and b.

The mask materials used for this work are W and SiO_x with their respective precursors, tungsten hexacarbonyl, and tetra-ethyl-ortho-silicate. W and SiO_x exhibit a higher energy-dependent Ar^+ sputter threshold than Au [101] enabling an efficient mask transfer into the Au and good contrast since the unmasked Au etches faster than the mask. The use of W leads to the highest mask resolution but complicates the subsequent milling process and W residues can compromise the plasmon resonance. SiO_x masks reveal slightly less resolution, yet the subsequent milling process is more reliable.

Using 20 kV primary electron energy for the SEM achieves the best compromise between mask resolution and mask deposition efficiency for both W and SiO_x precursor. The primary electron energy determines the depth of the main isotropic SE generation (plume) [102]. Based on the limited mean free path of these electrons, the deeper the plume, the fewer SEs reach the surface, and the closer to the incoming primary electrons, the SE emerge. Hence, by increasing the primary electron energy, one can achieve mask deposition resolutions of the order of the primary electron beam resolution [99].

A pattern generator moves the electron beam in incremental steps of 1 nm, drawing the desired form—in this work a nanobowtie with a 10 nm gap. Each pixel is exposed for a certain time (dwell time), depositing a specific amount of mask material. Together with the number of subsequent writing passages the final thickness of the mask is determined. The goal is to achieve masks between 10 and 15 nm thick for the efficient

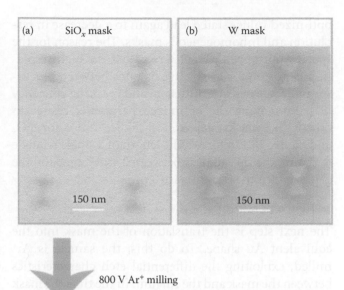

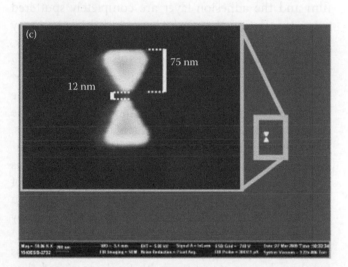

FIGURE 18.13 SEM image of FEBID-deposited SiO_x masks (a) and W masks (b) on a 20 nm Au film. (c) Depicts the structure after Ar^+ milling resulting into a well-defined Au optical bowtie antenna. (From Weber-Bargioni, A. et al., Functional plasmonic antenna scanning probes fabricated by induced-deposition mask lithography. *Nanotechnology*, 2010. **21**(6): 065306. With permission.)

transfer of the mask shape into the Au via Ar^+ milling based on the differential milling efficiencies of the mask and Au. The optimized dwell time and number of writing passages for a high-resolution 10-nm-thick SiO_x mask (measured via AFM) were found to be 1.5 μs dwell time and 12 passages, whereas to obtain the same height for a high-resolution W masks, 0.5 μs dwell time and 95 passages are optimal. Increasing the dwell time to write fewer passages will lead to a decrease in resolution, specifically increasing the radius of curvature of the triangle tips. Decreasing the dwell time and increasing the number of writing passages from the optimized values stated lead again to a decrease in resolution and inhomogeneous masks. The reason for the specific optimized parameters is that the high-resolution deposition is a sensitive balance between precursor molecule density, diffusion, and cracking efficiency [97]. Employing the optimized parameter results in bowtie-like masks, depicted in Figure 18.13a for SiO_x and Figure 18.13b for W. The length of a single triangle is 75 nm, the tip radius is between 2 and 3 nm, and the gap between the triangles is 10 nm.

18.3.2.3   Ar⁺ Etching

The next step is the translation of the mask into the equivalent Au shape. To do this, the sample is Ar^+ milled, exploiting the differential etch characteristics between the mask and the Au film. In the time the mask material is removed, the surrounding 20-nm-thick Au film and the adhesion layer are completely sputtered away, translating the mask shape with high accuracy into the underlying Au (cf. Figure 18.13c). For this work, the sample was sputtered under a 10° angle while being rotated to achieve more vertical sidewalls. The Ar^+ energy used is 0.8 keV and the Ar^+ current is 10 µA. The etch time is approximately 3 min and dependent on the Au film thickness and quality. Using these parameters, it was possible to obtain reproducible Au bowties composed of equilateral 75 nm high triangles, 5 nm radius of triangle tip curvature, and reproducible gap sizes of 12 nm. No impurities were detected in the resulting Au structures once surface contamination was removed by mild Ar sputter cleaning.

The quality/reproducibility of the milling process for these bowtie structures with final gap sizes below 15 nm is dependent on (1) the constitution of the mask, (2) the Au film quality, and (3) the thickness of the adhesion layer.

Masks above 15 nm height display a skirt of deposited material surrounding them. This can be removed by increasing the Ar^+ energy to 1.5 keV for the initial 20 s of milling time to sputter the skirt. However, this results generally into less well-defined edges, corners, and especially gaps of the bowtie structure.

The Au quality is essential for the high-resolution translation of the masks into the Au. One can observe a correlation between the surface roughness, and Au grain size and the roughness of the edges. Specifically, the surface roughness determines the reproducibility of the final gap size substantially. For the samples presented here, sputter-coated Au films were employed with a surface roughness below 1 nm. Ideally, one would employ either very flat (RMS < 1 nm) polycrystalline

films with grain sizes below 2 nm or flat single-crystalline films.

Finally, the thickness of the Cr or Ti adhesion layer influences the outcome of the milling and the performance of coupled plasmonic antennae since it is crucial that the triangles are electrically isolated from each other. In principle, the adhesion layer should be as thin as possible, due to the higher milling threshold of Cr (same for Ti) compared to Au. The time it takes to mill through an adhesion layer of more than 3 nm can lead to overmilling of the actual structure. Alternatively an attempt was made to remove the Cr adhesion layer using a chemical wet etch but that resulted in large amounts of residues on the structures.

In summary, using mask materials with a substantially larger sputtering threshold than Au such as W, SiO_x, or deposited diamond-like carbon, Au with a low surface roughness and adhesion layers below 2 nm provide the parameters to fabricate any desired form of optical antenna with reproducible resolutions close to 10 nm.

This method enables the precise placement of well-defined and reproducible shapes on a larger variety of substrates, and the employment of pure materials including single crystals, which are not contaminated during the fabrication process. This approach has the potential of feature resolutions comparable with state-of-the-art e-beam lithography process. In the next section, the functionality of the optical antennae produced here is demonstrated.

18.3.3   Optical Characterization of the Fabricated Bowtie Antennae

Optical or plasmonic antennae represent a cavity for localized surface plasmons, exhibiting a specific resonance that depends sensitively on the shape, the size of the cavity [103], the dielectric function of the antenna material and the surrounding medium [104], and, for coupled-particle plasmonic antennae, the gap size [95]. In the case of dipolar antennae or coupled plasmonic devices consisting of two or more nanoparticles, the resonance depends also on the polarization of the incoming light [52].

Here, the functionality of the plasmonic antennae fabricated via IDML is demonstrated as well as this technique's flexibility by placing these antennae on any desirable object such as the apex of an AFM tip.

First, darkfield spectroscopy is utilized to determine the resonance frequency of bowtie antennae on silica coverslips.

Second, a bowtie antenna fabricated on the apex of a silicon AFM tip is characterized by measuring the polarization-dependent nearfield enhancement of the AFM tip's Si Raman signal.

18.3.3.1 Determining the Resonance Frequency via Darkfield Spectroscopy

Darkfield spectroscopy allows determining the surface plasmon resonance of metal nanoparticles [105]. The sample is illuminated with polarized white light and reveals the wavelength at which the nanoparticles have the largest scattering cross section correlated to the surface plasmon. Figure 18.14a depicts the experimental setup. Bowtie antennae of varying sizes are placed on top of a silica coverslip via the novel fabrication method. A high numerical aperture condenser focuses white light on a bowtie array, where the light is polarized along the main axis of the bowtie structure. In the case of asymmetric and/or coupled plasmonic devices, the resonance, the scattering intensity, and the nearfield enhancement depended on the polarization of the incoming light [52]. The scattered light of the bowties was collected by a microscope objective and coupled to a spectrometer with a CCD detector. The spectra were corrected for source intensity and spectrometer response variations.

Figure 18.14b shows the spectra of three different bowtie antennae, assembled each of two equilateral triangles. The single triangle height (cf. inset Figure 18.14b) of the three bowties is 75, 90, and 100 nm, respectively and the gap between the triangles was measured to be 13–15 nm. All three spectra illustrate two peaks, a minor peak for smaller wavelength and a major one for longer wavelength.

The major peak is attributed to the dipole plasmon mode resonance of a bowtie antenna [52,95], which reveals for the 75 nm antenna (blue spectra) a maximum at 710 nm. For the 90 nm bowtie (green spectrum), the major peak is redshifted to 760 nm due to the longer resonance cavity for the surface plasmon. The 100 nm bowtie is shifted further to 790 nm but shows a smaller scattering intensity, which has to be investigated further. The smaller peak is attributed to the quadrupole mode of the antennae and is consistent with the calculations by Zhao et al. [106].

The peak positions are consistent with bowties produced via e-beam lithography [52] and indicate that the optical antennae produced with IDML represent effectively a cavity for localized surface plasmons.

Darkfield spectroscopy of the fabricated plasmonic antennae is very sensitive to small variations of the antenna structure and can be used as a measure of the fabrication quality. For example, overetching the sample and hence deteriorating the bowtie shape will lead to a featureless spectrum. On the other hand, it is crucial to etch sufficiently such that both Au

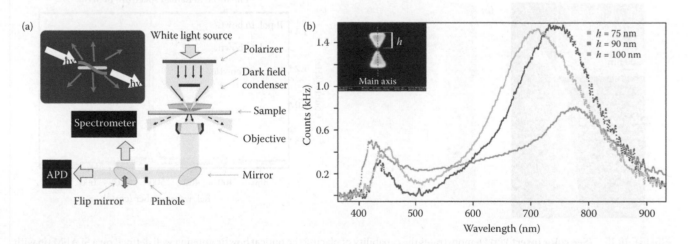

FIGURE 18.14 (**See color insert.**) (a) Diagram of the darkfield spectroscopy setup used to determine the resonant frequency of the plasmonic antennae. White light is focused onto the sample under an angle greater than what the objective would collect. The resonantly scattered light of the nanoparticle is collected by the objective and sent to an avalanche photo diode (APD) of spectrometer. (b) Inset: definition of the bowties main axes and the varying height parameter of the three bowties are characterized. Main: darkfield spectra of three differently sized bowtie antennae. Increasing the height of the triangle elongates the surface plasmon cavity length and therefore redshifts the major peak position. (From Weber-Bargioni, A. et al., Functional plasmonic antenna scanning probes fabricated by induced-deposition mask lithography. *Nanotechnology*, 2010. **21**(6): 065306. With permission.)

nanotriangles are electrically insulated by milling completely through the adhesion layer. If chromium or titanium can still be detected after the milling step around the bowtie (via Auger spectroscopy), it indicates the presence of chromium or titanium in the gap between the triangles. In this case, the plasmonic antennae represent very lossy cavities, which manifest in very broad spectra with very few distinct features. Rough edges will also widen the spectra or add additional resonance feature toward smaller wavelength from the main resonance peak.

18.3.3.2 Precise Placement and Nearfield Enhancement

Figure 18.15a illustrates the ability of this novel nanofabrication method to place these antennae reproducibly on the end of an AFM tip. The only prefabrication preparation is to cut the apex of the AFM tip to create a small 150–200-nm-wide flat plateau. This is done with an FIBM. Subsequently, the same fabrication method as described above is employed to place the optical antenna in the center of the plateau. The bowtie tip shown in this example was done using a SiO_x mask. The only modification to the recipe presented above is in the number of writing passages that is 18 instead of 12. The reduced writing efficiency is a result of the decreased precursor molecule density at the end of a tip compared to an extensive plane [97].

To prove that the optical antennae placed at the end of an AFM tip are functional, the effects of the plasmonic antennae's nearfield enhancement are shown. The nearfield enhancement is a crucial characterization since it is the motivation for the employment of the novel concept of optical antennae. One way, among others [88], of measuring the nearfield enhancement is to record the Raman signal of a material with and without a plasomic antenna in the proximity of the material (<10 nm). For the work presented here, the optical antenna is placed on a Si AFM tip, where Si itself has a large Raman cross section. Therefore, one would expect an increase in the Si Raman line for a bowtie plasmonic antenna placed on a Si AFM tip as long as the exciting incoming light is polarized along the main axis of the bowtie and therewith exciting the coupled surface plasmon mode that leads to strong nearfield enhancements in the bowtie gap.

The experimental setup used to investigate the Raman signal was an apertureless nearfield scanning optical microscope (NSOM) from Witec. The plasmonic antenna AFM tip is approached to a clean silica coverslip. The polarized laser light is aligned and focused from the bottom through the 300-μm-thick coverslip on the tip apex using a 100× oil objective (1.4 NA). Based on FDTD simulations, the higher Si dielectric function compared to that of silica will

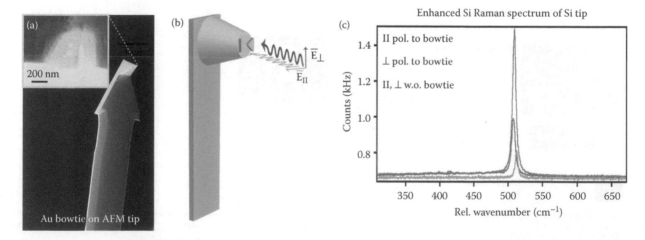

FIGURE 18.15 **(See color insert.)** (a) Demonstrates the capability of placing the optical bowtie antenna well defined on a Si AFM tip with this novel nanofabrication method. (b) Illuminating the bowtie tip at the antenna's resonance frequency with laser light polarized parallel to the bowtie's axis will create for a functional bowtie a nearfield enhancement in the bowtie gap. (c) Si Raman spectrum of the Si AFM tip. The blue spectrum shows the Si Raman peak for the bowtie tip illuminated parallel to the main axis and a 2.5-fold intensity increase can be recorded compared to the perpendicular polarization (red spectrum). The green spectra correspond to the equivalent Si tip without an Au bowtie. No polarization-dependent signal intensity can be observed. (From Weber-Bargioni, A. et al., Functional plasmonic antenna scanning probes fabricated by induced-deposition mask lithography. *Nanotechnology*, 2010. **21**(6): 065306. With permission.)

redshift the resonance of the 75 nm bowties toward 800 nm. Hence a 785 nm diode laser was employed to excite the bowtie antenna near resonance. The polarization is fixed with a rotatable λ-half-plate. No polarization-dependent power modulation was detected. The scattered light is collected by the same objective and passes through a beam splitter and a 790-nm-long-pass filter into a spectrometer where the Si Raman line is measured.

First, the laser light was polarized along the bowtie antenna's main axis (cf. Figure 18.15b) and a Raman spectrum is recorded observing a Si Raman peak at 508 wave numbers relative to the laser line. Subsequently, the polarization of the laser light is turned perpendicular to the bowtie's main axis and the Raman spectrum is recorded. Figure 18.15c illustrates the Si Raman spectrum in blue (red) for the parallel (perpendicular) polarization and a 2.5-fold increase for the parallel polarization can be recorded. The polarization dependence for the same tip was repeated five times reproducing the spectra represented here.

To ensure that the observed intensity difference is not due to a slight focus modification when the polarization is changed, the focus was optimized by maximizing the Raman signal for each polarization. Rotating back 90° reproduced the parallel optimized signal intensity.

In all, six tips were tested to determine the fabrication reproducibility. For four out of these six tips, an intensity increase between 1.7 and 2.5 of the Si Raman peak was detected. The other two tips showed no polarization-dependent peak in—or decrease due to insufficient etching—leaving both triangles connected.

To ensure that the Raman peak intensity modification does not relate to the Si crystal orientation, an identical FIB-flattened Si tip without a bowtie was placed into the NSOM. Figure 18.15c shows the resulting Si Raman spectrum in green, which is independent of the polarization. The lower peak intensity of the green spectrum compared to the blue and red spectra does not correlate to the absence of a plasmonic antenna based on alignment, tip, and laser intensity variations. Interestingly, the peak position for the Si tip with and without the plasmonic antenna is shifted from 508 to 512 relative wave numbers. Currently, the cause for the shift is not understood. A possible hypothesis is the strain on the Si due to the fabrication, which will be investigated. However, these results further support that the intensity increase in the Si Raman

peak originates from the nearfield enhancement of the bowtie antenna.

18.3.3.3 Estimating the Nearfield Enhancement

Based on the observed intensity increase of 2.5 for the parallel versus perpendicular excitation polarization, it is possible to make an estimate of the nearfield enhancement in the bowtie gap. There are two contributions to the Raman signal, one due to Raman scattering from the tip material in the (unenhanced) confocal volume, and another due to the strongly enhanced field within the zeptoliter volume near the bowtie gap [88]. The ratio of these volumes times the polarization ratio gives an estimate of the enhancement in the bowtie gap. For the volume of the confocal spot V_c, a lateral Gaussian spot size with an FWHM of 280 nm is assumed, based on the diffraction limit ($\lambda = 785$ nm, NA = 1.4) and a depth-of-focus of 820 nm, corresponding to the axial resolution of a confocal spot [107]. The volume contributing to the nearfield-enhanced signal V_g is based on a conservative estimate of 20 nm gap size resulting in a 400 nm^2 area times the 20 nm penetration depth of the evanescent wave into the silicon (details in supplementary information). The 2.5-fold signal intensity increase (150%) and the effective volume ratio lead to a Raman enhancement estimate of $1.5 \times V_c/V_g \approx 3.9 \times 10^4$. The total Raman enhancement is approximately proportional to the fourth power of the nanoantenna's electric field enhancement [108]. Therefore, the square root of the Raman enhancement gives an optical intensity enhancement of ~200 for the bowtie.

18.3.4 Conclusion

This work introduces a novel nanofabrication method, IDML, which enables the placement of nanostructures such as the Au bowtie with great flexibility and accuracy on insulating, semiconducting, or conducting surfaces with various topographies, including AFM tips. This novel fabrication technique can integrate genuinely functional nanostructures within larger-scale devices and frameworks; that is,—it is ideal for the development of novel scanning probes, as shown here, or, more generally, for the implementation of specific nanostructures in complex assemblies. The capabilities of this technique were demonstrated by fabricating and characterizing optical antennae on scanning AFM probe tips, showing size-dependent resonance effects and optical intensity enhancement comparable with good electron beam lithography (EBL) antennae.

References

1. Prasad, P.N., *Introduction. Nanophotonics.* 2004, Hoboken, New Jersey: John Wiley & Sons, Inc. 1–8.
2. Shalaev, V.M., Optical negative-index metamaterials. *Nature Photonics*, 2007. **1**(1): 41–48.
3. Pendry, J.B., Negative refraction makes a perfect lens. *Physical Review Letters*, 2000. **85**(18): 3966.
4. Leonhardt, U., Optical conformal mapping. *Science*, 2006. **312**(5781): 1777–1780.
5. Pendry, J.B., D. Schurig, and D.R. Smith, Controlling electromagnetic fields. *Science*, 2006. **312**(5781): 1780–1782.
6. Tsakmakidis, K.L., A.D. Boardman, and O. Hess, 'Trapped rainbow' storage of light in metamaterials. *Nature*, 2007. **450**(7168): 397–401.
7. Notomi, M., Theory of light propagation in strongly modulated photonic crystals: Refraction like behavior in the vicinity of the photonic band gap. *Physical Review B*, 2000. **62**(16): 10696.
8. Notomi, M., Negative refraction in photonic crystals. *Optical and Quantum Electronics*, 2002. **34**(1): 133–143.
9. Berrier, A. et al., Negative refraction at infrared wavelengths in a two-dimensional photonic crystal. *Physical Review Letters*, 2004. **93**(7): 073902.
10. Chatterjee, R. et al., Achieving subdiffraction imaging through bound surface states in negative refraction photonic crystals in the near-infrared range. *Physical Review Letters*, 2008. **100**(18): 187401.
11. Pendry, J.B. and S.A. Ramakrishna, Focusing light using negative refraction. *Journal of Physics: Condensed Matter*, 2003. **15**(37): 6345.
12. Pendry, J.B., Negative refraction. *Contemporary Physics*, 2004. **45**(3): 191–202.
13. Ramakrishna, S.A., Physics of negative refractive index materials. *Reports on Progress in Physics*, 2005. **68**(2): 449.
14. Shamonina, E. et al., Imaging, compression and Poynting vector streamlines for negative permittivity materials. *Electronics Letters*, 2001. **37**(20): 1243–1244.
15. Smolyaninov, I.I., Y.-J. Hung, and C.C. Davis, Magnifying superlens in the visible frequency range. *Science*, 2007. **315**(5819): 1699–1701.
16. Li, J. et al., Photonic band gap from a stack of positive and negative index materials. *Physical Review Letters*, 2003. **90**(8): 083901.
17. Panoiu, N.C. et al., Zero-n⁻ bandgap in photonic crystal superlattices. *Journal of the Optical Society of America B*, 2006. **23**(3): 506–513.
18. Mocella, V. et al., Self-collimation of light over millimeter-scale distance in a quasi-zero-average-index metamaterial. *Physical Review Letters*, 2009. **102**(13): 4.
19. Olynick, D.L., J.A. Liddle, and I.W. Rangelow, Profile evolution of Cr masked features undergoing HBr-inductively coupled plasma etching for use in 25 nm silicon nanoimprint templates. *Journal of Vacuum Science & Technology B*, 2005. **23**(5): 2073–2077.
20. MacDonald, K.F. and N.I. Zheludev, Active plasmonics: Current status. *Laser & Photonics Reviews*, 2010. **4**(4): 562–567.
21. Cao, L. and M.L. Brongersma, Ultrafast developments. *Nature Photonics*, 2009. **3**(1): 12–13.
22. Stockman, M.I., Nanoplasmonics: The physics behind the applications. *Physics Today*, 2011. **64**(2): 39–44.
23. Atwater, H.A., The promise of plasmonics. *Scientific American*, 2007. **296**(4): 56–63.
24. Schuck, P.J. et al., Improving the mismatch between light and nanoscale objects with gold bowtie nanoantennas. *Physical Review Letters*, 2005. **94**(1): 017402–017404.
25. Muhlschlegel, P. et al., Resonant optical antennas. *Science*, 2005. **308**(5728): 1607–1609.
26. Kuhn, S. et al., Enhancement of single-molecule fluorescence using a gold nanoparticle as an optical nanoantenna. *Physical Review Letters*, 2006. **97**(1): 017402–017404.
27. Taminiau, T.H. et al., Lambda/4 resonance of an optical monopole antenna probed by single molecule fluorescence. *Nano Letters*, 2007. **7**(1): 28–33.
28. Novotny, L., Effective wavelength scaling for optical antennas. *Physical Review Letters*, 2007. **98**(26): 266802.
29. Cubukcu, E. et al., Plasmonic laser antenna. *Applied Physics Letters*, 2006. **89**(9): 093120–093123.
30. Wang, H. et al., Symmetry breaking in individual plasmonic nanoparticles. *Proceedings of the National Academy of Sciences, USA*, 2006. **103**(29): 10856–10860.
31. Behr, N. and M.B. Raschke, Optical antenna properties of scanning probe tips: Plasmonic light scattering, tip–sample coupling, and near-field enhancement. *Journal of Physical Chemistry C*, 2008. **112**(10): 3766–3773.
32. Schnell, M. et al., Controlling the near-field oscillations of loaded plasmonic nanoantennas. *Nature Photonics*, 2009. **3**(5): 287–291.
33. Kang, J.H., D.S. Kim, and Q.H. Park, Local capacitor model for plasmonic electric field enhancement. *Physical Review Letters*, 2009. **102**(9): 093906.
34. Conway, J.A., Efficient optical coupling to the nanoscale, in *Electrical Engineering*. 2006, Los Angeles: University of California.
35. Li, K.R. et al., Surface plasmon amplification by stimulated emission in nanolenses. *Physical Review B*, 2005. **71**(11): 115409.
36. Ghenuche, P. et al., Cavity resonances in finite plasmonic chains. *Applied Physics Letters*, 2007. **90**(4): 041109.
37. Malyshev, A.V., V.A. Malyshev, and J. Knoester, Frequency-controlled localization of optical signals in graded plasmonic chains. *Nano Letters*, 2008. **8**(8): 2369–2372.
38. Laux, E. et al., Plasmonic photon sorters for spectral and polarimetric imaging. *Nature Photonics*, 2008. **2**(3): 161–164.
39. Kang, Z. and G.P. Wang, Coupled metal gap waveguides as plasmonic wavelength sorters. *Optics Express*, 2008. **16**(11): 7680–7685.
40. Choi, S.B. et al., Directional control of surface plasmon polariton waves propagating through an asymmetric bragg resonator. *Applied Physics Letters*, 2009. **94**(6): 063115.
41. Hao, F. et al., Symmetry breaking in plasmonic nanocavities: Subradiant LSPR sensing and a tunable Fano resonance. *Nano Letters*, 2008. **8**(11): 3983–3988.
42. Zhang, Z. et al., Manipulating nanoscale light fields with the asymmetric bowtie nano-colorsorter. *Nano Letters*, 2009. **9**(12): 4505–4509.
43. Sundaramurthy, A. et al., Field enhancement and gap-dependent resonance in a system of two opposing tip-to-tip Au nanotriangles. *Physical Review B: Condensed Matter and Materials Physics*, 2005. **72**(16): 165409–165416.

44. Palik, E.D., *Handbook of Optical Constants*. 1985. San Diego, California: Academic Press.
45. Tang, L. et al., Nanometre-scale germanium photodetector enhanced by a near-infrared dipole antenna. *Nature Photonics*, 2008. **2**(4): 226–229.
46. Farahani, J.N. et al., Single quantum dot coupled to a scanning optical antenna: A tunable superemitter. *Physical Review Letters*, 2005. **95**(1): 017402.
47. Weber-Bargioni, A. et al., Functional plasmonic antenna scanning probes fabricated by induced deposition mask lithography. *Nanotechnology*, 2010. **21**: 065306.
48. Weber-Bargioni, A. et al., Hyperspectral nanoscale imaging on dielectric substrates with coaxial optical antenna scan probes. *Nano Letters*, 2011. **11**: 1201–1207.
49. Chang, S.W. and S.L. Chuang, Normal modes for plasmonic nanolasers with dispersive and inhomogeneous media. *Optics Letters*, 2009. **34**(1): 91–93.
50. Hill, M.T. et al., Lasing in metal-insulator-metal sub-wavelength plasmonic waveguides. *Optics Express*, 2009. **17**(13): 11107–11112.
51. Biagioni, P. et al., Cross resonant optical antenna. *Physical Review Letters*, 2009. **102**(25): 256801.
52. Fromm, D.P. et al., Gap-dependent optical coupling of single "Bowtie" nanoantennas resonant in the visible. *Nano Letters*, 2004. **4**(5): 957–961.
53. Sundaramurthy, A. et al., Toward nanometer-scale optical photolithography: Utilizing the near-field of bowtie optical nanoantennas. *Nano Letters*, 2006. **6**(3): 355–360.
54. Prodan, E. et al., A hybridization model for the plasmon response of complex nanostructures. *Science*, 2003. **302**: 419–422.
55. Hao, F. et al., Tunability of subradiant dipolar and Fano-type plasmon resonances in metallic ring/disk cavities: Implications for nanoscale optical sensing. *ACS Nano*, 2009. **3**(3): 643–652.
56. Verellen, N. et al., Fano resonances in individual coherent plasmonic nanocavities. *Nano Letters*, 2009. **9**(4): 1663–1667.
57. Christ, A. et al., Controlling the Fano interference in a plasmonic lattice. *Physical Review B*, 2007. **76**(20): 201405.
58. Bachelier, G. et al., Fano profiles induced by near-field coupling in heterogeneous dimers of gold and silver nanoparticles. *Physical Review Letters*, 2008. **101**(19): 197401.
59. Liu, N., S. Kaiser, and H. Giessen, Magnetoinductive and electroinductive coupling in plasmonic metamaterial molecules. *Advanced Materials*, 2008. **20**(23): 4521–4525.
60. Fedotov, V.A. et al., Sharp trapped-mode resonances in planar metamaterials with a broken structural symmetry. *Physical Review Letters*, 2007. **99**(14): 147401.
61. Tang, L. et al., C-shaped nanoaperture-enhanced germanium photodetector. *Optics Letters*, 2006. **31**(10): 1519–1521.
62. Piner, R.D. et al., "Dip-pen" nanolithography. *Science*, 1999. **283**(5402): 661–663.
63. McLeod, A. et al., Nonperturbative visualization of nanoscale plasmonic field distributions via photon localization microscopy. *Physical Review Letters*, 2011. **106**(3): 037402.
64. Boyd, G.T., Z.H. Yu, and Y.R. Shen, Photoinduced luminescence from the noble-metals and its enhancement on roughened surfaces. *Physical Review B*, 1986. **33**(12): 7923–7936.
65. Beversluis, M.R., A. Bouhelier, and L. Novotny, Continuum generation from single gold nanostructures through near-field mediated intraband transitions. *Phys. Rev. B*, 2003. **68**: 115433.
66. Biagioni, P. et al., Dependence of the two-photon photoluminescence yield of gold nanostructures on the laser pulse duration. *Physical Review B*, 2009. **80**(4): 045411.
67. ten Bloemendal, D. et al., Local field spectroscopy of metal dimers by TPL microscopy. *Plasmonics*, 2006. **1**(1): 41–44.
68. Sondergaard, T. et al., Resonant plasmon nanofocusing by closed tapered gaps. *Nano Letters*, 2010. **10**(1): 291–295.
69. Ghenuche, P., S. Cherukulappurath, and R. Quidant, Mode mapping of plasmonic stars using TPL microscopy. *New Journal of Physics*, 2008. **10**: 105013.
70. Ghenuche, P. et al., Spectroscopic mode mapping of resonant plasmon nanoantennas. *Physical Review Letters*, 2008. **101**(11): 116805.
71. Huang, J.-S. et al., Mode imaging and selection in strongly coupled nanoantennas. *Nano Letters*, 2010. **10**(6): 2105–2110.
72. Ambrose, W.P., T. Basche, and W.E. Moerner, Detection and spectroscopy of single pentacene molecules in a para-terphenyl crystal by means of fluorescence excitation. *Journal of Chemical Physics*, 1991. **95**(10): 7150–7163.
73. Betzig, E. et al., Imaging intracellular fluorescent proteins at nanometer resolution. *Science*, 2006. **313**(5793): 1642–1645.
74. Feke, G.D. et al., Method for measuring profiles of photoacid patterns in chemically amplified photoresist. *MRS Proceedings*, 2001. **636**: D6.7.1.
75. Hess, S.T., T.P.K. Girirajan, and M.D. Mason, Ultra-high resolution imaging by fluorescence photoactivation localization microscopy. *Biophysical Journal*, 2006. **91**(11): 4258–4272.
76. Rust, M.J., M. Bates, and X.W. Zhuang, Sub-diffraction-limit imaging by stochastic optical reconstruction microscopy (STORM). *Nature Methods*, 2006. **3**(10): 793–795.
77. Sharonov, A. and R.M. Hochstrasser, Wide-field subdiffraction imaging by accumulated binding of diffusing probes. *Proceedings of the National Academy of Sciences of the United States of America*, 2006. **103**(50): 18911–18916.
78. Wu, D.M. et al., Super-resolution imaging by random adsorbed molecule probes. *Nano Letters*, 2008. **8**(4): 1159–1162.
79. Yildiz, A. et al., Myosin V walks hand-over-hand: Single fluorophore imaging with 1.5-nm localization. *Science*, 2003. **300**(5628): 2061–2065.
80. Moerner, W.E., New directions in single-molecule imaging and analysis. *Proceedings of the National Academy of Sciences*, 2007. **104**(31): 12596–12602.
81. Ober, R.J., S. Ram, and E.S. Ward, Localization accuracy in single-molecule microscopy. *Biophysical Journal*, 2004. **86**(2): 1185–1200.
82. Thompson, R.E., D.R. Larson, and W.W. Webb, Precise nanometer localization analysis for individual fluorescent probes. *Biophysical Journal*, 2002. **82**(5): 2775–2783.
83. Soljacic, M. and J.D. Joannopoulos, Enhancement of nonlinear effects using photonic crystals. *Nature Materials*, 2004. **3**(4): 211–219.
84. Wurtz, G.A., R. Pollard, and A.V. Zayats, Optical bistability in nonlinear surface-plasmon polaritonic crystals. *Physical Review Letters*, 2006. **97**(5): 057402.
85. Bozhevolnyi, S.I. et al., Channel plasmon subwavelength waveguide components including interferometers and ring resonators. *Nature*, 2006. **440**(7083): 508–511.
86. Mocella, V. et al., Self-collimation of light over millimeter-scale distance in a quasi-zero-average-index metamaterial. *Physical Review Letters*, 2009. **102**(13): 133902.
87. Valentine, J. et al., Three-dimensional optical metamaterial with a negative refractive index. *Nature*, 2008. **455**(7211): 376-U32.

Chapter 18

88. Schuck, P.J. et al., Improving the mismatch between light and nanoscale objects with gold bowtie nanoantennas. *Physical Review Letters*, 2005. **94**(1): 017402.

89. De Angelis, F. et al., Nanoscale chemical mapping using three-dimensional adiabatic compression of surface plasmon polaritons. *Nature Nanotechnology*, 2010. **5**(1): 67–72.

90. Hoppener, C. and L. Novotny, Antenna-based optical imaging of single Ca²⁺ transmembrane proteins in liquids. *Nano Letters*, 2008. **8**(2): 642–646.

91. Pettinger, B. et al., Nanoscale probing of adsorbed species by tip-enhanced Raman spectroscopy. *Physical Review Letters*, 2004. **92**(9): 096101 (4 pages).

92. Mock, J.J. et al., Shape effects in plasmon resonance of individual colloidal silver nanoparticles. *Journal of Chemical Physics*, 2002. **116**(15): 6755–6759: 065306.

93. Weber-Bargioni, A. et al., Functional plasmonic antenna scanning probes fabricated by induced-deposition mask lithography. *Nanotechnology*, 2010. **21**(6): 065306.

94. Sun, Y.G. and Y.N. Xia, Shape-controlled synthesis of gold and silver nanoparticles. *Science*, 2002. **298**(5601): 2176–2179.

95. Merlein, J. et al., Nanomechanical control of an optical antenna. *Nature Photonics*, 2008. **2**(4): 230–233.

96. Cubukcu, E. et al., Plasmonic laser antenna. *Applied Physics Letters*, 2006. **89**(9): 093120.

97. Utke, I., P. Hoffmann, and J. Melngailis, Gas-assisted focused electron beam and ion beam processing and fabrication. *Journal of Vacuum Science & Technology B*, 2008. **26**(4): 1197–1276.

98. Utke, I. et al., Focused electron beam induced deposition of gold. *Journal of Vacuum Science & Technology B*, 2000. **18**(6): 3168–3171.

99. Utke, I. et al., Resolution in focused electron- and ion-beam induced processing. *Journal of Vacuum Science & Technology B*, 2007. **25**(6): 2219–2223.

100. Mitsuishi, K. et al., Electron-beam-induced deposition using a subnanometer-sized probe of high-energy electrons. *Applied Physics Letters*, 2003. **83**(10): 2064–2066.

101. Yoshitake, M., Y. Yamauchi, and C. Bose, Sputtering rate measurements of some transition metal silicides and comparison with those of the elements. *Surface and Interface Analysis*, 2004. **36**(8): 801–804.

102. Silvis-Cividjian, N. et al., The role of secondary electrons in electron-beam-induced-deposition spatial resolution. *Microelectronic Engineering*, 2002. **61–2**: 693–699.

103. Kuwata, H. et al., Resonant light scattering from metal nanoparticles: Practical analysis beyond Rayleigh approximation. *Applied Physics Letters*, 2003. **83**(22): 4625–4627.

104. Barnes, W.L., A. Dereux, and T.W. Ebbesen, Surface plasmon subwavelength optics. *Nature*, 2003. **424**(6950): 824–830.

105. McFarland, A.D. and R.P. Van Duyne, Single silver nanoparticles as real-time optical sensors with zeptomole sensitivity. *Nano Letters*, 2003. **3**(8): 1057–1062.

106. Kelly, K.L. et al., The optical properties of metal nanoparticles: The influence of size, shape, and dielectric environment. *Journal of Physical Chemistry B*, 2003. **107**(3): 668–677.

107. Diaspro, A., *Confocal and Two-Photon Microscopy*, ed. Diaspro, A. 2002, New York: Wiley-Liss, Inc. pp. 110.

108. Schatz, G.C. and R.P. Van Duyne, Electromagnetic mechanism of surface-enhanced Raman spectroscopy, in *Handbook of Vibrational Spectroscopy*, eds, Chalmers, J.M. and P.R. Griffiths, 2002, Chichester: John Wiley & Sons Ltd.

19. Patterning Magnetic Nanostructures with Ions

Dan Kercher

Hitachi GST Research, San Jose, California

19.1 Introduction

Nanofabricated magnetic devices are extremely commonplace, but their fabrication processes are not so well understood by most. People tend to relate nanofabrication to silicon integrated circuit technology. However, computer hard disk drives (HDD) also rely on nanofabricated magnetic structures. In the 1980s, the discovery of interlayer exchange coupling in magnetic materials (Grunberg et al. 1986) and the giant magneto resistive (GMR) effect (Baibich et al. 1988, Binasch et al. 1989) drove an explosion of magnetic nanostructure fabrication research. Today, millions of hard drives with GMR read heads are sold every year. Recent research on magnetic devices also includes magnetic logic (Lmre et al. 2006), magnetic random access memory (Tehrani et al. 1999), racetrack memory (Parkin et al. 2008), and

patterned media (Terris 2009). These magnetic nanostructure technologies may enable higher data storage density and low-energy computing for future electronic devices.

Sputter deposition processes for thin-film magnetic materials have been well characterized due to their use in HDD read-and-write heads, and recording medium (Wood 2009). The real challenge for magnetic materials is the realization of isolated structures at the nanoscale. Anisotropic reactive ion etching has been the backbone of top-down nanofabrication. Unfortunately, the elements common in magnetic systems, such as Co, Ni, Fe, Pt, and Cr, do not easily form volatile reaction products. These materials can be etched with wet chemistries but the isotropic etch front from wet etching is not always desirable when working at nanoscale geometries.

Ion beam fabrication techniques are an effective method for top-down nanoscale patterning of magnetic materials. Electron ionization of atoms in a vacuum chamber is the basic process used for generating ion

Nanofabrication Handbook. Edited by Stefano Cabrini and Satoshi Kawata © 2012 CRC Press / Taylor & Francis Group, LLC. ISBN: 978-1-4200-9052-9

Chapter 19

beams. Once ionized, the positively charged atoms can be manipulated via electrostatic and Lorentz forces. With low ion beam acceleration voltages, typically <1 keV, the ion impact will result in material removal via sputtering. Higher beam accelerating voltages tend to implant the ions directly into the material, resulting in atomic dislocations and elemental composition changes. These two processes are referred to as ion beam etching and ion implantation. The combination of high-resolution lithography with ion beam exposure is a powerful tool in the nanofabrication toolbox. This chapter will provide an overview of ion beam etch and ion implantation processes for the realization of magnetic nanostructures.

19.2 Ion Beam Etching for Magnetic Materials

Ion beam etching, sometimes known as sputter etching or ion milling, is typically thought of as sand blasting at the atomic level. This technique has been used for years in the fabrication of microelectronics and is particularly useful in patterning of magnetic materials. Ions are generated in a plasma source and accelerated toward the sample. The impact of ions is strong enough to sputter material from the sample surface, etching away unwanted material.

19.2.1 Ion Beam Etch Sources

Ion beams for etching can be generated with a variety of different sources, including duoplasmotrons, Kaufman ion sources, Hall accelerators, Penning sources, saddle-field ion sources, radio frequency ion sources, electron resonance ion sources, and others. Some ion sources, such as the Kaufman type, use a grid for collimating the beam. Other sources, such as the end-Hall type, are gridless. When using a gridded ion beam source, it is important to keep in mind that the grids have a finite life from constant ion bombardment and etched material redeposition. In manufacturing environments, where high production throughput is required, the replacement cost for grids can be high.

An example of a gridded ion beam etch station is shown in Figure 19.1. A plasma is generated in the discharge source chamber by electron bombardment in a low-pressure gas with excitation from a direct current (DC) or radio frequency (RF) power source. Noble gas plasmas are used for ion beam etching of magnetic materials where no reactive gas chemistries are useful.

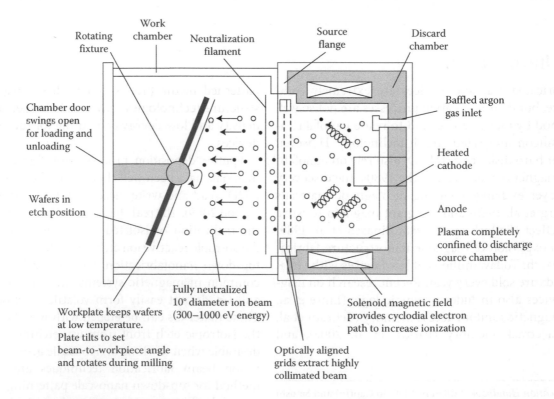

FIGURE 19.1 A gridded ion beam etching system. (Adapted from Ion Beam Milling, Inc. 2010: http://www.ionbeammilling.com/ABOUT_THE_ION_ MILLING_PROCESS.)

The ion beam is extracted from the discharge chamber through optically aligned hollow grids. These grids also provide collimation for the beam. A thermionic filament is used for beam neutralization. The substrate is mounted on a fixture capable of tilt and rotation.

The most important process parameters when using an ion source for etching are the potential of the beam, the ion current density and its distribution over the volume of the beam, the ion energy and its spatial distribution, and the divergence of the beam (Kaufman et al. 1982).

From a practical point of view, the two process parameters of most importance are the beam potential and beam current. Both these parameters can be measured with a Langmuir probe. The ion beam current has a large effect on the material etch rate. The ion beam potential directly impacts the selectivity between hard mask and substrate. For a given beam potential, the ion current can only be increased so high before the discharge plasma is extinguished.

A useful tool for ion beam etching is end point detection. Secondary ion mass spectrometry (SIMS) is a method of monitoring material composition of sputtered films. SIMS requires a low operating pressure, making it compatible with typical ion beam etching tools. The sensitivity of SIMS allows for very accurate end point control, detecting when each layer of material is etched away and a new material exposed.

There are commercially available ion beam etching tools on the market. If an ion beam etch chamber is not available, the etch processes described in this chapter can be approximated with noble gas in typical plasma chambers (RIE, ICP-RIE, Helicon, etc.) by controlling the plasma current density and DC bias. The advantages of a dedicated ion beam etch tool include lower operating pressure, a neutralized ion beam, and directional control.

19.2.2 The Sputter Etch Process

During ion beam etching, momentum transfer from ions that strike the substrate surface sets off a series of collisions. Energy is transferred to both atoms (nuclear energy loss) and to electrons (electronic energy loss). For low-energy ion beam etching (<10 keV), the process is dominated by nuclear energy loss (Smentkoski 2000). Ion collisions with the surface rattle atoms, and if an atom on the substrate surface is impacted with energy higher than its binding energy, then that atom is sputtered off. Along with sputtered material, secondary electrons are also emitted. The sputtering yield of a material

is the average number of atoms that are removed per ion striking the surface, and is proportional to

1. Energy of the incoming ion
2. Mass of the incoming ion
3. Mass of the atoms being etched
4. Surface binding energy of atoms being etched
5. Incident angle of the ion beam

When impacting a surface with an ion beam, there is a lower limit to the beam energy, below which no sputtering will occur.

The minimum energy required for material removal is typically around 10–100 eV. Below the sputtering threshold, ions will be reflected off the surface or implant into the material with minimal damage. Figure 19.2 shows the curve for nickel sputtering yield as a function of krypton ion beam potential. When selecting an ion beam potential for etching magnetic materials, a value on the curve with a low slope will give a larger process stability margin. However, lower beam potential does provide better selectivity and less damage to magnetic structures (McMorran et al. 2010).

Some sputter yield values for normally incident Ar ions on typical hard mask materials and magnetic materials are given in Table 19.1. Comparing these sputter yield values, one can observe the etch selectivity for ion beam etch processes. Carbon has a high surface binding energy and low sputter yield, making it an excellent etch mask material. To achieve the best resistance to ion beam etch, the carbon film should be deposited to optimize the number of sp^3 bonds, making the film more diamond-like. Another advantage of carbon hard masks is their ease of patterning with plasma etching. Proper plasma etch conditions can

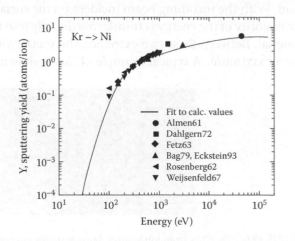

FIGURE 19.2 Ni sputtering yield versus incoming Kr^+ ion energy. (Adapted from Eckstein, W. 2008. *Vacuum*, 82, 930–934.)

Table 19.1 Sputter Yield Values for Normally Incident 500 eV and 1 keV Ar Ion Beams on Typical Hard Mask and Magnetic Materials

Material	Atomic Z	Sputter Yield at 500 eV	Sputter Yield at 1 keV
Carbon	6	0.197	0.449
Titanium	22	0.577	0.901
Chromium	24	1.131	1.723
Tantalum	73	0.664	1.083
Iron	26	1.157	1.767
Cobalt	27	1.248	1.905
Nickel	28	1.283	1.963
Palladium	46	2.092	3.127
Ruthenium	48	1.304	2.028
Platinum	78	1.265	1.989

Source: NPL United Kingdom, 2005. http://resource.npl.co.uk/docs/science_technology/nanotechnology/sputter_yield_values/arsputtertable.pdf

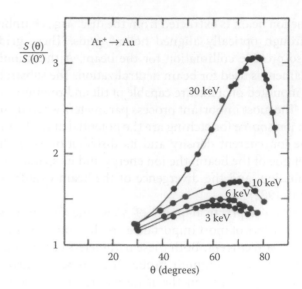

FIGURE 19.3 Sputtering yield versus ion beam angle for various Ar[+] beam energies. (Reprinted from Figure 7 with permission from American Physical Society. Olivia-Florio, A. et al. *Phys. Rev. B*, 35(5), 2198–2204, 1987. Copyright 1987 by the American Physical Society.)

produce carbon hard masks with high aspect ratio sidewalls that are important for ion beam etching.

As mentioned earlier, the ion beam potential has a large effect on the etch selectivity. Take, for example, ion beam etching cobalt with a carbon hard mask. In Table 19.1, increasing the ion beam potential from 500 to 1 keV increases the rate of removal for both materials. However, doubling the beam potential drops the media/mask selectivity from 6:1 to 4:1.

Above the minimum sputtering energy, there is a curious relationship between the ion beam angle and the sputtering yield. When the ion beam impacts at a shallow angle to the substrate, the affected atoms tend to slide across the material surface, reducing the sputter yield. With the incoming beam incident on the surface, the majority of the energy is transferred directly into the material. Between these two extremes, the sputter yield has a maximum. A typical example of this is shown in

Figure 19.3. This plot shows the sputtering yield of gold as a function of angle for various ion beam potentials. For a 3 keV Ar[+] ion beam, the maximum sputter yield is achieved around 40° from normal incidence. For further information, a full empirical formula to calculate angular dependent sputtering yield values has been published in the journal *Vacuum* (Eckstein 2008).

Sputtering yield that is a function of ion beam angle will cause faceting on corners (Neureuther et al. 1979), a common problem for ion beam etching. During the etch process, incoming ions preferentially eject material at an angle where the sputtering yield is maximized. Initially, square corners can quickly become damaged. A simulation of faceting damage on photoresist, by Zhang et al. (2002), and an experimental image with the same conditions can be seen in Figure 19.4. This faceting problem is typical and can also be seen in reactive ion etch processes where the DC bias voltage is particularly high.

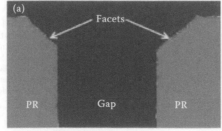
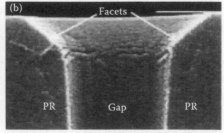

FIGURE 19.4 Faceting from ion sputter damage on the corners of a photoresist mask: (a) simulation and (b) SEM image. (From Zhang, D., S. Rauf, and T. Sparks. 2002. Modeling of photoresist erosion in plasma etching processes. *IEEE Trans. Plasma Sci.*, 30(1), 114–115, © (2002) IEEE.)

Combining the etch rate information from Table 19.1 and faceting, as depicted in Figure 19.4, one can see that patterning of magnetic nanostructures with ion beam etching is a difficult task. The selectivity of magnetic material to hard mask, in the bulk, is typically <8:1. Looking closely at the etch-front progression in Figure 19.4, faceting can significantly reduce etch selectivity near the corners of a feature. When etching magnetic nanostructures, it is not uncommon to see selectivity numbers as low as 2:1. This is much lower than etch selectivity values typically seen in reactive plasma etching.

19.2.3 Pattern Transfer

When using ion beam etching to fabricate magnetic nanostructures, some type of patterning must be used. Initial patterning can be done with techniques including photolithography, e-beam lithography, nano-imprinting, and block-copolymer formation. More information on these patterning techniques can be found elsewhere in this book. With a properly chosen hard mask and ion beam etch conditions, pattern transfer can be achieved with very high resolution and high feature density (Shaw et al. 2008). Figure 19.5 shows a scanning electron microscopy (SEM) image of a 30-nm-wide GMR spin valve fabricated with e-beam lithography and ion beam etching (Childress et al. 2008).

Most polymers used for lithography have poor ion beam etch resistance, so it is a good practice to transfer the initial polymer pattern into a hard mask such as carbon, silicon nitride, tantalum, or alumina. Anisotropic hard mask etch processes should be used to achieve the vertical sidewall angles that are desirable for ion beam etching. As pointed out by Peng et al. (2010), the initial mask sidewall angle can have a large effect on the etched feature sidewall. If you neglect faceting, the initial sidewall angle sets an

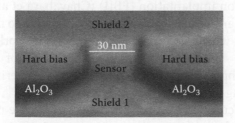

FIGURE 19.5 An ion beam etched giant magneto resistive spin valve sensor with 30-nm-wide IrMn, CoFe, Ru, and NiFe layers. (From Childress, J. R. et al. 2008. All-metal current-perpendicular-to-plane giant magnetoresistance sensors for narrow-track magnetic recording. *IEEE Trans. Magn.*, 44(1), 90–94, © (2008) IEEE.)

FIGURE 19.6 Depiction of ion beam etched material wall angle propagation.

upper limit to the achievable sidewall angle in the etched magnetic material.

Figure 19.6 depicts the propagation of the initial hard mask wall angle into the etched material. An exercise in trigonometry allows one to derive a formula to calculate an approximate sidewall angle α for the final etched material, based on hard mask to media selectivity, S, and the initial hard mask sidewall angle, θ.

$$\alpha = \arctan (S * \tan \theta) \tag{19.1}$$

Looking over Equation 19.1, one can see that with the best-case selectivity, $S = \infty$, the etched sidewall angle will be vertical. If the selectivity is unity, a more realistic number, the final sidewall angle can only be as good as the initial hard mask profile. This is a good approximation for incident angle ion beam etching when features are large enough to ignore faceting problems.

19.2.4 Redeposition of Material

Ion beam etching is useful for patterning materials with nonvolatile reaction products, but it is not immune to the problems associated with nonvolatility. As the ion beam sputters material from the surface of the substrate, the ejected material must have an unobstructed path for escape. If so, the etched material will travel off and be redeposited on the chamber walls and shields. If the etched material does not have a clear path and collides with a surface, for example, the etching hard mask, it has a probability of redepositing there. Redeposited material on the hard mask is often referred to as "fencing" or "redep." After stripping the hard mask, fencing often remains at the top of the etched features, marking the outline of the hard mask. This problem is particularly pronounced when a thick hard mask is required (Walsh et al. 2000).

There are methods to reduce fencing during etching of magnetic material. One method is the use of off-axis milling. To maintain a straight etched sidewall profile, a normally incident ion beam etch process is important. Normally incident ion beam etching generates the most fencing on the hard mask. After the bulk material removal is completed, an off-axis milling process, with substrate rotation, can be used to etch away the fencing material. With this approach, the fencing material can help support the etch mask from deformation during the etching process (Peng et al. 2010). A 60° off-axis etching process is typical for fencing clean up.

Alternatively, a slightly off-axis ion beam, around 5–10°, can be used for the entire etching process. By etching off axis for the entire process, the beam is continuously etching on the side of the hard mask. This process ensures that no material is redeposited on the hard mask sidewalls. Figure 19.7 shows CoPd multilayer bits for magnetic data storage, etched with this method by the author.

An angled hard mask as well as mask erosion will also suppress fencing problems. If the hard mask has a sidewall angle that is <90°, a normally incident ion beam will also be impinging on the hard mask sidewall, removing redeposited material. The problem of faceting can be advantageous for this purpose. If the etching process is stopped just as the hard mask facet meets the etched film, fencing problems can be reduced.

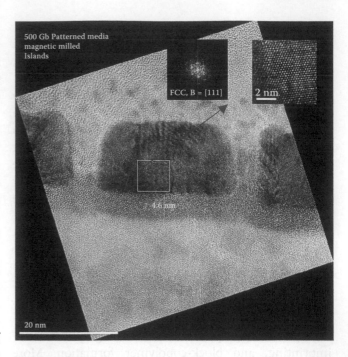

FIGURE 19.7 A TEM image of 7-nm-wide grooves etched between CoPd multilayer bits for magnetic data storage; inset box shows a diffraction pattern from the atomic lattice.

19.3 Ion Implantation for Magnetic Patterning

Ion implantation, historically common to the semiconductor industry, is another useful technique for patterning magnetic materials. This fabrication method is essentially an extension of ion beam etching. Ion implantation also relies on a plasma source to generate ions of a type desirable to the user. The ions are accelerated toward the substrate and impact with considerably more energy than those used for ion beam etching, around 10–500 keV. At these high energies, ion implantation occurs with minimal sputter etch damage to the substrate surface. When implanting ions, the two mechanisms for magnetic property modification are atomic/structural dislocations and compositional change. Exposing magnetic films through a lithographically defined hard mask can create high-resolution magnetic patterns.

19.3.1 Ion Implantation Tools

Ion implantation for semiconductor doping was invented by William Shockley in 1956. The traditional scanning ion implantation tool uses an ion source, accelerator, separator, and sweep controller. The ion source is a vacuum chamber with an RF power supply. Electrodes extract the beam from the plasma chamber

and accelerate it to a desired voltage. A magnetic field is used to separate ions by mass. The resulting ion beam is then focused, and x–y axis deflection plates are used to scan the beam on the substrate for uniform implantation. For example, BF_3 plasma can be used to generate B^+ ions that are filtered and focused for implantation. Some gas sources for ions can be quite dangerous, such as AsH_3 or PH_3.

Plasma source ion implantation (PSII) (Conrad et al. 1987) and plasma doping (PLAD) (Goeckener et al. 1999) use plasmas to generate the desired ions directly above the substrate. Ions are accelerated across the plasma sheath and impact the substrate. Plasma-based ion implantation (PBII) (Chayahara et al. 2004) is similar but relies on an arc-like discharge to melt the anode, which is ionized in front of the work piece. Plasma processing has the advantage of higher throughput but lacks any beam filtering, steering, and focusing capability.

A focused ion beam (FIB) tool can also be used for ion implantation on magnetic materials. This type of equipment is fairly common in analytical laboratories. The FIB implantation process has a very low throughput although it can be useful for research and process development.

19.3.2   The Implantation Process

When ions are accelerated and impact a material, collisions occur in a statistical process that modifies the substrate as the ions come to rest. Similar to ion beam etching, there are two contributors to the incoming ion's energy loss: losses from collisions with nuclei and losses from electrons. The "stopping power" of a material is defined as the loss per unit length of the ion's path. A simplified expression for stopping power is shown in Equation 19.2.

$$S = (dE/dx)_{\text{nuclear}} + (dE/dx)_{\text{electronic}} \qquad (19.2)$$

The values for the nuclear and electronic stopping power are not easily calculated. A more detailed description of stopping power can be found in Smith (1997). The nuclear component of stopping power is more important for lighter ions and lower ion energies. The electronic component is dominant for heavier ions and higher ion energies.

Ions have their highest energy at impact, and eventually the losses slow down and stop the ions within the substrate. When the ion energy is low enough for nuclear interaction, a collisional cascade is set off that dislocates atoms and can change the crystalline structure of the substrate. Figure 19.8 shows a possible path of an ion and its collisions as it travels through a solid material (Nordlund 2010). With high initial energy, electronic stopping forces drag down the ion speed. Approaching lower energy, nuclear stopping power dominates driving atomic collisions. With high-mass and low-energy ions, nuclear collisions near the substrate surface lead to sputter damage. This process is considered ion beam etching, which is described earlier in this chapter. With higher energy, the ions can penetrate deep enough into the substrate to reduce sputter damage. The eventual stopping point of ions results in a depth profile that is a Gaussian distribution.

The depth of the implantation process is a function of the energy of the ion beam, the mass of the ions, and the stopping power of the substrate. Equation 19.3 is the mathematical expression for the implanted material depth profile.

$$N(x) = N_p \exp[-(x - R_p)^2 / 2\Delta R_p^2] \qquad (19.3)$$

In Equation 19.3, the concentration $N(x)$ has a maximum when the depth is equal to the projected range (R_p). Lateral deviations of the incoming ions (ΔR_p) are known as the straggle. The projected range of ions is a function

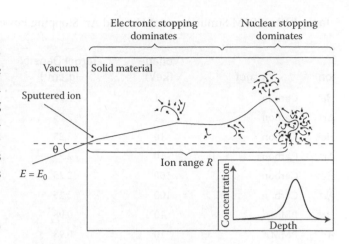

FIGURE 19.8   Ion interactions with a solid after impact. (From Nordlund, K. 2010. Personal communications, University of Helsinki.)

of incoming ion energy and the mass of both the ions and the substrate atoms. The values of R_p can be calculated or looked up in tables. Higher ion energy will result in a larger R_p. For magnetic materials, this is typically 10–1000 nm below the surface (Komvopoulos et al. 1994).

19.3.3   Ion Implantation Simulation

The plethora of information from the semiconductor industry makes it easy to locate the values of R_p and ΔR_p for ion implantation into silicon with impurities such as boron or arsenic. When implanting ions into magnetic materials, the pertinent information for R_p and ΔR_p cannot be easily located in the literature. Given this lack of experimental information, Monte Carlo simulators are a quick way to approximate these values. There are commercial simulation packages on the market, but one of the most widely used software applications for modeling the implantation process was written by Ziegler et al. (2009) and is available for free on Ziegler's web site (Ziegler et al. 1989). The program, stopping and range of ions in matter (SRIM), relies on Monte Carlo algorithms to efficiently calculate the R_p and ΔR_p for ions in energy as high as 10 GeV/amu. The statistical calculations in SRIM use velocity-dependent ion charge, Coulombic collisions, and long-range interactions within the material for accurate results. This simulation technique is a valuable tool for determining the ion implantation profile for a given process.

Table 19.2 is a list of ion implantation profiles, generated with SRIM, for He+ and Ar+ ions into hard mask and magnetic materials, typical for nanofabrication.

Chapter 19

Table 19.2   SRIM Simulations of He⁺ and Ar⁺ Stopping Power, Range, and Straggle for Various Ions and Targets

Ion	Target	Voltage (keV)	Target Density (g/cm³)	Stopping Power (eV/A)	Projected Range (A)	Longitudinal Straggle (A)	Lateral Straggle (A)
He⁺	Carbon	1	2.25	6.716	90	60	45
Ar⁺	Carbon	1	2.25	55.072	26	9	7
He⁺	Carbon	10	2.25	11.529	808	279	251
Ar⁺	Carbon	10	2.25	100.031	117	33	26
He⁺	Carbon	100	2.25	27.817	4901	589	718
Ar⁺	Carbon	100	2.25	121.612	819	159	134
He⁺	PMMA	10	0.98	4.708	1893	629	568
Ar⁺	PMMA	10	0.98	46.793	253	68	54
He⁺	SiO₂	10	2.32	6.376	1031	524	443
Ar⁺	SiO₂	10	2.32	74.508	139	54	40
He⁺	Ta	10	16.6	12.145	246	322	241
Ar⁺	Ta	10	16.6	80.576	51	59	45
He⁺	CoPtCr (70/15/15)	10	10.5	11.068	341	351	270
Ar⁺	CoPtCr (70/15/15)	10	10.5	118.629	56	42	31
He⁺	NiFe (80/20)	10	8.07	11.133	381	330	259
Ar⁺	NiFe (80/20)	10	8.07	129.822	59	38	28
He⁺	FePt (45/55)	1	15.34	4.698	36	72	52
Ar⁺	FePt (45/55)	1	15.34	44.806	12	14	10
He⁺	FePt (45/55)	10	15.34	11.133	266	337	253
Ar⁺	FePt (45/55)	10	15.34	100.38	50	48	37
He⁺	FePt (45/55)	100	15.34	34.479	2351	1268	1112
Ar⁺	FePt (45/55)	100	15.34	151.253	337	261	195

This reference chart is useful for visualizing how the stopping power is affected by ion mass and accelerating voltage.

The SRIM simulation software also has a package for simulating kinetic phenomena from the ion implantation process: TRIM (transport of ions in matter). The TRIM simulator can generate 3D profiles of implanted ions as well as calculate sputter damage, ionization, and phonon production. There are some limitations to the TRIM simulation process (Fassbender et al. 2004), including the fact that implantations are always calculated in virgin material and no implant accumulation is considered. Even with these limitations, it is an excellent method for modeling ion implantation of magnetic materials.

19.3.4   Magnetic Property Modification

Exposure to a high-energy ion beam can have different effects on magnetic materials depending on the process conditions. Collisions lead to disorder and compositional changes that can be exploited for modifying magnetic anisotropy and moment. With certain conditions, ion implantation can even be used to generate order in a magnetic system (Ravelosona et al. 2002, Yang et al. 2003).

Magnetic multilayers exhibit square hysteresis loops and high coercive fields in the presence of a perpendicular field. Their perpendicular magnetic anisotropy (PMA) is engineered with highly ordered layers. The dependence of PMA on film order makes multilayers ideal candidates for patterning with ion implantation. The implantation process is capable of breaking the order between layers and mixing the elements. Disorder in the multilayer system will result in reduced magnetization and coercive field (Weller et al. 2000). Perpendicular anisotropy modification can be achieved with disorder only and does not necessarily require a compositional modification from the implanted elements. Figure 19.7 is a diagram from

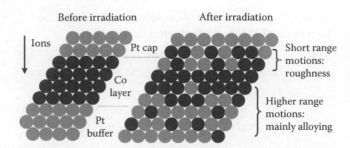

FIGURE 19.9 Atomic dislocations in CoPt magnetic multilayers resulting in loss of PMA. (Adapted from Fassbender, J., D. Ravelosona, and Y. Samson. 2004. *J. Phys. D: Appl. Phys.*, 37, R179–R196.)

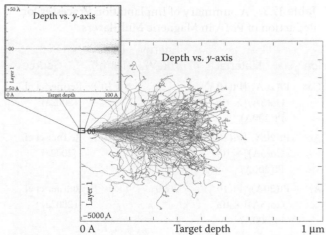

FIGURE 19.10 TRIM simulation of two hundred 700-keV N⁺ ions implanted in CoPt films.

Fassbender et al. (2004) showing the damaging effects of high-energy ion implantation.

The diagram in Figure 19.9 illustrates how back-scattered material from collisions results in short-range disorder above the multilayers, and longer-range disorder below the layers from forward scattering. Extended x-ray absorption fine-structure measurements can be done to verify these changes to multilayer structures (Devolder et al. 2001).

The first work using ion implantation to modify magnetic properties was done by Traverse et al. (1989). An FIB tool was used to implant NiAu magnetic multilayers with 30 keV He⁺ ions. It was shown that ion-induced damage to magnetic films could reduce their magnetic anisotropy and coercivity. This work also demonstrated that the ability of an ion beam to induce elemental mixing in multilayers can be improved if the elements in the film are miscible. In this case, the positive heat of mixing between elements, ΔH_{m}, contributed to the disorder from ballistic collisions.

When implanting to disorder a multilayer system, low ion mass and high ion energy are preferred. Figure 19.10 is a TRIM simulation of the ion 700 keV N⁺ implantation process similar to the work from Weller et al. (2000) on reducing PMA. These conditions push the ion deposition hundreds of nm below the surface of the CoPt film, thus achieving intermixing without doping. When implanting into thin magnetic layers with high-energy low-mass ion beams, very little ion scattering occurs in the thin film itself and atomic collisions are concentrated deep into the substrate. As seen in the inset picture of Figure 19.10, the first 100 A of the magnetic film shows lateral straggle that is <10 A. This type of process will leave initial grain size and structure intact and not disrupt the surface roughness of the film (Fassbender et al. 2004). The difficulty is finding a hard mask structure that can successfully block ions in areas of interest.

The reduction of PMA in magnetic multilayers can be increased with the addition of chemical processes. Gonzales et al. have shown that phosphorus and arsenic can be used to generate Co–P and Co–As bonds. This chemically reactive process of covalently bonding to cobalt ties up a valence electron and reduces the PMA for the film.

The dose, or fluence, required for reduction of PMA in multilayered magnetic films is very case dependent. When implanting N⁺ ions into CoPt multilayers, it was shown that 3–4 impinging ions were needed to separate one Co–Pt pair from a multilayered film, and essentially all pairs must be broken to completely reduce the PMA (Weller et al. 2000). The typical high-energy ion implantation dose is 1×10^{15} ion/cm². A review of ion implantation conditions for reduction of PMA is shown in Table 19.3.

High-energy and low-mass ion beams are useful for patterning ordered magnetic multilayers but are not as useful for magnetic alloy films. With alloy films, compositional changes to the material are required for lowering the magnetic moment. For this case, the implantation process should have the beam voltage tuned to implant the dose at the center of the magnetic film thickness. Because of this, when implanting to achieve a compositional change in alloy films, ion beam voltages are lower relative to those used for patterning magnetic multilayers. Ideally, the longitudinal spread of the ions should cover the thickness of the film.

The choice of ions to reduce the moment properties of alloy films is not so clear-cut. It has been shown that Ga⁺ ions implanted into 20-nm-thick $Co_{70}Cr_{18}Pt_{12}$ could poison magnetic properties (Rettner et al. 2002b). Here,

Chapter 19

Table 19.3 A Summary of Implantation Processes for Reduction of PMA in Magnetic Multilayers

Ion	Material	Energy (keV)	Dose (1/cm²)	Source
Ga⁺	Pt(20A)/[Pt(10A)/ Co(3A)] × 10/ Pt(200A)	30	4×10^{13}	Rettner et al. (2002a)
Ar⁺	Pt(20A)/[Pt(10A)/ Co(3A)] × 10/ Pt(200A)	20	8×10^{13}	Rettner et al. (2002a)
Ar⁺	Pt(20A)/[Pt(10A)/ Co(3A)] × 10/ Pt(200A)	2000	4×10^{14}	Rettner et al. (2002a)
He⁺	Pt(20A)/[Pt(10A)/ Co(3A)] × 10/ Pt(200A)	20	3×10^{15}	Rettner et al. (2002a)
He⁺	Pt(20A)/[Pt(10A)/ Co(3A)] × 10/ Pt(200A)	2000	3×10^{16}	Rettner et al. (2002)
N⁺	Pd(150A)/ [Co(2.8A)/ Pd(9.3A)] × 15	30	2×10^{16}	Martin-Gonzalez et al. (2010)
N⁺	Pt(200A)/[Co(3.5A)/ Pt(10A)] × 10/ Pt(20A)	700	3×10^{15}	Weller et al. (2000)
Ar⁺⁺	Pt(200A)/[Co(4.5A)/ Pt(10A)]	40	1×10^{15}	Bonder et al. (2003)
P⁺	Pd(150A)/ [Co(2.8A)/ Pd(9.3A)] × 15	20	2×10^{16}	Martin-Gonzalez et al. (2010)
As⁺	Pd(150A)/ [Co(2.8A)/ Pd(9.3A)] × 15	30	2×10^{16}	Martin-Gonzalez et al. (2010)

Table 19.4 A Summary of Implantation Processes for Reduction of Magnetic Properties in Alloy Films

Ion	Material	Energy (keV)	Dose (1/cm²)	Source
Ga⁺	$Ni_{80}Fe_{20}$(155A)	30	1×10^{16}	Kaminsky et al. (2001)
He⁺	$Co_{50}Pd_{50}$(500A)	40	6×10^{16}	Abes et al. (2004)
Ga⁺	$Co_{70}Cr_{18}Pt_{12}$(200A)	30	2×10^{16}	Rettner et al. (2001)
Cr⁺	CoCrPt(200A)	20	1×10^{16}	Hinoue et al. (2010)
N⁺	CoCrPt(200A)	17.5	2×10^{16}	Aniya et al. (2010)
Ga⁺	$Ni_{80}Fe_{20}$(300A)	30	1×10^{16}	Ozkaya et al. (2002)
Cr⁺	$Ni_{80}Fe_{20}$(500A)	40	5×10^{16}	Folks et al. (2003)

field, produces changes that are similar to results obtained with thermal annealing. Similarly, Dasgupta et al. (2006) have implanted rare earth elements into NiFe to modify the precessional dynamics of the film.

19.3.5 Patterning Magnetic Nanostructures

To pattern magnetic films, it is important to have precise control of where the ions are implanted. Selective placement for implantation can be achieved by direct implantation from an FIB tool, by ion exposure through a stencil mask (Terris et al. 1999), or with ion projection direct structuring (IPDS) (Dietzel et al. 2002). Direct writing of patterns onto a magnetic surface lacks the throughput needed for manufacturing and stencil masks limit pattern resolution. Ion implantation can also be performed through a patterned mask to generate isolated magnetic nanostructures (Rettner et al. 2002a,b, Martin-Gonzalez et al. 2010, Fassbender and McCord 2008).

Large surface area ion beam exposures through a mask can increase fabrication throughput but require the additional patterning steps. The critical parameters for masks used to block ion exposure are the ion stopping power and sputtering yield. Refer to Table 19.3 to compare the stopping power of various masking materials. Looking over the presented values of R_p and ΔR_p (longitudinal) reveals the importance of material density on stopping ions. For example,

the implantation process was sufficient to alloy the top 10 nm of the film with 10% Ga by volume. Similarly, Cr⁺, He⁺, and N⁺ ions have also been used for reduction of saturation magnetization. Table 19.4 summarizes the literature on the ion implantation process for the reduction of moment in magnetic alloy films. The ion doses used for these examples increase the material volume by 5–10%.

The main focus so far has been on implantation for the reduction of magnetic properties; however, adding atoms to a film via ion implantation can modify magnetic film properties in other ways.

McCord et al. (2009) have demonstrated that adding Co⁺ ions to FeCoSiB can change the amplitude and direction of its anisotropy field. In this case, ion implantation, with the sample immersed in a magnetic a 10 keV Ar⁺ beam is required to implant 5.6 nm deep in CoPtCr alloy. Stopping a 10 keV Ar⁺ beam can be achieved with a 39-nm-thick poly(methyl methacrylate) (PMMA) mask. The same beam can be stopped with only 17 nm of tantalum (Equation 19.4).

$$\text{Mask thickness} \geq R_p + 2(\Delta R_p) \qquad (19.4)$$

High-density films such as tantalum, molybdenum, and tungsten are excellent mask candidates because they can be patterned with fluorine plasma reactive ion etching.

The TRIM simulation results, by Martin-Gonzalez et al. (2010), in Figure 19.11 show the Gaussian distribution of ion ranges for 20 keV P⁺ ions implanted into CoPd multilayers and PMMA. For this simulation, 110 nm of PMMA is sufficient thickness to stop implantation. Fabrication of high aspect ratio implantation masks is difficult for nanoscale features, so reducing the mask thickness is desirable.

Ion implantation will roughen the surface of a material. The two processes that impact the surface roughness are sputter removal of material and local volumetric expansion from implantation. Sputter damage, from atomic collisions near the material surface, is more severe for lower-energy beams and high-mass ions that are necessary to achieve a shallow dopant profile. For example, Hinoue et al. have reported 4 nm of material

removed when implanting 20 keV Cr⁺ ions into CoCrPt films. The target implantation depth for this process was 10 nm. Devolder et al. (2000) have shown that the substrate surface roughness increases with implantation fluence. At the same time that the ions are sputtering away material from the surface, they are also implanting and increasing the volume of the substrate. Once again, volumetric expansion is more pronounced on the surface with lower ion energies. Typically, sputter damage is the dominant of these two roughening processes, resulting in a net reduction of material where the hard mask is opened. If planarity is not required in the final magnetic device, it is often advantageous to have some sputtered material removal for feature separation.

19.3.6 High-Resolution Patterning

When considering ion implantation, as is the case for other nanopatterning techniques, the question of ultimate resolution always arises. Nuclear collisions and scattering events in the material are the main contributors to loss of resolution. Referring again to Table 19.3, the values for longitudinal and lateral straggle highlight this problem. It is particularly challenging when fabricating densely packed nanostructures.

Reduction of straggle is not an easy task. The longitudinal and lateral straggles are related to the ion energy, ion mass, and target mass. Increasing the ion

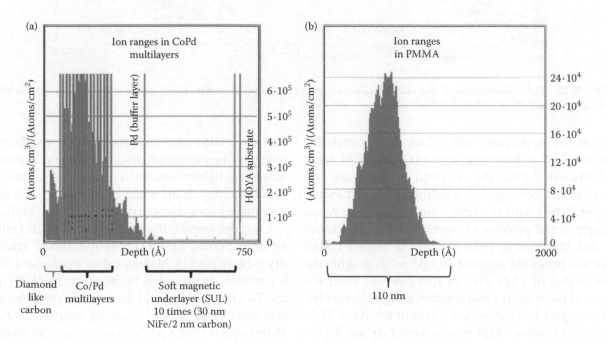

FIGURE 19.11 TRIM simulations from Martin-Gonzalez et al. of 20 keV P⁺ ions in CoPd and PMMA. (Martin-Gonzalez, M. S. et al. 2010. Nano-patterning of perpendicular magnetic recording media by low-energy implantation of chemically reactive ions. *J. Magn. Magn. Mater.*, 322, 2762–2768, © (2010) IEEE.)

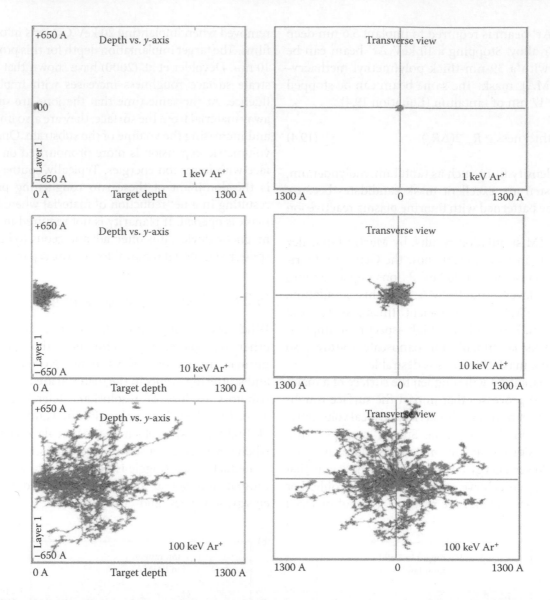

FIGURE 19.12   TRIM simulations of ions and collisional cascades that show the implant depth profile and transverse profile of 50 Ar⁺ ions implanted in FePt at 1, 10, and 100 keV.

mass or the target mass will reduce the ion mobility. This will lead to less straggle at the expense of lower implantation depth. The selection of ion mass and energy to reduce straggle is a balancing act that should be approached with the help from simulations. As an example of the problem's complexity, take the case of Ar⁺ and He⁺ ions in FePt. Because of argon's higher mass, the projected range of 10 keV Ar⁺ is roughly the same as that of 1 keV He⁺. In this case, the lower ion mass and lower energy still result in a slightly larger lateral straggle: 1 keV He⁺ = 52 A versus 10 keV Ar⁺ = 37 A. Figure 19.12 shows TRIM simulations of Ar⁺ ion deposited into FePt for 1, 10, and 100 keV energies. Comparing the longitudinal to lateral straggle, it is apparent that

the ratio is fairly constant, around 1.1, across the energy range. In the case of longitudinal straggle, it should be noted that lighter ions, such as B⁺, tend to have more backscattering events while heavier ions, such as Ar⁺, have more forward scattering events (Jager 2001).

Consider implanting into an 8-nm-thick FePt film for bit patterned media (BPM) at 1 Tb/in.² areal density (square lattice). Magnetic data storage at 1 Tb/in.² is currently an attractive number for the HDD industry. To achieve this areal density, the patterning process must be capable of resolving magnetic dots with 25 nm square lattice pitch. Then, assume an ideal hard mask for the implantation process. Referring to Table 19.4, SRIM simulations predict that a 10 keV Ar⁺

beam will implant to the center of the film thickness and the longitudinal straggle covers almost all the FePt film thickness. A 10 keV Ar^+ implantation into FePt gives a lateral straggle of 3.7 nm. The total lateral ion spread of 7.4 nm is a third of the entire feature pitch, driving the need for extremely narrow lithographic patterning of the hard mask. Although Ar^+ might not be the best choice of ion for BPM fabrication, it clearly demonstrates how lateral straggle limits the ultimate resolution of ion implantation. Implanting ions that complement the magnetic film, rather than degrading it, can use lateral straggle for feature growth and signal gain. Complementary implantation processes, described previously (McCord et al. 2009, Dasgupta et al. 2006), could be beneficial to highly dense packing of magnetic structures.

Experimentally, ion implantation has been used by Rettner et al. (2001) to pattern high-density BPM samples (Figure 19.13). A direct-write focused ion beam tool was used to implant 30 keV Ga^+ ions into 20-nm-thick $Co_{70}Cr_{18}Pt_{12}$ at density close to 200 Gb/in.² (60 nm

FIGURE 19.13 AFM (a) and MFM (b) images by M. Best showing 70 nm islands on a 100 nm pitch, created by ion implantation into CoCrPt. (Rettner, C. T., M. E. Best, and B. D. Terris. 2001. Patterning of granular magnetic media with a focused ion beam to produce single-domain islands at >140 Gb/in2. *IEEE Trans. Magn.*, 37(4), 1649–1651, © (2001) IEEE.)

pitch). The implantation process is easier when patterning isolated features. Ravelosona et al. (2001) were successful in generating 30-nm-wide magnetic lines by implanting 30 keV He ions at 2×10^{15} ions/cm² into CoPt magnetic multilayers.

19.4 Summary

The growing field of nanomagnetic devices relies on the pattern transfer into materials that do not readily form volatile reaction products. This limitation makes top-down patterning of magnetic materials challenging. Ion beams can be used either for physical removal of magnetic materials or for implantation of dopant atoms to modify magnetic properties. Combining either of these techniques with high-resolution lithography is an effective way to generate magnetic nanostructures.

Low-energy ion beams, particularly with heavy noble gasses, will remove nonvolatile materials via a sputtering process. Ion beam etching has some drawbacks that complicate its use for nanofabrication. Etch selectivity between material and hard mask is low. The angular dependence of sputtering yield leads to faceting that further reduces the etch selectivity. Also, redeposition of material can lead to fencing problems. Carbon has the lowest sputtering yield and should be used as an etching hard mask whenever

possible. With a properly chosen hard mask and appropriate ion beam etching conditions, the challenges associated with ion beam etching can be overcome to sculpt out nanostructures from magnetic films.

High-energy ion beams will physically implant material into magnetic films. Ion implanting through a hard mask can be used to create isolated magnetic features. The implantation process can modify magnetic properties by dislocating atoms in well-organized multilayers or by locally changing the elemental composition of a film. It is important to select a hard mask with enough stopping power to block unwanted ions. Selectively exposing magnetic material through an opened hard mask can create nanoscale magnetic structures with very little surface topography. Ultimately, feature resolution and packing density are limited by the Gaussian distribution of implanted ions.

References

Abes, M., J. Venuat, D. Muller et al. 2004. Magnetic patterning using ion irradiation for highly ordered CoPt alloys with perpendicular anisotropy, *J. Appl. Phys.*, 96(12), 7420–7423.

Aniya, M., A. Shimada, Y. Sonobe et al. 2010. Magnetization reversal process of hard/soft nano-composite structures formed by ion irradiation. *IEEE Trans. Magn.*, 46(6), 2132–2134.

Chapter 19

Baibich, M. N., J. M. Broto, A. Fert et al. 1988. Giant magnetoresistance of (001)Fe/(001)Cr magnetic superlattices. *Phys. Rev. Lett.*, 61(21), 2472–2475.

Binasch, G., P. Grunberg, F. Saurenbach, and W. Zinn, 1989. Enhanced magnetoresistance in layered magnetic structures with antiferromagnetic interlayer exchange. *Phys. Rev. B*, 39, 4828–4830.

Bonder, M. J., N. D. Telling, P. J. Grundy, C. A. Faunce, T. Shen, and V. M. Vishnyakov. 2003. Ion irradiation of Co/Pt multilayer films. *J. Appl. Phys.*, 93, 7226–7228.

Chayahara, A., Y. Mokuno, A. Kinomura, N. Tsubouchi, C. Heck, and Y. Horino. 2004. Metal plasma source for PBII using arclike discharge with a hot cathode. *Surface Coatings Technol.*, 186, 157–159.

Childress, J. R., M. J. Carey, S. Maat et al. 2008. All-metal current-perpendicular-to-plane giant magnetoresistance sensors for narrow-track magnetic recording. *IEEE Trans. Magn.*, 44(1), 90–94.

Conrad, J. R., J. L. Radtke, R. A. Dodd, F. J. Worzala, and N. C. Tran. 1987. Plasma source ion-implantation technique for surface modification of materials. *J. Appl. Phys.*, 62, 4591–4596.

Dasgupta, V., N. Litombe, W. E. Bailey, and H. Bakhru. 2006. Ion implantation of rare-earth dopants in ferromagnetic thin films. *J. App. Phys.*, 99, 08G312.

Devolder, T. 2000. Light ion irradiation of Co/Pt systems: Structural origin of the decrease in magnetic anisotropy. *Phys. Rev. B*, 62(9), 5794–5892.

Devolder, T., S. Pizzini, J. Vogle et al. 2001. X-ray absorption analysis of sputter-grown Co/Pt stackings before and after he irradiation. *Eur. Phys. J.*, B22(2), 193–201.

Dietzel, A., R. Berger, H. Grimm et al. 2002. Ion projection direct structuring for patterning of magnetic media. *IEEE Trans. Magn.*, 38(5), 1952–1954.

Eckstein, W. 2008. Sputtering yields. *Vacuum*, 82, 930–934.

Fassbender, J., and J. McCord. 2008. Magnetic patterning by means of ion irradiation and implantation. *J. Magn. Magn. Mater.*, 320 (3–4), 579–596.

Fassbender, J., D. Ravelosona, and Y. Samson. 2004. Tailoring magnetism by light-ion irradiation. *J. Phys. D: Appl. Phys.*, 37, R179–R196.

Folks, L., R. E. Fontana, B. A. Gurney et al. 2003. Localized magnetic modification of permalloy using Cr+ ion implantation. *J. Phys. D: Appl. Phys.*, 36, 2601–2604.

Goeckener, M. J., S. B. Felch, Z. Fang et al. 1999. Plasma doping for shallow junctions. *J. Vac. Sci. Technol. B*, 17(5), 2290–2293.

Grunberg, P., R. Schreiber, Y. Pang, M. B. Brodsky, and H. Sowers. 1986. Layered magnetic structures: Evidence for antiferromagnetic coupling of Fe layers across Cr interlayers. *Phys. Rev. Lett.*, 57(19), 2442–2445.

Hinoue, T., T. Ono, H. Inaba, T. Iwane, H. Akushiji, and A. Chayahara. 2010. Fabrication of discrete track media by Cr ion implantation, *IEEE Trans. Magn.*, 46(6), 1584–1586.

Ion Beam Milling, Inc. 2010.: http://www.ionbeammilling.com/ABOUT_THE_ION_MILLING_PROCESS

Jager, R. C. 2001. *Introduction to Microelectronic Fabrication: Volume 5 of Modular Series on Solid State Devices* (2nd Edition). Reading, MA: Addison-Wesley Publishing Company.

Kaminsky, W. M., G. A. C. Jones, N. K. Patel et al. 2001. Patterning ferromagnetism in Ni$_{80}$Fe$_{20}$ films via Ga+ ion irradiation. *Appl. Phys. Lett.*, 78, 1589–1591.

Kaufman, H. R., J. J. Cuomo, and J. M. E. Harper. 1982. Technology and applications of broad beam ion sources used in sputtering. Part I. Ion source technology. *J. Vac. Sci. Technol.*, 21(3), 725–736.

Komvopoulos K., B. Wei, S. Anders, A. Anders, and I. G. Brown. 1994. Surface modification of magnetic recording heads by plasma immersion ion implantation and deposition. *J. Appl. Phys.*, 76(3), 1656–1664.

Lmre, A., G. Csaba, L. Ji, A. Orlov, G. H. Bernstein, and W. Porod. 2006. Majority logic gate for magnetic quantum-dot cellular automata. *Science*, 311, 205–208.

Martin-Gonzalez, M. S., F. Briones, J. M. Garcia-Martin et al. 2010. Nano-patterning of perpendicular magnetic recording media by low-energy implantation of chemically reactive ions. *J. Magn. Magn. Mater.*, 322, 2762–2768.

McCord, J., I. Monch, J. Fassbender, A. Gerber, and E. Quandt. 2009. Local setting of magnetic anisotropy in amorphous films by co ion implantation. *J. Phys. D: Appl. Phys.*, 42, 055006.

McMorran, B. J., A. C. Cochran, R. K. Dumas et al. 2010. Measuring the effects of low energy ion milling on the magnetization of Co/Pd multilayers using scanning electron microscopy with polarization analysis. *J. App. Phys.*, 107, 09D305.

Neureuther, A. R., C. Y. Liu, and C. H. Ting. 1979. Modeling ion milling. *JVST*, 16(6), 1767–1771.

Nordlund, K. 2010. Personal communications, University of Helsinki.

NPL United Kingdom, 2005. http://resource.npl.co.uk/docs/science_technology/nanotechnology/sputter_yield_values/arsputtertable.pdf

Olivia-Florio, A., R. A. Baragiola, M. M. Jakas, E. V. Alonso, and J. Ferron. 1987. Noble-gas ion sputtering yield of gold and copper: Dependence on the energy and angle of incidence of the projectiles. *Phys. Rev. B*, 35(5), 2198–2204.

Ozkaya, D., R. M. Langford, W. L. Chan, and A. K. Petford-Long. 2002. Effect of Ga implantation on the magnetic properties of permalloy thin films, *J. Appl. Phys.*, 91(12), 9937–9942.

Parkin S. S. P., M. Hayashi, and L. Thomas. 2008. Magnetic domain-wall racetrack memory. *Science*, 320(5873), 190–194.

Peng, X., Z. Wang, Y. Lu, B. Lafferty, T. McLaughlin, and M. Ostrowski. 2010. On the geometry control of magnetic devices: Impact of photo-resist profile, shadowing effect, and material properties. *Vacuum*, 84, 1075–1079.

Ravelosona, D., C. Chappert, H. Bernas, D. Halley, Y. Samson, and A. Marty. 2002. Chemical ordering at low temperatures in FePd films. *J. Appl. Phys.*, 91, 8082–8084.

Ravelosona, D., T. Devolder, C. Chappert et al. 2001. Irradiation-induced magnetic patterning in magnetic multilayers. *Mater. Sci. Eng. C*, 15(1–2), 53–58.

Rettner, C. T., S. Anders, J. E. E. Baglin, T. Thomson, and B. D. Terris. 2002a. Characterization of the magnetic modifications of Co/Pt multilayer films by He+ Ar+ and Ga+ ion irradiation. *Appl. Phys. Lett.*, 80(2), 279–281.

Rettner, C. T., S. Anders, T. Thomson et al. 2002b. Magnetic characterization and recording properties of patterned Co$_{70}$Cr$_{18}$Pt$_{12}$ perpendicular media. *IEEE Trans. Magn.*, 38(4), 1725–1730.

Rettner, C. T., M. E. Best, and B. D. Terris. 2001. Patterning of granular magnetic media with a focused ion beam to produce single-domain islands at >140 Gb/in². *IEEE Trans. Magn.*, 37(4), 1649–1651.

Shaw, J. M., S. E. Russek, T. Thomson et al. 2008. Reversal mechanisms in perpendicularly magnetized nanostructures. *Phys. Rev. B*, 78, 024414.

Smentkoski, V. 2000. Trends in sputtering. *Prog. Surface Sci.*, 64, 1–58.

Smith, R. (ed.). 1997. *Atomic & Ion Collisions in Solids and at Surfaces: Theory, Simulations and Applications*. Cambridge, UK: Cambridge University Press.

Tehrani, S., J. M. Slaughter, E. Chen, M. Duriam, J. Shi, and M. DeHerrera. 1999. Progress and outlook for MRAM technology. *IEEE Trans. Magn.*, 35(5), 2814–2819.

Terris B. D. 2009 Fabrication challenges for patterned recording media. *J. Magn. Mater.*, 321, 512–517.

Terris, B. D., L. Folks, and D. Weller. 1999. Ion-beam patterning of magnetic films using stencil masks. *Appl. Phys. Lett.*, 35(3), 403–405.

Traverse, A., M. G. LeBoite, and G. Martin. 1989. Quantitative description of mixing with light ions. *Europhys. Lett.*, 8(7), 633–637.

Walsh, M. E., Y. Hao, C. A. Ross, and H. I. Smith. 2000. Optimization of a lithographic and ion beam etching process for nanostructuring magnetoresistive thin film stacks. *J. Vac. Sci. Technol. B*, 18(6), 3593–3597.

Weller, D., J. E. E. Baglin, A. J. Kellock et al. 2000. Ion induced magnetization reorientation in Co/Pt multilayers for patterned media. *J. Appl. Phys.*, 87(9), 5786–5788.

Wood, R. 2009. Future hard disk drive systems. *J. Magn. Magn. Mater.*, 321, 555–561.

Yang, C. H., C-H Lai, and S. Mao. 2003. Reversing exchange fields in CoFe/PtMn and CoFe/Ir/Mn bilayers by carbon field irradiation. *J. Appl. Phys.*, 93, 6596–6598.

Zhang, D., S. Rauf, and T. Sparks. 2002. Modeling of photoresist erosion in plasma etching processes. *IEEE Trans. Plasma Sci.*, 30(1), 114–115.

Ziegler, J. F., J. P. Biersack, and U. Littmark. 2009. *The Stopping and Range of Ions in Solids*. New York: Pergamon Press.

Ziegler, J. F., J. P. Biersack, and L. Haggmark. 1989. TRIM and SRIM, code at http://www.srim.org

Chapter 19

20. Plasmonic Metamaterials

Takuo Tanaka

Metamaterials Laboratory, RIKEN Advanced Science Institute, Hirosawa, Wako, Saitama, Japan

20.1 Introduction

Plasmonic metamaterial is an artificially designed material that consists of nanoscale metal structures (Figure 20.1). The most interesting feature of the plasmonic metamaterials is that their electromagnetic properties are not only because of their composition but also because of their subwavelength-engineered metallic structures. By engineering such materials, we can create materials exhibiting desired electromagnetic properties not attainable with naturally occurring materials. The creation of the magnetically active material is one of the most important and interesting applications of the metamaterial because all materials in nature lose magnetic response and their μ in the THz frequency region is fixed at unity (Landau et al. 1984). The magnetic metamaterial with μ ≠ 1 produces a great number of novel materials in the optical

Nanofabrication Handbook. Edited by Stefano Cabrini and Satoshi Kawata © 2012 CRC Press / Taylor & Francis Group, LLC. ISBN: 978-1-4200-9052-9

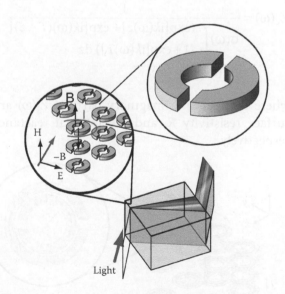

FIGURE 20.1 Schematic of plasmonic metamaterials. Plasmonic metamaterials are designed artificial materials with a controlled optical response. Both the size and the periodicity of the integrated resonant elements are significantly smaller than the wavelength. Therefore, the existence of the integrated elements is not sensed by the light and metamaterials act as homogeneous materials for the light.

Chapter 20

frequency region, which enables us to manipulate light freely (Cai et al. 2007). For example, "cloaking effect," which renders the object invisible, was proposed (Dolling et al. 2007; Alù and Engheta 2005) and experimentally investigated at a microwave region (Schurig et al. 2006). The concept of the metamaterial is introducing a new paradigm in the research field from the microwave region to the optical region. In this chapter, the theoretical background

of the plasmonic metamaterials with a discussion about the appropriate materials and structures to gain the magnetism in the visible light region are described, and then some nanofabrication techniques utilized for metamaterials are reviewed. The chapter closes with a discussion on the application of the metamaterials for optical devices using its unprecedented optical properties.

20.2   Structure of Plasmonic Metamaterials from THz to Visible Light Region

Theoretical investigation about the magnetic response of plasmonic metamaterials in the visible frequency region was carried out by Ishikawa et al. (2005), Ishikawa and Tanaka (2006a), Ishikawa et al. (2007). Figure 20.2 shows the split-ring-resonator (SRR) model used in the calculations. In Figure 20.2, r is the radius of the ring, w is the width of the ring, d is the gap distance between two rings of SRR, J is the induction current, a is the unit-cell dimension in the xy plane, and l is the distance between adjacent planes of the SRRs along the z-axis.

In order to describe the dispersion properties of metals, the internal impedance for a unit length and a unit width of the plane conductor ($Z_s(\omega)$) was introduced as

$$Z_s(\omega) = \cfrac{1}{\sigma(\omega)\displaystyle\int_0^T \cfrac{(\exp[ik(\omega)z] + \exp[ik(\omega)(T-z)])}{/1 + \exp[ik(\omega)T])} \, dz}$$
$$= R_s(\omega) + iX_s(\omega), \qquad (20.1)$$

where the real and imaginary parts of $Z_s(\omega)$ are the surface resistivity R_s and the internal reactance X_s, respectively.

Based on the dispersive properties of metals described by Equation 20.1, the frequency dependence of the magnetic responses of the metallic SRRs in the optical frequency region was calculated, and the effective permeability (μ_{eff}) of the SRRs was derived as

$$\mu_{eff} = \mu_{Re} + i\mu_{Im} = 1 - \frac{F\omega^2}{\omega^2 - ((1/CL) + i(Z(\omega)\omega/L))}, \quad (20.2)$$

where C and L are the geometrical capacitance and inductance, and F and $Z(\omega)$ are the filling factor and the ring metal impedance defined by

$$F = \frac{\pi r^2}{a^2}, \qquad (20.3)$$

$$C = \frac{2\pi r}{3}\varepsilon_0\varepsilon_r \frac{K[(1-t^2)^{1/2}]}{K(t)}, \quad t = \frac{g}{2w+g}, \qquad (20.4)$$

$$L = \frac{\mu_0 \pi r^2}{l}, \qquad (20.5)$$

$$Z(\omega) = \frac{2\pi r Z_s(\omega)}{w}, \qquad (20.6)$$

respectively. When deriving the geometrical capacitance, Gupta's formula was used to estimate the capacitance from the distance between two rings per unit length (Gupta et al. 1996).

By using Equations 20.1 and 20.2, and taking into account the empirical values of the damping constants (γ) and the plasma frequency (ω_p) of silver, gold, and copper ($\omega_p = 14.0 \times 10^{15}$ s^{-1} and $\gamma = 32.3 \times 10^{12}$ s^{-1} for silver, $\omega_p = 13.8 \times 10^{15}$ s^{-1} and $\gamma = 107.5 \times 10^{12}$ s^{-1} for

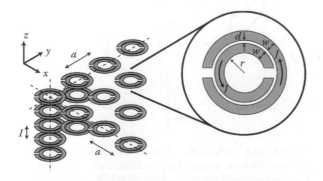

FIGURE 20.2 Sketch of the structure of plasmonic metamaterials and an SRR that is for artificially controlled magnetic permeability.

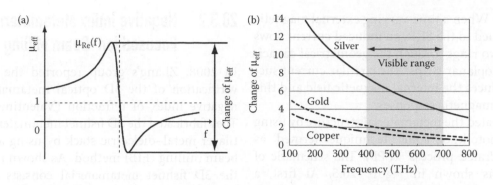

FIGURE 20.3 Frequency dependence of the change of μ_{eff} of the SRRs made of silver, gold, and copper.

gold, and $\omega_p = 13.4 \times 10^{15}$ s^{-1} and $\gamma = 144.9 \times 10^{12}$ s^{-1} for copper), the frequency dispersions of μ_{eff} from 100 to 800 THz covering the entire visible light frequency region were calculated (Johnson and Christy 1972). As shown in Figure 20.3a, at the resonant frequency of the SRR, μ_{eff} changes positively and negatively and it takes max μ_{eff} and min μ_{eff}. In Figure 20.3b, the calculation results of the change of μ_{eff}, which is defined by the difference between max μ_{eff} and min μ_{eff}, for each metal SRRs array were plotted. From these results, it was clarified that a three-dimensional (3D) array of SRRs made of silver can give a strong magnetic response in the visible light frequency region. As also shown in Figure 20.3b, silver SRRs exhibit μ_{eff} changes exceeding 2.0 in the

entire visible range, which means μ_{eff} can become a negative value, while the responses of gold and copper SRRs do not exceed 2.0 in the visible light region. The following is a brief summary of the design strategy of the plasmonic metamaterials that can reveal the magnetic response. In the lower-frequency region <100 THz, the original shape of SRR proposed by Pendry is appropriate, because it can provide large capacitance and wide metal rings. On the other hand, at a higher-frequency region, the single ring with a number of cuts is an appropriate structure. In this configuration, the capacitance of the resonant structure is easily decreased by increasing the number of cuts. This is advantageous to prevent the effect of $X_s(\omega)$.

20.3 Nanofabrication Techniques for Plasmonic Metamaterials

In this section, we discuss about the nanofabrication techniques used for creating plasmonic metamaterials. Due to the limitations of space, only the significant research results related to the nanofabrication techniques are reviewed. About the details of each research result of the metamaterial properties, we refer the reader to each original paper, and the details of each fabrication technique are already seen in Part I of this Handbook.

20.3.1 Metamaterials in THz Region Fabricated by Photolithography

Yen et al. (2004) reported the experimental result and theoretical verification of the magnetic metamaterials that act at THz frequencies. They fabricated an array of nonmagnetic and conducting SRRs, as shown in Figure 20.4. An SRR consists of two concentric square rings with a gap situated oppositely. The gaps are introduced to govern the direction of an AC current around

FIGURE 20.4 A secondary ion image of an array of SRR fabricated by PPPs. The split gap in SRR, the corresponding gap between the inner and outer square ring, the width of the metal lines, and the length of the outer ring were 2, 2, 4, and 26 μm, respectively. The periodicity of SRRs was 36 μm. (From Yen, T. J. et al. 2004. Terahertz magnetic response from artificial materials. *Science* 303:1494–1496. Copyright 2004, With permission of AAAS.)

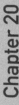

the two rings. When a time-varying external magnetic field H is applied to the SRR, an induced current flows around the two rings through the geometrical capacitance of the coplanar strips. The circular current successively produces the internal magnetic field and this results in the magnetic responses.

They fabricated the metamaterial structure by using a special photolithographic technique termed as "photo-proliferated process" (PPP). The schematic of the process is shown in Figure 20.5. At first, a 5-μm-thick negative photoresist layer was spun onto a quartz substrate, and then the designed SRR pattern was transferred using a contact photolithographic method (Figure 20.5a). After the photoresist patterning, 100-nm-thick chromium and 1-μm-thick copper layers were deposited by using an electron-beam evaporator (Figure 20.5b). Lift-off process using an acetone rinsed in an ultrasonic bath was employed to remove the photoresist to preserve the copper SRR pattern (Figure 20.5c). The photoresist layer is spin-coated again on the SRR pattern for the second lithographic patterning process. In the second lithography process, ultraviolet (UV) light was exposed from the backside of the quartz substrate using the copper SRR pattern fabricated in the first lithography as a photomask (Figure 20.5d). After the second UV exposure, evaporation of copper layer and lift-off process are repeated to increase the metal thickness of the SRR structures (Figure 20.5e). In the experiment, they successfully demonstrated the magnetic response from the fabricated structure at 1 THz.

20.3.2 Negative Index Metamaterials by Focused Ion–Beam Milling

In 2008, Zhang's group reported the experimental verification of the 3D optical metamaterials with a negative index of refraction (Valentine et al. 2008). They fabricated the 3D fishnet metamaterials on a multilayer metal–dielectric stack by using a focused ion-beam milling (FIB) method. As shown in Figure 20.6, the 3D fishnet metamaterial consists of multilayer metal–dielectric film stack. The multilayer stack was deposited by electron-beam evaporation of alternating layers of 30 nm silver (Ag) and 50 nm magnesium fluoride (MgF_2) thin films. A total of 21 layers with a total thickness of 830 nm were stacked on a transparent substrate. The fishnet pattern of 22×22 in-plane cells of 860 nm periodicity ($a = 565$ nm and $b = 265$ nm) was fabricated on the 21 layers by using the FIB milling techniques. This sample was used for the characterization of the transmittance, and they concluded that the experimentally measured figure of merit of their sample was 3.5 at $\lambda = 1775$ nm. They made another type of sample consisting of a prism fabricated on the multilayer stack with the number of functional layers ranging from 1 on one side to 10 on the other side (Figure 20.7). The prism with 10×10

FIGURE 20.6 An SEM image of the 21-layer fishnet structure with the side etched for appearance of the cross section. The structure consists of alternating layers of 30 nm silver and 50 nm magnesium fluoride. The inset shows a cross section of the pattern taken at a 45° angle. The sidewall angle is 4.3°. (Reprinted by permission from Macmillan Publishers Ltd., *Nature*, Valentine, J. et al. Three-dimensional optical metamaterial with a negative refractive index. 455:376–379. Copyright 2008.)

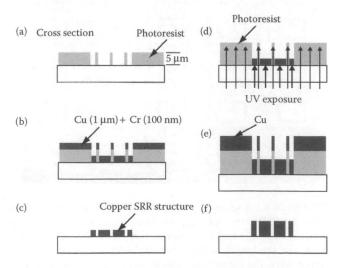

FIGURE 20.5 The process flow of PPP. (From Yen, T. J. et al. 2004. Terahertz magnetic response from artificial materials. *Science* 303:1494–1496. Copyright 2004, With permission of AAAS.)

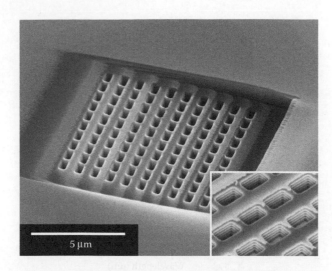

FIGURE 20.7 An SEM image of the fabricated 3D fishnet negative index materials (NIM) prism. A prism was fabricated on the multilayer stack with 10 functional layers using FIB. Subsequently, a 10 × 10 fishnet pattern was milled in the prism using FIB. The unit cell size is identical to that shown in Figure 20.6. The inset shows a magnified view with the film layers visible in each hole. (Reprinted by permission from Macmillan Publishers Ltd., *Nature*, Valentine, J. et al. Three-dimensional optical metamaterial with a negative refractive index. 455:376–379. Copyright 2008.)

fishnet pattern was formed by FIB etching the film. This sample was used to measure the effective refractive index of the fishnet structure. By measuring the absolute angle of refraction, they concluded that the refractive index varies from $n = 0.63 \pm 0.05$ at $\lambda = 1200$ nm to $n = -1.23 \pm 0.34$ at $\lambda = 1775$ nm.

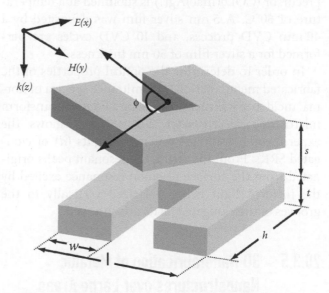

FIGURE 20.8 Schematic diagram of the stereo-SRR dimer metamaterials. The geometrical parameters l, h, w, t, and s are 230, 230 90, 50, and 100 nm, respectively. The period of SRR dimer in a lateral direction is 700 nm. (Reprinted by permission from Macmillan Publishers Ltd., *Nat. Photon.*, Liu, N. et al. Stereometamaterials. 3:157–162. Copyright 2009.)

20.3.3 Stereo Metamaterials Formed by Electron–Beam Lithography and Ion Beam Etching Technique

In 2009, Giessen's group in Universität Stuttgart reported the metamaterial structure that consists of an array of a stack of two identical SRRs (Figure 20.8) (Liu et al. 2009). They fabricated three types of stereo-SRR dimer metamaterials with specific twist angles $\varphi = 0°$, $90°$, and $180°$ (see Figure 20.9).

Fabrication process of stereo metamaterials is as follows. Three gold alignment marks (size 4×100 µm) with a gold thickness of 250 nm were fabricated using a lift-off process on a quartz substrate. The substrate

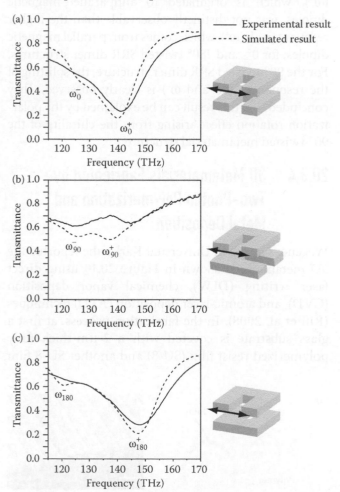

FIGURE 20.9 Transmittance spectra for twisted gold SRR dimer metamaterials. (a), (b), and (c) are for 0°-, 90°-, and 180°- twisted gold SRR dimer metamaterials. The structures were fabricated on a glass substrate. The gold SRR dimer was fabricated by electron-beam lithography and lift-off process. The SRRs were embedded in a photopolymer (PC403), which served as the dielectric spacer between SRRs. (Figure similar to that in Reprinted by permission from Macmillan Publishers Ltd., *Nat. Photon.*, Liu, N. et al. Stereometamaterials. 3:157–162. Copyright 2009.)

Chapter 20

was covered with another 50 nm gold film using electron-beam evaporation, and then gold SRR structures were fabricated using electron-beam lithography and Ar$^+$ beam etching processes. On the SRR patterns, a spacer layer of photopolymer (PC403, 120 nm in thickness) was spin-coated. Subsequently, a second SRR structure was fabricated on the sample using gold film evaporation and electron-beam lithography. The total area of the fabricated structures was 200 × 200 μm.

Transmittance spectra were measured with a Fourier-transform infrared spectrometer combined with an infrared microscope. From the transmittance spectra of three types of stereo-SRR dimer metamaterials, they demonstrated that the lower-frequency resonant peak (ω^-), which is originated to antiparallel magnetic dipoles, was not distinctly observable than the higher resonant peak (ω^+), which comes from parallel magnetic dipoles, for 0°- and 180°-twisted SRR dimer structures. For the 90°-twisted SRR dimer structure, the splitting of the resonances (ω^- and ω^+) is clearly observed. They concluded that this result can be explained by the polarization rotation effect arising from the chirality of the 90°-twisted metamaterial structure.

20.3.4 3D Metamaterials Fabricated by Two-Photon Polymerization and Metal Deposition

Wegener's group in Universität Karlsruhe reported the 3D metamaterial shown in Figure 20.10 using direct laser writing (DLW), chemical vapor deposition (CVD), and atomic-layer deposition (ALD) techniques (Rill et al. 2008). In the fabrication process, at first a glass substrate is covered with a 2-μm-thick fully polymerized resist film (SU-8) and another SU-8 film

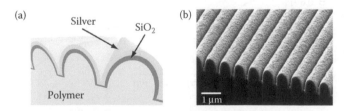

FIGURE 20.10 Photonic metamaterials fabricated by DLW and silver chemical vapor deposition. (a) Schematic diagram of a planar lattice of elongated and all connected SRRs. (b) Electron-beam micrograph of fabricated metamaterial corresponding to the design shown in (a). The FIB cut reveals the SiO$_2$ layer between the SU-8 polymer template and the silver coating. (Reprinted by permission from Macmillan Publishers Ltd., *Nat. Mater.*, Rill, M. et al. Photonic metamaterials by direct laser writing and silver chemical vapour deposition. 7:543–546. Copyright 2008.)

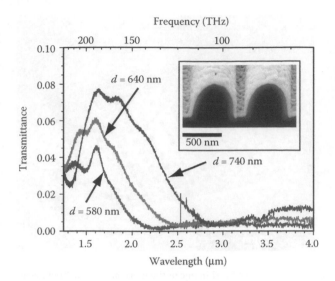

FIGURE 20.11 Normal incidence optical transmittance spectra. The incident linear polarization is perpendicular to the grooves. The insets show an electron micrograph of a sample structure of $d = 740$ nm. (Reprinted by permission from Macmillan Publishers Ltd., *Nat. Mater.*, Rill, M. et al. Photonic metamaterials by direct laser writing and silver chemical vapour deposition. 7:543–546. Copyright 2008.)

is spun-on and exposed using DLW. After the postbaking and developing processes, the SU-8 template is coated with a SiO$_2$ thin layer of a few 10s of nanometers in thickness using ALD. The SiO$_2$ surface is exposed to O$_2$ plasma for 15 min to activate the surface for the subsequent silver CVD process. The metal-organic precursor (COD)(hfac)Ag(I) is sublimed at a temperature of 60°C. A 5 nm silver film was deposited by a 40 min CVD process, and 10 CVD cycles are performed for a silver film of 50 nm thickness.

In order to determine the optical properties of the fabricated metamaterial, transmittance spectra of normal incidence was measured using a Fourier-transform microscope spectrometer. Figure 20.11 shows the experimental results of different heights (d) of elongated SRRs. From the result, the resonant peaks originated from the surface plasmon resonance excited by the incident light being polarized vertically to the grooves of the elongated SRRs are observed.

20.3.5 3D Nanofabrication of Metallic Nanostructures over Large Areas by Two-Photon Polymerization

Formanek et al. also reported the 3D fabrication of metallic nanostructures over large areas using a combination of two-photo polymerization and electroless plating techniques (Formanek et al. 2006a,b). One of the disadvantages of the single-beam laser direct drawing is that it is time consuming and thus this technique is unlikely to be adopted for mass manufacturing. The fabrication of numerous metallic 3D structures for plasmonic metamaterials would require a parallel approach. To solve this problem, multiple laser beam spots created by a microlens array were introduced to the DLW (Kato et al. 2005). Moreover, in order to deposit a thin metal film over polymer structures, electroless plating, which is a chemical process and can be effectively realized at ambient conditions, was employed. Electroless plating is suitable for metal deposition onto insulating samples since it allows uniform coating over large areas and even structures with complex shapes and occluded parts can be metal coated (Mallory and Hajdu 1990; Antipov et al. 2002; Takeyasu et al. 2005). However, polymers are naturally hydrophobic materials and they do not adhere well to metal films due to differences in surface energies (Gerenser 1990). For this reason, the chemical modification to the photopolymerizable resin was applied before the fabrication of the 3D polymer structures. Then, a pretreatment using $SnCl_2$ is applied before metal deposition to improve silver nucleation and adhesion on the polymer surface (Gray et al. 2005). On the other hand, in order to avoid unwanted metal deposition onto the glass substrate, a hydrophobic coating on the glass slides was applied using dimethyldichlorosilane (DMDCS, $(CH_3)_2SiCl_2$) (Guan et al. 2005).

Figure 20.12a shows an SEM image of a large 78×58 μm^2 fabricated area of 3D metalized polymer structures. Figure 20.12b shows an individual structure before coating by electroless plating, composed of a cube (2 μm in size), holding a self-standing spring (height 2.2 μm, inner diameter 700 nm).

To overcoat fine structures, the plating had to be optimized to reduce the thickness of the metal film as much as possible. To do so, they focused on the formation of very small silver particles, with diameters of 20 nm or less, by using a different reductant, changing the concentrations of the reagents and realizing two consecutive steps. The optimized plating solution was composed of a 0.3 mol/L $AgNO_3$ solution, mixed with a saturated solution of 2;5-dihydroxybenzoic acid ($C_7H_6O_4$) 300× diluted in water as reductant agent, in a 1:1 volume ratio. The reaction was performed at 37°C and stopped after 2 min. A second short plating (~10 s) was realized with the same $AgNO_3$ solution, mixed with ammonia (0.2 mol/L) and 10× diluted benzoic acid. Figure 20.12c shows an

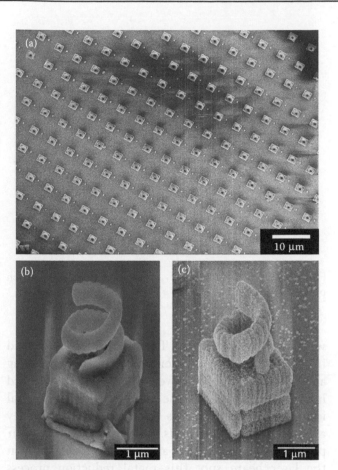

FIGURE 20.12 SEM image of 3D periodic metallic structure fabricated by two-photon-polymerization with a microlens array. (a) 78×58 μm^2 SEM image of a 3D periodic structure fabricated on a hydrophobic-coated glass surface. (b) Oblique magnified view of an individual uncoated polymer structure composed of a cube (2 μm in size) holding up a spring (2.2 μm in height, 1 μm in inner diameter). (c) SEM image of an individual silver-coated structure after electroless plating.

individual structure after metallization of the sample. Only a few silver particles adhered on the substrate and the polymer structure was uniformly overcoated by a 50 nm silver film.

20.3.6 Two-Photon-Induced Photoreduction Method

Again to create a plasmonic metamaterial structure, the fabrication technique requires the ability to make arbitrary 3D metallic structures. To satisfy this requirement, in 2006, Tanaka proposed a new fabrication technique that uses two-photon-induced reduction of metallic complex ions (Tanaka et al. 2006a; Ishikawa et al. 2006b).

Figure 20.13 shows a schematic of this two-photon reduction technique. A mode-locked Ti:Sapphire laser

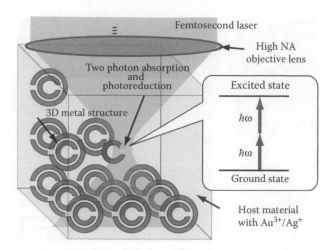

FIGURE 20.13 Schematic of two-photon-induced reduction process.

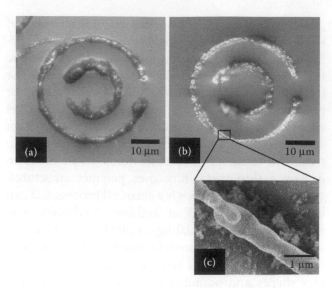

FIGURE 20.14 Double-ring structure fabricated on the glass substrate by two-photon-induced metal-ion reduction. (a) Optical microscope image of a silver double-ring pattern made by reduction of an $AgNO_3$ aqueous solution. (b) A gold pattern made by reduction of an $HAuCl_4$ solution. (c) Scanning electron micrograph of a magnified portion of the gold ring indicated by the rectangle in (b). (Reprinted with permission from Tanaka, T., Ishikawa, A., and Kawata, S. 2006a. Two-photon-induced reduction of metal ions for fabricating three-dimensional electrically conductive metallic microstructure. *Appl. Phys. Lett.* 88:81107. Copyright 2006, American Institute of Physics.)

was used as a light source. The laser beam was focused in the material that contains metal ions using an oil-immersion objective lens. When the focused laser beam illuminates the metal-ion solution, metal ions absorb two photons simultaneously and they are reduced to the metals. Owing to the nonlinear properties of two-photon absorption process, only at the laser beam spot this metal reduction process occurs and tiny metal particles are created in the 3D space. The focused laser beam was scanned in three dimensions (x–y–z scanning) to draw the metal structures.

Figure 20.14 shows two-dimensional (2D) metallic structures fabricated on the glass substrates. Figure 20.14a shows a silver ring structure, and Figure 20.14b shows a gold ring structure with the same pattern. These photographs were taken using a reflection optical microscope. Figure 20.14c is a scanning electron micrograph of a magnified portion of the gold ring structure, indicated by the rectangle in Figure 20.14b. The width of the gold line was about 0.7 μm; this value was almost the same as the diameter of the diffraction-limited focused laser beam spot.

The advantage of this technique is that it can create the highly electric conductive metal structures regardless of the micro/nanometer scale. To verify the electrical continuity of the metal structure, the resistivity of the fabricated metal wires was measured and the average of resistivity was determined as 5.30×10^{-8} Ωm. This value is only 3.3 times larger than that of bulk silver (1.62×10^{-8} Ωm), and this indicates the high conductivity of the fabricated silver wires.

Figure 20.15 is the scanning electron micrograph of 3D self-standing silver structures. Figure 20.15a shows silver gate microstructure on a glass substrate, whose width, height, and linewidth of the 3D silver gate are 12, 16, and 1.5 μm, respectively. Figure 20.15b shows a silver-tilted rod, whose length and the angle relative to the substrate were 34.64 μm and 60°, respectively. Figure 20.15c is a top-heavy silver cup. The height and the top and bottom diameter of the silver cup were 26 20, and 5 μm, respectively.

In this method, the major problem that inhibited the nanoscale size was the unwanted growth of the metal nanoparticles during laser irradiation, and the main issue to gain the nanometer scale depends on a way to avoid this unwanted metal particle growth and produce smaller nanoparticles to serve as building blocks. Cao et al. (2009a,b) presented a means to gain small feature sizes in the fabrication of metallic structures with the aid of a surfactant as a metal growth inhibitor. By using *n*-decanoylsarcosine sodium (NDSS) as the surfactant, they demonstrated the silver structure whose minimum size was finer than the diffraction limit of light. Figure 20.16 shows an SEM image of silver line whose linewidth was 120 nm. Figure 20.17 shows SEM images

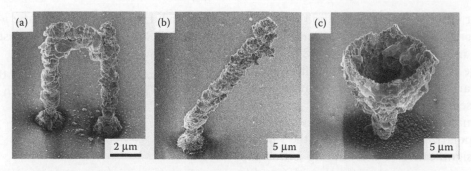

FIGURE 20.15 SEM images of 3D silver structures. (a) A microsized 3D silver gate structure standing on a glass substrate without any support. The width, height, and linewidth were 12, 16, and 2 μm, respectively. (b) A silver-tilted rod and (c) top-heavy silver cup on a substrate. The length of the rod and the angle relative to the substrate were 34.64 μm and 60°, respectively. The height and the top and bottom diameters of the cup were 26, 20, and 5 μm, respectively. (Reprinted with permission from Tanaka, T., Ishikawa, A., and Kawata, S. 2006a. Two-photon-induced reduction of metal ions for fabricating three-dimensional electrically conductive metallic microstructure. *Appl. Phys. Lett.* 88:81107. Copyright 2006, American Institute of Physics.)

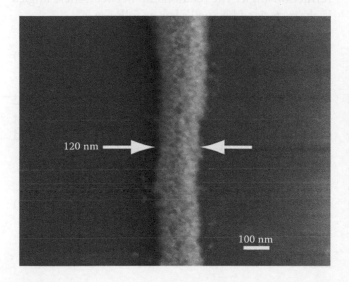

FIGURE 20.16 An SEM image of a silver line fabricated using a surfactant-assisted two-photon-induced reduction. The linewidth of the structure was 120 nm, which is finer than the diffraction limit of light ~700 nm at λ = 800 nm and NA = 1.42. (From Y. Cao et al. 2009b. Copyright 2009, Wiley-VCH Verlag GmbH & Co. KGaA, Weinheim.)

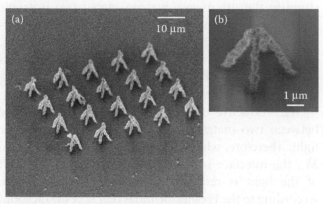

FIGURE 20.17 An SEM image of the free-standing silver pyramids taken at an observation angle of 45°. These silver pyramids were strong enough to resist the surface tension in the washing process, which demonstrates that the silver particles were closely combined. (b) The detail of the silver pyramid. From the image, the height was 5 μm and the angle for each edge relative to the substrate was 60°. (From Y. Cao et al. 2009b. Copyright 2009, Wiley-VCH Verlag GmbH & Co. KGaA, Weinheim.)

of the truly free-standing silver pyramids. The close-up view of the silver pyramid shown in Figure 20.17b reveals that the height of the pyramid was 5 μm and the angle for each edge relative to the substrate was 60°.

These silver pyramids structures were strong enough to resist the surface tension in the washing process, which demonstrates that the silver particles were closely combined.

20.4 Application of Metamaterials to the Optical Functional Devices

To date, a number of proposals about the application of plasmonic metamaterials have been proposed. One of the most popular challenges of the metamaterial application is creating the negative index materials at optical frequency as a way to realize the subwavelength imaging termed with "perfect lens" (Pendry 2000). For details about negative index materials and perfect lens, we refer the reader to specialized reviews (Smith et al. 2004). In this section, we discuss about the application of the plasmonic metamaterials whose permeability was changed to the positive (over one) direction.

The Brewster effect is used as a method to prevent unwanted reflection occurring at the material boundary of different indices (Born and Wolf 1980). A common knowledge in optics, the Brewster effect occurs only for *p*-polarized light. However, Tanaka et al. (2006b) proposed that the magnetically responsive metamaterial enables us to realize the Brewster effect not only for *p*-polarized light but also for the *s*-polarized one. The significance of this finding is that the metamaterial can interconnect materials with two different indices while eliminating the reflection arising from the index mismatch, and it can solve the problem of polarization dependence seen in conventional optical components based on the Brewster effect.

Here, we present the fundamental idea to produce the Brewster effect also for *s*-polarized light by suitably controlling the magnetic permeability of a material. To simplify the discussion, we considered two isotropic and homogeneous materials, material 1 (M_1) and material 2 (M_2) with different optical constants ε_1 and μ_1, and ε_2 and μ_2, as shown in Figure 20.18. The constants ε and μ represent the relative permittivity and the relative permeability, respectively.

The refractive index mismatch at the interface between two materials acts as a potential barrier for light. Therefore, when incident light passes from M_1 to M_2, the interface inhibits light transmission and part of the light is reflected, as shown in Figure 20.18. According to the Fresnel formulas that take into account

both ε and μ, the reflectance for the *p*- and *s*-polarization, R^p and R^s, respectively, can be described as

$$R^p = \left(\frac{-\mu_2 \sin\theta' \cos\theta' + \mu_1 \sin\theta \cos\theta}{\mu_2 \sin\theta' \cos\theta' + \mu_1 \sin\theta \cos\theta} \right)^2 \tag{20.7}$$

and

$$R^s = \left(\frac{\mu_2 \tan\theta' - \mu_1 \tan\theta}{\mu_2 \tan\theta' + \mu_1 \tan\theta} \right)^2, \tag{20.8}$$

respectively, where θ and θ' are the incident and the refraction angle. The relationship between these angles is characterized by Snell's law as follows:

$$\sqrt{\varepsilon_1}\sqrt{\mu_1}\sin\theta = \sqrt{\varepsilon_2}\sqrt{\mu_2}\sin\theta'. \tag{20.9}$$

Assuming the numerators of Equations 20.7 and 20.8 to be zero under the condition that the product $\varepsilon_1\mu_1$ is not equal to $\varepsilon_2\mu_2$, Brewster's angles for *p*- and *s*-polarized light θ_B^p and θ_B^s are found to be

$$\theta_B^p = \tan^{-1}\left[\sqrt{\frac{\varepsilon_2(\varepsilon_1\mu_2 - \varepsilon_2\mu_1)}{\varepsilon_1(\varepsilon_1\mu_1 - \varepsilon_2\mu_2)}} \right] \tag{20.10}$$

and

$$\theta_B^s = \tan^{-1}\left[\sqrt{\frac{\mu_2(\varepsilon_2\mu_1 - \varepsilon_1\mu_2)}{\mu_1(\varepsilon_1\mu_1 - \varepsilon_2\mu_2)}} \right]. \tag{20.11}$$

Figure 20.19a shows the incident angle dependencies of the reflectance R^p and R^s calculated under the condition that M_1 is vacuum ($\varepsilon_1 = \mu_1 = 1.0$) and the M_2 is glass ($\varepsilon_2 = 2.25$ and $\mu_2 = 1.0$). The zero reflectance exists for *p*-polarized light at $\theta = 56.3°$ and this angle is the standard Brewster's angle. On the other hand, there is no zero-reflectance angle for *s*-polarized light, indicating that there is no Brewster's angle for *s*-polarization. Figure 20.19b shows another result calculated under the condition that the M_1 is the same as in Figure 20.19a ($\varepsilon_1 = \mu_1 = 1.0$), but M_2 is a magnetic material with $\varepsilon_2 = 1.0$ and $\mu_2 = 2.25$. Under this condition, we can see that the reflectance falls to zero for *s*-polarization at the same angle $\theta = 56.3°$ as in Figure 20.19a. This is also known as the Brewster effect

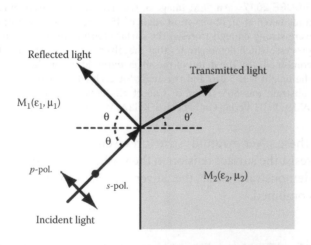

FIGURE 20.18 Analytical model of the Brewster effect for *p*- and *s*-polarized light. An incident plane wave reaches a boundary between two homogeneous materials (M_1 and M_2) at the incident angle of θ. The relative permittivity and permeability of M_1 and M_2 are ε_1 and μ_1, and ε_2 and μ_2, respectively. A reflected wave propagates back into the first material at the reflection angle of θ, which is the same as the incident angle. A transmitted wave is refracted and goes into the second material at the refraction angle of θ'.

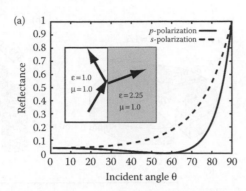

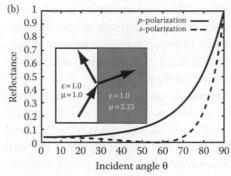

FIGURE 20.19 Incident angle dependence of the reflectance calculated under the condition that (a) M_1 is vacuum ($\varepsilon_1 = \mu_1 = 1.0$) and M_2 is glass ($\varepsilon_2 = 2.25$ and $\mu_2 = 1.0$) and (b) M_1 is vacuum and M_2 is a magnetic material ($\varepsilon_2 = 1.0$ and $\mu_2 = 2.25$). The Brewster effect occurs also for s-polarized light at $\theta = 56.3°$ when the light enters a magnetically active material.

shown in Figure 20.19a, but for s-polarization, not for p-polarization. These results can be explained by the symmetry between ε and μ in Equations 20.10 and 20.11, which resulted from the symmetric boundary conditions for p- and s-polarization, as discussed above. It was believed previously that the Brewster effect occurs only for p-polarized light because μ of most materials in nature is approximately unity in the optical frequency region. However, these results prove that controlling magnetic activity of the material enables us to realize the Brewster effect also for s-polarized light. Actually, this Brewster effect for s-polarization has already been experimentally investigated in the microwave region of 2.6 GHz by employing a 2D SRR array (Tamayama et al. 2006).

As discussed above, the magnetic material realizes the Brewster effect for s-polarized light, which is an unusual electromagnetic phenomenon in nature. This new finding also has the significant consequence that if we could produce the Brewster effect for both p- and s-polarized light simultaneously, the light could propagate through the material interface without any reflection at all. The realization of the Brewster for p- and s-polarized light simultaneously is the fundamental idea in realizing unattenuated transmission of light across the material boundary. However, Equations 20.10 and 20.11 tell us that the Brewster conditions for each polarization cannot be realized simultaneously.

To overcome this conflict, the idea of a uniaxial magnetic metamaterial whose ε and μ values depend on the direction of the material, which is analogous to a uniaxial crystal, is introduced (Figure 20.20). M_2 is a uniaxial magnetically active metamaterial that acts as an εbuffer layer for realizing the perfect light trans-

mission from M_1 to M_3. As shown in Figure 20.20, since M_2 consists of layer-stacked arrays of the SRRs lying only in the x–y plane, it responds to a magnetic field oscillating along the z-direction (H_z) and thus changes only μ_2^s. The basic concept of an anisotropic left-handed materials (LHM) was first introduced by Grzegorczyk et al. (2005) and they reported inversion of the critical angle and the Brewster's angle in such a material.

We proposed the practical application of the Brewster window for both p- and s-polarization in which light can propagate through the interface between two materials of different refractive indices

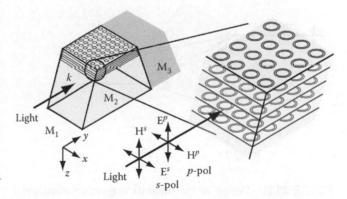

FIGURE 20.20 Uniaxial magnetic metamaterial exhibiting the Brewster effect for both p- and s-polarized light and its calculation model. (a) Incident light whose wave vector is k goes from M_1 through M_2 and is then transmitted to M_3. The magnetic metamaterial consists of layer-stacked 2D arrays of the SRRs lying in x–y plane. Because the SRRs react only to the magnetic field oscillating along the z-direction, only μ_2^s is controlled. (From Tanaka, T., Ishikawa, A., and Kawata, S. 2006b. Unattenuated light transmission through the interface between two materials with different indices of refraction using magnetic metamaterials. *Phys. Rev. B* 73:125423. Copyright 2006, The American Physical Society.)

without any reflection. This metamaterial-based device has a strong impact on optical technologies because the inherent problem of polarization dependence seen in the conventional optical component based on the Brewster effect is completely solved.

Figure 20.21 shows an example of the calculation results under the condition that M_1 was a vacuum ($\varepsilon_1 = 1.0$ and $\mu_1 = 1.0$) and M_3 was glass ($\varepsilon_3 = 2.25$ and $\mu_3 = 1.0$) with the additional constraint that the exit angles of p- and s-polarized light to M_3 are identical to zero ($\theta_{ex}^p = \theta_{ex}^s = 0.0$, i.e., the light was transmitted straight through from M_1 to M_3). The solution converged at the point $\varepsilon_2^p = 1.5$, $\mu_2^p = 1.0$, $\varepsilon_2^s = 1.5$, and $\mu_2^s = 3.29$. Under this condition, the Brewster angles for p- and s-polarized light at the interface between M_1 and M_2 were identically 50.8°. This result proved that the light completely passed through both interfaces between two different materials without any reflection losses.

We also investigated the design of the uniaxial magnetic metamaterial consisting of an SRR array. Figure 20.22a and b show an example of the designed layer-stacked 2D arrays of the s-SRRs made of silver. The inner radius (r), the ring width (w), the ring thickness (t), the gap distance of the division (g), the unit-cell dimension (a) in the x–y plane, and the distance between adjacent planes of the s-SRRs along the z-axis (l) were 50, 50, 50, 20, 300, and 50 nm, respectively,

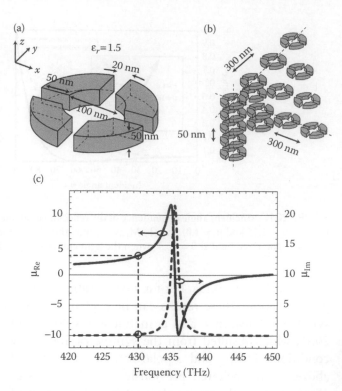

(a) (b)

(c)

FIGURE 20.22 (a, b) Design of layer-stacked 2D arrays of the silver s-SRRs with the relative permeability $\mu_{Re} = 3.29$ at 429.905 THz (697.829 nm in wavelength). For each s-SRR, the inner radius, the ring width, the ring thickness, and the gap distance of the division were 50, 50, 50, and 20 nm, respectively. The unit-cell dimensions were 300×300 nm^2 in the x–y plane and 50 nm along the z-axis, and these lattice constants produce the filling factor of 8.73%. (c) The numerically simulated real and imaginary parts of the effective permeability (μ_{Re} and μ_{Im}) of the s-SRR array. The resonant frequency was 435.215 THz (689.314 nm in wavelength) and the $|\mu_{Im}|/|\mu_{Re}|$ ratio, which is related to the transmission absorption loss, was only 0.076 at 429.905 THz at which $\mu_{Re} = 3.29$.

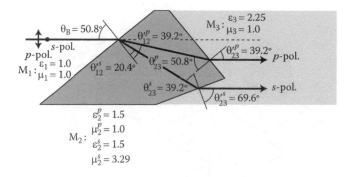

FIGURE 20.21 Design of the uniaxial magnetic metamaterial with the calculated optical constants ($\varepsilon_2^p = 1.5$, $\mu_2^p = 1.0$, $\varepsilon_2^s = 1.5$, and $\mu_2^s = 3.29$). Incident light enters the boundary between M_1 and M_2 at the Brewster angles $\theta_B = 50.8°$ and splits into two rays depending on the polarization. At the boundary between M_2 and M_3, the rays are refracted again and go into M_3 with the propagation direction identical to that of the incident light. The incident angle for p- and s-polarized light at the boundary between M_2 and M_3 were 50.8° and 39.2°, which were also the Brewster angles for the respective polarizations. (Reprinted with permission from Tanaka, T., Ishikawa, A., and Kawata, S. 2006b. Unattenuated light transmission through the interface between two materials with different indices of refraction using magnetic metamaterials. *Phys. Rev. B* 73:125423. Copyright 2006, The American Physical Society.)

and the corresponding filling factor (F) was 8.73%. Figure 20.22c shows the numerically simulated dispersion curves of the real and imaginary parts of the effective permeability μ_{Re} and μ_{Im} of the SRR array. The real part of the effective permeability $\mu_{Re} = 3.29$ was realized at 429.905 THz (697.829 nm in wavelength). In addition, the resonant frequency was 435.215 THz (689.314 nm in wavelength), and these operating parameters can be tuned by redesigning the structure of the s-SRRs.

This technology will find application in a huge variety of optical components, such as laser cavities and optical fiber communication system. Since the polarization dependence of the conventional Brewster window is completely eliminated, lasing of randomly polarized or circularly polarized light could be realized with reduced reflection losses. The situation of the light transmission from air ($n_1 = 1.0$) to glass ($n_2 = 1.5$), which we considered in Figure 20.22, can be easily found around us and this also reproduces the situation occurred at the junction between optical fibers. At the junction of optical fibers, light must go through the interface between air and the core of optical fibers and is reflected many times at every junction in the network. This reflection at every junction of optical components is a serious problem from the viewpoint of energy loss. However, as shown in Figure 20.23, embedding the metamaterial-based device at the end faces of optical fibers or other optical components enables us to completely eliminate the reflection loss at the interfaces.

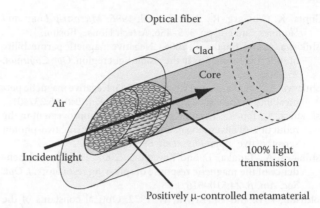

FIGURE 20.23 End face of the optical fiber equipped with the Brewster window made of a uniaxial magnetic metamaterial to eliminate the reflection loss completely. Standard silica-based optical fiber always has 4%-reflection loss at the interface between air and the core of optical fiber, and this is a serious problem from the viewpoint of energy loss for an optical fiber communication system.

20.5 Summary

We conclude this chapter by pointing out again the potential of the metamaterials technology. The metamaterial concept of creating composites with desired optical properties gives us the opportunity to engineer and specify the electromagnetic responses: the electric permittivity (ε) and the magnetic permeability (μ). As described here, controlling electromagnetic material parameters, such as ε and μ, will open the door for exotic physical phenomena and their applications. From the technical side, since the magnetic metamaterials should consist of 3D metallic micro/nanostructures, the realization of the metamaterials is strongly supported by the recent progress of nanofabrication technology, but it is still difficult to fabricate them. The important issues left to the future are three-dimensionality and mass productivity. Since metamaterials need the tremendously large amount of nanoscale resonant structures in it, the breakthrough technique in nanofabrication that can create the well-designed nanoscale metal structures in short time and integrate them into the appropriate host materials is strongly desired.

References

Alù, A. and Engheta, N. 2005. Achieving transparency with plasmonic and metamaterial coatings. *Phys. Rev. E* 72:016623-1–016623-9.

Antipov, A. A., Sukhorukov, G. B., Fedutik, Y. A. et al. 2002. Fabrication of a novel type of metallized colloids and hollow capsules. *Langmuir* 18:6687–6693.

Born, M. and Wolf, E. 1980. *Principles of Optics.* 6th ed. Pergamon Press, Oxford.

Cai, W., Chettiar, U. K., Yuan, H.-K. et al. 2007. Metamagnetics with rainbow colors. *Opt. Express* 15:3333–3341.

Cao, Y., Dong, X., Takeyasu, N. et al. 2009a. Morphology and size dependences of silver microstructures on fatty salts-assisted multiphoton photoreduction microfabrication. *Appl. Phys. A* 96:453–458.

Cao, Y., Takeyasu, N., Tanaka, T. et al. 2009b. 3D Metallic nanostructure fabrication by surfactant-assisted multi-photon-induced reduction. *Small* 5:1144–1148.

Dolling, G., Wegener, M., Soukoulis, C. M. et al. 2007. Negative-index metamaterial at 780 nm wavelength. *Opt. Lett.* 32:53–55.

Formanek, F., Takeyasu, N., Chiyoda, K. et al. 2006a. Selective electroless plating to fabricate complex three-dimensional metallic micro/nanostructures. *Appl. Phys. Lett.* 88:83110.

Formanek, F., Takeyasu, N., Tanaka, T. et al. 2006b. Three-dimensional fabrication of metallic nanostructures over large areas by two-photon polymerization. *Opt. Express* 14:800–809.

Gerenser, L. J. 1990. Photoemission investigation of silver/poly(ethylene terephthalate) interfacial chemistry: The effect of oxygen–plasma treatment. *J. Vac. Sci. Technol. A* 8:3682–3691.

Gray, J. E., Norton, P. R., and Griffiths, K. 2005. Mechanism of adhesion of electroless-deposited silver on poly(ether urethane). *Thin Solid Films* 484:196–207.

Grzegorczyk, T. M., Thomas, Z. M., and Kong, J. A. 2005. Inversion of critical angle and Brewster angle in anisotropic left-handed metamaterials. *Appl. Phys. Lett.* 86:251909.

Guan, F., Chen M., Yang, W. et al. 2005. Fabrication of patterned gold microstructure by selective electroless plating. *Appl. Surf. Sci.* 240:24–27.

Chapter 20

Gupta, K. C., Garg, R., Bahl, I. et al. 1996. *Microstrip Lines and Slotlines*. 2nd ed., pp. 375–456. Artech House, Boston.

Ishikawa, A. and Tanaka, T., 2006a. Negative magnetic permeability of split ring resonators in the visible light region. *Opt. Commun.* 258:300–305.

Ishikawa, A., Tanaka, T., and Kawata, S. 2005. Negative magnetic permeability in the visible light region. *Phys. Rev. Lett.* 95:237401.

Ishikawa, A., Tanaka, T., and Kawata, S. 2006b. Improvement in the reduction of silver ions in aqueous solution using two-photon sensitive dye. *Appl. Phys. Lett.* 89:113102.

Ishikawa, A., Tanaka, T., and Kawata, S., 2007. Frequency dependence of the magnetic response of split-ring resonators. *J. Opt. Soc. Am. B.* 24:510–515.

Johnson, P. B. and Christy, R. W. 1972. Optical constants of the noble metals. *Phys. Rev. B* 6:4370–4379.

Kato, J., Takeyasu, N., Adachi, Y., et al. 2005. Multiple-spot parallel processing for laser micronanofabrication. *Appl. Phys. Lett.* 86:44102.

Landau, L. D., Liftshitz, E. M., and Pitaevskii, L. P. 1984. *Electrodynamics of Continuous Media* 2nd ed. Chap. 79. Pergamon Press, Oxford.

Liu, N., Liu, H., Zhu, S. et al. 2009. Stereometamaterials. *Nat. Photon.* 3:157–162.

Mallory, G. O. and Hajdu, J. B. 1990. *Electroless Plating: Fundamentals and Applications*. American Electroplaters and Surface Finishers Society, Orlando, FL.

Pendry J. 2000. Negative refraction makes a perfect lens. *Phys. Rev. Lett.* 85:3966–3969.

Rill, M., Plet, C., Thiel, M. et al. 2008. Photonic metamaterials by direct laser writing and silver chemical vapour deposition. *Nat. Mater.* 7:543–546.

Schurig, D., Mock, J. J., Justice, B. J. et al. 2006. Metamaterial electromagnetic cloak at microwave frequencies. *Science* 314:977–980.

Smith, D., Pendry, J., and Wilshire, M. 2004. Metamaterials and negative refractive index. *Science* 305:788–792.

Takeyasu, N., Tanaka, T., and Kawata, S. 2005. Metal deposition deep into microstructure by electroless plating. *Jpn. J. Appl. Phys.* 44:1134–1137.

Tamayama, Y., Nakanishi, T., Sugiyama, K. et al. 2006. Observation of Brewster's effect for transverse-electric electromagnetic waves in metamaterials: Experiment and theory. *Phys. Rev. B* 73:193104.

Tanaka, T., Ishikawa, A., and Kawata, S. 2006a. Two-photon-induced reduction of metal ions for fabricating three-dimensional electrically conductive metallic microstructure. *Appl. Phys. Lett.* 88:81107.

Tanaka, T., Ishikawa, A., and Kawata, S. 2006b. Unattenuated light transmission through the interface between two materials with different indices of refraction using magnetic metamaterials. *Phys. Rev. B* 73:125423.

Valentine, J., Zhang, S., Zentgraf, T. et al. 2008. Three-dimensional optical metamaterial with a negative refractive index. *Nature* 455:376–379.

Yen, T. J., Padilla, W. J., Fang, N. et al. 2004. Terahertz magnetic response from artificial materials. *Science* 303:1494–1496.

21. Nanoelectronics

Koji Ishibashi

Advanced Device Laboratory, RIKEN Advanced Science Institute, Saitama, Japan

Nanofabrication Handbook. Edited by Stefano Cabrini and Satoshi Kawata © 2012 CRC Press / Taylor & Francis Group, LLC. ISBN: 978-1-4200-9052-9

Chapter 21

21.1 Introduction

Recent development of lithography techniques has made it possible to fabricate patterns with a few tens of nanometers. Accordingly, a gate length of the advanced CMOS (complementary metal oxide semiconductor) transistors could be <10 nm. The semiconductor roadmap suggests that the limit of the size miniaturization of the transistors would come in the near future. The main problem of the present CMOS transistor is an increase of power consumption due to the leakage current in addition to the difficulty of the size miniaturization. Various attempts have been carried out to overcome the problems without changing the operation principle of the transistor. This strategy is the so-called More Moore. There are two other approaches to overcome the problem. One is the so-called More than Moore, in which conventional integrated circuits (ICs) are combined with sensors and other functionalities to enhance the function of the chip. The other is the so-called Beyond CMOS, in which completely new devices will be explored. "Nanoelectronics" means different things to different people. In this chapter, I will limit myself to "Beyond CMOS." But, the "Beyond CMOS" may still include many things, and sometimes a concrete image of the devices is not established. For example, "Spintronics" is often used as concepts for one of the "Beyond CMOS" possibilities. In general, it uses "spin" instead of "charge." However, spintronic devices that have a concrete device image may be TMR (tunneling magnetoresistance) and GMR (giant magnetoresistance) devices, MRAM (magnetic random access memory), and a spin transistor (Waser 2005). Some of these need to be in "nano" scale, but others are not necessarily in nanoscale. Besides the device aspect, the field is attracting increasing interests from the basic physics point of view. Spin currents, the spin Hall effect, and topological currents are among those (Takahashi and Maekawa 2008). Because of the limitation of pages, I will focus on some specific topics that are essential in nanoscale.

When the device size becomes small, quantum effects start to emerge. To control electrons in a quantum mechanical manner goes back to the idea of "superlattices and quantum wells" proposed by Leo Esaki in the late 1960s (Esaki and Tsu 1970). In the quantum well, electrons are confined in one direction (growth direction) and now, it is possible to confine electrons in three dimensions (quantum dots). In the superlattices, he tried to realize the minigap with which electron oscillations in the k-space were

explored (Bloch oscillations). This route to confine electrons in a small space has been pursued with the idea that device characteristics of transistors and lasers would be improved due to the modification of the density of states (Arakawa and Sakaki 1982; Sakaki 1982).

One of the important effects that could be used for nanodevice application is the single-electron charging effect, or very often called "Coulomb blockade" (Meirav and Foxman 1996). This is an effect that occurs in quantum dots and small tunnel junctions, and with this mechanism, the number of electrons is controlled in the single-electron level by the gate voltage of the single-electron transistor (SET). The effect has attracted researchers to realize completely new devices (single-electron devices) that may produce ultralow power consumption. It is not clear as to whether the single-electron devices are really useful when considered in a circuit level, but it is worth being pursued. The basic physics of the single-electron tunneling will be presented later.

If quantum coherence is included in the single-electron effect, the concept of quantum computing devices is possible even in the normal electron systems. The quantum bit (qubit) is a basic element of the quantum computer, and the physical object of the qubit is a two-level system. These qubits may have to be in nanoscale, generally because the smaller the device size the larger the quantum effect. Superconducting qubits have been widely studied, and have various forms as a qubit. Probably because the macroscopic coherence is maintained in the superconductor, the superconducting qubits have been very successful, and a new physics where qubits as an artificial atom are placed in the electromagnetic cavity has opened up.

Until now, I have mentioned single-electron devices (Coulomb blockade) and quantum computing. Other attractive nanodevices may exist, but I do not have an ability to cover all of them. I believe the above effects continue to be a basic physics behind most of nanodevices for some time.

This chapter is organized as follows. After the introduction, fabrication processes that are widely used to study nanophysics are summarized. Then, the basic physics of the single-electron effect and the concept of quantum computing are described, followed by the description of nanodevices that are carbon nanotube transistors, single-electron devices, and quantum computing devices. The nanodevices covered in this chapter are only a fraction of emerging nanoelectronics, but I hope that the basic physics and

devices described here will be helpful to understand emerging nanodevices that come about in daily research.

21.2 Fabrication and Materials

21.2.1 Standard Techniques

In this section, fabrication processes and materials that have been used to study nanoscale physics are described. Most of the nanostructures studied so far are quantum dots, quantum wires, and small tunnel junctions. These structures have been fabricated in various materials. In the early stage of nanoscale physics, quantum dots or (classical) dots were commonly fabricated with Al tunnel junctions and the surface gate techniques applied to GaAs/AlGaAs two-dimensional electron gas (2DEG).

21.2.1.1 Shadow Evaporation Technique for Small Metallic Junctions

Figure 21.1 shows the fabrication process of the metallic SET, which is a basic device to study electronic properties of confined electrons in a small space (dot). In the SET, the dot in which electrons are confined is connected to source and drain electrodes through small tunnel (or Josephson) junctions. The commonly used method to fabricate the tunnel junctions was first proposed by Fulton and Dolan (1987). In this method, the shadow evaporation technique is used, as shown in Figure 21.1a. The first evaporation of Al is done with a shadow angle to the resist structure with the air bridge. After the first evaporation, the sample is transferred to air or to a different chamber where the surface oxidation is promoted. Then, the sample is moved back to the main chamber to evaporate the second Al from the opposite direction, as shown in Figure 21.1a. The

tunnel junctions with an Al surface oxide layer is completed (Figure 21.1b). An example of the Al SET fabricated in this method is shown in Figure 21.1c, where the tunnel junctions are formed in the dotted circles. The junction size is typically 0.1 μm^2. The resistance of the tunnel barrier is controlled by the oxidation condition. In this way, the normal tunnel junction with a large resistance and the Josephson junction with a low resistance can be fabricated. The control of the junction resistance is important to design superconducting qubits described later in this chapter.

21.2.1.2 Surface Gate Techniques for Semiconductor Nanostructures

Quantum dots have been commonly fabricated by the surface gate technique (Thornton 1995). In this technique, the metallic gates are deposited on top of the GaAs/AlGaAs 2DEG. By applying the negative gate voltage, electrons underneath the gate are depleted (Figure 21.2a). The quantum point contact (QPC), which is a constriction for current flow, is an important structure to form the controlled tunnel barrier. The QPC has a saddle potential, the shape and the height of which can be changed by the gate voltage (Beenakker and van Houten 1991). The important characteristic of the QPC is that the channel (mode) for electron waves is well defined, so that the conductance of the QPC is quantized in a unit of $2e^2/h$, which is a conductance for one channel (van Wees et al. 1988; Wharam et al. 1988). e is an elementary charge and h is the Planck's constant. Figure 21.2b shows a typical gate

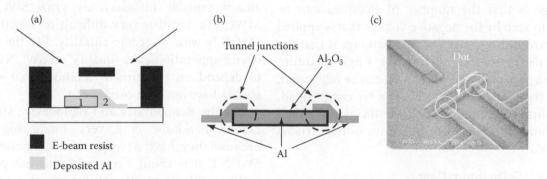

FIGURE 21.1   Shadow evaporation technique to fabricate metallic tunnel junctions. (a) Cross-sectional picture of the resist pattern and evaporated metals from two different directions. (b) Cross-sectional picture of the metallic SET. (c) Scanning electron microscope image of the SET made of Al. (Courtesy of A. Kanda.)

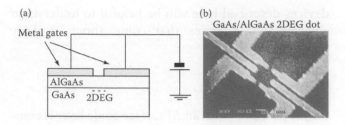

FIGURE 21.2 Surface gate technique for fabricating semiconductor quantum dots. (a) Cross-sectional picture of the wafer structure and the metallic gate. (b) SEM pattern of the single quantum dots.

pattern to form a quantum dot. The dot is connected to outside electrodes through the two QPCs. In this structure, the potential of the quantum dot can be changed by the plunger gates. The advantage of this technique is that the arbitrary structures can be fabricated simply by designing the gate pattern, and the disadvantage is that all the gates are capacitively coupled, so that changing a voltage on one gate affects the condition of other gates. For example, to keep the potential landscape of the dot constant after the voltage on one gate is changed, voltages on all the other gates may have to be adjusted. In the recent gate structures, the number of electrons can be reduced to zero by applying practical gate voltage (Ciorga et al. 2000).

21.2.1.3 Vertical Quantum Dots

Vertical dots have been fabricated in some groups (Reed et al. 1988; Tarucha et al. 1996). Starting from a wafer with a quantum well and tunnel barrier layers, the reactive ion etching technique is employed to form the narrow vertical channel. In this technique, very narrow columns can be fabricated. Besides, the shape of the tunnel barrier is not affected by the voltages on the gates, so that one does not need to worry about adjusting gate voltages, as is the case for the dot fabricated with the surface gate technique. Another advantage is that the number of electrons can be reduced to zero by the negative voltage that is applied to the surrounding gate. The disadvantage is that the only simple structures, such as the single quantum dot and the coupled quantum dots, can be fabricated. Besides, the fabrication process is a bit complicated. But the first observation of the artificial atom shell filling showed this kind of quantum dots (Tarucha et al. 1996).

21.2.1.4 Si Quantum Dots

Si quantum dots can be fabricated simply by patterning the thin-film Si crystal on a buried oxide layer (BOX Si) (Manoharan et al. 2008). In this structure, oxygen ions are implanted into the Si substrate, and the implanted layer is annealed to form an insulating layer in the Si substrate. The advantage of this method is that it is a simple process, and that the patterned dots can be further reduced by the surface oxidation. With this method, the coupled quantum dots as a charge qubit have been fabricated (Gorman et al. 2005).

Another unique method to fabricate extremely small quantum dots is the so-called PADOX (pattern-dependent oxidation) method developed in the NTT group (Takahashi et al. 1995). In this case, the PADOX mechanism is used. For example, to fabricate a single quantum dot, a simple constriction is patterned and thermally oxidized. The oxidation occurs more preferably at the edge of the constrictions to form the tunnel barrier. The technique is applied to fabricate more complicated quantum dot devices (Nishiguchi et al. 2006), and the dot size can be so small that the SET fabricated in this method works at a relatively high temperature.

21.2.1.5 Carbon Nanotubes

Since the discovery of carbon nanotubes (Iijima and Ichihashi 1993), they have been attractive building blocks for electronic and photonic device applications. Depending on the number of cylinders, there are two types of nanotubes, which are single-wall carbon nanotubes (SWCNTs) and multiwall carbon nanotubes (MWCNTs). It is well known that the SWCNT becomes semiconducting or metallic, depending on how it is rolled up from a graphene sheet (chirality). Carbon nanotubes are fabricated by various methods that include laser ablation, arc-discharge, catalytic chemical vapor deposition (CVD), and so on. Depending on the methods, the grown nanotubes are SWCNTs or MWCNTs, or sometimes double-walled carbon nanotubes (DWCNTs). At the moment, it is almost possible to selectively grow SWCNTs and MWCNTs, but it is very difficult to selectively grow SWCNTs with a specific chirality. For the electronic device applications, the "quality" of SWCNTs appears to depend on the growth methods, but systematic studies have not been carried out.

For the nanophysics and nanodevice studies, the carbon nanotube is a very interesting material because they have a very small diameter. For the SWCNT, it is about 1 nm, which is not possible to realize with conventional lithography techniques. Therefore, it is a natural choice for the study of the low-dimensional electron transport. To fabricate a sample to study electronic transport (Figure 21.3a),

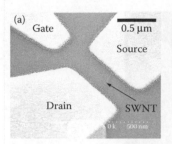

FIGURE 21.3 (a) Individual carbon nanotube with electrical contacts and the side gate. (b) Aligned carbon nanotubes grown on the quartz substrate.

source and drain contacts are fabricated in an individual carbon nanotube. A metallic gate near the nanotube or a heavily doped Si substrate is commonly used for the gate voltage application. The top gate can be fabricated with an insulating oxide grown by the atomic layer deposition technique, but the semiconducting SWCNT transistors fabricated with the method usually shows a large hysteresis in the gate voltage sweep (Kamimura and Matsumoto 2004). A primitive method is to disperse the SWCNTs from the solution onto the substrate on which large pads for wiring and alignment marks for later e-beam lithography are fabricated (Suzuki et al. 2001). Then, the SWCNTs are observed with the atomic force microscope (AFM) or scanning electron microscope (SEM) to record the position with respect to the alignment marks. Based on the position data, the e-beam lithography technique is used to fabricate the electrical contacts to an individual nanotube. But, it should be noted that the low-energy electron beam irradiation may damage the SWCNTs (Vijayaraghavan et al. 2007), so that the SEM observation should be avoided before devices are measured. Another method is to grow SWCNTs with the patterned catalyst, and a similar technique may be used to fabricate electrical contacts to an individual SWCNT. With the methods, metallic contacts are on top of the SWCNT.

The opposite method is sometimes used, and, in fact, for the very first measurement by the Delft group (Tans et al. 1998), the SWCNTs are dispersed onto the predefined electrical contacts for the current flow. In this case, the SWCNT is on top of the metallic contacts. The difference in terms of the transport characteristics is not clear between the two methods (Ishibashi et al. 2000), but in the former method, the electron beam is irradiated, and in the latter method, no electron beam is irradiated in the entire fabrication process. It should be noted that the damage due to the electron beam irradiation has been reported (Vijayaraghavan et al. 2007).

Another important difference is that in the latter method, no device fabrication processes are done on the nanotube so that the contamination that may be included in the processes is maximally avoided.

For the integrated SWCNT device fabrication, the aligned growth technique, as well as the chirality control, is necessary. Although the latter is still a big challenge, the former can be realized to some extent. It was found that the SWCNTs grew in a preferential direction on the quartz and sapphire substrates (Ago et al. 2005). Figure 21.3b shows an example of the aligned SWCNT growth on the quartz substrate. It has been reported that the aligned SWCNTs on the substrate can be transferred to another substrate, such as a surface-oxidized Si substrate that is suitable for device fabrication (Meitl et al. 2004), and serious damages do not appear to be included in the mechanical process (Tabata et al. 2009).

21.2.3 Graphene

A monolayer graphite (graphene) has been attracting much attention from the viewpoint of basic physics and device applications since it was shown that graphene was obtained simply by peeling off the graphite with a scotch tape (Novoselov et al. 2005). Because of the unique band structures with a linear dispersion, graphene is an interesting material for study in physics (Novoselov et al. 2005). On the other hand, it may also be attractive as a channel material for the field effect transistor (FET) because it may have a large mobility at room temperature (Novoselov et al. 2005). The disadvantage of using graphene for the channel material is that it does not have a band gap. The conduction band and the valence band touch at the "Dirac point." To open the gap, one needs to fabricate a double-layer graphene, or fabricate a graphene ribbon. In the latter case, the electronic property is predicted to strongly depend on the microscopic structure of the edge of the ribbon (Areshkin et al. 2007). Experimentally, the band gap obtained from the transport measurement increases as the width of the ribbon is decreased (Han et al. 2007). Because of the complexity of the edge, the ribbon may not be used for practical application. In the double-layer graphene, the band gap opens when an electric field is applied perpendicular to the layer (Ohta et al. 2006).

The peeling-off technique is obviously not suitable for practical application. A lot of effort has been made to grow a large-area graphene on a useful substrate. It has been known that a few-layer graphene can be grown on the SiC substrate by heating it up to >1000 (van Bommel et al. 1975). In this case, it is not easy to control the number of layers. An attempt is made to grow the few-layer graphene on the SiC layers that are in turn grown on the Si substrate (Fukidome et al. 2010), to avoid the use of expensive SiC substrate. Another useful method is to grow it on a metal substrate such as Ni (Reina et al. 2009). The grown layer should be transferred to a useful substrate (Geim 2009). The FETs with a few-layer graphene (Wu et al. 2008) as well as quantum dots (Stampfer et al. 2008; Moriyama et al. 2009) with a few-layer graphene have been demonstrated. It is theoretically predicted that the graphene could be used for the laser application in a THz range (Ryzhii 2006).

21.2.4 Emerging Techniques

In the previous sections, commonly used techniques and materials to fabricate nanostructures, mainly quantum dots and quantum wires, are described. The above methods rely on the lithography techniques, except in the case of carbon nanotubes, so that the device size is, in practice, in a submicron scale. There are attempts to go beyond the limitation. The use of carbon nanotubes is one of them. Here, two methods are introduced among many other attractive methods.

21.2.4.1 Semiconductor Nanowires

One of the promising attempts is to use semiconductor nanowires that are grown by catalytic CVD techniques. These semiconductor nanowires have a diameter of several tens of nanometers, but some of them could have a diameter <10 nm. Until now, Si and Ge nanowires (Fukata 2009; Morales and Lieber 1998) as well as various compound semiconductor nanowires (Dick et al. 2005; Tateno et al. 2004) have been grown. The advantage of the semiconductor nanowires is that heterostructures can be fabricated in the length direction. By controlling the growth condition, the core-shell structures can be fabricated (Lauhon et al. 2002; Sköld et al. 2005), which increases the variation of the structures. Figure 21.4 shows "one-dimensional heterostructures" fabricated in the compound semiconductor nanowire. The combination of the self-assembled nanowires, including carbon nanotubes, with lithography techniques has made it possible, to some extent, to fabricate nanostructures that are not easy only with the lithography techniques.

21.2.4.2 Molecular Lithography Technique

There are attempts to fabricate electrical contacts to a single molecule (Reed 2004), which is essentially

FIGURE 21.4 (**See color insert.**) Transmission microscope image of the heterostructures in the catalytically grown compound semiconductor nanowires. (Courtesy of L. Samuelson. From Dick, K. A. et al. 2005. *Nano Lett.* 5: 761. With permission.)

important to molecular electronics. Nanogaps with metallic electrodes are a widely used technique, and various techniques have been proposed to realize the nanogap electrodes with a distance <10 nm (Reed et al. 1997; Park et al. 1999; Liu et al. 2005; Chen et al. 2005). There exist large difficulties and problems on this route. In most cases, the nanogap electrodes are fabricated by electron beam lithography plus some tricks to realize the sub-10 nm gap. But the width of the gap is usually not well controlled, and the gate effect is very small because the electric field by the gate is strongly shielded. When the gap size does not fit the molecular size, it may make the molecule–metal contact unstable. Besides, to put a single molecule in the nanogap is not easy. One of the promising methods to fabricate the controlled nanogap may be the molecular ruler technique (Hatzor and Weiss 2001), in which self-assembled molecules (SAMs) are deposited with the controlled number of layers simply by counting the number of dipping times. Figure 21.5 shows the fabrication process of the nanogaps with a controlled gap size, which can be used for the electrical transport measurement (Negishi et al. 2006). In this technique, the SAMs are used as a resist to carry out a lift-off process. Recently, the problem of the small gate effect has been solved by fabricating a local back gate underneath the nanogap electrodes with an insulating layer (Nishino et al. 2010).

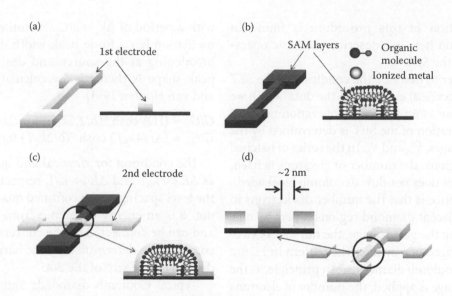

FIGURE 21.5 Nanogap fabrication process. (a) First electrode fabrication. (b) Coating SAM layers. The thickness of the SAM layers is determined by the number of layers that can be controlled accurately. One layer corresponds to ~2 nm. (c) Second electrode fabrication with e-beam lithography and metal lift-off. (d) Removal of the SAM layers.

21.3 Quantum Dots and Artificial Atom

21.3.1 Single-Electron Transistor

An equivalent circuit of the SET is shown in Figure 21.6a. A dot is connected to the source and drain electrodes through small tunnel junctions, and is connected to the gate electrode through the classical capacitor. The classical capacitor means a standard capacitor through which electrons cannot tunnel. It should be noted that the dot is floating electrically from the external circuit. In the single junction, the Coulomb blockade effect was difficult to observe because the voltage across the junction is fixed due to the existence of parasitic capacitances that shunt the junction in parallel (voltage biased). Therefore, the current-biased condition, necessary for the observation of the effect in the single junction, is difficult to realize in practice. On the other hand, in the SET with double junctions, the dot is electrically floating, so that the potential of the dot can fluctuate and the voltage bias does not kill the effect. Besides, the potential of the dot can be controlled by the gate voltage. Since the dot is isolated, the number of electrons in the dot can be fixed. In the SET, the number of electrons is controlled on the single-electron level by using the gate voltage.

A simple analysis of the SET is carried by considering the whole circuit as a capacitance circuit, and by finding the condition for an electron to tunnel through the source and drain junctions. It is the enthalpy that is important to analyze the circuit. It is given by the summation of the electrostatic energy that is stored in each capacitance, which is subtracted by the works that are done by the two voltage sources in the circuit, in order to make electrons tunnel through two junctions. The

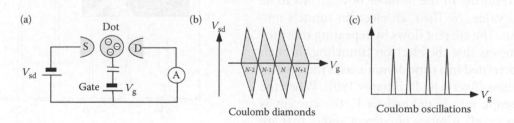

FIGURE 21.6 (a) Equivalent circuit of the SET. (b) Conductance as functions of the gate voltage (V_g) and the source-drain voltage (V_{sd}). In the diamonds with a gray color, the current does not flow (Coulomb blockade). The series of the diamonds are called Coulomb diamonds. (c) The current as a function of the gate voltage in the liner response regime (Coulomb oscillations).

excellent derivation of this procedure is found in Tucker (1992), and here, we describe the basic operation principle of the SET.

A good number to describe the condition of the SET is the number of (excess) electrons in the dots, which we define as N. Figure 21.6b shows the operation mode of the SET. The operation of the SET is determined by the two external voltages, V_{sd} and V_g. In the series of hatched diamond-like regions, the number of electrons is fixed, so that the current does not flow (Coulomb blockaded). An important notice is that the number of electrons in the dot in the adjacent diamond regions differs by one. Then, by changing the gate voltage, the number of electrons can be changed one by one. The pattern in Figure 21.6b is called "Coulomb diamonds." In principle, as the negative gate voltage is applied, the number of electrons in the dot can be zero. In practice, it is not possible to reduce the number of electrons to zero for metallic dots with a number of electrons. As is discussed later, it is possible in some semiconductor quantum dots and semiconductor carbon nanotube quantum dots. An important energy scale for the SET is the charging energy for a single electron, E_c, which is expressed as $E_c = e^2/C_\Sigma$, where $C_\Sigma = C_s + C_d + C_g$, the self-capacitance of the dot with C_s being the capacitance at the source, C_d the capacitance at the drain, and C_g the capacitance at the gate. The size of the diamond gives E_c, and the period of the diamonds gives C_g with $C_g\Delta V_g = e$. Since the dot is connected to the ground through three parallel capacitors, the effectiveness of the gate voltage to modulate the potential of the dot is defined as a conversion factor, κ, and is expressed as $\kappa = eC_g/C\Sigma$.

In the linear response regime with $V_{sd} \sim 0$, the current flows at the gate voltages where two diamonds meet. Consider the gate voltage value where the two diamonds that have N and $N + 1$ electrons meet. Since two states are degenerated at the gate voltage value, an electron can tunnel into the dot that has originally N electrons, resulting in the total number of electrons to be $N + 1$. After a trapping time, an electron tunnels out of the dot, resulting in the number of electrons to be the original value, N. Then, an electron tunnels into the dot again. The current flows by repeating this process. This means that the electron tunneling through the dot is correlated in a time domain, as is the case for single junctions (Averin and Likharev 1991). When the averaged escape rate is defined as Γ, the current is expressed as $I = e\Gamma$. A series of current spikes that are observed as the gate voltage is swept is called "Coulomb oscillations." In the simple capacitive model (orthodox theory), the oscillations are periodic in the gate voltage

with a period of $\Delta V_g = e/C_g$. At finite temperature, the oscillation has a finite peak width due to the thermal broadening at the source and drain electrodes. The peak shape is theoretically calculated as (Beenakker and van Houten 1991)

$$G/G_\infty \approx (1/2)\cosh^{-2}(\delta/2.5k_BT) \text{ for classical dot}$$
$$G/G_\infty \approx (\Delta E/4k_BT)\cosh^{-2}(\delta/2k_BT) \text{ for quantum dot}$$

The condition for classical and quantum is defined as $\Delta E \ll k_BT$ and $\Delta E \gg k_BT$, respectively, where ΔE is the level spacing of the confined quantum states in the dot. δ is an energy difference from the peak position and can be connected by the conversion factor. G_∞ is a conductance determined by the barriers and does not depend on the size of the dot.

Typical Coulomb diamonds and Coulomb oscillations measured in the surface-gated GaAs/AlGaAs 2DEG are shown in Figure 21.7. To be more precise, the device in the figure is a "quantum" dot, but the main features predicted in the orthodox model (Figure 21.6b and c) are reproduced. There are two major differences in the Coulomb diamonds and the Coulomb oscillations in the quantum dot, compared with those of the classical dot. One is the complicated features outside the

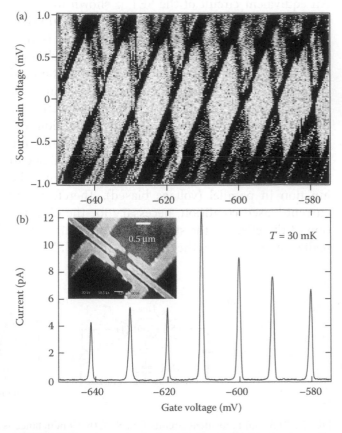

FIGURE 21.7 (a) Coulomb diamonds and (b) Coulomb oscillations, measured for the surface-gated GaAs/AlGaAs quantum dots.

Coulomb blockade diamonds, which originate from the effect of zero-dimensional confined levels. The other is that the height of the Coulomb peaks appears to be random. It is the effect of the energy dependence of the density of states in the quantum dot. In the classical dot, it is almost constant with a number of electrons. We do not discuss the quantum effect further in this section. For the SET in Figure 21.7, $E_c \sim 0.7$ meV.

21.3.2 Quantum Dot: Artificial Atom

The quantum dot resembles a natural atom in terms of the fact that electrons are confined in a small space. In Table 21.1, the well-known semiconductor artificial atom and the carbon nanotube artificial atom are compared with the natural atom. The main characteristics of the artificial atom, the shell structures, for example, are determined by the confinement potential. In the natural atom, electrons are confined in the three-dimensional Coulomb potential. The energy spectrum or the shell structures are analytically calculated in the text books of quantum mechanics. The energy levels are determined by the principal quantum number, n, and in the parenthesis, the degree of degeneracy is shown. When ΔE is not ignorable, compared with E_c, the Coulomb oscillations are no longer periodic, and the peak distances between adjacent peaks, which is referred to the addition energy (E_{add}), rather than E_c, which is usually constant in the orthodox model.

The well-known vertical semiconductor quantum dots confine electrons in the two-dimensional parabolic potential. The energy-level spacing is approximately constant and the degeneracy of each level is given by $2n$, where n is the level number. It should be noted that the number of degeneracy increases as the number of electrons increases, as is the case for the natural atom. The shell structures are observable by measuring E_{add} as a function of the number of electrons (gate voltage). The first experiment that showed the shell structures in the semiconductor artificial atom was carried out by Tarucha et al. (1996), and their results are shown in Figure 21.8. As seen in the figure, the peak distances are not constant, and the addition energy, calculated from them as a number of electrons, shows peaks at a certain number of electrons. These peaks indicate that the shell is closed, and the next electron needs the larger addition energy of the order of ΔE. Reflecting the shell structures determined by the confinement potential, the addition energy is larger at $N = 2, 4, 6, 8, 12$, and so on, where N is the number of electrons.

The SWCNT quantum dots (QDs) confine electrons in the one-dimensional hard-wall potential. Since the diameter of the SWCNT is about 1 nm, which is much smaller than the practical gap between two electrodes (~200 nm), only ground states of the quantum states in a circumference direction need to be considered. Therefore, the realistic confinement potential is a one-dimensional hard-wall potential. The SWCNT has two equivalent bands, so that by adding the twofold spin degeneracy, each level has a fourfold degeneracy. The important notice is that the degeneracy is constant, independent of the number of electrons in the dot. This is the reason why the shell structures are observable in the SWCNT quantum dot, irrespective of the number of electrons, which should be in contrast to the semiconductor quantum dots where the shell structure is observable only when the number of electrons is small (few electron quantum dots). As the quantum number increases in the semiconductor QDs, the degeneracy increases, resulting in the complicated shell filling. In fact, the peak spacing of the Coulomb oscillations in the standard semiconductor QDs with a number of electrons is constant, which is well explained by the orthodox model of the SET.

The four electron shell structures were observed in various groups (Liang et al. 2002), and experimental data by Moriyama et al. (2005) are shown in Figure 21.9. The Coulomb oscillations are observed in a wide gate voltage range (Figure 21.9a), and E_{add} calculated from the data is shown in Figure 21.9b. As expected from the

Table 21.1 Comparison among Artificial Atoms and a Natural Atom

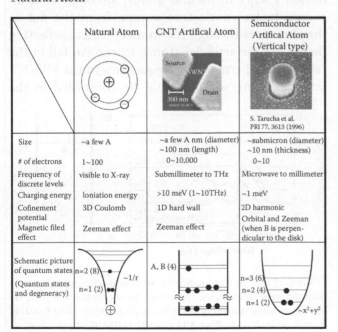

	Natural Atom	CNT Artifical Atom	Semiconductor Artifical Atom (Vertical type)
			S. Tarucha et al. PRl 77, 3613 (1996)
Size	~a few A	~a few A nm (diameter) ~100 nm (length)	~submicron (diameter) ~10 nm (thickness)
# of electrons	1~100	0~10,000	0~10
Frequency of discrete levels	visible to X-ray	Submillimeter to THz	Microwave to millimeter
Charging energy	Ionization energy	>10 meV (1~10THz)	~1 meV
Cofinement potential	3D Coulomb	1D hard wall	2D harmonic
Magnetic filed effect	Zeeman effect	Zeeman effect	Orbital and Zeeman (when B is perpendicular to the disk)
Schematic picture of quantum states (Quantum states and degeneracy)	n=2 (8) ~1/r n=1 (2)	A, B (4)	n=3 (6) n=2 (4) n=1 (2) ~x²+y²

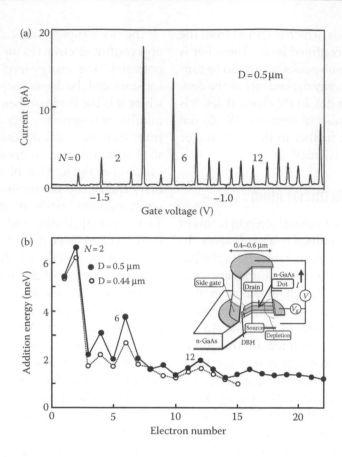

FIGURE 21.8 (a) Coulomb oscillations and (b) the addition energy as a function of the number of electrons in the vertical quantum dots. (Courtesy of S. Tarucha. From Tarucha, S. et al. 1996. *Rev. Lett.* 77: 3613. With permission.)

above arguments, E_{add} has a peak in every four electrons in one gate voltage range, while it has a two-electron periodicity in another gate voltage range. Experimentally, this kind of behavior is sometimes observed in SWCNT quantum dots. In the two-electron periodicity regime, twofold band degeneracy appears to be lifted for some reason that may destroy the symmetry of the bands (Oreg et al. 2000).

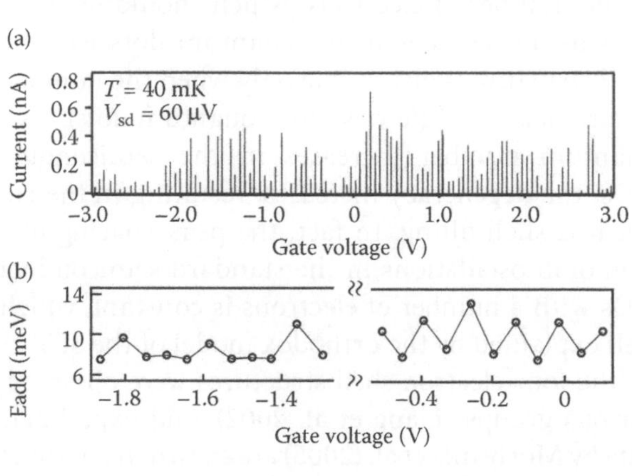

Important energy scales that characterize the artificial atom are the charging energy for the single electron (E_c) and the level spacing of confined quantum states (ΔE). The former could correspond to the ionization energy in natural atoms, although E_c is an energy to let an electron escape from the dot to the electrodes through the tunnel barrier. Reflecting the size of the artificial atoms, E_c and ΔE fall in the range of terahertz (THz) frequencies in the SWCNT artificial atom, which should be contrasted to the

FIGURE 21.9 (a) Coulomb oscillations and (b) the addition energy as a function of the gate voltage (the number of electrons) for the carbon nanotube quantum dots.

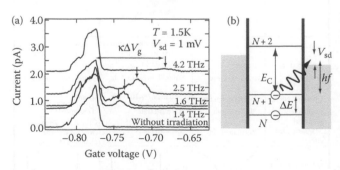

FIGURE 21.10 (a) Coulomb peak with and without THz waves with different frequencies. (b) Schematic mechanism of the THz photon-assisted tunneling (THz PAT). (Permission from Kawano, Y. et al. 2008. *J. Appl. Phys.* **103**: 034307-1–034307-4.)

semiconductor artificial atom where the energies fall in the range of microwave frequencies. This fact can be used for the quantum detection of the THz wave by the SWCNT artificial atom, as shown next (Fuse et al. 2007).

Figure 21.10 shows the Coulomb peaks with and without THz irradiation of various frequencies. Without THz irradiation (bottom), a Coulomb peak (main peak) is observed. When the THz wave is irradiated, the new peaks (side peaks) appear on the right-hand side of the main peak. The distance between the main peak and the side peaks increases linearly as the frequency is increased. This experimental observation clearly demonstrates the THz photon-assisted tunneling (THz PAT) of an electron in the dot to the drain bias window (Figure 21.10b) (Kawano et al. 2008). The simple PAT process shown in Figure 21.10b is possible when $E_c \gg hf \gg k_B T$, which is in fact satisfied in the experimental condition at $T = 1.5$ K and $E_c \sim 24$ meV for the measured dot.

21.4 Nanoelectronic Devices

There are many nanoelectronic devices. Some are experimentally demonstrated, some are at the level of theoretical proposal, and some are just concept without concrete device images. It is impossible to cover all of these, and so we cover only those nanodevices that have concrete device concept and need nanofabrication techniques. Most of them are based on the physics described in the previous sections.

21.4.1 Single–Electron Devices

21.4.1.1 Coupled Dots

21.4.1.1.1 Coulomb Blockade in the Coupled Dot System Coupled quantum dots are straightforward extension of the single quantum dots (SET), and it is important to understand their basic transport mechanism because they are used as building blocks of more complicated single-electron devices.

Figure 21.11 shows (a) an equivalent circuit and (b) the charge stability diagram of the coupled quantum dots that corresponds to the Coulomb diamonds for the single quantum dot. The double dots are connected through a coupling tunnel junction, and are connected to the source and drain electrodes through the small tunnel junctions. The device is controlled by three external voltages, V_{sd}, V_{gL}, and V_{gR}, which should be compared with the single quantum dot where the device is controlled by the two external voltages, V_{sd} and V_g. V_{gL} and V_{gR} modify the potential of the left dot and the right dot, respectively. In the coupled quantum dots connected in series, the charge state is defined by the number of electrons in each dot (m,n). When there is a finite coupling capacitance, the charge state shows a honeycomb pattern, as shown in Figure 21.11b for V_{sd} being close to zero. Inside each hexagon, the charge in each dot is fixed, so current does not flow (Coulomb blockade). The current flows at gate voltages where three hexagons meet, like at A and B in Figure 21.11b, where the three charge states are degenerated. For example, at the gate voltages, A, the charge states of (m,n), $(m + 1,n)$, and $(m,n + 1)$ meet. Starting from the original state (m,n), an electron moves $(m,n) \rightarrow (m + 1, n) \rightarrow (m,n + 1) \rightarrow (m,n)$. The current flows by repeating the process. Again, the current flows by electrons passing one by one, as is the case for the SET. Figure 21.12 shows a gray-scale plot of the current for the surface-gated coupled quantum dots (see inset of the SEM image) as functions of the gate voltages for each dot (V_{gL}, V_{gR}). Black color indicates a large current and white color indicates the zero current, showing the stability diagram of the coupled quantum dots. The distance between A and B corresponds to the strength of the capacitive coupling. When the tunnel coupling is included, the shape of the hexagon at A and B may be rounded due to the coherent coupling of the two dots (Blick et al. 1998).

The coupled quantum dots are basic structures for the charge qubit, where a single electron sits either on the left dot or on the right dot, and any superposition state is possible due to the tunnel coupling between the two dots. We will discuss this later.

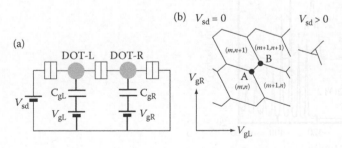

FIGURE 21.11 (a) Equivalent circuit of the coupled quantum dots. (b) Charge stability diagram for the coupled quantum dots for $V_{sd} \sim 0$. At the gate voltages of A and B, the current flows.

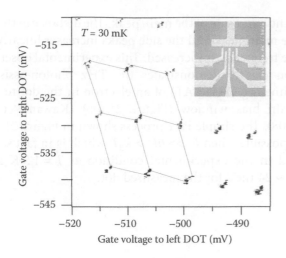

FIGURE 21.12 Experimental charge stability diagram measured for the GaAs/AlGaAs coupled quantum dots. The black color indicates the finite current while the white color indicates no current.

21.4.1.1.2 Electrometer Application

The coupled quantum dots arranged in parallel are used for detecting charges in a direct manner, and can be used for an electrometer that can measure extremely small current.

The basic idea is shown in Figure 21.13, where the two dots are coupled in parallel (Ishibashi et al. 1999). The SEM image of the device fabricated with the surface-gated GaAs system and its equivalent circuit are shown in Figure 21.13a and b, respectively. Figure 21.13c shows the current that flows through each dot, measured simultaneously. In this setup, DOT1 works as the electrometer that measures the change of charges in DOT2. As shown in the figure, even after the current is too small to measure in DOT2 due to the pinch-off the QPC gate, the small structures seen in the Coulomb peaks of the DOT1, which correspond to the change of the number of electrons in the DOT2, are still observable. If the gate voltage was set at some appropriate position, the electrometer current would change in time as an electron passes through the DOT2. In fact, very small current $<10^{-18}$ A has been measured by directly counting the number of electrons that passed in unit time (Fujisawa et al. 2004).

21.4.1.1.3 Spin Blockade

In the above discussions of the coupled dots, the spin has not been taken into account. In the coupled quantum dots where quantum

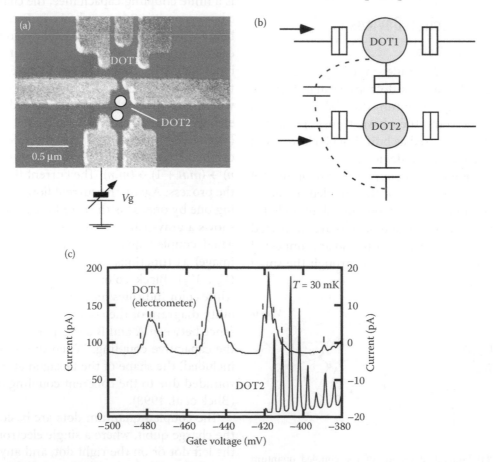

FIGURE 21.13 (a) SEM image of the parallel quantum dots. (b) Equivalent circuit of the device (a). (c) Currents measured simultaneously in the DOT1 and dots. (From Ishibashi, K. et al. 1999. *Microelectron. Eng.* 47: 185–187. With permisssion.)

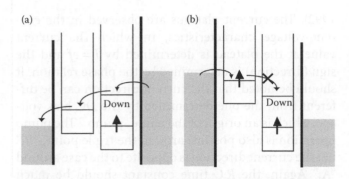

(a)   (b)

FIGURE 21.14 Spin blockade mechanism. (a) In this voltage bias direction, the current flows. (b) The current is blocked when the up-spin electron comes into the left dot.

levels become important, we cannot ignore effects of spin. The spin blockade effect is an important mechanism for the readout of the spin states in the spin qubit that is discussed later. Here, a basic mechanism of the spin blockade is introduced.

Suppose the coupled quantum dot in Figure 21.14 where one electron exists in the right dot and has an up-spin. In the bias condition of (a), a current can flow. Note that the excited level in the right dot has to be a down-spin state that is separated from the ground state by the amount of the on-site Coulomb energy. In contrast to the situation, suppose the opposite bias condition shown in Figure 21.14b. In this case, an electron with an arbitrary spin direction comes into the left dot, but only an electron with a down-spin can pass through the coupled dots. When an up-spin electron comes into the left dot, it cannot go into the right dot because of the Pauli principle. Then, the current flowing process stacks. This is called the spin blockade or Pauli blockade.

In the practical situation with GaAs quantum dots, there is a small leakage current even in the spin blockade condition (Ono et al. 2002). The up-spin electron can flip its spin direction and can pass through the right dot. The spin-flip may occur by the scattering with nuclear spins host atoms have. The nuclear spin plays an important role in the quantum dots in compound semiconductor quantum dots, and could be one of the spin decoherence sources in that material system.

21.4.1.2 Single-Electron Turnstile and Pump: Current Standard

21.4.1.2.1 Turnstile
By using the controllability of the number of electrons in the dot with a gate voltage, the current standard devices have been demonstrated. There are two different types for the current standard, which are the turnstile (Geerligs et al. 1990) and the

pump (Pothier et al. 1992). In both devices, the current is determined by $I = ef$, where f is the frequency that is applied to the gates. The Al-based junction technique was used for device fabrication.

Figure 21.15a shows the equivalent circuit of the turnstile device (Geerligs et al. 1990). The central dot is connected with two tunnel junctions on both sides, and has a gate on which the AC voltage is applied. In the figure, the symmetric source-drain bias is applied, but may not be essential. Figure 21.15b shows current–voltage curves with different AC frequencies applied to the gate. The current plateaus are observed, the height of which is determined by $I = ef$. In the experiment, the radio frequency (RF) frequencies were used from 4 to 20 MHz in 4 MHz steps (a through e), and should be much smaller than the R_tC time constant, which corresponds to 5 GHz.

In the analysis of the device based on the electrostatic energy, they found that an electron tunnels from the source to the central dot and stays there in the half period of the AC voltage. In the other half period of the AC voltage, the electron in the central dot tunnels from there to the drain electrode. The next electron does not tunnel until the second period of the AC voltage is applied. With this mechanism, a single electron is transferred in one period of the AC voltage.

A qualitative explanation of the turnstile device may be possible with the idea of the single-electron box that

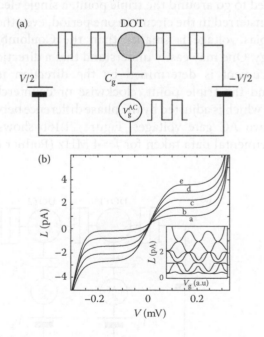

FIGURE 21.15 (a) Equivalent circuit of the turnstile device. (b) Current voltage characteristics with different frequencies. (Courtesy of J. M. Mooij. From Geerligs, L. J. et al. 1990. *Phys. Rev. Lett.* 64: 2691. With permission.)

is introduced in the next section. Seen from the central dot, the multiple junctions are connected to the source and the drain. One may note that the central dot corresponds to the memory node in the single-electron box. Since the single-electron box has a hysteresis characteristic, an electron that tunnels into the node (central dot) can stay there until the next half of the AC signal comes. Following the analogy with the single-electron box, the multiple tunnel junctions are necessary on both sides of the central dot for the turnstile operation.

The turnstile device was also demonstrated by the single quantum dots fabricated by the surface gate technique in 2DEG in GaAs/AlGaAs, where two tunnel barriers at the source and drain were modulated periodically out of phase (Kouwenhoven et al. 1991).

21.4.1.2.2 Pump

Another current standard device is called "pump device," and its equivalent circuit is shown in Figure 21.16a (Pothier et al. 1992). The device structure is similar to the coupled quantum dots in series, shown in Figure 21.11a, except that the AC voltages are applied to each gate superimposed on the DC voltages. Again, the symmetric bias is applied in the experiment, but may not be an essential requirement for the pump operation as long as the DC bias is small. In the pump device, the DC gate voltages on each dot set the device around the triple point, "A," for example, in Figure 21.11b. When the gate voltages are applied to go around the triple point, a single electron is transferred in the circuit in one period, even though the bias voltage is smaller than the Coulomb gap energy. One may easily understand that a direction of the current is determined by the direction to go around the triple point, clockwise or counterclockwise, which is adjusted by the phase difference between the two AC gate voltages. Figure 21.16b shows the experimental data taken for $f = 4$ MHz (Pothier et al.

1992). The current plateaus are observed in the current–voltage characteristics, in which the current value at the plateau is determined by $I = ef$ and the sign (direction) is determined by the phase relation. It should be noted that the current direction can be different from the one determined by the DC bias voltage, which is an origin of the name "pump." The pump operation is also possible around the triple point, "B," but the current direction is opposite to the case around "A." Again, the $R_t C$ time constant should be much smaller than f^{-1} for proper pump operation. For both turnstile and pump, the electrons move adiabatically with the AC voltage.

21.4.1.3 Single-Electron Logic and Memory

21.4.1.3.1 Single-Electron Logic

SETs could be used to realize some logic devices. The single-electron inverter, first proposed theoretically by Tucker (1992), has been demonstrated with Si-SETs (Ono et al. 2000), Al-SETs (Heij et al. 2001), and carbon nanotube SETs (Ishibashi et al. 2003). The complementary inverter is composed of two switches connected in series that work complementally. When one switch is off, the other is on, and vice versa. In the complementary-type single-electron inverter, which is commonly called CMOS-type single-electron inverter, the SET is used as a switch. The Coulomb blockade condition of the SET corresponds to the "OFF" state, and it is in the "ON" state when the Coulomb blockade is lifted. The equivalent circuit of the CMOS-type single-electron inverter is shown in Figure 21.17a. The input is applied to both the SETs through each gate capacitance, and the output is obtained in between the two SETs. The SEM image of the device fabricated in an individual multiwall carbon nanotube (MWNT) is shown in Figure 21.17b. In the case of MWCNT, the efficient tunnel barriers to confine electrons are not formed

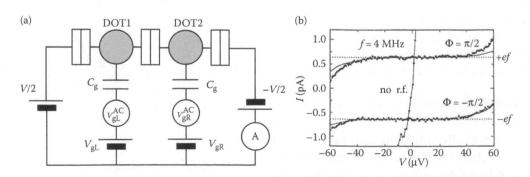

FIGURE 21.16 (a) Equivalent circuit of the current pump. (b) Current–voltage characteristics of the pump operating at 4 MHz and two different phase relations between the two AC gate voltages. (Courtesy of P. Pothier. From Pothier, H. et al. 1992. *Europhys. Lett.* 17: 249.)

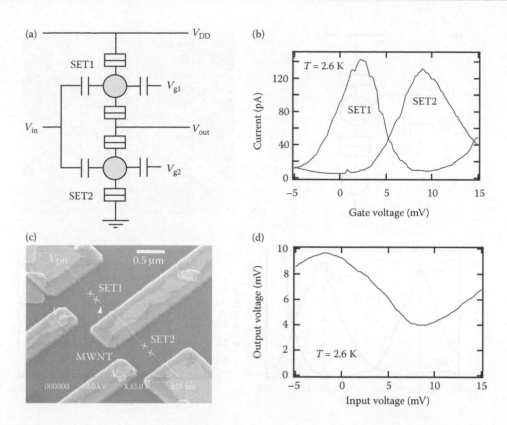

FIGURE 21.17 CMOS-type SET inverter fabricated with MWCNT. (a) Equivalent circuit. (b) SEM image. (c) Adjusted Coulomb peaks of each SET. (d) Transfer curve. (From Ishibashi, K. et al. 2003. *Appl. Phys. Lett.* 82: 3307. With permission.)

simply by depositing metallic contacts on top of it. To overcome the problem, the local ion beam irradiation technique was developed to form the controlled tunnel barriers (Suzuki et al. 2002). It seems that the host carbon atoms are kicked off by the ion beam irradiation to produce damages that may work as the tunnel barriers. The cross marks in Figure 21.17b indicate the tunnel barriers fabricated in this method. The heavily doped substrate was used for the common input gate.

To adjust the Coulomb peaks in each SET in such a way that one is "ON" when the other is "OFF," the peak positions were shifted by the adjacent gates that is located close to each SET (V_{g1}, V_{g2}). The shifted Coulomb peaks for each SET is shown in Figure 21.17c, ready for the inverter operation. The transfer characteristic is shown in Figure 21.17d, and the inverter-like transfer curve is obtained. But, it is not satisfactory in two ways. First, the voltage gain, which is the slope of the curve, is not larger than 1 (0.7 in this case). Second, the voltage swing is not large. These features are due to the small gate voltage gain of the SET, which is a general problem of the SET (Likharev 1987), and the nonideal SET operation. The former problem is solved by optimizing the device design, and in fact, the

gain larger than unity has been realized in the devices fabricated with the Si-SET and Al-SET.

It may not be clear that the logic circuits based on the CMOS-type single-electron inverter has advantages over the present CMOS transistor logic. First, the current level in the single-electron inverter is very small, typically in a range from 1 pA to 1 nA due to the large tunnel resistance to ensure the SET operation. The small current level is disadvantageous for driving the next stages in the cascading connection scheme. The large output resistance makes the RC time constant large when the load capacitor is charged or discharged. The second problem may be more serious: the background charge problem. SETs are very sensitive to any charge that may exist randomly near them, so all the SETs have to be tuned to work as the inverter. This is not practical, and seems to be impossible for integrated circuits that include many SETs. A room-temperature operation of the single SET may not be a problem, and in fact, SETs, fabricated with carbon nanotubes with a special technique to form tunnel barriers, worked at room temperature (Postma et al. 2001; Matsumoto et al. 2003).

The unique characteristic of the SET is a periodic current modulation by the gate voltage. In the CMOS-type

Chapter 21

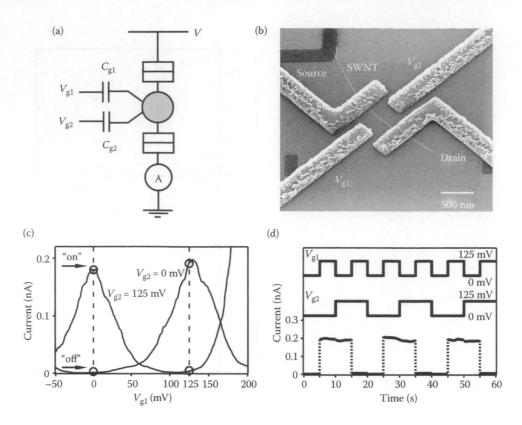

FIGURE 21.18 (a) Equivalent circuit of the SET XOR gate. (b) SEM image of the device fabricated with SWCNT. (c) Explanation of the operation principle of the XOR gate for the device. (c) XOR operation at the liquid He temperature. (From Tsuya, D. et al. 2005. *Appl. Phys. Lett.* 87: 153101. With permission.)

single-electron inverter, the periodic gate voltage dependence of the current is not used. It was pointed out that the XOR logic is simply realized with an SET that has two equivalent gates to the dot, by using the periodic characteristics (Tsuya et al. 2005). This fact may be advantageous over the conventional XOR gate that uses several transistors, although the offset charge problem still remains.

An equivalent circuit of the single-electron XOR gate is shown in Figure 21.18a. The two gate capacitances attached to the dot should have the same value, and the SEM image of the device fabricated with an individual SWCNT is shown in Figure 21.18b (Dresselhaus et al. 1994). Figure 21.18c shows the operation principle of the device. By applying the gate voltage on V_{g2} with a value that is half of the Coulomb oscillation period (=125 mV in this case), the Coulomb peak as a function of V_{g1} is shifted by half of the period, compared with the peak with $V_{g2} = 0$. Two Coulomb peaks as a function of V_{g1} for $V_{g2} = 0$ and 125 mV are shown in Figure 21.18c. When one looks at the figure carefully, one may notice that the XOR operation is realized for the current output. For example, when $V_{g1} = 0$ (LOW) and $V_{g2} = 125$ mV (HIGH), the current

has a peak, meaning that the output is "HIGH." When both inputs are zero ($V_{g1} = V_{g2} = $ LOW), the current does not flow, meaning that the output is "LOW." The XOR operation is demonstrated in Figure 21.18d. The XOR operation with a dual gate SET was also demonstrated with a Si-SET (Tsuya et al. 2005).

21.4.1.3.2 Single-Electron Memory An exact definition of the single-electron memory is not clear, but the configuration in Figure 21.19a is considered to be the simplest form of the single-electron memory (the authors call this a single-electron trap in their original paper; Dresselhaus et al. 1994). This configuration is also called a single-electron box. Electrons are accumulated one by one in the memory node through multiple small tunnel junctions by applying a positive node voltage (V_n). For the device to work as a memory, the number of electrons as a function of V_n should have hysteresis (two values). It is found that the hysteresis is realized when the multijunctions are used for the electron supply. As the number of junctions increases, the node voltage width increases where the number of electrons takes two values, and the trapping time for an electron to stay in the node increases. For the single-electron memory, the "1"

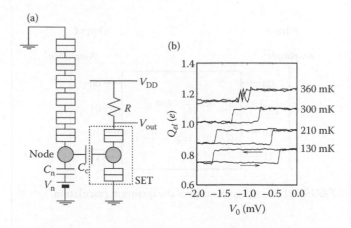

FIGURE 21.19 (a) Equivalent circuit of the single-electron memory (trap). (b) Operation of the device at three different temperatures. The device was fabricated with Al tunnel junctions. (Courtesy of K. K. Likharev. From Dresselhaus, P. D. et al. 1994. *Phys. Rev. Lett.* 72: 3226. With permission.)

state corresponds to the situation where the number of excess electrons in the node is one, and the "0" state corresponds to the value of zero. To read the memory state, which is the number of electrons in the node, the capacitively coupled SET is used, as shown in Figure 21.19a. The SET could be current biased with a large resistance, R, and the voltage could be read as an output.

The single-electron memory or trap was demonstrated with Al junctions, and the main result is shown in Figure 21.19b for different temperatures (Dresselhaus et al. 1994). Because of the large size of the junctions (small E_c), the device worked at low temperatures. As the temperature is increased, the cotunneling process takes place increasingly, leading to a narrower voltage width for the memory. The cotunneling is a higher-order tunneling process, and may occur even though junctions are in the Coulomb blockade condition when temperature is increased. In the experiment in Figure 21.19b, the hysteresis is not observable at $T = 360$ mK, and the charge in the node fluctuates in time between two states. It should be noted that the single-electron trap geometry is a part of the single-electron turnstile seen from the central dot. As mentioned in the turnstile section, the hysteresis behavior ensures the operation of the turnstile.

There is another type of the quantum-dot-based memory, in which the number of electrons in the dots is not always one, and is likely to be many (not known). The basic structure of this type is a small MOS transistor where metallic dots are embedded in the oxide layer. The pinch-off characteristic of the device shows the hysteresis, indicating the memory operation. We do not discuss the device further, and refer to Tiwari et al. (1996) and Huang et al. (2003). It should be noted

that this type of floating dot memories works at higher temperatures because the dot size can be made very small (<10 nm) easily with self-assembled techniques.

21.4.1.4 Quantum Dot Cellular Automata

The problem of the single-electron logic, mentioned above, comes from the small voltage gain of the SET. There are some new logic proposals that do not need the voltage gain. One of these is the quantum dot cellular automata (QCA), proposed by C. Lent (Lent et al. 1993). In the QCA, an elementary unit is composed of four (or five) dots that are coupled both capacitively and quantum mechanically, and so the tunneling of electrons is possible. The four dots are arranged like in Figure 21.20, and two electrons are put in the unit. Because of the Coulomb repulsion between the two electrons, the electrons prefer to stay in two possible diagonal configurations. The two states are defined as "0" and "1," and logic circuits are realized by arranging the units in an appropriate manner. Some logic examples are shown in Figure 21.20. The unique feature of the QCA is that the charges do not move to produce a current, but they change arrangements in each unit. The transition occurs when two states are in the same energy states. Because of these facts, there is a possibility that the QCA may not produce much heat consumption. The experimental demonstration of the simple device was demonstrated with Al junctions (Amlani et al. 1999).

21.4.2 Quantum Computing Devices

21.4.2.1 Basic Concept

Quantum computing takes the advantage of superposition of a number of quantum states (quantum

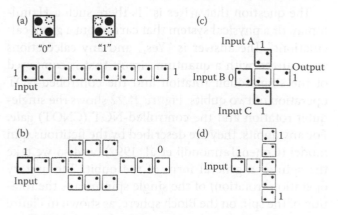

FIGURE 21.20 Quantum dot cellular automata. (a) Transfer. (b) Inverter. (c) Majority gate. (d) Branch. (From Lent, C. S. et al. 1993. *Nanotechnology* 4: 49. With permission.)

bit as defined later), and these quantum states can be manipulated simultaneously as a single wave function based on the time evolution of the Schrodinger equation. It is not our aim to describe the quantum computing (Williams and Clearwater 1997; Nielsen and Chuang 2000; Nakahara and Ohmi 2008), but in this section, we provide the basic concept to understand how it could be realized with solid-state devices.

The basic element of the quantum computing is a qubit, which is nothing more than a quantum two-level system. The qubit state is defined by a superposition of $|0>$ and $|1>$, such that $|\Phi> = a|0> + b|1>$ with $|a|^2 + |b|^2 = 1$. If the number of qubits is n, the number of quantum states that forms the superposition is 2^n (quantum parallelism). This can be a big number that is treated simultaneously, which could make the quantum computer a powerful machine to perform some calculations.

To make the story simple, suppose a two-qubit system, where four states are available, and we can consider the Hilbert space spanned by the four orthogonal states. If the calculation is considered a quantum transition from the INPUT to the OUTPUT in the Hilbert space, the transition could be realized by the unitary transform that can be done by the time evolution of the Schrodinger equation. In Figure 21.21, the situation is depicted, where the input is $|01>$ and the answer (OUTPUT) we want would be $|11>$. We prepare the INPUT state such that $|\Phi> = |01>$, and let the system evolve in time, and after some period of time, $|\Phi>$ would be, in general, the superposition of the four states.

$$|\Phi(t_1)> = C_1(t_1)|00> + C_2(t_1)|01> + C_3(t_1)|10> + C_4(t_1)|11>$$

If the answer we want is $|11>$, the ideal superposition is $|C_4(t_1)|^2 = 1$ and $|C_1(t_1)|^2 = |C_2(t_1)|^2 = |C_3(t_1)|^2 = 0$, because we would have a right answer when we measure the OUTPUT state.

The question that arises is "Is there such a Hamiltonian or a physical system that carries out a given calculation?" The answer is "Yes," and any calculations are realized with a quantum circuit that is composed of the single-qubit rotation and the controlled-NOT operation of two qubits. Figure 21.22 shows the single-qubit rotation and the controlled-NOT (CNOT) gate. For any qubits, they are described by the fictitious spin model (Cohen-Tannoudji et al. 1992), and so we take the spin as a general form of the qubit. The unitary operation (rotation) of the single spin means the rotation of the spin on the Bloch sphere, as shown in Figure 21.22a. The coherent rotation of the spin could be done by using the Lamor precession under the static magnetic field. The magnetic field pulses with two

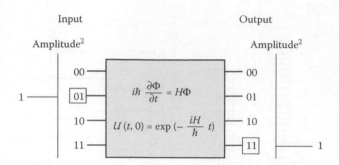

FIGURE 21.21 Basic concept of the quantum calculation.

directions, x and z, are necessary to produce the arbitrary spin direction that is defined by θ and ϕ.

The other necessary gate is the CNOT gate, as shown in Figure 21.22b. It is composed of the control bit and the target bit, and when the control bit is $|1>$, the target bit is flipped. The operation appears to be similar to the classical XOR operation. But, there are two main differences between the CNOT gate and the classical XOR. First, the CNOT gate has two inputs and two outputs to ensure the reversibility in the quantum mechanics, while the classical XOR has two inputs and one output. Second, the CNOT gate can create entanglement states by inputting the superposition state in the controlled bit. For example, when the controlled bit is $a|0> + b|1>$ and the target bit is $|0>$ for the input, the output state is $a|00> + b|11>$, an entangled state.

To illustrate the physical image of the CNOT gate, one composed of the coupled quantum dots is shown in Figure 21.23 (Ishibashi et al. 2002). In this case, the qubit is made of weakly coupled quantum dots, which have a confined state. To discriminate each qubit, the resonant energy could be tuned by a gate voltage that could be located near each qubit. The CNOT gate is realized by facing two qubits that are capacitively coupled, as shown in Figure 21.23a. Each bit may be

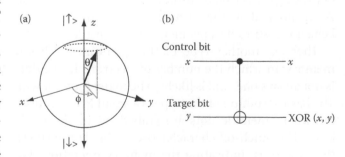

FIGURE 21.22 (a) Geometrical expression of the one-qubit operation. (b) Circuit diagram of the CNOT gate.

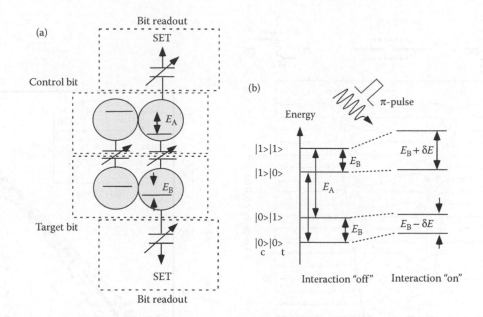

FIGURE 21.23 (a) Schematic picture of the CNOT gate with coupled quantum dots. The qubit is composed of the coupled quantum dot. (b) Energy diagram of the CNT gate and its operation principle. (From Ishibashi, K. et al. 2002. *Superlattices Microstruct.* 31: 141–149. With permission.)

manipulated for the single qubit rotation, in which case the capacitive qubit coupling has to be disconnected. For the two-qubit manipulation, like the CNOT, the capacitive coupling has to be switched on. The energy diagram of the two-qubit system is shown in Figure 21.23b. Because of the capacitive coupling between the two qubits, an electron in each qubit tends to have a diagonal arrangement. The two-qubit energy diagram when the coupling is switched on differs from the one when the coupling is switched off. By applying the microwave pulse with the frequency on resonance to |10> and |11>, the CNOT gate operation is realized. Note that the two levels |00> and the |01> do not respond, because they are in off-resonance. The qubit readout could be done with the SET. In the experiments to read the charge, the QPC set in the tunneling regime has been demonstrated (Vandersypen et al. 2004), as shown later. The experimental effort to demonstrate this type of the charge qubit is discussed later.

The quantum computing described in this section is simply about a basic principle, but for the practical implementation with solid-state devices, the most important problem would be to maintain the quantum coherence. Especially in solid-state qubits, there are many decoherence sources, which are the coupling of the quantum system to the environment. It is this problem that has to be overcome for the realization of the quantum computing. The decoherence sources are in many cases not clear, and should be considered in each case.

21.4.2.2 Quantum-Dot-Based Qubits

21.4.2.2.1 Charge Qubit
The coupled quantum dots with a single electron in the system can be seen as a charge qubit. The quantum state of the electron is a superposition of the electron being in the left dot and in the right dot. It is not obvious that the superposition state, or the molecular state, is really formed because there could be many decoherence sources. To see the coherent coupling between the two quantum states, the microwave spectroscopy has been carried out in the coupled quantum dots fabricated in the surface-gated GaAs/AlGaAs 2DEG (Oosterkamp et al. 1998). Figure 21.24a shows the microwave photon-assisted peaks as a function of an energy difference between the two states. It should be noted that the measurements were carried out with $V_{sd} \sim 0$. In Figure 21.24b, a distance between the two peaks is plotted as a function of applied frequency. The bending from the linear relation indicates the coherent coupling between the two states (formation of the molecular states). The main decoherence source would be the coupling to the electrodes that works as a reservoir, and so the coupling was made as small as possible. The Rabi (coherent) oscillations have been observed, as shown in Figure 21.25 (Hayashi et al. 2003).

The basic measurements to observe the Rabi (coherent) oscillations are to apply the pulsed high-frequency field that is on resonance to the energy of the two-level system, or to align two levels for a period of time, and

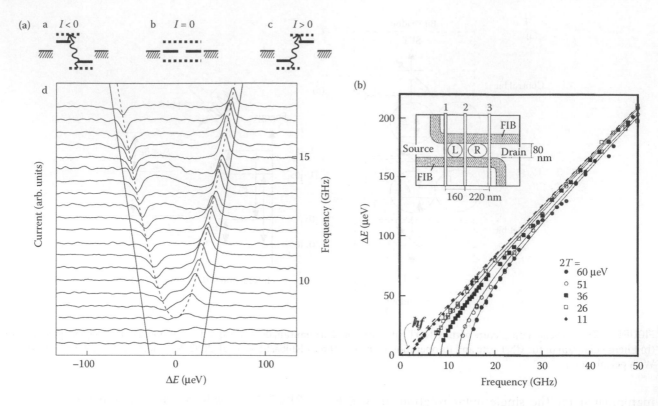

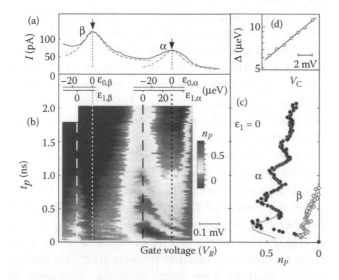

FIGURE 21.24 Coherent coupling of the coupled quantum dots in the GaAs/AlGaAs 2DEG. (a) Microwave photon-assisted peaks. (b) Energy spectroscopy results for different coupling strengths. (Courtesy of L. Kouvenhoven. From Oosterkamp, T. H. et al. 1998. *Nature* 395: 873–876. With permission.)

simultaneously to measure the steady-state current (repeated measurement), as schematically shown in Figure 21.26. In this measurement scheme, an electron

FIGURE 21.25 Rabi oscillations in the coupled quantum dots. (a) Resonant peaks. (b) Color scale plot of the oscillations in time domain with changing gate voltage. (c) Coupling strength as a function of the coupling gate. (d) Oscillations of the two peaks, α and β. (Courtesy of T. Fujisawa. From Hayashi, T. et al. 2003. *Phys. Rev. Lett.* 91: 226804. With permission.)

contributes to the current maximally when it finishes the oscillations at the right dot, but it contributes minimally when it finishes the oscillations at the left dot. The readout may not be correct when the electron that finishes the oscillations at the right dot tunnels to the left instead of tunneling into the drain electrodes. For this process not to occur, the central barrier may have to be thicker than the barrier between the right dot and the drain, but it may produce decoherence during the oscillations when the electron tunnels out of the system to the drain. When the current is measured, repeated measurement is necessary to obtain the measurable current. In contrast, the single-shot measurement is possible, where the single measurement is done after the manipulation. The single-shot readout of the spin is described in the next section. Recently, the conditional Rabi (coherent) oscillations have been observed in the two coupled-dot geometry (Shinkai et al. 2009).

21.4.2.2.2 Spin Qubit The electron spin is a two-level system that automatically forms a qubit. The spin qubit could be realized by putting absolutely one electron in the dot, or an unpaired electron in the shell. The former can be done in the quantum dots fabricated in the GaAs/AlGaAs 2DEG (Koppens et al. 2006), and the latter could

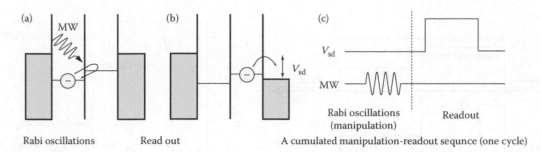

FIGURE 21.26 (a, b) Operation and readout principle of the Rabi oscillations in coupled quantum dots. (c) Sequence of V_{sd} and microwave irradiation.

be done in the carbon nanotube quantum dots (Moriyama et al. 2005). In the latter, the preparation and the initialization of the single qubit have been realized, but the manipulation has not been realized. In the former, the ESR (electron spin resonance)-type manipulation has not been realized. But, instead, the electrical manipulation has been demonstrated with the help of the spin–orbit interaction (Nowack et al. 2007). In compound semiconductors, the spin–orbit interaction is large, which means that the spin is not a good quantum state. This fact as well as the interaction with nuclei of the host material (Ga and As) limits the coherence time of the spin. In this aspect, Si, carbon nanotube, and graphene quantum dots may have a longer coherence time. To access the single spin is not easy experimentally, and so the technique to use the spin–orbit interaction as an electrical access tool may be an advantage for the material.

In the previous measurement, the repeated measurement was used for the Rabi oscillation measurement (Moriyama et al. 2005). But the single-shot measurement of the single spin has also been demonstrated, as shown in Figure 21.27 (Elzerman et al. 2004). The sample is a surface-gated single quantum dot to which the QPC is connected for the charge sensor. The idea is the conversion from the spin to the charge that can be

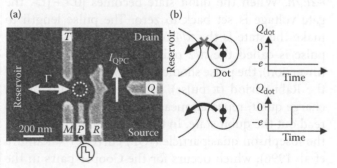

FIGURE 21.27 (a) SEM image of the single quantum dot for the spin readout. (b) Spin-dependent electron transfer to the reservoir. Spin-charge conversion. (Courtesy of L. Vandersypen. From Koppens, F. H. L. et al. 2006. *Nature* 442: 766. With permission.)

measured by the QPC (Figure 21.27b). As shown in Figure 21.28a, the pulse gate voltage is applied such that the Zeeman-splitted two levels ($E_\uparrow$, $E_\downarrow$) go below the Fermi level in the electrode. After some period of time (inject and wait time in Figure 21.28a), an electron may be injected either in $E_\uparrow$ or in $E_\downarrow$ in the dot. Then, the gate voltage is set such that the Fermi level in the electrode is located in between $E_\uparrow$ and $E_\downarrow$. When the electron has been injected in $E_\uparrow$, it can tunnel out to the electrode, while it would stay there when it is injected in $E_\downarrow$ (Figure 21.28b). This means that the number of charges in the dot changes, depending on the spin direction. The corresponding current through the QPC may behave, as shown in Figure 21.28a, and in fact, this behavior has been observed in the experiments in Figure 21.28c.

21.4.2.3 Superconducting Qubits

Basically, three types of superconducting qubits are considered. An excellent review on these is found in Devoret et al. (2004). The three types are summarized in Table 21.2. A superconducting system is in itself coherent, so the coherence might be realized in a macroscale (macroscopic quantum coherence, MQC). In fact, in the Flux and Phase qubit, the relatively large currents of the order of μA are flowing, and they can be superpositioned. This means that access to the qubits is much easier than those of microscopic states (ex. spin qubit), but at the same time, they are more sensitive to decoherence due to fluctuations of voltages and magnetic fields. The qubits are categorized based on the relation between E_c (charging energy for a single Cooper pair) and E_J (Josephson energy). Most of the devices are usually fabricated by the shadow evaporation technique with Al.

21.4.2.3.1 Charge (Cooper Pair) Qubit
The coherent oscillations of the charge (Cooper pair) qubit in Table 21.2 was first demonstrated (Nakamura et al. 1999). In this case, the qubit is composed of an excess number of Cooper pairs in the single-electron box (|0>

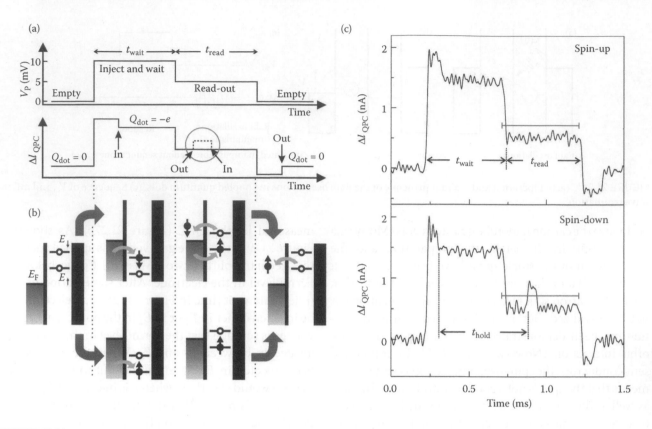

FIGURE 21.28 Single spin readout. (a) Pulse sequence of the central gate and the corresponding charge in the dot. (b) Schematic picture of the electron spin in the dot as a function of time. (c) Experimental observation that corresponds to (a). (Courtesy of L. Vandersypen. From Koppens, F. H. L. et al. 2006. *Nature* 442: 766. With permission.)

or $|1>$) and they couple by means of the Josephson coupling. For the charge state to be used, the charging energy for single Cooper pair (E_c) should be larger than the Josephson energy (E_J) ($E_c \gg E_J$). The qubit state is controlled by the gate voltage. The coupling effect is largest at the "sweet spot" where the two levels meet ($C_gV_g/e = 1$). The energy spectrum of the qubit in the table was measured with the microwave spectroscopy

Table 21.2   Three Types of Superconducting Qubits

	Charge	Flux	Phase
Basic element	C_g V_g	Φ_{ext}	I_b
Energy or potential	E 0 1 2 C_gV_g/e	E -1 0 1 Φ/Φ_0	E -1 -0.5 0 0.5 $\theta/2\pi$

technique (Nakamura et al. 1997), as was used for the coupled quantum dots to demonstrate the molecular state. To manipulate the qubit, the pulse voltage was applied to the gate. For example, to make the superposition state of $|0> + |1>$ from the initial state, $|0>$ ($C_gV_g/e = 0$), the gate pulse gate voltage is applied to set $C_gV_g/e = 1$, during which time the coherent oscillations (Rabi oscillations) occurs between $|0>$ and $|1>$. The oscillation frequency (Rabi frequency) is determined $\sim 2E_J/h$. When the qubit state becomes $|0> + |1>$, the gate voltage is set back to zero. The pulse length to make the state is the $\pi/4$ of the Rabi period, and the pulse is called the $\pi/2$ pulse. To make $|1>$ from $|0>$ (inversion), the pulse should be given during the half of the Rabi period (π pulse). In the experiments of the charge qubit, repeated measurement was employed to read out the qubit state. In this particular experiment, the Josephson quasiparticle (JQP) current (Nakamura et al. 1996), which occurs for the Cooper pairs in the dot to tunnel out as two quasiparticles, was used for the readout of the qubit. This JQP process is not controlled, and in that sense, the measurement itself is a cause of decoherence.

The other device for quantum computing, the CNOT gate, has been demonstrated with the use of capacitive coupling between the qubits (Yamamoto et al. 2003). When the first qubit was demonstrated with the charge (Cooper pair) qubit, it was found that the coherent oscillations did not last so long as one first assumed. The disadvantage of the charge qubit is that it is very sensitive to any voltage fluctuations, and so the coherence may not be long. A modified version of the Cooper pair qubit has been demonstrated with a longer coherence time (Quantronium) (Vion et al. 2002).

21.4.2.3.2 Flux Qubit

The second qubit in the table is the Flux qubit. A basic structure of the Flux qubit is a superconducting loop in which a Josephson junction is embedded (RF-SQUID). In this case, the current that flows in clockwise and counterclockwise directions can form the qubit. The relation $E_J > E_c$ is met for the qubit. The qubit potential is a superposition of the quadratic component that comes from the inductive energy and the Josephson energy with a cosine shape. It should be noted that one of the important roles of the Josephson junction is to induce the nonlinearity to the harmonic potential, and so the level spacing is no longer equal, and the qubit two levels are isolated from excited states. The overall potential landscape is controlled by the magnetic field applied to the loop by the current that couples to the loop through mutual inductance. The potential in Table 21.2 shows a case for the "sweet spot" of the qubit. The discrete energy levels and a coherent superposition between them have been demonstrated in the RF-SQUID geometry (Friedman et al. 2000).

More feasible qubit with three Josephson junctions has been proposed by Mooij et al. (1999), as shown in Figure 21.29, and experimentally demonstrated (van der Wal et al. 2000). In this particular Flux qubit, its state is read out by the SQUID that is connected with the qubit loop. In the other geometry, the qubit can be surrounded by the SQUID loop. To read out the qubit state, the bias current of the SQUID is set just below the switching current (I_{sw}), and depending on the qubit states that correspond to the current flowing in the clockwise and counterclockwise direction, the SQUID switches to the voltage state that is measured. To manipulate the qubit, the external pulse magnetic field (Φ_{ext}) is applied with a current that flows around the loop, and the manipulation can be done with a pulse sequence basically in a similar way with that in the charge qubit. The readout is done by the single-shot measurement, meaning that qubit readout is done just

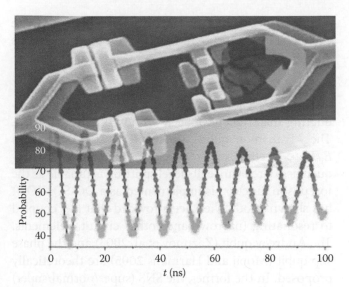

FIGURE 21.29 (See color insert.) SEM image of the flux qubit with SQUID for readout and Rabi oscillations. (Courtesy of J. E. Mooij.)

after the qubit manipulation is finished. The example of the demonstrated Rabi oscillations is shown in Figure 21.29. Other one-qubit manipulations such as the Ramsey interference and the spin echo have also been demonstrated (Chiorescu et al. 2003), and the CNOT operation has been demonstrated (Plantenberg et al. 2007).

There are advantages of the flux qubit over other qubits that make use of the microscopic states. First, the access to the qubit is easy because the qubit is composed of relatively large current and the size of the qubit is relatively large. Second, the flux qubit is considered to maintain more coherence than the charge qubit because magnetic noise is much smaller than electric noise. But the first advantage could also be a disadvantage.

21.4.2.3.3 Phase Qubit

A current-biased large Josephson junction is modeled with the resistively shunted junction (RSJ) model (Tinkham 2004). The model makes it possible to consider the system as a particle in the tilted washboard potential. The qubit is formed with the ground state and the first excited state of the potential well. The second excited state could be used for the readout. For more flexibility, the third excited state could be used for the readout (Martinis et al. 2002). In this case, $E_J \gg E_c$ is usually satisfied. The manipulation of the qubit is performed with the microwave pulse, the frequency of which is on resonance to the qubit energy. The readout is done after the manipulation by increasing the current, so the particle could tunnel out of the potential when it is in the excited state, resulting in the junction to switch to the voltage state.

21.4.2.3.4 Alternative Qubits and Quantum Circuits

Over the past few years, other qubits with superconducting materials have been proposed and developed. The "Transmon qubit" (Koch et al. 2007), although $E_J/E_C \gg 1$, effectively acts as a charge qubit for the frequencies of relevance. Its level splitting is tunable as the Josephson junction is implemented as a DC-SQUID. It has shown good coherence, provided that it is coupled to resonating (narrow band) other circuital elements. The Andreev qubit (Zazunov et al. 2003) and the phase slip qubit (Mooij and Harmans 2005) are theoretically proposed. In the former, the SNS (super/normal/super) structure with a normal material being a QPC is embedded in the RF-SQUID geometry. In the latter, a very narrow superconducting wire that allows superconducting fluctuations is embedded in the RF-SQUID geometry. Since the qubit loop is made with the same material, it may not suffer from charge fluctuations in the insulating layer of the Josephson junction.

In the superconducting qubit system, a number of sophisticated experiments have been demonstrated with the analogy of the quantum optics. A central aspect that has emerged is that, in order to optimally benefit from the intrinsic coherence of (superconducting) qubits, these quantum elements need to be mounted in a spectral environment that is strongly nonhomogeneous (Wallraff et al. 2004). By only allowing the environment to couple to the qubit at very well defined and restricted frequencies, the detrimental effects of wide-band noise can be suppressed, and in this way coherence could be increased. This leads to the widespread introduction of resonators or cavities, into/onto which qubits are attached. This area, recently called the cavity-QED and the circuit-QED, which also allows very strong qubit–cavity interaction, is shown to be extremely fertile. It is very interesting, but may be out of the scope of the book.

21.4.2.4 Other Qubits

There are other proposals and experiments on qubits. Among them, an interesting and unique proposal is by Kane (1998). In his proposal, schematically shown in Figure 21.30, the qubit is composed of the nuclear spin of a P atom embedded in the Si-MOSFET (metal oxide semiconductor field effect transistor). The nuclear spins are expected to have a long coherence because they do not interact with decoherence sources

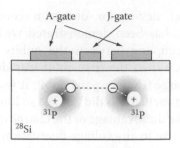

FIGURE 21.30 Schematic picture of the nuclear spin qubit. (From Kane, B. E. 1998. *Nature* **393**: 133.)

of environment. This means, again, that it is difficult to access them. Besides, an interaction between different nuclear spins, which is required for the CNOT gate, is very weak.

In his proposal, the manipulation of the nuclear spin qubit is performed by the RF field, and the original idea to realize the interaction between two qubit is to use an electron supplied from the donor ion. This electron-mediated nuclear spin interaction is switched on and off by using the J-gate. The readout of the qubit is by reading out the electron spin state, which may be basically possible with the spin-charge transformation. The A-gate is used to separate the qubit from the electron.

The experimental realization of the device is obviously not easy, but it is attractive also in terms of the compatibility with the present sophisticated Si technology. The nanofabrication to fabricate the qubit needs single-atom manipulations. The single-atom implantation would be a useful technique (Schenkel et al. 2003). The use of the scanning probe microscope to manipulate an atom would also be useful (Strocio and Eigler 1991). The effort to realize the device is going on (Rueβ et al. 2007).

21.4.3 Carbon Nanotube FET

Semiconducting single-walled carbon nanotubes are expected to be used for a channel of the field effect semiconductor (FET). The research for this direction could be considered as one of the possible solution for "More Moore." The operation principle of the CNTFET is different from the conventional Si MOSFETs. The simplest understanding of the operation mechanism of the CNTFET is the Schottky barrier transistor, where the Schottky barriers are formed at the source and drain electrodes. In this case, the carrier type is determined by work function difference between the metal and the CNT. The injection of the carriers occurs as shown in the top panel of Figure 21.31, either to the

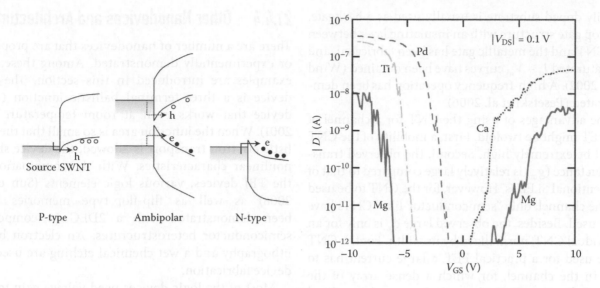

FIGURE 21.31   (a) Schematic band diagram of the metal–nanotube junction for the p-type carrier injection, ambipolar injection, and n-type injection. (b) Experimental observation of the gate voltage dependence of the current for various metals. (Courtesy of Nosho, Y. et al. 2006. *Nanotechnology* 17: 3412. With permission.)

conduction band for electrons or to the valence band for holes. This basically explains whether the intrinsic semiconducting CNT behaves as a p-type or n-type. A unique characteristic of the CNTFET is the ambipolar behavior where the carrier type changes between the n-type and the p-type, depending on the gate voltage range (negative or positive), with the pinch-off region in between. The basic behaviors of the carrier type of the semiconducting FET are demonstrated in the lower panel of Figure 21.31, where the different transfer characteristics (gate voltage dependence of the current) are shown for FETs with different metals (Nosho et al.

2006). In practice, the semiconducting CNT usually shows a p-type behavior when it is measured in air. The reason for this is believed to be oxygen molecules that are absorbed on the CNT surface. The controlled doping can also determine the carrier type.

The CMOS inverter has been realized with the p-type region and the n-type region in the same SWCNT, as shown in Figure 21.32, and the transfer characteristic with a larger voltage gain has been obtained (Derycke et al. 2001). With the arrays of inverters, a ring oscillator has been demonstrated (Chen et al. 2006). In most experimental CNTFETs, a

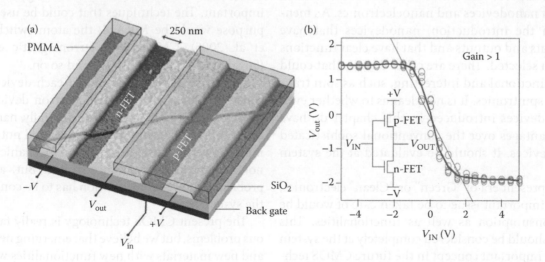

FIGURE 21.32   (a) AFM image of the CMOS inverter fabricated from SWCNT. (b) Transfer curve. (Courtesy of Derycke, V. et al. 2001. *Nano Lett.* 1: 453. With permission.)

Chapter 21

heavily doped substrate is usually used as a backgate. The top gate structure with an insulating layer between the CNT and the metallic gate has been fabricated, and the saturated $I_d - V_{sd}$ curves have been obtained (Wind et al. 2002). A high-frequency operation has been demonstrated (Pesetski et al. 2006).

The advantages of using the CNT for a channel of the FET might be twofold. First, a mobility of the CNT could be extremely high. Second, the observed transconductance (g_m) is relatively large compared to that of conventional Si FETs. However, for the CNT to be used for the channel, only "semiconductor-like" CNTs have to be used. Besides, the observed large g_m is only for an individual CNT, normalized by its width. For the CNT to be used for a practical FET, a large current has to flow in the channel, for which a dense array of the semiconducting CNT channels have to be realized (Cao and Rogers 2009).

More practical application of the CNTs would be a channel for flexible transistors. In this case, the CNT mat is used for the FET channel, and by adjusting an appropriate distance between the source and drain electrodes, current paths only connected by metallic CNTs rarely exist, which prevent unwanted shortage with metallic CNTs. In this application, the large on/off ratio is not required, and the mobility of the CNT mat is usually larger than that of organic semiconductors (Xiao et al. 2003).

21.4.4 Other Nanodevices and Architectures

There are a number of nanodevices that are proposed or experimentally demonstrated. Among these, two examples are introduced in this section. The first device is a three-terminal ballistic junction (TBJ) device that works even at room temperature (Xu 2001). When the junction area is so small that the ballistic electron transport is allowed, the device shows nonlinear characteristics. With the combination of the TBJ devices, various logic elements (Sun et al. 2008) as well as flip-flop-type memories have been demonstrated with a 2DEG in compound semiconductor heterostructures. An electron beam lithography and a wet chemical etching are used for device fabrication.

Most of the logic devices need voltage gain to use them for practical circuits. But, in the binary decision diagram (BDD) logic, the voltage gain is not needed, and the logic operation is carried out through the current path specific to the logic. Basically, the SET, which has a small voltage gain, could be used for the logic (Asahi et al. 1997). It may realize a very low-power-consuming logic. The BDD logic has been demonstrated with Schottky lap-gate FET with a compound semiconductor 2DEG (Hasegawa and Kasai 2001). By selecting the final destination of the current path, various logic gates are shown to be realized.

21.5 Conclusions

In this chapter, some nanoelectronic devices have been introduced. They however do not cover all aspects of nanodevices and nanoelectronics. As mentioned in the introduction, nanodevices that have clear inputs and outputs and that have clear functions have been selected. There are other devices that could be very functional and interesting, such as spin transistors or spintronics. It is not clear as to whether even the nanodevices introduced in this chapter do have clear advantages over the conventional sophisticated CMOS devices. It should be evaluated at the system level.

In the present era of "Green" or "Clean" electronics, the most important issue to be taken care of would be power consumption as well as functionalities. This problem should be considered completely at the system level. The important concept in the future CMOS technology would be that devices that do not need to operate should be switched off to reduce stand-by current.

To control the devices that should work at the given time, the nonvolatile memory technology may be very important. The techniques that could be used for this purpose would be MRAM, the atom switch (Terabe et al. 2005), CMOS with ferromagnetic electrodes (Sugahara and Tanaka 2005), and so on.

Of course, it would be nice if each device did not consume much power. Single-electron devices would consume less power, and more essentially, nanodevices based on the quantum mechanics would not consume much power because the quantum mechanics itself do not include dissipation mechanism. But, again, the problem of power consumption has to be considered at the system level.

The present CMOS technology is really facing serious problems, but we believe that emerging new devices and new materials with new functionalities would play an important role to solve the problems, based on the established Si technology.

Acknowledgment

The author would like to thank Professor C. J. P. M. Harmans of Delft University of Technology for reading the manuscript and providing critical comments on the superconducting qubit section.

References

Ago, H., Nakamura, K., Ikeda, K., Uehara, N., Ishigami, N., and Tsuji, M. 2005. Aligned growth of isolated single-walled carbon nanotubes programmed by atomic arrangement of substrate surface. *Chem. Phys. Lett.* **408**: 433.

Amlani, I., Orlov, A. O., Toth, G., Bernstein, G. H., Lent, C. S., and Gregory L. 1999. Digital logic gate using quantum-dot cellular automata. *Science* **284**: 289.

Arakawa, Y. and Sakaki, H. 1982. Multidimensional quantum well laser and temperature dependence of its threshold current. *Appl. Phys. Lett.* **40**: 939.

Areshkin, D. A., Gunlycke, D., and White, C. T. 2007. Ballistic transport in graphene nanostrips in the presence of disorder: Importance of edge effects. *Nano Lett.* **7**: 204.

Asahi, N., Akazawa, M., and Amemiya, Y. 1997. Single-electron logic device based on the binary decision diagram. *IEEE Trans. Electron Devices* **44**: 1109.

Averin, D. V. and Likharev, K. K. 1991. Single electronics: A correlated transfer of single electrons and copper pairs in systems of small tunnel junctions. In *Mesoscopic Phenomena in Solids*, (eds.) Altshuler, B., Lee, P. A., and Webb, R. A. Amsterdam: Elsevier Science Publishers, Chapter 6.

Beenakker, C. W. J. and van Houten, H. 1991. Quantum transport in semiconductor nanostructures. *Solid State Phys.* **44**: 1.

Blick, R. H., Pfannkuche, D., Haug, R.J., von Klitzing, K., and Eberl, K. 1998. Formation of a coherent mode in a double quantum dot. *Phys. Rev. Lett.* **80**(18): 4032–4035.

Cao, Q. and Rogers, J. A. 2009. Ultrathin films of single-walled carbon nanotubes for electronics and sensors: A review of fundamental and applied aspects. *Adv. Mater.* **21**: 29.

Chen, Z., Appenzeller, J., Lin, Y. et al. 2006. An integrated logic circuit assembled on a single carbon nanotube. *Science* **311**: 1735.

Chen, F., Qing, Q., Ren, L., Wu, Z., and Liu, Z. 2005. Electrochemical approach for fabricating nanogap electrodes with well controllable separation. *Appl. Phys. Lett.* **86**: 123105.

Chiorescu, I., Nakamura, Y., Harmans, C. J. P. M., and Mooij, J. E. 2003. Coherent quantum dynamics of a superconducting flux qubit. *Science* **299**: 1869.

Ciorga, M., Sachrajda, A. S., Hawrylak, P. et al. 2000. Addition spectrum of a lateral dot from Coulomb and spin-blockade spectroscopy. *Phys. Rev. B* **61**: R16315.

Cohen-Tannoudji, C., Diu, B., and Laloe, F. 1992. *Quantum Mechanics*. Weinheim: Wiley Interscience.

Derycke, V., Martel, R., Appenzeller, J., and Avouris, Ph. 2001. Carbon nanotube inter- and intramolecular logic gates. *Nano Lett.* **1**: 453.

Devoret, M. H., Wallraff, A., and Martinis, J. M. 2004. Superconducting qubits: A short review. Arxiv preprint cond-mat/0411174. Available at http://arxiv.org.

Dick, K. A., Deppert, K., Mårtensson, T., Mandl, B., Samuelson, L., and Seifert, W. 2005. Failure of the vapor–liquid–solid mechanism in Au-assisted MOVPE growth of InAs nanowires. *Nano Lett.* **5**: 761.

Dresselhaus, P. D., Ji, L., Han, S., Lukens, J. E., and Likharev, K. K. 1994. Measurement of single electron lifetimes in a multijunction trap. *Phys. Rev. Lett.* **72**: 3226.

Elzerman, J. M., Hanson, R., van Beveren, L. H. W., Witkamp, B., Vandersypen, L. M. K., and Kouwenhoven, L. P. 2004. Single-shot read-out of an individual electron spin in a quantum dot. *Nature* **430**: 431.

Esaki, L. and Tsu, R. 1970. Superlattice and negative differential conductivity in semiconductors. *IBM J. Res. Dev.* **14**: 61.

Friedman, J. R., Patel, V., Chen, W., Tolpygo, S. K., and Lukens, J. E. 2000. Quantum superposition of distinct macroscopic states. *Nature* **406**: 43.

Fujisawa, T., Hayashi, T., Hirayama, Y., Cheong, H. D., and Jeong, Y. H. 2004. Electron counting of single-electron tunneling current. *Appl. Phys. Lett.* **84**: 2343.

Fukata, N. 2009. Impurity doping in silicon nanowires. *Adv. Mater.* **21**: 2829.

Fukidome, H., Miyamoto, Y., Handa, H., Saito, E., and Suemitsu, M. 2010. Epitaxial growth processes of graphene on silicon substrates. *Jpn. J. Appl. Phys.* **49**: 01AH03.

Fulton, T. A. and Dolan, G. J. 1987. Observation of single-electron charging effects in small tunnel junctions. *Phys. Rev. Lett.* **59**: 109.

Fuse, T., Kawano, Y., Yamaguchi, T., Aoyagi, Y., and Ishibashi, K. 2007. http://iopscience.iop.org/0957-4484/18/4/044001, Quantum response of carbon nanotube quantum dots to terahertz wave irradiation. *Nanotechnology* **18**: 044001.

Geerligs, L. J., Anderegg, V. F., Holweg, P. A. M. et al. 1990. Frequency-locked turnstile device for single electrons. *Phys. Rev. Lett.* **64**: 2691.

Geim, A. K. 2009. Graphene: Status and prospects. *Science* **324**: 1530.

Gorman, J., Hasko, D. G., and Williams, D. A. 2005. Charge-qubit operation of an isolated double quantum dot. *Phys. Rev. Lett.* **95**: 090502.

Han, M. Y., Özyilmaz, B., Zhang, Y., and Kim, P. 2007. Energy band-gap engineering of graphene nanoribbons. *Phys. Rev. Lett.* **98**: 206805.

Hasegawa, H. and Kasai, S. 2001. Hexagonal binary decision diagram quantum logic circuits using Schottky in-plane and wrap-gate control of GaAs and InGaAs nanowires. *Physica E* **11**: 149–154.

Hatzor, A. and Weiss, P. S. 2001. Molecular rulers for scaling down nanostructures. *Science* **291**:1019.

Hayashi, T., Fujisawa, T., Cheong, H. D., Jeong, Y. H., and Hirayama, Y. 2003. Coherent manipulation of electronic states in a double quantum dot. *Phys. Rev. Lett.* **91**: 226804.

Heij, C. P., Hadley, P., and Mooij, J. E. 2001. Single-electron inverter. *Appl. Phys. Lett.* **78**: 1140.

Huang, S., Banerjee, S., Tung, R. T., and Oda, S. 2003. Electron trapping, storing, and emission in nanocrystalline Si dots by capacitance–voltage and conductance–voltage measurements. *J. Appl. Phys.* **93**: 576.

Iijima, S. and Ichihashi, T. 1993. Single-shell carbon nanotubes of 1-nm diameter. *Nature* **363**: 603.

Ishibashi, K., Ida, T., Kotani, H., Ochiai, Y., Sugano, T., and Aoyagi, Y. 1999. Characterization of SET electrometer coupled to the quantum dot in GaAs/AlGaAs 2DEG. *Microelectron. Eng.* **47**: 185–187.

Ishibashi, K., Suzuki, M., Ida, T., Tsukagoshi, K., and Aoyagi, Y. 2000. Quantum dots in carbon nanotubes. *Jpn. J. Appl. Phys.* **39**: 7053.

Ishibashi, K., Suzuki, M., Moriyama, S., Ida, T., and Aoyagi, Y. 2002. Single and coupled quantum dots in single-wall carbon nanotubes. *Superlattices Microstruct.* **31**: 141–149.

Ishibashi, K., Tsuya, D., Suzuki, M., and Aoyagi, Y. 2003. Fabrication of a single-electron inverter in multiwall carbon nanotubes. *Appl. Phys. Lett.* **82**: 3307.

Kamimura, T. and Matsumoto, K. 2004. Reduction of hysteresis characteristics in carbon nanotube field-effect transistors by refining process. *IEICE Trans. Electron. (Inst. Electron. Inf. Commun. Eng.)* **E87-C**: 1795.

Kane, B. E. 1998. A silicon-based nuclear spin quantum computer. *Nature* **393**: 133.

Kawano, Y., Fuse, T., Toyokawa, S., Uchida, T., and Ishibashi, K. 2008. Terahertz photon-assisted tunneling in carbon nanotube quantum dots. *J. Appl. Phys.* **103**: 034307-1–034307-4.

Koch, J., Yu, T. M., Gambetta, J. et al. 2007. Charge-insensitive qubit design derived from the Cooper pair box. *Phys. Rev. A* **76**: 042319.

Koppens, F. H. L., Buizert, C., Tielrooij, K. J. et al. 2006. Driven coherent oscillations of a single electron spin in a quantum dot. *Nature* **442**: 766.

Kouwenhoven, L. P., Johnson, A. T., van der Vaart, N. C., Harmans, C. J. P. M., and Foxon, C. T. 1991. Quantized current in a quantum-dot turnstile using oscillating tunnel barriers. *Phys. Rev. Lett.* **67**: 1626–1629.

Lauhon, L., Gudiksen, M. S., Wang, D., and Lieber, C. M. 2002. Epitaxial core-shell and core-multishell nanowire heterostructures. *Nature.* **420**, 57.

Lent. C. S., Tougaw, P. D., Porod, W., and Bernstein G. H. 1993. Quantum cellular automata. *Nanotechnology* **4**: 49.

Liang, W., Bockrath, M., and Park, H. 2002. Shell filling and exchange coupling in metallic single-walled carbon nanotubes. *Phys. Rev. Lett.* **88**: 126801.

Likharev, K. K. 1987. Single-electron transistors: Electrostatic analogs of the DC SQUIDS. *IEEE Trans. Magn.* **23**: 1142.

Liu, S., Tok, J. B.-H., and Bao, Z. 2005. Nanowire lithography: Fabricating controllable electrode gaps using Au–Ag–Au nanowires. *Nano Lett.*, 5 (6), 1071–1076.

Manoharan, M., Oda, S., and Mizuta, H. 2008. Impact of channel constrictions on the formation of multiple tunnel junctions in heavily doped silicon single electron transistors. *Appl. Phys. Lett.* **93**: 112107.

Martinis, J. M., Nam, S., Aumentado, J., and Urbina, C. 2002. Rabi oscillations in a large Josephson-junction qubit. *Phys. Rev. Lett.* **89**: 117901.

Matsumoto, K., Kinoshita, S., Gotoh, Y. et al. 2003. Single-electron transistor with ultra-high Coulomb energy of 5000 K using position controlled grown carbon nanotube as channel. *Jpn. J. Appl. Phys.* **42**: 2415.

Meirav, U. and Foxman, E. B. 1996. Single-electron phenomena in semiconductors. *Semicon. Sci. Technol.* **22**: 255.

Meitl, M. A., Zhou, Y., Gaur, A. et al. 2004. Solution casting and transfer printing single-walled carbon nanotube films. *Nano Lett.* **4**: 1643.

Mooij, J. E., Orlando, T. P., Levitov, L., Tian, L., van der Wal, C. H., and Lloyd, S. 1999. Josephson persistent-current qubit. *Science* **285**: 1036.

Mooij, J. E. and Harmans, C. J. P. M. 2005. Phase-slip flux qubits. *New J. Phys.* **7**: 219.

Morales, A. M. and Lieber, C. M. 1998. A laser ablation method for the synthesis of crystalline semiconductor nanowires. *Science* **279**: 208.

Moriyama, S., Fuse, T., Aoyagi, Y., and Ishibashi, K. 2005. Excitation spectroscopy of two-electron shell structures in carbon nanotube quantum dots in magnetic fields. *Appl. Phys. Lett.* **87**: 073103.

Moriyama, S., Tsuya, D., Watanabe, E. et al. 2009. Coupled quantum dots in a graphene-based two-dimensional semimetal. *Nano Lett.* **9**: 2891.

Nakahara, M. and Ohmi T. 2008. *Quantum Computing: From Linear Algebra to Physical Realizations.* Inst. of Physics Pub. Inc. CRC Press Boce Raton FL.

Nakamura, Y., Chen, C. D., and Tsai J. S. 1996. Quantitative analysis of Josephson-quasiparticle current in superconducting single-electron transistors. *Phys. Rev. B: Condens. Matter* **53**: 8234.

Nakamura, Y., Chen, C. D., and Tsai, J. S. 1997. Spectroscopy of energy-level splitting between two macroscopic quantum states of charge coherently superposed by Josephson coupling. *Phys. Rev. Lett.* **79**: 2328.

Nakamura, Y., Pashkin, Yu. A., and Tsai, J. S. 1999. Coherent control of macroscopic quantum states in a single-Cooper-pair box. *Nature* **398**: 786.

Negishi, R., Hasegawa, T., Terabe, K., Aono, M., Ebihara, T., Tanaka, H., and Ogawa, T. 2006. Fabrication of nanoscale gaps using a combination of self-assembled molecular and electron beam lithographic techniques. *Appl. Phys. Lett.* **88**: 223111.

Nielsen, M. A. and Chuang, I. 2000. *Quantum Computation and Quantum Information.* Cambridge: Cambridge University Press.

Nishiguchi, K., Fujiwara, A., Ono, Y., Inokawa, H., and Takahashi, Y. 2006. Room-temperature-operating data processing circuit based on single-electron transfer and detection with metal-oxide-semiconductor field-effect transistor technology. *Appl. Phys. Lett.* **88**: 183101.

Nishino, T., Negishi, R., Kawao, M., Nagata, T., Ozawa, H., and Ishibashi, K. 2010. The fabrication and single electron transport of Au nano-particles placed between Nb nanogap electrodes. *Nanotechnology* **21**: 225301.

Nosho, Y., Ohno, Y., Kishimoto, S., and Mizutani, T. 2006. Relation between conduction property and work function of contact metal in carbon nanotube field-effect transistors. *Nanotechnology* **17**: 3412.

Novoselov, K. S., Geim, A. K., Morozov, S. V. et al. 2005. Two-dimensional gas of massless Dirac fermions in graphene. *Nature* **438**: 197.

Nowack, K. C., Koppens, F. H. L., Nazarov, Yu. V., and Vandersypen, L. M. K. 2007. Coherent control of a single electron spin with electric fields. *Science* **318**: 1430.

Ohta, T., Bostwick, A., Seyller, T., Horn, K., and Rotenberg, E. 2006. Controlling the electronic structure of bilayer graphene. *Science* **313**: 951.

Ono, K., Austing, D. G., Tokura, Y., and Tarucha, S. 2002. Current rectification by Pauli exclusion in a weakly coupled double quantum dot system. **297**: 5585, 1313–1317.

Ono, Y., Takahashi, Y., Yamazaki, K., Nagase, M., Namatsu, H., Kurihara, K., and Murase, K. 2000. Si complementary single-electron inverter with voltage gain. *Appl. Phys. Lett.* **76**: 3121.

Oosterkamp, T. H., Fujisawa, T., van der Wiel, W. G. et al. 1998. Microwave spectroscopy of a quantum-dot molecule. *Nature* **395**: 873–876.

Oreg, Y., Byczuk, K., and Halperin, B.I. 2000. Spin configurations of a carbon nanotube in a nonuniform external potential. *Phys. Rev. Lett.* **85**: 365.

Park, H., Lim, A. K. L., Alivisatos, A. P., Park, J., and McEuen, P. L. 1999. Fabrication of metallic electrodes with nanometer separation by electromigration. *Appl. Phys. Lett.* **75**: 301.

Pesetski, A. A., Baumgardner, J. E., Folk, E., Przybysz, J. X., Adam, J. D., and Zhang, H. 2006. Carbon nanotube field-effect transistor operation at microwave frequencies. *Appl. Phys. Lett.* **88**: 113103.

Plantenberg, H., de Groot, P. C., Harmans, C. J. P. M., and Mooij, J. E. 2007. Demonstration of controlled-NOT quantum gates on a pair of superconducting quantum bits. *Nature* **447**: 836.

Postma, H. W. Ch., Teepen, T., Yao, Z., Grifoni, M., and Dekker, C. 2001. Carbon nanotube single-electron transistors at room temperature. *Science* **293**: 76.

Pothier, H., Lafarge, P., Urbina, C., Estive, D., and Devoret, M. H. 1992. Single-electron pump based on charging effects. *Europhys. Lett.* **17**: 249.

Reed, M. A. 2004. Molecular electronics: Back under control. *Nature Mater.* **3**: 286.

Reed, M. A., Randall, J. N., Aggarwal, R. J., Matyi, R. J., Moore, T. M., and Wessel, A. E. 1988. Observation of discrete electronic states in a zero-dimensional semiconductor nanostructure. *Phys. Rev. Lett.* **60**: 535.

Reed, M. A., Zhou, C., Muller, C. J., Burgin, T. P., and Tour, G. M. 1997. Conductance of a molecular junction. *Science* **278**: 252.

Reina, A., Jia, X., Ho, J. et al. 2009. Large area, few-layer graphene films on arbitrary substrates by chemical vapor deposition. *Nano Lett.* **9**: 30.

Rueß, F. J., Pok, W., Reusch, T. C. G. et al. 2007. Realization of atomically controlled dopant devices in silicon. *Small* **3**: 563–567.

Ryzhii, V. 2006. Terahertz plasma waves in gated graphene heterostructures. *Jpn. J. Appl. Phys.* **45**: L923.

Sakaki, H. 1982. Velocity-modulation transistor (VMT)—A new field-effect transistor concept. *Jpn. J. Appl. Phys.* **21**: L381.

Schenkel, T., Persaud, A., Park, S. J. et al. 2003. Solid state quantum computer development in silicon with single ion implantation. *J. Appl. Phys.* **94**: 7017.

Shinkai, G., Hayashi, T., Ota, T., and Fujisawa, T. 2009. Correlated coherent oscillations in coupled semiconductor charge qubits. *Phys. Rev. Lett.* **103**: 056802.

Sköld, N., Karlsson, L. S., Larsson, M. W., Pistol, M.-E., Seifert, W., Trägårdh, J., and Samuelson, L. 2005. Growth and Optical Properties of Strained GaAs–Ga$_x$In$_{1-x}$P Core–Shell Nanowires. *Nano Lett.* 5, 1943.

Stampfer, C., Güttinger, J., Molitor, F., Graf, D., Ihn, T., and Ensslin, K. 2008. Tunable Coulomb blockade in nanostructured graphene. *Appl. Phys. Lett.* **92**: 012102.

Strocio, J. A. and Eigler, D. M. 1991. Atomic and molecular manipulation with the scanning tunneling microscope. *Science* **254**: 1319.

Sugahara, S. and Tanaka, M. 2005. A spin metal-oxide-semiconductor field-effect transistor (spin MOSFET) with a ferromagnetic semiconductor for the channel. *J. Appl. Phys.* **97**: 10D503.

Sun, J., Wallin, D., Brusheim, P., Maximov, I., and Xu, H. Q. 2008. Novel room-temperature functional analogue and digital nanoelectronic circuits based on three-terminal ballistic junctions and planar quantum-wire transistors. *J. Phys. Conf. Ser.* **100**: 052073.

Suzuki, M., Ishibashi, K., Ida, T., Tsuya, D., Toratani, K., and Aoyagi, Y. 2001. Fabrication of single and coupled quantum dots in single-wall carbon nanotubes. *J. Vac. Sci. Technol. B* **19**: 2770.

Suzuki, M., Ishibashi, K., Toratani, T., Tsuya, D., and Aoyagi, Y. 2002. Tunnel barrier formation using argon-ion irradiation and single quantum dots in multiwall carbon nanotubes. *Appl. Phys. Lett.* **81**: 2273.

Tabata, H., Shimizu, M., and Ishibashi, K. 2009. Fabrication of single electron transistors using transfer-printed aligned single walled carbon nanotubes arrays. *Appl. Phys. Lett.* **95**: 113107.

Takahashi, Y., Nagase, M., Namatsu, H. et al. 1995. Fabrication technique for Si single-electron transistor operating at room temperature. *Electron. Lett.* **31**: 136.

Takahashi, S. and Maekawa, S. 2008. Spin current, spin accumulation and spin Hall effect. *Sci. Technol. Adv. Mater.* **9**: 014105.

Tans, S. J., Verschuren, A. R. M., and Dekker, C. 1998. Room-temperature transistor based on a single carbon nanotube. *Nature* **393**: 49.

Tateno, K., Goto, H., and Watanabe, Y. 2004. GaAs/AlGaAs nanowires capped with AlGaAs layers on GaAs (311)B substrates. *Appl. Phys. Lett.* **85** (10): 1808.

Tarucha, S., Austing, D. G., Honda, T., van der Hage, R. J., and Kouwenhiven, L. P. 1996. Shell filling and spin effects in a few electron quantum dot. *Phys. Rev. Lett.* **77**: 3613.

Terabe, K., Hasegawa, T., Nakayama, T., and Aono, M. 2005. Quantized conductance atomic switch. *Nature* **433**: 47.

Thornton, T. J. 1995. Mesoscopic devices. *Rep. Prog. Phys.* **58**: 311.

Tinkham, M. 2004. *Introduction to Superconductivity*. 2nd ed. Mineola, NY: Dover Publications.

Tiwari, S., Rana, F., Hanafi, H., Hartstein, A., Crabbé, E. F., and Chan, K. 1996. A silicon nanocrystals based memory. *Appl. Phys. Lett.* **68**: 1377.

Tsuya, D., Suzuki, M., Aoyagi, Y., and Ishibashi, K. 2005. Exclusive-OR gate using a two-input single-electron transistor in single-wall carbon nanotubes. *Appl. Phys. Lett.* **87**: 153101.

Tucker, J. R. 1992. Complementary digital logic based on Coulomb blockade. *J. Appl. Phys.* **72**: 4399.

van Bommel, A. J., Crombeen, J. E., and van Tooren, A. 1975. LEED and Auger electron observations of the SiC(0001) surface. *Surf. Sci.* **48**: 463.

van der Wal, C. H., ter Haar, A. C. J., Wilhelm, F. K. et al. 2000. Quantum superposition of macroscopic persistent-current states. *Science* **290**: 773–777.

van Wees, B. J., van Houten, H., Beenakker, C. W. J. et al. 1988. Quantized conductance of point contacts in a two-dimensional electron gas. *Phys. Rev. Lett.* **60**: 848.

Vandersypen, L. M. K., Elzerman, J. M., Schouten, R. N., van Beveren, L. H. W., Hanson, R., and Kouwenhoven, L. P. 2004. Real-time detection of single-electron tunneling using a quantum point contact. *Appl. Phys. Lett.* **85**: 4394.

Vijayaraghavan, A., Kanzaki, K., Suzuki, S. et al. 2007. Transition of single-walled carbon nanotubes from metallic to semiconducting in field-effect transistors by hydrogen plasma treatment. *Nano Lett.* **7**: 1622.

Vion, J. S. D., Aassime, A., Cottet, A. et al. 2002. Manipulating the quantum state of an electrical circuit. *Science* **296**: 886.

Wallraff, A., Schuster, D. I., Blais, A. et al. 2004. Strong coupling of a single photon to a superconducting qubit using circuit quantum electrodynamics. *Nature* **431**: 162.

Waser, R. ed., 2005. *Nanoelectronics and Information Technology*. Verlag, Berlin: Wiley-VCH.

Wharam, D. A., Thornton, T. J., Newbury, R. et al. 1988. One-dimensional transport and the quantisation of the ballistic resistance. *J. Phys. C: Solid State Phys.* **21**: L209.

Chapter 21

Williams, C. P. and Clearwater, S. H. 1997. *Explorations of Quantum Computing*. Berlin: Springer.

Wind, S. J., Appenzeller, J., Martel, R., Derycke, V., and Avouris, Ph. 2002. Vertical scaling of carbon nanotube field-effect transistors using top gate electrodes. *Appl. Phys. Lett.* **80**: 3817.

Wu, Y. Q., Ye, P. D., Capano, M. A. et al. 2008. Top-gated graphene field-effect-transistors formed by decomposition of SiC. *Appl. Phys. Lett.* **92**: 092102.

Xiao, K., Liu, Y., Hu, P., Yu, G., Wang, X., and Zhu, D. 2003. High-mobility thin-film transistors based on aligned carbon nanotubes. *Appl. Phys. Lett.* **83**: 150.

Xu, X. Q. 2001. Electrical properties of three-terminal ballistic junctions. *Appl. Phys. Lett.* **78**: 2064.

Yamamoto, T., Pashkin, Yu. A., Astafiev, O., Nakamura, Y., and Tsai, J. S. 2003. Demonstration of conditional gate operation using superconducting charge qubits. *Nature* **425**: 941–944.

Zazunov, A., Shumeiko, V. S., Bratus', E. N., Lantz, J., and Wendin, G. 2003. Andreev level qubit. *Phys. Rev. Lett.* **90**: 087003.

22. Manipulation and Nanostructuring for Biological Applications

Gobind Das and Carlo Liberale

Fondazione Istituto Italiano di Tecnologia (IIT), NanoBioScience Department, Genova, Italy

Francesco De Angelis, Maria Laura Coluccio, and Enzo Di Fabrizio

Fondazione Istituto Italiano di Tecnologia (IIT), NanoBioScience Department, Genova, Italy
BIONEM Lab, University of Magna Graecia, Campus S. Venuta, Germaneto, viale Europa, Catanzaro, Italy

Patrizio Candeloro

BIONEM Lab, University of Magna Graecia, Campus S. Venuta, Germaneto, viale Europa, Catanzaro, Italy

22.1   Introduction

The advent of micro- and nanotechnologies in life science is strongly affecting the biomedical field both in diagnostics and therapy. A comprehensive review of this field is practically impossible due to the incredible fast progress. One of the major issues in medicine is the search of novel tools for early diagnosis. Discovery and detection biomarkers from biological fluids appear challenging due to the tremendous number of biomolecular species with differences of many orders of

magnitude in their relative abundance. Several approaches have been proposed in the last few years; among them, the study of the proteome seems to hold the greatest potentials. The basic aim of proteomic analysis is the identification of specific protein patterns from cells, tissues, and biological fluids related to physiological or pathological conditions. Unfortunately, the most informative biofluid such as human serum or blood are also the most complex ones, and therefore new tools for proteome investigation and discovery of novel biomarker are strongly required.

Here we report three different tools, which, even though they are from different fields of nanotechnology

Nanofabrication Handbook. Edited by Stefano Cabrini and Satoshi Kawata © 2012 CRC Press / Taylor & Francis Group, LLC. ISBN: 978-1-4200-9052-9

Chapter 22

and experienced an independent development, they can actually converge in one unique tool: (1) single-cell analysis through microinjection, that considering the pipette size and volume involved could be also called nanoinjection; (2) nanoporous nanoparticles (NPs) (originating from bottom-up nanofabrication approach) acting as nanosponges to investigate biological fluids; (3) single/few molecule analysis by means of Raman scattering and surface-enhanced Raman scattering (SERS) devices (from the top-down nanofabrication approach).

Here, a brief introduction of the mentioned technique and their applications follows.

Microinjection is the direct-pressure injection of a solution into a cell through a glass-capillary tip. It constitutes a reproducible and reliable method for introducing (or taking) different kinds of biomaterials into cultured cells. However, the microinjection has been considered for long time as a method only available to skillful experimenters, because it requires a good ability and gentleness in working with micromanipulators under a microscope. Moreover, it is not well suited for standard biochemical analysis, since only a limited number of cells from a cultured population can be injected. Despite these drawbacks, the recent demand for investigation at the single-cell level and the improvements in fluorescence microscopy, as a tool for single-cell analysis, have generated a renewed interest in this manipulation technique. Indeed, microinjection intrinsically deals with single cells, and compared with other cell inclusion techniques, such as cell permeabilization via electroporation or lipofection, does not alter intensely the state of the cells and enables a more controlled quantification of the introduced (or taken) materials. Exploiting microinjection, small volumes of fluids can also be taken from a single or few cells or in close proximity to them, and then investigated. Two of the major challenges involved in such an investigation are filtration of the raw fluids and analysis with single/few molecule sensitivity. For instance, 90% of human serum is composed of albumin and immunoglobulin (molecular weight >30 kDa). Therefore, filtration is an unavoidable first step of many research protocols.

Discovered over 40 years ago, porous silicon (PSi) has attracted growing attention in many fields of research for its interesting features, in particular, its photoluminescence at room temperature, together with the demonstration in 1995 of its biodegradability in physiological environment, opened a new perspective for biomedical application. Nanoporous Si nanoparticles can be fabricated from PSi films, and employed as nanosponges to selectively capture low-molecular-weight (LMW) species from complex solutions. The filtered solution can be recovered and analyzed through proper investigation methods, without the crosstalk of the heavier ones.

Fluorescence-based spectroscopic methods, gel-electrophoresis, and mass spectrometry are the most common and successful investigation tools in molecular medicine. In recent years, Raman spectroscopy is emerging, and it is going to play an important role in identification and characterization of biomolecules. Unfortunately, Raman signal for diluted solutions is much lower with respect to florescence, limiting the identification of the material with low concentration. SERS is a technique that permits the detection of adsorbed molecules on noble metal surfaces, such as Au, Ag, Cu, and so on at subpicomolar concentration. It is a phenomenon resulting in strongly increased Raman signals when molecules are in close proximity to nanometer-sized metallic structures,[*] allowing detection of few molecules,[†] and, in some cases, singlemolecule detection.[‡]

22.2 Microinjection for Nanobiology

In recent years, microinjection has been used for introducing several bioactive compounds, such as peptides, RNAs, plasmids, antibodies, and diffusion markers [1–6] into single cells. In some cases, artificial microcarriers containing chemical and biological compounds are introduced into the cells and tissues [7,8]. Following the injection, cell responses can be monitored by means of microscopy techniques (fluorescence, confocal microscopy, and two-photon microscopy) or microspectroscopic techniques (micro-Raman, micro-Fourier-transform infrared (FTIR)).

When dealing with cellular microinjections, one of the first points to be addressed is the rate of successful injections and their reproducibility. One cell has to be considered successfully injected when the

---

[*] Haynes, C.L., McFarland, A.D., Van Duyne, R.P. 2005. Surface-enhanced Raman spectroscopy. *Anal. Chem.* 77:A–338–A–346.

[†] Das, G., Mecarini, F., Gentile, F., et al. 2009. Nano-patterned SERS substrate: Application for protein analysis vs. temperature. *Biosens. Bioelectron.* 24:1693–1699.

[‡] Otto, A. 1984. SERS 'classical' and 'chemical' origins. In *Light Scattering in Solids*, eds, M. Cardona, and G. Guentherodt, Springer, Berlin, p. 289.

desired amount of material is introduced, and the cell damage (due to the mechanical stress of the injection) is negligible for its survival. If fluorescent media can be included in the injected solutions, fluorescence methods can be used to monitor the injections and to observe the transfer of solution from the glass tip to the cell itself. Obviously this is not always possible; as for example, when more fluorescent dyes are injected as cellular markers, the use of a fluorescent medium could compromise the afterward observation of the dyes. In order to avoid this problem, dyes and medium have to be carefully chosen and their excitation and emission spectra should not overlap. Furthermore, the evidence of solution transferred from the glass tip to the cell does not provide any information about the cell damage (except in the case of a highly damaged cell, where the integrity of the cell is compromised and clear modifications of its phenotype are observed).

In order to practice with microinjection and also to estimate the success rate of established protocols, temporary-transfection experiments can be carried out with GFP–plasmid solutions (GFP, green fluorescent protein). More in details, the cells of interest are microinjected with a solution containing a GFP–plasmid and analyzed by fluorescence microscopy after an incubation time of 24 h. The presence of the GFP–plasmid inside the cells will induce the GFP synthesis during the incubation time, thus making the cells fluorescent, which are successfully microinjected. It is worth noting that GFP–plasmid erroneously released outside the cells will not undergo any uptake process, and consequently will not alter the fluorescence analysis results. Furthermore, among the microinjected cells, only the healthy ones will be able to carry out the protein synthesis and to produce the GFP protein (Figure 22.1). As a consequence, the fluorescence analysis will provide a direct picture of the successfully microinjected cells.

In the following, two applications of microinjections related to biological issues of nanofabrication are reported. In the first case, we explore the potentiality of microinjections combined with micro-Raman for investigating the interaction between single cells and few NPs. The interaction of living cells with NPs (and more generally with nanomaterials) has recently attracted the interest of the toxicology community, because of the fast-developing nanotechnologies. It is estimated that engineered NPs, nanofibers, or similar materials are already present in more than 800 different commercial products [9]. Today, it is possible to

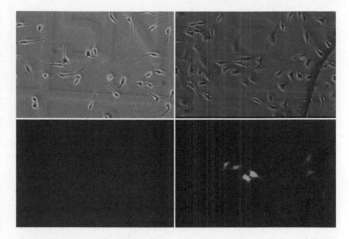

FIGURE 22.1   (See color insert.) Hela cell GFP–plasmid transfection experiment through microinjection technique. White-light (top row) and fluorescent (bottom row) microscopy images of injected cells, respectively, soon after injection (on the left) and 24 h of incubation later (on the right). During the incubation time, only microinjected cells in a healthy state can carry out the GFP synthesis, thus becoming fluorescent green.

produce particles and structures with sizes as small as that of proteins (i.e., at the nm level) and smaller than cell membrane sensors. The potential impact of these novel materials (not present in the nature before) on human health and environment is obviously under investigation. While large particles (size above 100 nm) are still recognized by the cell sensors as foreign entities, smaller particles (below 50 nm) can penetrate cell membranes and survive harsh endosome/lysosome pathways. The potential toxicological interaction between those nanostructured materials and the living matter is the subject studied by the so-called "nanotoxicology" [10].

Microinjection techniques offer the unique possibility to achieve the interaction of single cells with few NPs, without any decoration of the NPs from the biological environment (e.g., from the culturing medium used for cells growth). When NPs are incubated with the cells in the culturing dishes, they may undergo different processes, such as clusterization, in some cases dissolution, and very often surface decoration from the proteins inside the medium. If many NPs clusterize before interacting with the cell membrane, they could increase the overall size above 100 nm. At this point they should not be considered as NPs anymore, they cannot penetrate the cell membrane and they would experience the same process pathways as the larger microparticles. The effects of surface decoration are more subtle [11,12]: once the NP surface is totally decorated, it will be recognized by the cell membrane receptors according to the decorating molecules (and not according to the NP material) and different interaction mechanisms could arise. These effects could lead to an increase or a decrease of the cellular uptake, but without any a priori control of the process. Effectively, once NPs are incubated with the cells, there would be no confidence about the degree of cellular uptake, because of the variety of all the processes that NPs can undergo in a biological environment. In the worst case, there could be no interaction at all between NPs and cells. Conversely, microinjection techniques avoid all these problems related to the cellular uptake, since the nanomaterials of interest are directly injected into the cells. Furthermore, the amount of nanomaterials interacting with the cells can be reproduced with a good degree of confidence by means of a careful control of the injected volumes together with the concentration of the injected solutions.

In the present microinjection experiments, gold (Au), silver (Ag), and iron-oxide (Fe_3O_4) NPs of 20 nm diameter are microinjected into Hela cells. The NPs are produced exploiting different chemical reactions at the Catalan Institute of Nanotechnologies (ICN). For example, the gold NPs are synthesized reducing gold salts with different amounts of sodium citrate. Varying the ratio of gold salt to sodium citrate, different NPs sizes are obtained [13]. Before starting the investigation of single cell/few NPs interactions, a microinjection protocol has been validated by means of environmental scanning-electron microscopy (ESEM). In details, glass capillary tips with an external diameter of 0.3 μm (from WPI Inc.) are used to microinject a solution of Au NPs into Hela cells. The solvent of Au NPs is an aqueous solution of sodium citrate (2.2 mM) and former cytotoxicity experiments showed that this solvent has no influences on the cell viability. The NPs concentration in the starting solution is 2.39×10^{11} NPs/mL and the estimated injected solution is no larger than 50 fl, thus providing approximately 10 NPs for each injection. The microinjected cells are then fixated and analyzed through ESEM. Electron microscopy showed the presence of several Au-NPs clusters correspondingly to the injected cells, whereas outside the injection area no Au particles are revealed (Figure 22.2). Moreover, microchemical x-rays analysis performed inside the ESEM chamber confirmed that the observed clusters are purely made of gold. After this validation of the microinjection parameters, detection of single cell/ few NPs interaction is carried out by means of microspectroscopic techniques (micro-Raman in the

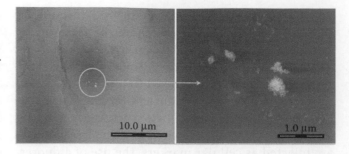

FIGURE 22.2 ESEM microscopy of Hela cells after gold NPs multiple-microinjections. (Recorded at the Biomaterials Laboratory of the University of Modena and Reggio Emilia, Italy.)

present case) combined with principal component analysis (PCA) for the interpretation of the spectra.

Raman spectroscopy provides a powerful tool for investigating small changes in the cellular biochemistry, thanks to its high sensitivity, noninvasive sampling capabilities, label-free investigation, and high spatial resolution (in the case of micro-Raman). Recently, micro-Raman has been widely employed in biological studies [14–16], since a typical Raman spectrum from a cell (Figure 22.3) represents a rich "fingerprint" of the global biochemical cell composition. However, because of this information richness, univariate methods for Raman spectra analysis (like the calculation of typical peak areas) could neglect a large amount of information. Besides few clear and pronounced peaks, Raman spectra of living cells are composed of complex bands coming from many overlapping peaks corresponding to various chemical species. Therefore, a lot of assumptions concerning peaks position, width, and shape should be made for a univariate analysis. On the other hand, multivariate statistical methods, such as PCA, do not require such assumptions to be made a priori and constitute more refined tools for Raman spectra analysis. Recently, PCA analysis of Raman spectra has been used for the discrimination of various cancerous tissues, identification of microorganisms, and tissue chemical imaging.

The combination of microinjection techniques with micro-Raman spectroscopy for investigation of living cells has been achieved, thanks to the development of home-made Petri dishes with very low Raman signal. Standard Petri dishes and microscope coverslips have large Raman signal which overcome the small signal coming from the cells. The solution very often reported in the literature is the use of MgF_2 or CaF_2 coverslips (which have a small Raman signal) for cell growth, put directly inside the Petri dish during incubation. This solution has two main disadvantages for

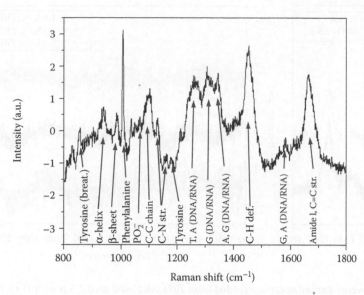

Peak (cm^{-1})	Assignment
854	Tyrosine (proteins)
1005	Phenylalanine (proteins)
1060–1095	PO$_2^-$ Chain C–C (lipids)
1128	C–N (proteins) C–O (carbohydrates)
1220–1284	T, A (DNA/RNA) Amide III (proteins) =CH bend (lipids)
1320	G (DNA/RNA) CH def (proteins)
1342	A, G (DNA/RNA) CH def (proteins)
1420–1480	G, A (DNA/RNA) CH def (proteins) CH def (lipids) CH def (carbohydrates)
1578	G, A (DNA/RNA)
1655–1680	Amide I (proteins) C=C str. (lipids)

FIGURE 22.3 Typical Raman spectrum from a cell: the richness of peaks reflects an equal richness in biochemical information.

microinjection: (1) MgF$_2$ or CaF$_2$ coverslips have too large thickness (at least 1 mm, due to technical reasons for their fabrication) and are not compatible with the small working distances (<0.5 mm) of high-magnification objectives on inverted microscopes, which is a must for live monitoring of the microinjection process; (2) MgF$_2$ or CaF$_2$ coverslips would not be fixed inside the Petri dish, and even small movements (in the μm range) due to their floating condition could break the fragile glass tips used for cell microinjection. In order to make cell microinjection fully compatible with Raman spectroscopy, we use bottom-glass Petri dish with a thin metal layer to prevent the Raman signal coming from the bottom-glass. At the same time, the metal layer is thin enough to be transparent and fully compatible with high-magnification inverted optical microscopy.

In order to test the Raman sensitivity to small changes induced by external stress in the cellular biochemistry, oxidative stress has been induced on Hela cells via H$_2$O$_2$ incubation. More in details, 9 mM H$_2$O$_2$ has been incubated for 1.0, 2.5, and 5.0 h before Raman measurements; for each incubation time a corresponding control Petri dish (i.e., no H$_2$O$_2$ incubation) has been prepared and measured. Moreover, Trypan blue viability assays are carried out to test the degree of stress induced on the cells. The results of these latter tests (Figure 22.4) show that the cell viability drops from 91% for 1.0 h of incubation time to 53% for the 5.0 h incubated cells. The oxidative damage of the surviving cells is correlated to the cell viability, so that 91% of living cells indicates a small oxidative stress and 53%

a very large one. For each incubation time (and relative control), Raman spectra are recorded on several cells. Further, the Raman signal outside the cells is acquired, in order to be subtracted later before data processing for the PCA analysis. Figure 22.5 (left panel) reports the average of all the spectra recorded for the 2.5 h incubation and the average of the spectra from the relative control. At first glance, the spectra look quite similar to each other and no large differences can be noted regarding the intensities and widths of the major peaks. To make the small differences more evident, the differential spectra are shown in Figure 22.5 (right panel) also.

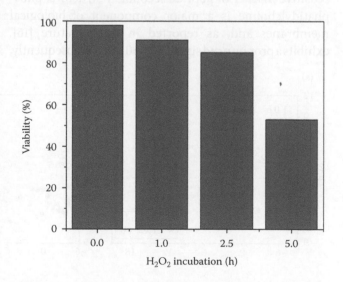

FIGURE 22.4 Cells viability (Trypan blue assay) of Hela cells incubated with H$_2$O$_2$ 9 mM for different incubation times.

Chapter 22

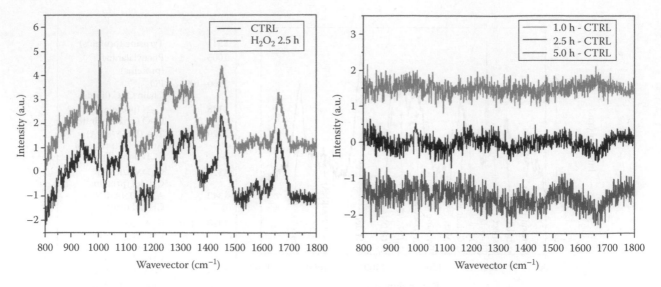

FIGURE 22.5 **(See color insert.)** On the right, average Raman spectra recorded from Hela cells incubated 2.5 h with H_2O_2 9 mM (top line) and from control cell (bottom line); on the left: difference between Raman spectra from incubated cells and control cells, for Hela incubated 1.0, 2.5, and 5.0 h with 9 mM H_2O_2.

Each differential spectrum is the difference between the H_2O_2-incubated cells average spectrum and the one of the corresponding control cells. As we can see the differential spectrum for the 1.0 h incubation is nearly flat, thus meaning that there is no evident difference between the average spectra of H_2O_2-incubated and control cells. As the incubation time increases, some differences are clearly observable. In particular, the peak located at around 1660 cm^{-1} slightly increases its magnitude in the differential spectra. In this spectral band, there is the overlapping of the peak coming from the amide I band of peptides and proteins with the one from phosphatidylcholine. While the amide I band is a sensitive marker of peptide secondary structure, phosphatidylcholine is a major component of biological membranes and, as reported in the literature [16], exhibits a pronounced signal at 1660 cm^{-1}. Consequently,

the differences observed in the differential peaks can be associated mainly with membrane damage due to the incubation with H_2O_2, and as expected the damage increases with the incubation time.

So far we only noticed the differences that a simple univariate analysis can also make evident. But PCA analysis is much more sensitive to the small variations of many correlated features, and constitutes a higher quality tool for discriminating the different cells groups. Figure 22.6 shows the PCA results along the first two principal components for the three incubation times and relative controls. In the graphs, each point represents the spectrum from one single cell, and for all the three incubation times two distinct analytical groups are formed corresponding to the control cells and to the H_2O_2-incubated ones. It is worth noting that also for the 1.0-h-incubated cells, where the differential analysis

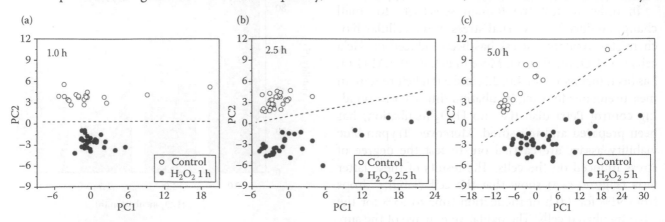

FIGURE 22.6 PCA analysis of Hela cells incubated with H_2O_2 9 mM for 1.0 h (a), 2.5 h (b), and 5.0 h (c). Data from incubated and control cells are separated into two groups for all the incubation times.

could not highlight any discrepancy, the PCA analysis is able to discriminate between the incubated and the control cells. This confirms the high potentialities of micro-Raman techniques combined with PCA analysis.

Coming back to the single cell/few NPs interactions, special Petri dishes with gridded bottom-glass are used for the microinjection experiments. The grid on the bottom glass provides a numbered reference system for an easy localization of the injected cells during the micro-Raman acquisition. Moreover, since not all the cells on the Petri dish are injected, all the remaining cells constitute a valid control which has undergone exactly the same conditions than the injected ones. All the NPs solutions are diluted so that their final concentration is 10^{11} NPs/mL = 0.1 NPs/fl (i.e., each cell is expected to be injected with 1–10 NPs). Cells are incubated for 3 h after microinjection and then Raman spectra are collected on the living cells (i.e., without any fixation and directly in nude culturing medium). Raman spectra are collected on several injected cells as well as control cells from the same Petri dish. The average spectra from control cells and from all the kinds of injected NPs are reported in Figure 22.7. At first glance, the spectra acquired on the microinjected cells do not exhibit large differences compared with the cells that are not injected (control cells). But a further multivariate statistical analysis of the data (PCA analysis) clearly discriminates between the control cells and the injected ones. As a first approach, PCA

analysis is performed separately for the different NPs microinjected into Hela cells (Figure 22.8). In all cases, microinjected cells and control ones are discriminated along the 1st principal component (PC1). Subsequently, the collected spectra are PCA-analyzed all together, even if they are from experiments carried out separately (one microinjected Petri dish for one kind of NPs). The results are shown in Figure 22.9 and a clear separation between microinjected and control cells is still observable. The majority of the differences are collected by the 2nd principal component (PC2). In order to gain a deeper understanding of NP's effects, decomposition of the PC2 is carried out, that is, we analyzed how the different Raman shifts contribute to the PC2. The so-called loadings of PC2 are then shown in Figure 22.10, together with the difference between the average spectrum from all the injected cells and the average spectrum from all the control cells. The two curves present the same behavior, thus confirming that PC2 is a good parameter to take into account the differences between injected and control cells. Moreover, it is evident that the main contribution to the PC2 is provided by the Raman band around 1070 cm^{-1}, corresponding to phosphate groups and C–C bonds in the lipids chains. However, the lipid bonds usually present contributions in other Raman bands (like the one around 1660 cm^{-1}) and in this case we do not observe any significant peak in the PC2 loadings from these other bands. Similarly, the main contribution to PC2, and then to the differences between NPs-injected cells and control cells, is from the phosphate groups of the cell. Figure 22.10 also shows the average spectra from microinjected cells and control ones: the 1070 cm^{-1} Raman band has clearly a lower intensity.

In order to ensure that the differences observed between microinjected cells and not microinjected cells are not due to the mechanical stress of the microinjection itself and/or the NPs solvents, microinjection experiments have been carried out only with the NPs solvents. The used solvents are sodium citrate 2.2 mM for both the Au and Ag NPs, while TMAOH 5.0 mM is used for Fe_3O_4 NPs. Hela cells are microinjected only with the NPs solvents and again after 3 h of incubation Raman spectra are recorded on microinjected and not microinjected cells. After performing PCA analysis (Figure 22.11), no difference can be detected between microinjected and control cells in the case of sodium citrate. Concerning TMAOH, a small difference is instead observable between the solvents injected and the control cells. But PCA analysis of TMAOH spectra together with Fe_3O_4 NPs spectra clearly shows that the solvents effects are much smaller compared with the Fe_3O_4 NPs effects.

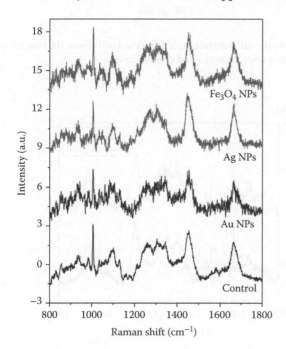

FIGURE 22.7 Average Raman spectra from Hela cells microinjected with iron oxide NPs (first curve from the top), silver NPs (second curve), gold NPs (third curve) and from control cells (last curve).

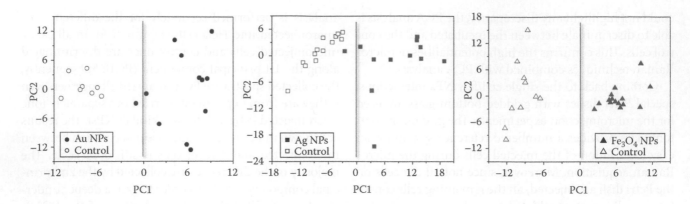

FIGURE 22.8 PCA analysis performed separately for the different NPs microinjected into Hela cells. In all cases, microinjected cells and control ones are separated along the PC1.

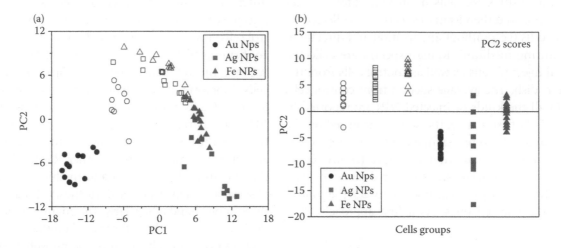

FIGURE 22.9 (a) PCA performed over all the Raman spectra simultaneously still separates microinjected cells from the control ones. (b) Principal component PC2 takes into account for the differences between microinjected and control cells.

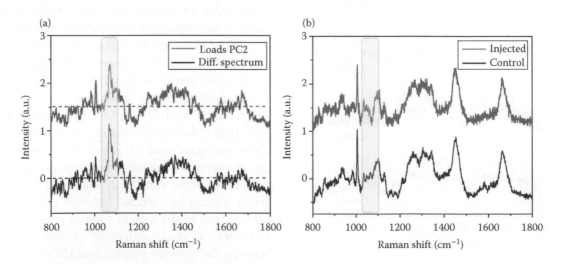

FIGURE 22.10 (a) PC2 loadings and difference Raman spectrum: the largest contribution to PC2 comes from 1070 cm^{-1} Raman band, corresponding to PO$_2^-$ groups and C.C chains in lipids. (b) Average spectra from injected and control cells: the 1070 cm^{-1} Raman band has a lower intensity in the average spectrum of the injected cells.

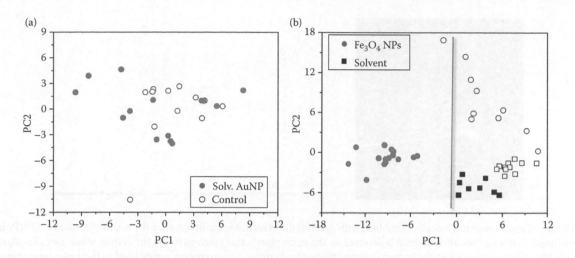

FIGURE 22.11 (a) PCA analysis over Raman spectra from Au (and Ag) NPs solvent (sodium citrate 2.2 mM) and from control cells: no difference is observable. (b) PCA analysis performed over iron-oxide NPs-injected cells, solvent (TMAOH 5.0 mM)-injected cells, and control cells: the difference between solvent-injected cells (black squares) and control cells (open symbols) is much smaller compared with the difference between the NPs-injected cells (gray circles) and the control ones.

Another application of microinjection for nanobiology concerns the use of nanoporous Si nanoparticles (NPNPs) for clinical purposes. As reported later in this chapter, NPNPs have attracted a lot of interest as potential carriers in drug-delivery applications. One interesting aspect that has not been widely investigated so far is the dissolution properties of the NPs inside the cellular environment. As already known, NPNPs have different dissolution behaviors according to the pH of the surrounding medium, and it would be interesting to follow the dissolution mechanisms of NPNPs directly inside the cells. Undoubtedly, microinjections offer the possibility to produce systematically the interaction of single cells with NPNPs, but a high-sensitive detection system combined with a high spatial resolution is then required for a direct monitoring of the activated processes. For this purpose, two-photon fluorescence microscopy is a good choice as monitoring system [17–19] and, in the following, we shortly report on the first results achieved combining microinjection and two-photon techniques.

In order to test the drug-delivery properties of NPs, NPNPs are incubated with rhodamine and subsequently microinjected in Hela cells. The detection of the microinjected particles via optical techniques is a challenging task because the volume delivered to the cell is a very small one, which in turn produces very small signals. For this purpose, two-photon fluorescence microscopy has been combined with a spectral detection method. While the two-photon mechanism is a very confined one and allows for a high optical resolution, the spectral acquisition provides a powerful tool to discriminate the chemical origin of the

fluorescence signal coming from the sample. Images are recorded just before and just after the microinjection. With standard fluorescence detection (i.e., without using the spectral acquisition), a strong rhodamine signal is clearly observable only at the micropipette tip (Figure 22.12, right picture). Concerning the cell, it is instead difficult to extrapolate information from the standard fluorescence, due to the wide autofluorescence signal that could be detected from the cell before the microinjection. However, spectral acquisition provides a better insight into the origin of the fluorescence from the sample. A clear rhodamine peak around 550 nm can be detected at the micropipette tip and a similar peak, but much weaker, is observable in the cell membrane only around the microinjected region (Figure 22.13).

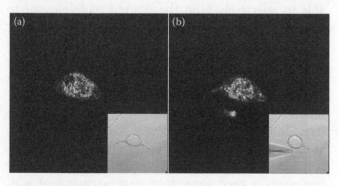

FIGURE 22.12 **(See color insert.)** Nanoporous Si nanoparticles microinjection into Hela cell: just before (a) and just after (b) the microinjection. Pictures are recorded via two-photon microscopy combined with standard fluorescence. On the right, a strong rhodamine signal coming from the micropipette tip is observable.

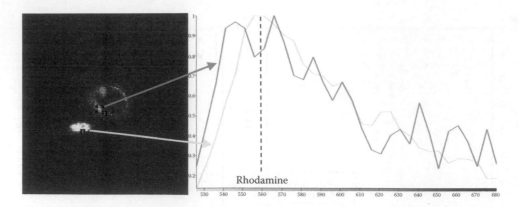

FIGURE 22.13 Two-photon microscopy combined with spectral fluorescence acquisition, after the microinjection of NPNPs incubated with rhodamine. A strong rhodamine signal is detected on the micropipette tip (green curve on the graph), while a smaller signal can be observed in the cell membrane close to the area directly injected (red curve). The curves are normalized to their maxima, respectively.

This result represents a clear evidence of the injected NPNPs directly inside the cellular environment. Since the microinjected volume is very small as well as the number of microinjected NPNPs, this technique reveals to be very appealing to achieve the detection of the single cell–single NP interaction. Moreover, further developments of the acquisition system could provide in the future the real-time detection of diffusing species inside the cell environment. Obviously, such an achievement would also constitute a novel powerful tool for investigating the dynamics of intracellular mechanisms activated as a response to external stimuli.

22.3 Nanoporous Nanoparticles as Nanosponges for Human Serum Filtration

In the most basic sense, PSi is a network of air holes within an interconnected crystalline silicon matrix. Pore size and distribution range from a few nanometers to a few microns and can be finely controlled during the fabrication. The well-known functionalization processes of the porous surface provide a further control of bioreactivity, hydrophobicity, and optical properties [20–27]. Various surface derivatization approaches can be found in the literature and hundreds of different cross-linking agents are now available to selectively bind the desired molecules, or to change the surface from hydrophobic to hydrophilic and vice versa. Also, the dissolution rate in water, or physiological environment, can be adjusted through the accurate control of the morphology, pore size, and pH of the medium. The typical dissolution rate in alkaline conditions varies from a few minutes up to a few days, and for pH values below 5, PSi does not dissolve at room temperature. The free volume inside pores can reach up to 80% of total volume of the material, and represents a very interesting nanoenvironment suitable for many applications.

Considering the incredible number of studies performed on PSi (more than 10,000 papers published over past 10 years), it is quite impossible to summarize all the applications proposed. After a brief description of the fabrication process, a novel and interesting application of PSi in biomedical field, such as molecular sieving of complex biofluid, is reported.

As introduced above, PSi is a spongy material with a well-defined pore size and distribution. It can be obtained from a silicon bulk sample trough and electrochemical reaction involving fluoride acid, methanol (or other alcohol), and water. A schematic representation of the process is reported in Figure 22.14: the starting silicon wafer is placed at the bottom of an electrolytic cell and works as the anode, whereas the cathode is typically made of platinum. The composition of

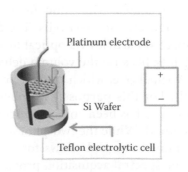

FIGURE 22.14 Electrolytic cell scheme.

the electrolytic solution strongly affects the morphology of the final porous film together with several other parameters that are extensively reported in many books and publications. Silicon doping type and concentration, applied voltage and current, illumination, time, and temperature can be exploited to rule the electrochemical reactions, and to obtain dissolution of the silicon film, electropolishing, deep vertical channel, and pore formation (see also Table 22.1).

Concerning porosity, four categories can be distinguished with respect to pore diameter and distance between pores [28–30]:

- Nanopores, with pore diameters of 2.3 nm, and pore distance of 5–6 nm
- Micropores, with geometries <10 nm
- Mesopores, with geometries in the 10–50 nm scale
- Macropores with geometries >50 nm

To understand the PSi formation it is necessary to deal with the dissolution chemistry of silicon when anodically biased in hydrofluoric acid. As depicted in Figure 22.15, when a hole (h+) reaches the silicon–solution interface, it enables a nucleophilic attack of fluoride ions on Si–H bonds, and a Si–F bond is established (step 1). Due to the polarizing influence of the bonded F, another F^- ion can attack, while an electron is injected into the electrode (step 2) and one hydrogen molecule is produced (step 3). Due to the polarization induced by the Si–F groups, the electron density of the Si–Si backbonds is lowered and these weakened bonds will now be attacked by HF or H_2O (steps 4 and 5) in a way that the silicon surface atoms remain bonded to hydrogen (step 5).

Table 22.1   Effect of Anodization Parameters on PSi Formation

An increase of … yields to	Porosity	Etching Rate	Critical Current
HF concentration	Decrease	Decrease	Increase
Current density	Increase	Increase	—
Anodization time	Increase	Almost constant	—
Temperature	—	—	Increase
Wafer doping (p-type)	Decrease	Increase	Increase
Wafer doping (n-type)	Increase	Increase	—

The electrochemical reaction goes on changing the morphology of silicon surface in many different ways depending on the overall conditions adopted (see Table 22.1). In particular, HF concentration directly affects the porosity: the higher the concentration, the lower the pore sizes and the porosity. It was found that the introduction of ethanol eliminates hydrogen and ensures complete infiltration of HF solution within the pores. Subsequently, uniform distribution of porosity and thickness is improved. Longer etching times lead to thicker layers, but induce anisotropy of the morphology along the depth. The applied current density determines the depletion width and the carrier injection rate, and then it affects both morphology and etching rate. In fact, porosity, thickness, pore diameter, and microstructure of PSi depend on the anodization conditions.

In Figure 22.16, the characteristic $i–V$ curves for n- and p-type-doped Si in aqueous HF are shown. Both n and p-type silicon is stable under cathodic polarization. Dissolution of silicon occurs only under anodic polarization. For high anodic overpotentials, the Si surface goes under electropolishing. On the contrary, for low anodic potentials, the surface morphology is dominated by a dense array of channels penetrating deeply into the bulk of the Si. The pore formation occurs only during the initial rising part of $i–V$ curve, for a potential value below the potential of the small sharp peak. For simplicity, $i–V$ curves shown in Figure 22.16 can be divided into four distinct regions depending on the sign of the applied potential and whether n- or p-type material is used. Table 22.2 summarizes the salient electrochemical features that are predominant in each region. For n-type substrates, this typical $i–V$ behavior is observed only under illumination because supply of holes is needed.

The growth process is obtained either by controlling the anodic current or the potential. The practice of working with constant current is widely diffused, because it allows a better control of the porosity, thickness, and reproducibility of the PSi layer.

22.3.1   Nanoporous Nanoparticles Fabrication

Nanoporous Si nanoparticles (NPNPs) can be fabricated by ultrasonication of a thin film of nanoporous silicon. Here, typical protocol for the fabrication of nanoporous NPs starting from bulk silicon is described.

PSi is obtained by anodization of boron-doped silicon wafer (resistivity 5–10 Ωcm) of [100] crystal orientation, using an electrolyte binary mixture of hydrofluoric acid (25%), water (25%), and ethanol (50%). Applied

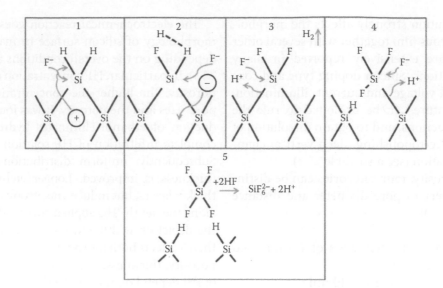

FIGURE 22.15 Proposed dissolution mechanism of silicon electrodes in hydrofluoric (HF) acid associated with PSi formation. Sketches from 1 to 5 represent the sequence of reaction steps involving F, H, Si, holes, and electrons.

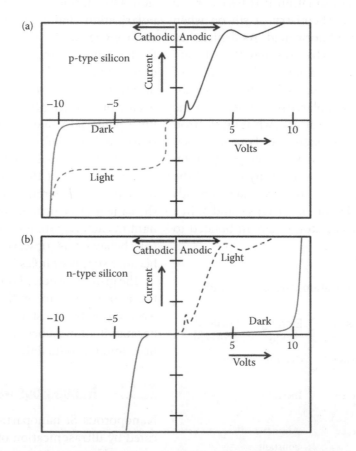

FIGURE 22.16 (a) "Typical" current–voltage relationships for p-type silicon. The solid line indicates the dark response, whereas the dashed line shows the response under illumination. The first (lower) current peak corresponds to the formation of a surface anodic oxide formed during electropolishing. The second (higher-) current peak marks the beginning of stable current (potential) oscillations with the possible formation of a second type of anodic oxide. (b) "Typical" current–voltage relationships for n-type silicon. The solid line indicates the dark response and the dashed line shows a response under illumination. The first (lower) current peak corresponds to the formation of a surface anodic oxide formed during electropolishing. The second (higher) current peak marks the beginning of stable current (potential) oscillations with the possible formation of a second type of anodic oxide.

Table 22.2 Breakdown of the Different Electrochemical Regions for the Silicon/HF System Summarizing the Important Features for Each Region

Cathodic Overpotentials	Anodic Overpotentials
p-type Si	

1. No silicon dissolution-inert	1. Silicon dissolution
	  a. Pore formation at low potentials
	  b. Electropolishing at high potentials

2. H₂ gas evolution	2. Forward-biased Schottky
	  i-V curves exponential
	  Tafel slope 60 mV
	  No apparent illumination effects

3. High hydrogen overpotential	3. Two current "peaks"
	  a. Lower potential, anodic electropolishing oxide
	  b. Oscillations at higher potential "peak"

4. Reverse-biased Schottky	
5. Photogenerated currents proportional to light intensity	
n-type Si	

1. No silicon dissolution-inert	1. Silicon dissolution
	  a. Pore formation at low potentials
	  b. Electropolishing at high potentials

2. H₂ gas evolution	2. Reverse-biased Schottky
	  High breakdown voltage before significant pore formation
	  Photogenerated currents, function of intensity and voltage

3. Low hydrogen overpotential	3. Single current peak without illumination a. Anodic oxide for electropolishing
4. Forward-biased Schottky	4. Second current peak with illumination a. Oscillations
5. No apparent illumination effects	

Note: In the table above, the chemical formula for hydrogen gas is H_2 and current density refers to $i\text{-}V$ curves.

constant current density: 10 mA/cm² for 10 min at 25°C. Samples are rinsed in deionized water, then in ethanol, and later in pentane. To obtain NPNPs, the PSi films are sonicated in H₂O/DMF for about 60 min and then ultra-

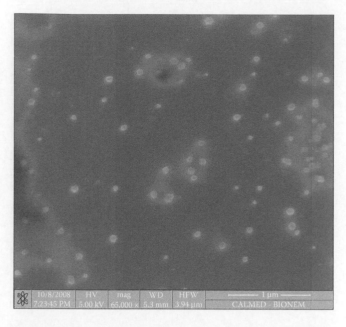

FIGURE 22.17 Electron scanning image of NPNPs with an average diameter of about 50–100 nm.

sonicated (5W) in water for 10 min at a constant temperature of 4°C. Finally, the fabricated NPNPs were filtered to eliminate impurities larger than desired medium size. Electron scanning image of the fabricated NPNPs is reported in Figure 22.17. The NPNPs exhibit an average diameter of about 50–100 nm, whereas their pore size is too small to be evidenced with electron scanning microscope. Nevertheless, their characteristic luminescence, with an emission spectrum centered at around 620 nm, confirms a pore size of a few nanometres (Figure 22.18). The average diameter of NPNPs can be adjusted modifying the time and/or the power of ultrasonication. For instance, decreasing the time of ultrasonication from 10 to 1 min, the average diameter increases from 50 to 200 nm (see Figure 22.19).

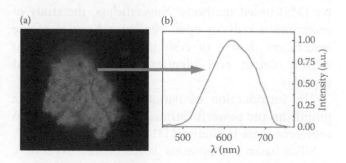

FIGURE 22.18 (a) Optical image of luminescent NPNPs (UV illumination); (b) NPNPs emission spectra. (Reprinted from ref. [37] Pujia, A. et al. 2010. Highly efficient human serum filtration with water-soluble nanoporous nanoparticles. *Int. J. Nanomed.* 5, 1005–1015. With permission from Dove Medical Press Ltd.)

Chapter 22

FIGURE 22.19 Electron scanning image of NPNPs with an average diameter of about 200 nm.

22.3.2 Nanoporous NPs as Nanosponges

Nowadays, the LMW fraction of human serum is the most informative source of biomarkers, but their study and identification are very difficult due to the incredible complexity of the raw human serum. Biomarkers discovery from biological fluids—and blood in particular—appears a challenging task, due to the tremendous number of biomolecular species, with differences of many orders of magnitude in their relative abundance. In the last few years, different approaches have been proposed for the study of the proteome, and some of them seem to hold the greatest potentials. The most common are two-dimensional-gel electrophoresis (2D-PAGE) and mass spectrometry (MS)-based methods. Nevertheless, the study of proteome is still an open challenge because of the insufficient degree of resolution and sensitivity to reliably detect and identify LMW low-abundant peptides.

The introduction of nanotechnologies appears a promising and powerful strategy that can be applied to overcome these limitations [31–35].

Silica-based nanoporous surfaces and NPs were developed in order to capture LMW peptides from human plasma. Human plasma was directly applied to a silica-based surface for selective peptide enrichment. After washing the surface, adsorbed molecules were extracted using classical matrix-assisted laser desorption/ionization (MALDI) matrix solutions. The method proved effective at isolating LMW peptides from this complex biological fluid [36], but recent advance shows that further improvement can be achieved exploiting NPs.

In fact, from this point of view, NPNPs exhibit three important properties [37,38]: (a) their ability to absorb small molecules depending on nanopores size; (b) their separability from solution through an efficient centrifugation (the NP density is higher than the solvent); and (c) their capability to be dissolved in water.

These three properties enable a fast protocol for the harvesting of small molecules from complex solution. NPNPs can be incubated with the desired biofluid, to let LMW fraction be absorbed into the nanopores. Later, the NPNPs can be separated from the starting fluid by means of centrifugation, and then resuspended in water where they dissolve releasing the harvested molecules. Two simple experiments are described to show the harvesting capability of the NPNPs.

Experiment 1. Interaction with one component solution: small dyes of different molecular weights.

NPNPs were incubated with two solutions of fluorescent polymer of different molecular weights (6 and 14 kDa dextran-FITC 10 μg, NPNP 5 μg, water 10 μL, 1 h).

After the incubation, the NPNPs were separated from supernatant (centrifugation); the recovered NPNPs can be dissolved in water with a rate dependent on the temperature and medium acidity. At pH = 8 and 90°C the dissolution takes a few minutes, but it can be also carried out at room temperature in a few hours if nondenaturing conditions are needed. For pH <5, the NPNPs do not dissolve allowing long-term storage in water at room temperature.

After the centrifugation, NPNPs were dropped on slide, dried, and analyzed with fluorescence microscopy.

In Figure 22.20, optical and fluorescence images of the NPNPs after incubation with two fluorescent polymers of different MW are reported: dextran-fitc 14 kDa (panels a and b), and dextran-fitc 6 kDa (panels c and d). The absorption of the lighter polymer is clearly indicated by the green fluorescence emitted from the NPs (panel d). On the contrary, in the case of heavier polymer no green fluorescence can be observed (the blue one is from salt residues), confirming the molecular weight cutoff. The amount of harvested fraction of dextran-FITC (6 kDa MW) can be evaluated comparing the total fluorescence intensity of two split

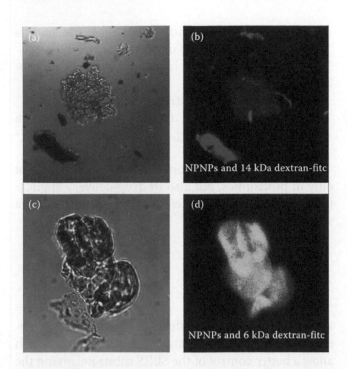

FIGURE 22.20 **(See color insert.)** Optical and fluorescence images of the NPNPs after incubation with two fluorescent polymers of different MW: dextran-fitc 14 kDa (panels a and b), and dextran-fitc 6 kDa (panels c and d). The adsorption of the lighter polymer is clearly indicated by the green fluorescence emitted from the NPs (panel d). On the contrary, in the case of heavier polymer no green fluorescence can be observed (the blue one is from salt residues), confirming the molecular weight cutoff. (Reprinted from Pujia, A. et al., 2010. *J.R. Soc. Interface* 1, 79; Pujia, A. et al. 2010. *Int. J. Nanomed.* 5, 1005–1015. With permission from Dove Medical Press Ltd.)

solution (harvested vs. supernatant). They exhibit very similar values, indicating that, under the experimental conditions described above, 50% of the molecules were harvested, while the remaining 50% was left in the solution, meaning a very high loading capacity of about 1 µg of harvested molecules for each µg of NPNPs (rough estimation).

Experiment 2. Interaction with a complex mixture of proteins with a wide range of molecular weights simulating a biological fluid.

The ability of NPNPs to enrich the LMW fraction of a complex mixture was tested with an "ad hoc" protein mixture. The protein mixture was prepared by mixing 50% (v/v) human serum albumin (MW 66,000 Da, Sigma Aldrich); 30% (v/v) bovine plasma gamma globulin (heavy-chain MW 45,000 Da, light-chain MW 30,000 Da, Biorad); 20% (v/v) aprotinin (6500 Da, Sigma Aldrich). All proteins were dissolved at the concentration of 1 mg/mL in 100 mM sodium phosphate buffer, 9% (w/v) sodium chloride, pH 7.4, and (PBC) to reproduce physiological conditions.

A volume of 200 µL of protein mixture was incubated with 5 µg of NPNPs (about 10^{15} particles) at room temperature for 1 h. NPNPs were then separated from the supernatant by centrifugation, and washed once with the phosphate buffer saline (PBS) buffer. The corresponding SDS-PAGE in Figure 22.21a clearly shows that NPs selectively retain small molecules as

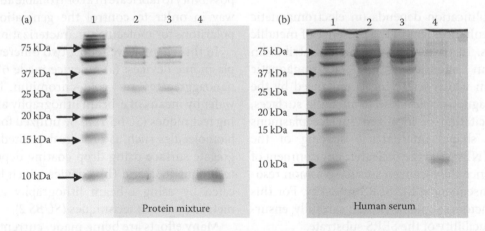

FIGURE 22.21   (a) The sodium dodecyl sulfate polyacrylamide gel electrophoresis (SDS-PAGE) analysis of protein mixture before and after incubation with nanoporous Si particles. A quantity of 100 mL of protein mixture (see text and ref. 37 for details) was subjected to incubation with NPs. Aliquots of the mixture before and after incubation were subjected to tris-tricine SDS-PAGE and stained with Coomassie Brilliant Blue. *Lane 1 A*: molecular weight markers; *lane 2A*: protein mixture before incubation; *lane 3A*: protein mixture following NP incubation (supernatant); *lane 4A*, LMW protein fraction enriched (pellet). (b) SDS-PAGE analysis of human serum before and after incubation with nanoporous silicon particles (see text and ref. 37 for details). Serum was diluted to 1:2 with 100 mM sodium phosphate buffer, 9% (p/v) sodium chloride, ph 7.4, and incubated with NPs. Aliquots of serum before and after incubation were subjected to tristricine SDS-PAGE and stained with Coomassie Brilliant Blue. *Lane 1B*, molecular weight markers; *lane 2B*, crude human serum; *lane 3B*, human serum following NP incubation (supernatant); *lane 4B*, LMW serum fraction enriched using (nondenaturing conditions) nanoporous silicon particles (pellet). (Reprinted from ref. 37 Pujia, A. et al. 2010. Highly efficient human serum filtration with water-soluble nanoporous nanoparticles. *Int. J. Nanomed.* 5, 1005–1015. With permission from Dove Medical Press Ltd.)

aprotinin (MW = 6500 Da), whereas proteins with higher MW are completely excluded from the nanopores. Finally, similar protocol can be directly applied to raw human serum [37,38], allowing to selectively enrich the LMW fraction (see Figure 22.21b).

The above experiments compared with the data in the literature show many remarkable characteristics of NPNPs:

- Cheap and easy of production.
- Controllable pore size and well-known surface chemistry.

- Capability of absorbing small molecule with a tuneable weight cutoff in molecular weight.
- Biocompatibility and biodegradability in physiological solution.
- Easy to recover by centrifugation.
- Wet storing for long term (water solution with pH < 5), or dry storing.

We note that such properties play vital roles not only in proteomics, but also in drug-delivery applications, where the development of novel nanocarriers is a subject of interest.

22.4 Single-Molecule Detection

Detection of single molecules has a great interest in scientific world for its utilities in medicine, pharmacology, chemistry, biology, and environmental science. Raman spectroscopy plays an important role in identification and characterization of molecules, but the Raman signal for diluted solutions is much lower with respect to florescence, limiting the identification of the material with low concentration [39–40]. SERS is a technique that permits the detection of adsorbed molecules on noble metal surfaces, such as Au, Ag, Cu, and so on at subpicomolar concentration.

Raman amplification depends on electromagnetic and chemical enhancement in the presence of metallic nanostructures. In particular, the main contribution to the Raman intensity is the electromagnetic phenomenon in which scattering is enhanced by the local electromagnetic field close to metallic surfaces, due to the excitation of localized surface plasmons [41–48]. Size, shape, interparticle spacing of the nanoparticles (NPs), and the dielectric environment of material influence the localized surface plasmon resonance and, consequently, the SERS intensity. For this reason, these factors should be chosen carefully, ensuring the reproducibility of the SERS substrate.

Many early SERS substrates used a random roughening of the surface so that only small uncontrolled areas of the total metal surface would have the correct geometry for Raman enhancement. Most traditional way of preparing an SERS surface is the chemical way; preparing Ag colloids and attaching the protein/peptides to these colloids [49,50]. SERS devices are also fabricated by means of electrochemically modified electrodes, island films [51,52], and regular particle arrays [53,54]. Recently, efforts are being made to fabricate well-defined nanolithographic structures which allow a better control of the SERS substrate, giving the possibility to fabricate controllable and reproducible devices [55–62].

Most well-defined SERS substrates are obtained by e-beam lithography technique and metal deposition (electroplating or electroless deposition [53]), in particular, using silver or gold, which allow a good enhancement of the Raman signal. The choice of e-beam lithography technique is because of the better possibility to fabricate in a controllable and reproducible way in order to control the generation of plasmon polaritons for molecular characterization.

In this chapter, we report two different novel nanoplasmonic devices: (a) a device made of array of gold nanoaggregates on gold–chromium base-plated Si wafer by means of e-beam lithography and electroplating techniques (*SUBS 1*) to be utilized for detecting the biomolecules such as protein deposited on a nanoaggregate surface using drop coating deposition (DCD) technique [63], and (b) a device which has been fabricated by using e-beam lithography and electroless metal deposition techniques (*SUBS 2*).

Many efforts are being made, currently, to fabricate reproducible periodic nanostructured SERS devices using e-beam and nanosphere lithography to have better control on generation of surface plasmon polariton (SPP). Kahl et al. [53] fabricated SERS devices based on Ag metal, which illustrates Raman enhancement for an organic compound (thiobenzene). However, ought to the inherent properties of Ag surface oxidation in air, the device durability as an SERS substrate is limited. The SERS device (*SUBS 1*) presented here for the generation of SPP is made of an array of gold nanoaggregates on gold–chromium base-plated Si wafer in order to detect the myoglobin (Mb) protein of low concentration. The presence of gold nanoaggregates array and e-beam patterning down to 10 nm of interspatial distance between two nanoaggregates are the major fabrication novelty of this device for SERS application. Mb protein is deposited on the nanostructure using the DCDR method [63]. Various SERS measurements are carried out for Mb deposited by varying the concentration and by changing the position on the nanostructure. The advantage of gold nanoaggregates over silver nanoaggregates is its chemical inactivity, having similar dielectric constant in the near-infrared region (NIR) (830 nm). The NIR laser is used due to the fact that the fluorescence background can be strongly quenched, giving an increase in 2–3 order of Raman intensity with respect to the fluorescence intensity because in a given time the Raman photon vibrational relaxation time is much shorter than the electronic relaxation time [42,64]. In this chapter, we show the detection of very small concentration of Mb down to attomole using this gold nanoaggregate SERS substrate.

Since there are critical fabrication complexities regarding matter growth using electroplating technique, an active SERS substrate (*SUBS 2*) is realized assembling site-selective silver NPs, by electroless technique, on the nanopatterned Si substrate with different shapes and sizes. [65]. In this chapter, the nanopattern-based substrate, obtained by e-beam lithography, has the function of controlling the diffusion of the metal particles on the surface. On the other hand, the deposition of metallic (silver or gold) nanospheres permits plasmonic enhancement. The electroless deposition on a substrate is based on an autocatalytic or a chemical reduction of aqueous metal ions. This process consists of an electron exchange between metal ions and a reducing agent [66]. In some cases the substrate is the catalyst for the reaction, instead, in the study of Coluccio et al. [65], the Si substrate itself is the reducing agent. Recently, a simple silver deposition method has just been developed, based on a fluoridric acid (HF) solution containing silver nitrate (AgNO$_3$) [67–71] where Ag is reduced to metal form by the Si substrate oxidation. However, Coluccio et al. reported the fabrication of well-reproducible SERS substrates without dendritic structures, typical of the electroless deposition technique, for the detection of rhodamine 6G (R6G) and benzenethiol with a concentration of 10^{-20} M.

22.4.1 Experimental

22.4.1.1 Device Fabrication

22.4.1.1.1 Periodic Gold Nanoaggregate Structure (*SUBS 1*) Electron beam lithography (EBL) and electroplating techniques are used for the fabrication of SERS device. Silicon wafer with a base-plating of a bilayer of chromium (10 nm) and gold (20 nm) is used as substrate. The wafer is prebaked on a covered hot plate for 30 min at 170°C. Upon the deposition (spin time 30 s at 1500 rpm) of a 20 nm layer of Omnicoat (MicoChem) to improve the adhesion of gold, a high-resolution positive electron resist (ZEP—520A) is spin-coated for 60 s at 2000 rpm to obtain a 80-nm-thick layer of ZEP and thus a 100-nm-thick final layer (ZEP—80 nm, Omnicoat—20 nm). Prior the EBL exposure, the sample is prebaked at 170°C for 3 min to remove the solvent from the resist. A pattern of 5×5 µm dots matrices is written on the sample using an EBL system (CRESTEC), employing a 50 keV acceleration voltage. Disk-like structures are obtained from a nominal pattern comprising arrays of concentric squares due to proximity effects. A 330 µC/cm^2 dose and 12 nm^2 pixels scanning resolution are set for the exposure. The sample is then immersed for 35 s in ZEP developer (ZED–N50®) to selectively remove the exposed resist. Si wafer is then processed using reactive ion etching in oxygen for 40 s to remove any residual resist from the disks. Arrays of empty disks with 80 nm of diameter and an interspatial gap ranging from 10 to 40 nm are finally obtained (Figure 22.22a).

The successive gold growth is accomplished in an electrolytic system using a solution of gold–potassium cyanide, keeping a density current of 10 mA/cm^2 for 8 s. Due to resist confinement, the gold grains could grow solely inside the holes. The grains assemblies are of 100 nm diameter and of 80 nm thickness with an interspatial gap of 10–30 nm. A scanning electron microscopy (SEM) micrograph of the grain assemblies are taken to assess uniformity and reproducibility (Figure 22.22b). The residual resist around the disk is finally stripped using a solvent. Figure 22.22c shows the optical image of the SERS substrate: positions "X" and "Y" show the location with and without nanospheres.

22.4.1.1.2 Site-Selective Metal NP Deposition (*SUBS 2*) The nanostructures are written by means of an ultrahigh-resolution e-beam lithography (CRESTEC CABL–9000C 50 KeV acceleration voltage) onto a clean silicon wafer (100) spin-coated by a

Chapter 22

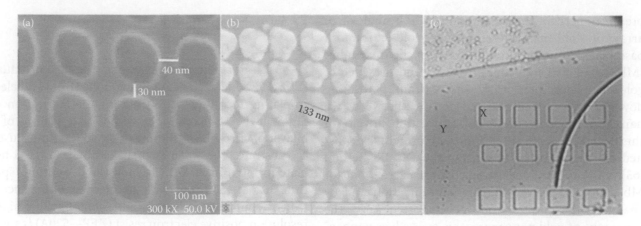

FIGURE 22.22 (a) Array of empty disk ready for successive gold growth; (b) SEM images of gold nanostructures; (c) optical image of Mb DCD spot. In figure position (Y) is referred to the place where no nanostructures are placed, and position (X) is the place where nanostructures have been developed. (Panels a,b reprinted from *Biosensors and Bioelectronics*, 24, Das, G. et al., Nano–patterned SERS substrate: Application for protein analysis vs.temperature, 1693–1699, Copyright (2009), with permission from Elsevier; panel c reprinted from *Microelectronic Eng.*, 85(5–6), Das, G. et al., Attomole (amol) myoglobin Raman detection from plasmonic nanostructures, 1282–1285, Copyright (2008), with permission from Elsevier.)

50-nm-thick ZEP layer. Substrates of different shapes are prepared with a maximum dimension size in the range of 50–1000 nm. Silver nanospheres are deposited on the substrate by means of an electroless method, in which the patterned silicon wafer is dipped in a 0.15 M HF (hydrofluoric acid) solution containing 1 mM silver nitrate for different times (10–60 s). The samples with 60 s deposition time are used for Raman analysis as these samples are showing better Raman response.

The driving force in this process is the difference between redox potentials of the two half-reactions, which depends on solution temperature, concentration, and pH. Consequently, these parameters influence the particles size and density.

Substrates with bimetallic electroless deposition (silver on gold and vice versa) [72,73] and substrates with gold deposition are also prepared to compare their SERS activity with Ag-deposited samples. Au is electroless deposited on nanopatterned silicon from a solution of 1 mM Au salt and 0.15 M HF, for 3 min. The bimetallic deposition is obtained with the same solution utilized in two steps: in the case of the Ag/Au-deposited substrate, the first step is silver electroless deposition for 20 s and the last step is Au electroless deposition for 30 s; on the contrary, for Au/Ag sample, the first step is Au electroless deposition for 20 s and the last step is Ag electroless deposition for 30 s. The R6G monolayer is deposited by soaking the substrate in an aqueous solution at known concentration ranging from 10^{-4} to 10^{-20} M for 30 min and then the sample is rinsed with water to remove the excess molecules not attached directly to the metal surfaces and finally dried with N_2.

SEM images of the Si substrate with lithography and after silver deposition are shown in Figure 22.23.

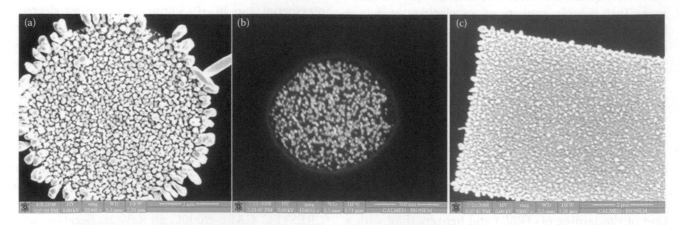

FIGURE 22.23 NPs deposition on microstructure on Si wafer: (a) Ag NPs; (b) Au NPs; and (c) Au/Ag NPs.

The Ag nanospheres array grew into the holes started from the wet Si surface by $AgNO_3$ solution. When the deposition size geometry is 1–2 μm regular, the silver nanospheres were distributed on the Si surface. Increasing the growth time, the process of aggregation generates silver dendritic structures formation (not shown in the figure).

22.4.1.2   Sample Preparation

22.4.1.2.1   *SUBS 1*   Mb, dissolved in water with different concentration (7, 30, 60, and 120 μM), of 2 μL volume was microdeposited, following the DCDR method, on the best gold substrate. To be noted, Zhang et al. [51] carried out many measurements for the aqueous samples and their corresponding DCDR samples, and observed that the spectra resemble each other. Moreover, after each deposition, it was ensured through the optical image that the ring (inner diameter—3.5 mm and width—0.08 mm) of adsorbed protein remains on the nanograin aggregates array, as shown in Figure 22.22c. The Raman measurements have been carefully performed at two positions: (1) on nanograin aggregate "X," (2) on no nanostructures "Y," by paying attention that at both position protein concentration remains equivalent.

22.4.1.2.2   *SUBS 2*   The R6G monolayer is deposited by soaking the substrate in an aqueous solution at known concentration ranging from 10^{-4} to 10^{-20} M for 30 min and then the sample is rinsed with water and dried with N_2. The laser power is always kept minimum so that the change in molecular bonding due to laser heating can be avoided. Benzenethiol monolayer is also deposited on the *SUBS 2* substrate following the same procedure as for R6G.

22.4.1.3   Characterization Technique

Microprobe Raman spectra were excited by visible laser with NIR laser with 830 nm (*SUBS 1*) from diode laser and by 514 nm (*SUBS 2*) laser line from Ar+ laser in backscattering geometry for both cases through 50X long-range objective (NA = 0.50, *SUBS 1*) and 50X (NA–0.75, *SUBS 2*) with resolution of about 1.1 cm⁻¹. The laser power was fixed to 6 mW (830 nm) and 0.018 mw (514 nm) whereas the accumulation time is kept to 100 s for *SUBS 1* and 60 s for *SUBS 2*. The MB protein, in case of *SUBS 1*, was always verified through optical image before and after the measurements to ensure that there is no protein damage through laser radiation. The Mb Raman spectrum on SERS substrate is shown in Figure 22.24. The SERS substrate Raman spectrum is shown in inset of Figure 22.24.

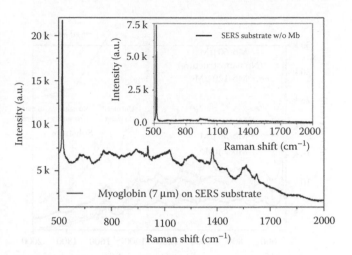

FIGURE 22.24   SERS substrate spectrum of Mb protein with a concentration of 7 μM. In the inset, the Raman spectrum of SERS background is shown. (Reprinted from *Biosensors and Bioelectronics*, 24, Das, G. et al., Nano–patterned SERS substrate: Application for protein analysis vs. temperature, 1693–1699, Copyright (2009), with permission from Elsevier.)

22.4.2   Results and Discussions

22.4.2.1   Mb Protein on Gold Nanoaggregate (*SUBS 1*)

SERS measurements are carried out for Mb deposited on a nanostructure using the drop coating deposition Raman (DCDR) technique [53,63] by varying concentration and by changing the position on SERS substrate. It should be noted that the dried but hydrated protein using DCDR is capable of detecting the biomolecules with shorter integration times with respect to the conventional Raman protein measurements. Gold nanograin aggregates show advantages over silver nanograin aggregates, having similar dielectric constant in NIR, because of its chemical inactivity. NIR laser (830 nm) excitation light source is used because the fluorescence background of Raman signal can be strongly quenched and the probability of sample damaging is very low. Raman effect in this case contributes total enhancement from the SERS substrate.

Raman spectroscopy provides the structural and chemical information of substances. Because of very low scattering cross section of water, this technique is very much suitable for biological substances. Mb protein, as in this case, of ~4.9 nm diameter and of an almost spherical molecular structure contains one planar Fe–protoporphyrin prosthetic heme group implanted in polypeptide chains, containing a very high amount of α-helix (75%). The major spectroscopic properties of Mb contribute from π–π* transitions within these heme groups. The absorption band of the

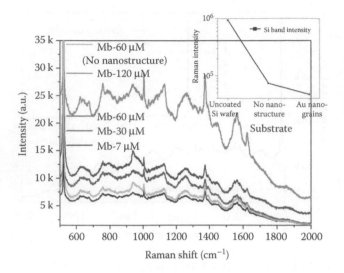

FIGURE 22.25 **(See color insert.)** Raman spectra for Mb protein by varying the concentration but by keeping the laser spot centered at nanostructures. The inset graph shows the background Raman signal of the Silicon for different substrate topologies. (Reprinted from *Microelectronic Eng.*, 85(5–6), Das, G. et al., Attomole (amol) myoglobin Raman detection from plasmonic nanostructures, 1282–1285, Copyright (2008), with permission from Elsevier.)

stretching vibrations of ring π–bands falls in the 1000–1700 cm^{-1} region. SERS spectra, carried out on the gold nanoaggregates plasmonic nanostructure substrate, for Mb with fixed concentration are illustrated in Figure 22.25 in the 500.1800 cm^{-1} region. Figure 22.25, shows the well-known vibrational band for Mb centered at 1126, 1373, and 1560 cm^{-1} which are attributed to the C–N stretching [74], an oxidation marker band of heme iron [75] and C–C vibrational band, respectively. A sharp peak at about 520 cm^{-1} in Raman spectra, as shown by an arrow in Figure 22.25, arises from Si wafer over which the SERS substrate is fabricated. Raman protein spectra show some unique intense peaks at 760, 1011, 1365, and 1554 cm^{-1}, and could be attributed to the different vibrational modes of tryptophan (Trp), whereas the bands around 1005 and 1034 cm^{-1} could be relied to the phenylalanine (Phe) residues. Figure 22.25 shows that with a decrease in Mb concentration there is a decrease in intensity which is obvious because a higher number of Mb could be attached to the gold nanograins with the increase in protein concentration.

The measurement has also been performed for Mb deposited on c.Si wafer. A faint peak of Phe residue is observed, whereas rest of the protein Raman is covered by strong Si Raman band (not shown here). The intensity of Si band (520 cm^{-1}) for various measurements carried out for Mb on Si wafer, nanograin aggregates, and on a base plate without nanostructures is illustrated in the

inset of Figure 22.25. Noticeably, protein Raman is not observed at positions farther away from the ring. Here, all the calculations are being carried out for 7 120 µM protein (concentration of our protein) to estimate the quantity of proteins being detected. If we suppose that all the protein is located on the ring of DCD mark, then the protein surface density will be approximately 15–270 pmol/mm^2 in the ring and if the proportion is 1/10 or 1/3, then the protein surface density would be 8–235 pmol/mm^2 or 3–100 pmol/mm^2, respectively. Realistically, the actual protein surface density could be around 5–10 pmol/mm^2 and 140–230 pmol/mm^2 for 7 and 120 µM protein, respectively. By assuming the size of one Mb molecule to be 20 nm^2, the minimum amount of Mb within the focal spot of laser (radius: 1 µm) could, therefore, be estimated around 10–240 attomole.

22.4.2.2 Site-Selective Metal NP Deposition (*SUBS 2*)

R6G is deposited by immersing the Ag-, Au-, Ag/Au- and Au/Ag-based SERS substrates into the R6G aqueous solution for 30 min. In Figure 22.26a, Raman spectra of R6G molecules, obtained after 60 s silver deposition time, are shown for fixed concentrations (10^{-5}, 10^{-16}, and 10^{-20} M). There are various bands associated to R6G molecules at about 1650, 1509, and 1361 cm^{-1}, attributed to the xanthenes ring stretching of C–C vibrations, 1575 cm^{-1} attributed to C=O stretching; while the band at 1,183 cm^{-1} is related to the C–H bending and N–H bending vibration of xanthenes ring. As can be clearly observed in Figure 22.26a, the Raman intensity decreases with decreasing concentration of R6G, as expected. Although the concentration is very low, various bands are clearly visible even for R6G molecules with a concentration of 10^{-20} M, keeping the measurement parameters invariable.

As it is well known that

a. Silver-based SERS substrate shows very high Raman scattering enhancement, while the silver metal exhibits the surface oxidation in air due to its inherent properties. The device durability as a SERS substrate is limited.

b. Gold-based SERS substrates show better durability but exhibit lower SERS enhancement and so bimetallic substrates are also fabricated. In order to estimate SERS behavior, Raman measurements have been carried out on Ag, Au, Au–core/Ag–shell, and Ag–core/Au–shell metallic SERS substrates.

SERS response of the silver-deposited substrates for the R6G monolayer is compared with that of gold, gold/

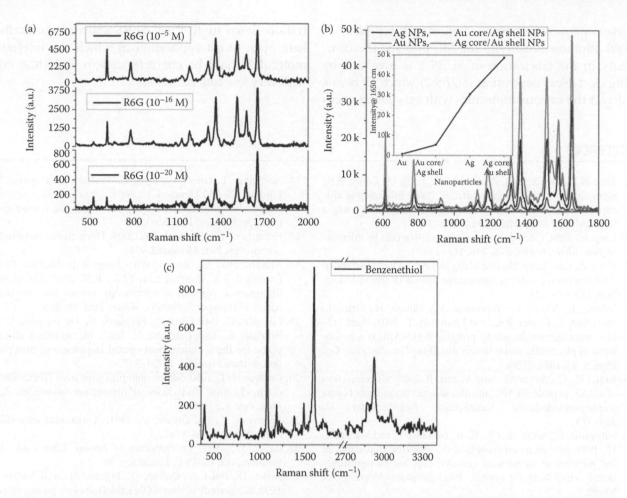

FIGURE 22.26 **(See color insert.)** Raman spectra of (a) R6G at various concentrations on Ag-based *SUBS 2*; (b) R6G with a concentration of 10^{12} M, absorbed on the silver, gold, Ag/Au- and Au/Ag-based *SUBS 2*; (c) benzenethiol [10 mM] absorbed on Ag-based *SUBS 2*. (Reprinted from *Microelectronic Eng.*, 86(4–6), Coluccio, M.L. et al., Silver–based surface enhanced Raman scattering (SERS) substrate fabrication using nanolithography and site selective electroless deposition, 1085–1088, Copyright (2009), with permission from Elsevier.)

silver, and silver/gold-deposited substrates. Raman spectra of 10^{-12} M R6G, deposited on various kinds of above-described substrates are shown in Figure 22.26b. Raman spectra show similar vibrational bands for individual and mixed metallic nanospheres-based substrates. It is clearly evident from Figure 22.26b that Raman signal intensity is increasing, respectively, for Au, Au/Ag, Ag, and Ag/Au samples. As it is shown in the inset of Figure 22.26b, the Raman intensity of reference vibrational band centered at 1,649 cm^{-1} shows around 800 counts for gold metal and around 30,000 counts for Ag nanometals, whereas about 48,700 counts for Ag–core/Au–shell mixed nanometallic structures. In this last substrate, it seems that gold has a protection function against the Ag oxidation and sulfur passivation, thus allowing a better SERS activity.

Benzenethiol molecule is also investigated with an Ag-based SERS substrate. Figure 22.26c shows the baseline-corrected Raman spectrum of benzenethiol. Various bands are observed at about 390, 1078, 1593 cm^{-1} and in the range of 2820–3120 cm^{-1} which are attributed to the C–C rocking, ring breathing, ring stretching, and C–Hx stretching vibrations, respectively.

22.4.3 Conclusions

We have fabricated two different controllable and reproducible SERS substrates: (a) nanograin aggregates array with internanoaggregate space down to 10 nm by means of electroplating and e-beam lithography techniques deposited with Mb using the DCD method, (b) site-selective nanocrystalline metal deposition using e-beam lithography and electroless metal deposition technique, deposited with R6G, and benzenthiol molecules using the immersion technique. SERS measurements on plasmonic nanostructures (*SUBS 1*) reveal attomole sensitivity for Mb protein. The complexity in the fabrication of *SUBS 1* is overcome by using electroless deposition (*SUBS 2*) which allows us to detect the various molecules with very low concentration down to 10^{-20} M. SERS substrates described here open broad application in which the interested molecules could be characterized in real time even with very low concentration.

References

1. Tran, N.D., Liu, X., Yan, Z., Abbote, D., Jiang, Q., Kmiec, E.B., Sigmund, C.D., and Engelhardt, J.F., 2003. Efficiency of chimeraplast gene targeting by direct nuclear injection using a GFP recovery assay. *Mol. Ther.* 7, 248–253.

2. King, R., 2004. Gene delivery to mammalian cells by microinjection. *Methods Mol. Biol.* 245, 167–174.

3. Zhao, Z., Cao, Y., Li, M., and Meng, A., 2001. Double-stranded RNA injection produces nonspecific defects in zebrafish. *Dev. Biol.* 229, 215–223.

4. Yokota, E., Vidali, L., Tominaga, M., Tahara, H., Orii, H., Morizane, Y., Hepler, P.K., and Shimmen, T., 2003. Plant 115-kDa actinfilament bundling protein, P-115-ABP, is a homologue of plant villin and is widely distributed in cells. *Plant Cell Physiol.* 44, 1088–1099.

5. Kim, J.H., Creekmore, E., and Vezina, P., 2003. Microinjection of CART peptide 55-102 into the nucleus accumbens blocks amphetamineinduced locomotion. *Neuropeptides* 37, 369–373.

6. Sotoyama, H., Saito, M., Oh, K.-B., Nemoto, Y., and Matsuoka, H., 1998. In vivo measurement of the electrical impedance of cell membranes of tobacco cultured cells with a multifunctional microelectrode system. *Bioelectrochem. Bioenerg.* 45, 83–92.

7. McAllister, D.V., Wang, P.M., Davis, S.P., Park, J.H., Canatella, P.J., Allen, M.G., and Prausnitz, M.R., 2003. Microfabricated needles for transdermal delivery of macromolecules and nanoparticles: fabrication methods and transport studies. *Proc. Natl. Acad. Sci. U.S.A.* 100, 13755–13760.

8. Perennes, F., Marmiroli, B., Matteucci, M., Tormen, M., Vaccari, L., and Di Fabrizio, E., 2006. Sharp beveled tip hollow microneedle arrays fabricated by LIGA and 3D soft lithography with polyvinyl alcohol. *J. Micromech. Microeng.* 16, 473–479.

9. Service, F., 2008. Report faults U.S. strategy for nanotoxicological research. *Science* 322, 1779.

10. Fischer, H.C. and Chan, W.C.W., 2007. Nanotoxicity: The growing need for in vivo study. *Curr. Opin. Biotechnol.* 18, 565–571.

11. Lynch, I. and Dawson, K.A., 2008. Protein–nanoparticle interactions. *Nanotoday* 3(1–2), 40.

12. Jiang, W., Kim, B.Y.S., Rutka, J.T., and Chan, W.C.W., 2008. Nanoparticle-mediated cellular response is size-dependent. *Nat. Nanotechnol.* 3, 145.

13. Puntes, V.F., Krishnan, K.M., and Alivisatos, A.P., 2001. Colloidal nanocrystal shape and size control: The case of cobalt. *Science* 291, 2115.

14. McAnally, G.D., Everall, N.J., Chalmers, J.M., and Smith, W.E., 2003. Analysis of thin film coatings on poly(ethylene terephtalate) by confocal raman microscopy and surface-enhanced Raman scattering. *Appl. Spectrosc.* 57, 44.

15. Kneipp, K., Kneipp, H., Itzkan, I., Dassari, R.R., and Feld, M.S., 2002. Surface enhanced raman scattering and biophysics. *J. Phys. Condens. Matter.* 14, R598.

16. Notingher, I., Green, C., Dyer, C., Perkins, E., Hopkins, N., Lindsay, C., and Hench, L.L., 2004. Discrimination between ricin and sulphur mustard toxicity in vitro using Raman spectroscopy. *J. R. Soc. Interface* 1, 79.

17. Helmchen, F., and Denk, W., 2005. Deep tissue two-photon microscopy. *Nat. Methods* 2, 932.

18. Alexandrakis, G., Brown, E.B., Tong, R.T., McKee, T., D., Campbell, R.B., Boucher Y., and Jain, R.K., 2004. Two-photon fluorescence correlation microscopy reveals the two-phase nature of transport intumors. *Nature Med.* 10, 203.

19. Liberale, C., Minzioni, P., Bragheri, F., De Angelis, F., Di Fabrizio, E., and Cristiani, I., 2007. Miniaturized all-fibre probe for three-dimensional optical trapping and manipulation. *Nature Photonics* 1, 723–727.

20. Canham, L.T., 1990. Silicon quantum wire array fabrication by electrochemical and chemical dissolution of wafers. *Appl. Phys. Lett.* 57, 1046.

21. Lehmann, V., and Gijsele, U., 1991. A quantum wire effect. *Appl. Phys. Lett.* 58, 8568.

22. Bellet, D., 1997. In *Properties of Porous Silicon*, ed., L.T. Canham, IEE INSPEC, London, p. 38.

23. Bellet, D., Billat, S., Dolino, G., Ligeon, M., and Muller, F., 1993. X-ray study of the anodic oxidation of p+ porous silicon. *Solid State Commun.* 86, 51.

24. Perez, J.M., Billabo, J., McNeill, P., Prasad, J., Cheek, R., Kelber, J., Stevens, J.P., and Glosser, R., 1992. Direct evidence for the amorphous silicon phase in visible photoluminescent porous silicon. *Appl. Phys. Lett.* 61, 563.

25. Schupper, S., Friedman, S.L., Marcus, M.A., Adler, D.L., Xie, Y.H., Ross, F.M., Harris, T.D. et al., 1994. Dimensions of luminescent oxidized and porous silicon structures. *Phys. Rev. Lett.* 72, 2648.

26. Vaccari, L., Canton, D., Zaffaroni, N., Villa, R., Tormen, M., and Di Fabrizio, E., 2006. Porous silicon as drug carrier for controlled delivery of doxorubicin anticancer agent. *Microelectronic Eng.* 83, 1598–1601.

27. Cullis, A.G. and Canham, L.T., 1991. Visible light emission due to quantum size effects in highly porous crystalline silicon. *Nature*, 353, 335.

28. Föll, H., Christophersen, M., Carstensen, J., and Hasse, G., 2002. Formation and application of porous silicon. *Mater. Sci. Eng.: R: Reports*, 39, 93–141.

29. Lehmann, V. and Gijsele, U., 1991. A quantum wire effect. *Appl. Phys. Lett.* 58, 8568.

30. Smith, R. L. and Collins, S. D., 1992. Porous silicon formation mechanisms. *J. Appl. Phys.* 71, R1–R22.

31. Fu, J., Mao, P., and Han, J., 2008. Artificial molecular sieves and filters: A new paradigm for biomolecule Separation. *Trends Biotechnol.* 26, 311–320.

32. Luchini, A., Geho, D.H., Bishop, B., Tran, D., Xia, C., Dufour, R.L., Jones, C.D. et al., 2008. Smart hydrogel particles: Biomarker harvesting: One-step affinity purification, size

exclusion, and protection against degradation. *Nano Lett.* 8, 350–361.

33. Cheng, M.M., Cuda, G., Bunimovich, Y.L., Gaspari, M., Heath, J.R., Hill, H.D., Mirkin, C.A. et al., 2006. Nanotechnologies for biomolecular detection and medical diagnostics. *Curr. Opin. Chem. Biol.* 10, 11–19.

34. Luo, C., Fu, Q., Li, H., Xu, L., Sun, M., Ouyang, Q., Chen, Y., and Ji, H., 2005. PDMS microfludic device for optical detection of protein immunoassay using gold nanoparticles. *Lab Chip.* 5, 726–729.

35. Cojoc, D., Difato, F., Ferrari, E., Shahapure, R.B., Laishram, J., Righi, M., Di Fabrizio, E.M., and Torre, V., 2007. Properties of the force exerted by filopodia and lamellipodia and the involvement of cytoskeletal components. *PLOS ONE*, issue 10, e1072, doi: 10.1371/journal.pone.0001072.

36. Terracciano, R., Gaspari, M., Testa, F., Pasqua, L., Tagliaferri, P., Cheng, M.M., Nijdam A.J. et al., 2006. Selective binding and enrichment for low molecular weight biomarker molecules in human plasma after exposure to nanoporous silica particles. *Proteomics.* 6, 3243–3250.

37. Pujia, A., De Angelis, F., Scumaci, D., Gaspari, M., Liberale, C., Candeloro, P., Cuda, G., and Di Fabrizio, E., 2010. Highly efficient human serum filtration with water-soluble nanoporous nanoparticles. *Int. J. Nanomed.* 5, 1005–1015.

38. De Angelis, F., Pujia, A., Falcone, C., Iaccino, E., Palmieri, C., Liberale, C., Mecarini, F. et al., 2010. Water soluble nanoporous nanoparticle for in vivo targeted drug delivery and controlled release in B cells tumor context. *Nanoscale.* 2, 2230–2236.

39. Kneipp, K., Kneipp, H., Itzkan, I., Dasari, R.R., and Feld, M.S., 2002. Surface-enhanced raman scattering and biophysics. *J. Phys. Condens. Matter.* 14, R597–R624.

40. De Angelis, F., Gentile F., Mecarini, F., Das, G., Moretti, M., Candeloro, P., Coluccio, M.L. et al. 2011. Breaking the diffusion limit with super-hydrophobic delivery of molecules to plasmonic nanofocusing SERS structures. *Nature Photonics*, doi:10.1038/nphoton.2011.222

41. Chang, R.K. and Furtak, T.E., 1982. Surface-enhanced Raman-scattering. New York: Plenum Press.

42. Moskovits, M., 1985. Surface-enhanced spectroscopy. *Rev. Mod. Phys.* 57, 783–826.

43. Aroca, R., 2006. Surface enhanced vibrational spectroscopy. Chichester: Wiley.

44. Tourrel, G. and Corset, J., 1996. Raman microscopy: developments and applications. Elsevier Academic Press, Malta.

45. De Angelis, F., Patrini, M., Das, G. Maksymov, I., Galli, M., Businaro, L., Andreani, L.C., and Di Fabrizio, E., 2008. A hybrid plasmonic-photonic nanodevice for label-free detection of few molecules. *Nano Lett.* 8(8), 2321–2327.

46. De Angelis F., Das G., Candeloro P. et al. 2010. Nanoscale chemical mapping using three-dimensional adiabatic compression of surface plasmon polaritons. *Nature Nanotech.* 5(1): 67–72.

47. Accardo, A., Gentile, F., Mecarini, F. et al. 2010. *In situ* x-ray scattering studies of protein solution droplets drying on micro-and nanopatterned superhydrophobic PMMA surfaces. *Langmuir*, 26(18): 15057–15064.

48. Gentile, F., Das, G., Coluccio, M.L. et al. 2010. Ultra low concentrated molecular detection using super hydrophobic surface based biophotonic devices. *Microelectronics Eng.* 87(5–8) 798–801.

49. Kneipp, K., Kneipp, H., Abdali, S. et al. 2004. Single molecule raman detection of enkephalin on silver colloidal particles. *Spectroscopy.* 18, 433–440.

50. Xu, H., Bjerneld, E.J., Kall, M. et al. 1999. Spectroscopy of single hemoglobin molecules by surface enhanced Raman scattering. *Phys. Rev. Lett.* 83: 4357–4360.

51. Constantino, C.J.L., Lemma, T., Antunes, P.A. et al. 2001. Single-molecule detection using surface-enhanced resonance Raman scattering and Langmuir–sBlodgett monolayers. *Anal. Chem.* 73(15): 3674–3678.

52. Weitz, D.A., Garoff, S., Gersten, J.I. et al. 1983. The enhancement of Raman scattering, resonance Raman scattering and fluorescence from molecules adsorbed on a rough silver surface. *J. Chem. Phys.* 78: 5324–5338.

53. Kahl, M., Voges, E., Kostrewa, S. et al. 1998. Periodically structured metallic substrates for SERS. *Sens. Actuators B: Chem.* 51: 285–291.

54. Haes, A.J., Zhao, J., Zou, S. et al. 2005. Solution–phase, triangular Ag nanotriangles fabricated by nanosphere lithography. *J. Phys. Chem. B* 109(22): 11158–11162.

55. Zhang, X., Yonzon, C.R., and Duyne, R.P.V., 2006. Nanosphere lithography fabricated plasmonic materials and their applications. *J. Mater. Res.* 21: 1083–1092.

56. Perney, N., Baumberg, J., Zoorob, M. et al. 2006. Tuning localized plasmons in nanostructured substrates for surface enhanced Raman scattering. *Optical Exp.* 14(2): 847–857.

57. Le Ru, E.C., Etchegoin, P.G., Grand, J. et al. 2008. Surface enhanced Raman spectroscopy on nanolithography–prepared substrates. *Curr. Appl. Phys.* 8: 467–470.

58. Cabrini, S., Carpentiero, A., Kumar, R., Businaro, L., Candeloro, P., Prasciolu, M., Gosparini, A. et al. 2005. Focused ion beam lithography for two dimensional array structures for photonic applications. *Microelectronic Eng.* 78–79, 11–15.

59. Tormen, M., Businaro, M., Altissimo, M., Romanato, F., Cabrini, S., Perennes, F., Proietti, R., Hong-Bo Sunc, Satoshi Kawata, and Di Fabrizio, E., 2004. 3D patterning by means of nanoimprinting, x-ray and two-photon lithography. *Microelectronic Engineering*, 73–74, 535–541.

60. De Angelis, F., Liberale, C., Coluccio, M.L., Cojoc, G., and Di Fabrizio, E., 2011. Emerging fabrication techniques for 3D nano-structuring in plasmonics and single molecule studies *Nanoscale*, 3, 2689–2696.

61. Das, G., Mecarini, F., Gentile, F. et al. 2009. Nano-patterned SERS substrate: Application for protein analysis vs. temperature. *Biosensors and Bioelectronics*, 24: 1693–1699.

62. Das, G., Mecarini, F., De Angelis, F. et al. 2008. Attomole (amol) myoglobin Raman detection from plasmonic nanostructures. *Microelectronic Eng.* 85(5–6): 1282–1285.

63. Zhang, D., Xie, Y., Mrozek, M.F. et al. 2003. Raman detection of proteomic analytes. *Anal. Chem.* 75(21): 5703–5709.

64. Kneipp, K., Wang, Y., Kneipp, H. et al. 1997. Single molecule detection using surface–enhanced Raman scattering (SERS). *Phys. Rev. Lett.* 78(9): 1667–1670.

65. Coluccio, M.L., Das, G., Mecarini, F. et al. 2009. Silver-based surface enhanced Raman scattering (SERS) substrate fabrication using nanolithography and site selective electroless deposition. *Microelectronic Eng.* 86(4–6): 1085–1088.

66. Qiu, T. and Chu, P.K., 2008. Self–selective electroless plating: An approach for fabrication of functional 1D nanomaterials. *Mater. Sci. Eng. R* 61: 59–77.

67. Peng, K., Yan, Y., Gao, S. et al. 2002. Synthesis of large–area silicon nanowire arrays via self-assembling nanochemistry. *Adv. Mater.* 14: 1164–1167.

68. Qiu, T., Wu, X.L., Mei, Y.F. et al. 2005. Self-organized synthesis of silver dendritic nanostructures via an electroless metal deposition method. *Appl. Phys. A* 81: 669–671.

Chapter 22

69. Ye, W., Shen, C., Tian, J. et al. 2008. Self–assembled synthesis of SERS-active silver dendrites and photoluminescence properties of a thin porous silicon layer. *Electrochem. Commun.* 10(4): 625–629.

70. Yae, S., Nasu, N., Matsumoto, K. et al. 2007. Nucleation behavior in electroless displacement deposition of metals on silicon from hydrofluoric acid solution. *Electrochim. Acta.* 53(1): 35–41.

71. Goia, Dan V. and Matijevic, E., 1998. Preparation of mono-dispersed metal particles. *New J. Chem.* 22: 1203–1215.

72. Yang, Y., Shi, J., Hawamura, G. et al. 2008. Preparation of Au–Ag, Ag–Au core–shell bimetallic nanoparticles for surface-enhanced Raman scattering. *Scr. Mater.* 58: 862–865.

73. Bulovas, A., Talaikyte, Z., Niaura, G. et al. 2007. Double-layered Ag/Au electrode for SERS spectroscopy: preparation and application for adsorption studies of chromophoric compounds, *CHEMJA* 10(4): 9–15.

74. Spiro, T.G. 1985. Resonance Raman spectroscopy as a probe of heme protein structure and dynamics. *Adv. Protein Chem.* 27: 111–159.

75. Sato, H., Chiba, H., Tashiro, H. et al. 2001. Excitation wavelength-dependent changes in Raman spectra of whole blood and hemoglobin: comparison of the spectra with 514.5-, 720-, and 1064-nm excitation. *J. Biomed., Optics.*, 6(3): 366–370.

Index

T - #0117 - 071024 - C548 - 279/216/29 - PB - 9780367381653 - Gloss Lamination